TABLE 3.5-1   Useful Formulas for the Linear Second-Order Model

Model:	$m\ddot{x} + c\dot{x} + kx = 0$
	m, c, k constant
1. Roots:	$s = \dfrac{-c \pm \sqrt{c^2 - 4mk}}{2m}$
2. Stability property:	Stable if and only if both roots have negative real parts. This occurs if and only if m, c, and k have the same sign.
3. Damping ratio or damping factor:	$\zeta = \dfrac{c}{2\sqrt{mk}}$
4. Undamped natural frequency:	$\omega_n = \sqrt{\dfrac{k}{m}}$
5. Damped natural frequency:	$\omega_d = \omega_n \sqrt{1 - \zeta^2}$
6. Time constant:	$\tau = 2m/c = 1/\zeta\omega_n$ if $\zeta \leq 1$
7. Logarithmic decrement:	$\delta \equiv \dfrac{1}{n} \ln \left(\dfrac{B_i}{B_{i+n}} \right)$
	$\delta = \dfrac{2\pi\zeta}{\sqrt{1 - \zeta^2}}$
	$\zeta = \dfrac{\delta}{\sqrt{(2\pi)^2 + \delta^2}}$

TABLE 4.1-2   Steady-State Sine Response of a Second-Order Model

Model: $m\ddot{x} + c\dot{x} + kx = F_o \sin \omega t$

$$\zeta = \frac{c}{2\sqrt{mk}} \qquad \omega_n = \sqrt{\frac{k}{m}} \qquad r = \frac{\omega}{\omega_n}$$

Steady-state response:

$$x_{ss}(t) = \frac{F_o}{k} \frac{1}{\sqrt{(1 - r^2)^2 + (2\zeta r)^2}} \sin(\omega t + \phi)$$

$$\phi = -\tan^{-1} \frac{2\zeta r}{1 - r^2}$$

Third quadrant if $1 - r^2 < 0$
Fourth quadrant if $1 - r^2 > 0$
$\phi = -90°$ if $r^2 = 1$

Magnitude ratio: $M = \dfrac{|x_{ss}|}{F_o} = \dfrac{1}{\sqrt{(k - m\omega^2)^2 + (c\omega)^2}} = \dfrac{1}{k} \dfrac{1}{\sqrt{(1 - r^2)^2 + (2\zeta r)^2}}$

Dimensionless magnitude ratio: $\dfrac{X}{\delta_{st}} = kM \qquad \delta_{st} = \dfrac{F_o}{k}$

Peak frequency: $\omega_p = \omega_n \sqrt{1 - 2\zeta^2} \qquad 0 \leq \zeta \leq \dfrac{1}{\sqrt{2}}$

Peak response: $X_p = \delta_{st} \dfrac{1}{2\zeta\sqrt{1 - \zeta^2}} \qquad 0 \leq \zeta \leq \dfrac{1}{\sqrt{2}}$

(continued at back of book)

Mechanical Vibration

Mechanical Vibration

William J. Palm III

University of Rhode Island

John Wiley & Sons, Inc.

To my wife Mary Louise and
our children: Aileene, Bill, and Andy.

Publisher	*Bruce Spatz*
Associate Publisher	*Daniel Sayre*
Acquisitions Editor	*Joseph Hayton*
Production Editor	*Nicole Repasky*
Marketing Manager	*Frank Lyman*
Designer	*Hope Miller*
Media Editor	*Stefanie Liebman*
Editorial Assistant	*Mary Moran*
Production Services Management	*Ingrao Associates*

This book was set in 10/12 pt. Times Roman by Thomson Digital, India and printed and bound by R.R. Donnelley/Willard Division. The cover was printed by *Phoenix Color.*

This book is printed on acid-free paper. ∞

To order books or for customer service, please call 1-800-CALL WILEY (225-5945).

Library of Congress Cataloging-in-Publication Data:
Palm, William J. (William John), 1944–
   Mechanical vibrations/William J. Palm III.
      p. cm.
   ISBN-13: 978-0-471-34555-8 (cloth)
   ISBN-10: 0-471-34555-5 (cloth)
   1. Vibration. 2. Damping (Mechanics). I. Title

   TA355.P25 2006
   620.3–dc22   2005058117

Printed in the United States of America

10 9 8 7 6 5 4 3 2

Preface

This text is an introduction to the analysis and design of vibrating systems. Such a course is normally offered in the junior or senior year in mechanical engineering, engineering mechanics, and aeronautical/aerospace engineering.

It is assumed that the student has a background in calculus and dynamics. Any other required material in mathematics, such as differential equations and matrix algebra, is developed in the text.

Text Objectives

This text provides a foundation in the ***modeling, analysis, and design*** of vibratory systems through text discussions, worked examples, applications, and use of modern computer tools.

Vibrating systems are broad and varied – from the vibration of a baseball bat upon impact with a ball to the ground motion caused by earthquakes. Understanding vibrations enables engineers to design products and systems that eliminate unwanted vibrations, and to harness vibrations for useful purposes. Vibration analysis must start with a **mathematical model** of the dynamics of the system under study, and the model must be simple enough to be amenable to **analysis** yet detailed enough to describe the pertinent behavior of the system. This requires a firm foundation in dynamics, and practice in developing system models. Chapter 2 provides background in understanding and describing the dynamics of vibrating systems. The remaining chapters systematically cover systems of increasing complexity, providing extensive sets of examples, applications, and homework problems to practice modeling and analyzing a wide variety of vibratory systems.

Computational tools are useful in modeling and analyzing mechanical vibration, and ABET encourages use of modern engineering tools in the curriculum. This text uses MATLAB®, the most commonly used engineering program, and the related MATLAB product Simulink®, for simulation and analysis[1]. Most chapters end with a MATLAB or Simulink section showing how to use these tools to solve situations not easily amenable to analytic solutions.

Features

The text:

1. Applies knowledge of math and science to solve engineering vibration problems. Chapter 2 reviews the necessary dynamics background needed for the remaining chapters. Differential equations and matrix algebra are discussed in the text in context where they are needed to solve related vibration problems.

2. Provides practice in identifying, formulating, and solving vibration problems. An extensive set of worked examples and homework problems offer the

---

[1] MATLAB and Simulink are registered trademarks of The MathWorks, Inc. This text is based on Release 14 of the software (version 7.0 of MATLAB, version 6.0 of Simulink, and version 6.0 of the Control System Toolbox).

opportunity to apply concepts discussed in the book to model, analyze, and solve a variety of problems.

3. Uses modern engineering tools:

   a. Separate **MATLAB** sections at the end of most chapters show how to use the most recent features of this standard engineering tool, in the context of solving vibration problems. Appendix A provides a self-contained introduction to MATLAB for readers who are not familiar with the language.

   b. Chapters 4, 5, and 8 cover MATLAB functions in the Control Systems toolbox that are useful for vibration applications.

   c. A unique feature is the introduction of **Simulink** into the vibration course. Having a graphical user interface that many consider easier to use than the command line programming in MATLAB, Simulink is used where solutions may be difficult to program in MATLAB, such as modeling Coulomb friction effects and simulating systems that contain non-linearities.

   d. Chapter 11 covers the **finite element method,** which provides a more accurate system description than that used to develop a lumped-parameter model, but that for complex systems is easier to solve than partial differential equations.

4. Includes information on designing and conducting experiments, and analyzing the data. In particular,

   a. Chapter 1 provides unique coverage of the use of the **least squares method** to obtain stiffness and damping values from data.

   b. Chapter 9 introduces vibration measurement and testing, and it discusses the hardware available to produce vibration of the system under test and to measure the response. Knowledge of the algorithms and software available for processing the data is also necessary. The chapter outlines the equipment requirements and the procedures that are available for these tasks. The chapter also shows how to perform a transform analysis of signal data, including use of the Fast Fourier transform with MATLAB, to estimate natural frequencies.

Additional features of the text include

5. Learning objectives stated at the beginning of each chapter.

6. A summary at the end of each chapter.

7. A special section dealing with active vibration control in sports equipment. (chapter 7).

Organization

Chapter 1 introduces basic vibration terminology and the concepts of stiffness and damping. A novel feature is coverage of the least squares method for obtaining stiffness and damping values from data. This topic has traditionally not been given much attention in vibration texts, but the methods are often needed by engineers working with real devices.

Many of the concepts and methods of vibration analysis can be explained with models of systems having a single degree of freedom. Many such models are developed in

Chapter 2. The description may be in the form of a differential equation of motion, derived directly from Newton's laws, but other methods that do not require the equation of motion, such as work-energy methods and impulse-momentum methods, are also treated in Chapter 2. The chapter concludes with a discussion of equivalent mass and equivalent inertia. These concepts simplify the modeling of systems containing both translating and rotating parts, whose motion is coupled, and of systems containing distributed mass.

Chapter 3 covers the free response of damped and undamped systems having a single degree of freedom. The chapter contains numerous examples of systems having viscous damping and Coulomb damping, and it includes two sections dealing with computer methods for analyzing and simulating system response.

Chapter 4 treats the harmonic response of systems having one degree of freedom, and it includes two important topics associated with such systems: resonance and bandwidth. Damping sources other than linear viscous damping, such as hysteresis, fluid drag, and Coulomb friction, can introduce nonlinear effects, and the chapter contains a discussion of the effects of nonlinearities on system response. The chapter introduces several MATLAB functions that are especially intended for harmonic response analysis, and it covers several Simulink features useful for simulating systems containing nonlinearities that are difficult to program in MATLAB or other languages.

Chapter 5 shows how to obtain the response of single-degree-of-freedom systems to non-harmonic forcing functions. The Fourier series representation of a periodic function as a series of sines and cosines enables us to use the results of Chapter 4 to obtain the response. Then the Laplace transform method is introduced. This method provides a means of obtaining the response of a linear system, of any order, for most of the commonly found forcing functions. The chapter shows how MATLAB can be used to perform some of the algebra required to use the Laplace transform method. The final section shows how to use Simulink to obtain the general forced response.

Many practical vibration applications require a model having more than one degree of freedom in order to describe the important features of the system response. Chapter 6 treats models that have two degrees of freedom, and it introduces Lagrange's equations as an alternative to deriving models from Newton's laws. The final three sections of Chapter 6 treat applications of MATLAB and Simulink to the chapter's topics. MATLAB has powerful capabilities for analyzing systems with more than one degree of freedom. One of these, the root locus plot, was developed for control system design and has useful applications in vibration, but has been ignored in the vibration literature. In addition, MATLAB can be used not only to solve the differential equations numerically, but also to perform some of the algebra required to obtain closed form solutions. Simulink is useful in applications where the stiffness and/or the damping elements are nonlinear, and where the input is not a simple function of time.

Chapter 7 considers how to design systems to eliminate or at least reduce the effects of unwanted vibration. In order to determine how much the vibration should be reduced, we need to know what levels of vibration are harmful, or at least disagreeable. To reduce vibration, it is often important to understand the vibration source, and these two topics are discussed in the beginning of the chapter. The chapter then treats the design of vibration isolators, which consist of a stiffness element and perhaps a damping element, and which are placed between the vibration source and the surrounding environment. The chapter also treats the design of vibration absorbers. The chapter concludes with a discussion of active vibration control, which uses a power source such as a hydraulic cylinder or an electric motor to provide forces needed to counteract the forces producing the unwanted vibration.

The algebra required to analyze systems having more than two degrees of freedom can become very complicated. For such systems it is more convenient to use matrix representation of the equations of motion and matrix methods to do the analysis. This is the topic of Chapter 8, which begins by showing how to represent the equations of motion in compact matrix form. Then systematic procedures for analyzing the modal response are developed. Besides providing a compact form for representing the equations of motion and performing the analysis, matrix methods also form the basis for several powerful MATLAB and Simulink functions that provide useful tools for modal analysis and simulation. These are described at the end of the chapter.

There are applications in which it is difficult to develop a differential equation model of the system from basic principles such as Newton's laws. In such cases we must resort to using measurements of the system response. Chapter 9 introduces vibration measurement and testing, and it discusses the hardware available to produce vibration of the system under test and to measure the response.

Chapter 10 treats the vibration of systems that cannot be described adequately with lumped-parameter models consisting of ordinary differential equations. The chapter begins by considering how to model the simplest distributed system, a cable or string under tension. Torsional and longitudinal vibration of rods are also described by such a model. The partial differential equation model – the wave equation – is second-order, and we introduce two methods for solving such an equation. The second solution method, separation of variables, is also useful for solving the fourth-order model that describes beam vibration. This model is also derived in the chapter. Examples are then provided to show how to use MATLAB to solve the resulting transcendental equations for the natural frequencies.

In Chapter 11 we introduce the finite element method, which provides a more accurate system description than that used to develop a lumped-parameter model, but that for complex systems is easier to solve than partial differential equations. The method is particularly useful for irregular geometries, such as bars that have variable cross sections, and for systems such as trusses that are made up of several bars. The chapter concludes with applications of MATLAB to finite element analysis.

The text has three appendices. Appendix A is a self-contained introduction to MATLAB. Appendix B is an introduction to numerical methods for solving differential equations, and it focuses on the Runge-Kutta family of algorithms. Appendix C contains physical property data for common materials.

Typical Syllabi

The first seven chapters constitute a basic course in mechanical vibration. The remaining four chapters can be used to provide coverage of additional topics at the discretion of the instructor. Some examples of such coverage are as follows.

1. Cover Chapter 8 for treatment of matrix methods and modal analysis.

2. Cover Chapter 9 for an introduction to experimental methods, including spectral analysis and the Fast Fourier transform.

3. Cover Chapter 10 for distributed parameter models.

4. Cover both Chapters 10 and 11 for finite element analysis.

Chapters 8 through 11 can be covered in any order, with the exception that Chapter 10 should be covered before Chapter 11.

Supplements

An online solutions manual, containing typeset solutions to all the chapter problems, is available at http://www.wiley.com/college/palm for instructors who have adopted the text for their course.

Here you can also find the text illustrations, available as both Powerpoint and HTML, as well as MATLAB files to accompany the text.

Acknowledgments

The publisher, John Wiley and Sons, provided the author with much support during the development of this text by surveying many instructors and by commissioning extensive reviews of the chapters. This feedback significantly influenced the final result. The author wishes to express his appreciation to Joe Hayton for this support, and to the rest of the editorial staff for their expert help during the production process.

The author is grateful to the following reviewers for providing especially useful comments in this project.

David Bridges – Mississippi State University
Ali Mohammadzadeh – Grand Valley State University
Fred Barez – San Jose State University
Lawrence Bergman – University of Illinois – Urbana Champaign

The author appreciates the support of the University of Rhode Island, which approved a sabbatical leave for this project. The Department of Mechanical Engineering and Applied Mechanics at URI has always encouraged teaching excellence via textbook writing, and the author wishes to acknowledge particularly the support of the department chair, Professor Arun Shukla.

The final acknowledgement goes to the author's wife, Mary Louise, and their children, Aileene, Bill, and Andy, for their encouragement and understanding during the creation of this text.

Contents

Appendix A   Introduction to MATLAB

Appendix A can be found on the following website: http://www.wiley.com/college/palm.

Appendix B   Numerical Solution Methods   685

Appendix C   Mechanical Properties of Common Materials   693

CHAPTER 1

INTRODUCTION TO MECHANICAL VIBRATION

CHAPTER OUTLINE

THIS CHAPTER introduces the basic terminology used in the study of mechanical vibration, as well as two important elements found in vibration models. These are the *spring* element, which produces a restoring force or moment as a function of the displacement of the mass element, and the *damping* element, which produces a restoring force or moment as a function of the velocity of the mass element. In this chapter we establish the basic principles for developing mathematical models of these elements and apply these principles to several commonly found examples.

To be useful, numerical values eventually must be assigned to the parameters of a mathematical model of a spring or a damping element, in order to make predictions about the behavior of the physical device. Therefore, this chapter includes two sections that show how to obtain numerical parameter values from data by using the least-squares method. This topic has traditionally not been given much attention in vibration texts, but the methods are often needed by engineers working with real devices.

The programming language MATLAB has several useful functions that enable you to apply these methods easily. These functions are presented in a separate section that may be skipped by readers who will not be using MATLAB. If you are such a reader, however, you are strongly encouraged to consider learning MATLAB, because it will be very useful to you in the future.

LEARNING OBJECTIVES

After you have finished this chapter, you should be able to do the following:

- Develop models of spring elements from the basic principles of mechanics of materials.
- Develop models of damping elements from the basic principles of fluid mechanics.
- Apply the least-squares method to obtain numerical parameter values for a spring or damping element model when given the appropriate data.
- Use MATLAB to implement the least-squares method.

1.1 INTRODUCTION

This text is an introduction to the study of mechanical vibration. Although the terms *good vibrations* and *bad vibrations* (or "vibes") have entered popular culture with a non-engineering sense, we as engineers must be concerned about good and bad vibrations in our designs. An example of "good" engineering vibration is an electric shaver, in which the cutter is caused to oscillate by an electromagnet. Other examples are a guitar string, vibratory finishers, vibratory conveyors, and vibratory sieves for sorting objects by size. An example of "bad" engineering vibration is the oscillation endured by a passenger as a car rides over a bumpy road. Perhaps the premier example of bad vibration is the building motion caused by an earthquake. In an engineering sense, good vibration is vibration that is useful, and bad vibration is vibration that causes discomfort or damage.

In this text you will learn how to design systems that make use of vibration and how to design systems that reduce or protect against vibration. The material is a necessary prerequisite to related but more specialized courses in finite element analysis, acoustics, modal analysis, active vibration control, and fatigue failure.

Inherent in the study of vibration is *oscillation*. Electrical circuits can have oscillatory voltages or currents, but these are not called vibratory systems, and their study is not called vibration. Air pressure oscillation is called *sound*, but the study of sound is called *acoustics*, not *vibration*. The term *vibration* is usually used to describe the motion of mechanical objects that oscillate or have the potential to oscillate. Common examples include a pendulum, a seesaw, and a child's swing (which is really a pendulum). For now, let us not dwell too much on what is meant by "potential to oscillate"; the point is that some "vibratory" systems may not oscillate. For example, if we place a pendulum in molasses and give it a push, the pendulum will not oscillate because the fluid is so sticky, but the pendulum itself has the potential to oscillate under the influence of gravity—were it not for the molasses.

Oscillation, or vibration, of a mechanical object is caused by a force or moment that tries to return the object to an equilibrium, or rest, position. Such a force is called a *restoring* force or moment. Restoring forces and moments are usually functions of displacement, such as the gravitational moment acting on the pendulum. A common theme in mechanical vibration is the interplay between restoring forces and frictional forces, which are constant and oppose motion, or between restoring forces and *damping* forces, which also oppose motion but are velocity dependent (like the force exerted on the pendulum by the molasses).

The study of mechanical vibration has a long tradition. We will not present a detailed history of the subject but merely mention several of the most important investigators. Pythagoras, who was born in 582 B.C., studied the music produced by the vibration of strings. Galileo, born in A.D. 1564, also investigated stringed musical instruments, as well as pendulum oscillations. Of course, Newton, born in A.D. 1642, established the laws of motion that are the basis for deriving the equations of motion of vibrating systems. Many famous mathematicians and physicists contributed to our understanding of vibration. Some of these are Taylor (who developed the Taylor series), Leonhard Euler, Daniel Bernoulli, D'Alembert, Fourier, Lagrange, Poisson, and Coulomb. One of the most famous names in vibration is Rayleigh, who in 1877 published his classic work, *The Theory of Sound*. Major contributors in the twentieth century include Timoshenko, Stodola, De Laval, Frahm, Minorsky, Duffing, and van der Pol. Many illuminating stories concerning the application of vibration theory to practical problems are given by Den Hartog [Den Hartog, 1985].

Knowledge of vibration is important for the modern engineer. One reason is that so many engineering devices contain or are powered by engines or motors. These often produce oscillatory forces that cause mechanical vibration which can result in unwanted noise, uncomfortable motion, or structural failure. Examples where the resulting vibration can have serious consequences include coolant pumps in submarines (because of the noise they generate), fuel pumps in aircraft and rocket engines, helicopter rotors, turbines, and electrical generating machinery. Often the motion of the object itself produces vibration. Examples are air turbulence acting on an aircraft and road forces acting on a vehicle suspension. The environment can produce vibration, as with earthquakes or wind forces acting on a structure. Temperature gradients within a satellite structure, caused by solar radiation, are time varying because the satellite moves in and out of the earth's shadow. These gradients can cause structural vibration that interferes with the operation of the instruments, as was the case early on with the Hubble space telescope until the problem was corrected. In fact, satellite structures are particularly susceptible to vibration because they must be light in weight for their size.

1.2 DESCRIPTIONS OF VIBRATIONAL MOTION

The sine function $\sin \omega t$ and its related function, the cosine function $\cos \omega t$, often appear as solutions to vibration models. The two functions are related as $\cos \omega t = \sin(\omega t + \pi/2)$. The parameter ω is called the *radian frequency*. It is the frequency of oscillation of the function expressed as radians per unit time, for example, as radians per second. The related frequency is cycles per unit time and is often denoted by f. The two frequencies are related by $\omega = 2\pi f$. When expressed as cycles per second, the SI unit for frequency is the *hertz*, abbreviated as Hz. Thus 1 Hz is one cycle per second.

The sine function $A \sin \omega t$ is plotted in Figure 1.2-1. The *amplitude A* is shown in the figure. The function oscillates between the minimum value, $-A$, and the maximum value, A; thus its range is $2A$. The *period P* is the time between two adjacent peaks and is thus the time required for the oscillation to repeat. The period is related to the frequencies as $P = 1/f = 2\pi/\omega$.

Simple Harmonic Motion

We will see that the equation of motion of many oscillatory systems has a solution of the form

$$y(t) = A \sin(\omega t + \phi)$$

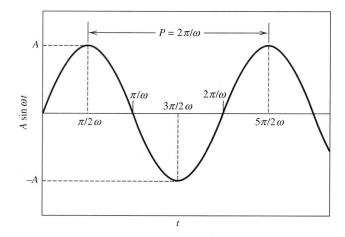

FIGURE 1.2-1   Graph of the function $y(t) = A \sin \omega t$.

where $y(t)$ is the displacement of a mass. The radian frequency is ω and the period is $P = 2\pi/\omega$. Expressions for the velocity and acceleration are obtained by differentiating $y(t)$:

$$\dot{y}(t) = A\omega \cos(\omega t + \phi) = A\omega \sin\left(\omega t + \phi + \frac{\pi}{2}\right)$$

$$\ddot{y}(t) = -A\omega^2 \sin(\omega t + \phi) = A\omega^2 \sin(\omega t + \phi + \pi)$$

The displacement, velocity, and acceleration all oscillate with the same frequency ω but have different amplitudes. The velocity is $\pi/2$ rad or $90°$ out of phase with the displacement, and the acceleration is π rad or $180°$ out of phase with the displacement. Thus the velocity is zero when the displacement and acceleration are at a minimum or a maximum. The acceleration is at a minimum when the displacement is at a maximum. These functions are plotted in Figure 1.2-2 for the case where $y(t) = \sin(2t + \pi/2)$.

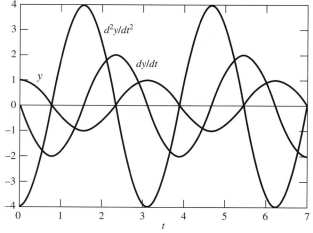

FIGURE 1.2-2   Displacement, velocity, and acceleration for simple harmonic motion. The specific function plotted is $y(t) = \sin(2t + \pi/2)$.

This type of motion, where the acceleration is proportional to the displacement but opposite in sign, is called *simple harmonic motion*. It occurs when the force acting on the mass is a restoring force that is proportional to the displacement.

Forms of Harmonic Functions

When the sine and cosine functions appear with a *phase angle* or *phase shift* ϕ, as $A \sin(\omega t + \phi)$, we can reduce this expression to simple sines and cosines by using the trigonometric identity

$$A \sin(\omega t + \phi) = A \cos \phi \sin \omega t + A \sin \phi \cos \omega t \qquad (1.2\text{-}1)$$

Sometimes we will be given an expression like

$$y(t) = B \sin \omega t + C \cos \omega t \qquad (1.2\text{-}2)$$

and we will need to express $y(t)$ as $A \sin(\omega t + \phi)$. To find A and ϕ, given B and C, compare Equations 1.2-1 and 1.2-2 and obtain

$$A \cos \phi = B \qquad \text{and} \qquad A \sin \phi = C \qquad (1.2\text{-}3)$$

If we square both equations and add, we obtain

$$B^2 + C^2 = (A \cos \phi)^2 + (A \sin \phi)^2 = A^2 \left(\cos^2 \phi + \sin^2 \phi \right) = A^2$$

because of the identity $\cos^2 \phi + \sin^2 \phi = 1$. Thus there are two solutions for A:

$$A = \pm \sqrt{B^2 + C^2} \qquad (1.2\text{-}4)$$

Choosing the positive solution for A, we can then solve Equation 1.2-3 for $\cos \phi$ and $\sin \phi$ as follows:

$$\cos \phi = \frac{B}{A} \qquad \text{and} \qquad \sin \phi = \frac{C}{A} \qquad (1.2\text{-}5)$$

These equations uniquely specify the angle ϕ. Although there are two solutions to $\phi = \cos^{-1}(B/A)$ and two solutions to $\phi = \sin^{-1}(C/A)$, the quadrant of ϕ is uniquely determined by the signs of $\cos \phi$ and $\sin \phi$, or equivalently by the signs of B/A and C/A.

For example, suppose we are given $y(t) = -10 \sin 5t + 6 \cos 5t$. Then

$$A = +\sqrt{(-10)^2 + 6^2} = \sqrt{136} = 2\sqrt{34}$$

and

$$\cos \phi = \frac{-10}{2\sqrt{34}} = \frac{-5}{\sqrt{34}} \qquad \text{and} \qquad \sin \phi = \frac{6}{2\sqrt{34}} = \frac{3}{\sqrt{34}}$$

Because $\cos \phi < 0$ and $\sin \phi > 0$, ϕ lies in the second quadrant ($\pi/2 \leq \phi \leq \pi$) and is given by $\phi = \cos^{-1}(-5/\sqrt{34}) = 2.601$ rad. Thus $y(t) = 2\sqrt{34} \sin(5t + 2.601)$.

Oscillations with Exponential Amplitude

A very commonly occurring function is the product of an exponential and a sine or cosine; for example,

$$y(t) = A e^{-t/\tau} \sin(\omega t + \phi) \qquad (1.2\text{-}6)$$

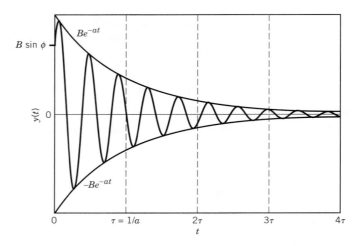

FIGURE 1.2-3   The function $y(t) = Ae^{-t/\tau}\sin(\omega t + \phi)$. The envelopes of the oscillations are given by $Ae^{-t/\tau}$ and $-Ae^{-t/\tau}$.

This function is illustrated in Figure 1.2-3.

   The exponential function is $e^{-t/\tau}$. If $\tau > 0$, it is called the *time constant*. This function is shown in Figure 1.2-4. As $t \to \infty$, $e^{-t/\tau} \to 0$. In most engineering applications we simply need to know how long it takes for the function to decay to some small value, so the following values, which are correct to two decimal places, will be very useful to us:

- At $t = \tau$, $e^{-t/\tau} = e^{-1} = 0.37$
- At $t = 4\tau$, $e^{-t/\tau} = e^{-4} = 0.02$
- At $t = 5\tau$, $e^{-t/\tau} = e^{-5} = 0.01$

Noting that $e^0 = 1$, we see that the exponential is 37% of its initial value at $t = \tau$, 2% at $t = 4\tau$, and 1% at $t = 5\tau$.

   The oscillation amplitudes of Equation 1.2-6 decay with time because of the decaying exponential, and thus we can use the time constant τ to estimate how long it will

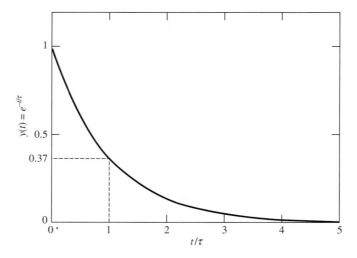

FIGURE 1.2-4   Graph of the function $y(t) = e^{-t/\tau}$.

TABLE 1.2-1 The Exponential Function

Taylor series

$$e^x = 1 + x + \frac{x^2}{2} + \frac{x^3}{6} + \cdots + \frac{x^n}{n!} + \cdots$$

Euler's identities

$$e^{i\theta} = \cos\theta + i\sin\theta$$

$$e^{-i\theta} = \cos\theta - i\sin\theta$$

Limits

$$\lim_{x\to\infty} xe^{-x} = 0 \qquad \text{if } x \text{ is real}$$

$$\lim_{t\to\infty} e^{-st} = 0 \qquad \text{if the real part of } s \text{ is positive}$$

If τ is real and positive,

$$e^{-t/\tau} < 0.02 \text{ if } t > 4\tau$$

$$e^{-t/\tau} < 0.01 \text{ if } t > 5\tau$$

take for the oscillations to disappear (after $t = 4\tau$, the amplitudes will be less than 2% of the largest amplitude).

A displacement represented by Equation 1.2-6 does not undergo simple harmonic motion because the acceleration is not proportional to the displacement. This can be verified by deriving the expression for $\ddot{y}$ from Equation 1.2-6.

The exponential is not always a decaying exponential. For example, we will see solutions of the form

$$y(t) = Ae^{rt}\sin(\omega t + \phi)$$

where $r > 0$. In this case the oscillation amplitude grows indefinitely with time.

Table 1.2-1 summarizes some important formulas and properties related to the exponential function.

Euler's Formula and Complex Numbers

We will find it convenient to use the exponential function to solve equations of motion even when the motion is oscillatory. This convenience arises from *Euler's identity*, which we now derive.

The Taylor series expansion of e^x about $x = 0$ is

$$e^x = 1 + x + \frac{x^2}{2!} + \frac{x^3}{3!} + \frac{x^4}{4!} + \frac{x^5}{5!} + \cdots$$

Let $x = i\theta$ and use the fact that $i^2 = -1$, $i^3 = -i$, $i^4 = 1$, and $i^5 = i$ to obtain

$$e^{i\theta} = 1 + i\theta - \frac{\theta^2}{2!} - i\frac{\theta^3}{3!} + \frac{\theta^4}{4!} + i\frac{\theta^5}{5!} + \cdots$$

Collecting the real and imaginary parts gives

$$e^{i\theta} = \left(1 - \frac{\theta^2}{2!} + \frac{\theta^4}{4!}\right) + i\left(\theta - \frac{\theta^3}{3!} + \frac{\theta^5}{5!}\right) + \cdots$$

The expansions for the sine and cosine function about $\theta = 0$ are

$$\sin\theta = \theta - \frac{\theta^3}{3!} + \frac{\theta^5}{5!} - \frac{\theta^7}{7!} + \cdots$$

$$\cos\theta = 1 - \frac{\theta^2}{2!} + \frac{\theta^4}{4!} - \frac{\theta^6}{6!} + \cdots$$

Comparing these with the expansion for $e^{i\theta}$, we see that

$$e^{i\theta} = \cos\theta + i\sin\theta$$

This is Euler's identity. Its companion form can be obtained by using the fact that $\sin(-\theta) = -\sin\theta$ and $\cos(-\theta) = \cos\theta$:

$$e^{-i\theta} = \cos\theta - i\sin\theta$$

We will find that some equations of motion have the following solution form:

$$y(t) = D_1 e^{i\theta} + D_2 e^{-i\theta}$$

where $\theta = \omega t$. This solution form is difficult to interpret, so we will use Euler's identities to express the solution in terms of sine and cosine functions. Substituting these two identities and collecting terms, we find that $y(t)$ has the form

$$y(t) = (D_1 + D_2)\cos\theta + i(D_1 - D_2)\sin\theta$$

Because $y(t)$ is a displacement and must therefore be real, D_1 and D_2 must be complex conjugates. Thus this equation reduces to

$$y(t) = B_1\cos\theta + B_2\sin\theta = B_1\cos\omega t + B_2\sin\omega t$$

where B_1 and B_2 are real and are given by $B_1 = D_1 + D_2$ and $B_2 = i(D_1 - D_2)$. This form can be converted into the form $y(t) = A\sin(\omega t + \phi)$, as shown earlier.

Roots and Complex Numbers

We will often need to solve the quadratic equation $as^2 + bs + c = 0$ in order to solve an equation of motion. Its solution is given in Table 1.2-2. Sometimes the quadratic roots are complex numbers, and in this case it is often more convenient to express the quadratic factor as shown in the table.

We will often need to work with algebraic expressions containing complex numbers. The properties of complex numbers that we will find useful are listed in Table 1.2-3. Note that a complex number z may be represented in several ways, two of which are based on Figure 1.2-5. The rectangular form is $z = x + yi$, and the polar or vector form is $z = |z|\angle\theta$, where $|z| = \sqrt{x^2 + y^2}$ and $\theta = \angle z = \tan^{-1}(y/x)$. The exponential form is based on Euler's identity and is $z = |z|e^{i\theta}$.

TABLE 1.2-2   The Quadratic Equation

The roots of $as^2 + bs + c = 0$ are given by

$$s = \frac{-b \pm \sqrt{b^2 - 4ac}}{2a}$$

For complex roots, $s = \sigma \pm i\omega$, the quadratic can be expressed as

$$as^2 + bs + c = a[(s + \sigma)^2 + \omega^2] = 0$$

TABLE 1.2-3   Complex Numbers

Rectangular representation:

$$z = x + iy, \quad i = \sqrt{-1}$$

Magnitude and angle:

$$|z| = \sqrt{x^2 + y^2} \qquad \theta = \angle z = \tan^{-1}\frac{y}{x}$$

Polar and exponential representation:

$$z = |z|\angle\theta = |z|e^{i\theta}$$

Equality: If $z_1 = x_1 + iy_1$ and $z_2 = x_2 + iy_2$, then

$$z_1 = z_2 \text{ if } x_1 = x_2 \text{ and } y_1 = y_2$$

Addition:

$$z_1 + z_2 = (x_1 + x_2) + i(y_1 + y_2)$$

Multiplication:

$$z_1 z_2 = |z_1||z_2|\angle(\theta_1 + \theta_2)$$
$$z_1 z_2 = (x_1 x_2 - y_1 y_2) + i(x_1 y_2 + x_2 y_1)$$

Complex-conjugate multiplication:

$$(x + iy)(x - iy) = x^2 + y^2$$

Division:

$$\frac{1}{z} = \frac{1}{x + yi} = \frac{x + iy}{x^2 + y^2}$$
$$\frac{z_1}{z_2} = \frac{|z_1|}{|z_2|}\angle(\theta_1 - \theta_2)$$
$$\frac{z_1}{z_2} = \frac{x_1 + iy_1}{x_2 + iy_2} = \frac{x_1 + iy_1}{x_2 + iy_2}\frac{x_2 - iy_2}{x_2 - iy_2} = \frac{(x_1 + iy_1)(x_2 - iy_2)}{x_2^2 + y_2^2}$$

Units

None of the modeling and analysis techniques we will develop depend on a specific system of units. However, in order to make quantitative statements based on the resulting models, a set of units must be employed. In this book we use two systems of units, the

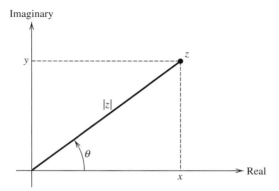

FIGURE 1.2-5   Graphical representation of the complex number z.

FPS (foot-pound-second) system and the metric Système International d'Unités (SI) system. A common system of units in business and industry in English-speaking countries has been the foot-pound-second (FPS) system. The FPS system is also known as the *U.S. customary system* or the *British engineering system.*

The FPS system is a *gravitational* system. This means that the primary variable is force and that the unit of mass is derived from Newton's second law. The *pound* is selected as the unit of force and the *foot* and *second* as units of length and time, respectively. From Newton's second law of motion, force equals mass × acceleration, or

$$f = ma \qquad\qquad (1.2\text{-}7)$$

where f is the net force acting on the mass m and producing an acceleration a. Thus the unit of mass must be

$$\text{mass} = \frac{\text{force}}{\text{acceleration}} = \frac{\text{pound}}{\text{foot}/(\text{second})^2}$$

This mass unit is named the *slug*.

Through Newton's second law, the weight W of an object is related to the object mass m and the acceleration due to gravity, denoted by g, as follows: $W = mg$. At the surface of the earth, the standard value of g in FPS units is $g = 32.2$ ft/sec^2.

Energy has the dimensions of mechanical work; namely, force × displacement. Therefore, the unit of energy in this system is the *foot pound* (ft-lb). Another energy unit in common use for historical reasons is the *British thermal unit* (Btu). The relationship between the two is given in Table 1.2-4. Power is the rate of change of energy with time, and a common unit is *horsepower*.

The SI metric system is an *absolute* system, which means that the mass is chosen as the primary variable and the force unit is derived from Newton's law. The *meter* and the second are selected as the length and time units, and the *kilogram* is chosen as the mass unit. The derived force unit is called the *newton*. In SI units the common energy unit is the newton-meter, also called the *joule*, while the power unit is the joule/second, or *watt*. At the surface of the earth, the standard value of g in SI units is $g = 9.81$ m/s^2.

Table 1.2-5 gives the most commonly needed factors for converting between the FPS and the SI systems.

TABLE 1.2-4   SI and FPS Units

Quantity	Unit name and abbreviation	
	SI unit	**FPS unit**
Time	second (s)	second (sec)
Length	meter (m)	foot (ft)
Force	newton (N)	pound (lb)
Mass	kilogram (kg)	slug
Energy	joule (J)	foot-pound (ft-lb), Btu
Power	watt (W)	ft-lb/sec, horsepower (hp)

TABLE 1.2-5   Unit Conversion Factors

Length	1 m = 3.281 ft	1 ft = 0.3048 m
	1 mile = 5280 ft	
Speed	1 ft/sec = 0.6818 mi/hr	1 mi/hr = 1.467 ft/sec
	1 m/s = 3.6 km/h	1 km/h = 0.2778 m/s
	1 km/h = 0.6214 mi/hr	1 mi/hr = 1.609 km/h
Force	1 N = 0.2248 lb	1 lb = 4.4482 N
Mass	1 kg = 0.06852 slug	1 slug = 14.594 kg
Energy	1 J = 0.7376 ft-lb	1 ft-lb = 1.3557 J
Power	1 hp = 550 ft-lb/sec	1 hp = 745.7 W
	1 W = 1.341 $\times 10^{-3}$ hp	

1.3   SPRING ELEMENTS

When a body's deformation due to externally applied forces is small, we use a rigid-body model. However, mechanical systems are often made up of rigid bodies connected with elastic elements. The elastic element can be intentionally designed into the system, as with a helical-coil spring in a car's suspension. Sometimes the element is not intended to be elastic but deforms anyway because it is subjected to large forces or torques. This can be the case with a drive shaft that transmits the motor torque to the driven object. Unintended deformation may also occur in the boom and cables of a crane lifting a heavy load.

The most familiar spring is probably the helical-coil spring, such as those used in vehicle suspensions and those found in retractable pens and pencils. The purpose of the spring in both applications is to provide a restoring force. However, considerably more engineering analysis is required for the vehicle spring application because the spring can cause undesirable motion of the wheel and chassis, such as vibration. Because the pen motion is constrained and cannot vibrate, we do not need as sophisticated an analysis to see if the spring will work.

Many engineering applications involving elastic elements, however, do not contain coil springs but rather involve the deformation of beams, cables, rods, and other mechanical members. In this section we develop the basic elastic properties of many of these common elements, which are called *spring elements.*

Force-Deflection Relations

A coil spring has a *free length*, denoted by L in Figure 1.3-1. The free length is the length of the spring when no tensile or compressive forces are applied to it. When a spring is

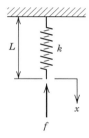

FIGURE 1.3-1   Symbol and nomenclature for a spring element. The free length is *L*.

compressed or stretched, it exerts a restoring force that opposes the compression or extension. The general term for the spring's compression or extension is *deflection*. The greater the deflection (compression or extension), the greater the restoring force. The simplest model of this behavior is the so-called *linear force-deflection model*,

$$f = kx \qquad (1.3\text{-}1)$$

where f is the restoring force, x is the compression or extension distance (the change in length from the free length), and k is the *spring constant*, which is defined to be always positive. Typical units for k are lb/ft and N/m. Some references, particularly in the automotive industry, use the term *spring rate* or *stiffness* instead of *spring constant*.

When $x = 0$, the spring is at its free length. We must decide whether extension is represented by positive or negative values of x. This choice depends on the particular application. If $x > 0$ corresponds to extension, then a positive value of f represents the force of the spring pulling against whatever is causing the extension. Conversely, because of Newton's law of action-reaction, the force causing the extension has the same magnitude as f but is in the opposite direction.

Tensile Test of a Rod

A plot of the data for a tension test on a rod is given in Figure 1.3-2. The elongation is the change in the rod's length due to the tension force applied by the testing machine. As the tension force was increased, the elongation followed the curve labeled "Increasing." The behavior of the elongation under decreasing tension is shown by the curve labeled "Decreasing." The rod was stretched beyond its *elastic limit*, so that a permanent elongation remained after the tension force was removed.

For the smaller elongations the "Increasing" curve is close to a straight line with a slope of 3500 pounds per one-thousandth of an inch, or 3.5×10^6 lb/in. This line is labeled "Linear" in the plot. If we let x represent the elongation in inches and f the tension force in pounds, then the model $f = 3.5 \times 10^6 x$ represents the elastic behavior of the rod. We thus

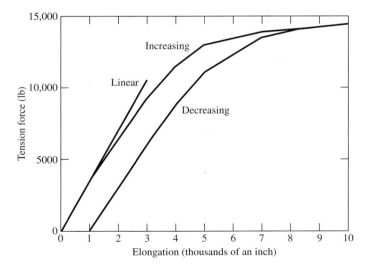

FIGURE 1.3-2 Plot of tension-test data.

see that the rod's spring constant is 3.5×10^6 lb/in. Converting this to SI units, we find that the spring constant is $k = 3.5 \times 10^6 (4.4482)(12)/0.3048 = 6.126 \times 10^8$ N/m.

This experiment could have been repeated using compressive instead of tensile force. For small compressive deformations x, we would find that the deformations are related to the compressive force f by $f = kx$, where k would have the same value as before. This example indicates that mechanical elements can be described by the linear law $f = kx$, for both compression and extension, as long as the deformations are not too large, that is, deformations not beyond the elastic limit. Note that the larger the deformation, the greater the error that results from using the linear model.

Formulas for Spring Constants

Table 1.3-1 lists the expressions for the spring constants of several common elements. Some of the expressions in the table contain the constants E and G, which are the modulus of elasticity and the shear modulus of elasticity of the material. The formula for the spring constant of a coil spring is derived in references on machine design. Other mechanical elements that have elasticity, such as beams, rods, and rubber mounts, are usually represented pictorially as a coil spring. Most of the expressions in the table can

TABLE 1.3-1   Spring Constants for Common Elements

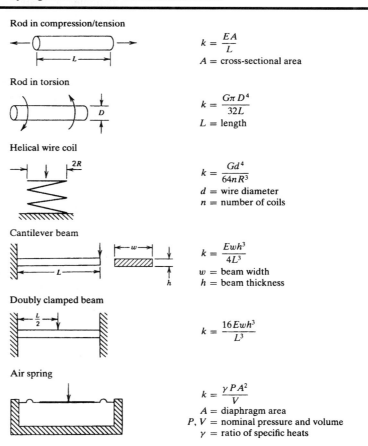

Rod in compression/tension

$$k = \frac{EA}{L}$$

A = cross-sectional area

Rod in torsion

$$k = \frac{G\pi D^4}{32L}$$

L = length

Helical wire coil

$$k = \frac{Gd^4}{64nR^3}$$

d = wire diameter
n = number of coils

Cantilever beam

$$k = \frac{Ewh^3}{4L^3}$$

w = beam width
h = beam thickness

Doubly clamped beam

$$k = \frac{16Ewh^3}{L^3}$$

Air spring

$$k = \frac{\gamma P A^2}{V}$$

A = diaphragm area
P, V = nominal pressure and volume
γ = ratio of specific heats

be derived from the force-deflection formulas given in texts dealing with the mechanics of materials. An exception is the *air spring*, which is used in pneumatic suspension systems. Its spring constant must be derived from the thermodynamic relations for a gas.

Analytical Determination of the Spring Constant

In much engineering design work we do not have the elements available for testing, and thus we must be able to calculate their spring constant from the geometry and material properties. To do this we can use results from the study of the strength of materials. The following examples show how this is accomplished.

EXAMPLE 1.3-1 *Rod with Axial* *Loading*	Derive the spring constant expression for a cylindrical rod subjected to an axial force (either tensile or compressive). The rod length is L and its area is A.
Solution	From strength-of-materials references—for example [Roark, 2001]—we obtain the force-deflection relation of a cylindrical rod:

$$x = \frac{L}{EA}f = \frac{4L}{\pi E D^2}f$$

where x is the axial deformation of the rod, f is the applied axial force, A is the cross-sectional area, and D is the diameter. Rewrite this equation as

$$f = \frac{EA}{L}x = \frac{\pi E D^2}{4L}x$$

Thus we see that the spring constant is $k = EA/L = \pi E D^2/4L$.

The modulus of elasticity of steel is approximately $E = 2 \times 10^{11}$ N/m². Thus a steel rod 0.5 m long and 4.42 cm in diameter would have a spring constant of 6.129×10^8 N/m, the same as the rod whose curve is plotted in Figure 1.3-2. ∎

Beams used to support objects can act like springs when subjected to large forces. Beams can have a variety of shapes and can be supported in a number of ways. The beam geometry, beam material, and the method of support determine its spring constant.

EXAMPLE 1.3-2 *Spring Constant* *for a Cantilever* *Beam*	Derive the spring constant expression for a cantilever beam of length L, thickness h, and width w, assuming that the force f and deflection x are at the end of the beam (see Table 1.3-1).
Solution	The force-deflection relation of a cantilever beam is

$$x = \frac{L^3}{3EI_A}f$$

where x is the deflection at the end of the beam, P is the applied force at the end, and I_A is the *area* moment of inertia about the beam's longitudinal axis [Roark, 2001]. The area moment I_A is computed with an integral similar to the mass moment equation, except that an area element dA is used instead of a mass element dm:

$$I_A = \int r^2 \, dA$$

Formulas for the area moments are available in most engineering mechanics texts. For a beam having a rectangular cross section with a width w and thickness h, the area moment is

$$I_A = \frac{wh^3}{12}$$

Thus the force-deflection relation reduces to

$$x = \frac{12L^3}{3Ewh^3}f = \frac{4L^3}{Ewh^3}f$$

The spring constant is the ratio of the applied force f to the resulting deflection x, or

$$k = \frac{f}{x} = \frac{Ewh^3}{4L^3} \tag{1}$$

∎

Note that two beams of identical shape and material, one a cantilever and one fixed-end, have spring constants that differ by a factor of 64. The fixed-end beam is thus 64 times "stiffer" than the cantilever beam. This illustrates the effect of the support arrangement on the spring constant.

A leaf spring in a vehicle suspension is shown in Figure 1.3-3. Leaf springs are constructed by strapping together several beams. The value of the total spring constant depends not only on the spring constants of the individual beams but also on the how they are strapped together, the method of attachment to the axle and chassis, and whether any material to reduce friction has been placed between the layers. Some formulas are available in the automotive literature. See, for example, [Bosch, 1986].

Torsional Spring Elements

Table 1.3-1 shows a rod in torsion. This is an example of a *torsional* spring, which resists with an opposing torque when twisted. For a torsional spring element we will use the "curly" symbol shown in Figure 1.3-4. The spring relation for a torsional spring is usually written as

$$T = k\theta \tag{1.3-2}$$

where θ is the net angular twist in the element, T is the opposing torque, and k is the *torsional spring constant*. We assign $\theta = 0$ at the spring position where there is no torque

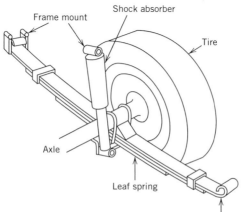

Frame mount   **FIGURE 1.3-3**   A leaf-spring suspension.

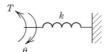

FIGURE 1.3-4   Symbol and nomenclature for a torsional spring element.

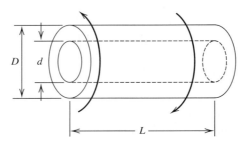

FIGURE 1.3-5   A hollow cylindrical shaft.

in the spring. This is analogous to the free length position of a translational spring. Although the same symbol k is used for the spring constant, note that the units of the torsional and translational spring constants are not the same. FPS units for k are lb-ft/rad; SI units are N·m/rad.

For a hollow cylindrical shaft, Figure 1.3-5, the formula for the torsional spring constant is

$$k = \frac{\pi G(D_o^4 - D_i^4)}{32L} \tag{1.3-3}$$

Note that a solid rod or a hollow shaft can be designed for axial or torsional loading. Thus there are two spring constants for a solid rod or a hollow shaft, a translational constant and a torsional constant. Figure 1.3-6 shows an example of a torsion-bar

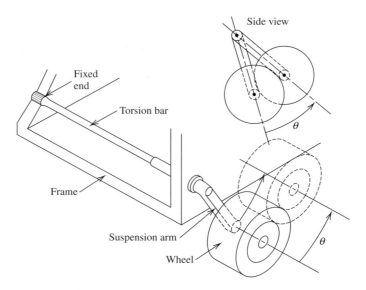

FIGURE 1.3-6   A torsion-bar suspension.

suspension, which was invented by Dr. Fernand Porsche in the 1930s. As the ground motion pushes the wheel up, the torsion bar twists and resists the motion.

A coil spring can also be designed for axial or torsional loading. Thus there are two spring constants for coil springs: a translational constant, which is given in Table 1.3-1, and a torsional constant, which is given by the following formula for a spring made from round wire ([Den Hartog, 1985]).

$$k = \frac{Ed^4}{64nD} \tag{1.3-4}$$

Parallel and Series Spring Elements

In many applications, multiple spring elements are used, and in such cases we must obtain the single equivalent spring constant of the combined elements. When two springs are connected side-by-side, as in Figure 1.3-7a, we can determine their equivalent spring constant as follows. Assuming that the force f is applied so that both springs have the same deflection x but different forces f_1 and f_2, then

$$x = \frac{f_1}{k_1} = \frac{f_2}{k_2}$$

If the system is in static equilibrium, then

$$f = f_1 + f_2 = k_1 x + k_2 x = (k_1 + k_2)x$$

For the equivalent system with a single spring, $f = kx$; thus we see that its equivalent spring constant is given by $k = k_1 + k_2$. This formula can be extended as follows to the case of n springs experiencing the same deflection:

$$k = \sum_{i=1}^{n} k_i \tag{1.3-5}$$

When two springs are connected end-to-end, as in Figure 1.3-7b, so that they experience the same force, we can determine their equivalent spring constant as follows. Assuming both springs are in static equilibrium, then both springs are subjected to the same force f, but their deflections f/k_1 and f/k_2 will not be the same unless their spring constants are equal. The total deflection x of the system is obtained from

$$x = \frac{f}{k_1} + \frac{f}{k_2} = \left(\frac{1}{k_1} + \frac{1}{k_2} \right) f$$

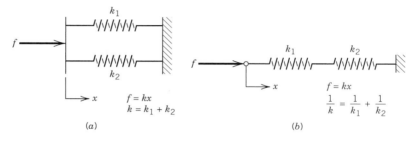

FIGURE 1.3-7 Parallel and series spring elements.

For the equivalent system with a single spring, $f = kx$; thus we see that its equivalent spring constant is given by

$$\frac{1}{k} = \frac{1}{k_1} + \frac{1}{k_2}$$

This formula can be extended as follows to the case of n springs experiencing the same force:

$$\frac{1}{k} = \sum_{i=1}^{n} \frac{1}{k_i} \tag{1.3-6}$$

The derivations of Equations 1.3-5 and 1.3-6 assumed that the product of the spring mass times its acceleration is zero, which means that either the system is in static equilibrium or the spring masses are very small compared to the other masses in the system.

The symbols for springs connected end-to-end look like the symbols for electrical resistors similarly connected. Such resistors are said to be in *series*, and therefore springs connected end-to-end are sometimes said to be in series. However, the equivalent electrical resistance is the sum of the individual resistances, whereas series springs obey the reciprocal rule of Equation 1.3-6. This similarity in appearance of the symbols often leads people to mistakenly add the spring constants of springs connected end-to-end, just as series resistances are added. Springs connected side-by-side are sometimes called *parallel* springs, and their spring constants should be added.

According to this usage, then, parallel springs have the same deflection; series springs experience the same force or torque.

EXAMPLE 1.3-3
Torsional Spring Constant of a Stepped Shaft

Determine the expression for the equivalent torsional spring constant for the stepped shaft shown in Figure 1.3-8a.

Solution

Assuming that the shafts are rigidly connected, we see that $\phi_2 = \phi_3$ and that each shaft sustains the same torque T but has a different twist angle. The net twist for the left-hand shaft is $\theta_1 = \phi_1 - \phi_2$, and the twist for the other shaft is $\theta_2 = \phi_3 - \phi_4 = \phi_2 - \phi_4$. Therefore, $T = k_1\theta_1 = k_2\theta_2$, and the equivalent spring constant is given by

$$\frac{1}{k_e} = \frac{1}{k_1} + \frac{1}{k_2}$$

where k_1 and k_2 are given in Table 1.3-1 as

$$k_i = \frac{G\pi D_i^4}{32L_i} \qquad i = 1,2$$

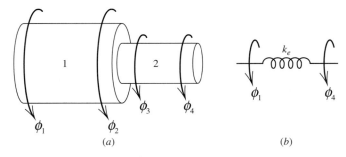

FIGURE 1.3-8   Torsional spring constant of a stepped shaft.

Thus

$$k_e = \frac{k_1 k_2}{k_1 + k_2}$$

and the original system is equivalent to that shown in Figure 1.3-8*b*, for which $T = k_e(\phi_1 - \phi_4)$. ∎

Linearization

In order to obtain the equations of motion of vibratory systems, we will need a mathematical description of the forces and moments involved, as functions of displacement or velocity, for example. As we will see in later chapters, solution of vibration models to predict system behavior requires solution of differential equations. We will see that differential equations based on linear models of the forces and moments are much easier to solve than ones based on nonlinear models. A *linear* model of a force *f* as a function of displacement *x* has the form $f = mx + b$. A *nonlinear* model involves *x* raised to some power or as the argument of a transcendental function. Examples of nonlinear functions are $f = 5x^2$, $f = 4\sqrt{x}$, $f = 3e^{6x}$, and $f = 5 \sin x$.

We therefore try to obtain a linear model whenever possible. Sometimes the use of a linear model results in a loss of accuracy, and the engineer must weigh this disadvantage with advantages gained by using a linear model. If the model is nonlinear, we can obtain a linear model that is an accurate approximation over a limited range of the independent variable. The next example illustrates this technique, which is called *linearization*.

EXAMPLE 1.3-4
Linearization of the Sine and Cosine

Many of our applications will involve the sine and cosine. Consider the pendulum shown in Figure 1.3-9, which contains elements of many vibration applications. Determine its angular velocity $\dot{\theta}$ as a function of θ, assuming that θ is small.

Solution

From the geometry of the circle, we see that the arc length swept out by the mass *m* as it rotates through the angle θ is $L_1\theta$, and thus its velocity is $L_1\dot{\theta}$. This relation involves no approximation. So the kinetic energy of the mass is $KE = m(L_1\dot{\theta})^2/2$.

Recall the Taylor series expansion of the sine and cosine about $\theta = 0$.

$$\sin \theta = \theta - \frac{\theta^3}{3!} + \frac{\theta^5}{5!} - \frac{\theta^7}{7!} + \cdots$$

$$\cos \theta = 1 - \frac{\theta^2}{2!} + \frac{\theta^4}{4!} - \frac{\theta^6}{6!} + \cdots$$

If θ is "small," then we can neglect the higher-order terms. So the common approximation for the sine is $\sin \theta = \theta$, which is linear. If the spring in Figure 1.3-9 is constrained to move only horizontally, then the displacement of its left-hand end is $L_2 \sin \theta$, which is approximately $L_2\theta$ for small angles. So the potential energy V_s due to the compression of the spring is approximately $V_s = k(L_2\theta)^2/2$.

The cosine function requires more care. If we measure the gravitational potential energy V_g of the mass *m* from its lowest point (where $\theta = 0$), its expression is

$$V_g = -mgh = -mgL_1(1 - \cos \theta)$$

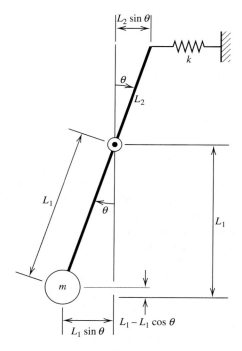

FIGURE 1.3-9   A pendulum and spring.

It is often tempting for beginners to follow the sine example and drop all the higher-order terms to obtain $\cos\theta = 1$ and thus $V_g = 0$, which is not a function of θ at all. If, however, we keep the next term in the series, we obtain $\cos\theta = 1 - \theta^2/2$ and $V_g = -mgL_1\theta^2/2$, which is nonlinear. This does not pose any difficulties because the potential energy expression does not appear in the differential equation of motion. It appears only in the integrated form, which is stated as conservation of mechanical energy. This principle states that the sum of the kinetic energy K and the potential energy V is a constant, which is zero here since we take V_g to be zero at $\theta = 0$. Thus

$$K + V = \frac{1}{2}m\left(L_1\dot{\theta}\right)^2 + \frac{1}{2}k(L_2\theta)^2 - \frac{1}{2}mgL_1\theta^2 = 0$$

This gives

$$\dot{\theta} = \sqrt{\frac{kL_2^2 - mgL_1}{mL_1^2}}\,\theta$$

which is a solution of the differential equation of motion. ∎

EXAMPLE 1.3-5
Spring Constant
of a Lever System

Figure 1.3-10 shows a horizontal force f acting on a lever that is attached to two springs. Assume that the resulting motion is small enough to be only horizontal, and determine the expression for the equivalent spring constant that relates the applied force f to the resulting displacement x.

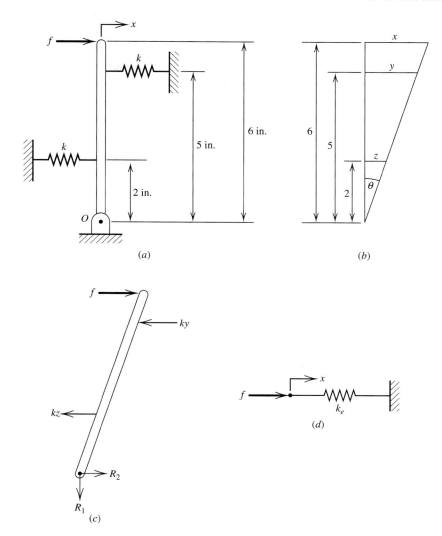

FIGURE 1.3-10   Equivalent spring constant of a lever system.

Solution

From the triangle shown in Figure 1.3-10*b*, for small angles θ, the upper spring deflection is $y = 5x/6$ and the deflection of the lower spring is $z = 2x/6 = x/3$. Thus the free-body diagram is as shown in Figure 1.3-10*c*. For static equilibrium, the net moment about point O must be zero. This gives

$$6f - 5(ky) - 2(kz) = 0$$

or

$$6f - 5k\frac{5x}{6} - 2k\frac{x}{3} = 0$$

Therefore

$$f = \left(\frac{29}{36}k\right)x$$

and the equivalent spring constant is $k_e = 29k/36$. Thus the original system is equivalent to the system shown in Figure 1.3-10d. That is, the force f will cause the same displacement x in both systems.

Note that although these two springs appear to be connected side-by-side, they are not in parallel, because they do not have the same deflection. Thus their equivalent spring constant is not given by the sum, $2k$. ■

Analytical Approach to Linearization

In the previous example we used a graphical approach to develop the linear approximation. The linear approximation may also be developed with an analytical approach based on the Taylor series. Taylor's theorem states that a function $f(\theta)$ can be represented in the vicinity of $\theta = \theta_r$ by the expansion

$$f(\theta) = f(\theta_r) + \left(\frac{df}{d\theta}\right)_{\theta=\theta_r}(\theta - \theta_r) + \frac{1}{2}\left(\frac{d^2f}{d\theta^2}\right)_{\theta=\theta_r}(\theta - \theta_r)^2 + \cdots$$

$$+ \frac{1}{k!}\left(\frac{d^kf}{d\theta^k}\right)_{\theta=\theta_r}(\theta - \theta_r)^k + \cdots + R_n \tag{1.3-7}$$

The term R_n is the remainder and is given by

$$R_n = \frac{1}{n!}\left(\frac{d^nf}{d\theta^n}\right)_{\theta=b}(\theta - \theta_r)^n \tag{1.3-8}$$

where b lies between θ_r and θ.

These results hold if $f(\theta)$ has continuous derivatives through order n. If R_n approaches zero for large n, the expansion is called the Taylor series. If $\theta_r = 0$ the series is sometimes called the Maclaurin series.

Consider the nonlinear function $f(\theta)$, which is sketched in Figure 1.3-11. Let $[f(\theta_r), \theta_r]$ denote the reference operating condition of the system. A model that is approximately linear near this reference point can be obtained by expanding $f(\theta)$ in a Taylor series near this point and truncating the series beyond the first-order term. If θ is

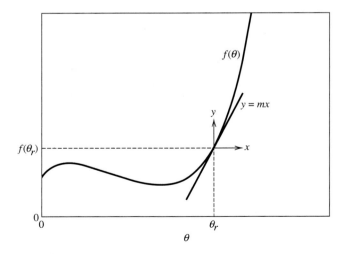

FIGURE 1.3-11 Linearization of a nonlinear function.

"close enough" to θ_r, the terms involving $(\theta - \theta_r)^i$ for $i \geq 2$ are small compared to the first two terms in the series. The result is

$$f(\theta) = f(\theta_r) + \left(\frac{df}{d\theta}\right)_r (\theta - \theta_r) \tag{1.3-9}$$

where the subscript r on the derivative means that it is evaluated at the reference point. This is a linear relation. To put it into a simpler form, let m denote the slope at the reference point:

$$m = \left(\frac{df}{d\theta}\right)_r \tag{1.3-10}$$

Let y denote the difference between $f(\theta)$ and the reference value $f(\theta_r)$:

$$y = f(\theta) - f(\theta_r) \tag{1.3-11}$$

Let x denote the difference between θ and the reference value θ_r:

$$x = \theta - \theta_r \tag{1.3-12}$$

Then Equation 1.3-9 becomes

$$y \approx mx \tag{1.3-13}$$

The geometric interpretation of this result is shown in Figure 1.3-11. We have replaced the original function $f(\theta)$ with a straight line passing through the point $f(\theta_r), \theta_r$ and having a slope equal to the slope of $f(\theta)$ at the reference point. Using the (y, x) coordinates gives a zero intercept and simplifies the relation.

Nonlinear Spring Relations

The linear spring relation $f = kx$ becomes less accurate with increasing deflection (for either compression or extension). In such cases it is often replaced by

$$f = k_1 x + k_2 x^3 \tag{1.3-14}$$

where $k_1 > 0$ and x is the spring displacement from its free length. The spring element is said to be *hardening* or *hard* if $k_2 > 0$ and *softening* or *soft* if $k_2 < 0$. These cases are shown in Figure 1.3-12. A hard spring is used to limit excessively large deflections.

The spring stiffness k is the slope of the force-deflection curve and is constant for the linear spring element. A nonlinear spring element does not have a single stiffness

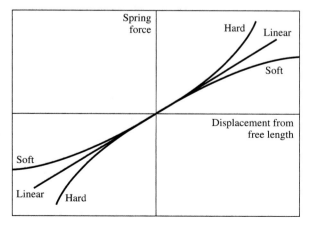

FIGURE 1.3-12 Hard, soft, and linear spring functions.

value because its slope is variable. For a hard spring, its slope and thus its stiffness increase with deflection. The stiffness of a soft spring decreases with deflection.

We therefore try to obtain a linear spring model whenever possible because the resulting equation of motion is much easier to solve. The following example illustrates how a linear model can be obtained from a nonlinear description.

EXAMPLE 1.3-6
Linearizing A Nonlinear Spring Model

Suppose a particular nonlinear spring is described by

$$f(y) = k_1 y + k_2 y^3$$

where y is the stretch in the spring from its free length, $k_1 = 120$ lb/ft, and $k_2 = 360$ lb/ft^3. Determine its equivalent linear spring constant for the two equilibrium positions corresponding to the following applied forces:
(a) $f_r = 10$ lb
(b) $f_r = 35$ lb

Solution

The first step is to determine the spring stretch caused by each force. At equilibrium, the spring force must equal the applied force. Thus

$$120 y_r + 360 y_r^3 = f_r$$

where y_r is the spring stretch from free length and $f_r = f(y_r)$ is the force causing that stretch. For $f_r = 10$,

$$120 y_r + 360 y_r^3 = 10$$

The only real root of this cubic equation is $y_r = 0.082$ ft. For $f_r = 35$,

$$120 y_r + 360 y_r^3 = 35$$

The only real root of this cubic equation is $y_r = 0.247$ ft.

The truncated Taylor series for this function is

$$f(y) = f(y_r) + \left(\frac{df}{dy}\right)_r (y - y_r) \tag{1}$$

Note that

$$\left(\frac{df}{dy}\right)_r = k_1 + 3k_2 y_r^2 = 120 + 1080 y_r^2$$

Thus Equation 1 gives

$$f(y) = f_r + (120 + 1080 y_r^2)(y - y_r) \tag{2}$$

This equation may be expressed as

$$\Delta f = (120 + 1080 y_r^2)\Delta y = k_e \Delta y$$

where $\Delta f = f(y) - f_r$, $\Delta y = y - y_r$, and k_e is the equivalent linearized spring constant. This relation says that for small displacements from the reference length y_r, the change in spring force is proportional to the change in spring length.

Equation 2 is the linearized approximation to the nonlinear spring force function for the given values of k_1 and k_2. It approximately describes the nonlinear function near the specified equilibrium point. For the equilibrium at $y_r = 0.082$,

$$f(y) = 10 + 127.3(y - 0.082)$$

For the equilibrium at $y_r = 0.247$,

$$f(y) = 35 + 185.9(y - 0.247)$$

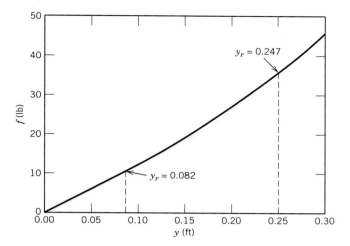

FIGURE 1.3-13 Plot of the spring force relation $f = 120y + 360y^3$. The short straight-line segments are the linear approximations to the curve near the reference points $y_r = 0.082$ and $y_r = 0.247$ ft.

These equations describe the straight lines shown for each equilibrium in Figure 1.3-13 by the short straight-line segments passing through the two equilibrium points.

The equivalent linear spring constant k_e is the slope of the curve $f(y)$ at the specific equilibrium point. This slope is given by $k_e = 120 + 1080y_r^2$. Thus the linearized spring constants for each equilibrium are:

(a) For $y_r = 0.082$, $k_e = 120 + 1080(0.082)^2 = 127.3$ lb/ft
(b) For $y_r = 0.247$, $k_e = 120 + 1080(0.247)^2 = 185.9$ lb/ft

In Chapter 3 you will learn how to compute the frequency of oscillation of a mass attached to such a spring. ∎

Sometimes a nonlinear model is unavoidable. This is the case when a system is designed to utilize two or more spring elements to achieve a spring constant that varies with the applied load. Even if each spring element is linear, the combined system will be nonlinear.

An example of such as system is shown in Figure 1.3-14a. This is a representation of systems used for packaging and in vehicle suspensions, for example. The two side springs provide additional stiffness when the weight W is too heavy for the center spring.

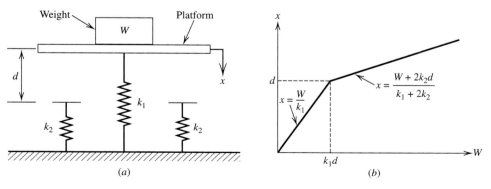

FIGURE 1.3-14 (a) A nonlinear spring arrangement. (b) The plot of platform displacement versus applied weight.

EXAMPLE 1.3-7
Deflection of a Nonlinear System

Obtain the deflection of the system model shown in Figure 1.3-14a as a function of the weight W. Assume that each spring exerts a force that is proportional to its compression.

Solution

When the weight W is gently placed, it moves through a distance x before coming to rest. From statics, we know that the weight force must balance the spring forces at this new position. Thus

$$W = k_1 x \qquad \text{if } x < d$$

and

$$W = k_1 x + 2k_2(x - d) \qquad \text{if } x \geq d$$

We can solve these relations for x as follows:

$$x = \frac{W}{k_1} \qquad \text{if } x < d$$

$$x = \frac{W + 2k_2 d}{k_1 + 2k_2} \qquad \text{if } x \geq d$$

These relations can be used to generate the plot of x versus W, shown in Figure 1.3-14b. ∎

1.4 DAMPING ELEMENTS

The general term for fluid resistance force is *damping*. It is produced when a fluid layer moves relative to an object. Such is the case, for example, when the block in Figure 1.4-1 slides on a lubricated surface. The fluid's *viscosity* produces a shear stress that exerts a resisting force on the block and on the surface. Viscosity is an indication of the "stickiness" of the fluid; molasses and oil have greater viscosities than water, for example. This resisting force is called the damping force, and its magnitude depends on the relative velocity between the fluid and the surface. This dependence can be quite complicated, but we usually model it as a linear function of the relative velocity. This approach allows us to obtain equations of motion that are easier to solve, without ignoring the effect of the velocity-dependent damping force.

Engineering systems can contain damping as an unwanted effect, such as with bearings and other surfaces lubricated to prevent wear. On the other hand, damping elements can be deliberately included as part of the design. Such is the case with shock absorbers, fluid couplings, and torque converters.

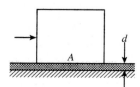

FIGURE 1.4-1 A mass sliding on a lubricated surface.

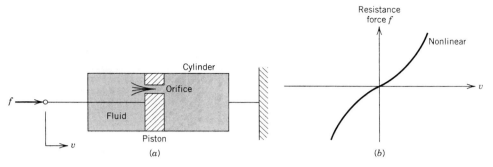

FIGURE 1.4-2 A fluid damper.

A Damper

A spring element exerts a reaction force in response to a *displacement*, either compression or extension, of the element. On the other hand, a *damping* element is an element that resists relative *velocity* across it. A simple way to achieve damping is with a *dashpot* or *damper*, which is the basis of a shock absorber. It consists of a piston moving inside a cylinder that is sealed and filled with a viscous fluid (Figure 1.4-2a). The piston has a hole or orifice through which the fluid can flow when the piston moves relative to the cylinder, but the fluid's viscosity resists this motion (the "stickier" the fluid, the greater the viscosity and the resistance).

   If we hold the piston rod in one hand and the cylinder in the other hand, and move the piston and the cylinder at the same velocity, we will feel no reaction force. However, if we move the piston and the cylinder at different velocities, we will feel a resisting force that is caused by the fluid moving through the holes from one side of the piston to the other. From this example we can see that the resisting force in a damper is caused by fluid friction (in this case, friction between the fluid and the walls of the piston holes) and that the force depends on the relative velocity of the piston and the cylinder. The faster we move the piston relative to the cylinder, the greater the resisting force.

   Let v be this relative velocity. If the cylinder is fixed, then v is the velocity of the piston stem. The expression for the force f on the piston required to keep it moving at velocity v is often taken to be the linear model:

$$f = cv \qquad (1.4\text{-}1)$$

where c is the *damping coefficient*. However, in general the damping force is a nonlinear function of v (Figure 1.4-2b).

Damper Symbols

The symbol shown in Figure 1.4-3a is the general symbol for a damping element, even when the damping is produced by something other than a piston and cylinder. The units of c are [force/speed]; for example, N·s/m or lb-sec/ft.

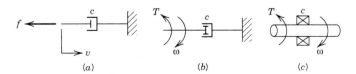

FIGURE 1.4-3 Symbols for dampers. (*a*) Translational damper. (*b*) Rotational damper. (*c*) Rotational damping in bearings.

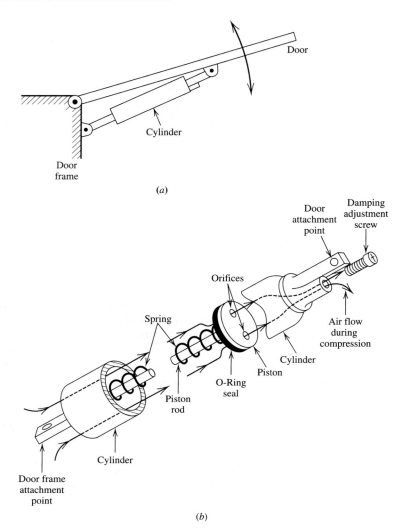

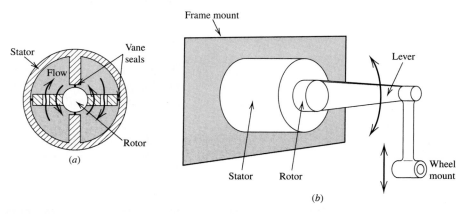

FIGURE 1.4-4 A pneumatic door closer.

FIGURE 1.4-5 A rotary damper.

The simplest type of axle bearing is a *journal* bearing, in which the axle is supported by passing it through an opening in a support and lubricating the contact area between the support and axle. This element is an example of a *torsional damper* or *rotational damper*, in which the resistance is a torque rather than a force.

Torsional dampers are represented by a slightly different symbol, shown in Figure 1.4-3b. When rotational resistance is due to viscous friction in bearings, the bearing symbol shown in Figure 1.4-3c is often used with the symbol c to represent the damping. Although the symbol c represents damping in both translational and rotational systems, its units are different for each type. For torsional damping, the units of c are [torque/angular velocity]; for example, N·m·s/rad or lb-ft-sec/rad.

A Door Closer

An example from everyday life of a device that contains a damping element as well as a spring element is the door closer (Figure 1.4-4). In some models the working fluid is air, while others use a hydraulic fluid. The cylinder is attached to the door, and the piston rod is fixed to the door frame. As the door is closed, the air is forced both through the piston holes and out past the adjustment screw, which can be used to adjust the amount of damping resistance (a smaller passageway provides more resistance to the flow and thus more damping force). The purpose of the spring is to close the door; if there were no spring, the door would remain stationary because the damper does not exert any force unless its endpoints are moving relative to each other. The purpose of the damper is to exert a force that prevents the door from being opened or closed too quickly (such as may happen due to a gust of wind). If you have such a door closer, try closing the door at different speeds and notice the change in resisting force.

A common example of a rotary damper is the vane-type damper shown in Figure 1.4-5a. The rotating part (the rotor) has vanes with holes through which the fluid can flow. The stator is the stationary housing. This device is the basis of some door closers. It is also used to provide damping of wheel motion in some vehicle suspensions (Figure 1.4-5b).

Shock Absorbers

The telescopic shock absorber is used in many vehicles. A cutaway view of a typical shock absorber is very complex, but the basic principle of its operation is the damper concept illustrated in Figure 1.4-6. The damping resistance can be designed to be dependent on the sign of the relative velocity. For example, Figure 1.4-7 shows a piston containing spring-loaded valves that partially block the piston passageways. If the two spring constants are different or if the two valves have different shapes, then the flow resistance will be dependent on the direction of motion. This design results in a force-versus-velocity curve like that shown in Figure 1.4-8. During *jounce* (compression), when the wheel hits a road bump, the resisting force is different than during *rebound* (extension), when the wheel is forced back to its neutral position by the suspension spring. The resisting force during jounce should be small in order to prevent a large force from being transmitted to the passenger compartment, whereas during rebound the resisting force should be large to prevent wheel oscillation. An aircraft application of a shock absorber is the *oleo strut*, shown in Figure 1.4-9.

By combining the characteristics of several valves, the damping can be designed to be *progressive*, which means that the slope of the damping force-versus-velocity curve increases with velocity, or *degressive*, where the slope decreases with velocity (Figure 1.4-10).

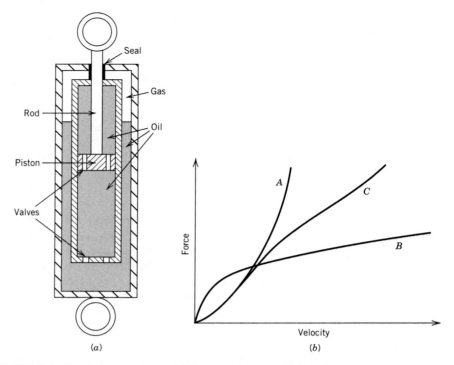

FIGURE 1.4-6 Shock absorber design. (*a*) Cross-section view. (*b*) Damping curves.

Other examples of damping include aerodynamic drag and hydrodynamic drag. Engineering systems can exhibit damping in bearings and other surfaces lubricated to prevent wear. Damping can also be caused by nonfluid effects, such as the energy loss that occurs due to internal friction in solid but flexible materials.

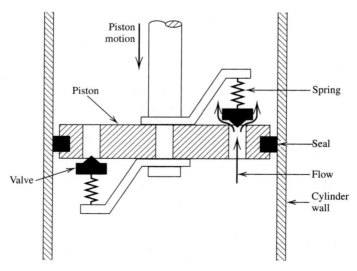

FIGURE 1.4-7 Shock absorber with spring-loaded valves.

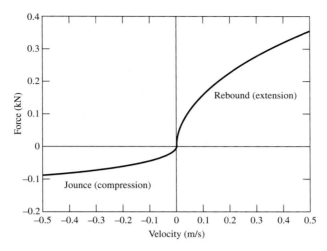

FIGURE 1.4-8   Shock absorber damping curves for rebound and jounce.

Viscous Damping Elements

For a so-called *Newtonian fluid*, the shear stress τ is proportional to the velocity gradient, as

$$\tau = \mu \frac{du}{dy} \tag{1.4-2}$$

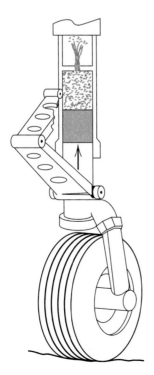

FIGURE 1.4-9   An oleo strut.

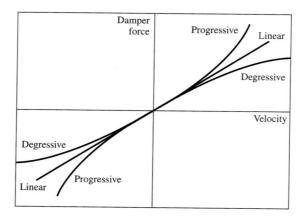

FIGURE 1.4-10 Progressive and degressive damping curves.

where μ is the viscosity and $u(y)$ is the fluid velocity as a function of distance y from some reference point. This relation may be used to develop expressions for the damping constant c in some applications.

EXAMPLE 1.4-1
Viscous Damping between Two Flat Surfaces

Derive the expression for the viscous damping coefficient c for two flat surfaces separated by a distance d and moving with a relative velocity v.

Solution

We may think of this situation as a flat plate moving with a velocity v relative to a fixed surface, as shown in Figure 1.4-11. Let A be the plate area that is in contact with the fluid. The fluid next to the plate moves with the velocity v, and the fluid next to the fixed surface has zero velocity. Assuming that the fluid velocity varies linearly with distance from the fixed surface, we see that

$$u = \frac{y}{d}v \tag{1}$$

and thus $du/dy = v/d$. The force f of the fluid acting on the plate is the shear stress times the area A. Thus, from Equation 1.4-2,

$$f = A\tau = A\mu\frac{du}{dy} = A\mu\frac{v}{d}$$

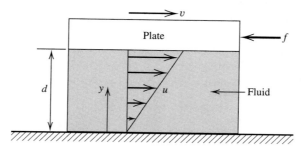

FIGURE 1.4-11 Velocity profile in a fluid layer.

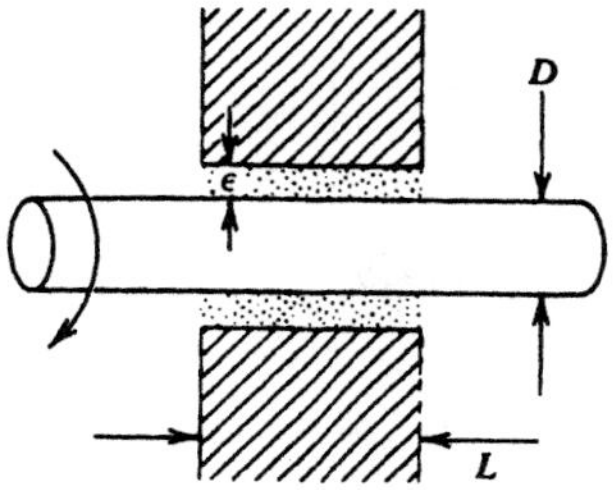

FIGURE 1.4-12 A journal bearing.

From the definition of the damping coefficient, we have

$$c = \frac{f}{v} = \frac{\mu A}{d} \tag{2}$$

As a practical matter, it is usually difficult to determine the distance d, but Equation 2 is useful nonetheless because it tells us that c is inversely proportional to d, a result that might not have been intuitively obvious. ∎

An application where the clearance distance is easier to determine is a *journal bearing*, in which a shaft is supported around its circumference by a lubricating fluid enclosed by a housing.

EXAMPLE 1.4-2
Damping Coefficient of a Journal Bearing

A journal bearing is illustrated in Figure 1.4-12. Derive the expression for its rotational damping coefficient.

Solution

A cross section is shown in Figure 1.4-13, where D is the shaft diameter and ϵ is the clearance between the shaft and the housing. The fluid next to the rotating shaft moves with the velocity $V = D\omega/2$, where ω is the shaft angular velocity. The fluid next to the stationary housing has zero velocity. Assuming that the fluid velocity varies linearly with radial distance r from the fixed surface, we see that

$$u = \frac{r}{\epsilon} V \tag{1}$$

and thus $du/dr = V/\epsilon$. The shear stress acting on the shaft is

$$\tau = \mu \frac{du}{dr} = \mu \frac{V}{\epsilon}$$

The torque T on the shaft due to the shear stress acting over the surface area A is

$$T = (A\tau)\frac{D}{2} = \left(A\mu \frac{V}{\epsilon}\right)\frac{D}{2}$$

The area A is the surface area of the shaft exposed to the lubricating fluid. Thus

$$A = 2\pi L \frac{D}{2} = \pi L D$$

and

$$T = (\pi L D)\left(\mu \frac{V}{\epsilon}\right)\frac{D}{2} = \frac{\pi \mu L D^3}{4\epsilon}\omega$$

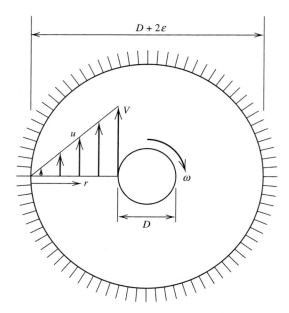

FIGURE 1.4-13   Velocity profile in a journal bearing.

From the definition of the rotational damping coefficient, $c = T/\omega$, we have

$$c = \frac{\pi \mu L D^3}{4\epsilon} \tag{2}$$

This is known as *Petrov's equation*, and it has been widely used to design journal bearings. In this application the designer can specify the clearance ϵ, so we may expect this formula for c to give more accurate results than the flat plate formula derived in the previous example. ∎

From fluid mechanics it is known that the mass flow rate q of fluid through an orifice is proportional to the square root of the pressure difference Δp across the orifice; that is,

$$q = \frac{1}{R} \sqrt{\Delta p} \tag{1.4-3}$$

where R is a proportionality constant that is usually determined from experimental data. For devices that produce damping with fluid flow through an orifice, this relation can be used to determine the damping force as a function of velocity.

EXAMPLE 1.4-3
An Orifice
Damper

A damper may be constructed using a piston within a cylinder filled with a fluid, as shown in Figure 1.4-2a. The piston has a hole (an orifice) that allows the fluid to flow from one side of the piston to the other, depending on the piston motion. Derive the expression for the force f as a function of the piston velocity v, assuming that the piston moves slowly or that the piston mass is small.

Solution

If the fluid is incompressible with mass density ρ, then the volume rate q/ρ of fluid passing through the orifice must equal the rate at which volume is swept out by the piston's motion. Thus

$$\frac{q}{\rho} = Av \tag{1}$$

where A is the area of the piston surface. Combining Equation 1 with Equation 1.4-3 we obtain

$$v = \frac{1}{\rho A R} \sqrt{\Delta p} \qquad (2)$$

If the piston moves slowly or if the piston mass is small, we have a situation that is essentially static, which means that the force f is equal and opposite to the force on the piston caused by the pressure difference. Thus

$$A \Delta p = f \qquad (3)$$

Solving Equation 3 for Δp and substituting in Equation 2, we obtain

$$v = \frac{1}{\rho A R} \sqrt{\frac{f}{A}}$$

which may be solved for f as follows:

$$f = A^3 \rho^2 R^2 v^2$$

We note immediately that the force is proportional to the *square* of the velocity. Thus the force-velocity relation for this damper is nonlinear and progressive. ∎

Series and Parallel Damper Elements

In many applications, multiple damper elements are used, and in such cases we must obtain the single equivalent damping constant of the combined elements. When two dampers are connected side-by-side, as in Figure 1.4-14a, we can determine their equivalent damping constant as follows. Assuming that the force f is applied so that both dampers have the same velocity difference v across them but different forces f_1 and f_2, then

$$v = \frac{f_1}{c_1} = \frac{f_2}{c_2}$$

If the system is in static equilibrium, then

$$f = f_1 + f_2 = c_1 v + c_2 v = (c_1 + c_2) v$$

For the equivalent system with one damper, $f = cv$; thus we see that its equivalent damping constant is given by $c = c_1 + c_2$. This formula can be extended as follows to the case of n dampers experiencing the same velocity difference:

$$c = \sum_{i=1}^{n} c_i \qquad (1.4\text{-}4)$$

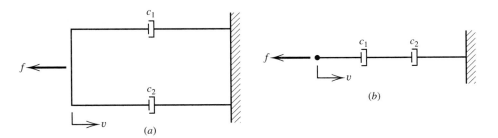

FIGURE 1.4-14 (a) Parallel dampers. (b) Series dampers.

When two dampers are connected end-to-end, as in Figure 1.4-14b, so that they experience the same force, we can determine their equivalent damping constant as follows. Assuming both dampers are in static equilibrium, then both are subjected to the same force f, but their velocity differences f/c_1 and f/c_2 will not be the same unless their damping constants are equal. The total velocity difference v across the system is obtained from

$$v = \frac{f}{c_1} + \frac{f}{c_2} = \left(\frac{1}{c_1} + \frac{1}{c_2}\right) f$$

For the equivalent system with one damper, $f = cv$; thus we see that its equivalent damping constant is given by

$$\frac{1}{c} = \frac{1}{c_1} + \frac{1}{c_2}$$

This formula can be extended as follows to the case of n dampers experiencing the same force:

$$\frac{1}{c} = \sum_{i=1}^{n} \frac{1}{c_i} \tag{1.4-5}$$

These derivations assumed that the product of the damper mass times its acceleration is zero, which means that either the system is in static equilibrium or the damper masses are very small compared to the other masses in the system.

According to standard usage, parallel dampers have the same velocity difference; series dampers experience the same force or torque.

1.5 PARAMETER ESTIMATION

Although there are many applications where we can derive an expression for the spring force as a function of deflection or the damping force as a function of velocity, in practice we must often determine these relations from measured data. The most common model forms for the spring relation are the linear model

$$f(x) = kx$$

and the cubic model

$$f(x) = k_1 x + k_2 x^3$$

in which the constant term and the x^2 term cannot appear if the force function is antisymmetric about the origin; that is, if $f(-x) = -f(x)$.

The most common model forms for the damping relation are:

1. The linear model:
$$f(v) = cv$$

2. The square-law model:
$$f(v) = cv^2$$

3. The general *progressive* model (of which the square law is a special case):
$$f(v) = cv^n \qquad n > 1$$

4. The *degressive* model:

$$f(v) = cv^m \qquad n < 1$$

The difference between the progressive and the degressive models is that the slope of the progressive curve increases with v, whereas the slope of the degressive curve decreases with v. The slope of the linear model is constant, and its line lies between the progressive and the degressive curves.

Identifying Functions from Data

When trying to find a function to describe a set of data, we try to find a set of axes on which the data will plot as a straight line. We look for a straight line on the plot because it is the one most easily recognized by eye, and therefore we can easily tell if the function will fit the data well. Each of the following functions gives a straight line when plotted using a specific set of axes:

1. The linear function $y = mx + b$ gives a straight line when plotted on rectilinear axes. Its slope is m and its intercept is b.

2. The power function $y = bx^m$ gives a straight line when plotted on log-log axes.

3. The exponential function $y = be^{mx}$ and its equivalent form $y = b(10)^{mx}$ give a straight line when plotted on a semilog plot whose y axis is logarithmic.

We may prove these properties as follows. Using the following properties of base-10 logarithms, which are shared with natural logarithms, we have:

$$\log(ab) = \log a + \log b$$

$$\log(a^m) = m \log a$$

Take the logarithm of both sides of the power equation $y = bx^m$ to obtain

$$\log y = \log(bx^m) = \log b + m \log x$$

This has the form $Y = B + mX$ if we let $Y = \log y$, $X = \log x$, and $B = \log b$. Thus, if the data can be described by the power function, they will form a straight line when plotted on log-log axes.

Taking the logarithm of both sides of the exponential form $y = be^{mx}$ gives

$$\log y = \log(be^{mx}) = \log b + mx \log e$$

This has the form $Y = B + Mx$ if we let $Y = \log y$, $B = \log b$, and $M = m \log e$. Thus, if the data can be described by the exponential function, they will form a straight line when plotted on semilog axes (with the log axis used for the abscissa).

Once we have identified the proper function form, we may obtain the function's coefficients by drawing a straight line that passes near most of the points. For example, for the exponential function $y = be^{mx}$, the slope and intercept of this straight line give the values of B and M, from which we may calculate $b = 10^B$ and $m = M/\log e$.

EXAMPLE 1.5-1
A Cantilever Beam Deflection Model

The deflection of a cantilever beam is the distance its end moves in response to a force applied at the end (Figure 1.5-1). The following table gives the measured deflection x that was produced in a particular beam by the given applied force f. Plot the data to see whether a linear relation exists between f and x, and determine the spring constant.

Force f (lb)	0	100	200	300	400	500	600	700	800
Deflection x (in.)	0	0.15	0.23	0.35	0.37	0.5	0.57	0.68	0.77

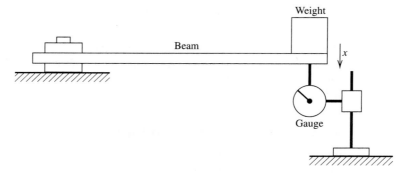

FIGURE 1.5-1 Measurement of beam deflection.

Solution

The plot is shown in Figure 1.5-2. Common sense tells us that there must be zero beam deflection if there is no applied force, so the curve describing the data must pass through the origin. The straight line shown was drawn by aligning a ruler so that it passes through the origin and near most of the data points (note that this line is subjective; another person might draw a different line). The data points lie close to a straight line, so we may use the linear function $f = kx$ to describe the relation. The value of the constant k can be determined from the slope of the line, which is

$$k = \frac{800 - 0}{0.78 - 0} = 1026 \text{ lb/in.}$$ ∎

Once we have discovered a functional relation that describes the data, we can use it to make predictions for conditions that lie *within* the range of the original data. This process is called *interpolation*. For example, we can use the beam model to estimate the deflection when the applied force is 550 lb. We can be fairly confident of this prediction because we have data below and above 550 lb and we have seen that our model describes this data very well.

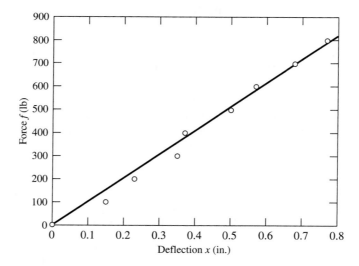

FIGURE 1.5-2 Plot for Example 1.5-1.

Extrapolation is the process of using the model to make predictions for conditions that lie *outside* the original data range. Extrapolation might be used in the beam application to predict how much force would be required to bend the beam, say, 1.2 in. We must be careful when using extrapolation, because we usually have no reason to believe that the mathematical model is valid beyond the range of the original data. For example, if we continue to bend the beam, eventually the force is no longer proportional to the deflection, and it becomes much greater than that predicted by the linear model. Extrapolation has a use in making tentative predictions, which must be backed up later on by testing.

Most of time, however, our data will not fall close enough to a straight line to enable us to identify a unique line. If we ask two people to draw a straight line passing as close as possible to all the data points, we will probably get two different answers. In such cases we need to use a more objective method for identifying the parameter values. The most commonly used method is the least-squares method, which can also be applied to functions other than the three listed previously. The remainder of this section illustrates how to use the least-squares method to fit a variety of functions. When the number of data points is large, the required calculations are tedious and best done with a computer. Section 1.6 shows how to use MATLAB for this purpose.

The Least-Squares Method

A systematic and objective way of obtaining a functional description of a set of data is the *least-squares* method. Suppose we want to find the coefficients of the straight line $y = mx + b$ that best fits the following data:

x	0	5	10
y	2	6	11

According to the least-squares criterion, the line that gives the best fit is the one that minimizes the sum of the squares of the vertical differences between the line and the data points (Figure 1.5-3). We denote this sum by J. These differences are called the

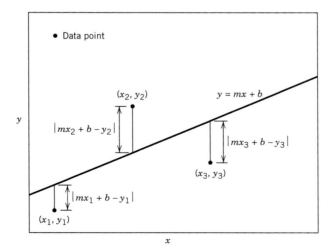

FIGURE 1.5-3   Least-squares example.

residuals. Here there are three data points, and J is given by

$$J = \sum_{i=1}^{3} (mx_i + b - y_i)^2$$

Substituting the data values (x_i, y_i) given in the table, we obtain

$$J = (0m + b - 2)^2 + (5m + b - 6)^2 + (10m + b - 11)^2$$

The values of m and b that minimize J can be found from $\partial J / \partial m = 0$ and $\partial J / \partial b = 0$. These conditions give the following equations that must be solved for the two unknowns m and b:

$$250m + 30b = 280$$

$$30m + 6b = 38$$

The solution is $m = 0.9$ and $b = 11/6$. The best straight line in the least-squares sense is $y = 0.9x + 11/6$. If we evaluate this equation at the data values $x = 0$, 5, and 10, we obtain the values $y = 1.833$, 6.333, and 10.8333. These values are different from the given data values $y = 2$, 6, and 11 because the line is not a perfect fit to the data. The value of J is $J = (1.833 - 2)^2 + (6.333 - 6)^2 + (10.8333 - 11)^2 = 0.1666$. No other straight line will give a lower value of J for this data.

The General Linear Case

We can generalize the above results to obtain formulas for the coefficients m and b in the linear equation $y = mx + b$. Note that for n data points,

$$J = \sum_{i=1}^{n} (mx_i + b - y_i)^2$$

The values of m and b that minimize J are found from $\partial J / \partial m = 0$ and $\partial J / \partial b = 0$. These conditions give the following equations that must be solved for m and b:

$$\partial J / \partial m = 2 \sum_{i=1}^{n} (mx_i + b - y_i)x_i = 2 \sum_{i=1}^{n} mx_i^2 + 2 \sum_{i=1}^{n} bx_i - 2 \sum_{i=1}^{n} y_i x_i = 0$$

$$\partial J / \partial b = 2 \sum_{i=1}^{n} (mx_i + b - y_i) = 2 \sum_{i=1}^{n} mx_i + 2 \sum_{i=1}^{n} b - 2 \sum_{i=1}^{n} y_i = 0$$

Noting that

$$\sum_{i=1}^{n} b = nb$$

we see that these equations become

$$m \sum_{i=1}^{n} x_i^2 + b \sum_{i=1}^{n} x_i = \sum_{i=1}^{n} y_i x_i \qquad (1.5\text{-}1)$$

$$m \sum_{i=1}^{n} x_i + bn = \sum_{i=1}^{n} y_i \qquad (1.5\text{-}2)$$

These are two linear equations in terms of m and b.

EXAMPLE 1.5-2
Fitting the Linear Spring Function

Consider the spring data given in Example 1.5-1. Use the least-squares method to estimate the spring constant k for the linear model $f = kx$.

Solution

Here our model is $f = mx + b$, so f plays the role of y in Equations 1.5-1 and 1.5-2, and $k = m$. The required terms for these equations are $n = 9$ and

$$\sum_{i=1}^{n} x_i^2 = 1.965 \qquad \sum_{i=1}^{n} x_i = 3.62$$

$$\sum_{i=1}^{n} y_i x_i = \sum_{i=1}^{n} f_i x_i = 1998$$

$$\sum_{i=1}^{n} y_i = \sum_{i=1}^{n} f_i = 3600$$

Thus Equations 1.5-1 and 1.5-2 become

$$1.965m + 3.62b = 1998$$
$$3.62m + 9b = 3600$$

Their solution is $m = 1081$ and $b = -0.0347$. Thus $k = m = 1081$ lb/in., and the fitted function is $f = 1081x - 0.0347$. The plot is shown in Figure 1.5-4. Note that it does not pass through the origin. In a later example we will show how to modify the least-squares method to obtain a function that passes through the origin. ∎

The degressive damping model is the power function $f = cv^m$, where $m < 1$. If we can perform an experiment to measure the force f as a function of velocity v, we can estimate the value of the damping coefficient c as follows.

EXAMPLE 1.5-3
Fitting the Power Function

Find a functional description of the following data, where f is force and v is velocity.

v (m/s)	0.5	1	1.5	2	2.5	3	3.5	4
f (N)	35	59	83	104	125	145	163	182

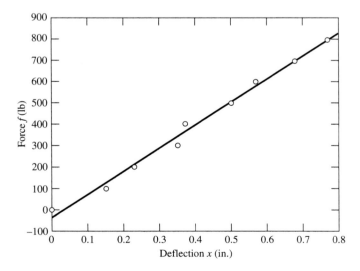

FIGURE 1.5-4 Plot for Example 1.5-2.

Solution

These data lie close to a straight line when plotted on log-log axes, but not when plotted on linear or semilog axes. Thus a power function $f = cv^m$ can describe the data. Because $\log f = \log c + m \log v$, we use the transformations $V = \log v$ and $F = \log f$, and obtain the new data table:

$V = \log v$	−0.301	0	0.1761	0.301	0.3979	0.4771	0.5441	0.6021
$F = \log f$	1.5441	1.7709	1.9191	2.017	2.0969	2.1614	2.2122	2.2601

From this table we obtain

$$\sum_{i=1}^{4} V_i^2 = 1.2567 \qquad \sum_{i=1}^{4} V_i = 2.1973$$

$$\sum_{i=1}^{4} V_i F_i = 4.9103 \qquad \sum_{i=1}^{4} F_i = 15.9816$$

Using V, F, and $B = \log c$ instead of x, y, and b in Equations 1.5-1 and 1.5-2, we obtain

$$1.2567m + 2.1973B = 4.9103$$

$$2.1973m + 8B = 15.9816$$

The solution is $m = 0.7972$ and $B = 1.7787$. This gives $c = 10^B = 60.0801$. Thus the desired function is $f = 60.0801v^{0.7972}$, and the the estimate of the damping constant is $c = 60.0801$.

■

As we will see in a later chapter, the velocity of an object of mass m acted on by a linear damping force cv is the exponential function $v(t) = v(0)e^{-ct/m}$. If we can perform an experiment to measure $v(t)$, we can estimate the value of the coefficient c/m as follows.

EXAMPLE 1.5-4
Fitting Data with
the Exponential
Function

Find a functional description of the following data, where t is time and v is velocity.

t (s)	0	1	2	3	4	5	6	7	8
v (m/s)	5.1	3.3	1.8	1.1	0.68	0.4	0.25	0.15	0.09

Solution

These data lie close to a straight line only when plotted on semilog axes (with v plotted on the log axis). Thus an exponential function can describe the data. Choosing the exponential form $v(t) = be^{at}$, we note that $\ln v = \ln b + at$ and use the transformation $V = \ln v$ to obtain the new data table:

t	0	1	2	3	4	5	6	7	8
$V = \ln v$	1.6292	1.1939	0.5878	0.0953	−0.3857	−0.9163	−1.3863	−1.8971	−2.4079

From this table we obtain

$$\sum_{i=1}^{4} t_i^2 = 204 \qquad \sum_{i=1}^{4} V_i = 36$$

$$\sum_{i=1}^{4} V_i t_i = -44.3298 \qquad \sum_{i=1}^{4} y_i = -3.4871$$

Using V, t, a, and $B = \ln b$ instead of x, y, m, and b in Equations 1.5-1 and 1.5-2, we obtain

$$204a + 36B = -44.3298$$
$$36a + 9B = -3.4871$$

The solution is $a = -0.5064$ and $B = 1.638$. Thus gives $b = e^B = 5.1448$. Thus the desired function is $v = 5.1448e^{-0.5064t}$, and we see that $v(0) = 5.144$ m/s and $c/m = 0.5064$ 1/s. Note that this value of $v(0)$ does not equal the first data value ($v = 5.1$). If we know the mass m, we can compute c from $c = 0.5064m$.

■

Constraining Models to Pass through a Given Point

Many applications require a model whose form is dictated by physical principles. For example, the force-deflection model of a spring must pass through the origin $(0, 0)$ because the spring exerts no force when it is not stretched. Thus a linear model $y = mx + b$ sometimes must have a zero value for b. However, in general the least-squares method will give a nonzero value for b because of the scatter or measurement error that is usually present in the data.

To obtain a zero-intercept model of the form $y = mx$, we must derive the equation for m from basic principles. The sum of the squared residuals in this case is

$$J = \sum_{i=1}^{n}(mx_i - y_i)^2$$

Computing the derivative $\partial J / \partial m$ and setting it equal to zero gives the result

$$m \sum_{i=1}^{n} x_i^2 = \sum_{i=1}^{n} x_i y_i \tag{1.5-3}$$

which can be easily solved for m.

EXAMPLE 1.5-5
A Constrained Linear Spring Model

For the data given in Example 1.5-1, estimate the value of the spring constant k for the model $f = kx$, which is constrained to pass through the origin.

Solution

The solution is given by Equation 1.5-3 with $m = k$ and the force data f_i replacing y_i:

$$k = \frac{\sum_{i=1}^{n} x_i f_i}{\sum_{i=1}^{n} x_i^2} = 1017 \text{ lb/in.}$$

Compare this value with $k = 1026$ and $k = 1081$ from Examples 1.5-1 and 1.5-2. ∎

Fitting a Function Having a Specified Power

Sometimes we know from physical theory that the data can be described by a power function with a specified power. For example, the orifice damping relation states that $f = cv^2$. In this case we know that the relation is a power function with an exponent of 2, and we need to estimate the value of the coefficient c. In such cases we can modify the least-squares method to find the best-fit function of the form $f = cv^2$.

To fit the power function $f = cv^m$ with a known value of m, the least-squares criterion is

$$J = \sum_{i=1}^{n}(cv^m - f_i)^2$$

To obtain the value of c that minimizes J, we must solve $\partial J / \partial c = 0$:

$$\frac{\partial J}{\partial c} = 2 \sum_{i=1}^{n} v_i^m (cv_i^m - f_i) = 0$$

This gives

$$\sum_{i=1}^{n} v_i^{2m} c - \sum_{i=1}^{n} v_i^m f_i = 0$$

or

$$c = \frac{\sum_{i=1}^{n} v_i^m f_i}{\sum_{i=1}^{n} v_i^{2m}} \tag{1.5-4}$$

EXAMPLE 1.5-6
Fitting a
Progressive
Damping Function

Use the least-squares method to estimate the value of c for the progressive damping model $f = cv^2$. The data are

v (m/s)	0	0.05	0.13	0.26	0.46
f (kN)	0	0.05	0.14	0.45	1.1

Solution

Here the power is $m = 2$, so Equation 1.5-4 becomes

$$c = \frac{\sum_{i=1}^{5} v_i^2 f_i}{\sum_{i=1}^{5} v_i^4} = \frac{0.265671}{0.04963618} = 5.3524$$

The best-fit value of c is 5.3524. ∎

The Quality of a Curve Fit

In general, if the arbitrary function $y = f(x)$ is used to represent the data, then the error in the representation is given by $e_i = f(x_i) - y_i$, for $i = 1, 2, 3, \ldots, n$. The error e_i is the difference between the data value y_i and the value of y obtained from the function; that is, $f(x_i)$. The least-squares criterion used to fit a function $f(x)$ is the sum of the squares of the residuals, J. It is defined as

$$J = \sum_{i=1}^{n} [f(x_i) - y_i]^2 \tag{1.5-5}$$

We can use this criterion to compare the quality of the curve fit for two or more functions used to describe the same data. The function that gives the smallest J value gives the best fit.

We denote the sum of the squares of the deviation of the y values from their mean $\bar{y}$ by S, which can be computed from

$$S = \sum_{i=1}^{n} [y_i - \bar{y}]^2 \tag{1.5-6}$$

This formula can be used to compute another measure of the quality of the curve fit, the *coefficient of determination*, also known as the *r-squared value*. It is defined as

$$r^2 = 1 - \frac{J}{S} \tag{1.5-7}$$

For a perfect fit, $J = 0$ and thus $r^2 = 1$. Thus, the closer r^2 is to 1, the better the fit. The largest r^2 can be is 1. The value of S is an indication of how much the data are spread around the mean, and the value of J indicates how much of the data spread is left unaccounted for by the model. Thus the ratio J/S indicates the fractional variation left unaccounted for by the model. It is possible for J to be larger than S, and thus it is possible for r^2 to be negative. Such cases, however, are indicative of a very poor model that should not be used. As a rule of thumb, a very good fit accounts for at least 99% of the data variation. This corresponds to $r^2 \geq 0.99$.

For example, the function $y = 0.9x + 11/6$ derived at the beginning of this section has the values $S = 40.6667$, $J = 0.1666$, and $r^2 = 0.9959$, which indicates a very good fit.

1.6 MATLAB APPLICATIONS

Appendix A is an introduction to the MATLAB programming language. If you are unfamiliar with MATLAB, you should study Appendix A before proceeding. In this section we show how to use MATLAB to fit functions to data.

When the least-squares method is applied to fit quadratic and higher-order polynomials, the resulting equations for the coefficients are linear algebraic equations that are easily solved. Their solution forms the basis for the algorithm contained in the MATLAB polyfit function. Its syntax is p = polyfit(x,y,n). The function fits a polynomial of degree n to data described by the arrays x and y, where x is the independent variable. The result p is the row array of length $n + 1$ that contains the polynomial coefficients in order of descending powers.

When we type p = polyfit(z,w,1), MATLAB will fit a linear function $w = p_1 z + p_2$. The coefficients p_1 and p_2 are the first and second elements in the array p; that is, p will be $[p_1, p_2]$. With a suitable transformation, the power and exponential functions can be transformed into a linear function, but the polynomial $w = p_1 z + p_2$ has a different interpretation in each of the three cases:

The Linear Function: $y = mx + b$. In this case the variables w and z in the polynomial $w = p_1 z + p_2$ are the original data variables, and we can find the linear function that fits the data by typing p = polyfit(x,y,1). The first element p_1 of the array p will be m, and the second element p_2 will be b.

The Power Function: $y = bx^m$. In this case log $y = m$ log $x + $ log b, which has the form $w = p_1 z + p_2$, where the polynomial variables w and z are related to the original data variables x and y by $w = $ log y and $z = $ log x. Thus we can find the power function that fits the data by typing p = polyfit(log10(x),log10(y),1). The first element p_1 of the array p will be m, and the second element p_2 will be log b. We can find b from $b = 10^{p_2}$.

The Exponential Function: $y = be^{mx}$. In this case ln $y = mx + $ ln b, which has the form $w = p_1 z + p_2$, where the polynomial variables w and z are related to the original data variables x and y by $w = $ ln y and $z = x$. Thus we can find the exponential function that fits the data by typing p=polyfit(x,log(y),1). The first element p_1 of the array p will be m, and the second element p_2 will be ln b. We can find b from $b = e^{p_2}$.

EXAMPLE 1.6-1
Fitting a Linear Model

Consider the spring deflection data used in Examples 1.5-1 and 1.5-2. Use MATLAB to fit a linear model to the data.

Force f (lb)	0	100	200	300	400	500	600	700	800
Deflection x (in.)	0	0.15	0.23	0.35	0.37	0.5	0.57	0.68	0.77

Solution

The MATLAB script file is:

```
x=[0,0.15,023,0.35,0.37,0.5,0.57,0.68,0.78];
f=[0:100:800];
p=polyfit(x,f,1)
```

The array of computed coefficients is p = [1081, -0.0347]. Thus the fitted model is $f = 1081x - 0.0347$. Note that it does not pass through the origin. ∎

EXAMPLE 1.6-2
Fitting a Power Function

Consider the force-velocity data used in Example 1.5-3. Use MATLAB to fit a power function to the data.

v (m/s)	0.5	1	1.5	2	2.5	3	3.5	4
f (N)	35	59	83	104	125	145	163	182

Solution

For the power function $f = cv^m$, we have $\log f = \log c + m \log v$. So we transform the data with the transformations $V = \log v$ and $F = \log f$, then apply the polyfit function. The MATLAB script file is:

```
v =[0.5,1,1.5,2,2.5,3,3.5,4];
f = [35,59,83,104,125,145,163,182];
V = log10(v);F = log10(f);
p = polyfit(V,F,1)
m = p(1)
c = 10^p(2)
```

The array of computed coefficients is p = [0.7972, 1.7787], where $m = p_1 = 0.7972$ and $c = 10^{p_2} = 10^{1.7787} = 60.0801$. Thus the fitted model is $f = 60.0801v^{0.7972}$. ∎

EXAMPLE 1.6-3
Fitting an Exponential Function

Consider the velocity-versus-time data used in Example 1.5-4. Use MATLAB to fit an exponential function to the data.

t (s)	0	1	2	3	4	5	6	7	8
v (m/s)	5.1	3.3	1.8	1.1	0.68	0.4	0.25	0.15	0.09

Solution

For the exponential function $v = be^{mt}$, we note that $\ln v = \ln b + mt$, and so use the transformation $V = \ln v$ on the data before applying the polyfit function. The MATLAB script file is:

```
v = [5.1,3.3,1.8,1.1,0.68,0.4,0.25,0.15,0.09];
t = [0:8];
V = log10(v);
p = polyfit(V,f,1)
m = p(1)
c = 10^p(2)
```

The array of computed coefficients is p = [-0.5064, 1.6380], where $m = p_1 = -0.5064$ and $b = e^{p_2} = e^{1.6380} = 5.1448$. Thus the fitted model is $v = 5.1448e^{-0.5064t}$. ∎

Fitting the Cubic Spring Model

For the cubic model of spring force $f = k_1x + k_2x^3$, there is no constant term and no x^2 term. However, using the MATLAB polyfit function polyfit(x,f,3) will give a function of the form $f = a_0 + a_1x + a_2x^2 + a_3x^3$, in which a_0 and a_2 are nonzero in general. So we cannot use the polyfit function without modification to compute k_1 and k_2. We may derive expressions for k_1 and k_2 using the least-squares method, but noting that $f(-x) = -f(x)$ for the cubic spring function, an easier way is to use the polyfit function with an antisymmetric data set created from the original data. The following example shows how this is done.

EXAMPLE 1.6-4
Fitting the Cubic Spring Model

Use the least-squares method to estimate the values of k_1 and k_2 for the following cubic model of spring force: $f = k_1 x + k_2 x^3$. The data are

x (m)	0	0.015	0.030	0.045	0.060	0.075	0.090
f (N)	0	22	62	80	129	160	214

Solution

To the original data set (x, f), we append their negative values $(-x, -f)$. The following MATLAB program performs the calculations:

```
x1= [15:15:90]*0.001;
x = [-x1,0,x1]
f1 = [22, 62, 80, 129, 160, 214]
f = [-f1,0,f1]
p = polyfit(x,f,3)
xp = [0:0.1:90]*0.001;
fp = polyval(p,xp)
plot(xp,fp,x1,f1,'o')
```

The computed coefficients are $p = [7.2175 \times 10^4, 0, 1.7815 \times 10^3, 0]$, and thus $k_1 = 1781.5$ and $k_2 = 7.2175 \times 10^4$. ∎

EXAMPLE 1.6-5
Fitting a Progressive Damping Function

Use the least-squares method to estimate the value of c for the progressive damping model $f = cv^2$. The data are

v (m/s)	0	0.05	0.13	0.26	0.46
f (kN)	0	0.05	0.14	0.45	1.1

Solution

Here the power is specified to be $m = 2$. The MATLAB program to implement Equation 1.5-4 is as follows:

```
v=[0,0.05,0.13,0.26,0.46];
f = [0,0.050,0.14,0.45,1.1];
vp = [0:.01:.5];
c = sum(v.^2.*f)/sum(v.^4)
```

The best-fit value of c is 5.3524. ∎

EXAMPLE 1.6-6
Evaluating the Quality of Fit with MATLAB

Use MATLAB to compute from Equations 1.5-5, 1.5-6, and 1.5-7 the quality-of-fit indicators J, S, and r^2 for the function derived in Example 1.6-5.

Solution

Continue the script file given in Example 1.6-5 as follows:

```
mu = mean(f);
f_fit = c*v.^2;
J = sum((f_fit-f).^2)
S = sum((f -mu).^2)
r2 = 1-J/S
```

The results are $J = 0.0126$, $S = 0.8291$, and $r^2 = 0.9848$, which indicates a good fit. ∎

1.7 CHAPTER REVIEW

This chapter introduced the basic terminology used in the study of mechanical vibration and analyzed two important elements found in vibration models. These are the spring element, which produces a restoring force or moment as a function of the displacement of the mass element, and the damping element, which produces a restoring force or moment as a function of the velocity of the mass element. In this chapter we established the basic principles for developing mathematical models of these elements and applied these principles to several commonly found examples.

To be useful, numerical values eventually must be assigned to the parameters of a mathematical model of a spring or a damping element in order to make predictions about the behavior of the physical device. Therefore, this chapter included two sections that show how to obtain numerical parameter values from data by using the least-squares method. This topic has traditionally not been given much attention in vibration texts, but the methods are often needed by engineers working with real devices.

The programming language MATLAB has several useful functions that enable you to apply these methods easily. These functions were presented in a separate section that may be skipped by readers who will not be using MATLAB. However, if you are such a reader, you are strongly encouraged to consider learning MATLAB, because it will be very useful to you in the future.

Now that you have finished this chapter, you should be able to do the following:

1. Develop models of spring elements from the basic principles of mechanics of materials.

2. Develop models of damping elements from the basic principles of fluid mechanics.

3. Apply the least-squares method to obtain numerical parameter values for a spring or damping element model when given the appropriate data.

4. Use MATLAB to implement the least-squares method.

PROBLEMS

SECTION 1.1 INTRODUCTION

1.1 Name two examples of useful vibration and two examples of harmful vibration.

1.2 What are the three basic elements in a vibratory system?

SECTION 1.2 DESCRIPTIONS OF VIBRATIONAL MOTION

1.3 A particular function has the form $y(t) = Ae^{-t/\tau}$ and has the values $y(1) = 4.912$ and $y(3) = 3.293$. Compute the values of A and τ.

1.4 If you were to sketch the plot of the function $y(t) = 7e^{-t/3}$, what would be appropriate ranges for y and t?

1.5 If you were to sketch the plot of the function $y(t) = 6e^{-t/2} + 12e^{-t/5}$, what would be appropriate ranges for y and for t?

1.6 A particular motor rotates at 3000 revolutions per minute (rpm). What is its speed in rad/sec, and how many seconds does it take to make one revolution?

1.7 The displacement of a certain object is described by $y(t) = 23 \sin 5t$, where t is measured in seconds. Compute its period and its oscillation frequency in rad/sec and in Hz.

1.8   What values must B and C have for the following to be true?

$$10 \sin(3t + 2) = B \sin 3t + C \cos 3t$$

1.9   What values must A and ϕ have for the following to be true?

$$A \sin(6t + \phi) = -6.062 \sin 6t - 3.5 \cos 6t$$

1.10   The displacement of a certain object in meters is described by $y(t) = 0.5 \sin(7t + 4)$, where t is measured in seconds. Obtain the expesssions for its velocity and acceleration as functions of time. What is the amplitude of the velocity and of the acceleration?

1.11   The displacement of a certain object in feet is described by $y(t) = 0.08 \sin 3t + 0.05 \cos 3t$, where t is measured in seconds. Obtain the expressions for its velocity and acceleration as functions of time. What is the amplitude of the displacement, of the velocity, and of the acceleration?

1.12   The displacement of a certain object in meters is described by $y(t) = 0.07 \sin(5t + \phi)$, where t is measured in seconds. It is known that the object's velocity at $t = 0$ is 0.041 m/s. Compute the value of the phase angle ϕ.

1.13   The displacement of a certain object in meters is described by $y(t) = A \sin(5t + \phi)$, where t is measured in seconds. It is known that the object's displacement at $t = 0$ is 0.048 m and that its velocity at $t = 0$ is 0.062 m/s. Compute the values of A and ϕ.

1.14   As measured by an accelerometer, the amplitude of acceleration of a certain vibrating object was $0.7g$ with a frequency of 10 Hz. Compute the amplitudes of the object's displacement and velocity, assuming that the motion is simple harmonic. Use $g = 9.81$ m/s^2.

1.15   The amplitudes of the displacement and acceleration of an unbalanced motor were measured to be 0.15 mm and $0.6g$, respectively. Determine the speed of the motor. Use $g = 9.81$ m/s^2.

1.16   If you were to sketch the plot of the function $y(t) = 6e^{-t/3} \sin 0.25\pi t$, what would be appropriate ranges for y and t?

1.17   Derive the expressions for $\dot{y}(t)$ and $\ddot{y}(t)$, where $y(t)$ is given by Equation 1.2-6.

SECTION 1.3   SPRING ELEMENTS

1.18   The distance a spring stretches from its "free length" is a function of how much tension force is applied to it. The following table gives the spring length y that was produced in a particular spring by the given applied force f. The spring's free length is 0.12 m. Find a functional relation between f and x, the extension from the free length ($x = y - 0.12$).

Force f (N)	Spring length y (m)
0	0.120
2.09	0.183
5.12	0.269
7.30	0.328

1.19   Suppose you have two springs to use to increase the overall stiffness of a system. How should you connect the springs, end-to-end or side-by-side?

1.20   Derive the spring constant expression of a doubly clamped beam of length L, thickness h, and width w, assuming that the force f and deflection x are at the center of the beam (see Table 1.3-1). The force-deflection relation is

$$x = \frac{L^3}{192EI_A} f$$

where the area moment is $I_A = wh^3/12$.

1.21 Figure P1.21 shows a horizontal force f acting on a lever that is attached to two springs. Assume that the resulting motion is small and thus essentially horizontal, and determine the expression for the equivalent spring constant that relates the applied force f to the resulting displacement x.

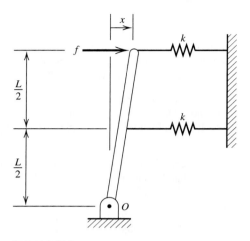

FIGURE P1.21

1.22 Compute the translational spring constant of a particular steel helical-coil spring, of the type used in automotive suspensions. The coil has six turns. The coil diameter is 0.1 m, and the wire diameter is 0.013 m. For the shear modulus, use $G = 8 \times 10^{10}$ N/m^2.

1.23 In the spring arrangement shown in Figure P1.23, the displacement x is caused by the applied force f. Assuming the system is in static equilibrium, sketch the plot of f versus x. Determine the equivalent spring constant k_e for this arrangement, where $f = k_e x$.

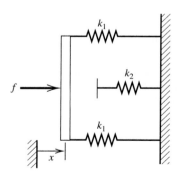

FIGURE P1.23

1.24 In the arrangement shown in Figure P1.24, a cable is attached to the end of a cantilever beam. We will model the cable as a rod. Denote the translational spring constant of the beam by k_b and the translational spring constant of the cable by k_c. The displacement x is caused by the applied force f.

(a) Are the two springs in series or in parallel?

(b) What is the equivalent spring constant for this arrangement?

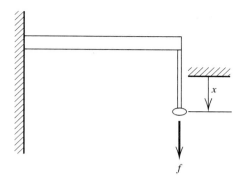

FIGURE P1.24

1.25 The two stepped solid cylinders in Figure P1.25 consist of the same material and have an axial force f applied to them. Determine the equivalent translational spring constant for this arrangement. (*Hint*: Are the two springs in series or in parallel?)

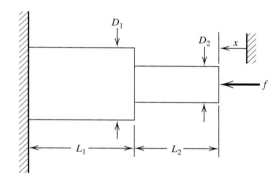

FIGURE P1.25

1.26 A table with four identical legs supports a vertical force. The solid cylindrical legs are identical and made of metal with $E = 2 \times 10^{11}$ N/m². The legs are 1 m in length and 0.03 m in diameter. Compute the equivalent spring constant due to the legs, assuming the table top is rigid.

1.27 The beam shown in Figure P1.27 has been stiffened by the addition of a spring support. The steel beam is 0.91 m long, 0.03 m thick, and 0.3 m wide, and its mass is 55 kg. The mass m is 580 kg.

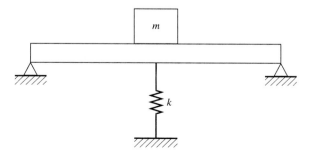

FIGURE P1.27

Neglecting the mass of the beam, compute the spring constant k necessary to reduce the static deflection to one-half its original value before the spring k was added.

1.28 Determine the equivalent spring constant of the arrangement shown in Figure P1.28. All the springs have the same spring constant k.

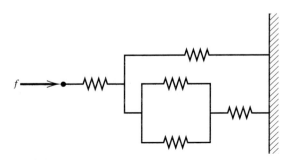

FIGURE P1.28

1.29 Compute the equivalent torsional spring constant of the stepped shaft arrangement shown in Figure P1.29. For the shaft material, $G = 8 \times 10^{10}$ N/m^2.

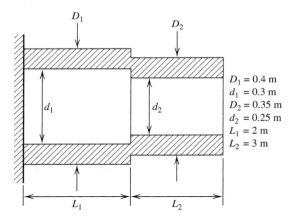

$D_1 = 0.4$ m
$d_1 = 0.3$ m
$D_2 = 0.35$ m
$d_2 = 0.25$ m
$L_1 = 2$ m
$L_2 = 3$ m

FIGURE P1.29

1.30 The following "small-angle" approximation for the sine is used in many engineering applications to obtain a simpler model that is easier to understand and analyze. This approximation states that $\sin x \approx x$ where x must be in radians. Investigate the accuracy of this approximation by creating three plots. For the first plot, plot $\sin x$ and x versus x for $0 \le x \le 1$. For the second plot, plot the approximation error $\sin(x) - x$ versus x for $0 \le x \le 1$. For the third plot, plot the percent error $[\sin(x) - x]/\sin(x)$ versus x for $0 \le x \le 1$. How small must x be for the approximation to be accurate within 5%?

1.31 Obtain two linear approximations of the function $f(\theta) = \sin \theta$, one valid near $\theta = \pi/4$ rad and the other valid near $\theta = 3\pi/4$ rad.

1.32 Obtain two linear approximations of the function $f(\theta) = \cos\theta$, one valid near $\theta = \pi/3$ rad and the other valid near $\theta = 2\pi/3$ rad.

1.33 Obtain two linear approximations of the function $f(v) = 6v^2$, one valid near $v = 5$ and the other valid near $v = 10$.

1.34 Obtain a linear approximation of the function $f(v) = \sqrt{v}$, valid near $v = 16$. Noting that $f(v) \geq 0$, what is the value of v below which the linearized model loses its meaning?

1.35 In the spring arrangement shown in Figure P1.35, the displacement x is caused by the applied force f. Assuming the system is in static equilibrium and that the angle θ is small, determine the equivalent spring constant k_e for this arrangement, where $f = k_e x$.

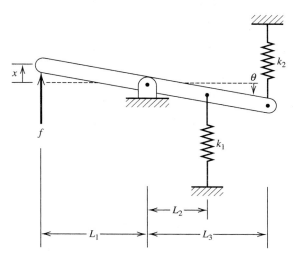

FIGURE P1.35

1.36 Plot the spring force felt by the mass shown in Figure P1.36 as a function of the displacement x. When $x = 0$, spring 1 is at its free length. Spring 2 is at its free length in the configuration shown.

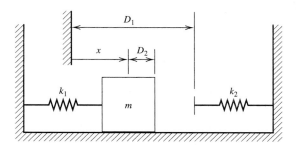

FIGURE P1.36

1.37 Suppose a particular nonlinear spring is described by

$$f(y) = k_1 y + k_2 y^3$$

where y is the stretch in the spring from its free length, $k_1 = 1000$ N/m, and $k_2 = 3000$ N/m^3. Determine its equivalent linear spring constant for the two equilibrium positions corresponding to the following applied forces:

(a) $f_r = 100$ N

(b) $f_r = 500$ N

1.38 Gravity in certain situations acts like a spring by applying a restoring moment that is a function of displacement. Figure P1.38 shows a pendulum with a "gravity spring." Derive the relation between the restoring torque T acting on the mass due to gravity and the angle θ.

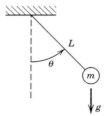

FIGURE P1.38

1.39 Buoyancy in certain situations acts like a spring by applying a restoring force that is a function of displacement. Figure P1.39 shows a cylinder acted on by a "buoyancy spring." Derive the relation between the restoring force f due to buoyancy and the displacement x from the rest position of the cylinder. (*Hint*: Use Archimedes' principle, which states that the buoyancy force equals the weight of displaced liquid.)

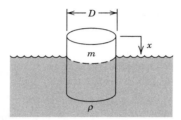

FIGURE P1.39

1.40 The wire of length $2L$ shown in Figure P1.40 has an initial tension T. The mass m in the middle moves a distance x horizontally. Derive the relation between the horizontal restoring force f and the displacement x, assuming that x is small.

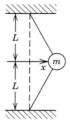

FIGURE P1.40

1.41 For each case shown in Figure P1.41, plot the force exerted *on* the mass by the springs as a function of the displacement x. Neglect friction.

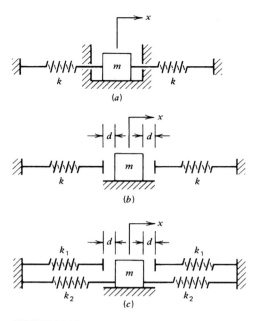

FIGURE P1.41

SECTION 1.4 DAMPING ELEMENTS

1.42 A flat plate slides over a horizontal flat surface, which is lubricated with an oil whose viscosity is $\mu = 0.9$ N·s/m^2. The contact area is 0.16 m^2.

(a) Assuming that the oil film is 1 mm thick, compute the damping coefficient c.

(b) How much force is required to keep the plate moving at 0.5 m/s?

(c) Suppose the estimate of the film thickness is off by $\pm 50\%$. Compute the uncertainty in the value of the damping coefficient.

1.43 The rotational damper shown in Figure P1.43 consists of a cylinder of diameter D rotating in a cylindrical housing filled with liquid of viscosity μ. The damper resists with a torque T as a function

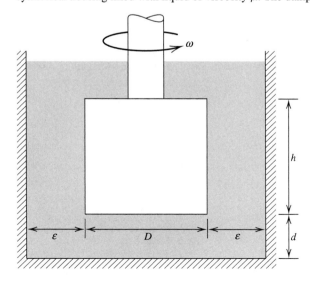

FIGURE P1.43

of its constant rotational velocity ω. Derive the expression for its damping constant c in terms of D, ϵ, d, h, and μ, where $T = c\omega$.

1.44 The damper shown in Figure P1.44 consists of a piston of diameter D translating in a cylindrical housing filled with liquid of viscosity μ. The damper resists with a force T as a function of the piston velocity v. Derive the expression for its damping constant c in terms of D, ϵ, L, and μ, where $f = cv$.

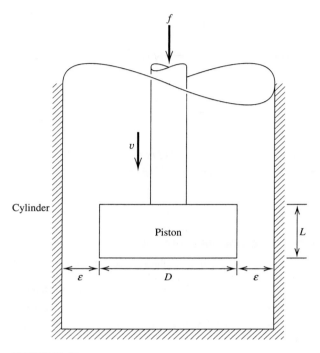

FIGURE P1.44

1.45 The *viscometer* shown in Figure P1.45 is used to measure the viscosity of the liquid contained in the cylindrical pan. A thin disk of diameter D, suspended by a wire, is in contact with the surface of the liquid, whose thickness is h. The wire acts like a torsional spring and the liquid acts like a torsional

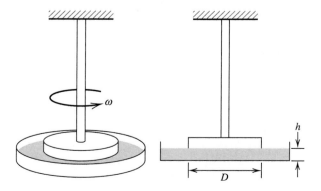

FIGURE P1.45

damper, providing a resisting torque $T = c\omega$, where ω is the angular velocity of the disk. In later chapters we will derive an expression for the damping coefficient c in terms of the observed oscillation frequencies of the disk when in and out of contact with the liquid. To compute the viscosity μ, we need an expression for μ as a function of c, D, and h. Derive this expression.

1.46 Compute the minimum and maximum values of the torsional damping constant c of a journal bearing lubricated with SAE 30 oil, as the temperature of the oil changes from 50°F to 150°F. The oil viscosity varies from $\mu = 120 \times 10^{-6}$ reyn at 50°F to $\mu = 3 \times 10^{-6}$ reyn at 150°F (1 reyn = 1 lb-sec/in.²). The shaft diameter is $D = 2$ in., the bearing length is $L = 2$ in., and the bearing clearance is $\epsilon = 0.001$ in.

1.47 (a) Obtain the expression for the equivalent damping coefficient c_e for Figure P1.47a, where $f = c_e v$. Are the dampers in series or in parallel?

(b) Obtain the expression for the equivalent damping coefficient c_e for Figure P1.47b, where $f = c_e v$.

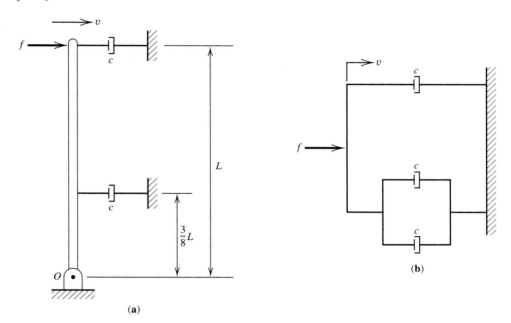

FIGURE P1.47

1.48 Derive the expression for the linearized damping coefficient c, valid near the reference velocity v_r, for the orifice damper whose force-velocity relation is $f = A^3 \rho^2 R^2 v^2$. The linearized expression has the form $\Delta f = c \Delta v$, where $\Delta f = f - f_r$, $\Delta v = v - v_r$, and $f_r = A^3 \rho^2 R^2 v_r^2$.

SECTION 1.5 PARAMETER ESTIMATION

1.49 The deflection of a cantilever beam is the distance its end moves in response to a force applied at the end (Figure 1.5-1). The following table gives the measured deflection x that was produced in a particular beam by the given applied force f. Plot the data to see whether a linear relation exists between f and x, and determine the spring constant by drawing a straight line through the data by eye.

Force f (N)	0	200	400	600	800	1000	1200	1400	1600
Deflection x (cm)	0	0.3	0.46	0.7	0.75	1.05	1.14	1.36	1.55

1.50 Use the least-squares method to find the coefficients of the straight line $y = mx + b$ that best fits the following data:

x	0	10	20	30
y	22	87	144	201

1.51 Consider the spring data given in Problem 1.49. Use the least-squares method to estimate the spring constant k for the linear model $f = kx + b$. Discuss the significance of the intercept b.

1.52 Find a functional description of the following data, where f is force and v is velocity:

v (m/s)	1	2	3	4	5	6	7	8
f (N)	9	48	124	245	401	620	880	1200

1.53 Find a functional description of the following data, where t is time and v is velocity.

t (s)	0	2	4	6	8	10	12	14	16
v (m/s)	8.5	3.9	2.2	1.05	0.6	0.3	0.15	0.07	0.04

1.54 For the data given in Problem 1.49, use the least-squares method to compute the value of the spring constant k for the model $f = kx$, which is constrained to pass through the origin.

1.55 Use the least-squares method to estimate the value of c for the progressive damping model $f = cv^2$. The data are

v (m/s)	0	0.1	0.2	0.3	0.4	0.5
f (kN)	0	0.11	0.42	0.88	1.63	2.45

SECTION 1.6   MATLAB APPLICATIONS

1.56 Solve Problem 1.51 using MATLAB, and evaluate the quality-of-fit parameters J, S, and r^2.

1.57 Solve Problem 1.54 using MATLAB, and evaluate the quality-of-fit parameters J, S, and r^2.

1.58 Solve Problem 1.55 using MATLAB, and evaluate the quality-of-fit parameters J, S, and r^2.

1.59 Use the least-squares method to estimate the values of k_1 and k_2 for the following cubic model of spring force: $f = k_1 x + k_2 x^3$. The data are

x (m)	0	0.03	0.06	0.09	0.12	0.15	0.18
f (N)	0	45	102	165	245	360	512

MODELS WITH ONE DEGREE OF FREEDOM

DYNAMICS DEALS with the motion of bodies under the action of forces. The subject consists of *kinematics*, the study of motion without regard to the forces causing the motion, and *kinetics*, the relationship between the forces and the resulting motion.

The single most important step in analyzing a vibratory system is to derive the correct description of the dynamics of the system. If this description is incorrect, then any analysis based on that description will also be incorrect. The description may be in the form of a differential equation of motion, but there are analysis methods that do not require the equation of motion, such as the work-energy methods of Section 2.3, the impulse-momentum methods of Section 2.4, and Rayleigh's method, to be treated in Chapter 3. These methods, however, are based on the principles of dynamics covered in this chapter.

The chapter begins with a discussion of kinematics principles. Vibrating systems are often categorized by their number of *degrees of freedom*, abbreviated DOF; this concept is developed in Section 2.1. This chapter focuses on the simplest systems: those having one degree of freedom.

The kinetics of rigid bodies is covered in Section 2.2. Section 2.3 covers the work-energy method, which is based on an integration of the equation of motion over displacement. Section 2.4 covers the impulse-momentum method, which is based on an integration of the equation of motion over time. Because these methods are based on integrations of the equation of motion, the mathematics associated with them is easier to apply to solve some types of problems.

Section 2.5 discusses equivalent mass and equivalent inertia. These concepts use kinetic energy equivalence to simplify the process of obtaining descriptions of systems containing both translating and rotating parts, systems containing distributed mass, and systems consisting of multiple masses whose motion is coupled.

Section 2.6 contains many examples showing how to obtain the differential equation of motion for single-degree-of-freedom systems containing spring elements.

Although the chapter focuses on motion with one degree of freedom, its principles can be easily extended to obtain models of systems having two or more degrees of freedom. This is done in Chapters 6 and 8.

LEARNING OBJECTIVES

After you have finished this chapter, when presented with a vibratory system, you should be able to do the following:

- Identify the type of motion, apply the appropriate kinematic equations, and choose an appropriate set of coordinates to describe the motion.
- Identify the degrees of freedom.
- Apply Newton's laws of motion if the object can be modeled as a particle.
- Identify the type of plane motion, and apply the appropriate form of Newton's laws if the object can be modeled as a rigid body in plane motion.
- Apply work-energy methods for problems involving force, displacement, and velocity, but not time.
- Apply impulse-momentum methods for problems in which the applied forces act either over very short times, such as happens during an impact, or over a specified time interval.
- Apply the principle of kinetic energy equivalence to simplify the process of obtaining a description of a system containing both translating and rotating parts, or one containing distributed mass, or one consisting of multiple masses whose motion is coupled.
- Obtain the differential equation of motion for single-degree-of-freedom systems containing spring elements.

2.1 KINEMATICS AND DEGREES OF FREEDOM

Kinematics is the study of motion without regard for the masses that are in motion or the forces that produce the motion. Kinematics is primarily concerned with the relationship among displacement, velocity, and acceleration, and thus it provides the basis for describing the motion of the systems we will be studying. For the most part we will restrict our attention to motion in a plane. This means that the object can translate in two dimensions only and can rotate only about an axis that is perpendicular to the plane containing these two dimensions. The completely general motion case involves translation in three dimensions and rotation about three axes. This type of motion is considerably more complex to analyze. For such details, consult a reference on dynamics. However, many practical engineering problems can be handled with the plane motion methods covered here. We present a summary of the pertinent relations without derivation or extensive discussion.

Categories of Plane Motion

A *particle* is a mass of negligible dimensions. We may consider a body to be a particle if its dimensions are irrelevant for specifying its position and the forces acting on it. We may represent the position of a particle by a *position vector* **r** measured relative to any convenient coordinate system. It is important to understand that the validity of the laws of physics does not depend on the coordinate system used to express them; the choice of a coordinate system should be made entirely for convenience.

If, however, the dimensions of the body are not negligible, we must consider the possibility of rotational motion. There are four categories of plane motion. These are illustrated in Figure 2.1-1. In *translation* every line in the body remains parallel to its original position, and there is no rotation of any line in the body. *Rectilinear translation* occurs when all points in the body move in parallel straight lines, whereas *curvilinear translation* occurs when all points move on congruent curves. In both cases, all points on the body have the same motion, and thus the motion of the body may be completely described by describing the motion of any point on the body. Rectilinear translation is usually easily recognized, but curvilinear translation is not always obvious. For example, the swinging plate in Figure 2.1-2*a* does not rotate but undergoes curvilinear translation if the links are parallel.

The third category of plane motion is *rotation about a fixed axis*. In such motion all points in the body move in circular paths. The fourth category, *general plane motion*, consists of both translation and rotation. The connecting rod in Figure 2.1-2*b* undergoes general plane motion, whereas the piston displays rectilinear translation and the crank rotates about a fixed axis.

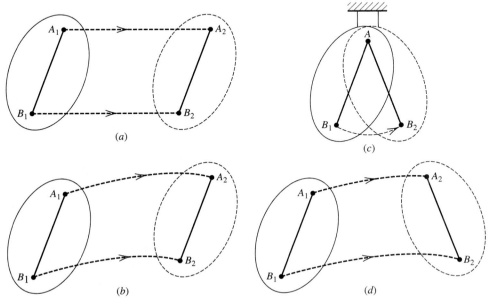

FIGURE 2.1-1 Types of rigid-body plane motion. (*a*) Rectilinear translation. (*b*) Curvilinear translation. (*c*) Fixed-axis rotation. (*d*) General plane motion.

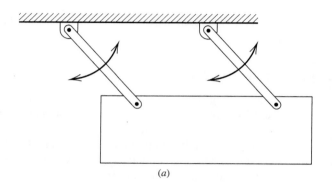

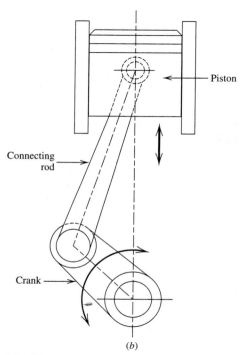

FIGURE 2.1-2 Examples of plane motion. (*a*) The plate shows curvilinear translation if the links are identical and parallel; each link rotates about a fixed axis. (*b*) The connecting rod has general plane motion, while the piston has rectilinear translation and the crank has fixed-axis rotation.

EXAMPLE 2.1-1
Rectangular versus Normal–Tangential Coordinates

Consider the pendulum shown in Figure 2.1-3. Obtain the description of the velocity and acceleration of point G as functions of θ, $\dot{\theta}$, and $\ddot{\theta}$ (**a**) in terms of normal and tangential coordinates, and (**b**) in terms of the rectangular coordinates x and y.

Solution

(**a**) Using the normal n and tangential t directions shown in the figure, we may express the velocity and acceleration as follows, noting that $v = L\dot{\theta}$ and $\dot{v} = L\ddot{\theta}$:

$$\mathbf{v} = v\mathbf{e}_t = L\dot{\theta}\mathbf{e}_t$$

$$\mathbf{a} = \frac{v^2}{L}\mathbf{e}_n + \dot{v}\mathbf{e}_t = L\dot{\theta}^2\mathbf{e}_n + L\ddot{\theta}\mathbf{e}_t$$

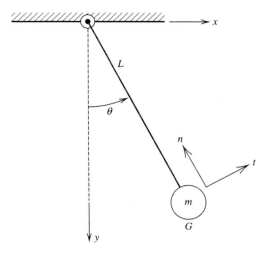

FIGURE 2.1-3 Rectangular and normal-tangential coordinates for a pendulum.

(**b**) Note that $x = L \sin \theta$ and $y = L \cos \theta$. The velocity in terms of the unit vectors **i** and **j** of the rectangular coordinates x and y is

$$\mathbf{v} = \dot{x}\mathbf{i} + \dot{y}\mathbf{j}$$

where

$$\dot{x} = L\dot{\theta} \cos \theta$$

$$\dot{y} = -L\dot{\theta} \sin \theta$$

The acceleration is

$$\mathbf{a} = \ddot{x}\mathbf{i} + \ddot{y}\mathbf{j}$$

where

$$\ddot{x} = L\ddot{\theta} \cos \theta - L\dot{\theta}^2 \sin \theta$$

$$\ddot{y} = -L\ddot{\theta}^2 \sin \theta - L\dot{\theta}^2 \cos \theta$$

For this example, note how much simpler the velocity and acceleration expressions are when described by normal and tangential coordinates. ∎

Coordinates and Degrees of Freedom

The term *coordinate* refers to an independent quantity used to specify position. You may use any convenient measure of displacement as a coordinate. The proper choice of coordinates can simplify the resulting expressions for position, velocity, or acceleration, and even a simple problem can be made difficult if you select the coordinates unwisely.

As we just saw, the motion of a simple pendulum (see Figure 2.1-3) can be described with the rectangular coordinates x and y or with just the single coordinate θ, since L is constant. So the constant length L acts as a *constraint* on the motion of the mass at the end of the pendulum. Not only do the rectangular coordinates result in more complicated expressions for velocity and acceleration, but the two coordinates are unnecessary because the pendulum rotating in a single plane has only one *degree of freedom*, abbreviated as DOF. The coordinates x and y are constrained by the fact that $x^2 + y^2 = L^2$. Thus only one of the two coordinates is independent.

The presence of constraints that restrict the motion of an object usually means that fewer coordinates are needed to describe the motion of the object. With this insight, we may now see that the number of degrees of freedom is the minimum number of coordinates required to completely describe the motion of the object.

Generalized coordinates are those that reflect the existence of constraints and yet still provide a complete description of the motion of the object. Referring again to the pendulum, the angle θ is a generalized coordinate, but the coordinates x and y are not. However, if a spring replaces the rigid rod, as in Figure 2.1-4*a*, then the system has two degrees of freedom, and either (r, θ) or (x, y) may be used as generalized coordinates.

If the base of the pendulum is free to slide, as in Figure 2.1-4*b*, the system has two degrees of freedom. The coordinates (x, θ) may be used as generalized coordinates.

Two pendula coupled as shown in Figure 2.1-4*c* have two degrees of freedom. The coordinates (θ_1, θ_2) may be used as generalized coordinates. Similarly, the system shown in Figure 2.1-4*d* has two degrees of freedom. The coordinates (x_1, x_2) may be used as generalized coordinates.

In Figure 2.1-4*e*, if the wheel does not slip on the surface, we may describe its motion with either the displacement x or the angle θ. The object has one degree of freedom because $\Delta x = R\Delta\theta$. Either x or θ may be used as a generalized coordinate.

A vibrating string or cable, such as shown in Figure 2.1-4*e*, has an infinite number of degrees of freedom because its vertical displacement y as a function of x is a continuous function. There are an infinite number of possible values of x; therefore there are an infinite number of possible values of y.

Finally, when interpreting the illustrations in this text, you should assume that the number of degrees of freedom is the smallest number that is consistent with the figure; that is, assume that the object has the simplest motion consistent with the figure, unless otherwise explicitly stated. For example, an object attached to a spring, as in Figure 2.1-5*a*, can in general translate in three directions as well as rotate about three axes. Thus it has six degrees of freedom. However, with the representation shown in Figure 2.1-5*b*, you may assume that the object has only one degree of freedom, that associated with the coordinate x, because no other coordinates or indications of other motions are given.

EXAMPLE 2.1-2
Pulley-Cable Kinematics

Pulleys can be used to change the direction of an applied force or to amplify forces. In our examples we will assume that the cords, ropes, chains, and cables that drive the pulleys do so without slipping and are inextensible; if not, then they must be modeled as springs. Figure 2.1-6 shows a simple pulley system. Suppose the pulley is free to rotate and to translate vertically, but not horizontally. Let L be the length of the cable, which is attached to a support. Determine the relation between ω, $\dot{y}$, and $\dot{x}$.

Solution

Suppose the system is initially in the configuration shown in Figure 2.1-6*a*, where L_1 is the unwrapped length of the cable. Because the cable length wrapped around the pulley is one-half the pulley's circumference, or πR, then

$$L = L_1 + \pi R \tag{1}$$

Suppose we pull up on the cable at its endpoint at point A. Let x be the displacement of the endpoint and y be the displacement of the pulley center, as shown in Figure 2.1-6*b*. Therefore

$$L = (x - y) + \pi R + (L_1 - y)$$

Substituting for L_1 from Equation 1, we obtain

$$L = (x - y) + \pi R + (L - \pi R - y) = L + x - 2y$$

or

$$x = 2y \tag{2}$$

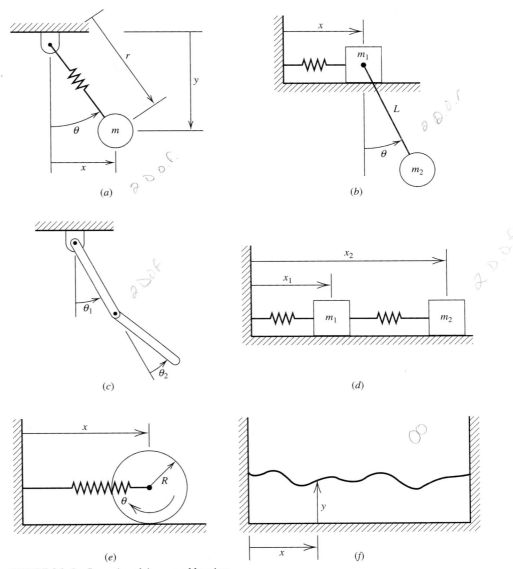

FIGURE 2.1-4 Examples of degrees of freedom.

Differentiating this equation with respect to time gives $\dot{x} = 2\dot{y}$. This says that the speed of the cable end A is twice the vertical translational speed of the pulley center.

To find the pulley's rotational speed ω as a function of the endpoint's speed $\dot{x}$, note that if the pulley center moves up a distance Δy, the pulley must rotate an angle $\Delta\theta = \Delta y/R$ if the cable does not slip. Dividing this relation by the time Δt gives

$$\frac{\Delta\theta}{\Delta t} = \frac{1}{R}\frac{\Delta y}{\Delta t}$$

or

$$\omega = \frac{1}{R}\dot{y} = \frac{1}{2R}\dot{x} \tag{3}$$

∎

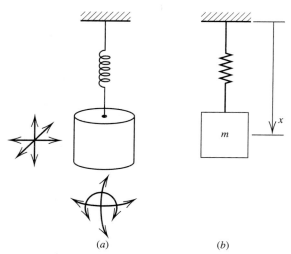

FIGURE 2.1-5 Representations of degrees of freedom. (*a*) The mass attached to a spring can move with six degrees of freedom in general (translation in three directions and rotation about three axes). (*b*) Representation of a mass constrained to have one degree of freedom, vertical translation.

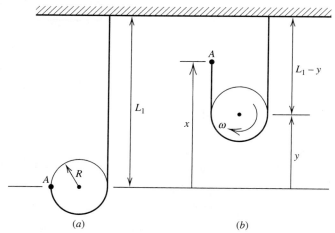

FIGURE 2.1-6 Pulley-cable kinematics.

EXAMPLE 2.1-3
A Pulley System

Consider the system shown in Figure 2.1-7.

(**a**) How many degrees of freedom does it have?

(**b**) Determine the velocity $\dot{x}_2$ and acceleration $\ddot{x}_2$ of mass 2 as functions of the velocity $\dot{x}_1$ and acceleration $\ddot{x}_1$ of mass 1.

Solution

(**a**) If the cable does not stretch, the system has one degree of freedom, because the motion of mass m_2 is not independent of the motion of mass m_1. To see why this is true, write the expression for the cable length. The total cable length is L, where

$$L = x_1 + \frac{1}{4}(2\pi r_1) + 2x_2 + \frac{1}{2}(2\pi r_2) + D = x_1 + \frac{\pi r_1}{2} + 2x_2 + \pi r_2 + D$$

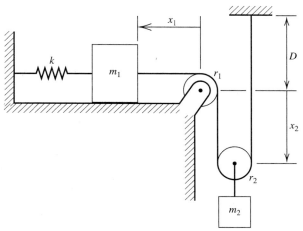

FIGURE 2.1-7 A system having one degree of freedom.

We can solve this for x_2 as follows:

$$x_2 = C - \frac{1}{2}x_1$$

where C is a constant that depends on D, L, r_1, and r_2.

(**b**) Differentiate this equation with respect to time t to obtain

$$\dot{x}_2 = -\frac{1}{2}\dot{x}_1$$

Differentiate again to obtain

$$\ddot{x}_2 = -\frac{1}{2}\ddot{x}_1$$

So the speed and acceleration of mass m_2 is one-half the speed and acceleration of mass m_1, and in the opposite direction. ■

If we need only the velocities or accelerations in a pulley system, it is inconvenient and unnecessary to determine the pulley kinematics by writing the equation for the total length of the cable. In Figure 2.1-7 note that D and the cable lengths wrapped around the pulleys are constant, so we can write $x_1 + 2x_2 = $ constant. Thus $\dot{x}_1 + 2\dot{x}_2 = 0$ and $\ddot{x}_1 + 2\ddot{x}_2 = 0$. As another example, consider Figure 2.1-8. Suppose we need to determine the relation between the velocities of mass m_A and mass m_B. Define x and y as shown from a common reference line attached to a fixed part of the system. Noting that the cable lengths wrapped around the pulleys are constant, we can write $x + 3y = $ constant. Thus $\dot{x} + 3\dot{y} = 0$. So the speed of point A is three times the speed of point B, and in the opposite direction.

2.2 PLANE MOTION OF A RIGID BODY

We now consider the case of plane motion of a rigid body. Such motion can involve both translation and rotation. A *particle* is a mass of negligible dimensions whose motion can consist of translation only. We may consider a body to be a particle if its dimensions are irrelevant for specifying its position and the forces acting on it. For example, we normally need not know the size of an earth satellite in order to describe is orbital path. *Newton's first law* states that a particle originally at rest, or moving in a straight line with a constant

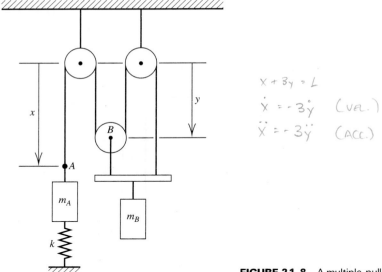

$$x + 3y = L$$
$$\dot{x} = -3\dot{y} \quad (\text{vel.})$$
$$\ddot{x} = -3\ddot{y} \quad (\text{acc.})$$

FIGURE 2.1-8 A multiple-pulley system.

speed, will remain that way as long as it is not acted upon by an unbalanced external force. *Newton's second law* states that the acceleration of a mass particle is proportional to the vector resultant force acting on it and is in the direction of this force. *Newton's third law* states that the forces of action and reaction between interacting bodies are equal in magnitude, opposite in direction, and collinear.

The second law implies that the vector sum of external forces acting on a body of mass m must equal the time rate of change of momentum. Thus

$$\frac{d(m\mathbf{v})}{dt} = \mathbf{f} \tag{2.2-1}$$

where $\mathbf{f}$ is the three-dimensional force vector and $\mathbf{v}$ is the vector velocity of the body's center of mass. For an object treated as a particle of mass m, if the mass is constant, the second law can be expressed as

$$m\frac{d\mathbf{v}}{dt} = m\mathbf{a} = \mathbf{f} \tag{2.2-2}$$

where $\mathbf{a}$ is the acceleration vector of the mass center and $\mathbf{f}$ is the net force vector acting on the mass. Note that the acceleration vector and the force vector lie on the same line. If the mass is constrained to move in only one direction, say along the direction of the coordinate x, then the equation of motion is the scalar equation

$$ma = m\frac{dv}{dt} = f \tag{2.2-3}$$

In Equation 2.2-1, the vector $\mathbf{f}$ is the vector sum of *all* the forces acting on the particle. The free-body diagram provides a reliable way of accounting for all these forces. To draw the diagram, you mentally isolate the particle from all contacting and influencing bodies and replace the bodies so removed by the forces they exert on the particle. The free-body diagram must contain all identifiable forces acting on the particle, both known and unknown (whose magnitude and/or direction is unknown). For any unknown forces, when drawing the free-body diagram, you should assume a direction for the force and denote its magnitude with a variable.

If we assume that the object is a rigid body and we neglect the force distribution within the object, we can treat the object as a particle, as if its mass were concentrated at its mass center. This is the *point-mass* assumption, which makes it easier to obtain the translational equations of motion, because the object's dimensions can be ignored and all external forces can be treated as if they acted through the mass center.

If the object can rotate, then the translational equations must be supplemented with the rotational equations of motion, which are treated in this section. The simplest case in this category of motion occurs when the body is constrained to rotate about a fixed axis. Slightly more complicated cases occur when the object is free to rotate about an axis passing through the center of mass, which may be accelerating, or when the object is constrained to rotate about an axis passing through an arbitrary but accelerating point.

Fixed-Axis Rotation

For plane motion, which means that the object can translate in two dimensions and can rotate only about an axis that is perpendicular to the plane, Newton's second law can be used to show that

$$\dot{\mathbf{H}} = \mathbf{M} \tag{2.2-4}$$

where $\mathbf{H}$ is the *angular momentum* vector of the object about an axis through a nonaccelerating point fixed to the body and $\mathbf{M}$ is the vector sum of the moments acting on the body about that point. The magnitude of the angular momentum can be expressed as

$$H = I\omega \tag{2.2-5}$$

where I is the body's *mass moment of inertia* about the axis and ω is the angular velocity about that axis. Thus, Equation 2.2-4 becomes

$$I\dot{\omega} = M \tag{2.2-6}$$

This situation is illustrated by the free-body diagram in the Figure 2.2-1. The angular displacement is θ, and $\dot{\theta} = \omega$.

The term *torque* and the symbol T are often used instead of *moment* and M. Also, when the context is unambiguous, we use the term *inertia* as an abbreviation for "mass moment of inertia."

Calculating Inertia

The mass moment of inertia I about a specified reference axis is defined as

$$I = \int r^2 \, dm \tag{2.2-7}$$

where r is the distance from the reference axis to the mass element dm. The expressions for I for some common shapes are given in Table 2.2-1. More extensive tables can be found in handbooks and engineering mechanics texts (for example, see [Meriam, 2002]).

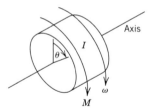

FIGURE 2.2-1 A body in rotation about a single axis.

TABLE 2.2-1   Mass Moment of Inertia

Definition: $I = \int r^2\, dm$ r = distance from reference axis to mass element dm

Conversion factor: $1\,\text{slug-ft}^2 = 1.356\,\text{kg} \cdot \text{m}^2$

Values for common elements	(m = element mass) (G = mass center)

Hollow Cylinder

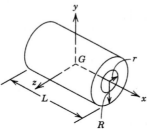

$$I_x = \frac{1}{2}m(R^2 + r^2)$$

$$I_y = I_z = \frac{1}{12}m(3R^2 + 3r^2 + L)$$

Rectangular Prism

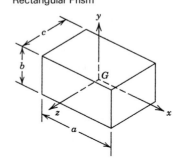

$$I_x = \frac{1}{12}m(b^2 + c^2)$$

$$I_y = \frac{1}{12}m(c^2 + a^2)$$

$$I_z = \frac{1}{12}m(a^2 + b^2)$$

Sphere

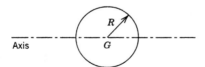

$$I = \frac{2}{5}mR^2$$

Revolving Point Mass

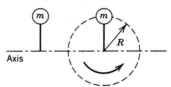

$$I = mR^2$$

Lead screw

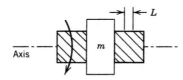

$$I = \frac{mL^2}{4\pi^2}$$

L = screw lead

If the rotation axis of a homogeneous rigid body does not coincide with the body's axis of symmetry but is parallel to it at a distance d, then the mass moment of inertia about the rotation axis is given by the *parallel-axis theorem* ([Meriam, 2002])

$$I = I_s + md^2 \tag{2.2-8}$$

where I_s is the inertia about the symmetry axis.

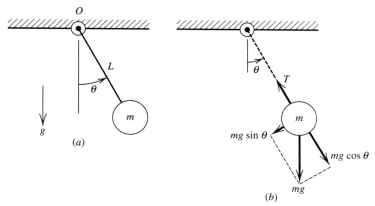

FIGURE 2.2-2 (*a*) A pendulum. (*b*) Free-body diagram showing the reaction force in the rod.

EXAMPLE 2.2-1
Equation of
Motion of a
Pendulum

Obtain the equation of motion of the pendulum shown in Figure 2.2-2*a*. The pendulum swings about the fixed pivot at point *O*. Assume that the rod mass is negligible compared to *m*.

Solution

We can use Equation 2.2-6 with $\omega = \ddot{\theta}$ as the angular acceleration. In the free-body diagram in Figure 2.2-2*b*, the tension force in the rod is T and the weight mg has been resolved into a component perpendicular to the rod and one parallel to the rod. Only the weight component perpendicular to the rod supplies a moment, so that $M = -mgL \sin\theta$ in Equation 2.2-6.

From Table 2.2-1, we see that $I = mL^2$, and thus the equation of motion is

$$mL^2\ddot{\theta} = -mgL \sin\theta$$

Canceling *m* and one *L*, we obtain

$$L\ddot{\theta} = -g \sin\theta \tag{1}$$

∎

Pulley Dynamics

Figure 2.2-3 provides an example of pulley dynamics. The pulley's center is fixed, and forces F_1 and F_2 are the tensions in the cord on either side of the pulley. From Equation 2.2-6,

$$I\dot{\omega} = F_1R - F_2R = (F_1 - F_2)R \tag{2.2-9}$$

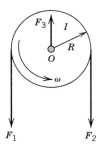

FIGURE 2.2-3 Dynamics of a single pulley.

An immediate result of practical significance is that the tension forces are approximately equal if $I\dot{\omega}$ is negligible. This condition is satisfied if either the pulley rotates at a constant speed (so that $\dot{\omega} = 0$) or the pulley inertia is negligible compared to the other inertias in the system. The pulley inertia will be negligible if either its mass or its radius is small. Thus, when we neglect the inertia of a pulley, the tension forces in the cable may be taken to be the same on both sides of the pulley. The force on the support at the pulley center is $F_3 = F_1 + F_2$. If $I\dot{\omega}$ is negligible, then the support force is approximately $2F_1$.

General Plane Motion

We now consider the case of an object undergoing general plane motion, that is, translation and rotation about an axis through an accelerating point. Assume that the object in question is a rigid body that moves in a plane passing through its mass center, and assume also that the object is symmetrical with respect to that plane. Thus it can be thought of as a slab with its motion confined to the plane of the slab. We assume that the mass center and all forces acting on the mass are in the plane of the slab.

We can describe the motion of such an object by its translational motion in the plane and by its rotational motion about an axis perpendicular to the plane. Two force equations describe the translational motion, and a moment equation is needed to describe the rotational motion.

Consider the slab shown in Figure 2.2-4, where we arbitrarily assume that three external forces f_1, f_2, and f_3 are acting on the slab. Define an x-y coordinate system as shown with its origin located at a nonaccelerating point. Then the two force equations can be written as

$$f_x = ma_{Gx} \tag{2.2-10}$$

$$f_y = ma_{Gy} \tag{2.2-11}$$

where f_x and f_y are the net forces acting on the mass m in the x and y directions respectively. The mass center is located at point G. The quantities a_{Gx} and a_{Gy} are the accelerations of the mass center in the x and y directions, relative to the fixed x-y coordinate system.

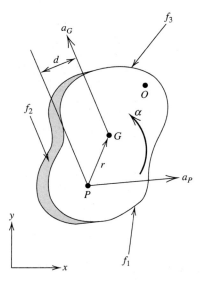

FIGURE 2.2-4 Plane motion of a slab.

We previously treated the case where an object is constrained to rotate about a nonaccelerating axis that passes through a point, say point O, fixed to the body. For this case we can apply the following moment equation:

$$M_O = I_O \alpha \qquad (2.2\text{-}12)$$

where α is the angular acceleration of the mass about an axis through a nonaccelerating point O fixed to the body, I_O is the mass moment of inertia of the body about the axis through point O, and M_O is the sum of the moments applied to the body about that axis.

The following moment equation applies regardless of whether the axis of rotation is constrained or not:

$$M_G = I_G \alpha \qquad (2.2\text{-}13)$$

where M_G is the net moment acting on the body about an axis that passes through the *mass center* G and is perpendicular to the plane of the slab. The terms I_G and α are the mass moment of inertia and angular acceleration of the body about this axis. The net moment M_G is caused by the action of the external forces $f_1, f_2, f_3, \ldots$ and any couples applied to the body. The positive direction of M_G is determined by the right-hand rule (counterclockwise if the x-y axes are chosen as shown).

Note that point G in the preceding equations must be the mass center of the object; no other point may be used. However, in many problems the acceleration of some point P is known, and sometimes it is more convenient to use this point rather than the mass center or a fixed point. The following moment equation applies for an accelerating point P, which need not be fixed to the body:

$$M_P = I_G \alpha + m a_G d \qquad (2.2\text{-}14)$$

where the moment M_P is the net moment acting on the body about an axis that passes through P and is perpendicular to the plane of the slab, a_G is the magnitude of the acceleration vector $\mathbf{a}_G$, and d is the distance between $\mathbf{a}_G$ and a parallel line through point P (see Figure 2.2-4).

An alternative form of this equation is

$$M_P = I_P \alpha + m r_x a_{Py} - m r_y a_{Px} \qquad (2.2\text{-}15)$$

where a_{Px} and a_{Py} are the x and y components of the acceleration of point P. The terms r_x and r_y are the x and y components of the location of G relative to P. I_P is the mass moment of inertia of the body about an axis through P. Note that, in general, M_P does not equal M_G and I_P does not equal I_G. If point P is fixed at some point O, then $a_{Px} = a_{Py} = 0$, and the moment equation, Equation 2.2-15, simplifies to Equation 2.2-12, because $M_O = M_P$ and $I_O = I_P$. Note that the angular acceleration α is the same regardless of whether point O, G, or P is used.

Curvilinear Translation

If the mass center moves along a curved path (curvilinear translation), the following forms of the force equations may be more useful. In terms of the mutually perpendicular *normal* and *tangential* coordinates n and t, the force equations are written as (Figure 2.2-5)

$$f_n = m a_n \qquad (2.2\text{-}16)$$

$$f_t = m a_t \qquad (2.2\text{-}17)$$

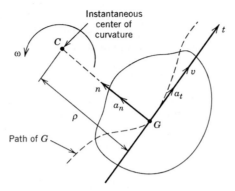

FIGURE 2.2-5 Normal and tangential coordinates.

where f_n and f_t are the net external forces in the n and t directions, and a_n and a_t are the acceleration components in those directions. Because the coordinates are perpendicular, the total acceleration magnitude a can be found from

$$a = \sqrt{a_n^2 + a_t^2} \tag{2.2-18}$$

The normal and tangential acceleration components depend on the radius of curvature ρ, the speed v, and the angular velocity ω as follows:

$$a_n = \frac{v^2}{\rho} = \rho\omega^2 \tag{2.2-19}$$

$$a_t = \dot{v} \tag{2.2-20}$$

Circular motion is a special case of curvilinear translation in which the radius of curvature ρ is a constant R, the radius of the circle. In this case the following relations are also true:

$$v = R\omega \tag{2.2-21}$$

$$a_t = R\alpha \tag{2.2-22}$$

where $\alpha = \dot{\omega} = \ddot{\theta}$ is the angular acceleration.

In summary, the equations of motion for a rigid body in plane motion are given by two force equations expressed in an appropriate coordinate system, and the moment equation—either Equation 2.2-12, 2.2-13, 2.2-14, or 2.2-15. All the equations need not be used for given problem; the appropriate choice depends on the nature of the problem.

The Pendulum Revisited

Many engineering devices rotate about a fixed point and are examples of a pendulum, although they might not appear to be so at first glance. In the pendulum analysis of Example 2.2-1, it was assumed that the rod mass was negligible and therefore that the center of mass of the system was at m at the end of the rod. This is not always the case, however, and the following example examines this situation.

EXAMPLE 2.2-2
A Compound Pendulum

An object of mass m with an arbitrary shape is attached at a fixed point O and is free to swing about that point (Figure 2.2-6a). Its mass moment of inertia about its mass center G is I_G. Its mass center is a distance L from point O. Neglect friction. Obtain its equation of motion in terms of the angle θ and determine the reaction forces on the support at point O, in terms of θ and $\dot{\theta}$.

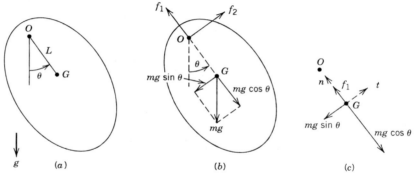

FIGURE 2.2-6 (*a*) A compound pendulum. (*b*) Free-body diagram. (*c*) Forces acting at point *G*.

Solution | Choosing the moment equation, Equation 2.2-12, about point O, we have $M_O = I_O \alpha$, where $\alpha = \ddot{\theta}$. The moment M_O is caused by the weight mg of the object acting through the mass center at G (Figure 2.2-6b). Thus $M_O = -(mg \sin \theta)L$ and the desired equation of motion is

$$I_O \ddot{\theta} = -mgL \sin \theta \tag{1}$$

From the parallel-axis theorem, $I_O = I_G + mL^2$. Because the mass center G does not translate in a straight line, it is more convenient to use normal-tangential coordinates to find the reaction forces. Figure 2.2-6c shows the forces at point G resolved along the normal-tangential directions. The reaction forces acting on the body at point O are f_1 and f_2. For the normal direction, Equations 2.2-16 and 2.2-19 give

$$ma_n = f_n = f_1 - mg \cos \theta$$

$$a_n = L\omega^2$$

Thus

$$f_1 = mL\omega^2 + mg \cos \theta \tag{2}$$

We can find f_2 by applying Equation 2.2-13:

$$I_G \ddot{\theta} = -f_2 L \tag{3}$$

Using Equation 1, we obtain

$$f_2 = -\frac{I_G}{L}\ddot{\theta} = \frac{I_G}{L}\left(\frac{mgL}{I_O}\sin \theta\right) = \frac{mgI_G}{I_G + mL^2}\sin \theta \tag{4}$$

In a later chapter we will discuss how to solve Equation 1 for $\theta(t)$ and $\dot{\theta}(t)$. We can then substitute $\theta(t)$ and $\dot{\theta}(t)$ into Equations 2 and 4 to obtain the reaction forces as functions of time. ∎

The moment equation about the mass center G is likely to give useful results when the object is not constrained about a fixed point and no information is given about the acceleration of any point on the object.

EXAMPLE 2.2-3 | Figure 2.2-7a shows the cross-section view of a ship undergoing rolling motion.
Rolling Motion | Archimedes' principle states that the buoyancy force B equals the weight of the displaced
of a Ship | liquid. In order to float, B must equal the ship's weight $W = mg$. Thus $B = W = mg$. The metacenter M is the intersection point of the line of action of the buoyancy force and the ship's centerline. The distance h of M from the mass center G is called the *metacentric height*. Obtain the equation describing the ship's rolling motion in terms of θ.

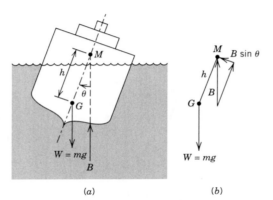

FIGURE 2.2-7 (a) Rolling motion of a ship. (b) Free-body diagram.

Solution

Use the moment equation about the mass center G and the force diagram in Figure 2.2-7b to obtain

$$I_G\ddot{\theta} = M_G = -(B\sin\theta)h$$

or, with $B = mg$,

$$I_G\ddot{\theta} + mgh\sin\theta = 0 \tag{1}$$

where I_G is the ship's moment of inertia about G. This has the same form as the pendulum equation, Equation 1 of Example 2.2-1. In a later chapter we will see how to solve this equation for $\theta(t)$. We can use this solution to determine the roll frequency of the ship. This model neglects drag on the ship's hull as it rolls, and in a later chapter we will see how to include the effects of drag. ■

The moment equations, Equation 2.2-14 and 2.2-15, are useful when information is given about the acceleration of a point on the object.

EXAMPLE 2.2-4
A Crane
Application

A crane with a movable trolley is shown in Figure 2.2-8a. Its hook and suspended load act like a pendulum whose base is moving. The hook pulley and load have a combined weight mg. The trolley is moving to the left with an acceleration a. Treat the pulley cable as a rigid rod of length L and negligible mass, and obtain the equation of motion in terms of θ.

Solution

This is an application for the moment equation, Equation 2.2-15, about an accelerating point P. Here $I_P = mL^2$, $a_{Px} = -a$, and $a_{Py} = 0$. From Figure 2.2-8b we see that $r_x = L\sin\theta$ and $r_y = -L\cos\theta$. Equation 2.2-15 gives:

$$mL^2\ddot{\theta} - mL\cos\theta a = M_P = -mgL\sin\theta$$

Dividing by m and rearranging, we obtain:

$$L\ddot{\theta} + g\sin\theta = a\cos\theta$$

In a later chapter we will see how to solve this equation for $\theta(t)$. ■

Sliding versus Rolling Motion

Wheels are common examples of systems undergoing general plane motion with both translation and rotation. The wheel shown in Figure 2.2-9a can have one of three possible motion types:

1. Pure rolling motion. This occurs when there is no slipping between the wheel and the surface. In this case, $v = R\omega$.

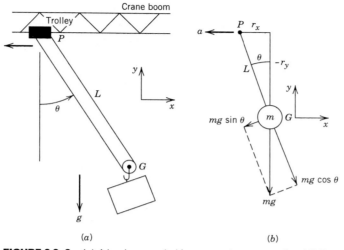

FIGURE 2.2-8 (*a*) A load suspended from a moving crane trolley. (*b*) Free-body diagram of the equivalent pendulum model.

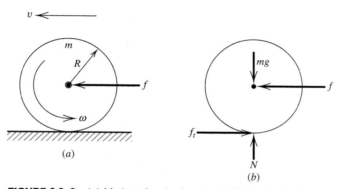

FIGURE 2.2-9 (*a*) Motion of a wheel and axle. (*b*) Free-body diagram.

2. Pure sliding motion. This occurs when the wheel is prevented from rotating (such as when a brake is applied). In this case $\omega = 0$ and $v \neq R\omega$.

3. Sliding and rolling motion. In this case $\omega \neq 0$. Because slipping occurs in this case, $v \neq R\omega$.

The wheel will roll without slipping (pure rolling) if the tangential force f_t is smaller than the static friction force $\mu_s N$, where N is the force of the wheel normal to the surface (Figure 2.2-9*b*). In this case the tangential force does no work because it does not act through a distance. If the static friction force is smaller than f_t, the wheel will slip. If slipping occurs, the tangential force is computed from the dynamic friction equation: $f_t = \mu_d N$.

2.3 WORK-ENERGY METHODS

Methods based on the concepts of work and energy are useful for solving problems involving force, displacement, and velocity, but not time. In the next section we will see that methods based on impulse and momentum enable us to solve problems involving force, velocity, and time, but not displacement. Work-energy methods are based on an integration of the equation of motion over displacement, whereas impulse-momentum methods are based on an integration over time.

Conservation of Mechanical Energy

Conservation of mechanical energy is a direct consequence of Newton's second law. Consider a scalar case in which the force f is a function of displacement x:

$$m\dot{v} = f(x)$$

Multiply both sides by $v\,dt$ and use the fact that $v = dx/dt$.

$$mv\,dv = vf(x)\,dt = \frac{dx}{dt}f(x)\,dt = f(x)\,dx$$

Integrate both sides to obtain

$$\int mv\,dv = \frac{mv^2}{2} = \int f(x)\,dx + C \tag{2.3-1}$$

where C is a constant of integration.

Work is force times displacement, so the integral on the right represents the total work done on the mass by the force $f(x)$. The term on the left is called the *kinetic energy* (*KE*), often denoted by the symbol T.

If the force $f(x)$ is derivable from a function $V(x)$ as follows,

$$f(x) = -\frac{dV}{dx} \tag{2.3-2}$$

then $f(x)$ is called a *conservative force*. If we integrate both sides of this equation, we obtain

$$V(x) = \int dV = -\int f(x)\,dx$$

or, from Equation 2.3-1,

$$\frac{mv^2}{2} + V(x) = C \tag{2.3-3}$$

This equation shows that $V(x)$ has the same units as kinetic energy. $V(x)$ is called the *potential energy function* (*PE*).

Equation 2.3-3 states that the sum of the kinetic and potential energies must be constant if no force other than the conservative force is applied. If v and x have the values v_0 and x_0 at the time t_0, then

$$\frac{mv_0^2}{2} + V(x_0) = C$$

Comparing this with Equation 2.3-3 gives

$$\frac{mv^2}{2} - \frac{mv_0^2}{2} + V(x) - V(x_0) = 0 \tag{2.3-4}$$

which may be expressed as

$$\Delta T + \Delta V = 0 \tag{2.3-5}$$

where the change in kinetic energy is $\Delta T = m(v^2 - v_0^2)/2$ and the change in potential energy is $\Delta V = V(x) - V(x_0)$. This equation expresses the principle of conservation of mechanical energy, which states that the change in kinetic energy plus the change in potential energy is zero. This principle is not valid if nonconservative forces do work on the object.

For some problems, the following forms of the principle are more convenient to use:

$$\frac{mv_0^2}{2} + V(x_0) = \frac{mv^2}{2} + V(x) \tag{2.3-6}$$

or

$$T_1 + V_1 = T_2 + V_2 \tag{2.3-7}$$

where the subscripts refer to two different states of the system.

Note that the potential energy has a relative value only. The choice of reference point for measuring x determines only the value of the constant C, which Equation 2.3-5 shows to be irrelevant.

Gravitational Potential Energy

Gravity is an example of a conservative force, for which $f = -mg$. Thus, if h represents vertical displacement (positive upward),

$$V(h) = mgh + \text{a constant}$$

where the value of the constant depends on our chosen reference height (called the "datum"). Thus

$$\frac{mv^2}{2} - \frac{mv_0^2}{2} + mg(h - h_0) = 0 \tag{2.3-8}$$

where $v = dh/dt$.

Elastic Potential Energy

An elastic element has a potential energy whenever it is stretched or compressed. For a linear element, the resisting force is given by $f = -kx$, where x is the displacement from free length. Thus the elastic potential function is given by

$$V(x) = -\int_0^x f(x)\, dx = \frac{1}{2}kx^2$$

The work done in deflecting the spring a distance x from its free length is thus $V(x) = kx^2/2$. Note that this work is the same regardless of whether the spring is compressed or stretched.

In many of our applications the potential energy consists of gravitational potential energy and the potential energy due to any spring elements. To distinguish them, we will denote the gravitational potential energy by V_g and the potential energy due to a spring element by V_s. Thus we may express conservation of mechanical energy as follows:

$$T_1 + V_{g1} + V_{s1} = T_2 + V_{g2} + V_{s2}$$

Energy and Rotational Motion

The principle of conservation of mechanical energy applies as well to systems of particles and to rigid bodies. The equation of motion of a body of inertia I not translating but in pure rotation about an axis through a fixed point or through the center of mass is

$$I\frac{d\omega}{dt} = M$$

where ω is the angular velocity and M is the applied moment. This equation has the same form as the equation of motion for an object in pure translation, $m\dot{v} = f(x)$. Following the development leading to Equation 2.3-1, we see that

$$\frac{1}{2}I\omega^2 = \int M\,d\theta + C$$

We thus see that the work done by the moment M produces the kinetic energy of rotation: $T = I\omega^2/2$.

The kinetic energy of a rigid body is the sum of the kinetic energy of translation and the kinetic energy of rotation. For general plane motion, the kinetic energy is

$$T = \frac{1}{2}mv_G^2 + \frac{1}{2}I_G\omega^2$$

where v_G is the velocity of the center of mass and I_G is the moment of inertia about the mass center.

The Work-Energy Principle

Let us look more closely at the concept of work. The work done on a particle by a force vector $\mathbf{F}$ acting through a differential displacement $d\mathbf{r}$ is

$$dW = \mathbf{F} \cdot d\mathbf{r} = F\,ds\,\cos\theta$$

where $ds = |d\mathbf{r}|$ and θ is the angle between $\mathbf{F}$ and $d\mathbf{r}$. Thus the work may be interpreted either as the component of $\mathbf{F}$, $F\cos\theta$, along the $d\mathbf{r}$ direction times the displacement ds or as the force F times the component of $d\mathbf{r}$, $ds\cos\theta$, along the $\mathbf{F}$ direction. The work is positive if $F\cos\theta$ is in the direction of the displacement. Note that the component of $\mathbf{F}$ normal to $d\mathbf{r}$, $F\sin\theta$, does no work.

Let T_1 be the kinetic energy of the particle at point 1 and T_2 the kinetic energy at point 2. Similarly, let V_1 and V_2 be the potential energy of the particle at point 1 and point 2. Let W_{1-2} denote the work done on the particle during displacement from point 1 to point 2 by all external forces other than gravity or spring forces. That is,

$$W_{1-2} = \int_{\text{state 1}}^{\text{state 2}} dW = \int_{s_1}^{s_2} F\cos\theta\,ds$$

The principle of work-energy may be stated as follows:

$$T_1 + V_1 + W_{1-2} = T_2 + V_2$$

For our applications, the potential energy consists of gravitational potential energy V_g and the potential energy V_s due to any spring elements. So we may express the work-energy principle as follows:

$$T_1 + V_{g1} + V_{s1} + W_{1-2} = T_2 + V_{g2} + V_{s2}$$

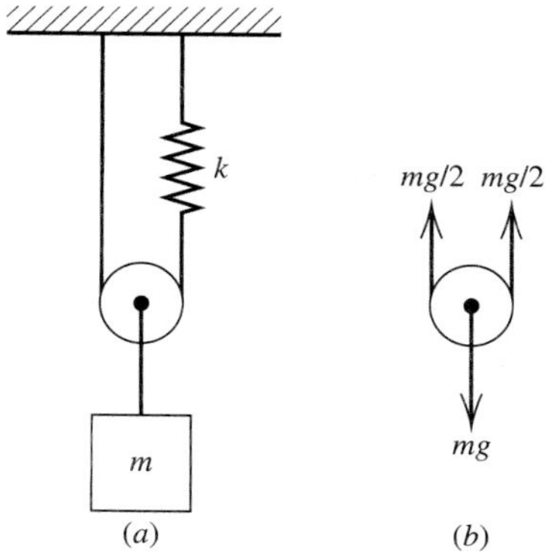

FIGURE 2.3-1 A mass-spring-pulley system.

EXAMPLE 2.3-1
Application of
Conservation of
Energy

In the system shown in Figure 2.3-1a the masses of the spring and pulley are negligible compared to the mass m. Use the values $m = 20$ kg and $k = 500$ N/m. (**a**) What is the static deflection in the spring when the mass m is stationary? (**b**) Suppose the spring is initially stretched 0.1 m when we release the mass m. Calculate the velocity of the mass m after it has dropped 0.05 m.

Solution

(**a**) From the equilibrium free-body diagram shown in Figure 2.3-1b, because we are neglecting the masses of the spring and pulley, the spring is subjected to the force $mg/2$, and thus the static deflection δ in the spring is given by $k\delta = mg/2$, or

$$\delta = \frac{mg}{2k} = \frac{20(9.81)}{2(500)} = 0.196 \text{ m}$$

(**b**) Since we are releasing the mass above its static equilibrium position, it will fall. From conservation of energy,

$$T_1 + V_{g1} + V_{s1} = T_2 + V_{g2} + V_{s2}$$

Selecting the initial position as the datum for gravitational potential energy, and noting that the spring deflection is twice that of the mass m, we obtain

$$0 + 0 + \frac{1}{2}500(0.1)^2 = \frac{1}{2}20v^2 - 20(9.81)(0.05) + \frac{1}{2}500[0.1 + 2(0.05)]^2$$

or

$$2.5 = 10v^2 - 9.81 + 10$$

which gives $v = 0.481$ m/s. ∎

EXAMPLE 2.3-2
Application of
the Work-Energy
Method

The mass m is released from rest a distance d above the spring as shown in Figure 2.3-2a. The coefficient of dynamic friction between the mass and the surface is μ. Using the values $m = 3$ kg, $d = 0.6$ m, $k = 1500$ N/m, $\theta = 60°$, and $\mu = 0.3$, (**a**) calculate the velocity of m when it first touches the spring and (**b**) calculate the resulting maximum spring deflection.

Solution

(**a**) The free-body diagram is shown in Figure 2.3-2b. From this we can see that the normal force is $N = 3(9.81) \cos 60° = 29.4$ N. The force doing work is the friction force μN. The spring is not involved in this phase, so from the work-energy principle,

$$T_1 + V_{g1} + W_{1-2} = T_2 + V_{g2}$$

Choosing the datum for V_g to be at the initial position of m, we obtain

$$0 + 0 - 0.3(29.4)0.6 = \frac{1}{2}(3)v^2 - 3(9.81)0.6 \sin 60°$$

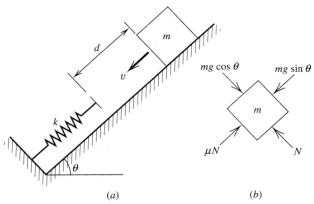

FIGURE 2.3-2 (*a*) A mass impacting a spring. (*b*) Free-body diagram before impact.

which gives $v = 3.16$ m/s. Note that the work done by the friction force on the mass m is *negative* because the friction force is in the opposite direction of the displacement. Note also that the normal force N does not do work and thus does not appear in the work-energy equation.

(**b**) Let δ be the maximum spring deflection. From the work-energy principle,

$$T_1 + V_{s1} + V_{g1} + W_{1-2} = T_2 + V_{s2} + V_{g2}$$

Choosing the datum for V_g to be at the point where m first touches the spring, we obtain

$$\frac{1}{2}3(3.16)^2 + 0 + 0 - [0.3(3)(9.81)\cos 60°]\delta = 0 + \frac{1}{2}1500\delta^2 - 3(9.81)\delta \sin 60°$$

or

$$750\delta^2 - 21.0726\delta - 14.9784 = 0$$

which gives $\delta = 0.156$ and $\delta = -0.128$ m. The desired answer is the positive solution, $\delta = 0.156$ m. ∎

2.4 IMPULSE-MOMENTUM METHODS

Methods based on impulse and momentum aid us in solving problems in which the applied forces act either over very short times, such as happens during an impact, or over a specified time interval.

The Linear Impulse-Momentum Principle

Newton's second law gives

$$m\frac{d\mathbf{v}}{dt} = \mathbf{F}$$

where $\mathbf{F}$ is the net force acting on the particle of mass m. This can be expressed as $\dot{\mathbf{G}} = \mathbf{F}$, where the *linear momentum* of the particle is $\mathbf{G} = m\mathbf{v}$. This equation says that the time rate of change of the linear momentum of a particle equals the resultant of all forces acting on the particle.

Integrating $\mathbf{G} = m\mathbf{v}$ over time, we obtain

$$\int_{t_1}^{t_2} \dot{\mathbf{G}}\, dt = \mathbf{G}_2 - \mathbf{G}_1 = \int_{t_1}^{t_2} \mathbf{F}\, dt \tag{2.4-1}$$

where $\mathbf{G}_1 = \mathbf{G}(t_1)$, $\mathbf{G}_2 = \mathbf{G}(t_2)$, and $\int_{t_1}^{t_2} \mathbf{F}\, dt$ is the *linear impulse* of the force $\mathbf{F}$. This equation expresses the linear impulse-momentum principle, which states that the linear impulse equals the change in linear momentum. If the linear impulse is zero, then $\mathbf{G}_1 = \mathbf{G}_2$, which states that the linear momentum is conserved.

Note that these results apply also to a system of particles. For example, in a system composed of two particles, if the two particles have equal but opposite forces applied to them, the vector sum of forces acting on the system is zero, and thus the momentum of the system is conserved.

EXAMPLE 2.4-1
Application of the
Linear Momentum
Principle

The mass m_1 is dropped from rest a distance h onto the mass m_2, which is initially resting on the spring support (Figure 2.4-1). Assume that the impact is inelastic so that m_1 sticks to m_2. Calculate the maximum spring deflection caused by the impact. The given values are $m_1 = 0.5$ kg, $m_2 = 4$ kg, $k = 400$ N/m, and $h = 2$ m.

Solution

We may think of this process as consisting of three phases: (1) the drop, (2) the impact, and (3) the motion immediately after the impact. Let v_1 be the velocity of m_1 just before impact. During the drop phase, conservation of energy gives

$$0.5(9.81)2 = \frac{1}{2}(0.5)v_1^2$$

or $v_1 = 6.264$ m/s.

Let v_2 be the velocity of the combined mass after impact. At impact, conservation of linear momentum gives

$$0.5(6.264) = (0.5 + 4)v_2$$

or $v_2 = 0.696$ m/s.

The initial spring deflection caused by the weight of m_2 is

$$\delta_1 = \frac{4(9.81)}{400} = 0.098 \text{ m}$$

We can use conservation of energy to compute the additional spring deflection δ_2 after the impact:

$$T_1 + V_{g1} + V_{s1} = T_2 + V_{g2} + V_{s2}$$

Selecting V_g to be zero at the initial position of m_2, we have

$$\frac{1}{2}(4.5)(0.696)^2 + 0 + \frac{1}{2}400(0.098)^2 = 0 - 4.5(9.81)\delta_2 + \frac{1}{2}400(0.098 + \delta_2)^2$$

This gives

$$200\delta_2^2 - 4.945\delta_2 - 1.09 = 0$$

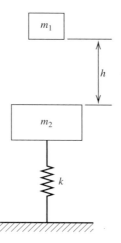

FIGURE 2.4-1 A falling object impacting a mass-spring system.

which has the solutions $\delta_2 = 0.087$ and $\delta_2 = -0.063$ m. Selecting the positive solution and adding it to the initial static deflection of 0.098 m, we obtain the total deflection: $0.087 + 0.098 = 0.185$ m. This is the lowest point reached by the combined mass after the impact. Of course, the mass will not remain at rest in this position but will oscillate.   ■

The Angular Impulse-Momentum Principle

The *angular momentum* $\mathbf{H}_O$ of the particle about a point O is the moment of the linear momentum of the particle about point O and is expressed as

$$\mathbf{H}_O = \mathbf{r} \times \mathbf{G} = \mathbf{r} \times m\mathbf{v}$$

where $\mathbf{r}$ is the position vector of the particle measured from the point O and $\mathbf{v} = \dot{\mathbf{r}}$. If the net external force acting on the particle is $\mathbf{F}$, its moment $\mathbf{M}_O$ about point O is

$$\mathbf{M}_O = \mathbf{r} \times \mathbf{F} = \mathbf{r} \times m\dot{\mathbf{v}}$$

if the mass is constant. Thus

$$\dot{\mathbf{H}}_O = \dot{\mathbf{r}} \times m\mathbf{v} + \mathbf{r} \times m\dot{\mathbf{v}} = \mathbf{v} \times m\mathbf{v} + \mathbf{r} \times m\dot{\mathbf{v}}$$

However, $\mathbf{v} \times m\mathbf{v} = 0$ because the vector $m\mathbf{v}$ is collinear with $\mathbf{v}$. Therefore

$$\dot{\mathbf{H}}_O = \mathbf{r} \times m\dot{\mathbf{v}} = \mathbf{M}_O$$

which states that the time rate of change of the angular momentum of m about a fixed point O equals the moment about O of all forces acting on m. This relation can be extended to a system of particles.

If we integrate the previous equation over time, we obtain

$$\int_{t_1}^{t_2} \dot{\mathbf{H}}_O \, dt = \mathbf{H}_{O2} - \mathbf{H}_{O1} = \int_{t_1}^{t_2} \mathbf{M}_O \, dt \qquad (2.4\text{-}2)$$

The integral $\int_{t_1}^{t_2} \mathbf{M}_O \, dt$ is the *angular impulse*. This equation expresses the principle of angular impulse and momentum, which states that the change in angular momentum of m about a fixed point O equals the total angular impulse acting on m about O. If the angular impulse is zero, then

$$\mathbf{H}_{O1} = \mathbf{H}_{O2}$$

which states that the angular momentum is conserved.

Impact

When one particle impacts another, the *coefficient of restitution e* is defined as

$$e = \frac{|\text{relative velocity of separation}|}{|\text{relative velocity of approach}|} \qquad (2.4\text{-}3)$$

The impact is *perfectly elastic* if $e = 1$. This means that the particles rebound from each other and no energy is lost. The impact is *perfectly plastic* or *inelastic* if $e = 0$. This means that the particles stick to one another; however, all the energy is not necessarily lost.

Application to Rigid-Body Motion

The linear impulse-momentum principle for a rigid body states that

$$\mathbf{G}_1 + \int_{t_1}^{t_2} \mathbf{F} \, dt = \mathbf{G}_2$$

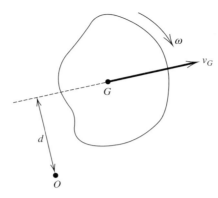

FIGURE 2.4-2 Illustration of angular momentum H_O about point O, where $H_O = I_G\omega + mv_Gd$.

where $\mathbf{G} = m\mathbf{v}_G$. For plane motion, the angular momentum and angular impulse vectors are normal to the plane of motion, and so vector notation is not needed. The angular impulse-momentum principle for this case states that

$$H_{G1} + \int_{t_1}^{t_2} M_G\, dt = H_{G2}$$

where $H_G = I_G\omega$. For any point O, fixed or moving, on or off the body,

$$H_O = I_G\omega + mv_Gd \tag{2.4-4}$$

where the distance d is defined in Figure 2.4-2.

If the body is constrained to rotate about an axis through a fixed point O, which may be on or off the body, then

$$H_O = I_O\omega \tag{2.4-5}$$

$$M_O = I_O\dot{\omega} \tag{2.4-6}$$

$$\int_{t_1}^{t_2} M_O\, dt = I_O(\omega_2 - \omega_1) \tag{2.4-7}$$

EXAMPLE 2.4-2
Pendulum
Equation of
Motion

Neglect the mass of the rod in the pendulum shown in Figure 2.4-3a, and derive the equation of motion (**a**) using the angular impulse-momentum principle and (**b**) using conservation of energy.

Solution

(**a**) The free-body diagram is shown in Figure 2.4-3b. The principle of angular impulse-momentum states that

$$M_O = \dot{H}_O$$

or

$$-(mg\,\sin\theta)L = \frac{d}{dt}\left(mL^2\dot{\theta}\right) = mL^2\ddot{\theta}$$

which gives the equation of motion

$$L\ddot{\theta} = -g\,\sin\theta$$

(**b**) As the pendulum moves from $\theta = 0$ to θ, conservation of energy gives

$$T_1 + V_{g1} = T_2 + V_{g2}$$

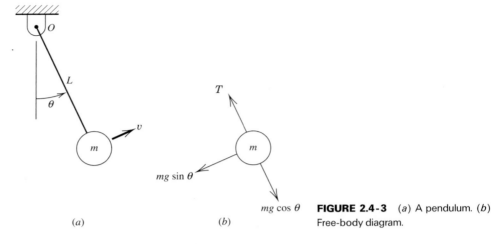

FIGURE 2.4-3 (a) A pendulum. (b) Free-body diagram.

(a)

(b)

Selecting the datum for potential energy to be at $\theta = 0$, we obtain

$$0 + 0 = \frac{1}{2}mv^2 + mgL(1 - \cos \theta)$$

Since $v = L\dot{\theta}$, we have

$$0 = \frac{1}{2}m(L\dot{\theta})^2 + mgL(1 - \cos \theta)$$

or

$$0 = \frac{1}{2}L\dot{\theta}^2 + g(1 - \cos \theta)$$

Differentiating both sides of this equation with respect to time t gives

$$0 = \frac{1}{2}L(2\dot{\theta}\ddot{\theta}) + g\dot{\theta} \sin \theta$$

Cancel $\dot{\theta}$ to obtain

$$L\ddot{\theta} + g \sin \theta = 0$$

■

EXAMPLE 2.4-3
Pendulum Impact

A mass m_1 moving with a velocity v_1 strikes and becomes embedded in the pendulum as shown in Figure 2.4-4. Assume that the pendulum is a slender rod of mass m_2. Use the values $m_1 = 10^{-3}$ kg, $m_2 = 20$ kg, $v_1 = 500$ m/s, and $L = 0.75$ m. **(a)** Determine the angular velocity ω of the pendulum immediately after the impact. **(b)** Determine the maximum angular displacement of the bar following impact.

Solution

(a) The angular momentum of the entire system consisting of the mass m_1 and the pendulum is conserved because the force of impact acting on the pendulum is equal but opposite to that acting on the mass m_1. Thus, taking counterclockwise to be the positive direction of rotation, we have

$$H_{O1} = H_{O2}$$

or

$$10^{-3}(500)2(0.75) = \left\{ I_O + 10^{-3}[2(0.75)]^2 \right\}\omega$$

where I_O is given by $I_O = I_G + m_2L^2$. For a slender rod,

$$I_G = \frac{1}{12}m_2(2L)^2 = \frac{1}{12}(20)[2(0.75)]^2 = 3.75 \ \ \text{kg} \cdot \text{m}^2$$

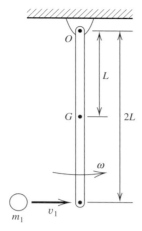

FIGURE 2.4-4   A mass impacting a pendulum.

Thus $I_O = 3.75 + 20(0.75)^2 = 15$ kg $\cdot$ m^2, and the angular velocity immediately after the impact is $\omega = 0.05$ rad/sec.

(b) After impact, the location of the mass center of the combined masses m_1 and m_2 changes, but only very slightly because $m_1 \ll m_2$. Thus we will take the mass center still to be located a distance L from the pivot point. The kinetic energy of the system just after impact is, with $v_G = L\omega$,

$$T = \frac{1}{2}(m_1 + m_2)v_G^2 + \frac{1}{2}I_G\omega^2 = \frac{1}{2}(10^{-3} + 20)[0.75(0.05)]^2 + \frac{1}{2}(3.75)(0.05)^2 = 0.0188$$

The change in gravitational potential energy when the bar reaches its maximum displacement θ is

$$\Delta V = (m_1 + m_2)gL(1 - \cos\theta) = 20(9.8)(0.75)(1 - \cos\theta) = 147(1 - \cos\theta)$$

Equating T with ΔV gives $\theta = 0.016$ rad, or $0.9°$.   ∎

2.5 EQUIVALENT MASS AND INERTIA

The concepts of equivalent mass and equivalent inertia use kinetic energy equivalence to simplify the process of obtaining descriptions of systems containing both translating and rotating parts, systems containing distributed mass, and systems consisting of multiple masses whose motion is coupled.

The system in Figure 2.5-1a consists of a cart having two wheels connected by a rigid, massless axle (only one wheel is shown). The cart body has a mass m_b, and each wheel has a mass m_w. The next section treats systems with springs, but for now let us concentrate on the cart dynamics.

One way to analyze the cart dynamics is to draw two free-body diagrams, one for the cart body and one for the wheel-axle system. To do this you must include the reaction forces between the axle and the cart body. These diagrams will produce three equations of motion for the wheel-axle system (one rotational and two translational equations) and two translational equations for the cart body. These five equations must be combined to eliminate the unknowns to obtain a single equation in terms of the variable x.

An easier way to solve this problem is to find the equivalent mass of the cart-wheel system and model the system as shown in Figure 2.5-1b. As we will see in the next

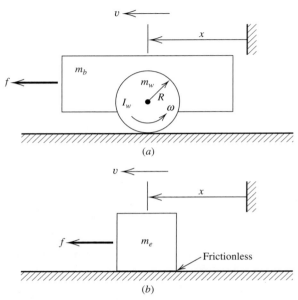

FIGURE 2.5-1 (*a*) A cart with two wheels. (*b*) Equivalent translating sytem.

section, it is much easier to obtain the equation of motion for this equivalent system because we need not deal with two separate masses and the reaction forces between them.

Equivalent Mass of a Wheel

To understand the equivalent-mass method, consider the equation of motion for a block with mass m moving without friction on a horizontal surface:

$$m\dot{v} = f \tag{2.5-1}$$

where f is an applied force (Figure 2.5-2*a*). Instead of a block, suppose we have a wheel of mass m_w, radius R, and moment of inertia I_w (Figure 2.5-2*b*). Suppose we push on the

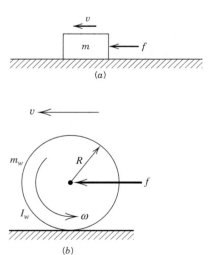

FIGURE 2.5-2 (*a*) Motion of a mass on a frictionless horizontal surface. (*b*) Motion of a wheel and axle.

axle with a force f and we want to find the equation of motion in terms of the axle's translational speed v. If the wheel does not slip, its rotational speed is related to the translational speed by $v = R\omega$.

One way to solve this problem is to find the *equivalent mass* of the wheel and use Equation 2.5-1. The equivalent mass is the fictitious mass value that makes the wheel have the same kinetic energy as the block if they are both moving at the same speed v. The equivalent mass can be interpreted as the mass value that a blindfolded person would feel while pushing on the axle, thinking he or she was pushing a block (assuming that the wheel and block are moving horizontally). The equivalent mass can be computed by equating the kinetic energies of the block and wheel, assuming they are moving at the same speed v:

$$\frac{1}{2}mv^2 = \frac{1}{2}m_wv^2 + \frac{1}{2}I_w\omega^2 = \frac{1}{2}\left(m_w + \frac{I_w}{R^2}\right)v^2$$

where $v = R\omega$. Thus the wheel is equivalent to a block having a mass

$$m = m_w + \frac{I_w}{R^2} \tag{2.5-2}$$

Using this value in Equation 2.5-1, we obtain

$$\left(m_w + \frac{I_w}{R^2}\right)\dot{v} = f \tag{2.5-3}$$

which is the equation of motion for the wheel.

Another way to compute the equivalent mass is to equate the force f required to give the block and the axle the same acceleration $\dot{v}$. If the wheel does not slip, then $v = R\omega$ and $\dot{v} = R\dot{\omega}$. The surface produces a force f_t tangential to the wheel at its rim (Figure 2.5-3). This force produces a torque Rf_t on the wheel. Thus the equations of motion are, for translation, $m_w\dot{v} = f - f_t$, and for rotation, $I_w\dot{\omega} = Rf_t$. Solve the second equation for f_t and substitute it into the first equation, using $\dot{v} = R\dot{\omega}$, to obtain Equation 2.5-3. This approach gives the same equivalent-mass value obtained with energy equivalence, but the energy-equivalence method is easier because it does not involve the tangential force f_t.

Equivalent Mass of a Cart

For the cart shown in Figure 2.5-1a, its kinetic energy is

$$KE = \frac{1}{2}m_bv^2 + \frac{1}{2}(2m_w)v^2 + \frac{1}{2}(2I_w)\omega^2$$

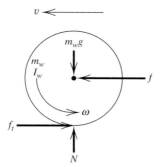

FIGURE 2.5-3 Free-body diagram of a wheel.

where I_w is the inertia of each wheel. Thus, since $v = R\omega$ if the wheels do not slip,

$$KE = \frac{1}{2}\left(m_b + 2m_w + \frac{2I_w}{R^2}\right)v^2$$

So the equivalent mass is

$$m_e = m_b + 2m_w + \frac{2I_w}{R^2}$$

We can now represent the cart system shown in Figure 2.5-1a as the simpler system shown in Figure 2.5-1b. Note that because the tangential wheel force does no work, we must treat the suface in Figure 2.5-1b as frictionless.

The equivalent mass expresses only the equivalent *inertial* resistance to changes in motion. It expresses only the kinetic energy, not the potential energy. Thus we must be careful using the equivalent mass when gravity affects the motion.

EXAMPLE 2.5-1
A Wheel on an Incline

Use the equivalent-mass concept to find the equation of motion for a wheel being pushed at its axle by a force on an incline of angle ϕ (Figure 2.5-4).

Solution

The equivalent mass is, from Equation 2.5-3,

$$m_e = m_w + \frac{I_w}{R^2}$$

Note that the gravity force is computed from the actual mass m_w, not the equivalent mass m_e. Thus the gravity force is $m_w g$. The equation of motion is

$$m_e \dot{v} = f - m_w g \sin \phi$$

■

Equivalent Inertia

Sometimes we want to visualize a system consisting of translating and rotating parts as a purely rotational one. In such cases the concept of *equivalent inertia* is useful.

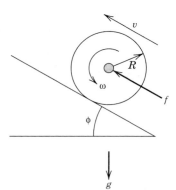

FIGURE 2.5-4 Motion of a wheel and axle on a inclined plane.

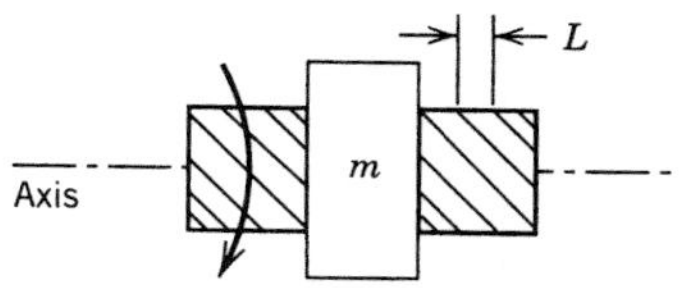

FIGURE 2.5-5   A lead screw.

EXAMPLE 2.5-2 *Equivalent Inertia* *of a Lead Screw*	Consider the lead screw given in Table 2.2-1 and repeated here in Figure 2.5-5. One finds such a device on lathes, for example, where it is used to translate the cutting tool. It consists of a rotating threaded rod and a translating mass. A motor supplies the torque T required to rotate the rod. Obtain the expression for its equivalent inertia and obtain its equation of motion.
Solution	The expression for the equivalent inertia of the lead screw is found by equating the kinetic energy of the real system (translation plus rotation) with the kinetic energy of rotation of the fictitious, purely rotational system. The mass m rides on a support having ball bearings aligned with the screw threads. When the screw rotates one revolution (2π radians), the mass translates the distance L, which is the distance between the threads. Thus if the screw rotates through θ radians, the mass translates a distance $x = L\theta/2\pi$. Thus the velocities are related as $v = L\dot{\theta}/2\pi$.

If the rod has an inertia I_r, the total kinetic energy of the system is

$$KE = \frac{1}{2}I_r\dot{\theta}^2 + \frac{1}{2}m\dot{v}^2 = \frac{1}{2}I_r\dot{\theta}^2 + \frac{1}{2}m\left(L\dot{\theta}/2\pi\right)^2$$

or

$$KE = \frac{1}{2}\left(I_r + m\frac{L^2}{4\pi^2}\right)\dot{\theta}^2 = \frac{1}{2}I_e\dot{\theta}^2$$

Thus the equivalent inertia is

$$I_e = I_r + m\frac{L^2}{4\pi^2}$$

and the equation of motion is

$$I_e\ddot{\theta} = T$$

■

Equivalent mass and equivalent inertia are complementary concepts. A system should be viewed as an equivalent mass if an applied force is specified and as an equivalent inertia if an applied torque is specified.

Equivalent Mass of Spring Elements

Kinetic energy equivalence is often used to determine the equivalent mass of the spring element, because kinetic energy is associated with the mass parameter. As spring elements are normally represented, it is assumed that the mass of the spring element either is negligible compared to the rest of the system's mass or has been included, or "lumped," in the mass attached to the spring. The following example shows how to include the mass of a spring.

EXAMPLE 2.5-3 *Equivalent Mass* *of a Cantilever* *Spring*	The static deflection of a cantilever beam is described by

$$x_y = \frac{P}{6EI_A}y^2(3L - y) \tag{1}$$

where P is the load applied at the end of the beam and x_y is the vertical deflection at a point a distance y from the support (Figure 2.5-6).

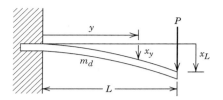

FIGURE 2.5-6   A cantilever beam of mass m_d.

Solution

The deflection x_L at the end of the beam, where $y = L$, is found from Equation 1 to be:

$$x_L = \frac{PL^3}{3EI_A} \tag{2}$$

Comparing the last two equations shows that

$$x_y = \frac{y^2(3L - y)}{2L^3} x_L \tag{3}$$

Differentiate this equation with respect to time for a fixed value of y to obtain

$$\dot{x}_y = \frac{y^2(3L - y)}{2L^3} \dot{x}_L \tag{4}$$

Because Equation 1 describes the *static* deflection, it does not account for inertia effects, and thus Equation 4 is not exactly true. However, lacking another reasonable expression for $\dot{x}_y$, we will use Equation 4.

The kinetic energy of a beam mass element dm at position y is $\dot{x}_y^2 \, dm/2$. Let v be the beam's mass per unit length. Then $dm = v\, dy$, and the total kinetic energy KE in the beam is

$$\begin{aligned}
KE &= \frac{1}{2} \int_0^L \dot{x}_y^2 \, dm \\
&= \frac{v}{8L^6} \dot{x}_L^2 \int_0^L y^4 (3L - y)^2 \, dy \\
&= \frac{v}{8L^6} \dot{x}_L^2 \frac{33L^7}{35} = \frac{33vL}{2(140)} \dot{x}_L^2
\end{aligned}$$

Because the beam mass $m_b = vL$,

$$KE = \frac{33 m_b}{2(140)} \dot{x}_L^2$$

If a mass m_e is located at the end of the beam and moves with a velocity $\dot{x}_L$, its kinetic energy is $m_e \dot{x}_L^2/2$. Comparing this with the beam's energy, we see that $m_e = (33/140)m_b \approx 0.23 m_b$. Thus the equivalent mass of the cantilever spring is 23% of the beam mass.   ∎

Equivalent Mass of Other Springs

The same approach can be applied to find the equivalent mass of other types of springs. The rod in tension/compression, shown in Table 2.5-1, has an equivalent mass equal to one-third its actual mass. This fraction is obtained by assuming that the velocity of a particle along the length of the rod is proportional to its distance from the support. At the support, the velocity is of course zero. At a point halfway from the support, the velocity is one-half the velocity at the end of the rod, and so on.

The equivalent inertia of rotational spring elements can be found using kinetic energy equivalence. The rod in torsion, shown in Table 2.5-1, has an equivalent inertia equal to one-third its actual inertia, assuming that the rotational velocity is proportional to the distance from the support.

TABLE 2.5-1   Equivalent Masses of Common Elements

m_c = concentrated mass
m_d = distributed mass
m_e = equivalent lumped mass

$$m_e \ddot{x} + kx = 0$$

Equivalent system

k Massless spring

m_e

Helical spring or rod in tension/compression

m_d

m_d

m_c

m_c

$$m_e = m_c + m_d/3$$

Cantilever beam

m_d

m_c

L

$$m_e = m_c + 0.23\, m_d$$

Fixed-end beam

m_d m_c

$L/2$ $L/2$

$$m_e = m_c + 0.375\, m_d$$

Simply supported beam

m_d m_c

$L/2$ $L/2$

$$m_e = m_c + 0.50\, m_d$$

Helical spring or rod in torsion

I_d

I_d

I_c

I_c

θ

θ

$$I_e = I_c + I_d/3$$

Equivalent system

k Inertialess spring

I_e

θ

$$I_e \ddot{\theta} + k\theta = 0$$

Mechanical Transformers

Mechanical devices such as gears, belts, levers, and pulleys transform an input motion and input force into another motion and force at the output; thus they are called *mechanical transformers*. They have dynamics of their own, due to their mass, elasticity, and damping properties. For example, a lever can flex if the forces it transmits are great enough. However, we try to use low-order models to describe these devices so that the total system model is not too complicated. For example, we treat a lever as a rigid rod whenever possible. Another simplification is to describe the effects of the transformer's mass, elasticity, and damping relative to a single reference location, for example, the location of the input motion. This gives a simpler model whose coordinates are those of the reference location.

Here we show how to use kinetic energy equivalence to facilitate this process. In a later chapter we will show how to use elastic potential energy and energy dissipation expressions to obtain equivalent spring and damping models.

Geared Systems

Several types are gears are used in mechanical systems. These include helical-spur gears, worm gears, bevel gears, and planetary gears. We now use the spur gear pair to demonstrate how to use kinetic energy equivalence to obtain a simpler model. This approach can be used to analyze other gear types and other drive types, such as belt or chain drives.

EXAMPLE 2.5-4
A Geared System

In the system shown in Figure 2.5-7a, a motor whose inertia is I_F rotates a pendulum through a gear pair. The gear inertias are I_C and I_D. The shaft inertias are I_B and I_E. The end view is shown in Figure 2.5-7b. Represent the system by the equivalent single-shaft, gearless system shown in Figure 2.5-7c, and obtain its equation of motion. The torque T_F acting on I_F is an externally applied torque from the motor. The gear ratio is $N = \omega_1/\omega_2$. Assume that the shaft torsional elasticity, gear tooth elasticity, and gear tooth friction are negligible.

Solution

We can represent the geared system shown in Figure 2.5-7a by the equivalent single-shaft, gearless system shown in Figure 2.5-7c by requiring that the equivalent system have the same expression for kinetic energy as the original system. The inertia of the pendulum about its pivot point is $I_A = mL^2$. The kinetic energy expression for the original system is

$$KE = \frac{1}{2}(I_A + I_B + I_C)\omega_1^2 + \frac{1}{2}(I_D + I_E + I_F)\omega_2^2 = \frac{1}{2}I_1\omega_1^2 + \frac{1}{2}I_2\omega_2^2$$

where

$$I_1 = I_A + I_B + I_C \qquad I_2 = I_D + I_E + I_F$$

Replacing ω_2 with ω_1/N gives

$$KE = \frac{1}{2}I_1\omega_1^2 + \frac{1}{2}I_2\left(\frac{\omega_1}{N}\right)^2 = \frac{1}{2}\left(I_1 + \frac{I_2}{N^2}\right)\omega_1^2$$

The expression for the kinetic energy of the equivalent single-shaft, gearless system shown in Figrue 2.5-7d is

$$KE = \frac{1}{2}I_e\omega_1^2$$

where I_e is the equivalent inertia of the new system.

Equating the energy expression of the original system to that of the equivalent system, we obtain

$$I_e = I_1 + \frac{I_2}{N^2} = mL^2 + I_B + I_C + \frac{1}{N^2}(I_D + I_E + I_F)$$

Remember that when an inertia is referenced to a slower shaft, its equivalent value must be greater than its true value so that its kinetic energy will remain the same. When referenced to a faster shaft, its equivalent value must be less than its true value.

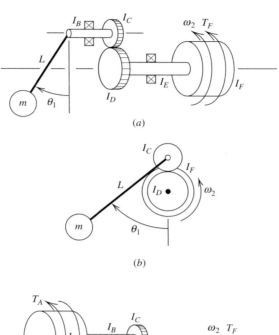

(a)

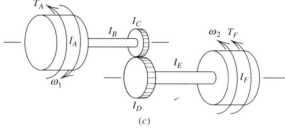

(b)

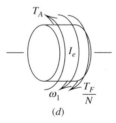

(c)

(d)

FIGURE 2.5-7 (a) A motor driving a pendulum through a gear pair. (b) End view. (c) Representation of the pendulum as an equivalent inertia. (d) Single inertia equivalent to the system shown in part (a).

We can now obtain the equation of motion for the equivalent system, by noting that the torque T_A is the torque about shaft AB due to the weight of the pendulum. It is $T_A = mgL \sin \theta_1$. From conservation of energy, we can show that the effect of the motor torque T_F felt in shaft 1 is T_F/N. Summing moments about shaft AB gives

$$I_e \frac{d\omega_1}{dt} = \frac{T_F}{N} - T_A = \frac{T_F}{N} - mgL \sin \theta_1$$

or

$$I_e \ddot{\theta}_1 + mgL \sin \theta_1 = \frac{T_F}{N} \tag{1}$$

Given T_F, this model can be solved to compute the angle θ_1 as a function of time, but it cannot be used to obtain information about gear tooth forces and gear backlash, or the effects of gear tooth elasticity or shaft elasticity. For this a more detailed model is required. We will develop such models in later chapters. ∎

2.6 SYSTEMS WITH SPRING ELEMENTS

In our examples thus far, we have derived the differential equation of motion for systems in which the restoring force was due to gravity. We now consider a number of examples in which the restoring force is due to spring elements. For now we restrict ourselves to systems with one degree of freedom. We will consider two or more degrees of freedom starting with Chapter 6.

Real versus Ideal Spring Elements

By their very nature, all real spring elements have mass and are not rigid bodies. Thus, because it is much easier to derive an equation of motion for a rigid body than for a distributed-mass, flexible system, the basic challenge in modeling mass-spring systems is to first decide whether and how the system can be modeled as a single rigid body.

If the system consists of an object attached to a spring, the simplest way to model the systems as a single rigid body is to neglect the spring mass relative to the mass of the object and take the mass center of the system to be located at the mass center of the object. With this assumption, we treat the spring as an *ideal* spring element, which is massless.

This assumption is accurate in many practical applications, but to be comfortable with this assumption, you should know the numerical values of the masses of the object and the spring element. In some of the chapter problems and some of the examples in this text, the numerical values are not given. In such cases, unless otherwise explicitly stated, you should assume that the spring mass can be neglected.

A real spring element can be represented as an ideal element either by neglecting its mass or by including it in another mass in the system, using the equivalent-mass methods developed in Section 2.5.

Effect of Spring Free Length and Object Geometry

Suppose we attach a cube of mass m and side length $2a$ to a linear spring of negligible mass, and we fix the other end of the spring to a rigid support, as shown in Figure 2.6-1a. We assume that the horizontal surface is frictionless. If the mass is homogeneous, its center of mass is at the geometric center G of the cube. The free length of the spring is L. The force f is the net horizontal force acting on the mass due to sources other than the spring. If the applied force f is zero, when we position the mass m so that the spring is at its free length, the mass will be in equilibrium. The equilibrium location of G is the point marked E. Figure 2.6-1b shows the mass displaced a distance x from the equilibrium position. In this position the spring has been stretched a distance x from its free length, and thus its force is kx. The free-body diagram of this situation is shown in Figure 2.6-1c. From this diagram we can obtain the following equation of motion:

$$m\ddot{x} = -kx + f \tag{2.6-1}$$

Note that neither the free length L nor the cube dimension a appears in the equation of motion. These two parameters need to be known only to locate the equilibrium position E of the mass center. Therefore we can represent the object as a point mass, as shown in Figure 2.6-1d.

Unless otherwise specified, you should assume that the objects in our diagrams can be treated as point masses and therefore their geometric dimensions need not be known to obtain the equation of motion. You should also assume that the location of the equilibrium

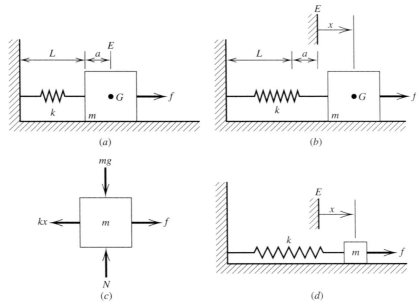

FIGURE 2.6-1 The point-mass approximation. (*a*) The block of side length 2*a* is in equilibrium at point *E* when the applied force *f* is zero. (*b*) The displacement *x* is measured from the equilibrium position of the mass center *G*. (*c*) Free-body diagram. (*d*) With the point-mass approximation, the entire object is taken to be at the mass center and its dimensions do not appear in the equation of motion, which is $m\ddot{x} = f - kx$.

position is known. The hatched-line symbol shown in Figure 2.6-1*a* is used to indicate a rigid support, such as the horizontal surface and the vertical wall, and also to indicate the location of a fixed coordinate origin, such as the origin of *x*.

The Equilibrium Position as Coordinate Reference

Figure 2.6-2*a* shows a spring at its free length (the spring is not stretched by gravity because we are neglecting its mass). Figure 2.6-2*b* shows a mass *m* attached to the spring.

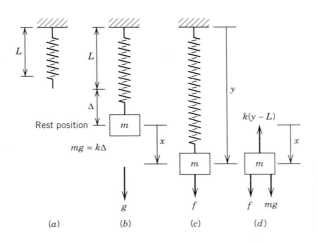

FIGURE 2.6-2 A mass-spring system in vertical motion. (*a*) The spring's free length is *L*. (*b*) The mass shown in its rest position. (*c*) General motion. (*d*) Free-body diagram.

The applied force f is due to some source other than gravity or the spring. Suppose for now that $f = 0$. When the mass m is attached and allowed to come to rest, the spring stretches an additional distance Δ, which is called the *static deflection* of the spring. The spring force at this new position is $k\Delta$, which must equal the gravity force mg if the mass remains at rest. Thus

$$mg = k\Delta \tag{2.6-2}$$

Suppose now that $f \neq 0$. When the mass is displaced a distance x from the equilibrium position, the spring force is $k(\Delta + x)$. The free-body diagram gives

$$m\ddot{x} = mg - k(\Delta + x) = f + mg - k\Delta - kx = f - kx$$

where we have used Equation 2.6-2 to cancel some terms. Thus the equation of motion is

$$m\ddot{x} + kx = f \tag{2.6-3}$$

When we use the equilibrium position as the reference point for measuring the displacement of the mass, the gravity force mg in the equation of motion is canceled by the equilibrium spring force $k\Delta$. This results in a simplified form of the equation of motion.

We could use any fixed point as the coordinate reference; for example, we can measure the displacement from the fixed support. Let y be this displacement. Then the spring force is $k(y - L)$ and Newton's law gives

$$m\ddot{y} = mg - k(y - L) + f$$

or

$$m\ddot{y} + ky = mg + kL + f \tag{2.6-4}$$

Both Equations 2.6-3 and 2.6-4 are correct, but the solution to Equation 2.6-3 can be expressed more compactly because it has fewer constants.

We have demonstrated a general principle of modeling systems containing spring elements:

> Any constant forces or moments will not appear
> in the equations of motion of a system containing
> a *linear* restoring force or moment if the mass
> displacements are measured from the equilibrium positions.

This is so because the combined restoring forces acting on each mass at its equilibrium position must cancel the gravity force on that mass. Unless otherwise stated, any displacement variables for such a system are assumed to be measured from the equilibrium position.

Thus the mass-spring systems shown in Figures 2.6-1 and 2.6-2 have the same equation of motion. Gravity's only effect in Figure 2.6-2 is to determine the location of the equilibrium position. For the system in Figure 2.6-1, $\Delta = 0$.

The advantages of choosing the equilibrium position as the coordinate origin are that (1) we need not specify the geometric dimensions of the mass and (2) this choice simplifies the equation of motion by eliminating the static forces.

Figure 2.6-3 shows three situations that have the same equation of motion, $m\ddot{x} = -kx + f$. If possible, it is good practice to select the positive direction of x to coincide with the positive direction of f. This choice will help to avoid mistakes in the free-body diagram.

It is important to understand that any forces acting on the mass, other than gravity and the spring force, are not to be included when determining the location of the

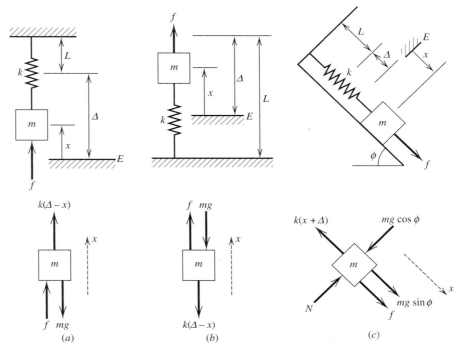

FIGURE 2.6-3 The effect of orientation, gravity, and static spring deflection Δ on the equation of motion. All three systems have the same equation of motion, which is $m\ddot{x} = f - kx$.

equilibrium position. For example, in Figure 2.6-3c a force f acts on the mass but the equilibrium position E is the location of the mass at which $k\Delta = mg \sin \phi$ when $f = 0$. The equation of motion is $m\ddot{x} = f - k(x + \Delta) + mg \sin \phi$, or $m\ddot{x} = f - kx$.

The previous analysis is based on a system model that contains a linear spring and a constant gravity force or moment. For nonlinear spring elements, the gravity terms may or may not be canceled by the static spring forces. Also note that the effect of gravity in some applications is not constant but acts like a spring, and thus gravity may appear in the equation of motion. For example, the equation of motion for a pendulum, derived in Section 2.3, is $mL^2\ddot{\theta} + mgL \sin \theta = 0$. The gravity term is not canceled out in this equation because the effect of gravity here is not a constant torque but rather is a torque $mgL \sin \theta$ that is a function of the coordinate θ.

EXAMPLE 2.6-1
A Mass with Two Springs

Obtain the equation of motion in terms of x for the system shown in Figure 2.6-4. The equilibrium position is at $x = 0$ when $f = 0$.

Solution

The two springs are connected end-to-end, so their spring constants combine as follows:

$$\frac{1}{k_e} = \frac{1}{k_1} + \frac{1}{k_2}$$

This gives the spring constant of a single equivalent spring:

$$k_e = \frac{k_1 k_2}{k_1 + k_2}$$

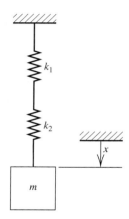

FIGURE 2.6-4 A mass with two springs in series.

Thus the equation of motion is $m\ddot{x} = -k_e x + f$, or

$$m\ddot{x} + \frac{k_1 k_2}{k_1 + k_2} x = f$$

∎

Displacement Inputs

Thus far the inputs to our systems have been specified forces. In some applications we are not given the force that is driving the system, but rather an input displacement.

EXAMPLE 2.6-2
A Cam-Driven System

Figure 2.6-5 shows a mass-spring system that is driven by a rotating noncircular cam. The displacement $y(t)$ is a given function, which is determined by the cam shape and its rotation speed. Derive the equation that describes the displacement x of the mass m.

Solution

Assuming that $y > x$, we obtain the free-body diagram shown and the following equation of motion:

$$m\ddot{x} = -k_2 x + k_1(y - x)$$

Rearranging gives

$$m\ddot{x} + (k_1 + k_2)x = k_1 y$$

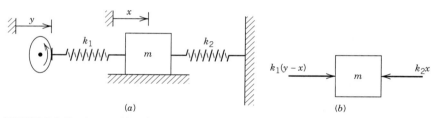

(a) (b)

FIGURE 2.6-5 A mass driven by cam motion.

Such a model might be used to find an expression for the force on the support, which is $k_2x(t)$. To compute this force we need $x(t)$. This can be found by solving the equation of motion.

Note that the assumption $y > x$ is completely arbitrary. We would have obtained the same equation of motion if we had made the assumption that $y < x$. In this case the magnitude of the spring force on the free-body diagram would be $k_1(x - y)$ and its direction would be to the left instead of to the right. You should become comfortable with making such assumptions about the relative motion of two points in a system. If you assign the force magnitude and direction to be consistent with the assumed relative motion, then you will obtain the correct equation of motion, regardless of your assumption. ■

EXAMPLE 2.6-3
A Model of a Torsion-Bar Suspension

Figure 2.6-6 shows a torsion-bar suspension of the type often used in vehicles to prevent excessive rotation. Figure 2.6-7*a* is a representation of a torsion-bar suspension. It shows an inertia I_1 connected to two shafts. Assume that the inertia of each shaft is negligible compared to I_1. The shafts' torsional spring constants are k_1 and k_2. The input variable is the given angular displacement θ_i. Develop the model for the angular displacement θ.

Solution

The equivalent model is shown in Figure 2.6-7*b*. When $\theta_i = 0$, the equilibrium is at $\theta = 0$. Assume that $\theta_i > \theta$. If so, then the resulting torque in spring 1 will accelerate I_1 in the positive θ direction, and the free body will be as shown in Figure 2.6-7*c*. The equation of motion is

$$I_1\ddot{\theta} = k_1(\theta_i - \theta) - k_2\theta$$

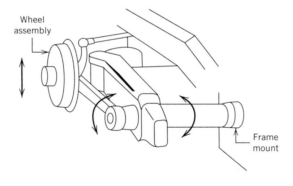

Wheel
assembly

Frame
mount

FIGURE 2.6-6 Torsion-bar suspensions are used in vehicles to prevent excessive wheel motion.

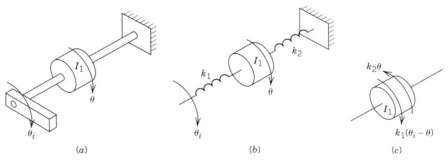

(a) (b) (c)

FIGURE 2.6-7 (a) Model of a torsion-bar suspension. (b) Equivalent representation. (c) Free-body diagram.

or

$$I_1\ddot{\theta} + (k_1 + k_2)\theta = k_1\theta_i$$

You should obtain the same equation of motion by drawing the free-body diagram using the assumption that $\theta_i < \theta$. ∎

2.7 CHAPTER REVIEW

This chapter reviewed the basic topics of dynamics required for studying vibrations. Dynamics deals with the motion of bodies under the action of forces. The subject consists of kinematics, the study of motion without regard to the forces causing the motion, and kinetics, the relationship between the forces and the resulting motion.

The single most important step in analyzing a vibratory system is to derive the correct model, or description, of the dynamics of the system because if this description is incorrect, then any analysis based on that description will also be incorrect. There are several ways to derive such a description. It may be a differential equation of motion derived from Newton's second law, or a work-energy description, or an impulse-momentum description.

The chapter began with a review of kinematics principles and coordinate systems for representing motion. Vibrating systems are often categorized by their number of degrees of freedom, and this concept was developed in Section 2.1.

The kinetics of rigid bodies was treated in Section 2.2. Section 2.3 covered the work-energy method, which is based on an integration of the equation of motion over displacement. Methods based on work-energy are useful for solving problems involving force, displacement, and velocity, but not time. Section 2.4 covered the impulse-momentum method, which is based on an integration of the equation of motion over time. Methods based on impulse and momentum aid us in solving problems in which the applied forces act either over very short times, such as happens during an impact, or over a specified time interval. Because the work-energy and impulse-momentum methods are based on integrations of the equation of motion, the mathematics associated with them is easier to apply to solve some types of problems.

Section 2.5 covered concepts of equivalent mass and equivalent inertia. These concepts use kinetic energy equivalence to simplify the process of obtaining descriptions of systems containing both translating and rotating parts, systems containing distributed mass, and systems consisting of multiple masses whose motion is coupled.

Section 2.6 developed many examples showing how to obtain the differential equation of motion for single-degree-of-freedom systems containing spring elements.

Now that you have finished this chapter, when presented with a vibratory system, you should be able to do the following:

1. Identify the type of motion, apply the appropriate kinematic equations, and choose an appropriate set of coordinates to describe the motion.
2. Identify the degrees of freedom.
3. Apply Newton's laws of motion if the object can be modeled as a particle.
4. Identify the type of plane motion, and apply the appropriate form of Newton's laws if the object can be modeled as a rigid body in plane motion.

5. Apply work-energy methods for problems involving force, displacement, and velocity, but not time.

6. Apply impulse-momentum methods for problems in which the applied forces act either over very short times, such as happens during an impact, or over a specified time interval.

7. Apply the principle of kinetic energy equivalence to simplify the process of obtaining a description of a system containing both translating and rotating parts, or one containing distributed mass, or one consisting of multiple masses whose motion is coupled.

8. Obtain the differential equation of motion for single-degree-of-freedom systems containing spring elements.

PROBLEMS

Section 2.1   Kinematics and Degrees of Freedom

2.1   A cylinder of radius R rolls on a semicircular track of radius r without slipping (Figure P2.1). Derive the relation between ϕ and θ.

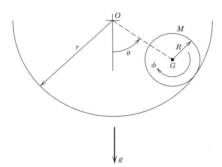

FIGURE P2.1

2.2   A cylinder of radius R rolls on a semicircular track of radius r without slipping (Figure P2.2). Its axle moves in a frictionless slot in a link. Derive the relation between ϕ and θ.

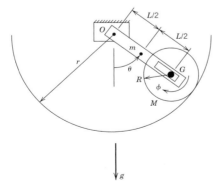

FIGURE P2.2

2.3   For the wheel rolling with angular velocity ω and angular acceleration α, shown in Figure P2.3, determine the expression for the acceleration of point A as a function of ω, α, and the radius R.

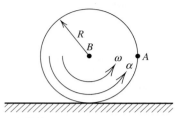

FIGURE P2.3

2.4   The vehicle shown in Figure P2.4 hoists the block A by moving to the left with a constant velocity $v = 0.2$ m/s. The cable length is 10 m. Assume the pulley diameter is small enough to be negligible. Compute the velocity and acceleration of the block A when it reaches the height of the platform.

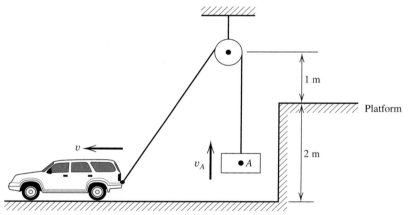

FIGURE P2.4

2.5   The vehicle shown in Figure P2.5 hoists the mass m by moving to the right with a velocity v_1. Obtain the expression for the velocity v_2 in terms of the velocity v_1.

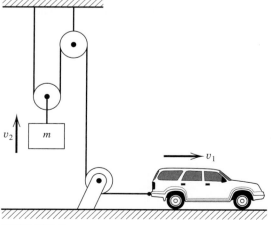

FIGURE P2.5

2.6 For the system shown in Figure P2.6, derive the relation between the velocities v_A and v_B.

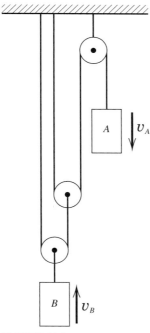

FIGURE P2.6

Section 2.2   Plane Motion of a Rigid Body

2.7 Find the equation of motion for the pendulum shown in Figure P2.7, which consists of a concentrated mass m_C a distance L_C from point O, attached to a rod of length L_R and inertia I_{RG} about its mass center. Discuss the case where the rod's mass is small compared to the concentrated mass.

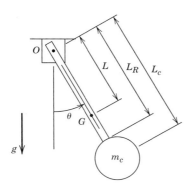

FIGURE P2.7

2.8 Consider the cylinder treated in Problem 2.1 (Figure P2.1). The cylinder has a radius R, a mass M, and an inertia I_G about its mass center. It rolls on a semicircular track of radius r without slipping. Derive its equation of motion.

2.9 Consider the cylinder treated in Problem 2.2 (Figure P2.2). The cylinder has a radius R, a mass M, and an inertia I_G about its mass center. It rolls on a semicircular track of radius r without slipping. Its axle moves in a frictionless slot in a link. The link has a mass m and inertia I_O about point O. Derive the equation of motion of the cylinder.

2.10 The pendulum shown in Figure P2.10 consists of a slender rod of mass 1.4 kg and a block of mass 4.5 kg.

(a) Determine the location of the center of mass.

(b) Derive the equation of motion in terms of θ.

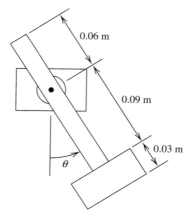

0.06 m

0.09 m

0.03 m

θ

FIGURE P2.10

2.11 A slender rod 1.4 m long and of mass 20 kg is attached to a wheel of radius 0.05 m and negligible mass, as shown in Figure P2.11. A horizontal force f is applied to the wheel axle. Derive the equation of motion in terms of θ. Assume the wheel does not slip.

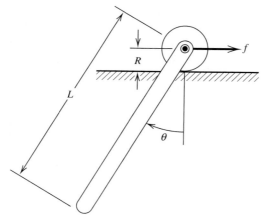

f

R

L

θ

FIGURE P2.11

Section 2.3   Work-Energy Methods

2.12 In the system shown in Figure P2.12 the masses of the spring and pulley are negligible compared to the mass m. Use the values $m = 30$ kg and $k = 600$ N/m.

(a) What is the static deflection in the spring when the mass m is stationary?

(b) Suppose the spring is initially stretched 0.06 m when we release the mass m. Calculate the velocity of the mass m after it has dropped 0.09 m.

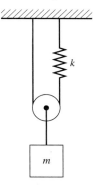

FIGURE P2.12

2.13 The mass m is released from rest a distance d above the spring as shown in Figure P2.13. The coefficient of kinetic friction between the mass and the surface is μ. Using the values $m = 3$ kg, $d = 0.5$ m, $k = 1000$ N/m, $\theta = 45°$, and $\mu = 0.4$, **(a)** calculate the velocity of m when it first touches the spring and **(b)** calculate the resulting maximum spring deflection.

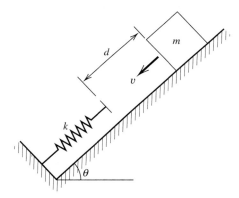

FIGURE P2.13

2.14 The block in Figure P2.14 has a mass of 11 kg and is initially at rest on the smooth inclined plane, with the spring stretched 0.5 m. A constant force of $f = 400$ N, parallel to the plane, pushes the block up the incline. Compute the work done by all the forces acting on the block as the block is pushed 2 m up the incline. The spring constant is $k = 44$ N/m.

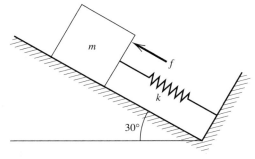

FIGURE P2.14

2.15 The disk shown in Figure P2.15 has a radius of 0.25 m, a mass of 10 kg, and an inertia of 0.4 kg·m² about its center. The free length L of the spring is 1 m and its stiffness is 25 N/m. The disk is released from rest in the position shown and rolls without slipping. Calculate its angular velocity when its center is directly below point O. Assume that $D_1 = 2.25$ m and $D_2 = 2$ m.

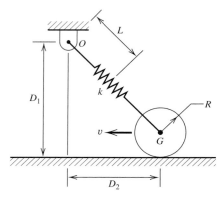

FIGURE P2.15

Section 2.4   Impulse-Momentum Methods

2.16 The 100-kg mass is released from rest at a height 0.75 m above the middle spring shown in Figure P2.16. The spring constants are $k_1 = 10$ kN/m and $k_2 = 8$ kN/m. Compute the maximum deflection of the middle spring.

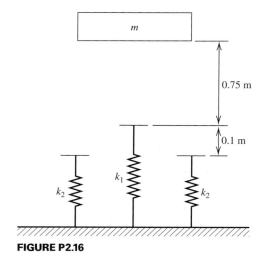

FIGURE P2.16

2.17 A mass m_1 moving with a velocity v_1 strikes and becomes embedded in the pendulum as shown in Figure P2.17. Assume that the pendulum is a slender rod of mass m_2. Determine the angular velocity of the pendulum immediately after the impact. Use the values $m_1 = 0.8$ kg, $m_2 = 4$ kg, and $v_1 = 10$ m/s.

2.18 A mass m_1 moving with a velocity v_1 strikes and rebounds from the pendulum as shown in Figure P2.18. The impact is perfectly elastic. Assume that the pendulum is a slender rod of mass m_2. Determine the angular velocity of the pendulum immediately after the impact. Use the values $m_1 = 0.8$ kg, $m_2 = 4$ kg, and $v_1 = 10$ m/s.

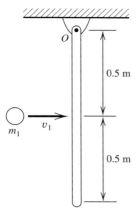

FIGURE P2.17

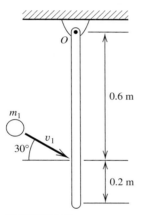

FIGURE P2.18

2.19 An object of mass 0.005 kg moving with a velocity 365 m/s strikes and becomes embedded in the pendulum as shown in Figure P2.19, where $L = 1.2$ m. Assume that the pendulum is a slender rod whose mass is 4.5 kg. Determine the angular velocity of the pendulum immediately after the impact.

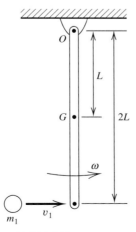

FIGURE P2.19

2.20 The pendulum shown in Figure P2.20 is a slender rod of mass m and is released from rest at $\theta = 0$. In terms of m and g, obtain the expression for the reaction force on the pin at point O when $\theta = 90°$.

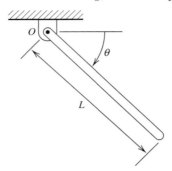

FIGURE P2.20

Section 2.5   Equivalent Mass and Inertia

2.21 Suppose we push on the axle of a wheel with a force of 400 N, as shown in Figure P2.21. The wheel's mass is 80 kg, and it has a radius of 0.3 m and an inertia of $3 \text{ kg} \cdot \text{m}^2$. The slope is 25°, and the wheel is initially at rest. Estimate the axle's speed and the wheel's rotational speed after 1 minute.

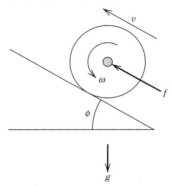

FIGURE P2.21

2.22 The vibration of a motor mounted on the end of a cantilever beam can be modeled as a mass-spring system. The motor mass is 30 kg, and the beam mass is 15 kg. When the motor is placed on the beam, it causes an additional static deflection of 0.003 m. Find the equivalent mass m and equivalent spring constant k.

2.23 The vibration of a motor mounted in the middle of a doubly clamped beam can be modeled as a mass-spring system. The motor mass is 10 kg, and the beam mass is 2 kg. When the motor is placed on the beam, it causes an additional static deflection of 0.02 m. Find the equivalent mass m and equivalent spring constant k.

2.24 A motor connected to a pinion gear of radius R drives a load of mass m on the rack (Figure P2.24). Neglect all masses except m. What is the energy-equivalent inertia due to m, as felt by the motor?

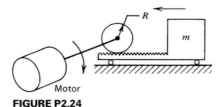

FIGURE P2.24

2.25 The geared system shown in the Figure P2.25 represents an elevator system. The motor has an inertia I_1 and supplies a torque T_1. Neglect the inertias of the gears, and assume that the cable does not slip on the pulley. Derive an expression for the equivalent inertia I_e felt on the input shaft (shaft 1). Then derive the dynamic model of the system in terms of the speed ω_1 and the applied torque T_1.

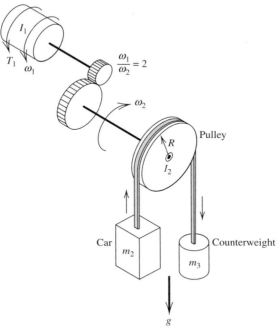

FIGURE P2.25

2.26 For the geared system shown in Figure P2.26, assume that shaft inertias and the gear inertias I_1, I_2, and I_3 are negligible. The motor and load inertias in kg·m² are

$$I_4 = 0.02 \qquad I_5 = 0.1$$

The speed ratios are:

$$\frac{\omega_1}{\omega_2} = \frac{\omega_2}{\omega_3} = 1.4$$

Derive the system model in terms of the speed ω_3, with the applied torque T as the input.

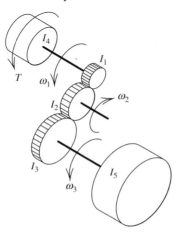

FIGURE P2.26

Section 2.6 Systems with Spring Elements

2.27 For the system shown in Figure P2.27, obtain the equation of motion in terms of x. Neglect the pulley mass.

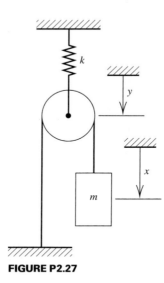

FIGURE P2.27

2.28 For the system shown in Figure P2.28, obtain the equation of motion in terms of x. Neglect the pulley mass.

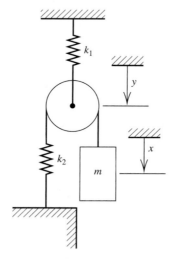

FIGURE P2.28

2.29 In the system shown in Figure P2.29, the cart wheels roll without slipping. Neglect the pulley masses. Obtain the equation of motion in terms of x.

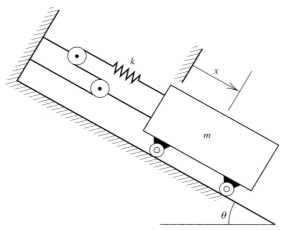

FIGURE P2.29

2.30 For the system shown in Figure P2.30, obtain the equation of motion in terms of x. Neglect the mass of the L-shaped arm.

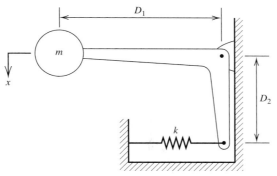

FIGURE P2.30

2.31 Derive the equation of motion for the system shown in Figure P2.31 for small motions. Neglect the pulley mass.

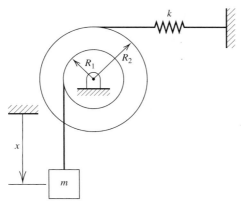

FIGURE P2.31

2.32 Derive the equation of motion for the system shown in Figure P2.32 for small motions. The applied force f is a given function of time. The lever inertia about the pivot is I.

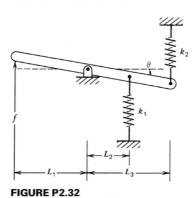

FIGURE P2.32

2.33 In the pulley system shown in Figure P2.33, assume that the cable is massless and inextensible, and assume that the pulley masses are negligible. The force f is a known function of time. Derive the system's equation of motion in terms of the displacement x_1.

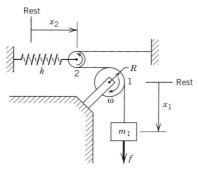

FIGURE P2.33

2.34 For the system shown in Figure P2.34, the solid cylinder of inertia I and mass m rolls without slipping. Neglect the pulley mass and obtain the equation of motion in terms of x.

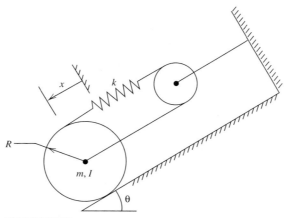

FIGURE P2.34

2.35 For the system shown in Figure 2.1-7, assume the cable is inextensible, neglect the pulley masses, and obtain the equation of motion in terms of x_2.

2.36 For the system shown in Figure 2.1-8, assume the cable is inextensible, neglect the pulley masses, and obtain the equation of motion in terms of x.

FREE RESPONSE WITH A SINGLE DEGREE OF FREEDOM

THIS CHAPTER covers the free response of damped and undamped systems having a single degree of freedom. The free response refers to the motion of a mass when no external forces act on the mass, other than spring forces, damping forces, or gravitational forces. In Chapter 4 we treat the forced response of single-degree-of-freedom systems having periodic forces acting on them.

In order to analyze the free response completely, we need to obtain the equation of motion. So far our examples have not had damping forces. Thus in this chapter we illustrate, with numerous examples, how to obtain the equation of motion for systems having damping forces. In Section 3.1 we start by treating the response of a mass with viscous damping but no spring force. From a mathematical point of view, such a system is simpler to analyze because the equation of motion is first order.

Sections 3.2 and 3.3 treat the mathematical analysis and modeling of undamped systems, whose equation of motion is second order. Sections 3.4 through 3.7 treat the analysis and modeling of systems having spring elements and viscous damping. Section

115

3.8 covers the free vibration of systems having another type of damping, one that is due to Coulomb or dry friction.

The chapter concludes with two sections dealing with computer methods for analyzing and simulating system response. Section 3.9 introduces the MATLAB functions for solving ordinary differential equations, the `ode` solvers. Section 3.10 introduces the Simulink program, which is based on MATLAB. It has an easy-to-use graphical interface that provides built-in functions for analyzing phenomena, such as Coulomb friction, that are difficult to treat analytically or numerically with the `ode` solvers.

LEARNING OBJECTIVES

After you have finished this chapter, you should be able to do the following:

- Obtain the equation of motion of a system containing spring elements, viscous damping, and/or Coulomb friction, and having a single degree of freedom.
- Linearize a nonlinear equation of motion.
- Solve a linear equation of motion for the free response.
- Compute the damped and undamped natural frequencies, the logarithmic decrement, the time constant, and the damping factor, and determine whether or not the system is stable.
- Use the chapter's methods to estimate values for the spring constant, damping coefficient, and friction coefficient.
- Use the MATLAB `ode` functions and Simulink to solve ordinary differential equations.

3.1 FIRST-ORDER RESPONSE

We now consider systems whose models are first-order differential equations. Such equations do not have an oscillatory solution unless the forcing function oscillates, and so such systems will not exhibit vibration. However, the models provide an easy introduction to some important concepts such as the *time constant*.

Consider a block sliding on a horizontal lubricated surface (Figure 3.1-1). The two horizontal forces acting on the mass are the resistance of the lubricated surface, modeled as viscous damping, and an arbitrary applied force $f(t)$. Assuming that the resistance is proportional to the speed v, the equation of motion can be obtained from the free-body diagram in Figure 3.1-1, which shows only the forces in the horizontal direction:

$$m\dot{v} = f(t) - cv \qquad (3.1\text{-}1)$$

This is a *first-order* differential equation because its highest derivative is first order.

Suppose that the force $f(t)$ acts on the mass until time $t = 0$, when it ceases to act. We denote the velocity so produced at $t = 0$ by $v(0)$. So for $t \geq 0$, the equation of motion becomes

$$m\dot{v} = -cv \qquad (3.1\text{-}2)$$

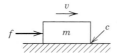

FIGURE 3.1-1   A mass with viscous surface friction.

It can be solved in several ways, but perhaps the simplest is to use a trial solution of the form

$$v(t) = Ae^{st}$$

where A and s are unknown constants. If we are given the initial velocity $v(0)$ of the block, then we can determine A by substituting $t = 0$ into the previous equation. The result is $v(0) = Ae^0 = A$. Thus the trial solution reduces to

$$v(t) = v(0)e^{st}$$

If we substitute this form into the differential equation and collect terms, we obtain

$$(ms + c)v(0)e^{st} = 0$$

This relation is valid for $v(0) \neq 0$ only if $ms + c = 0$, because e^{st} may not be zero if the solution is to be valid for any time t. Thus we see that $s = -c/m$, and the solution is

$$v(t) = v(0)e^{-ct/m} \qquad (3.1\text{-}3)$$

If $c/m > 0$, as it will be for problems of practical interest, this solution predicts that the velocity will decay to zero, as shown in Figure 3.1-2. It also predicts that the block will take an infinite time to stop, whereas in reality other effects such as Coulomb friction will stop the block in a finite time. If $c/m > 0$, the speed decays exponentially and the time constant τ, which has units of time, is often used to indicate the speed of the decay. It is defined as

$$\tau = \frac{m}{c} \qquad (3.1\text{-}4)$$

In terms of this parameter, the equation and its solution can be written as

$$\tau\dot{v} + v = 0 \qquad (3.1\text{-}5)$$

$$v(t) = v(0)e^{-t/\tau} \qquad (3.1\text{-}6)$$

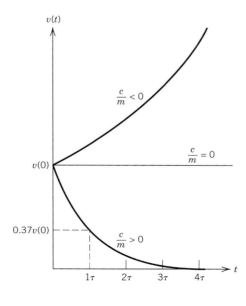

FIGURE 3.1-2 Free response of the model $m\dot{v} = f - cv$, illustrating the meaning of the time constant $\tau = m/c$.

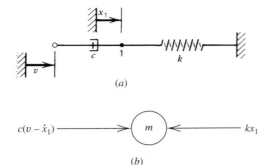

(a)

(b)

FIGURE 3.1-3 (a) A system with a velocity input variable. (b) Free-body diagram.

The Fictitious Mass Method

As an aid to organizing the appropriate equations of motion, it is sometimes useful to place a "fictitious" mass at an appropriate point in the system and obtain the free-body diagrams and equations of motion. Then set the fictitious mass equal to zero, and rearrange the equations in the desired form.

EXAMPLE 3.1-1
A Cam-Driven System

In the system shown in Figure 3.1-3a, a mechanism such as a cam drives the left-hand endpoint so that its velocity v is a known function of time. Derive the equation of motion of the system in terms of the displacement x_1 of point 1.

Solution

We place a fictitious mass m at point 1, the point corresponding to the desired displacement x_1. The free-body diagram is shown in Figure 3.1-3b. From Newton's law we have $m\ddot{x}_1 = c(v - \dot{x}_1) - kx_1$. Set $m = 0$ to obtain the equation of motion: $c\dot{x}_1 + kx_1 = cv$. The time constant is $\tau = c/k$. ∎

Systems in Pure Rotation

The next two examples illustrate how to obtain the equation of motion for damped systems undergoing pure rotation.

EXAMPLE 3.1-2
A Model of an Axle and Journal Bearing

Figure 3.1-4a shows an axle driven by a torque T and supported by a journal bearing having a rotational damping constant c. The axle's inertia about its axis of rotation is I. We can also represent this system abstractly as in Figure 3.1-4b, where the hatched symbol represents the stationary bearing support. Derive a model for the axle's rotational speed ω and determine the time constant.

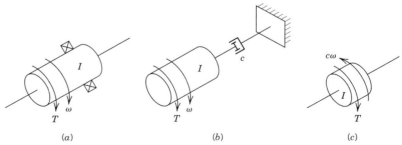

(a)   (b)   (c)

FIGURE 3.1-4 (a) A model of an axle with a journal bearing. (b) Equivalent representation. (c) Free-body diagram.

Solution | The free-body diagram is shown in Figure 3.1-4c. Applying the basic law for rotation about a fixed axis, we obtain

$$I\dot{\omega} = T - c\omega \tag{1}$$

This equation may be rearranged as

$$\frac{I}{c}\dot{\omega} + \omega = \frac{1}{c}T$$

from which the time constant can be identified as $\tau = I/c$.   ∎

Power Dissipation in Damping

Energy is dissipated in damping as heat. The rate of energy dissipation is the power consumed, and its expression can be derived as follows. Consider a rotational damper that obeys the linear relation $T = c\omega$, where T is the net torque across the damper. If the damper rotates an angle $d\theta$, the work done by the torque is $dW = T\,d\theta$. Dividing both sides by dt gives

$$\frac{dW}{dt} = T\frac{d\theta}{dt} = T\omega = (c\omega)\omega$$

Therefore the power dissipated in the damper is $c\omega^2$.

We can use the expression $c\omega^2$ for power dissipation to represent a system with a gear pair by an equivalent single-shaft, gearless system. We do this by requiring that the equivalent system have the same expressions for kinetic energy, potential energy, and energy dissipation as the original system.

EXAMPLE 3.1-3
Equivalent Damping in Geared Systems | Obtain the equation of motion of the system shown in Figure 3.1-5a with the torque T_1 as the input. Assume that there is negligible elasticity in the system and that the damping on the input shaft is also negligible.

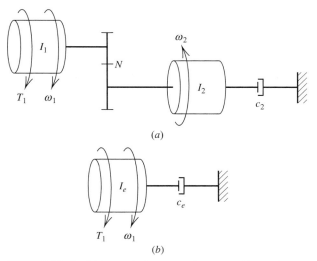

FIGURE 3.1-5   (a) An equivalent geared system with massless gears and massless shafts. (b) Equivalent gearless system.

Solution

Referring to Figure 3.1-5a, we arbitrarily choose the driving shaft (shaft 1) as the reference shaft. The expressions for kinetic energy KE and power dissipation P of the original system are

$$KE = \frac{1}{2}I_1\omega_1^2 + \frac{1}{2}I_2\omega_2^2$$

$$P = c_2\omega_2^2$$

The expressions for the energy terms in the equivalent single-shaft, gearless system shown in Figure 3.1-5b are

$$KE = \frac{1}{2}I_e\omega_1^2$$

$$P = c_e\omega_1^2$$

where I_e and c_e are the inertia and damping constant of the equivalent system.

Equating the energy expressions of the original system to those of the equivalent system, and writing the speed and displacement ratios in terms of N, we obtain

$$I_e = I_1 + I_2\left(\frac{\omega_2}{\omega_1}\right)^2 = I_1 + \frac{1}{N^2}I_2$$

$$c_e = c_2\left(\frac{\omega_2}{\omega_1}\right)^2 = \frac{1}{N^2}c_2$$

Thus the equation of motion is

$$I_e\dot{\omega}_1 = T_1 - c_e\omega_1$$

The time constant of this model is $\tau = I_e/c_e$. ∎

3.2   FREE UNDAMPED VIBRATION

In Chapter 2 we saw examples of systems whose equation of motion has the form $m\ddot{x} + kx = f(t)$. This is the form of the equation of motion of an undamped system with a single degree of freedom. We now develop the solution of this equation for the case where the force $f(t)$ is zero. This solution is called the *free response* because it describes the response of the system when it is free of, or not influenced by, the force $f(t)$.

The Natural Frequency ω_n

There are several ways to solve the equation

$$m\ddot{x} + kx = 0 \tag{3.2-1}$$

We know from physical insight that such a system will oscillate, so this suggests a solution in the form of a sine or cosine. In fact, both a sine function and a cosine function will solve the equation. To see this, substitute the function $x = A_1 \sin \omega t$ into the differential equation, using the fact that $\ddot{x} = -\omega^2 A_1 \sin \omega t$, to obtain

$$m(-\omega^2 A_1 \sin \omega t) + k(A_1 \sin \omega t) = 0$$

where A_1 and ω are unknown constants. Collect terms to obtain

$$(-m\omega^2 + k)(A_1 \sin \omega t) = 0$$

Now A_1 cannot be zero because this would give the trivial solution $x = 0$. For the same reason, $\sin \omega t$ cannot be zero for all values of t. This means that $-m\omega^2 + k$ must be zero, which gives $\omega = \sqrt{k/m}$. Thus $x = A_1 \sin \sqrt{k/m} t$ solves the differential equation $m\ddot{x} + kx = 0$. This solution shows that the mass m oscillates at the radian frequency $\sqrt{k/m}$. This is the frequency of the free response and is called the *natural frequency*, or sometimes the *undamped natural frequency* to emphasize that it is the oscillation frequency of a system having no damping. The symbol reserved for this frequency is ω_n. Thus for the equation of motion $m\ddot{x} + kx = 0$,

$$\omega_n = \sqrt{\frac{k}{m}} \tag{3.2-2}$$

General Solution for the Free Response

Using the same procedure, we find that the cosine function $x = A_2 \cos \omega t$ also solves the differential equation if $\omega = \omega_n$. Because the sine and cosine are independent functions, their linear combination

$$x = A_1 \sin \omega_n t + A_2 \cos \omega_n t \tag{3.2-3}$$

is also a solution of the differential equation. From the trigonometric identity

$$A \sin(\omega_n t + \phi) = A \sin \omega_n t \cos \phi + A \cos \omega_n t \sin \phi$$

we see that

$$x = A \sin(\omega_n t + \phi) \tag{3.2-4}$$

is also a solution. The constants A and ϕ are related to A_1 and A_2 as follows: $A_1 = A \cos \phi$ and $A_2 = A \sin \phi$.

Initial Conditions

The sine and cosine solutions show that the mass m oscillates with a radian frequency of $\sqrt{k/m}$. Often this is the only information we need to obtain. However, the solution to a vibration problem in general requires more than just finding a solution to the equation of motion. For example, we may need to compute the value of the displacement x or the velocity $\dot{x}$ at one or more values of time t. In this case we must be able to compute the values of the constants A_1 and A_2, or A and ϕ.

From basic dynamics we know that the displacement and velocity at some time $t > 0$ depend on the displacement and velocity at time $t = 0$ as well as on the force $f(t)$ for $t \leq 0$. If this force is zero (as we are assuming to obtain the free response), then $x(t)$ and $\dot{x}(t)$ depend only on $x(0)$ and $\dot{x}(0)$. These values are called the *initial conditions*.

For example, suppose that at time $t = 0$ we position the mass at x_0. Thus $x(0) = x_0$. If we release the mass from rest, then $\dot{x}(0) = 0$. The displacement $x(t)$ and the velocity $\dot{x}(t)$ so resulting will be different than if we give the mass some initial velocity v_0 when we release it. In this case, $\dot{x}(0) = v_0$.

In other situations we may not know the initial displacement or velocity but are given the values of the displacement or velocity at some other times. For example, we may know the initial displacement $x(0)$ and the displacement $x(t_1)$ at some time $t_1 \neq 0$. In fact, we must be given *two* values because there are two unknown constants in the solutions, Equations 3.2-3 and 3.2-4.

In most of our applications we will be given the initial displacement x_0 and the initial velocity v_0, so we will now express the solution in terms of them.

Solution Form $x = A_1 \sin \omega_n t + A_2 \cos \omega_n t$

Evaluating the solution form in Equation 3.2-3 and its derivative at $t = 0$, we obtain

$$x(0) = A_1 \sin 0 + A_2 \cos 0 = A_2$$

$$\dot{x}(0) = A_1 \omega_n \cos 0 - A_2 \omega_n \sin 0 = A_1 \omega_n$$

Thus $A_1 = \dot{x}(0)/\omega_n = v_0/\omega_n$ and $A_2 = x(0) = x_0$. So the solution of Equation 3.2-3 may be expressed as

$$x = \frac{v_0}{\omega_n} \sin \omega_n t + x_0 \cos \omega_n t \tag{3.2-5}$$

Note that if the initial velocity v_0 is zero, the response is cosinusoidal, $x = x_0 \cos \omega_n t$, and its amplitude is the initial displacement x_0.

Solution Form $x = A \sin(\omega_n t + \phi)$

Repeating this process for Equation 3.2-4 gives

$$x(0) = x_0 = A \sin \phi$$

$$\dot{x}(0) = v_0 = A \omega_n \cos \phi$$

or

$$\sin \phi = \frac{x_0}{A} \tag{3.2-6}$$

$$\cos \phi = \frac{v_0}{A \omega_n} \tag{3.2-7}$$

The phase angle ϕ can be computed from

$$\phi = \tan^{-1} \left(\frac{\sin \phi}{\cos \phi} \right) = \tan^{-1} \frac{x_0 \omega_n}{v_0} \tag{3.2-8}$$

Note that this equation will give two solutions for ϕ. The correct quadrant for ϕ can be found from Equations 3.2-6 and 3.2-7 once A is known.

Using Equations 3.2-6 and 3.2-7 with the identity $\sin^2 \phi + \cos^2 \phi = 1$, we obtain

$$\sin^2 \phi + \cos^2 \phi = \left(\frac{x_0}{A} \right)^2 + \left(\frac{v_0}{A \omega_n} \right)^2 = 1$$

We can solve this equation for A as follows. Because A represents the amplitude of motion, we choose the *positive* square root:

$$A = +\sqrt{x_0^2 + \frac{v_0^2}{\omega_n^2}}$$

Thus the response is

$$x = \sqrt{x_0^2 + \frac{v_0^2}{\omega_n^2}} \sin(\omega_n t + \phi) \tag{3.2-9}$$

where ϕ is computed from Equation 3.2-8 with the correct quadrant determined from Equations 3.2-6 and 3.2-7 with $A > 0$.

Note that if the initial velocity v_0 is zero, the amplitude of motion is $A = x_0$, and $\phi = \pi/2$ if $x(0) > 0$ and $\phi = -\pi/2$ if $x(0) < 0$. So if $x(0) > 0$, the response is $x = x_0 \sin(\omega_n t + \pi/2) = x_0 \cos \omega_n t$.

It is easier to compute the constants for Equation 3.2-5 than for Equation 3.2-9. However, Equation 3.2-9 has the advantage of giving the amplitude of motion directly and it is a more compact representation of the free response.

EXAMPLE 3.2-1 ***Applying the Initial Conditions***	Obtain the free response of $2\ddot{x} + 128x = f(t)$ **(a)** in the form $x = A_1 \sin \omega_n t + A_2 \cos \omega_n t$ and **(b)** in the form $x = A \sin(\omega_n t + \phi)$. The initial conditions are $x(0) = 0.05$ m and $\dot{x}(0) = -0.3$ m/s.
Solution	The natural frequency is $\omega_n = \sqrt{128/2} = 8$ rad/s.

(a) Note that $x_0 = 0.05$ and $v_0 = -0.3$. From Equation 3.2-5,

$$x = \frac{-0.3}{8} \sin 8t + 0.05 \cos 8t = -0.0375 \sin 8t + 0.05 \cos 8t \text{ m}$$

(b) From Equation 3.2-9,

$$A = +\sqrt{(0.05)^2 + \frac{(-0.3)^2}{8^2}} = 0.062 \text{ m}$$

From Equations 3.2-6 and 3.2-7,

$$\sin \phi = \frac{0.05}{0.062} = 0.806$$

$$\cos \phi = \frac{-0.3}{0.062(8)} = -0.605$$

Because $\sin \phi > 0$ and $\cos \phi < 0$, ϕ is in the second quadrant ($\pi/2 \leq \phi \leq \pi$). The angle is computed from Equation 3.2-8:

$$\phi = \tan^{-1} \frac{0.806}{-0.605} = \tan^{-1}(-1.333) = -0.927 \text{ or } -0.927 + \pi = 2.214 \text{ rad}$$

Choosing the solution in the second quadrant gives $\phi = 2.214$ rad. Thus the free response is

$$x = 0.062 \sin(8t + 2.214) \text{ m}$$

Note that the amplitude of motion (0.062 m) is greater than the initial displacement. This is because of the nonzero initial velocity. ∎

Simple Harmonic Motion

From the equation of motion $m\ddot{x} = -kx$, we can see that the acceleration is $\ddot{x} = -kx/m = -\omega_n^2 x$. This type of motion, where the acceleration is proportional to the displacement but opposite in sign, is called *simple harmonic motion*. It occurs when the restoring force— here, the spring force—is proportional to the displacement. It is helpful to understand the relation among the displacement, velocity, and acceleration in simple harmonic motion.

The free response of $m\ddot{x} + kx = f$ is $x(t) = A \sin(\omega_n t + \phi)$ where $\omega_n = \sqrt{k/m}$. The amplitude A and phase angle ϕ depend on the initial conditions.

Expressions for the velocity and acceleration are obtained by differentiating $x(t)$:

$$\dot{x} = A\omega_n \cos(\omega_n t + \phi) = A\omega_n \sin\left(\omega_n t + \phi + \frac{\pi}{2}\right)$$

$$\ddot{x} = -A\omega_n^2 \sin(\omega_n t + \phi)$$

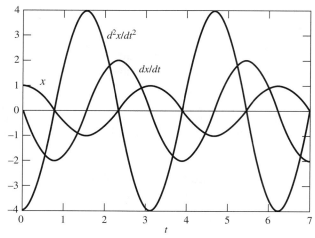

FIGURE 3.2-1 Displacement, velocity, and acceleration versus time, for simple harmonic motion.

The displacement, velocity, and acceleration all oscillate with the same frequency ω_n but have different amplitudes. The velocity is zero when the displacement is minimum or maximum. The sign of the acceleration is the opposite of that of the displacement, and the magnitude of the acceleration is ω_n^2 times the magnitude of the displacement. These functions are plotted in Figure 3.2-1 for the case where $x(0) = 1$, $\dot{x}(0) = 0$, and $\omega_n = 2$.

EXAMPLE 3.2-2
Energy Dynamics in Simple Harmonic Motion

Consider a mass-spring system described by $2\ddot{x} + 17x = 0$, where $x(0) = 2$ and $\dot{x}(0) = 0$. Plot the kinetic, potential, and total mechanical energy of the system versus time.

Solution

The free response is

$$x(t) = 2 \sin\left(\sqrt{\frac{17}{2}} + \frac{\pi}{2}\right)$$

Differentiate with respect to t to obtain the velocity:

$$\dot{x}(t) = 2\sqrt{\frac{17}{2}} \cos\left(\sqrt{\frac{17}{2}}t + \frac{\pi}{2}\right)$$

The position and velocity oscillate with the same radian frequency $\left(\sqrt{17/2}\right)$ but different amplitudes. The period of the oscillations is $2\pi/\sqrt{17/2} = 2.16$.

The expressions for kinetic, potential, and total mechanical energy are: $KE = m\dot{x}^2/2 = \dot{x}^2$, $PE = kx^2/2 = 17x^2/2$, and $Total = KE + PE$. Using the expressions for $x(t)$ and $\dot{x}(t)$, we can plot the energies versus time. These plots are shown in Figure 3.2-2 along with $x(t)$ for reference. Note that the total mechanical energy remains constant because there is no friction present to convert energy into heat. The maximum kinetic energy occurs when the potential energy is at a minimum. The system receives its energy initially in the form of potential energy from the initial displacement $x(0)$ of the mass. The dynamics of the system result from the energy flowing back and forth between kinetic and potential energy. ■

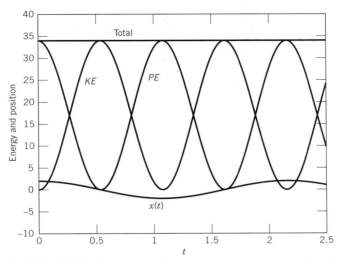

FIGURE 3.2-2 Kinetic, potential, and total energy versus time, for a mass-spring system, from Example 3.2-2.

EXAMPLE 3.2-3
A Cantilever
Support

Figure 3.2-3a shows a motor mounted on a cantilever beam support. It is impossible to balance the motor perfectly, and the imbalance will result in a vertical force f that oscillates at the same frequency as the motor's rotational speed. We will see in a later chapter that excessive support motion or even support failure can occur if the motor's speed is near the natural frequency of the support system (this condition is called *resonance*). Determine the expression for the natural frequency of the system.

Solution

We can model this system as if it were a single mass located at the end of the beam. The resulting equivalent system is shown in Figure 3.2-3b. The equivalent mass of the system is the motor mass plus the equivalent mass of the beam. From Table 2.5-1, we see that the beam's equivalent mass is $0.23m_d$. Thus the system's equivalent mass is $m_e = m_c + 0.23m_d$.

The equivalent spring constant of the beam is found from Table 1.4-1. It is

$$k = \frac{Ewh^3}{4L^3}$$

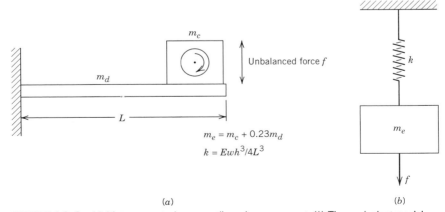

$$m_e = m_c + 0.23m_d$$
$$k = Ewh^3/4L^3$$

(a)   (b)

FIGURE 3.2-3 (a) Motor mounted on a cantilever beam support. (b) The equivalent model.

where h is the beam's thickness (height) and w is its width. Thus the system model is

$$m_e \ddot{x} + kx = f$$

where x is the vertical displacement of the beam end from its equilibrium position. The natural frequency is

$$\omega_n = \sqrt{\frac{k}{m_e}} = \sqrt{\frac{Ewh^3/4L^3}{m_c + 0.23 m_d}}$$

∎

Rotational Systems

The equivalent inertia of rotational spring elements can be found using kinetic energy equivalence. The rod in torsion shown in Table 2.5-1 has an equivalent inertia equal to one-third its actual inertia, assuming that the twist is proportional to the distance from the support.

EXAMPLE 3.2-4
A Torsional
Vibration Model

Figure 3.2-4a shows an inertia I_1 connected to a shaft with inertia I_2. The other end of the shaft is rigidly attached to the support. The applied torque is T.
(a) Develop the equation of motion.

(b) Calculate the system's natural frequency if I_1 is a cylinder 0.1 m in diameter and 0.05 m long and I_2 is a cylinder 0.05 m in diameter and 0.3 m long. Both are made of steel with a shear modulus $G = 8 \times 10^{10}$ N/m^2 and a density $\rho = 7800$ kg/m^3.

Solution

(a) Note that when $T = 0$, the equilibrium position is $\theta = 0$. As in Table 2.5-1, we lump one-third of the shaft's inertia into the inertia at the end of the shaft. Thus the equivalent inertia is

$$I_e = I_1 + \frac{1}{3} I_2$$

The equivalent representation is shown in Figure 3.2-4b, and the free-body diagram is shown in Figure 3.2-4c. From this we obtain the equation of motion:

$$I_e \ddot{\theta} = T - k\theta$$

where, from Table 1.4-1,

$$k = \frac{G\pi D^4}{32L}$$

(b) The value of k is

$$k = \frac{8 \times 10^{10} \pi (0.05)^4}{32(0.3)} = 1.64 \times 10^5 \text{ N·m/rad}$$

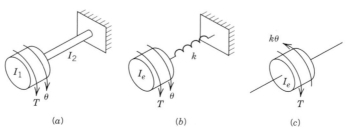

FIGURE 3.2-4   (a) A torsional vibration model. (b) Equivalent model. (c) Free-body diagram.

The moments of inertia are

$$I_1 = \frac{\pi(7800)}{2}(0.05)^4 0.05 = 3.83 \times 10^{-3} \text{ kg·m}^2$$

$$I_2 = \frac{\pi(7800)}{2}\left(\frac{0.05}{2}\right)^4 (0.3) = 1.44 \times 10^{-3} \text{ kg·m}^2$$

Thus

$$I_e = \left(3.83 + \frac{1}{3}1.44\right)10^{-3} = 4.31 \times 10^{-3} \text{ kg·m}^2$$

This system's natural frequency is $\sqrt{k/I_e} = 6169$ rad/sec, or $6169/2\pi = 982$ cycles per second. This gives a period of 0.001 s. ∎

Small-Angle Approximation

The equations of motion for several examples in Chapter 2 have the form

$$\ddot{\theta} + \omega_n^2 \sin \theta = 0 \tag{3.2-10}$$

These examples include:

1. A compound pendulum (Example 2.2-2), $I_O\ddot{\theta} + mgL \sin \theta = 0$
2. The rolling motion of a ship (Example 2.2-3), $I_G\ddot{\theta} + mgh \sin \theta = 0$

This equation can be solved in closed form, but the solution is cumbersome.

As we will see, quite often the displacement θ is small enough that we can use the *small-angle approximation*: $\sin \theta \approx \theta$. In this case the equation can be written as

$$\ddot{\theta} + \omega_n^2\theta = 0 \tag{3.2-11}$$

which is the equation for simple harmonic motion. Thus its solution is

$$\theta(t) = A \sin(\omega_n t + \phi)$$

where A and ϕ depend on the initial conditions. The period is $2\pi/\omega_n$. Thus the period of the compound pendulum is

$$P = \frac{2\pi}{\sqrt{mgL/I_O}} = 2\pi\sqrt{\frac{I_O}{mgL}}$$

The roll period of the ship is

$$P = \frac{2\pi}{\sqrt{mgh/I_G}} = 2\pi\sqrt{\frac{I_G}{mgh}}$$

General Linearization Method

The model $\ddot{\theta} + \omega_n^2 \sin \theta = 0$ is easily linearized because of the small-angle approximation and the fact that its equilibrium is at $\theta = 0$. However, a more general method of linearization is needed for other model forms that do not contain the sine function and whose equilibrium is nonzero. In Chapter 1 we discussed the process of obtaining a model that is "approximately" linear; that is, a linear equation that approximately describes the system's behavior under certain conditions.

Knowing the system's equilibrium position, we can replace any nonlinear functions in the equation of motion with linear functions that are approximately correct near the equilibrium. A systematic procedure for doing this is based on the Taylor series expansion. Consider the nonlinear function $w = f(y)$. Let (w_r, y_r) be the values of w and y at the reference equilibrium. A function that is approximately linear near the reference point (w_r, y_r) can be obtained by expanding $f(y)$ in a Taylor series near this point and truncating the series beyond the first-order term. The result is

$$w = f(y) = f(y_r) + \left(\frac{df}{dy}\right)_r (y - y_r) \tag{3.2-12}$$

where the subscript r on the derivatives means that they are evaluated at the reference equilibrium (w_r, y_r). If y is "close enough" to y_r, the terms involving $(y - y_r)^i$ for $i \geq 2$ are small compared to the first two terms in the series. This is a linear relation. We have replaced the original function with a straight line passing through the point (w_r, y_r) and having a slope equal to the slope of $f(y)$ at the reference point.

Linearization and Spring Constants

Elastic elements that are extended or compressed beyond their linear range are often described by the following nonlinear model:

$$f(y) = k_1 y + k_2 y^3 \tag{3.2-13}$$

where $f(y)$ is the spring force and y is the spring's displacement from its free length. If a mass m is attached to the spring, as shown in Figure 3.2-5, its equation of motion is

$$m\ddot{y} = -(k_1 y + k_2 y^3) + mg \tag{3.2-14}$$

When the mass is in equilibrium at the position $y = y_r$, then $\ddot{y}_r = 0$, and the previous equation becomes

$$k_1 y_r + k_2 y_r^3 - mg = 0 \tag{3.2-15}$$

This can be solved for y_r.

In general, if the force-deflection model of a spring is $f(y)$, the equation of motion is

$$m\ddot{y} = -f(y) + mg \tag{3.2-16}$$

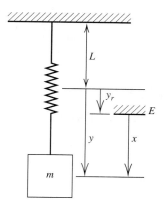

FIGURE 3.2-5 A mass-spring model. The equilibrium location is E.

Evaluating this equation at $y = y_r$, we obtain

$$m\ddot{y}_r = -f(y_r) + mg$$

Expand $f(y)$ in a Taylor series, keeping only the linear term, and subtract these two equations to obtain

$$m(\ddot{y} - \ddot{y}_r) = f(y_r) - f(y)$$

$$= f(y_r) - \left[f(y_r) + \left(\frac{df}{dy}\right)_r (y - y_r) \right]$$

$$= -\left(\frac{df}{dy}\right)_r (y - y_r)$$

Let $x = y - y_r$, where x denotes the displacement of the mass from its equilibrium position. Substituting $x = y - y_r$ and $\ddot{x} = \ddot{y} - \ddot{y}_r$ yields

$$m\ddot{x} = -\left(\frac{df}{dy}\right)_r x \qquad (3.2\text{-}17)$$

For the *linear* spring relation $f = kx$, the equation of motion is

$$m\ddot{x} = -kx \qquad (3.2\text{-}18)$$

Comparing Equations 3.2-17 and 3.2-18, we see that

$$k = \left(\frac{df}{dy}\right)_r \qquad (3.2\text{-}19)$$

The important result of this analysis is that *the value of the linearized spring constant k is the slope of the spring force curve $f(y)$ at the equilibrium position y_r.*

EXAMPLE 3.2-5
Linearizing a Nonlinear Spring Model

Suppose a particular nonlinear spring is described by Equation 3.2-13 with $k_1 = 120$ lb/ft and $k_2 = 360$ lb/ft^3. Determine its equivalent linear spring constant k for the two equilibrium positions corresponding to the following weights: **(a)** $mg = 10$ lb, **(b)** $mg = 35$ lb. In addition, find the linearized equations for motion for each equilibrium, and solve them.

Solution

The first step is to determine the equilibrium positions for each weight. From Equation 3.2-15,

$$120y_r + 360y_r^3 - mg = 0$$

For $mg = 10$,

$$120y_r + 360y_r^3 - 10 = 0$$

The only real root of this cubic equation is $y_r = 0.082$ ft.
  For $mg = 35$,

$$120y_r + 360y_r^3 - 35 = 0$$

The only real root of this cubic equation is $y_r = 0.247$ ft.
  Note that $f(y_r) = mg$ and that

$$\left(\frac{df}{dy}\right)_r = k_1 + 3k_2 y_r^2 = 120 + 1080y_r^2$$

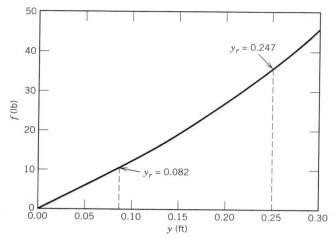

FIGURE 3.2-6 Plot of the spring force relation $f = 120y + 360y^3$. The short straight-line segments are the linear approximations to the curve near the equilibrium positions $y_r = 0.082$ and $y_r = 0.247$ ft.

Thus Equation 3.2-12 gives

$$f(y) = mg + (120 + 1080y_r^2)(y - y_r)$$

This is the linearized approximation to the nonlinear spring force function for the given values of k_1 and k_2. It approximately describes the nonlinear function near the specified equilibrium point. It is shown in Figure 3.2-6 by the short straight-line segments passing through the two equilibrium points.

The equivalent linear spring constant k is the slope m of the curve $f(y)$ near the specific equilibrium point. Thus the linearized spring constants for each equilibrium are:

(a) For $y_r = 0.082$, $k = 120 + 1080(0.082)^2 = 127.3$ lb/ft
(b) For $y_r = 0.247$, $k = 120 + 1080(0.247)^2 = 185.9$ lb/ft

The linearized equations of motion, which are accurate only near their respective equilibrium points, are found from Equation 3.2-17. They are

(a) For $y_r = 0.082$, $x = y - 0.082$ and $(10/32.2)\ddot{x} + 127.3x = 0$
(b) For $y_r = 0.247$, $x = y - 0.247$ and $(35/32.2)\ddot{x} + 185.9x = 0$

The solution of the equation $m\ddot{x} + kx = 0$ for $\dot{x}(0) = 0$ is

$$x(t) = x(0)\sin\left(\sqrt{\frac{k}{m}}t + \pi/2\right)$$

Using the initial conditions, $x(0) = 0.05$ and $\dot{x}(0) = 0$, the solutions for each equilibrium are

(a) For $y_r = 0.082$, $x(t) = 0.05\sin(20.25t + \pi/2)$
(b) For $y_r = 0.247$, $x(t) = 0.05\sin(13.08t + \pi/2)$

If we wish, we can find $y(t)$ from the relation $y(t) = x(t) + y_r$. These solutions are plotted in Figure 3.2-7. Note that in each case the mass oscillates about the specific equilibrium position. The oscillation frequency is different for each equilibrium because of the different values of the linearized spring constant k. ∎

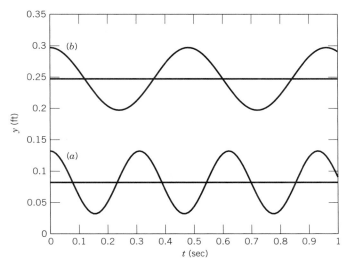

FIGURE 3.2-7 Free response of the mass for two equilibrium points, from Example 3.2-5.

3.3 RAYLEIGH'S METHOD

In linear mass-spring systems with negligible friction and damping undergoing simple harmonic motion, we can often determine the frequency of vibration without obtaining the equation of motion. The method for doing this is known as Rayleigh's method.

In simple harmonic motion, the principle of conservation of energy implies that the kinetic energy is maximum and the potential energy is minimum at the equilibrium position $x = 0$. When the displacement is maximum, the potential energy is maximum but the kinetic energy is zero. From conservation of energy,

$$T_{max} + V_{min} = T_{min} + V_{max}$$

Thus

$$T_{max} + V_{min} = 0 + V_{max}$$

or

$$T_{max} = V_{max} - V_{min} \tag{3.3-1}$$

For example, for the mass-spring system oscillating vertically as shown in Figure 3.3-1, $T = m\dot{x}^2/2$ and $V = k(x + \Delta)^2/2 - mgx$, and from Equation 3.3-1 we have

$$T_{max} = \frac{1}{2}m(\dot{x}_{max})^2 = V_{max} - V_{min} = \frac{1}{2}k(x_{max} + \Delta)^2 - mgx - \frac{1}{2}k\Delta^2$$

or

$$\frac{1}{2}m(\dot{x}_{max})^2 = \frac{1}{2}k(x_{max})^2$$

where we have used the fact that $k\Delta = mg$. In simple harmonic motion $|\dot{x}_{max}| = \omega_n|x_{max}|$, and thus

$$\frac{1}{2}m(\omega_n|x_{max}|)^2 = \frac{1}{2}k|x_{max}|^2$$

Cancel $|x_{max}|^2$ and solve for ω_n to obtain $\omega_n = \sqrt{k/m}$.

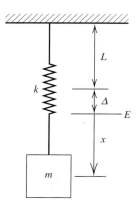

FIGURE 3.3-1 The static spring deflection is Δ and the equilibrium position of the mass m is at E.

In this simple example we merely obtained the expression for ω_n that we already knew. However, in other applications the expressions for T and V may be different, but if the motion is simple harmonic, we can directly determine the natural frequency by using the fact that $|\dot{x}_{\max}| = \omega_n |x_{\max}|$ to express $T_{\max}$ as a function of $|x_{\max}|$ and then equating $T_{\max}$ to $V_{\max} - V_{\min}$. This approach is the basis of Rayleigh's method.

EXAMPLE 3.3-1
Natural Frequency of a Pendulum

Apply Rayleigh's method to determine the natural frequency of the pendulum shown in Figure 3.3-2. Neglect the mass of the rod.

Solution

The kinetic energy expression is

$$T = \frac{1}{2}m\left(L\dot{\theta}\right)^2 = \frac{1}{2}mL^2\dot{\theta}^2$$

Choosing the gravitational potential energy to be zero at $\theta = 0$, from Figure 3.3-2 we see that the potential energy is given by

$$V = mg(L - L\cos\theta) = mgL(1 - \cos\theta)$$

Note that for small angles, if we use the approximation that $\cos\theta \approx 1$, then $V \approx 0$ and thus does not appear as a function of θ. To remedy this oversimplification and represent V as a function of θ, we keep an additional term in the Taylor series expansion for $\cos\theta$ and use the following approximation:

$$\cos\theta \approx 1 - \frac{\theta^2}{2} \tag{1}$$

Thus

$$V \approx mgL\left(1 - 1 + \frac{\theta^2}{2}\right) = mgL\frac{\theta^2}{2}$$

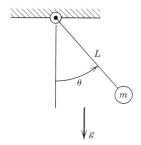

FIGURE 3.3-2 A pendulum.

Assuming simple harmonic motion, we have $\theta = A\sin(\omega_n t + \phi)$ and $\dot{\theta} = \omega_n A\cos(\omega_n t + \phi)$. Thus $\theta_{max} = A$, $\dot{\theta}_{max} = \omega_n A$, and

$$T_{max} = \frac{1}{2}mL^2(\omega_n A)^2$$

$$V_{max} = \frac{1}{2}mgLA^2$$

$$V_{min} = 0$$

From Rayleigh's method, $T_{max} = V_{max} - V_{min}$, we obtain

$$\frac{1}{2}mL^2(\omega_n A)^2 = \frac{1}{2}mgLA^2 - 0$$

This gives $L\omega_n^2 = g$, or

$$\omega_n = \sqrt{\frac{g}{L}}$$

∎

A common mistake made when applying Rayleigh's method is to assume that $V_{min} = 0$, but this is not always true, as shown by the next example.

EXAMPLE 3.3-2
*Natural
Frequency of a
Rolling Cylinder*

Apply Rayleigh's method to determine the natural frequency of the cylinder shown in Figure 3.3-3. Neglect the mass of the spring.

Solution

Let $x = 0$ denote the rest position of the cylinder and let $\theta = x/R$ and $\omega = \dot{\theta}$. Take the gravitational potential energy to be 0 at the rest position. The elastic potential energy at the rest position is $k\Delta^2/2$, where Δ is the static deflection of the spring. The total potential energy is

$$V = \frac{1}{2}k(x + \Delta)^2 - mgh$$

where $h = x\sin\phi$. Since $mg\sin\phi = k\Delta$ from statics, the expression for V becomes

$$V = \frac{1}{2}k(x^2 + 2x\Delta + \Delta^2) - mgx\sin\phi = \frac{1}{2}kx^2 + \frac{1}{2}k\Delta^2$$

Thus $V_{min} = k\Delta^2/2$.

The kinetic energy is

$$T = \frac{1}{2}m\dot{x}^2 + \frac{1}{2}I\dot{\theta}^2 = \frac{1}{2}m\dot{x}^2 + \frac{1}{2}I\frac{\dot{x}^2}{R^2} = \frac{1}{2}\left(m + \frac{I}{R^2}\right)\dot{x}^2$$

since $\dot{\theta} = \dot{x}/R$ if the cylinder rolls without slipping.

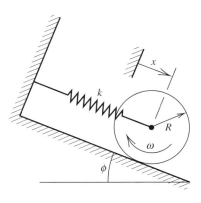

FIGURE 3.3-3   A cylinder-spring system.

Assuming simple harmonic motion, we obtain $x = A\sin(\omega_n t + \phi)$ and $\dot{x} = \omega_n A\cos(\omega_n t + \phi)$. Thus $x_{max} = A$, $\dot{x}_{max} = \omega_n A$, and

$$T_{max} = \frac{1}{2}\left(m + \frac{I}{R^2}\right)(\omega_n A)^2$$

$$V_{max} = \frac{1}{2}kA^2 + \frac{1}{2}k\Delta^2$$

From Rayleigh's method, $T_{max} = V_{max} - V_{min}$, we have

$$\frac{1}{2}\left(m + \frac{I}{R^2}\right)(\omega_n A)^2 = \frac{1}{2}kA^2 + \frac{1}{2}k\Delta^2 - \frac{1}{2}k\Delta^2 = \frac{1}{2}kA^2$$

This gives

$$\left(m + \frac{I}{R^2}\right)\omega_n^2 = k$$

or

$$\omega_n = \sqrt{\frac{k}{m + I/R^2}}$$

∎

Rayleigh's method is especially useful when it is difficult to derive the equation of motion. The following is an example of such an application.

EXAMPLE 3.3-3
Natural Frequency of a Suspension System

Figure 3.3-4 shows the suspension of one front wheel of a car in which $L_1 = 0.4$ m and $L_2 = 0.6$ m. The coil spring has a spring constant of $k = 36\,000$ N/m and the car weight associated with that wheel is 3500 N. Determine the suspension's natural frequency for vertical motion.

Solution

Imagine that the frame moves down by a distance A_f while the wheel remains stationary. Then, from similar triangles, the amplitude A_s of the spring deflection is related to the amplitude A_f of the frame motion by $A_s = L_1 A_f / L_2 = 0.4 A_f / 0.6 = 2A_f/3$.

Using the fact that $k\Delta = mg$, the change in potential energy can be written as

$$V_{max} - V_{min} = \frac{1}{2}kA_s^2 = \frac{1}{2}k\left(\frac{2}{3}A_f\right)^2$$

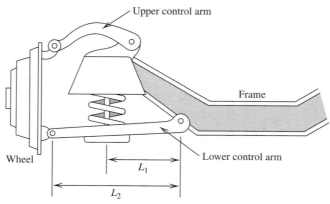

FIGURE 3.3-4 A vehicle suspension.

The amplitude of the velocity of the mass in simple harmonic motion is $\omega_n A_f$, and thus the maximum kinetic energy is

$$T_{\max} = \frac{1}{2} m \left(\omega_n A_f \right)^2$$

From Rayleigh's method, $T_{\max} = V_{\max} - V_{\min}$, we obtain

$$\frac{1}{2} m \left(\omega_n A_f \right)^2 = \frac{1}{2} k \left(\frac{2}{3} A_f \right)^2$$

Solving this for ω_n, we obtain

$$\omega_n = \frac{2}{3} \sqrt{\frac{k}{m}} = \frac{2}{3} \sqrt{\frac{36\ 000}{3500/9.8}} = 6.69 \text{ rad/s}$$

∎

3.4   FREE VIBRATION WITH VISCOUS DAMPING

Consider the system shown in Figure 3.4-1a. The surface is lubricated, so the mass is subjected to viscous damping. Using the linear damping model, we obtain the free-body diagram shown in Figure 3.4-1b, where x is the displacement of the mass from its equilibrium position. The equation of motion is

$$m\ddot{x} = f - c\dot{x} - kx$$

or

$$m\ddot{x} + c\dot{x} + kx = f \qquad (3.4\text{-}1)$$

Similarly, the system shown in Figure 3.4-2 has the same equation of motion because the static spring force $k\Delta$ is canceled by the weight mg. Note that the location of the equilibrium is unaffected by the presence of viscous damping because the damping force is a function of velocity, which is zero at equilibrium.

Equation 3.4-1 is the form of the equation of motion for a linear, single-degree-of-freedom system with viscous damping. In this section we will obtain the solution of this equation for the case where $f = 0$; namely, the free response. The simplest way to obtain this solution is to use the exponential function Ae^{st}, just as we did in Section 3.1 for the first-order model. Substituting this function into Equation 3.4-1 with $f = 0$ and collecting terms, we obtain

$$(ms^2 + cs + k)Ae^{st} = 0$$

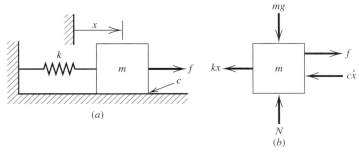

(a)

(b)

FIGURE 3.4-1   (a) Mass-spring system on a surface having viscous friction.   (b) Free-body diagram.

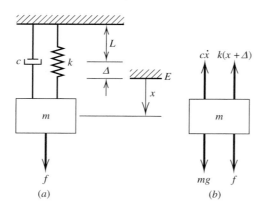

FIGURE 3.4-2   A mass-spring-damper system. (a) The equilibrium position E as the coordinate reference. (b) Free-body diagram.

As we saw in Section 3.1, a general solution is not possible if $Ae^{st} = 0$, so the previous equation requires that

$$ms^2 + cs + k = 0 \tag{3.4-2}$$

This is the *characteristic equation* of Equation 3.4-1, and its roots are the *characteristic roots*. The solution is

$$s = \frac{-c \pm \sqrt{c^2 - 4mk}}{2m} \tag{3.4-3}$$

The solution for the roots will be one of the following cases:

1. Two distinct, real roots, denoted r_1 and r_2. This case will occur if $c^2 - 4mk > 0$.
2. Two equal (repeated) roots, denoted r_1 and r_1. This case will occur if $c^2 - 4mk = 0$. These roots will be real.
3. Two complex conjugate roots, denoted $r + iq$ and $r - iq$, where we define $q > 0$. This case will occur if $c^2 - 4mk < 0$.

We can see that case 1 corresponds to a value of the damping constant c large enough to prevent the mass from oscillating. This case is called the *overdamped* case. Case 3 corresponds to a value of the damping constant small enough to allow the mass to oscillate. This is the *underdamped* case. Case 2 is the borderline case between cases 1 and 2. This is the *critically damped* case.

We may obtain the solution for the free response for each case as follows. Recall that we need two arbitrary constants in the solution so that it can satisfy the two initial conditions $x(0) = x_0$ and $\dot{x}(0) = v_0$.

Overdamped Case

For case 1, the solution is the linear combination of the exponential form Ae^{st} for each root. Thus

$$x = A_1 e^{r_1 t} + A_2 e^{r_2 t} \tag{3.4-4}$$

where the two roots are $s = r_1$ and $s = r_2$. Applying the initial conditions gives $x_0 = A_1 + A_2$ and $v_0 = r_1 A_1 + r_2 A_2$. These give

$$A_1 = \frac{v_0 - r_2 x_0}{r_1 - r_2}$$

$$A_2 = \frac{r_1 x_0 - v_0}{r_1 - r_2} = x_0 - A_1$$

Critically Damped Case

For case 2, the required second independent term in the solution is given by tAe^{st}. Thus

$$x = A_1 e^{r_1 t} + tA_2 e^{r_1 t} \qquad (3.4\text{-}5)$$

Applying the initial conditions gives $x_0 = A_1$ and $v_0 = r_1 A_1 + A_2$. These give

$$A_1 = x_0$$

$$A_2 = v_0 - r_1 x_0$$

Underdamped Case

For case 3, the solution is the linear combination of the exponential form Ae^{st} for each root. Thus

$$x = A_1 e^{(r+iq)t} + A_2 e^{(r-iq)t} = e^{rt}\left(A_1 e^{iqt} + A_2 e^{-iqt}\right)$$

As demonstrated in Section 1.2, this form is equivalent to

$$x = Be^{rt}\sin(qt + \phi) \qquad (3.4\text{-}6)$$

Applying the initial conditions gives $x_0 = B\sin\phi$ and $v_0 = rB\sin\phi + qB\cos\phi$. These give

$$\sin\phi = \frac{x_0}{B} \qquad (3.4\text{-}7)$$

$$\cos\phi = \frac{v_0 - rx_0}{qB} \qquad (3.4\text{-}8)$$

The phase angle ϕ can be computed from

$$\phi = \tan^{-1}\left(\frac{\sin\phi}{\cos\phi}\right) = \tan^{-1}\frac{qx_0}{v_0 - rx_0} \qquad (3.4\text{-}9)$$

Note that this equation will give two solutions for ϕ. The correct quadrant for ϕ can be found from Equations 3.4-7 and 3.4-8 once B is known.

Using Equations 3.4-7 and 3.4-8 with the identity $\sin^2\phi + \cos^2\phi = 1$, we obtain

$$\sin^2\phi + \cos^2\phi = \left(\frac{x_0}{B}\right)^2 + \left(\frac{v_0 - rx_0}{qB}\right)^2 = 1$$

We can solve this equation for B as follows. Because B represents the amplitude of motion, we choose the *positive* square root:

$$B = +\sqrt{x_0^2 + \left(\frac{v_0 - rx_0}{q}\right)^2} = +\frac{1}{q}\sqrt{(qx_0)^2 + (v_0 - rx_0)^2} \qquad (3.4\text{-}10)$$

The angle ϕ is computed from Equation 3.4-9 with the correct quadrant determined from the signs of $\sin\phi$ and $\cos\phi$ obtained from Equations 3.4-7 and 3.4-8 with $B > 0$.

The solutions for all three cases are summarized in Table 3.4-1.

The response for case 3 for $r < 0$ is illustrated in Figure 3.4-3, where we have defined $a = -r$, $a > 0$. The sinusoidal oscillation has a frequency q (radians/unit time). Because the amplitude decays to zero, the mass eventually returns to equilibrium, so this

Table 3.4-1   Free Response with Viscous Damping

Equation of motion:	$m\ddot{x} + c\dot{x} + kx = 0$
Initial conditions:	$x(0) = x_0 \qquad \dot{x}(0) = v_0$
Characteristic roots:	$s = \dfrac{-c \pm \sqrt{c^2 - 4mk}}{2m}$
Case 1: Distinct, real roots (overdamped case):	$r_1 \neq r_2$ $x(t) = A_1 e^{r_1 t} + A_2 e^{r_2 t}$ $A_1 = \dfrac{v_0 - r_2 x_0}{r_1 - r_2}$ $A_2 = \dfrac{r_1 x_0 - v_0}{r_1 - r_2} = x_0 - A_1$
Case 2: Repeated roots (critically damped case):	$r_1 = r_2$ $x(t) = (A_1 + A_2 t)e^{r_1 t}$ $A_1 = x_0$ $A_2 = v_0 - r_1 x_0$
Case 3: Complex conjugate roots (underdamped case):	$s = r \pm iq, q > 0$ $x(t) = B e^{rt} \sin(qt + \phi)$ $\phi = \tan^{-1} \dfrac{q x_0}{v_0 - r x_0}$ $B = +\dfrac{1}{q}\sqrt{(q x_0)^2 + (v_0 - r x_0)^2}$ $\sin\phi = \dfrac{x_0}{B}$ $\cos\phi = \dfrac{v_0 - r x_0}{qB}$

behavior is called *stable*. The amplitude of oscillation decays exponentially; that is, the oscillation is bracketed on top and bottom by envelopes that are proportional to e^{-at}. These envelopes have a time constant of $\tau = 1/a$. Thus the amplitude of the next oscillation occurring after $t = 1/a$ will be less than 37% of the peak amplitude. For $t > 4/a$, the amplitudes are less than 2% of the peak. So we can say that the response has disappeared after $t \approx 4/a$.

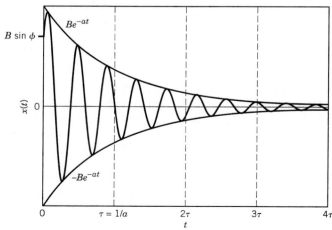

FIGURE 3.4-3   Free response of a second-order model having the complex roots $s = -a \pm iq$. The time constant is $\tau = 1/a$.

Note that if $r < 0$, we may express the roots as $s = -1/\tau \pm qi$. Therefore we see that for complex roots with negative real parts, the time constant is the negative reciprocal of the *real* part of the roots.

If $r > 0$, the oscillation amplitudes increase with time; this behavior is called *unstable*. If $r = 0$, the amplitudes remain constant. This corresponds to undamped free vibration and is the *neutrally stable* case.

Dominant-Root Approximation

We have seen that a time constant τ is a measure of the decay rate of an exponential $e^{-t/\tau}$. The time constant corresponds to the characteristic root $s = -1/\tau$. If a differential equation has several roots, all with negative real parts, then every root will have a time constant. The root having the *largest* time constant is the root whose exponential term dominates the response. This root is thus called the *dominant* root, and its time constant is the *dominant* time constant.

Consider a differential equation whose roots are $s = -2, -20$. Its free response has the form

$$x = A_1 e^{-2t} + A_2 e^{-20t}$$

The time constants are $\tau_1 = 1/2$ and $\tau_2 = 1/20$. Thus the second exponential will essentially disappear after $t = 4/20$, ten times faster than the first exponential. The dominant time constant is thus $\tau_1 = 1/2$. Because the secondary time constant is $1/20$, for $t > 4/20$, the response will essentially look like $A_1 e^{-2t}$. We cannot make exact predictions based on the dominant root because the initial conditions, which determine the values of A_1 and A_2, may be such that $A_2 \gg A_1$, so that the second exponential influences the response for longer than expected.

The dominant-root concept is only an approximation; it cannot be used to make exact predictions about system response. The farther away the dominant root is from the other roots (the "secondary" roots), the better the approximation. Nevertheless, the concept is a very useful one. For example, a useful rule of thumb is that the free response is essentially zero after $t = 4\tau_d$, where τ_d is the dominant time constant.

EXAMPLE 3.4-1 *Free Response* *with Distinct* *Roots*	Obtain the free response of the following model, $$2\ddot{x} + 10\dot{x} + 8x = 0$$ for the following three cases: **(a)** $x(0) = 2$, $\dot{x}(0) = 5$; **(b)** $x(0) = 2$, $\dot{x}(0) = -5$; **(c)** $x(0) = 2$, $\dot{x}(0) = 0$.
Solution	With $m = 2$, $c = 10$, and $k = 8$, Equation 3.4-3 gives the roots $s = -1$ and $s = -4$. Thus the free response has the form $$x(t) = A_1 e^{-t} + A_2 e^{-4t}$$ Using Table 3.4-1 with $r_1 = -1$ and $r_2 = -4$ to evaluate the coefficients A_1 and A_2 for each of the three cases, we obtain: **(a)** $x(0) = 2$, $\dot{x}(0) = 5$: $A_1 = 13/3$, $A_2 = -7/3$ **(b)** $x(0) = 2$, $\dot{x}(0) = -5$: $A_1 = 1$, $A_2 = 1$ **(c)** $x(0) = 2$, $\dot{x}(0) = 0$: $A_1 = 8/3$, $A_2 = -2/3$

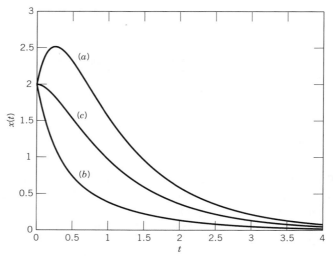

FIGURE 3.4-4 Free responses of second-order models with distinct roots, from Example 3.4-1.

The responses for the three cases are shown in Figure 3.4-4. The time constant corresponding to $s = -1$ is $\tau_1 = 1$, and the time constant corresponding to $s = -4$ is $\tau_2 = 1/4$. The largest time constant (the dominant time constant) determines how long it takes for the response to disappear. In this example, that time is $4\tau_1 = 4$, and this is confirmed for all three cases by Figure 3.4-4. ∎

EXAMPLE 3.4-2
Response of a
Crash Barrier

Highway crash barriers are designed to absorb a vehicle's kinetic energy without bringing the vehicle to such an abrupt stop that the occupants are injured. Figure 3.4-5 represents such a barrier. The barrier's materials and thickness are chosen to accomplish this. It can be modeled as the mass-spring-damping system shown in Figure 3.4-1. Knowledge of the barrier's materials provide the spring constant k and the damping coefficient c; the mass m is the vehicle mass. For this application, $t = 0$ denotes the time at which the moving vehicle contacts the barrier at $x = 0$; thus $\dot{x}(0)$ is the speed of the vehicle at the time of contact and $x(0) = 0$. The applied force f is zero. Most of the barrier's resistance is due to the term $c\dot{x}$, and it stops resisting after the vehicle comes to rest; so the barrier does not reverse the vehicle's motion.

A particular barrier's construction gives $k = 18\,000$ N/m and $c = 20\,000$ N·s/m. A 1800-kg vehicle strikes the barrier at 22 m/s. Determine how long it takes for the vehicle to come to rest, how far the vehicle compresses the barrier, and the maximum deceleration of the vehicle.

Solution

The equation of motion has the form of Equation 3.4-1. It is

$$1800\ddot{x} + 2\times10^4\dot{x} + 1.8\times10^4 x = 0$$

FIGURE 3.4-5 A highway crash barrier.

The initial conditions are $x(0) = 0$ and $\dot{x}(0) = 22$ m/s. The characteristic roots are $s = -0.998, -10.1$. Using the solution given by case 1 in Table 3.4-1, we obtain the vehicle's displacement as a function of time:

$$x = 2.417\left(e^{-0.998t} - e^{-10.1t}\right)$$

We can find the desired answers by plotting this function to find its maximum or we can solve the problem analytically by finding the value of t for which $\dot{x} = 0$. Using this approach, we obtain

$$\dot{x} = 2.417\left(-0.998e^{-0.998t} + 10.1e^{-10.1t}\right) = 0$$

This equation can be solved numerically or by plotting it. The answer is $t = 0.25$ s, which is the time required for the barrier to stop the vehicle. Substituting this value into the expression for x gives the maximum displacement of the barrier: 1.69 m.

The acceleration is found by differentiating x twice:

$$\ddot{x} = 2.417\left[(0.998)^2 e^{-0.998t} - (10.1)^2 e^{-10.1t}\right]$$

Its *minimum* value gives the maximum *deceleration*. It can be found numerically or by plotting. The results show that the maximum deceleration is -244 m/s^2, which is $-24.9g$. It occurs at $t = 0$.

∎

EXAMPLE 3.4-3
Free Response with Repeated Roots

Obtain the free response of the following equation for $x(0) = 3$, $\dot{x}(0) = 5$:

$$5\ddot{x} + 20\dot{x} + 20x = 0$$

Solution

With $m = 5$, $c = 20$, and $k = 20$, Equation 3.4-3 gives the roots $s = -2$ and $s = -2$. The free response form is

$$x = (A_1 + tA_2)e^{-2t}$$

From Table 3.4-1, $A_1 = 3$ and $A_2 = 11$. The response is shown in Figure 3.4-6. ∎

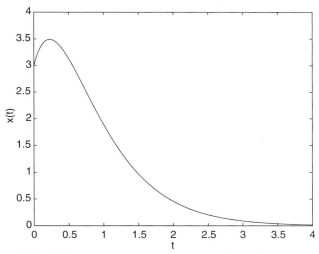

FIGURE 3.4-6   Response of the equation $5\ddot{x} + 20\dot{x} + 20x = 0$ for $x(0) = 3$ and $\dot{x}(0) = 5$.

EXAMPLE 3.4-4
Free Response
with Complex
Roots

Obtain the free response of the following equation

$$2\ddot{x} + 6\dot{x} + 17x = 0$$

for the following three cases: **(a)** $x(0) = 2$, $\dot{x}(0) = 5$; **(b)** $x(0) = 2$, $\dot{x}(0) = -5$; **(c)** $x(0) = 2$, $\dot{x}(0) = 0$.

Solution

The characteristic equation is $2s^2 + 6s + 17 = 0$, and the roots are $s = -1.5 \pm 2.5i$. In terms of the notation of case 3, Table 3.4-1, $r = -1.5$ and $q = 2.5$. Thus the form of the response is

$$x(t) = Be^{-1.5t}\sin(2.5t + \phi)$$

Use Table 3.4-1 to compute the constants B and ϕ for the given initial conditions. For all three cases, $x(0) = 2$ and thus, with $v_0 = \dot{x}(0)$,

$$B = +\frac{1}{2.5}\sqrt{5^2 + (v_0 + 3)^2}$$

$$\phi = \tan^{-1}\left(\frac{5}{v_0 + 3}\right)$$

$$\sin\phi = \frac{2}{B} > 0 \qquad \cos\phi = \frac{v_0 + 3}{2.5B}$$

The remaining results for each case are as follows:
(a) For $\dot{x}(0) = 5$,

$$B = +\frac{1}{2.5}\sqrt{5^2 + (5 + 3)^2} = 3.773$$

$$\phi = \tan^{-1}\left(\frac{5}{5 + 3}\right) = 0.559 \quad \text{or} \quad 0.559 + \pi$$

$$\sin\phi = \frac{2}{3.773} > 0 \qquad \cos\phi = \frac{5 + 3}{2.5(3.773)} > 0$$

Because $\sin\phi > 0$ and $\cos\phi > 0$, ϕ is in the first quadrant, so $\phi = 0.559$ rad. The response is

$$x(t) = 3.773e^{-1.5t}\sin(2.5t + 0.559)$$

(b) For $\dot{x}(0) = -5$,

$$B = +\frac{1}{2.5}\sqrt{5^2 + (-5 + 3)^2} = 2.154$$

$$\phi = \tan^{-1}\left(\frac{5}{-5 + 3}\right) = -1.19 \quad \text{or} \quad -1.19 + \pi = 1.951$$

$$\sin\phi = \frac{2}{2.154} > 0 \qquad \cos\phi = \frac{-5 + 3}{2.5(2.154)} = \frac{-2}{2.5(2.154)} < 0$$

Because $\sin\phi > 0$ but $\cos\phi < 0$, ϕ is in the second quadrant, so $\phi = 1.951$ rad. The response is

$$x(t) = 2.154e^{-1.5t}\sin(2.5t + 1.951)$$

(c) For $\dot{x}(0) = 0$,

$$B = +\frac{1}{2.5}\sqrt{5^2 + (0 + 3)^2} = 2.332$$

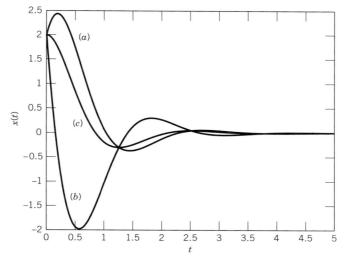

FIGURE 3.4-7 Response of the equation $2\ddot{x} + 6\dot{x} + 17x = 0$ for $x(0) = 2$ and (a) $\dot{x}(0) = 5$, (b) $\dot{x}(0) = -5$, and (c) $\dot{x}(0) = 0$.

$$\phi = \tan^{-1}\left(\frac{5}{0+3}\right) = 1.03 \quad\text{or}\quad 1.03 + \pi$$

$$\sin\phi = \frac{2}{2.332} > 0 \qquad \cos\phi = \frac{0+3}{2.5(2.154)} = \frac{3}{2.5(2.154)} > 0$$

Because $\sin\phi > 0$ and $\cos\phi > 0$, ϕ is in the first quadrant, so $\phi = 1.03$ rad. The response is

$$x(t) = 2.332e^{-1.5t}\sin(2.5t + 1.03)$$

In parts (a), (b) and (c), note that only the amplitude and phase angle are affected by the initial conditions. The frequency and time to decay depend only on the roots. The responses are plotted in Figure 3.4-7. For all three cases, the oscillation frequency is 2.5 rad/sec, if time is measured in seconds. The period is $1/0.398 = 2.51$ sec. The time constant for all three cases is $\tau = 1/1.5 = 2/3$ sec. Thus the oscillations will be essentially gone by $t = 4\tau = 8/3 = 2.67$ sec. Because the period is close to this value, we see only one oscillation before the response dies out. ∎

EXAMPLE 3.4-5
Energy Dynamics
with Viscous
Damping

Plot the kinetic and potential energy versus time for the model given in Example 3.4-4, part (c).

Solution

The expressions for kinetic, potential, and total mechanical energy are: $KE = m\dot{x}^2/2 = \dot{x}^2$, $PE = kx^2/2 = 4x^2$, and $Total = KE + PE$. The expression for $x(t)$ obtained in part (c) of Example 3.4-4 is

$$x(t) = 2.332e^{-1.5t}\sin(2.5t + 1.03)$$

Differentiate this to obtain the velocity expression:

$$x(t) = -1.5(2.332)e^{-1.5t}\sin(2.5t + 1.03) + 2.5(2.332)e^{-1.5t}\cos(2.5t + 1.03)$$

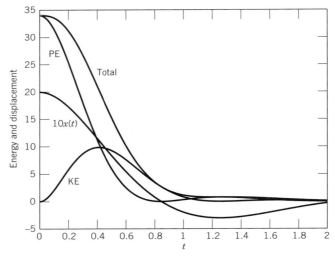

FIGURE 3.4-8   Kinetic and potential energy versus time, for a mass-spring system having viscous friction, from Example 3.4-5.

We can use these expressions to plot the energies versus time. These plots are shown in Figure 3.4-8, along with $10x(t)$ for reference [the factor of 10 was used to magnify the plot of $x(t)$]. The total mechanical energy is the sum of the kinetic and potential energies; it does not remain constant because the viscous damping converts energy into heat. Thus the total mechanical energy eventually becomes zero.

In simple harmonic motion, where there is no damping, the maximum kinetic energy occurs when the potential energy is at a minimum. However, when viscous damping is present, this is not true. With no viscous damping, if the mass is released from rest, its maximum speed occurs when it passes through the equilibrium position at $x = 0$, because the spring's resisting force kx is zero at $x = 0$. However, with viscous damping, the resisting force is due to both the spring and the damping, and is $kx + c\dot{x}$. The maximum resisting force occurs before the mass reaches the equilibrium position. This causes the mass to begin slowing down before it reaches the equilibrium position.   ■

3.5   ANALYSIS OF THE CHARACTERISTIC ROOTS

The characteristic equation plays a central role in understanding the behavior of the system represented by its equation of motion. Figure 3.5-1 shows how the location of the characteristic roots in the complex plane affects the free response. The real part of the root is plotted on the horizontal axis, and the imaginary part is plotted on the vertical axis. This plot is said to be the *s-plane plot*. Because complex roots are conjugate, we show only the upper root. Using the results we found in the previous section, we see that the following types of behavior can occur:

1. Unstable behavior occurs when the root lies to the right of the imaginary axis.

2. Neutrally stable behavior occurs when the root lies on the imaginary axis.

3. The response oscillates only when the root has a nonzero imaginary part.

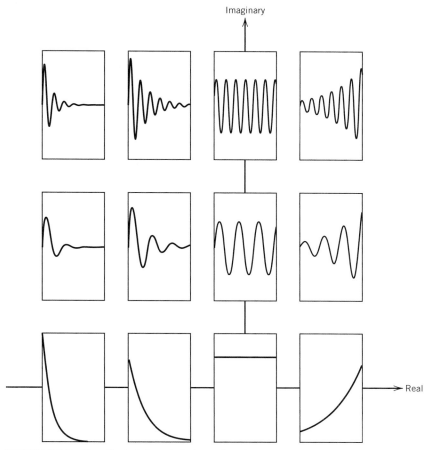

FIGURE 3.5-1 The effect of characteristic root location on system response.

4. The greater the imaginary part, the higher the frequency of the oscillation.

5. The farther to the left the root lies, the faster the response decays.

We now develop some techniques useful for predicting system behavior, including response time; oscillatory behavior; and frequency of oscillation, if any. All these are properties of the system itself and are thus determined by the characteristic roots, not by the initial conditions, which determine the amplitudes and phase angles. Many of these methods utilize a plot of the roots in the complex plane.

The Damping Ratio

A second-order system's free response for the stable case can be conveniently character-ized by the *damping ratio* ζ (sometimes called the *damping factor*).

For the characteristic equation, Equation 3.4-3, repeated here,

$$ms^2 + cs + k = 0 \tag{3.5-1}$$

the damping ratio is defined as

$$\zeta = \frac{c}{2\sqrt{mk}} \tag{3.5-2}$$

This definition is not arbitrary but is based on the way the roots change from real to complex as the value of c is changed; that is, from the root solution

$$s = \frac{-c \pm \sqrt{c^2 - 4mk}}{2m}$$

we see that three cases can occur:

1. *The critically damped case*: Repeated, real roots occur if $c^2 - 4mk = 0$; that is, if $c = 2\sqrt{mk}$. This value of the damping constant is the *critical damping constant* c_c; when c has this value, the system is said to be *critically damped*.
2. *The overdamped case*: If $c > c_c = 2\sqrt{mk}$, two distinct, real roots exist, and the system is *overdamped*.
3. *The underdamped case*: If $c < c_c = 2\sqrt{mk}$, complex conjugate roots occur, and the system is *underdamped*.

The damping ratio is thus seen to be the ratio of the actual damping constant c to the critical value c_c. Note the following:

1. For a critically damped system, $\zeta = 1$.
2. Exponential behavior occurs if $\zeta > 1$ (the overdamped case).
3. Oscillations exist when $\zeta < 1$ (the underdamped case).

For an unstable system, the damping ratio is meaningless and therefore not defined. For example, the equation $s^2 - 3s + 9 = 0$ is unstable, and we do not compute its damping ratio, which would be negative if you used Equation 3.5-2. If you obtain a *negative* damping ratio, you have made an error.

The damping ratio can be used as a quick check for oscillatory behavior. For example, the equation $s^2 + 3ds + d^2 = 0$ is stable if $d > 0$ and has the following damping ratio:

$$\zeta = \frac{3d}{2\sqrt{d^2}} = \frac{3}{2} > 1$$

Because $\zeta > 1$, no oscillations can occur in the system's free response regardless of the value of d and regardless of the initial conditions.

Natural and Damped Frequencies of Oscillation

When there is no damping, the characteristic roots are purely imaginary. The imaginary part, and therefore the frequency of oscillation, for this case is $\omega_n = \sqrt{k/m}$. This frequency is the undamped natural frequency. We can write the characteristic equation in terms of the parameters ζ and ω_n. First divide Equation 3.5-1 by m and use the fact that $2\zeta\omega_n = c/m$. The equation becomes

$$s^2 + 2\zeta\omega_n s + \omega_n^2 = 0 \tag{3.5-3}$$

and the roots are

$$s = -\zeta\omega_n \pm i\omega_n\sqrt{1 - \zeta^2} \tag{3.5-4}$$

Thus the time constant τ is

$$\tau = \frac{1}{\zeta\omega_n} \tag{3.5-5}$$

The imaginary part of the roots is frequency of oscillation, which is called the *damped natural frequency* ω_d to distinguish it from ω_n:

$$\omega_d = \omega_n \sqrt{1 - \zeta^2} \tag{3.5-6}$$

The frequencies ω_n and ω_d have physical meaning only for the underdamped case ($\zeta < 1$). For this case Equation 3.5-6 shows that $\omega_d < \omega_n$. Thus the damped frequency is always less than the undamped frequency.

We can express the free response of the underdamped second-order model in terms of the parameters ζ, τ, and ω_n as follows:

$$x(t) = Be^{-\zeta\omega_n t}\sin(\omega_d t + \phi) \tag{3.5-7}$$

where B and ϕ depend on the initial conditions $x(0)$ and $\dot{x}(0)$.

Graphical Interpretation

The preceding relations can be represented graphically by plotting the location of the roots (Equation 3.5-4) in the complex plane (Figure 3.5-2). The parameters ζ, ω_n, ω_d, and τ are normally used to describe stable systems only, and so we will assume for now that all the roots lie to the left of the imaginary axis (in the left half-plane).

The lengths of two sides of the right triangle shown in Figure 3.5-2 are $\zeta\omega_n$ and $\omega_n\sqrt{1 - \zeta^2}$. Thus the hypotenuse is of length ω_n. It makes an angle β with the negative real axis, and

$$\cos\beta = \zeta \tag{3.5-8}$$

Therefore all roots lying on the circumference of a given circle centered on the origin are associated with the same undamped natural frequency ω_n. Figure 3.5-3*a* illustrates this for two different frequencies, ω_{n1} and ω_{n2}. From Equation 3.5-8, we see that all roots lying on the same line passing through the origin are associated with the same damping ratio (Figure 3.5-3*b*). The limiting values of ζ correspond to the imaginary axis ($\zeta = 0$) and the negative real axis ($\zeta = 1$). Roots lying on a given line parallel to the real axis all give the same damped natural frequency (Figure 3.5-3*c*). All roots lying on a line parallel to the imaginary axis have the same time constant (Figure 3.5-3*d*). The farther to the left this line is, the smaller the time constant.

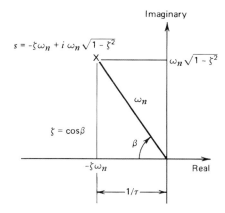

FIGURE 3.5-2 Location of the upper complex root in terms of the parameters ζ, τ, ω_n, and ω_d.

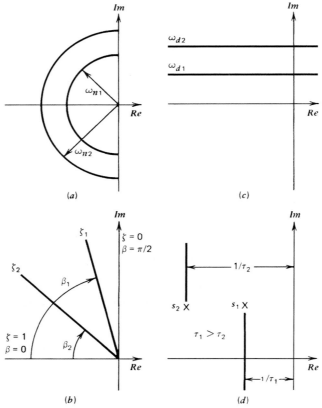

FIGURE 3.5-3 Graphical representation of the parameters ζ, τ, ω_n, and ω_d in the complex plane. (*a*) Roots with the same natural frequency ω_n lie on the same circle. (*b*) Roots with the same damping ratio ζ lie on the same line through the origin. (*c*) Roots with the same damped frequency ω_d lie on the same line parallel to the real axis. (*d*) Roots with the same time constant τ lie on the same line parallel to the imaginary axis. The root lying the farthest to the right is the dominant root.

EXAMPLE 3.5-1
A Viscometer

The *viscometer* shown in Figure 3.5-4 is used to measure the viscosity μ of the liquid contained in the cylindrical pan. A thin disk of diameter D, suspended by a wire, is in contact with the surface of the liquid, whose thickness is h. The wire acts like a torsional spring and the liquid acts like a torsional damper providing a resisting torque $T = c\omega$, where ω is the angular velocity of the disk. Using the methods of Section 1.4, we can derive the following expression for the damping coefficient as a function of μ, D, and h:

$$c = \frac{\mu\pi D^4}{32h} \tag{1}$$

Use this to derive an expression for μ in terms of h, D, the disk inertia I, and the observed oscillation frequencies ω_d and ω_n of the disk when in and out of contact with the liquid.

Solution

The equation of motion of the disk in rotation is

$$I\ddot{\theta} + c\dot{\theta} + k\theta = 0$$

This is of the same form as Equation 3.4-1, so we can use the formulas based on Equation 3.4-1 by substituting I for m. Thus

$$\zeta = \frac{c}{2\sqrt{Ik}} \tag{2}$$

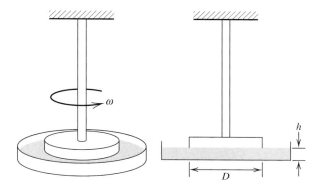

FIGURE 3.5-4   A viscometer.

and $\omega_d = \omega_n\sqrt{1-\zeta^2}$. Solve this for ζ and use Equation 2 to obtain

$$\zeta = \sqrt{1 - \left(\frac{\omega_d}{\omega_n}\right)^2} = \frac{c}{2\sqrt{Ik}} \tag{3}$$

But $k = I\omega_n^2$, so $\sqrt{Ik} = I\omega_n$. Using this and solving Equation 3 for c gives

$$c = 2I\omega_n\sqrt{1 - \left(\frac{\omega_d}{\omega_n}\right)^2}$$

Comparing this with Equation 1 and solving for μ, we obtain

$$\mu = \frac{32hc}{\pi D^4} = \frac{64hI}{\pi D^4}\omega_n\sqrt{1 - \left(\frac{\omega_d}{\omega_n}\right)^2}$$

■

The Logarithmic Decrement

Usually the damping coefficient c is the parameter most difficult to estimate. Mass m and stiffness k can be measured with static tests, but measuring damping requires a dynamic test. If the system exists and dynamic testing can be done with it, then the *logarithmic decrement* provides a good way to estimate the damping ratio ζ, from which we can compute c from the formula $c = 2\zeta\sqrt{mk}$. To see how this is done, use the form $s = -\zeta\omega_n \pm \omega_d i$ for the characteristic roots, and write the free response for the under-damped case as follows:

$$x(t) = Be^{-\zeta\omega_n t}\sin(\omega_d t + \phi) \tag{3.5-9}$$

The frequency of the oscillation is ω_d, and thus the period P is $P = 2\pi/\omega_d$. The logarithmic decrement δ is defined as the natural logarithm of the ratio of two successive amplitudes; that is,

$$\delta = \ln\frac{x(t)}{x(t+P)} \tag{3.5-10}$$

Using Equation 3.5-9, we see that this becomes

$$\delta = \ln\left[\frac{Be^{-\zeta\omega_n t}\sin(\omega_d t + \phi)}{Be^{-\zeta\omega_n(t+P)}\sin(\omega_d t + \omega_d P + \phi)}\right] \tag{3.5-11}$$

Note that $e^{-\zeta\omega_n(t+P)} = e^{-\zeta\omega_n t}e^{-\zeta\omega_n P}$. In addition, because $\omega_d P = 2\pi$ and $\sin(\theta + 2\pi) = \sin\theta$, $\sin(\omega_d t + \omega_d P + \phi) = \sin(\omega_d t + \phi)$, and Equation 3.5-11 becomes

$$\delta = \ln e^{\zeta\omega_n P} = \zeta\omega_n P$$

Because $P = 2\pi/\omega_d = 2\pi/\omega_n\sqrt{1 - \zeta^2}$, we have

$$\delta = \frac{2\pi\zeta}{\sqrt{1 - \zeta^2}} \tag{3.5-12}$$

We can solve this for ζ to obtain

$$\zeta = \frac{\delta}{\sqrt{4\pi^2 + \delta^2}} \tag{3.5-13}$$

If we have a plot of $x(t)$ from a test, we can measure two values x at two times t and $t + P$. These values can be measured at two successive peaks in x. The x values are then substituted into Equation 3.5-10 to compute δ. Equation 3.5-13 gives the value of ζ, from which we compute $c = 2\zeta\sqrt{mk}$.

The plot of $x(t)$ will contain some measurement error, and for this reason, the above method is usually modified to use measurements of two peaks n cycles apart (Figure 3.5-5). Let the peak values be denoted B_1, B_2, and so on, and note that

$$\ln\left(\frac{B_1}{B_2}\frac{B_2}{B_3}\frac{B_3}{B_4}\cdots\frac{B_n}{B_{n+1}}\right) = \ln\left(\frac{B_1}{B_{n+1}}\right)$$

or

$$\ln\frac{B_1}{B_2} + \ln\frac{B_2}{B_3} + \ln\frac{B_3}{B_4}\cdots\ln\frac{B_n}{B_{n+1}} = \ln\frac{B_1}{B_{n+1}}$$

Thus

$$\delta + \delta + \delta + \cdots + \delta = n\delta = \ln\frac{B_1}{B_{n+1}}$$

or

$$\delta = \frac{1}{n}\ln\frac{B_1}{B_{n+1}} \tag{3.5-14}$$

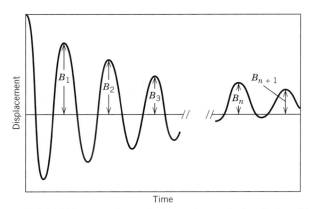

FIGURE 3.5-5   Use of nonadjacent peaks to calculate the logarithmic decrement.

We normally take the first peak to be B_1, because this is the highest peak and least subject to measurement error, but this is not required. The above formula applies to any two points n cycles apart.

EXAMPLE 3.5-2
Estimating
Damping and
Stiffness

Measurement of the free response of a certain system of mass 450 kg shows that after five cycles, the amplitude of the displacement is 10% of the first amplitude. Also, the time for these five cycles to occur was measured to be 20 s. Estimate the system's damping c and stiffness k.

Solution

From the given data, $n = 5$ and $B_6/B_1 = 0.1$. Thus from Equation 3.5-14,

$$\delta = \frac{1}{5} \ln \left(\frac{B_1}{B_6}\right) = \frac{1}{5} \ln 10 = \frac{2.302}{5} = 0.4605$$

From Equation 3.5-13,

$$\zeta = \frac{0.4605}{\sqrt{4\pi^2 + (0.4605)^2}} = 0.0731$$

Because the measured time for five cycles was 20 s, the period P is $P = 20/5 = 4$ s. Thus $\omega_d = 2\pi/P = 2\pi/4 = \pi/2$. The damped frequency is related to the undamped frequency by Equation 3.5-6:

$$\omega_d = \frac{\pi}{2} = \omega_n \sqrt{1 - \zeta^2} = \omega_n \sqrt{1 - (0.0731)^2}$$

Thus $\omega_n = 1.576$ and

$$k = m\omega_n^2 = 450(1.576)^2 = 1118 \text{ N/m}$$

The damping constant is calculated as follows:

$$c = 2\zeta\sqrt{mk} = 2(0.0731)\sqrt{450(1118)} = 104 \text{ N·s/m}$$ ∎

Table 3.5-1 summarizes the formulas developed thus far for the second-order model.

TABLE 3.5-1 Useful Formulas for the Linear Second-Order Model

Model:	$m\ddot{x} + c\dot{x} + kx = 0$
	m, c, k constant
1. Roots:	$s = \dfrac{-c \pm \sqrt{c^2 - 4mk}}{2m}$
2. Stability property:	Stable if and only if both roots have negative real parts. This occurs if and only if m, c, and k have the same sign.
3. Damping ratio or damping factor:	$\zeta = \dfrac{c}{2\sqrt{mk}}$
4. Undamped natural frequency:	$\omega_n = \sqrt{\dfrac{k}{m}}$
5. Damped natural frequency:	$\omega_d = \omega_n \sqrt{1 - \zeta^2}$
6. Time constant:	$\tau = 2m/c = 1/\zeta\omega_n$ if $\zeta \leq 1$
7. Logarithmic decrement:	$\delta \equiv \dfrac{1}{n} \ln \left(\dfrac{B_i}{B_{i+n}}\right)$
	$\delta = \dfrac{2\pi\zeta}{\sqrt{1 - \zeta^2}}$
	$\zeta = \dfrac{\delta}{\sqrt{(2\pi)^2 + \delta^2}}$

Graphical Representation of Root Migration

As shown in Figures 3.5-2 and 3.5-3, a graphical display of the characteristic root locations gives insight into the system response. A *root locus plot* is a plot of the location of the characteristic roots as a parameter value is varied. Such a plot shows how the roots migrate and thus how the response will change if the parameter value is changed. Root locus plots were used widely in engineering design well before digital computers became available, but the usefulness of the root locus has been enhanced by programs such as MATLAB, which can quickly generate the plots. In this section, we will present some simple plots that can be sketched by hand. In Chapter 6 we will show how to use MATLAB to generate the plots.

Varying the Spring Constant

In this section we will consider several examples where the characteristic equation has the form

$$ms^2 + cs + k = 0 \qquad (3.5\text{-}15)$$

We will consider, in order, the effects of varying first the spring constant k, then the damping constant c, and finally the mass m.

Suppose that $m = 2$ and $c = 8$ and that we wish to display the root locations as k varies. In this case the characteristic equation becomes $2s^2 + 8s + k = 0$. The roots are found from the quadratic formula

$$s = \frac{-8 \pm \sqrt{64 - 8k}}{4} \qquad (3.5\text{-}16)$$

It is easily seen that if $k < 8$, the roots are real and distinct. They are repeated if $k = 8$ and complex conjugates if $k > 8$. By repeatedly evaluating Equation 3.5-16 for various values of $k \geq 0$, the root locations can be plotted with dots in the s plane, as in Figure 3.5-6a. Noting the general trend of the dots, we can connect them with solid lines to produce the plot in Figure 3.5-6b. The spacing of the dots is uneven but corresponds to an even spacing in the k values. By convention, the root locations corresponding to $k = 0$ are

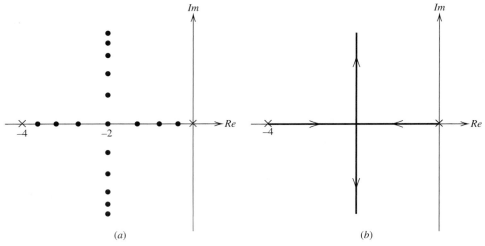

FIGURE 3.5-6 Plot of the roots of the equation $2s^2 + 8s + k = 0$ as k increases through positive values.

denoted by a cross ($\times$). These locations are $s = 0$ and $s = -4$. The arrows on the plot indicate the direction of root movement as k increases.

The plot shows that if $k \geq 0$, we cannot achieve a dominant time constant any smaller than $1/2$ by changing k. If $k < 8$, the dominant root lies between $s = 0$ and $s = -2$, and thus the dominant time constant is never less than $1/2$. If $k \geq 8$, the time constant is always $1/2$. This illustrates the type of insight that can be obtained from the root locus plot.

The root locus plot gives us a picture of the roots' behavior as one parameter is varied. It thus enables us to obtain a more general understanding of how the system's response changes as a result of a change in that parameter. The plot is useful in system design, where we need to select a value of the parameter to obtain a desired response.

Varying the Damping Constant

Suppose that $m = 2$ and $k = 8$, and that we wish to display the root locations as c varies. In this case the characteristic equation becomes $2s^2 + cs + 8 = 0$. The roots are found from the quadratic formula

$$s = \frac{-c \pm \sqrt{c^2 - 64}}{4} \tag{3.5-17}$$

When $c = 0$, the roots are $s = \pm 2i$, and we mark these locations on the plot with a $\times$. By evaluating and plotting the roots for many values of c and connecting the points with a solid line, we obtain the plot shown in Figure 3.5-7. We find that as c is increased through large values, one root moves to the left while the other root gradually approaches the origin $s = 0$. This can be proved analytically by taking the limit of Equation 3.5-17 as $c \to \infty$; we find that one root approaches 0 and the other root approaches $s = -\infty$. By convention, the root location corresponding to an infinite value of the parameter is denoted by a large circle ($\bigcirc$).

When $c = 8$, both roots are $s = -2$. Recall that ω_n is the radius of a circle centered at the origin. From the plot it is easily seen that for $c < 8$, the roots are complex and that $\omega_n = 2$ (because the plot is a circle of radius 2 centered at the origin). For $c > 8$, the roots are real and distinct, and the plot shows that the dominant root is always no less than -2,

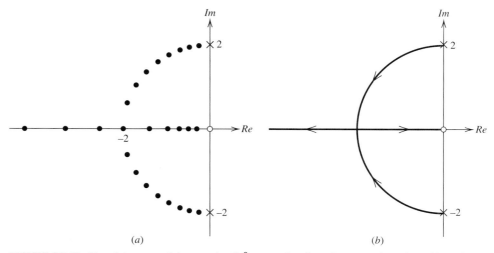

FIGURE 3.5-7 Plot of the roots of the equation $2s^2 + cs + 8 = 0$ as c increases through positive values.

and thus the dominant time constant is always $\geq 1/2$. Thus, as long as $c \geq 0$, we cannot reduce the dominant time constant below $1/2$ by changing c.

In this example the characteristic equation can be expressed as

$$s^2 + 4 + \frac{c}{2}s = 0$$

which is a special case of the more general form in terms of the variable parameter μ:

$$s^2 + \beta s + \gamma + \mu(s + \alpha) = 0 \tag{3.5-18}$$

where $\beta = 0$, $\gamma = 4$, $\alpha = 0$, and $\mu = c/2$. When $\mu = 0$, $s^2 + \beta s + \gamma = 0$, and thus the starting points of the root locus, denoted by crosses ($\times$), are given by

$$s = \frac{-\beta \pm \sqrt{\beta^2 - 4\gamma}}{2}$$

These may be real or complex numbers, depending on the values of β and γ. It can be shown that as $\mu \to \infty$, one root of Equation 3.5-18 approaches $s = -\alpha$ and the other root approaches $s = -\infty$.

A simple geometric analysis will show that off the real axis, the root locus of Equation 3.5-18 in terms of the variable parameter μ is a circle centered at $s = -\alpha$ and having a radius of

$$R = \sqrt{\alpha^2 + \gamma - \alpha\beta} \tag{3.5-19}$$

To prove this, substitute $s = x + iy$ into Equation 3.5-18 and separate the real and imaginary parts to obtain

$$x^2 - y^2 + \beta x + \gamma + \mu(x + \alpha) + iy(2x + \beta + \mu) = 0 + i0$$

Thus

$$x^2 - y^2 + \beta x + \gamma + \mu(x + \alpha) = 0$$

and

$$y(2x + \beta + \mu) = 0$$

Since $y \neq 0$ in general, the second equation gives $2x + \beta + \mu = 0$, or $\mu = -2x - \beta$. When this is substituted into the first equation, we obtain

$$(x + \alpha)^2 + y^2 = \alpha^2 + \gamma - \alpha\beta \tag{3.5-20}$$

which is the equation of the circle just described.

Varying the Mass

Now suppose that $c = 8$ and $k = 6$ and we want to investigate the effects of varying the mass m. The characteristic equation becomes $ms^2 + 8s + 6 = 0$, and the quadratic formula gives

$$s = \frac{-8 \pm \sqrt{64 - 24m}}{2m}$$

We immediately see that a problem arises if we try to examine the effects of varying the mass starting with $m = 0$, because m is in the denominator. In fact, if $m = 0$, the equation is no longer second order and the quadratic formula no longer applies. So we conclude

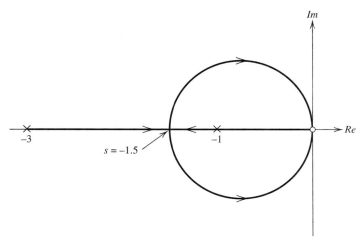

FIGURE 3.5-8 Plot of the roots of the equation $ms^2 + 8s + 6 = 0$ for $m \geq 2$.

that we must be careful when varying the leading coefficient in the characteristic equation because the equation order might change.

Instead, suppose we examine the effects of varying the mass for $m \geq 2$. For this case the equation always remains second order. When $m = 2$, the two roots are $s = -1$ and $s = -3$; these starting points are marked in Figure 3.5-8. As $m \to \infty$, the quadratic formula shows that both roots approach $s = 0$, and we mark this point with a small circle. By repeated evaluation of the quadratic formula for different m values, we obtain the plot shown in Figure 3.5-8. The locus off the real axis is a circle.

From the plot we see that as m is increased from $m = 2$, the two roots are real and approach one another, meeting at $s = -1.5$ when $m = 8/3$. For $m > 8/3$, the roots become complex and move to the right. Thus the smallest dominant time constant the model can have is $\tau = 1/1.5 = 2/3$ and the largest possible undamped natural frequency is $\omega_n = 1.5$, which occurs when $m = 8/3$.

Let μ be the relative deviation from its nominal value of 2; that is, let $\mu = (m - 2)/2$. Make the substitution $m = 2\mu + 2$ to obtain $(2\mu + 2)s^2 + 8s + 6 = 0$ or

$$s^2 + 4s + 3 + \mu s^2 = 0$$

where $\mu \geq 0$. This equation is a special case of the more general form

$$s^2 + \beta s + \gamma + \mu(s + \alpha)(s + \delta) = 0 \tag{3.5-21}$$

The root locus properties of this form can be developed as was done with Equation 3.5-18.

The examples in this section were second-order equations, for which a closed-form solution for the roots is available. For higher-order equations, however, a computer method is needed to obtain the root locus plot; a MATLAB method for doing so is presented in Chapter 6.

3.6 STABILITY

A model whose free response approaches ∞ as $t \to \infty$ is said to be *unstable*. If the free response approaches zero, the model is *stable*. The stability properties of a linear model are determined from its characteristic roots.

To understand the relationship between stability and the characteristic roots, we will consider some simple examples. The first-order model $\dot{x} = rx$ has the response

$$x(t) = x(0)e^{rt}$$

which approaches zero as $t \to \infty$ if the characteristic root $s = r$ is *negative*. The model is unstable if the root is *positive* because $x(t) \to \infty$ as $t \to \infty$. A borderline case, called *limited* stability or *neutral* stability, occurs if the root is zero. In this case $x(t)$ remains at $x(0)$. Neutral stability describes a situation where the free response does not approach ∞ but does not approach zero.

Now consider some second-order examples, all having the same initial conditions: $x(0) = 1$ and $\dot{x}(0) = 0$. Their responses are shown in Figure 3.6-1.

1. The model $\ddot{x} - x = 0$ has the roots $s = \pm 1$ and the response

$$x(t) = \frac{1}{2}(e^t + e^{-t})$$

2. The model $\ddot{x} - 2\dot{x} + 900x = 0$ has the roots $s = 1 \pm 30i$ and the response

$$x(t) = 1.001e^t \sin(t + 1.604)$$

3. The model $\ddot{x} + 1000x = 0$ has the roots $s = \pm 31.62i$ and the response

$$x(t) = \sin\left(31.62t + \frac{\pi}{2}\right)$$

From the plots we can see that none of the three models display stable behavior. The first and second models are unstable, while the third is neutrally stable. Thus the free response of a neutrally stable model can either approach a nonzero constant or settle down to a constant-amplitude oscillation.

The first model has a negative root and a positive root, while the roots of the second model have positive real parts. The roots of the neutrally stable model have real parts that are zero. The effect of the real part can be seen from the free response form for complex roots:

$$x(t) = Be^{rt} \sin(qt + \phi) \tag{3.6-1}$$

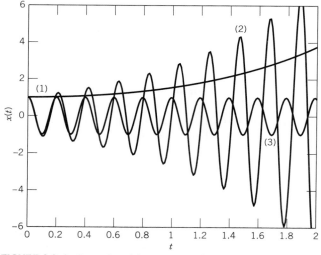

FIGURE 3.6-1   Examples of the response of unstable and neutrally stable models.

Clearly, if the real part is positive (that is, if $r > 0$), then the exponential e^{rt} will grow with time and so will the amplitude of the oscillations. This is an unstable case.

If the real part is zero (that is, $r = 0$), then the exponential becomes $e^{0t} = 1$ and the amplitude of the oscillations remains constant. This is the neutrally stable case. In both cases, the imaginary part of the root has no effect on the stability; it is the frequency of oscillation.

Model 1 is unstable because of the exponential e^t, which is due to the positive root $s = +1$. The negative root $s = -1$ has no effect on the stability because its exponential disappears in time.

If we realize that a real number is simply a special case of a complex number whose imaginary part is zero, then these examples show that a linear model is unstable if at least one root has a positive real part. The free response of any linear, constant-coefficient model, of any order, consists of a sum of terms, each multiplied by an exponential. Each exponential will approach ∞ as $t \to \infty$ if its corresponding root has a positive real part. Thus we can make the following general statement:

A constant-coefficient linear model is stable if and only if all of its characteristic roots have negative real parts. The model is neutrally stable if one or more roots have a zero real part and the remaining roots have negative real parts. The model is unstable if any root has a positive real part.

If a linear model is stable, then it is not possible to find a set of initial conditions for which $x(t) \to \infty$. However, if the model is unstable, there might still be certain initial conditions that result in a response that dies out. For example, the model $\ddot{x} - x = 0$ has the roots $s = \pm 1$ and thus is unstable. However, if the initial conditions are $x(0) = 1$, $\dot{x}(0) = -1$, then the free response is $x(t) = e^{-t}$, which certainly approaches zero as $t \to \infty$. Note that the exponential e^t corresponding to the root at $s = +1$ does not appear in the response because of the special nature of the initial conditions. We will shed more light on this phenomenon in Chapter 6.

EXAMPLE 3.6-1
Stability of an Inverted Pendulum

Figure 3.6-2a shows a representation of a simple robot arm driven by a motor at the base (point O). The arm is modeled as a rigid rod. The moment of inertia of the arm and its payload about point O is I; the mass is m. The motor supplies a torque T. Thus we can model the arm as an inverted pendulum, as in Figure 3.6-2b, whose equation of motion is

$$I\ddot{\theta} = T - mgL \sin \theta \tag{1}$$

We want the motor to hold the arm in equilibrium at five positions corresponding to the following angles: $\theta = 0°$, $45°$, $90°$, $135°$, and $180°$. Find the torque required and determine what will happen if something hits the arm and slightly alters its position.

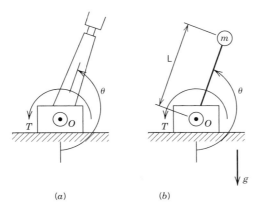

(a)   (b)

FIGURE 3.6-2 (a) A simple robot arm. (b) Pendulum model of the robot arm.

Solution

The torque T_e required to keep the arm in equilibrium at $\theta = \theta_e$ is found by setting $\ddot{\theta} = 0$ in Equation 1. This gives

$$T_e = mgL \sin \theta_e \qquad (2)$$

The torque required for each angle is

For $\theta_e = 0°$ and $180°$: $T_e = 0$

For $\theta_e = 45°$ and $135°$: $T_e = 0.707mgL$

For $\theta_e = 90°$: $T_e = mgL$

The linearized model can be obtained by replacing the nonlinear term $\sin \theta$ in Equation 1 with the Taylor series approximation $\sin \theta = \sin \theta_e + \cos \theta_e(\theta - \theta_e)$ and substituting the variables $x = \theta - \theta_e$ and $u = T - T_e$. It is

$$I\ddot{x} + (mgL \cos \theta_e)x = u \qquad (3)$$

We now need to investigate the stability of the arm for each position. The characteristic equation of the linearized model in Equation 3 is:

$$Is^2 + mgL \cos \theta_e = 0 \qquad (4)$$

which has the roots

$$s = \pm\sqrt{\frac{-mgL \cos \theta_e}{I}} \qquad (5)$$

The roots and stability property for each of the cases are as follows:

For $\theta_e = 0°$: $s = \pm i\sqrt{\dfrac{mgL}{I}}$ (both imaginary; thus neutrally stable)

For $\theta_e = 45°$: $s = \pm i\sqrt{\dfrac{0.707mgL}{I}}$ (both imaginary; thus neutrally stable)

For $\theta_e = 90°$: $s = 0, 0$ (both zero; thus neutrally stable)

For $\theta_e = 135°$: $s = \pm\sqrt{\dfrac{0.707mgL}{I}}$ (one negative, one positive; thus unstable)

For $\theta_e = 180°$: $s = \pm\sqrt{\dfrac{mgL}{I}}$ (one negative, one positive; thus unstable)

The torque T_e required to keep the pendulum at $\theta_e = 0°$ and at $\theta_e = 180°$ is zero in both cases. However, the equilibrium at $\theta_e = 0°$ is neutrally stable while the equilibrium at $\theta_e = 180°$ is unstable. This result is easy to understand with a little physical insight. If the arm is disturbed when hanging at $\theta = 0°$, it will swing back and forth continually (no friction is present in this model). However, when at $\theta = 180°$, any disturbance will cause it to fall.

However, the stability properties of the other cases are not as obvious, and mathematical analysis is needed to obtain the results. The same torque is required to keep the arm at $\theta = 45°$ and at $135°$, but the former position is neutrally stable while the latter is unstable. To understand why, recall that $T_e = mgL \sin \theta_e$. The slope $\partial T_e/\partial \theta_e = mgL \cos \theta_e$ is positive at $\theta_e = 45°$ and negative at $\theta_e = 135°$. At $\theta = 45°$, the motor is supplying a torque T_e that is not large enough to support the arm at a higher angle; thus if the arm is knocked upward, it will fall back. But the motor torque is larger than that required to keep the arm at an angle less than $45°$; thus the arm will move upward (back toward $45°$) if knocked downward. The opposite is true of the equilibrium at $\theta_e = 135°$ because $\partial T_e/\partial \theta_e$ is negative. The arm will not return to that position if it is knocked up or down.

In reality, the neutrally stable cases will actually be stable if a velocity-dependent damping torque is present to damp out the oscillations. However, the unstable cases will remain unstable.

The implication of these results for robot design is clear. It will be more difficult to achieve stable arm control for positions greater than 90°. If possible, the arm's base elevation should be designed to permit positioning objects at angles less than 90°. ■

Local and Global Stability

In the robot arm example, there are actually an infinite number of equilibrium positions, one for each value of T_e. Because some are unstable and others are stable (or at least neutrally stable), we might wonder what happens when the arm is displaced from an unstable equilibrium. Such questions are difficult to answer for nonlinear systems in general, and we leave it to the reader to think about what the arm will do.

Figure 3.6-3 shows a simpler situation. A ball rolls on a surface that has a valley and a hill. Clearly, the bottom of the valley is an equilibrium, and if we displace the ball slightly from this position, it will oscillate forever about the bottom if there is no friction (the neutrally stable case) or return to the bottom if friction is present (the stable case). However, if we displace the ball so much to the left that it lies outside the valley, it will never return. Thus, if friction is present, we say that the valley equilibrium is *locally stable* but *globally unstable*. An equilibrium is globally stable only if the system returns to it for *any* initial displacement.

The equilibrium on the hilltop is both locally and globally unstable. If displaced to the right, the ball will continue to roll to the right. If displaced to the left, it will end up at the bottom of the valley if it is moving slowly enough not to overshoot the valley's opposite rim. (Use this insight to think about the robot arm's behavior.)

For linear models, stability analysis using the characteristic roots gives global stability information. However, for nonlinear models, linearization about an equilibrium gives us only local stability information.

Accuracy of Linearized Models

The question that inevitably arises in a discussion of linearization is how small the deviation variables must be for the truncated Taylor series to be accurate. Often the linearized model not used to obtain a solution but to determine the system's stability in the presence of small perturbations moving the system slightly away from equilibrium. If the equilibrium is stable, then any slight displacement from equilibrium will die out and the system will return to the equilibrium.

In vibration applications, linearization is widely used to compute the equivalent spring constant. If the system oscillates with only a small amplitude near the equilibrium, then the linearization and the computed spring constant will be accurate. Linearization is also useful for estimating the speed of response of a nonlinear system, from the time constant of the linearized model.

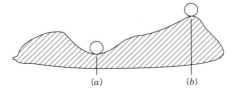

(a)   (b)

FIGURE 3.6-3 Illustration of stability: a ball rolling on a surface with a hill and a valley.

3.7 VIBRATION MODELS WITH VISCOUS DAMPING

The equation

$$m\ddot{x} + c\dot{x} + kx = f \tag{3.7-1}$$

is the general form of the equation of motion for a linear, single-degree-of-freedom system with viscous damping. In the two previous sections we obtained and analyzed the solution of this equation for the case where $f = 0$; namely, the free response. We will see in this section the following related form, which contains the derivative of the forcing function f:

$$m\ddot{x} + c\dot{x} + kx = af + b\dot{f} \tag{3.7-2}$$

Note that the characteristic equation is the same for both forms, namely,

$$ms^2 + cs + k = 0$$

Thus the conditions for stability and all the expressions based on the characteristic equation, such as the damping ratio, oscillation frequencies, and time constant, are the same for both forms. For example,

$$\zeta = \frac{c}{2\sqrt{mk}}$$

$$\omega_n = \sqrt{\frac{k}{m}} \qquad \omega_d = \omega_n\sqrt{1 - \zeta^2}$$

$$\tau = \frac{2m}{c} \qquad \text{if } \zeta \le 1$$

In this section we illustrate how Equations 3.7-1 and 3.7-2 arise in a variety of applications. In the examples to follow, pay special attention to the way viscous damping must be considered when drawing a free-body diagram. Remember that the force or moment exerted by a viscous damper depends on the *relative* velocity across its endpoints. When drawing the free-body diagram, you must make an assumption concerning the direction of the relative velocity. If you consistently apply this assumption, you will obtain the correct equation of motion.

EXAMPLE 3.7-1
A Single-Mass Model of a Car's Suspension

A simplified representation of a car's suspension system is shown in Figure 3.7-1. The spring constant k models the elasticity of both the tire and the suspension spring. The damping constant c models the shock absorber. The masses of the wheel, tire, and axle are neglected; here the mass m represents one-fourth of the car's mass. For this reason, this model is called the *quarter-car model*. As usual, $x = 0$ corresponds to the equilibrium position of m when $y = 0$. The road surface displacement $y(t)$ can be derived from the road surface profile and the car's speed. Derive the equation of motion of m in terms of the displacement x.

Solution

It is important to understand that the damper acts to eliminate any velocity difference across it. The force that it exerts has a magnitude $c|\dot{y} - \dot{x}|$ and acts in the direction that will reduce the speed difference. Figure 3.7-2 shows the free-body diagram, which is drawn for $\dot{y} > \dot{x}$. In this case the damper will try to increase the speed $\dot{x}$ of the mass so that it will equal $\dot{y}$. Thus the damper force has the direction shown. Assuming that $y > x$, the spring force acts in the direction shown. The variable x is the displacement of m from its equilibrium position when $y = 0$; that is, when $y = 0$, $x = 0$, and the spring force cancels the gravity force. From this free-body diagram we write

$$m\ddot{x} = c(\dot{y} - \dot{x}) - k(y - x)$$

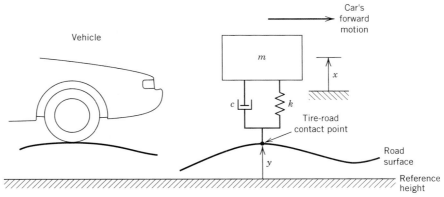

FIGURE 3.7-1   Single-mass representation of a car suspension.

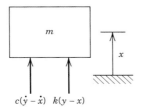

$c(\dot{y} - \dot{x})$ $k(y - x)$ FIGURE 3.7-2   Free-body diagram of a single-mass suspension model.

or

$$m\ddot{x} + c\dot{x} + kx = c\dot{y} + ky \qquad (1)$$

This corresponds to Equation 3.7-2 with $f = y$ ■

EXAMPLE 3.7-2
A Model with
Angular
Displacement
Input

Figure 3.7-3*a* represents a shaft supported by a bearing having damping. The shaft has a torsional spring constant k and inertia I about its axis of rotation. The angular displacements at each end of the shaft are θ and θ_i. Displacement θ_i is a specified function of time in this example. Derive the equation of motion for θ.

Solution

From the free-body diagram in Figure 3.7-3*b*, which is drawn for $\theta_i > \theta$, we obtain

$$I\dot{\omega} = k(\theta_i - \theta) - c\omega$$

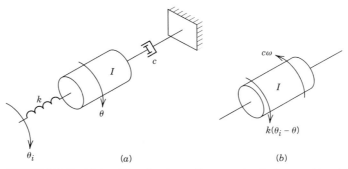

FIGURE 3.7-3   (*a*) A rotational system with an angular displacement input. (*b*) Free-body diagram.

Substituting $\dot{\theta}$ for ω gives

$$I\ddot{\theta} = k(\theta_i - \theta) - c\dot{\theta}$$

or

$$I\ddot{\theta} + c\dot{\theta} + k\theta = k\theta_i \tag{1}$$

This corresponds to Equation 3.7-1 with $x = \theta$ and $f = k\theta_i$.   ■

EXAMPLE 3.7-3
A Swing-Axle
Suspension

Figure 3.7-4a shows an automotive independent rear suspension of the swing-axle type. Power is transmitted from the engine to the axle through the differential gear at point A, which is fixed to the chassis. The half-axle pivots about the fixed point O. The shock absorber combines a spring and a damping element. Treat the masses of the absorber and the half-axle as negligible compared to the wheel-tire assembly mass m. The system's equilibrium position corresponds to the axle being horizontal at $\theta = 0$.

We wish to develop a model of the suspension's motion for the situation where the tire is not in contact with the road force. This model can then be used to investigate the phenomenon of "wheel hop." Derive the equation of motion of the wheel-tire assembly in terms of θ, for small values of θ.

Solution

We can model this system in simplified form as shown in Figure 3.7-4b. For small θ, the spring and damper displacement is $L_1 \sin\theta \approx L_1\theta$ and is perpendicular to the axle. The weight force perpendicular to the axle is $mg\cos\theta \approx mg$ and thus is constant. Therefore the weight force will not appear in the equation of motion because it is canceled by the static deflection spring force.

Applying the moment equation about the fixed point O, we obtain

$$mL^2\ddot{\theta} = -(kL_1\theta)L_1 - (cL_1\dot{\theta})L_1$$

or

$$mL^2\ddot{\theta} + cL_1^2\dot{\theta} + kL_1^2\theta = 0 \tag{1}$$

The damping ratio is

$$\zeta = \frac{cL_1^2}{2\sqrt{mL^2kL_1^2}} = \left(\frac{c}{2\sqrt{mk}}\right)\frac{L_1}{L}$$

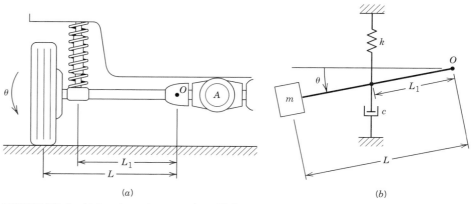

(a) (b)

FIGURE 3.7-4 (a) A swing-axle suspension. (b) Suspension model.

If $\zeta \le 1$, the time constant is given by

$$\tau = \frac{2m}{c}\left(\frac{L}{L_1}\right)^2$$

This value tells the designer how quickly the wheel oscillations will die out. ∎

EXAMPLE 3.7-4
A Cylinder in
Translation and
Rotation

A homogeneous cylinder of radius R and mass m is free to rotate about an axle that is connected to a support by a spring and damper (Figure 3.7-5a). Assume that the cylinder rolls without slipping or bouncing on the surface of inclination α. Find the equation of motion in terms of the displacement x_1, which is the displacement of the cylinder's center parallel to the inclined surface and is measured from the equilibrium position.

Solution

Let x_2 measure the displacement of the cylinder's center perpendicular to the inclined surface. Let θ measure the rotation of the cylinder from its equilibrium position. The assumption of no bouncing implies that $x_2 = 0$. The assumption of no slipping implies that $x_1 = R\theta$.

The free-body diagram is shown in Figure 3.7-5b. Note that the spring force that exists at rest is canceled by the weight component $mg \sin \alpha$ parallel to the surface. Thus the static spring force is not shown on the free-body diagram. The remaining spring force is kx_1. The forces f_n and f_t are the normal and tangential components of the reaction force at the surface. Newton's law in the x_1 direction gives

$$m\ddot{x}_1 = -c\dot{x}_1 - kx_1 - f_t \tag{1}$$

because $x_2 = 0$, $\ddot{x}_2 = 0$, and Newton's law in the x_2 direction gives

$$m\ddot{x}_2 = 0 = f_n - mg \cos \alpha$$

or $f_n = mg \cos \alpha$.

For the θ direction, Newton's law gives

$$I\ddot{\theta} = Rf_t$$

Solve this for f_t and substitute $\theta = x_1/R$ to obtain

$$f_t = \frac{I}{R^2}\ddot{x}_1$$

Substitute this into Equation 1 to obtain

$$\left(m + \frac{I}{R^2}\right)\ddot{x}_1 + c\dot{x}_1 + kx_1 = 0$$

Substituting the inertia of a solid cylinder $I = 0.5mR^2$ into the above equation gives the answer:

$$1.5m\ddot{x}_1 + c\dot{x}_1 + kx_1 = 0 \tag{2}$$

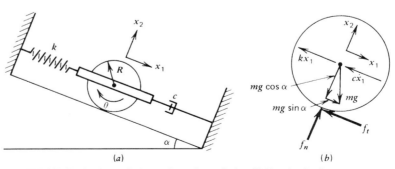

FIGURE 3.7-5 (a) A translating and rotating cylinder. (b) Free-body diagram.

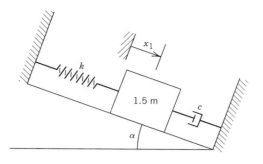

FIGURE 3.7-6   A translational system equivalent to the one shown in Figure 3.7-5.

This equation shows that the system has effective mass 50% greater than if the mass were to translate but not rotate. To see why, consider the system's kinetic energy expression, which is

$$KE = \frac{1}{2}m\dot{x}_1^2 + \frac{1}{2}I\dot{\theta}^2 = \frac{1}{2}m\dot{x}_1^2 + \frac{1}{2}mR^2\left(\frac{\dot{x}_1}{R}\right)^2 = \frac{1}{2}(1.5m)\dot{x}_1^2$$

Thus the system has the same kinetic energy as a translating mass of $1.5m$. The equivalent system is shown in Figure 3.7-6. Sometimes we can use this kinetic energy approach to expedite the development of a model of a system undergoing rotation and translation. By converting the system into an equivalent translational system, we can sometimes obtain the model more quickly. To do this, the translational and rotational motion of the system must be directly coupled. For example, $x_1 = R\theta$ in this example because of the assumption that the wheel does not slip.   ∎

EXAMPLE 3.7-5
A Rack-and-Pinion
Gear

In terms of the angular displacement θ, develop the equivalent rotational model of the rack-and-pinion gear shown in Figure 3.7-7. The applied torque T is a given function of time. Neglect any twist in the shaft.

Solution

The system's kinetic energy KE is

$$KE = \frac{1}{2}I_m\dot{\theta}^2 + \frac{1}{2}I_s\dot{\theta}^2 + \frac{1}{2}I_p\dot{\theta}^2 + \frac{1}{2}m_r\dot{x}^2$$

where

$$I_p = \frac{1}{2}m_pR^2$$

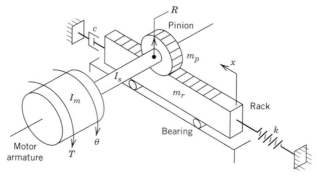

FIGURE 3.7-7   A system with a rack-and-pinion gear.

Using $\dot{x} = R\dot{\theta}$ and collecting terms gives

$$KE = \frac{1}{2}\left(I_m + I_s + \frac{m_p R^2}{2} + m_r R^2\right)\dot{\theta}^2$$

So the equivalent inertia is

$$I_e = I_m + I_s + \frac{m_p R^2}{2} + m_r R^2$$

The system's potential energy PE is

$$PE = \frac{1}{2}kx^2 = \frac{1}{2}kR^2\theta^2$$

So the equivalent torsional spring constant is $k_e = kR^2$. Finally, the system's power dissipation PD is

$$PD = c\dot{x}^2 = cR^2\dot{\theta}^2$$

and the equivalent torsional damping constant is $c_e = cR^2$. The model of the equivalent system is found from Newton's law:

$$I_e\ddot{\theta} = T - c_e\dot{\theta} - k_e\theta$$

or

$$I_e\ddot{\theta} + c_e\dot{\theta} + k_e\theta = T \tag{1}$$

∎

Levers are common mechanical elements. Figure 3.7-8 is a representation of a lever. For small angular displacements θ, the displacements of the endpoints are approximately $x = a\theta$ and $y = b\theta$. The lever is an ideal transformer (i.e., no power loss) if its mass, elasticity, and pivot friction are negligible. In this case, the work $f_1 x$ done by force f_1 equals the work $f_2 y$ done by force f_2. Thus $f_2 = af_1/b$. The lever ratio is a/b.

We can represent a lever system as a translational system lumped at coordinate x or at coordinate y, or as a rotational system lumped at coordinate θ. For example, we can represent the system shown in Figure 3.7-9a by any of the equivalent systems shown in Figure 3.7-9b. The equivalent parameter values are given in the figure. These values are obtained by equating the energy expressions for the original and equivalent systems.

EXAMPLE 3.7-6
A Lever System

Consider the lever system shown in Figure 3.7-10. (**a**) Derive its equation of motion in terms of the coordinate x using energy principles. (**b**) Derive its equation of motion in terms of the coordinate θ using Newton's law.

Solution

(**a**) The kinetic energy of the entire system is

$$KE = \frac{1}{2}m_a\dot{x}^2 + \frac{1}{2}m_b\dot{y}^2 = \frac{1}{2}m_a\dot{x}^2 + \frac{1}{2}m_b\left(\frac{b}{a}\dot{x}\right)^2 = \frac{1}{2}\left[m_a + m_b\left(\frac{b}{a}\right)^2\right]\dot{x}^2$$

Thus the equivalent mass at the coordinate x is $m_e = m_a + m_b(b/a)^2$.

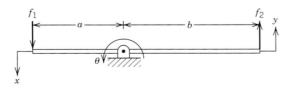

FIGURE 3.7-8 A lever.

Original levered system

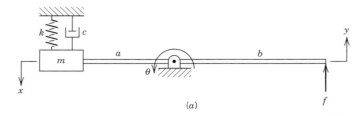

(a)

Equivalent Systems
(For small displacements and negligible level mass, lever elasticity, and pivot friction)

Lumped for coordinate x

$$m_e = m$$
$$k_e = k$$
$$c_e = c$$
$$f_e = f(b/a)$$

Lumped for coordinate y

$$m_e = m(a^2/b^2)$$
$$k_e = k(a^2/b^2)$$
$$c_e = c(a^2/b^2)$$
$$f_e = f$$

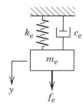

Lumped for coordinate θ

$$I_e = ma^2 \qquad k_e = ka^2$$
$$c_e = ca^2 \qquad T_e = fb$$

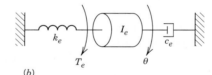

(b)

FIGURE 3.7-9   Three equivalent representations of a lever system.   (*a*) The original system.   (*b*) Equivalent systems in terms of the coordinates *x, y,* and θ.

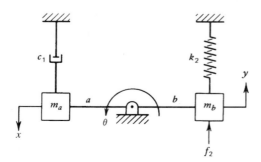

FIGURE 3.7-10   A lever system.

The potential energy of the entire system is due to the spring:

$$PE = \frac{1}{2}k_2 y^2 = \frac{1}{2}k_2 \left(\frac{b}{a}\right)^2$$

Thus the equivalent spring constant at the coordinate x is $k_e = k_2(b/a)^2$.

The power dissipation in the system is

$$PD = c_1 \dot{x}^2$$

Thus the equivalent damping constant is simply $c_e = c_1$. The equivalent applied force is due to the applied force f_2, and is $f_e = (b/a)f_2$.

The equivalent model for coordinate x is thus

$$m_e \ddot{x} + c_e \dot{x} + k_e x = f_e$$

or

$$\left[m_a + m_b \left(\frac{b}{a}\right)^2\right]\ddot{x} + c_1 \dot{x} + k_2 \left(\frac{b}{a}\right)^2 x = \left(\frac{b}{a}\right)f_2 \tag{1}$$

(**b**) The system's inertia relative to the pivot is

$$I_e = m_a a^2 + m_b b^2$$

The torques about the pivot are caused by the applied force f_2, the spring force $k_2 y$, and the damping force $c_1 \dot{x}$. Their moment arms are b, b, and a, respectively. Thus, from Newton's law, and using the fact that $x = a\theta$ and $y = b\theta$, we have

$$I_e \ddot{\theta} = f_2 b - (k_2 y)b - (c_1 \dot{x})a = f_2 b - (k_2 b\theta)b - (c_1 a\dot{\theta})a$$

or

$$\left(m_a a^2 + m_b b^2\right)\ddot{\theta} = f_2 b - k_2 b^2 \theta - c_1 a^2 \dot{\theta} \tag{2}$$

Equations 1 and 2 can be shown to be equivalent by substituting $x = a\theta$ into the latter equation. ∎

When you need to develop a model of a lever system, choose whatever coordinate system is most convenient to use, and then transform the answer into the desired coordinate by one of the following substitutions: $x = a\theta$, $y = b\theta$, or $y = (b/a)x$. You can use either the energy approach or Newton's law, whichever is easier for you.

3.8   FREE VIBRATION WITH COULOMB DAMPING

Coulomb friction will cause free vibrations to die out, so we may consider it to be a form of damping. Unlike viscous damping, which is velocity dependent, Coulomb damping is a constant force or torque as long as the normal force is constant. We now investigate the effects of Coulomb friction on the free vibration of a mass and a spring (Figure 3.8-1a).

Note that the friction force always opposes the motion. The free-body diagram in Figure 3.8-1b is for the case where motion is to the left. The equation of motion is

$$m\ddot{x} = -kx + \mu N$$

Its solution has the form

$$x = A_1 \sin \omega_n t + B_1 \cos \omega_n t + \frac{\mu N}{k} \tag{3.8-1}$$

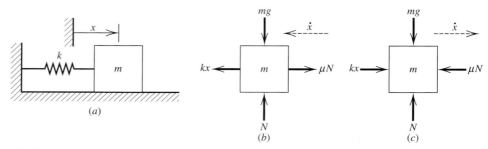

FIGURE 3.8-1 (*a*) A mass-spring system with Coulomb friction. (*b*) Free-body diagram for the case where the mass is moving to the left. (*c*) Free-body diagram for the case where the mass is moving to the right.

where $\omega_n = \sqrt{k/m}$. Figure 3.8-1*c* shows the free-body diagram for the case where the motion is to the right. The equation of motion is

$$m\ddot{x} = -kx - \mu N$$

and its solution has the form

$$x = A_2 \sin \omega_n t + B_2 \cos \omega_n t - \frac{\mu N}{k} \qquad (3.8\text{-}2)$$

The constants A_i and B_i depend on the values of the displacement x and the velocity $\dot{x}$ at the time the mass starts to move either to the left or to the right.

Suppose that at time $t = 0$, the mass is displaced to the right a distance $x(0) = x_0 > 0$ and released from rest so that $\dot{x}(0) = 0$. In this case the mass moves to the left and Equation 3.8-1 applies. We can show that $A_1 = 0$ and $B_1 = x_0 - \mu N/k$, so the solution is

$$x = \left(x_0 - \frac{\mu N}{k}\right) \cos \omega_n t + \frac{\mu N}{k} \qquad (3.8\text{-}3)$$

The response is a cosine of frequency ω_n shifted up by the amount $\mu N/k$. This solution holds until $\dot{x} = 0$, after which the mass begins moving to the right. The previous equation gives

$$\dot{x} = -\omega_n \left(x_0 - \frac{\mu N}{k}\right) \sin \omega_n t$$

from which we can find the time when $\dot{x} = 0$. It is $t = \pi/\omega_n$. At this time

$$x(\pi/\omega_n) = \frac{2\mu N}{k} - x_0$$

With the mass now moving to the right, we use Equation 3.8-2 and evaluate A_2 and B_2 using $x(\pi/\omega_n)$, given previously, and $\dot{x} = 0$. The solution is

$$x = \left(x_0 - \frac{3\mu N}{k}\right) \cos \omega_n t - \frac{\mu N}{k} \qquad (3.8\text{-}4)$$

This is a cosine shifted down by the amount $\mu N/k$, and it holds for $\pi/\omega_n \leq t \leq 2\pi/\omega_n$, after which time the mass reverses direction again and begins to move to the left.

At $t = 2\pi/\omega_n$, $x = x_0 - 4\mu N/k$. So after one complete cycle, $0 \leq t \leq 2\pi/\omega_n$, the displacement has been reduced by $4\mu N/k$.

If we were to continue this analysis, we would find that the amplitude of displacement is reduced by $4\mu N/k$ after each cycle. The motion will stop when the spring force kx is less than the friction force μN. This occurs at the end of the half-cycle in which the amplitude is less than $\mu N/k$. The mass does not come to rest at $x = 0$ as with viscous damping, but at $x = \mu N/k$.

The number n of *half* cycles required for the motion to stop is given by

$$x_0 - n\left(\frac{2\mu N}{k}\right) \leq \frac{\mu N}{k}$$

Thus n is the smallest integer satisfying the following relation:

$$n \geq \frac{kx_0}{2\mu N} - \frac{1}{2} \tag{3.8-5}$$

In summary, we note the following:

1. We thus have shown that the successive amplitudes of free vibration with Coulomb damping are related as follows:

$$X_{n+1} = X_n - \frac{4\mu N}{k}$$

Thus the envelope of the oscillations consists of a pair of *straight* lines having slopes of $\pm 2\mu N\omega_n/\pi k$. By contrast, the envelope for viscous damping is a pair of *exponentially decaying* curves.

2. The motion comes to a complete stop with Coulomb friction, whereas with viscous friction the motion continues forever (theoretically).

3. The frequency of free vibration with Coulomb friction is the same as the natural frequency ω_n, whereas the frequency with viscous damping is different from the natural frequency $(\omega_d = \omega_n\sqrt{1 - \zeta^2})$.

4. Free vibration with Coulomb damping is always periodic, whereas with viscous damping it may be nonperiodic if the damping is large enough (if $\zeta \geq 1$).

You should keep in mind that an accurate value of the friction coefficient μ in a given situation is very difficult to obtain, so you should treat any predictions based on μ to be approximate at best.

EXAMPLE 3.8-1
Estimating the Friction Coefficient

A spring-mass system on a horizontal surface was set in motion by displacing the mass 20 mm from its rest position and releasing it with zero velocity. The mass oscillated with a frequency of 10 Hz and came to a stop after 20 half-cycles. Estimate the value of the friction coefficient.

Solution

Because the mass is on a horizontal surface, the normal force is $N = mg$. Noting also that $k/m = \omega_n^2 = [2\pi(10)]^2$, we see that the inequality, Equation 3.8-5, can be written as

$$n \geq \frac{kx_0}{2\mu mg} - \frac{1}{2} = \frac{x_0\omega_n^2}{2\mu g} - \frac{1}{2}$$

Solve this for μ to obtain

$$\mu \geq \frac{x_0\omega_n^2}{2g(n + 0.5)} = \frac{0.02(20\pi)^2}{2(9.81)(20 + 0.5)} = 0.19$$

Thus we estimate that $\mu = 0.19$. ■

3.9 THE MATLAB ode SOLVERS

Numerical methods for solving ordinary differential equations (abbreviated as ODE) may be applied to both linear and nonlinear equations, although their primary application is to nonlinear equations because it is usually impossible to obtain a closed-form solution for such equations. A nonlinear ordinary differential equation can be recognized by the fact that the dependent variable or its derivatives appear raised to a power or in a transcendental function. For example, the following equations are nonlinear:

$$y\ddot{y} + 5\dot{y} + y = 0 \qquad \dot{y} + \sin y = 0 \qquad \dot{y} + \sqrt{y} = 0$$

Numerical methods require that the derivatives in the model be described by finite-difference expressions and that the resulting difference equations be solved in a step-by-step procedure. The issues related to these methods are as follows:

- What finite-difference expressions provide the best approximations for derivatives?
- What are the effects of the step size used to obtain the approximations?
- What are the effects of round-off error when solving the finite-difference equations?

These issues are more likely to be significant when the solution is rapidly changing with time, and errors may occur if the step size is not small compared to the smallest time constant of the system or the smallest oscillation period.

If you are unfamiliar with these concepts, Appendix B provides an introduction to numerical methods for solving ordinary differential equations. The predictor-corrector and Runge-Kutta algorithms presented there are simplified versions of the ones used by MATLAB and Simulink, and so an understanding of these methods will improve your understanding of these two software packages.

In addition to the many variations of the predictor-corrector and Runge-Kutta algorithms that have been been developed, there are more advanced algorithms that use a variable step size. These algorithms use larger step sizes when the solution is changing more slowly. MATLAB provides functions called *solvers* that implement Runge-Kutta methods with variable step size. Some of these are the ode23, ode45, and ode113 functions. The ode23 function uses a combination of second- and third-order Runge-Kutta methods, whereas the ode45 function uses a combination of fourth- and fifth-order methods. These solvers are classified as low- and medium-order respectively. The ode113 solver is based on a variable-order algorithm. MATLAB contains additional solvers. However, the ode23 and ode45 solvers are sufficient to solve the problems encountered in this text.

Solver Syntax

We begin our coverage with two examples of solving first-order equations. Solution of higher-order equations is covered later in this section. When used to solve the equation $\dot{y} = f(t, y)$, the basic syntax is (using ode45 as the example):

```
[t,y] = ode45('ydot',tspan,y0)
```

where ydot is the name of the function file whose inputs must be t and y and whose output must be a *column* vector representing dy/dt, that is, $f(t, y)$. The number of rows in this column vector must equal the order of the equation. The syntax for the other solvers is identical.

The vector tspan contains the starting and ending values of the independent variable t and, optionally, any intermediate values of t where the solution is desired. For example, if no intermediate values are specified, tspan is [t0, tf], where t0 and tf are the desired starting and ending values of the independent parameter t. As another example, using tspan = [0, 5, 10] tells MATLAB to find the solution at $t = 5$ and at $t = 10$. You can solve equations backward in time by specifying t0 to be greater than tf.

The parameter y0 is the initial value $y(t_0)$. The function file must have two input arguments, t and y, even for equations where $f(t, y)$ is *not* an explicit function of t. You need not use array operations in the function file because the ode solvers call the file with scalar values for the arguments.

As a first example of using a solver, let us solve an equation whose solution is known in closed form so that we can make sure we are using the method correctly.

EXAMPLE 3.9-1
MATLAB
Solution of
ẏ = -10y

Use the ode45 solver to solve the following equation:

$$\dot{y} = -10y \qquad y(0) = 2$$

The exact solution is $y(t) = 2e^{-10t}$.

Solution

Create the following function file. Note that the input arguments must be t and y in that order.

```
function ydot = eqn1(t,y)
ydot = -10*y;
```

The initial time is $t = 0$, so set t0 to be 0. Choosing tf to be 0.4 will show most of the response. The function is called in MATLAB as shown in the following script file, and the solution is plotted along with the exact solution y_exact.

```
[t,y] = ode45('eqn1',[0,0.4],2);
y_exact = 2*exp(-10*t);
plot(t,y,'o',t,y_exact),xlabel('t'),...
    ylabel('y')
```

Note that we need not generate the array t to evaluate y_exact because t is generated by the ode45 function. The plot is shown in Figure 3.9-1. The numerical solution is marked by the circles[1] and the exact solution is indicated by the solid line. Clearly the numerical solution gives an accurate answer. ∎

We now assess the performance of the ode45 solver when applied to find an oscillating solution.

EXAMPLE 3.9-2
MATLAB
Solution of
ẏ = sin t

Use the ode45 solver for the equation

$$\dot{y} = \sin t \qquad y(0) = 0$$

for $0 \le t \le 4\pi$. The exact solution is $y(t) = 1 - \cos t$.

Solution

Create the following function file:

```
function ydot = sinefn(t,y)
ydot = sin(t);
```

---

[1] The algorithm used by the ode functions to select the step size depends on the specific version of MATLAB, so you may obtain slightly different spacing, depending on which version you are using.

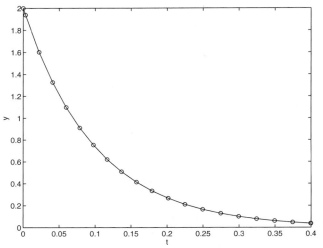

FIGURE 3.9-1 Analytical and Runge-Kutta numerical solutions of the equation $\dot{y} = -10y$ for $y(0) = 2$.

Use the following script file to compute the solution. Because of the oscillations, we will use many points to plot the exact solution:

```
t_exact = [0:0.01:4*pi];
[t,y] = ode45('sinefn',[0,4*pi],0);
y_exact = 1-cos(t_exact);
plot(t,y,'o',t_exact,y_exact),xlabel('t'),ylabel('y'),...
  axis([0 4*pi -0.5 2.5])
```

Figure 3.9-2 shows that the solution generated by `ode45` is correct. ∎

Extension to Higher-Order Equations

To use the `ode` solvers to solve an equation of order two or greater, you must first write the equation as a set of first-order equations (called the *Cauchy form* or *state-variable*

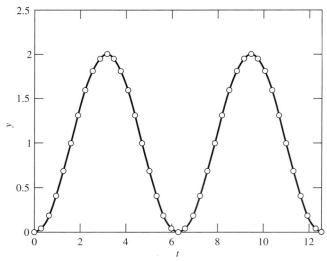

FIGURE 3.9-2 Analytical and Runge-Kutta numerical solutions of the equation $\dot{y} = \sin t$ for $y(0) = 0$.

form) and then create a function file that computes the derivatives of the state variables. Consider the second-order equation $2\ddot{y} + 10\dot{y} + 8y = f(t)$. Define the variables, $x_1 = y$ and $x_2 = \dot{y}$. The state-variable form is

$$\dot{x}_1 = x_2 \tag{3.9-1}$$

$$\dot{x}_2 = \frac{1}{2}[f(t) - 8x_1 - 10x_2] \tag{3.9-2}$$

Now write a function file that computes the values of $\dot{x}_1$ and $\dot{x}_2$ and stores them in a *column* vector. To do this, we must first have a function specified for $f(t)$. Suppose that $f(t) = \sin t$. Then the required file is

```
function xdot = example1(t,x)
% Computes derivatives of two equations
xdot(1) = x(2);
xdot(2) = (1/2)*(sin(t)-8*x(1)-10*x(2));
xdot = [xdot(1);xdot(2)];
```

Note that `xdot(1)` represents $\dot{x}_1$, `xdot(2)` represents $\dot{x}_2$, `x(1)` represents x_1, and `x(2)` represents x_2.

Suppose we want to solve Equations 3.9-1 and 3.9-2 for $0 \le t \le 3$ with the initial conditions $y(0) = x_1(0) = 2$, $\dot{y}(0) = x_2(0) = 5$. Then the initial condition for the *vector x* is `[2, 5]`. To use `ode45`, you type

```
[t,x] = ode45('example1',[0,3],[2,5]);
```

Each row in the matrix x corresponds to a time returned in the column vector `t`. If you type `plot(t,x)`, you will obtain a plot of both x_1 and x_2 versus t. Note that x is a matrix with two columns; the first column contains the values of x_1 at the various times generated by the solver. The second column contains the values of x_2. Thus, to plot only x_1, type `plot(t,x(:,1))`.

We do not as yet have the closed-form solution of this equation when $f(t) = \sin t$, but we do have the solution for $f = 0$. It is shown in Figure 3.4-4, curve (*a*). To obtain this solution, replace the fourth line in the function file `example1.m` with the following line:

```
xdot(2) = (1/2)*(-8*x(1)-10*x(2));
```

If you type `plot(t,x(:,1))`, you should obtain the curve shown in Figure 3.4-4.

So far we have used as examples only equations that have closed-form solutions so that we could confirm that we were using the numerical method correctly. However, of course, the main application of numerical methods is to solve equations for which a closed-form solution cannot be obtained. When solving nonlinear equations, sometimes it is possible to check the numerical results by using an approximation that reduces the equation to a linear one. The following example illustrates such an approach with a second-order equation.

EXAMPLE 3.9-3
A Nonlinear Pendulum Model

Consider a pendulum consisting of a concentrated mass m attached to a rod whose mass is small compared to m. The rod's length is L. The equation of motion for this pendulum is

$$\ddot{\theta} + \frac{g}{L}\sin\theta = 0 \tag{1}$$

Suppose that $L = 1$ m and $g = 9.81$ m/s². Use MATLAB to solve this equation for $\theta(t)$ for two cases: $\theta(0) = 0.5$ rad and $\theta(0) = 0.8\pi$ rad. In both cases $\dot{\theta}(0) = 0$. Discuss how to check the accuracy of the results.

Solution | If we use the small-angle approximation $\sin\theta \approx \theta$, the equation becomes

$$\ddot{\theta} + \frac{g}{L}\theta = 0 \tag{2}$$

which is linear and has the solution

$$\theta(t) = \theta(0)\cos\sqrt{\frac{g}{L}}\,t \tag{3}$$

Thus the amplitude of oscillation is $\theta(0)$ and the period is $P = 2\pi/\sqrt{g/L} = 2$ seconds. We can use this information to select a final time and to check our numerical results.

First, rewrite the pendulum equation, Equation 1, as two first-order equations. To do this, let $x_1 = \theta$ and $x_2 = \dot{\theta}$. Thus

$$\dot{x}_1 = \dot{\theta} = x_2$$

$$\dot{x}_2 = \ddot{\theta} = -\frac{g}{L}\sin x_1$$

The following function file is based on the last two equations. Remember that the output xdot must be a *column* vector.

```
function xdot = pendul(t,x)
global g L
xdot = [x(2);-(g/L)*sin(x(1))];
```

The function file is called as follows. The vectors ta and xa contain the results for the case where $\theta(0) = 0.5$. The vectors tb and xb contain the results for $\theta(0) = 0.8\pi$. Note how the use of the global statement enables the values of g and L to be used within the function file without being passed as arguments of the function:[2]

```
global g L
g = 9.81;L = 1;
[ta,xa] = ode45('pendul',[0,5],[0.5,0]);
[tb,xb] = ode45('pendul',[0,5],[0.8*pi,0]);
plot(ta,xa(:,1),tb,xb(:,1)),xlabel('Times (s)'),…
   ylabel('Angle(rad)'),gtext('Case 1'),gtext('Case2')
```

The results are shown in Figure 3.9-3. The amplitude remains constant, as predicted by the small-angle analysis, and the period for the case where $\theta(0) = 0.5$ is a little larger than 2 s, the value predicted by the small-angle analysis. So we can place some confidence in the numerical procedure. For the case where $\theta(0) = 0.8\pi$, the period of the numerical solution is about 3.3 s. This illustrates an important property of nonlinear differential equations. The free response of a linear equation has the same period for any initial conditions; however, the form of the free response of a nonlinear equation often depends on the particular values of the initial conditions. ■

3.10 INTRODUCTION TO SIMULINK

Simulink is built on top of MATLAB. It provides a graphical user interface that uses various types of elements called *blocks* to create a simulation of a dynamic system; that is, a system that can be modeled with differential or difference equations whose

---

[2] You can avoid the use of the global statement by passing parameters such as g and L within the ode45 function call. This, however, requires that a "function handle" be defined and used for the function pendul. Function handles are not that useful for our applications, so we do not cover them. See the MATLAB documentation for information on function handles and how they can be used with the ode functions.

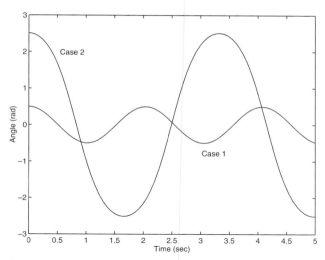

FIGURE 3.9-3 The pendulum angle as a function of time for two starting positions, from Example 3.9-3.

independent variable is time. For example, one block type is a multiplier, another performs a sum, and another is an integrator. The user interface enables you to position the blocks, resize them, label them, specify block parameters, and interconnect the blocks to describe complicated systems for simulation.

The advantages of Simulink include the graphical interface, which some users prefer to coding the MATLAB `ode` solvers, and its variety of special-purpose blocks that enable the user to work easily with piecewise-linear functions, discontinuous functions, and other functions that are difficult to program.

Type `simulink` in the MATLAB Command window to start Simulink. The Simulink Library Browser window opens. To create a new model, click on the icon that resembles a clean sheet of paper, or select **New** from the **File** menu in the Browser. A new **Untitled** window opens for you to create the model. To select a block from the Library Browser, double-click on the appropriate library category, and a list of blocks within that category then appears (see Figure 3.10-1). Click on the block name or icon, hold the mouse button down, drag the block to the new model window, and release the button. Note that when you click on the block name in the Library Browser, a brief description of the block's function appears at the top of the Browser. You can access help for that block by right-clicking on its name or icon and selecting **Help** from the drop-down menu.

Simulink model files have the extension `.mdl`. Use the **File** menu in the model window to Open, Close, and Save model files. To print the block diagram of the model, select **Print** on the **File** menu. Use the **Edit** menu to copy, cut, and paste blocks. You can also use the mouse for these operations. For example, to delete a block, click on it and press the **Delete** key.

Getting started with Simulink is best done through examples, which we now present.

Simulation Diagrams

You construct Simulink models by constructing a diagram that shows the elements of the problem to be solved. Such diagrams are called *simulation diagrams* or *block diagrams*. Consider the equation $\dot{y} = 10f(t)$. Its solution can be represented symbolically as

$$y(t) = \int 10f(t)\,dt$$

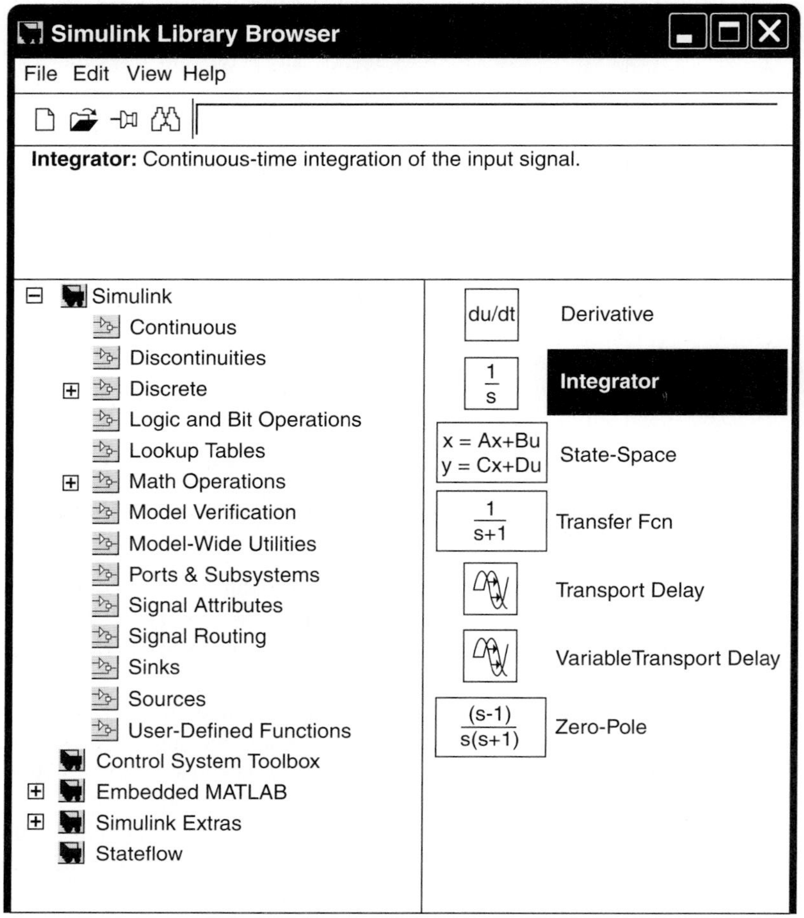

Simulink Library Browser [─][□][✕]

File Edit View Help

[□] [☞] [⊣◻] [⚔] [   ]

Integrator: Continuous-time integration of the input signal.

⊟ **Simulink**
    ▷ Continuous
    ▷ Discontinuities
  ⊞ ▷ Discrete
    ▷ Logic and Bit Operations
    ▷ Lookup Tables
  ⊞ ▷ Math Operations
    ▷ Model Verification
    ▷ Model-Wide Utilities
    ▷ Ports & Subsystems
    ▷ Signal Attributes
    ▷ Signal Routing
    ▷ Sinks
    ▷ Sources
    ▷ User-Defined Functions
    Control System Toolbox
  ⊞ Embedded MATLAB
  ⊞ Simulink Extras
    Stateflow

du/dt	Derivative
$\frac{1}{s}$	**Integrator**
x = Ax+Bu, y = Cx+Du	State-Space
$\frac{1}{s+1}$	Transfer Fcn
	Transport Delay
	VariableTransport Delay
$\frac{(s-1)}{s(s+1)}$	Zero-Pole

FIGURE 3.10-1 The Simulink Library Browser.

which can be thought of as two steps, using an intermediate variable x:

$$x(t) = 10f(t) \qquad \text{and} \qquad y(t) = \int x(t)\, dt$$

This solution can be represented graphically by the simulation diagram shown in Figure 3.10-2. The arrows represent the variables y, x, and f. The blocks represent cause-and-effect processes. Not all "blocks" in Simulink are rectangular. For example, a triangle represents a *multiplier* block, which is also called a *gain* block. Thus the block containing the number 10 represents the process $x(t) = 10f(t)$, where $f(t)$ is the cause (the *input*) and $x(t)$ represents the effect (the *output*).

    The block containing the fraction $1/s$ represents the integration process $y(t) = \int x(t)\, dt$, where $x(t)$ is the cause (the *input*) and $y(t)$ represents the effect (the

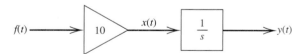

FIGURE 3.10-2 Block diagram for the equation $y(t) = \int 10f(t)\, dt$.

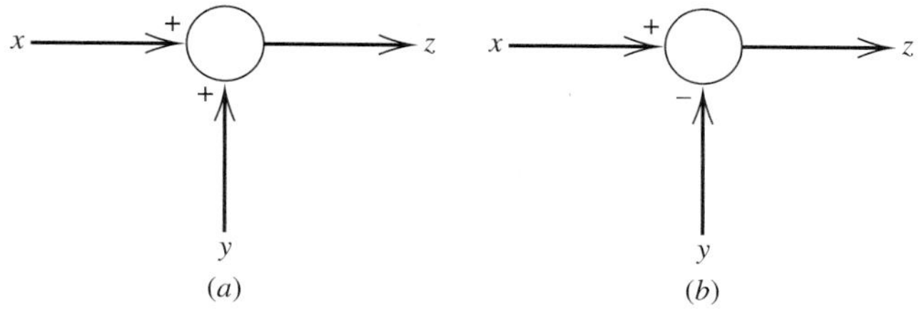

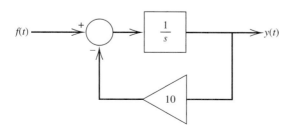

FIGURE 3.10-3 (a) Block diagram for the equation $z = x + y$. (b) Block diagram for the equation $z = x - y$.

output). This type of block is called an *integrator* block. The fraction $1/s$ may seem to be an unusual symbol, but it is based on the notation used with the Laplace transform, a mathematical method to be covered in Chapter 5.

Another element used in simulation diagrams is the *summer* that, despite its name, is used to subtract as well as to sum variables. Two versions of its symbol are shown in Figure 3.10-3. In Figure 3.10-3a the symbol represents the equation $z = x + y$. In Figure 3.10-3b it represents the equation $z = x - y$. Note that a plus or minus sign is required for each input arrow.

The summer symbol can be used to represent the equation $\dot{y} = f(t) - 10y$, which can be expressed as

$$y(t) = \int [f(t) - 10y]\, dt$$

You should study the simulation diagram shown in Figure 3.10-4 to confirm that it represents this equation.

Figure 3.10-2 forms the basis for developing a Simulink model to solve the equation $\dot{y} = 10f(t)$.

EXAMPLE 3.10-1
Simulink Solution of $\dot{y} = 10\sin t$

Let us use Simulink to solve the following problem for $0 \le t \le 13$:

$$\frac{dy}{dt} = 10\sin t \qquad y(0) = 0$$

Although the exact solution is easily found to be $y(t) = 10(1 - \cos t)$, we may use it to verify that we are using Simulink correctly.

Solution

To construct the simulation, perform the following steps. Refer to Figure 3.10-5.
1. Start Simulink and open a new model window as described previously.

2. Select and place in the new window the Sine block from the Sources category. Double-click on it to open the Block-Parameters window, and make sure the Amplitude is set to 1, the Frequency to 1, the Phase to 0, and the Sample time to 0. Then click **OK**.

FIGURE 3.10-4 Block diagram for the equation $y(t) = \int [f(t) - 10y(t)]\, dt$.

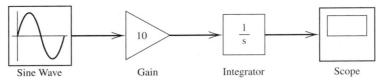

FIGURE 3.10-5 Simulink model of the equation $\dot{y} = 10 \sin t$.

3. Select and place the Gain block from the Math category, double-click on it, and set the Gain Value to 10 in the Block Parameters window. Then click **OK**. Note that the value 10 then appears in the triangle. To make the number more visible, click on the block and drag one of the corners to expand the block so that all the text is visible.

4. Select and place the Integrator block from the Continuous category, double-click on it to obtain the Block Parameters window, and set the Initial Condition to 0 [this is because $y(0) = 0$]. Then click **OK**.

5. Select and place the Scope block from the Sinks category.

6. Once the blocks have been placed as shown in Figure 3.10-5, connect the input port on each block to the outport port on the preceding block. To do this, move the cursor to an input port or an output port; the cursor will change to a cross. Hold the mouse button down, and drag the cursor to a port on another block. When you release the mouse button, Simulink will connect the two blocks with an arrow pointing at the input port. Your model should now look like that shown in Figure 3.10-5.

7. Click on the **Simulation** menu and click the **Simulation Parameters** item. Click on the Solver tab, and enter 13 for the Stop time. Make sure the Start time is 0. Then click **OK**.

8. Run the simulation by clicking on the **Simulation** menu and then clicking the **Start** item. You can also start the simulation by clicking on the **Start** icon on the toolbar (this is the black triangle).

9. You will hear a bell sound when the simulation is finished. Then double-click on the Scope block and then click on the binoculars icon in the Scope display to enable autoscaling. You should see a curve oscillating about the value 10 with an amplitude of 10 and a period of 2π. The independent variable in the Scope block is time t; the input to the block is the dependent variable y. This completes the simulation. ∎

Note that blocks have a Block Parameters window that opens when you double-click on the block. This window contains several items, the number and nature of which depend on the specific type of block. In general you can use the default values of these parameters, except where we have explicitly indicated that they should be changed. You can always click on **Help** within the Block Parameters window to obtain more information.

Note that most blocks have default labels. You can edit text associated with a block by clicking on the text and making the changes. You can save the Simulink model as an `.mdl` file by selecting **Save** from the **File** menu in Simulink. The model file can then be reloaded at a later time. You can also print the diagram by selecting **Print** on the **File** menu.

The Scope block is useful for examining the solution, but if you want to obtain a labeled and printed plot, you can use the To Workspace block, which is described in the next example.

EXAMPLE 3.10-2
Exporting to the MATLAB Workspace

We now demonstrate how to export the results of the simulation to the MATLAB workspace, where they can be plotted or analyzed with any of the MATLAB functions.

Solution

Modify the Simulink model constructed in Example 3.10-1 as follows. Refer to Figure 3.10-6.

1. Delete the arrow connecting the Scope block by clicking on it and pressing the **Delete** key. Delete the Scope block in the same way.

2. Select and place the To Workspace block from the Sinks category and the Clock block from the Sources category.

3. Select and place the Mux block from the Signals & Systems category, double-click on it, and set the Number of inputs to 2. Click **OK**. (The name *Mux* is an abbreviation for *multiplexer*, which is an electrical device for transmitting several signals.)

4. Connect the top input port of the Mux block to the output port of the Integrator block. Then use the same technique to connect the bottom input port of the Mux block to the outport port of the Clock block. Your model should now look like that shown in Figure 3.10-6.

5. Double-click on the To Workspace block. You can specify any variable name you want as the output; the default is simout. Change its name to y. The output variable y will have as many rows as there are simulation time steps and as many columns as there are inputs to the block. The second column in our simulation will be time because of the way we have connected the Clock to the second input port of the Mux. Specify the Save Format as Array. Use the default values for the other parameters (these should be inf, 1, and −1 for Maximum number of rows, Decimation, and Sample Time, respectively). Click on **OK**.

6. After running the simulation, you can use the MATLAB plotting commands from the Command window to plot the columns of y (or simout in general). To plot $y(t)$, type in the MATLAB Command window:

```
≫plot(y(:,2),y(:,1)),xlabel('t'),ylabel('y')
```
∎

Simulink can be configured to to put the time variable tout into the MATLAB workspace automatically when using the To Workspace block. This is done with the Data I/O tab under Configuration Parameters on the **Simulation** menu. The alternative is to use the Clock block to put tout into the workspace. The Clock block has one parameter,

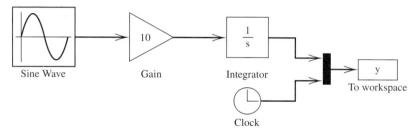

FIGURE 3.10-6 Exporting to the MATLAB workspace.

Decimation. If this parameter is set to 1, the Clock will output every time step; if set to 10, the Clock will output every 10 time steps, and so on.

EXAMPLE 3.10-3
Simulink Model for
$\dot{y} = -10y + f(t)$

Construct a Simulink model to solve

$$\dot{y} = -10y + f(t) \qquad y(0) = 1$$

where $f(t) = 2\sin 4t$, for $0 \leq t \leq 3$.

Solution

To construct the simulation, do the following steps.

1. You can use the model shown in Figure 3.10-6 by rearranging the blocks as shown in Figure 3.10-7. You will need to add a Sum block.

2. Select the Sum block from the Math library and place it as shown in the simulation diagram. Double-click on it and select "round" for the icon shape. Its default setting adds two input signals. To change this, double-click on the block, and in the List of Signs window, type |+−. The signs are ordered counterclockwise from the top. The symbol | is a spacer indicating here that the top port is to be empty.

3. To reverse the direction of the Gain block, right-click on the block, select **Format** from the pop-up menu, and select **Flip Block**.

4. When you connect the negative input port of the Sum block to the output port of the Gain block, Simulink will attempt to draw the shortest line. To obtain the more standard appearance shown in Figure 3.10-7, first extend the line vertically down from the Sum input port. Release the mouse button and then click on the end of the line and attach it to the Gain block. The result will be a line with a right angle. Do the same to connect the input of the Gain block to the arrow connecting the Integrator and the Scope. A small dot appears to indicate that the lines have been successfully connected. This point is called a *take-off point* because it takes the value of the variable represented by the arrow (here, the variable y) and makes that value available to another block.

5. Double-click on the Sine block, and set the Amplitude to 2, the Frequency to 4, the Phase to 0, and the Sample time to 0. Click **OK**.

6. Select **Simulation Parameters** from the **Simulation** menu, and set the Stop time to 3. Then click **OK**.

7. Run the simulation as before and observe the results in the Scope.   ∎

Unlike linear models, closed-form solutions are not available for most nonlinear differential equations, and we must therefore solve such equations numerically. This can

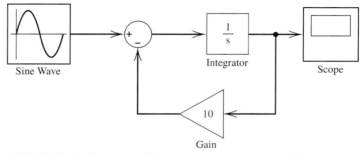

FIGURE 3.10-7   Simulink model of the equation $\dot{y} = -10y + f(t)$.

be done by coding using the MATLAB `ode` solvers or with Simulink (which uses the `ode` solvers in the integration block). In the **Simulation Parameters** submenu under the **Simulation** menu, you can select the `ode` solver to use. The default is `ode45`.

EXAMPLE 3.10-4
Simulink Model of a Pendulum

A pendulum has the following equation of motion if there is viscous friction in the pivot and if there is an applied moment $M(t)$ about the pivot.

$$I\ddot{\theta} + c\dot{\theta} + mgL \sin \theta = M(t) \tag{1}$$

where I is the mass moment of inertia about the pivot. Create a Simulink model for this system for the case where $I = 4$ kg·m^2, $mgL = 10$ N·m, $c = 0.8$ N·m·s, and $M(t)$ is a square wave with an amplitude of 3 and a frequency of 0.5 Hz. Assume that the initial conditions are $\theta(0) = \pi/4$ rad and $\dot{\theta}(0) = 0$.

Solution

To simulate this model in Simulink, define a set of variables that lets you rewrite the equation as two first-order equations. Thus let $\omega = \dot{\theta}$. Then the model can be written as

$$\dot{\theta} = \omega$$

$$\dot{\omega} = \frac{1}{I}[-c\omega - mgL \sin \theta + M(t)] = 0.25[-0.8\omega - 10 \sin \theta + M(t)]$$

Integrate both sides of each equation over time to obtain

$$\theta = \int \omega \, dt$$

$$\omega = 0.25 \int [-0.8\omega - 10 \sin \theta + M(t)] \, dt$$

Obtain a new model window and do the following:

1. Select and place in the new window the Integrator block from the Continuous category, and change its label to Velocity as shown in Figure 3.10-8. You can edit text associated with a block by clicking on the text and making the changes. Double-click on the block to obtain the Block Parameters window, and set the Initial condition to 0 [this is the initial condition $\dot{\theta}(0) = 0$]. Click **OK**.

2. Copy the Integrator block to the location shown and change its label to Displacement. Set its initial condition to $\pi/4$ by typing `pi/4` in the Block Parameters window. This is the initial condition $\theta(0) = \pi/4$.

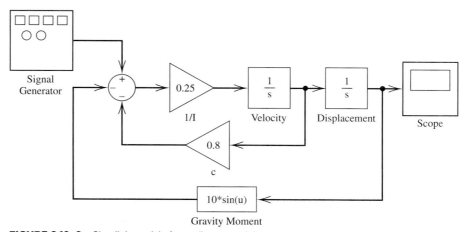

FIGURE 3.10-8 Simulink model of a nonlinear pendulum.

3. Select and place a Gain block from the Math category, double-click on it, and set the Gain value to 0.25. Click **OK**. Change its label to `1/I`. Then click on the block, and drag one of the corners to expand the box so that all the text is visible.

4. Copy the Gain box, change its label to `c`, and place it as shown in Figure 3.10-8. Double-click on it, and set the Gain value to 0.8. Click **OK**. To flip the box left to right, right-click on it, select **Format**, and select **Flip**.

5. Select and place the Scope block from the Sinks category.

6. For the term $10 \sin \theta$, we cannot use the Trig function block in the Math category because we need to multiply the $\sin \theta$ by 10. So we use the Fcn block under the Functions and Tables category (*Fcn* stands for *function*). Select and place this block as shown. Double-click on it, and type `10*sin(u)` in the expression window. This block uses the variable `u` to represent the input to the block. Change its label to Gravity Moment. Click **OK**. Then flip the block.

7. Select and place the Sum block from the Math category. Double-click on it, and select round for the Icon shape. In the List of Signs window, type $+--$. Click **OK**.

8. Select and place the Signal Generator block from the Sources category. Double-click on it, select **Square Wave** for the Wave form, 3 for the Amplitude, and 0.5 for the Frequency, and Hertz for the Units. Click **OK**.

9. Once the blocks have been placed, connect arrows as shown in the figure.

10. Set the Stop time to 10, run the simulation, and examine the plot of $\theta(t)$ in the Scope. This completes the simulation.   ∎

Piecewise-linear models are actually nonlinear, although they may appear to be linear. They are composed of linear models that take effect when certain conditions are satisfied. The effect of switching back and forth between these linear models makes the overall model nonlinear. An example of such a model is a mass attached to a spring and sliding on a horizontal surface with Coulomb friction. The model is

$$m\ddot{x} + kx = f(t) - \mu mg \quad \text{if } \dot{x} > 0$$

$$m\ddot{x} + kx = f(t) \quad \text{if } \dot{x} = 0$$

$$m\ddot{x} + kx = f(t) + \mu mg \quad \text{if } \dot{x} < 0$$

These three linear equations can be expressed using the sign function as the single, nonlinear equation

$$m\ddot{x} + kx = f(t) - \mu mg \, \text{sign}(\dot{x}) \quad \text{where} \quad \text{sign}(\dot{x}) = \begin{cases} +1 & \text{if } \dot{x} > 0 \\ 0 & \text{if } \dot{x} = 0 \\ -1 & \text{if } \dot{x} < 0. \end{cases} \tag{3.10-1}$$

Solution of linear or nonlinear models that contain piecewise-linear functions is very tedious to program. However, Simulink has built-in blocks that represent many of the commonly found functions such as Coulomb friction. Therefore Simulink is especially useful for such applications.

EXAMPLE 3.10-5
Coulomb Friction

Consider a 1-kg mass moving on a horizontal surface with $\mu = 0.4$ and attached to a spring of stiffness $k = 5$ N/m. Suppose the mass is released from rest after being displaced 10 m. Use Simulink to obtain the free response up to the time the mass comes to rest.

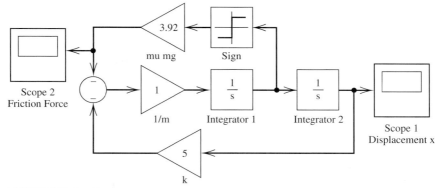

FIGURE 3.10-9 Simulink model of a mass-spring system with Coulomb damping.

Solution

Here the applied force $f(t)$ is zero, $\mu mg = 3.92$, and Equation 3.10-1 becomes

$$\ddot{x} + 5x = -3.92\,\text{sign}(\dot{x})$$

The Simulink model is shown in Figure 3.10-9. The only new feature in this model is the Sign block, which is in the Math category. It implements the sign function defined in Equation 3.10-1. The output of the Integrator 1 block is $\dot{x}$. Double-click on it and make sure the initial condition is set to 0. The output of the Integrator 2 block is x. Double-click on it and set the initial condition 10. (Simulink also has the Coulomb Friction block in the Nonlinear category, which, despite its name, implements viscous as well as Coulomb damping. Thus the Sign block is easier to use for this example.)

The Stop time is unknown here, so we must experiment with the model. The Sign block may have numerical difficulties when $\dot{x}$ is near 0, so try a small Stop time, and examine the output in Scope 1. The mass will stop when the spring force kx equals the friction force μmg; that is, when $x = \mu mg/k = 0.784$. If x is not near 0.784, increase the Stop time and run the simulation again. The required Stop time was found to be 8.4 s.

From the theory developed in Section 3.7, the mass should come to rest in three cycles, the decrement in x per cycle should be 3.1, and the period should be 2.8. The final simulation results are shown in Figure 3.10-10. This curve agrees with the theoretical results.

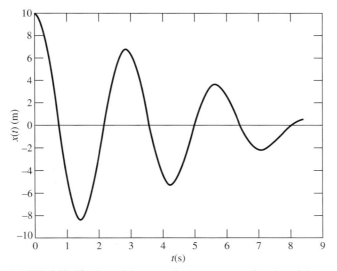

FIGURE 3.10-10 Plot of the mass displacement as a function of time.

This example was chosen because theoretical results are available to check the simulation. However, this model can be modified to study the effects of an applied force $f(t)$. In this case no general theoretical results are available, and we must rely on simulation to obtain the answers. ∎

Summary

Problems involving nonlinear functions such as the Sign block and discontinuous inputs like the square wave are much easier to solve with Simulink. In later chapters we will discover other nonlinear blocks and other advantages to using Simulink. There are menu items in the model window we have not discussed. However, the ones we have discussed are the most important ones for getting started. We have introduced just a few of the blocks available within Simulink, and we will introduce more in later chapters. In addition, some blocks have additional properties that we have not mentioned. However, the examples given here will help you get started in exploring the other features of Simulink. Consult the online help for information about these items.

3.11 CHAPTER REVIEW

This chapter treated the free response of damped and undamped systems having a single degree of freedom. The free response refers to the motion of a mass when no external forces act on the mass other than spring forces, damping forces, or gravitational forces. In this chapter we demonstrated how to obtain the equation of motion of such systems in order to analyze the free response. Some systems have a nonlinear equation of motion, and we demonstrated how to linearize such equations so that we can utilize analysis techniques based on linear equations.

We started by considering the response of a mass with viscous damping but no spring force. From a mathematical point of view, such a system is simpler to analyze because the equation of motion is first order.

We then studied the mathematical analysis and modeling of undamped mass-spring systems, whose equation of motion is second order. We saw that Rayleigh's method is an easier way to obtain the natural frequency of such systems for cases where the kinematics is complicated.

For systems having viscous damping, which is a velocity-dependent force, the free response may be oscillatory or nonoscillatory, depending on the value of the damping coefficient relative to the spring constant and the mass value. From the roots, or equivalently the coefficients, of the characteristic equation, we can compute the damped and undamped natural frequencies, the logarithmic decrement, the time constant, and the damping factor, and determine whether or not the system is stable.

Another type of damping is one due to Coulomb or dry friction, and we analyzed the response of a mass-spring system acted on by such a force.

The chapter concluded with two sections dealing with computer methods for calculating system response. In one section we covered the MATLAB `ode` functions, which are especially useful for solving nonlinear differential equations. In the next section we introduced the Simulink program, which is based on MATLAB. It provides an easy-to-use graphical interface that contains built-in functions for analyzing phenomena, such as Coulomb friction, that are difficult to treat analytically.

Now that you have finished this chapter, you should be able to do the following:

1. Obtain the equation of motion of a system containing spring elements, viscous damping, and/or Coulomb friction, and having a single degree of freedom.
2. Linearize a nonlinear equation of motion.
3. Solve a linear equation of motion for the free response.
4. Compute the damped and undamped natural frequencies, the logarithmic decrement, the time constant, and the damping factor, and determine whether or not the system is stable.
5. Use the chapter's methods to estimate values for the spring constant, damping coefficient, and friction coefficient.
6. Use the MATLAB `ode` functions and Simulink to solve ordinary differential equations.

PROBLEMS

SECTION 3.1   FIRST-ORDER RESPONSE

3.1   Obtain the free response of the following models, and determine the time constant, if any.
(a) $8\dot{y} + 7y = 0$, $y(0) = 6$
(b) $12\dot{y} + 5y = 15$, $y(0) = 3$
(c) $13\dot{y} + 6y = 0$, $y(0) = -2$
(d) $7\dot{y} - 5y = 0$, $y(0) = 9$

3.2   The equation of motion for an ascending rocket having a linear drag force is

$$m\dot{v} = T - mg - cv$$

where T is the rocket's thrust and g is the acceleration due to gravity. Obtain an expression for the vertical velocity $v(t)$ and the rocket's height $h(t)$ given that $h(0) = v(0) = 0$.

3.3   A rocket sled moving on a horizontal track has the following equation of motion: $2\dot{v} = 900 - 8v$. How long must the rocket fire before the sled travels 2500 m? The sled starts from rest.

3.4   A load inertia I_2 is driven through gears by a motor with inertia I_1 (Figure P3.4). The shaft inertias are I_3 and I_4; the gear inertias are I_5 and I_6. The gear ratio is 5:1 (the motor shaft has the greater speed). The motor torque is T_1, and the viscous damping coefficient is $c = 1.6$ N·m·s/rad. Neglect elasticity in the system, and use the following inertia values (in kg·m^2):

$$I_1 = 0.0136 \quad I_2 = 0.68$$
$$I_3 = 0.00136 \quad I_4 = 0.0068$$
$$I_5 = 0.0272 \quad I_6 = 0.041$$

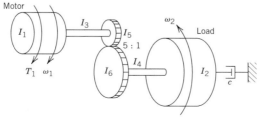

FIGURE P3.4

     (a) Derive the model for the motor shaft speed ω_1 with the torque T_1 a given function of time.

     (b) Derive the model for the load shaft speed ω_2.

3.5 A load inertia I_5 is driven through a double gear pair by a motor with inertia I_4 (Figure P3.5). The shaft inertias are negligible. The gear inertias are I_1, I_2, and I_3. The speed ratios are given in the figure. The motor torque is T_1, and the viscous damping coefficient is $c = 5.4$ N·m·s/rad. Neglect elasticity in the system, and use the following inertia values (in kg·m^2):

$$I_1 = 0.136 \quad I_2 = 0.272$$
$$I_3 = 0.544 \quad I_4 = 0.408$$
$$I_5 = 0.952$$

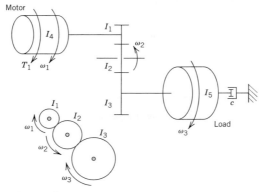

FIGURE P3.5

Derive the model for the motor shaft speed ω_1 with T_1 a given function of time.

3.6 In the damper system shown in Figure P3.6, the velocity v_1 of point 1 is a given function of time. The velocity v_2 is the velocity of the mass m at point 2. Derive the equation of motion in terms of v_2.

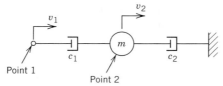

FIGURE P3.6

3.7 In the system shown in Figure P3.7, the displacement x is a given function of time. Derive the equation of motion in terms of x_1, which is the displacement of point 1.

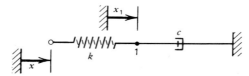

FIGURE P3.7

SECTION 3.2 FREE UNDAMPED VIBRATION

3.8 A certain mass-spring system has the equation of motion $2\ddot{x} + 1800x = 0$. If $x(0) = 0.05$ m and $\dot{x}(0) = 0.3$ m/s, determine the values of A, ω_n, and ϕ in the following solution:

$$x(t) = A \sin(\omega_n t + \phi)$$

3.9   Given that $x(t) = 10\sin(3t + \pi/4)$, find the values of $x(0)$ and $\dot{x}(0)$.

3.10   A certain mass-spring system has the equation of motion $20\ddot{x} + 1280x = 0$. If $x(0) = 0.05$ m and $\dot{x}(\pi/2) = -0.06$ m/s, determine the values of A, ω_n, and ϕ in the following solution:

$$x(t) = A\sin(\omega_n t + \phi)$$

3.11   For the system shown in Figure P3.11, obtain the equation of motion in terms of x and determine the natural frequency. Neglect the pulley mass.

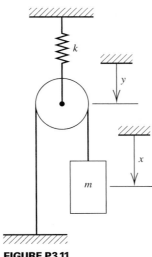

FIGURE P3.11

3.12   For the system shown in Figure P3.12, obtain the equation of motion in terms of x and determine the natural frequency. Neglect the pulley mass.

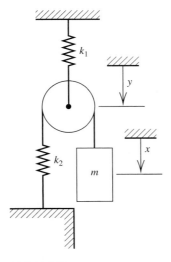

FIGURE P3.12

3.13 In the system shown in Figure P3.13, the cart wheels roll without slipping. Neglect the pulley masses. Obtain the equation of motion in terms of x and determine the natural frequency for **(a)** $\theta = 30°$ and **(b)** $\theta = 60°$.

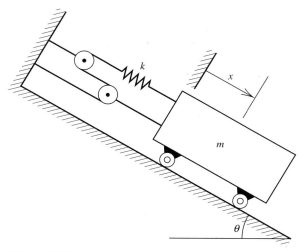

FIGURE P3.13

3.14 The connecting rod shown in Figure P3.14 has a mass of 3.6 kg. It oscillates with a frequency of 40 cycles per minute when supported as shown. Its center of mass is located 0.15 m below the support. Obtain the moment of inertia about the mass center.

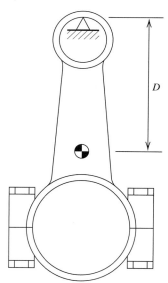

FIGURE P3.14

3.15 A certain 200-kg machine is mounted on a rubber pad to isolate it from the motion of the factory floor. The machine compresses the pad by 8 mm when it is placed on the pad. Determine the natural frequency of the system.

3.16 A certain machine weighing 10 000 N is to be mounted on four identical springs to isolate it from the motion of the factory floor. Determine the spring constant required to give a natural frequency between 1 Hz and 2 Hz.

3.17 Determine the natural frequency of the system shown in Figure P3.17. Neglect the mass of the rod.

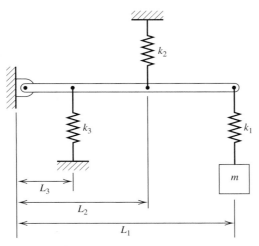

FIGURE P3.17

3.18 A vehicle of mass 1800 kg causes a static deflection of 0.05 m when placed on its suspension. Determine the natural frequency of the system for vertical motion.

3.19 A bungee cord of length 70 m and stiffness 1500 N/m is used by a jumper whose mass is 65 kg. If the person jumps from a bridge, determine the amplitude and frequency of the person's oscillation and the point about which the oscillation occurs.

3.20 Obtain the natural frequency of the system shown in Figure P3.20. Assume small motions and neglect the pulley mass.

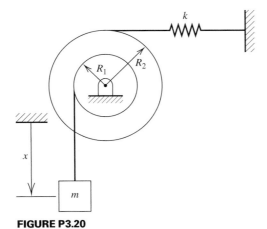

FIGURE P3.20

3.21 Obtain the natural frequency of the system shown in Figure P3.21 for small motions. The applied force f is a given function of time. The lever inertia about the pivot is I.

3.22 In the pulley system shown in Figure P3.22, assume that the cable is massless and inextensible, and assume that the pulley masses are negligible. The force f is a known function of time. Obtain the expression for the natural frequency.

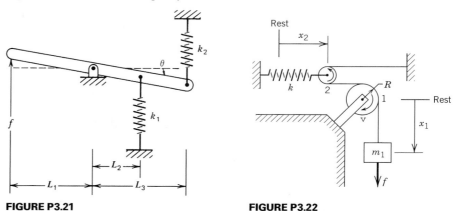

FIGURE P3.21   **FIGURE P3.22**

3.23 For the system shown in Figure P3.23, the solid cylinder of inertia I and mass m rolls without slipping. Neglect the pulley mass and determine the natural frequency.

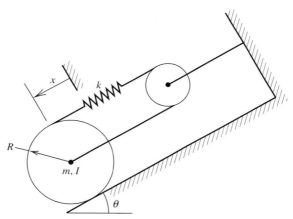

FIGURE P3.23

3.24 A rectangular street sign is mounted on top of a post (Figure P3.24). The sign is 0.3 m high by 0.6 m wide by 0.003 m thick. The post is a hollow cylinder 1.8 m high. Its inner radius is 0.016 m and its outer radius is 0.025 m. Both the post and the sign are made of steel, whose properties are as follows. Mass density: $\rho = 7800$ kg/m^2; modulus of elasticity: $E = 2 \times 10^{11}$ N/m^2; and shear modulus: $G = 8 \times 10^{10}$ N/m^2.

Consider two ways the sign may vibrate: in the transverse (horizontal) direction (as a cantilever beam) and about the vertical axis (as a torsional spring). The stiffness of the post in each of these directions is as follows:

1. Transverse:

$$k = \frac{3EI_A}{L^3} = \frac{3\pi E}{2L^3}\left(r_o^4 - r_i^4\right)$$

where r_i and r_o are the inner and outer radii of the post.

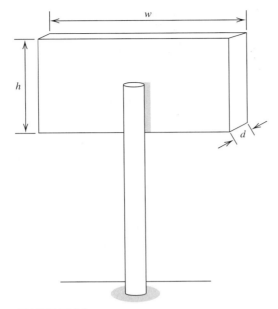

FIGURE P3.24

2. Torsional:

$$k = \frac{\pi G}{2L}\left(r_o^4 - r_i^4\right)$$

The mass moment of inertia of the sign about the vertical axis is

$$I_s = \frac{1}{12}m_s\left(d^2 + w^2\right)$$

where m_s is the sign mass, d is the thickness, and w is the width.

Compute the natural frequency for each of these modes of vibration, taking into account the mass of the post as well as that of the sign.

3.25 Do Problem 3.24 but use aluminum instead of steel for the post and sign. Use the following properties of aluminum: $\rho = 2.7 \times 10^3$ kg/m^3, $E = 7.1 \times 10^{10}$ N/m^2, and $G = 2.67 \times 10^{10}$ N/m^2.

3.26 The following model describes a mass supported by a nonlinear spring. The units are SI, so $g = 9.81$ m/s^2.

$$5\ddot{y} = 5g - \left(900y + 1700y^3\right)$$

(a) Find the equilibrium position y_r, obtain a linearized model using the equilibrium as the reference operating condition, and compute the oscillation frequency of the linearized model.

(b) Find the free response of the linearized model, in terms of $x(t) = y(t) - y_r$, for the initial conditions: $x(0) = 0.002$ m and $\dot{x}(0) = 0.005$ m/s.

3.27 Elastic elements such as springs that are extended or compressed beyond their linear range are often described by the nonlinear model

$$f(y) = k_1y + k_2y^3$$

where $f(y)$ is the spring force and y is the spring's displacement from its free length. If a mass m is hung from such a spring, its equation of motion is

$$m\ddot{y} = -(k_1y + k_2y^3) + mg$$

Because the system is nonlinear, its oscillation frequency depends on the equilibrium position, which is determined by m, g, k_1, and k_2.

Develop a program that a designer could use to compute the maximum weight mg that could be supported by a given spring, provided that its frequency of oscillation is less than a prescribed value ω_{max}. The given information is g, k_1, k_2, and ω_{max}.

3.28 The following equation of motion describes the rolling motion of a boat:

$$I_G\ddot{\theta} + mgh\sin\theta = 0$$

Measurements of a certain boat's roll period give a value of 5 s. The boat's metacentric height is $h = 0.3$ m. How much change in the metacentric height must be made for the roll period to be increased to 10 s?

SECTION 3.3   RAYLEIGH'S METHOD

3.29 Determine the natural frequency of the system shown in Figure P3.29 by using Rayleigh's method. Assume small angles of oscillation.

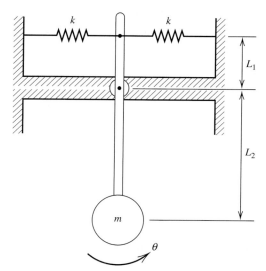

FIGURE P3.29

3.30 Determine the natural frequency of the system shown in Figure P3.30 using Rayleigh's method. Assume small angles of oscillation.

3.31 Determine the natural frequency of the system shown in Figure P3.31 using Rayleigh's method. Assume small angles of oscillation.

3.32 Determine the natural frequency of the system shown in Figure P3.17 using Rayleigh's method.

3.33 Figure 3.3-4 shows the suspension of one front wheel of a car in which $L_1 = 0.6$ m and $L_2 = 0.9$ m. The coil spring has a spring constant of $k = 60\,000$ N/m and the car mass associated with that wheel is 360 kg. Determine the suspension's natural frequency for vertical motion.

3.34 Use Rayleigh's method to derive the natural frequency of the system shown in Figure P3.20.

3.35 Use Rayleigh's method to derive the natural frequency of the system shown in Figure P3.17.

3.36 Determine the expression for the natural frequency of the liquid mass in the manometer shown in Figure P3.36. The total length of the liquid column is L, and the liquid mass density is ρ.

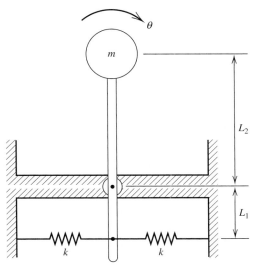

FIGURE P3.30

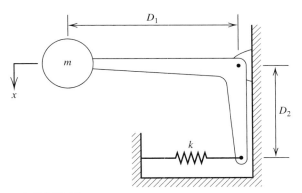

FIGURE P3.31

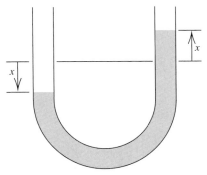

FIGURE P3.36

SECTION 3.4 FREE VIBRATION WITH VISCOUS DAMPING

3.37 Obtain the free response of the following models with the initial conditions $x(0) = 0$ and $\dot{x}(0) = 1$:
(a) $\ddot{x} + 4\dot{x} + 8x = 0$
(b) $\ddot{x} + 8\dot{x} + 12x = 0$
(c) $\ddot{x} + 4\dot{x} + 4x = 0$

3.38 Obtain the free response of the following models with the initial conditions $x(0) = 1$ and $\dot{x}(0) = -1$:
(a) $3\ddot{x} + 21\dot{x} + 30x = 0$
(b) $5\ddot{x} + 20\dot{x} + 20x = 0$
(c) $2\ddot{x} + 8\dot{x} + 58x = 0$

3.39 A 2.3-kg object is hung from a spring whose other end is attached to a rigid support. The object oscillates with a frequency of 250 cycles per minute in air and 240 cycles per minute when immersed in a liquid. Determine the spring constant k, the damping ratio ζ, and the damping constant c when the object is immersed in the liquid.

3.40 A highway crash barrier is described in Example 3.4-2. Develop a program that a designer could use to evaluate different barrier materials and designs. The given information consists of the constant's k and c, the vehicle weight, and the speed at which it strikes the barrier. The program should compute how long it takes for the vehicle to come to rest, how far the vehicle compresses the barrier, and the maximum deceleration of the vehicle. You can use the results of Example 3.4-2 to check your program.

3.41 In the vehicle suspension system shown in Figure P3.41, $m = 200$ kg, $k = 2500$ N/m, and $c = 500$ N·s/m. The spring is compressed 0.1 m when the tire hits a bump, which gives a velocity of $\dot{x}(0) = 1.2$ m/s. Assuming that the tire remains out of contact with the road, (a) determine the tire displacement $x(t)$ and (b) compute the maximum spring compression.

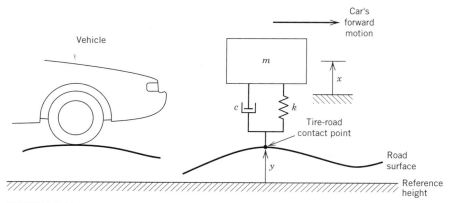

FIGURE P3.41

3.42 Figure P3.42 represents a drop forging process. The anvil mass is $m_1 = 1000$ kg and the hammer mass is $m_2 = 200$ kg. The support stiffness is $k = 10^7$ N/m, and the damping constant is $c = 2 \times 10^4$ N·s/m. The anvil is at rest when the hammer is dropped from a height of $h = 1$ m. Obtain the expression for the displacement of the anvil as a function of time after the impact. Do this for two values of the coefficient of restitution: (a) $e = 0$ and (b) $e = 1$.

3.43 A box car of mass 18 000 kg hits a shock absorber at the end of the track while moving at 1.3 m/s (Figure P3.43). The stiffness of the absorber is $k = 73\,000$ N/m and the damping is $c = 88\,000$ N·s/m. Determine (a) the maximum spring compression and (b) the time for the boxcar to stop.

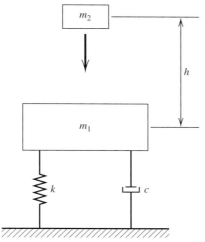

FIGURE P3.42

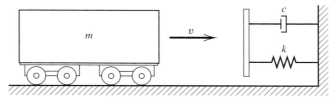

FIGURE P3.43

SECTION 3.5   ANALYSIS OF THE CHARACTERISTIC ROOTS

3.44 Compute the time constant and the frequency of oscillation, if any, for the following models. Estimate how long it will take for the free response to disappear.

(a) $3\ddot{x} + 21\dot{x} + 30x = 0$

(b) $5\ddot{x} + 20\dot{x} + 20x = 0$

(c) $2\ddot{x} + 8\dot{x} + 58x = 0$

3.45 Find the time constants of the model $2\ddot{x} + 26\dot{x} + 60x = 0$. Estimate how long it will take for the free response to disappear.

3.46 If applicable, compute ζ, τ, ω_n, and ω_d for the following roots, and find the corresponding characteristic polynomial:

(a) $s = -2 \pm 6i$

(b) $s = 1 \pm 5i$

(c) $s = -10, -10$

(d) $s = -10$

3.47 If applicable, compute ζ, τ, ω_n, and ω_d for the dominant root in each of the following sets of characteristic roots:

(a) $s = -2, -3 \pm i$

(b) $s = -3, -2 \pm 2i$

3.48 A certain fourth-order model has the roots:

$$s = -2 \pm 4i, -10 \pm 7i$$

Identify the dominant roots and use them to estimate the system's time constant, damping ratio, and oscillation frequency.

3.49 A mass-spring-damper system has a mass of 100 kg. In 60 s, its free response amplitude decays such that the amplitude of the 30th cycle is 20% of the amplitude of the 1st cycle. Estimate the damping constant c and the spring constant k.

3.50 A certain mass-spring system undergoes viscous damping. What is the ratio of successive amplitudes of free vibration if the damping ratio is (**a**) $\zeta = 0.5$ and (**b**) $\zeta = 0.707$?

3.51 Determine the logarithmic decrement δ and the damping ratio ζ if each successive amplitude of free vibration is $2/3$ of the previous amplitude.

3.52 A certain mass-spring system undergoes viscous damping. The mass is 11 kg and the spring constant is 15 000 N/m. The amplitudes in meters of the first through fourth cycle of free vibration are 0.098, 0.073, 0.055, and 0.041. Compute the damping constant c.

3.53 For a certain mass-spring system, $m = 100$ kg and $k = 8000$ N/m. Determine (**a**) the value of the critical damping constant c_c, and (**b**) the damped natural frequency ω_d and the logarithmic decrement δ if $c = c_c/3$.

3.54 Sketch the root locus plot of $3s^2 + 12s + k = 0$ for $k \geq 0$. What is the smallest possible dominant time constant, and what value of k gives this time constant?

3.55 Sketch the root locus plot of $3s^2 + cs + 12 = 0$ for $c \geq 0$. What is the smallest possible dominant time constant, and what value of c gives this time constant? What is the value of ω_n if $\zeta < 1$?

3.56 Sketch the root locus plot of $ms^2 + 12s + 10 = 0$ for $m \geq 2$. What is the smallest possible dominant time constant, and what value of m gives this time constant?

SECTION 3.6   STABILITY

3.57 Determine whether the following models are stable, neutrally stable, or unstable:
(**a**) $3\ddot{x} + 2\dot{x} + 30x = 0$
(**b**) $5\ddot{x} + 20x = 0$
(**c**) $2\ddot{x} - 8\dot{x} + 58x = 0$
(**d**) $5\ddot{x} + 4\dot{x} - 20x = 0$
(**e**) $2\ddot{x} - 58x = 0$

3.58 Given the model

$$\ddot{x} - (r + 2)\dot{x} + (2r + 5)x = 0$$

(**a**) Find the values of the parameter r for which the system is (i) stable, (ii) neutrally stable, (iii) unstable.
(**b**) For the stable case, for what values of r is the system (i) underdamped, (ii) overdamped?

3.59 Find the equilibrium solution (corresponding to $\dot{y} = 0$) of the following models, obtain a linearized model valid near the equilibrium, and compute the time constant if the linearized model is stable.
(**a**) $\dot{y} = 9 - \sqrt{y}$
(**b**) $\dot{y} = \sin y, \ 0 \leq y \leq 2\pi$
(**c**) $6\ddot{y} + 5\dot{y} + 3\sin y = 0, \ 0 \leq y \leq \pi$
(**d**) $6\ddot{y} + 5\dot{y} + 3\sin y = 2, \ 0 \leq y \leq \pi$

3.60 Consider the robot arm discussed in Example 3.6-1. Suppose there is a slight amount of friction in the joint. Discuss what happens if the arm is held in equilibrium at $\theta_e = 135°$ when it is knocked downward slightly. Does the arm eventually come to rest at $\theta = 0$, at $\theta = 45°$, or somewhere else?

3.61 For the system of Problem 3.30, shown in Figure P3.30, obtain the equation of motion in terms of θ for small angles. Neglect the rod mass. What relation among L_1, L_2, m, and k must be satisfied for the system to be stable?

3.62 Derive the equation of motion of the system shown in Figure P3.62. Assume small angles of oscillation and neglect the rod mass. What relation among L_1, L_2, m, and k must be satisfied for the system to be stable?

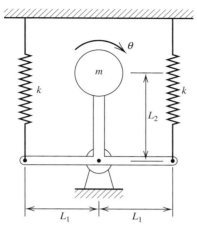

FIGURE P3.62

SECTION 3.7   VIBRATION MODELS WITH VISCOUS DAMPING

3.63 Obtain the equation of motion for the system shown in Figure P3.63. The equilibrium position corresponds to $x = y = 0$.

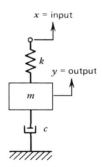

FIGURE P3.63

3.64 Obtain the equation of motion for the system shown in Figure P3.64. The equilibrium position corresponds to $x = y = 0$.

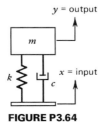

FIGURE P3.64

3.65 Obtain the equation of motion for the system shown in Figure P3.65. The equilibrium position corresponds to $\theta_i = \theta = 0$.

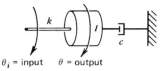

θ_i = input θ = output

FIGURE P3.65

3.66 Obtain the equation of motion for the system shown in Figure P3.66. Neglect the inertia of the gears. The number of gear teeth is n_1 and n_2. The equilibrium position corresponds to $\theta_i = \theta = 0$.

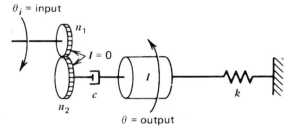

FIGURE P3.66

3.67 The solid door, whose top view is shown in Figure P3.67, has a mass of 40 kg and is 2.1 m high, 1.2 m wide, and 0.05 m thick. Its door closer has a torsional spring constant of 13.6 N·m/rad. The door will close the fastest without oscillating if the torsional damping coefficient c in the closer is set to the critical damping value corresponding to $\zeta = 1$. Determine this critical value of c.

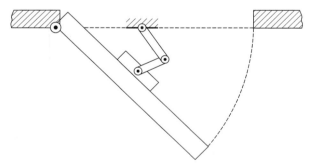

FIGURE P3.67

3.68 Determine a model for the system shown in Figure P3.68, with the force f as a given function of time. Assume small displacements. The lever has an inertia I relative to the pivot.

3.69 The mass m in Figure P3.69 is attached to a rigid lever having negligible mass and negligible pivot friction. The displacement x is a given function of time. When x and θ are zero, the spring is at its free length. Assuming that θ is small, derive the equation of motion for θ.

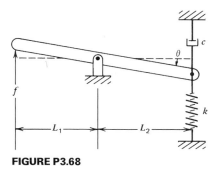

FIGURE P3.68

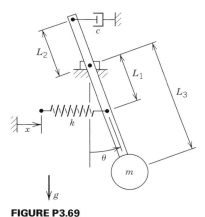

FIGURE P3.69

SECTION 3.8 FREE VIBRATION WITH COULOMB DAMPING

3.70 A spring-mass system on a horizontal surface was set in motion by displacing the mass 0.1 m from its rest position and releasing it with zero velocity. The spring constant is $k = 800$ N/m and the coefficient of kinetic friction is $\mu = 0.2$. Determine how many half-cycles are required for the mass to come to rest, and determine its position at that time.

3.71 A spring-mass system on a horizontal surface was set in motion by displacing the 5-kg mass 30 mm from its rest position and releasing it with zero velocity. The mass oscillated with a frequency of 20 Hz and came to a stop after 35 half-cycles. Estimate the value of the friction coefficient.

3.72 A spring-mass system on a horizontal surface was set in motion by displacing the mass 0.076 m from its rest position and releasing it with zero velocity. The mass oscillated with a period of 10 s, with an amplitude reduction of 0.005 m per cycle. Estimate the value of the friction coefficient, and determine how many half-cycles are required for the mass to come to rest.

SECTION 3.9 THE MATLAB ode SOLVERS

3.73 Use MATLAB to compute and plot the solution of the following problem

$$\dot{y} = 5\sin(3t + 2) \qquad y(0) = 4$$

for $0 \le t \le 8$. Compare the results with the exact solution, which can be obtained by integration.

3.74 Use MATLAB to compute and plot the solution of the following problem

$$3\ddot{x} + 8\dot{x} + 10x = 0 \qquad x(0) = 5 \qquad \dot{x}(0) = 3$$

for $0 \le t \le 3$. Compare the results with the exact solution.

3.75 Use MATLAB to compute and plot the solution of the following nonlinear equation of motion of a simple pendulum with viscous damping

$$10\ddot{\theta} + 15\dot{\theta} + 20 \sin \theta = 0 \qquad \theta(0) = 1.5 \text{ rad} \qquad \dot{\theta}(0) = 0$$

for $0 \le t \le 5$. Compare the results with the exact solution of the linearized equation.

3.76 A mass hung from a certain nonlinear spring has the equation of motion

$$m\ddot{y} = -(k_1 y + k_2 y^3) + mg$$

where $k_1 = 120$ lb/ft and $k_2 = 360$ lb/ft^3. In Example 3.2-5 the solution of the linearized equation was obtained for two cases: (a) $mg = 10$ lb and (b) $mg = 35$ lb. The equilibrium for each case was found to be $y_r = 0.082$ for case (a) and $y_r = 0.247$ for case (b). The initial conditions used were $\dot{y}(0) = 0$ for both cases: $y(0) = 0.05 + y_r = 0.05 + 0.082 = 0.132$ for case (a) and $y(0) = 0.05 + y_r = 0.05 + 0.247 = 0.297$ for case (b).

Use MATLAB to compute and plot the solution of the nonlinear equation of motion for each case, and compare the results with the linearized solution.

3.77 The following model describes a mass supported by a nonlinear spring. The units are SI. Use $g = 9.81$ m/s^2.

$$5\ddot{y} = 5g - (900y + 1700y^3)$$

(a) Suppose that $\dot{y}(0) = 0$. Use a numerical method to plot the solution for two initial conditions: (i) $y(0) = 0.06$ and (ii) $y(0) = 0.1$.

(b) Compare the results with those expected from the answers to Problem 3.26.

3.78 Van der Pol's equation is a nonlinear model for some oscillatory processes. It is

$$\ddot{y} - b(1 - y^2)\dot{y} + y = 0$$

(a) Find the equilibrium point and obtain a linearized model using the equilibrium as the reference operating condition. Analyze the stability of the equilibrium. Compute the oscillation frequency and the time constant of the linearized model if applicable.

(b) Use a numerical method to solve the nonlinear equation for the following cases:

(i) $b = 0.1$, $y(0) = \dot{y}(0) = 1$, $0 \le t \le 25$; (ii) $b = 0.1$, $y(0) = \dot{y}(0) = 3$, $0 \le t \le 25$; (iii) $b = 3$, $y(0) = \dot{y}(0) = 1$, $0 \le t \le 25$. Compare the numerical solution with the solution of the linearized model.

3.79 The equation of motion for the hanging load of an overhead crane whose base is accelerating horizontally with an acceleration $a(t)$ is (see Figure 2.4-14a)

$$L\ddot{\theta} + g \sin \theta = a(t)\cos \theta$$

Suppose that $g = 9.81$ m/s^2, $L = 1$ m, and $\dot{\theta}(0) = 0$. Use a numerical method to plot $\theta(t)$ for $0 \le t \le 10$ s for the following three cases:

(a) The acceleration is constant: $a = 5$ m/s^2, and $\theta(0) = 0.5$ rad.

(b) The acceleration is constant: $a = 5$ m/s^2, and $\theta(0) = 3$ rad.

(c) The acceleration is linear with time: $a = 0.5t$ m/s^2, and $\theta(0) = 3$ rad.

3.80 Van der Pol's equation is

$$\ddot{y} - b(1 - y^2)\dot{y} + y = 0$$

This equation can be challenging to solve numerically if the parameter b is large. Compare the performance of two or more numerical methods for this equation. Use $b = 1000$ and $0 \le t \le 3000$, with the initial conditions $y(0) = 2$, $\dot{y}(0) = 0$.

3.81 The equation of motion for the hanging load of an overhead crane whose base is accelerating horizontally with an acceleration $a(t)$ is

$$L\ddot{\theta} + g\sin\theta = a(t)\cos\theta$$

Suppose that $L = 9$ m, $\theta(0) = 0$, and $\dot{\theta}(0) = 0$. Find the maximum allowable base acceleration a required to keep the angle θ less than $20°$.

SECTION 3.10   INTRODUCTION TO SIMULINK

3.82 Use Simulink to compute the solution of the following problem for $0 \le t \le 8$:

$$\dot{y} = 5\sin(3t + 2) \qquad y(0) = 4$$

Export the results to the MATLAB workspace and plot the solution. Compare the results with the exact solution, which can be obtained by integration.

3.83 Use Simulink to compute the solution of the following problem for $0 \le t \le 3$:

$$3\ddot{x} + 8\dot{x} + 10x = 0 \qquad x(0) = 5 \qquad \dot{x}(0) = 3$$

Export the results to the MATLAB workspace and plot the solution. Compare the results with the exact solution.

3.84 Use Simulink to compute and plot the solution of the following nonlinear equation of motion of a simple pendulum with viscous damping for $0 \le t \le 5$:

$$10\ddot{\theta} + 15\dot{\theta} + 20\sin\theta = 0 \qquad \theta(0) = 1.5 \text{ rad} \qquad \dot{\theta}(0) = 0$$

Export the results to the MATLAB workspace and plot the solution. Compare the results with the exact solution of the linearized equation.

3.85 Use Simulink to compute and plot the solution for $0 \le t \le 5$ of the following nonlinear equation of motion of a simple pendulum with viscous damping and an applied moment $M(t)$, which is a square wave with an amplitude of 2 and a frequency of 1 Hz:

$$10\ddot{\theta} + 15\dot{\theta} + 20\sin\theta = M(t) \qquad \theta(0) = \dot{\theta}(0) = 0$$

Export the results to the MATLAB workspace and plot the solution.

3.86 Consider a mass-spring system in which the mass slides on a horizontal surface with a coefficient of kinetic friction of $\mu = 0.2$. The mass weighs 100 N and the spring constant is 1500 N/m. Suppose the mass is displaced 0.15 m from equilibrium and released from rest. Use Simulink to compute and plot the solution until the mass comes to rest. Compare the simulation results with the analytical results given in Section 3.8.

3.87 Consider a mass-spring system in which the mass slides on a horizontal surface with a coefficient of kinetic friction of $\mu = 0.2$. The mass weighs 100 N and the spring constant is 1500 N/m. Suppose the mass is displaced 0.15 m from equilibrium and released with a velocity of $+1.5$ m/s. Use Simulink to compute and plot the solution until the mass comes to rest. Since we do not have any analytical results for this case (because the initial velocity is not zero), how would you make a rough check of the validity of the simulation results?

3.88 Consider a mass-spring-damper system in which the mass slides on a horizontal surface with a coefficient of kinetic friction of $\mu = 0.2$. The equation of motion is

$$m\ddot{x} + c\dot{x} + kx = -\mu mg \, \text{sign}(\dot{x})$$

The mass weighs 100 N, the spring constant is 1500 N/m, and the viscous damping coefficient is $c = 120$ N·s/m. Suppose the mass is displaced 0.15 m from equilibrium and released from rest. Use Simulink to compute and plot the solution until the mass comes to rest. Since we do not have any analytical results for this case (because of the viscous damping), how would you make a rough check of the validity of the simulation results?

3.89 Consider the lever system whose equation of motion was derived in Example 3.7-6. If the applied force f_2 is zero and if a Coulomb friction torque T_f acts at the pivot, the equation of motion becomes

$$(m_a a^2 + m_b b^2)\ddot{\theta} + c_1 a^2 \dot{\theta} + k_2 b^2 \theta = -T_f \,\text{sign}\,(\dot{\theta})$$

The given parameter values are $m_a = 3$ kg, $m_b = 5$ kg, $a = 1$ m, $b = 0.8$ m, $c_1 = 60$ N·m/s, $k_1 = 1000$ N/m, and $T_f = 1000$ N·m. Suppose that $\theta(0) = 0.02$ rad and $\dot{\theta} = 0$. Use Simulink to compute and plot the solution until the system comes to rest. Compare the simulation results with the analytical results given in Section 3.8.

CHAPTER 4

HARMONIC RESPONSE WITH A SINGLE DEGREE OF FREEDOM

CHAPTER OUTLINE

A *harmonic* input is one that is either sinusoidal or cosinusoidal. This chapter treats harmonic response, which is the response of a system when subjected to a harmonic input. Many naturally occurring inputs are harmonic, but harmonic inputs are a special case of periodic inputs. In Chapter 5 we will see how to describe general periodic inputs in terms of sines and cosines, so the results of this chapter are especially useful.

This chapter treats the harmonic response of systems having one degree of freedom. In Section 4.1 we begin by obtaining the harmonic response of a system having one degree of freedom. Then in Section 4.2 we treat two important topics associated with such systems: resonance and bandwidth.

Ground motion caused by earthquakes and displacement inputs caused by vehicle motion are two important examples of base excitation, which is treated in Section 4.3. Two quantities of interest are the displacement of the mass and the force transmitted to the mass as a result of the base motion.

Rotating unbalanced machinery produces harmonic forces on the supporting structures, and this topic is covered in Section 4.4. The resulting motion of the mass and the force transmitted to the supporting structure are important in many applications.

An important application is the design of vibration isolators to minimize the transmission of vibration between a support and a machine. A related topic, critical speeds of rotating shafts, concerns the vibration of high-speed machinery such as a rotor-shaft system, like those found in jet engines and electrical generators. This is treated in Section 4.5.

Sections 4.6 and 4.7 deal with harmonic response in the presence of damping sources other than linear viscous damping. These sources introduce nonlinear effects and include hysteresis, drag, and Coulomb friction. A more general discussion of the effects of nonlinearities of system response is given in Section 4.8.

Coverage of computer methods concludes the chapter. Section 4.9 introduces several MATLAB functions that are especially intended for harmonic response analysis. Section 4.10 covers several Simulink features useful for simulating systems containing nonlinearities that are difficult to program in MATLAB or other languages.

LEARNING OBJECTIVES

After you have finished this chapter, you should be able to do the following:

- Obtain the harmonic response of systems having a single degree of freedom.
- Obtain the transfer function from the equation of motion.
- Obtain the frequency transfer function.
- Use the plots or the frequency transfer function to determine the steady-state output amplitude and phase that result from a sinusoidal input.
- Determine the resonance frequency, peak response, and bandwidth, and identify the system as a low-pass, high-pass, or band-pass filter.
- Analyze the displacement and transmitted force of systems having base excitation, rotating unbalance, or rotor-shaft vibration.
- Design simple vibration isolation systems.
- Model and analyze nonlinear sources of damping, such as hysteresis, drag, and Coulomb friction.
- Describe the differences between the response characteristics of linear and nonlinear systems.
- Apply MATLAB and Simulink to analyze harmonic response.

4.1 SOLUTION FOR THE HARMONIC RESPONSE

As we saw in Chapter 3, the *free vibration*, also called the *free response*, is the response caused by the initial conditions. When a vibratory system has an external force acting on it, such as from a gust of wind, an unbalanced motor, or road surface variation, its response depends not only on the initial conditions but also on the nature of the external force, which is called the forcing function. We call that part of the response due to the forcing function the *forced response*. If the system model consists of linear differential equations, then the *total* or *complete* response is the sum of the free and the forced responses.

In some applications the forcing function is not actually a force; it can be a moment or a displacement, for example. Sometimes the term *input* is used instead of *forcing function*, and *output* is used instead of *response*. The response of a system to either a sinusoidal or cosinusoidal forcing function is termed the *harmonic response*, and it is a special case of forced response.

We can use the substitution method to obtain the response where the forcing function is sinusoidal. This method may be familiar to you from the study of differential equations, where the terms *homogeneous solution* and *particular solution* may have been used. The homogeneous solution has the same form as the free response. The particular solution has the form of the forcing function plus derivatives of the forcing function, if necessary.

Consider the following equation of motion of a mass-spring-damper system:

$$m\ddot{x} + c\dot{x} + kx = f(t) \tag{4.1-1}$$

Suppose that the forcing function is $f(t) = F_o \sin \omega t$. Then the model becomes

$$m\ddot{x} + c\dot{x} + kx = F_o \sin \omega t \tag{4.1-2}$$

Free Response

As we saw in Chapter 3, the form of the free response depends on the nature of the roots of $ms^2 + cs + k = 0$. The solution for the roots will be one of the following cases:

1. Two distinct, real roots, denoted r_1 and r_2. This case will occur if $c^2 - 4mk > 0$. For this case the free response form is $A_1 e^{r_1 t} + A_2 e^{r_2 t}$.

2. Two equal (repeated) roots, denoted r_1 and r_1. This case will occur if $c^2 - 4mk = 0$. These roots will be real. For this case the free response form is $A_1 e^{r_1 t} + t A_2 e^{r_1 t}$.

3. Two complex conjugate roots, denoted $r + iq$ and $r - iq$, where

$$r = -\frac{c}{2m} \qquad q = \frac{\sqrt{4mk - c^2}}{2m}$$

This case will occur if $c^2 - 4mk < 0$. For this case the free response form is $Be^{rt} \sin(qt + \psi)$.

The solutions for the free response for these three cases are given in Table 3.5-1.

Sine Response with Complex Roots

To obtain the complete solution, we try the form

$$x(t) = \text{Free Response Form} + \text{Forcing Function Form}$$
$$+ \text{Forcing Function Derivative Form}$$

The general substitution method includes *all* the derivatives of the forcing function, but since the higher derivatives of the sine function repeat as sines and cosines, we need include only the forcing function and its first derivative.

So, for example, if the roots are complex, then we try a solution for the total response of the form

$$x(t) = Ae^{rt}\sin(qt + \psi) + B \sin \omega t + C \cos \omega t \tag{4.1-3}$$

If we use this form, however, we will not be able to separate the free and the forced responses, because the amplitude A and the phase angle ψ will depend on both the initial conditions and the forcing frequency ω. If, however, we use the following form, then we can separate the free and forced responses:

$$x(t) = Ae^{rt}\sin qt + Be^{rt}\cos qt + C\sin \omega t + D\cos \omega t \tag{4.1-4}$$

We need the first and second derivatives. After collecting terms, these are:

$$\dot{x} = (rA - Bq)e^{rt}\sin qt + (qA + Br)e^{rt}\cos qt + C\omega \cos \omega t - D\omega \sin \omega t \tag{4.1-5}$$

$$\begin{aligned}\ddot{x} = \left(r^2A - q^2A - 2rqB\right)e^{rt}\sin qt + \left(r^2B - q^2B + 2rqA\right)e^{rt}\cos qt \\ - C\omega^2 \cos \omega t - D\omega^2 \cos \omega t\end{aligned} \tag{4.1-6}$$

Substitute these into the differential equation and collect terms to obtain

$$\alpha e^{rt}\sin qt + \beta e^{rt}\cos qt + \gamma \sin \omega t + \delta \cos \omega t = 0$$

where

$$\alpha = (mr^2 - mq^2 + cr + k)\, A - q(2mr + c)B$$

$$\beta = q(2mr + c)A + \left(mr^2 - mq^2 + cr + k\right)B$$

$$\gamma = \left(k - m\omega^2\right)C - c\omega D - F_o$$

$$\delta = c\omega + \left(k - m\omega^2\right)D$$

Because the sine and cosine are independent functions, α, β, γ, and δ must all be zero. Because the roots are $s = r \pm qi$, we can show that

$$mr^2 - mq^2 + cr + k = 0 \quad \text{and} \quad 2mr + c = 0$$

Thus $\alpha = \beta = 0$ for any A and B (the values of A and B depend on the initial conditions).

Setting the expressions for γ and δ equal to zero gives two equations for the unknowns C and D. Their solution is

$$C = \frac{(k - m\omega^2)F_o}{(k - m\omega^2)^2 + (c\omega)^2} \tag{4.1-7}$$

$$D = -\frac{c\omega F_o}{(k - m\omega^2)^2 + (c\omega)^2} \tag{4.8-8}$$

The values of A and B can be found from the initial conditions. To obtain only the forced response, we set $x(0) = \dot{x}(0) = 0$. Thus, from Equations 4.1-4 and 4.1-5,

$$x(0) = B + D = 0$$

$$\dot{x}(0) = qA + Br + C\omega = 0$$

These have the solution

$$A = -\frac{Br + C\omega}{q}$$

$$= -\frac{\omega F_o}{q}\frac{rc + k - m\omega^2}{(k - m\omega^2)^2 + (c\omega)^2}$$

$$B = -D = \frac{c\omega F_o}{(k - m\omega^2)^2 + (c\omega)^2} \tag{4.1-9}$$

TABLE 4.1-1   Sine Response of an Underdamped Model

Model: $m\ddot{x} + c\dot{x} + kx = F_o \sin \omega t$

Roots: $s = r \pm qi$

$$r = -\frac{c}{2m} \qquad q = \frac{\sqrt{4mk - c^2}}{2m}$$

$$x(t) = \frac{F_o}{(k - m\omega^2)^2 + (c\omega)^2}\left[-\frac{\omega}{q}\left(rc + k - m\omega^2\right)e^{rt}\sin qt + c\omega\, e^{rt}\cos qt\right]$$
$$+ \frac{F_o}{(k - m\omega^2)^2 + (c\omega)^2}\left[\left(k - m\omega^2\right)\sin \omega t - c\omega \cos \omega t\right]$$

Thus, for the sinusoidal forcing function $F_o \sin \omega t$, the forced response, which is the total response for *zero* initial conditions, is

$$x(t) = \frac{F_o}{(k - m\omega^2)^2 + (c\omega)^2}\left[-\frac{\omega}{q}\left(rc + k - m\omega^2\right)e^{rt}\sin qt + c\omega e^{rt}\cos qt\right]$$
$$+ \frac{F_o}{(k - m\omega^2)^2 + (c\omega)^2}\left[(k - m\omega^2)\sin \omega t - c\omega \cos \omega t\right] \tag{4.1-10}$$

This solution is summarized in Table 4.1-1.

If the initial conditions are not zero, then the total response is found by adding the free response (given in Table 3.5-1) to the forced response. For the complex roots $s = r \pm iq$, the free response is

$$x(t) = Be^{rt}\sin(qt + \psi) \qquad B = \frac{1}{q}\sqrt{(qx_0)^2 + (v_0 - rx_0)^2} \tag{4.1-11}$$

$$\sin \psi = \frac{x_0}{B} \qquad \cos \psi = \frac{v_0 - rx_0}{qB} \tag{4.1-12}$$

Steady-State Response

The *steady-state response* is that part of the response which does not disappear as time goes on. The *transient response* is that part of the response which disappears. Note that the forcing function influences the transient response as well as the steady-state response, in general. For Equation 4.1-10, the transient response consists of the terms containing e^{rt} if $r < 0$. The steady-state response consists of the $\sin \omega t$ and $\cos \omega t$ terms.

For the complex roots $s = r \pm iq$, if the real part r is negative, then the terms in Equation 4.1-10 containing the exponential term e^{rt} disappear, and the steady-state response is

$$x(t) = \frac{F_o}{(k - m\omega^2)^2 + (c\omega)^2}\left[(k - m\omega^2)\sin \omega t - c\omega \cos \omega t\right] \tag{4.1-13}$$

Note that this is also the steady-state response for the real root cases, as long as both roots are negative. This is because the total response (which is *not* given by Equation 4.1-10 for the real root cases) contains decaying exponentials if the roots are negative. After these exponential terms die out, the total response is given by Equation 4.1-13.

The expression in Equation 4.1-13 can be written as a single sine function as follows:

$$x(t) = X(\omega) \sin[\omega t + \phi(\omega)] \tag{4.1-14}$$

where

$$X(\omega) = \frac{F_o}{\sqrt{(k - m\omega^2)^2 + (c\omega)^2}} \tag{4.1-15}$$

$$\phi(\omega) = \tan^{-1} \frac{c\omega}{m\omega^2 - k} \tag{4.1-16}$$

The angle ϕ will be in the third quadrant if $m\omega^2 - k > 0$, in the fourth quadrant if $m\omega^2 - k < 0$, and $\phi = -90°$ if $m\omega^2 - k = 0$ and $c \neq 0$. The case where $c = 0$ will be treated in Section 4.2.

EXAMPLE 4.1-1 *Response of* *an Overdamped* *System*	The equation of motion of a certain mass-spring-damper system is $$10\ddot{x} + 70\dot{x} + 100x = f(t)$$ Obtain its steady-state response when $f(t) = 307 \sin 4t$.
Solution	The characteristic roots are $s = -2$ and $s = -5$. Thus the transient response will contain the terms e^{-2t} and e^{-5t} and thus will disappear in time. From Equation 4.1-16, we note that $m\omega^2 - k = 10(16) - 100 = 60 > 0$, and thus ϕ is in the third quadrant: $$\phi = \tan^{-1} \frac{70(4)}{60} = 1.36 + \pi = 4.5 \text{ rad}$$ From Equations 4.1-14 through 4.1-16 with $F_o = 307$, $$x(t) = 1.072 \sin(4t + 4.5)$$ Note that the steady-state response is oscillatory even though the system is over-damped. ∎

EXAMPLE 4.1-2 *Response of an* *Underdamped* *System*	Obtain the forced response of the following model for two cases: **(a)** $\omega = 1$ and **(b)** $\omega = 2$. $$\ddot{x} + \dot{x} + x = \sin \omega t$$ Time is measured in seconds.
Solution	**(a)** Here $m = c = k = F_o = 1$. For $\omega = 1$, the formulas in Table 4.1-1 give $$x(t) = 0.5774 e^{-0.5t} \sin 0.866t + e^{-0.5t} \cos 0.866t - \cos t$$ The steady-state response is thus $x(t) = -\cos t$. From Equations 4.1-15 and 4.1-16, we obtain $X = 1$ and $\phi = -\pi/2$. This gives $x(t) = \sin(t - \pi/2) = -\cos t$, so the results are identical, as they should be. The roots are $s = -0.5 \pm 0.866i$, and the time constant is $\tau = 2$. Thus the transient response is essentially zero after $t = 4\tau = 8$ sec. The top graph in Figure 4.1-1 shows the response for $\omega = 1$. At steady state the response is a constant-amplitude harmonic with a radian frequency of 1 and an amplitude of 1, which is equal to the amplitude of the forcing function. From the graph, we measure

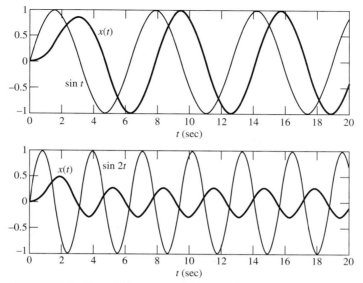

FIGURE 4.1-1 Sinusoidal response of the model $\ddot{x} + \dot{x} + x = \sin \omega t$ for $\omega = 1$ and $\omega = 2$.

that the peak in the steady-state response occurs 1.57 sec after the peak in the forcing function. We could have computed this delay from $|\phi|/\omega = \pi/2 = 1.57$ sec.

(b) For $\omega = 2$, we have

$$x(t) = 0.0769 \left(8.0829 e^{-0.5t} \sin 0.866t + 2e^{-0.5t} \cos 0.866t - 3 \sin 2t - 2 \cos 2t \right)$$

The steady-state response is thus $x(t) = 0.0769(-3 \sin 2t - 2 \cos 2t) = -0.231 \sin 2t - 0.154 \cos 2t$. From Equations 4.1-15 and 4.1-16, we obtain $X = 0.2774$ and $\phi = -2.554$. This gives $x(t) = 0.2774 \sin(2t - 2.554) = -0.231 \sin 2t - 0.154 \cos 2t$, so the results are identical, as they should be.

The bottom graph shows the response for $\omega = 2$. At steady state the response is a constant-amplitude harmonic with a radian frequency of 2 and a measured amplitude of approximately 0.28, which is 28% of the amplitude of the forcing function. From the graph, we estimate that the peak in the steady-state response occurs 1.28 sec after the peak in the forcing function. Computing this delay from $|\phi|/\omega$ gives $2.554/2 = 1.28$ sec to three significant figures. ∎

Magnitude Ratio and Frequency Transfer Function

The *magnitude ratio* is defined to be the magnitude of the steady-state response divided by the magnitude F_o of the forcing function. (Because the magnitude of a sine function is its amplitude, the term *amplitude ratio* is sometimes used instead of *magnitude ratio*.) The magnitude ratio represents the effect of the forcing function on the amplitude of the steady-state portion of the forced response.

We will use the symbol $M(\omega)$ to represent the magnitude ratio. Thus for the model $m\ddot{x} + c\dot{x} + kx = F_o \sin \omega t$, from Equation 4.1-15 the magnitude ratio is seen to be

$$M(\omega) = \frac{X(\omega)}{F_o} = \frac{1}{\sqrt{(k - m\omega^2)^2 + (c\omega)^2}} \qquad (4.1\text{-}17)$$

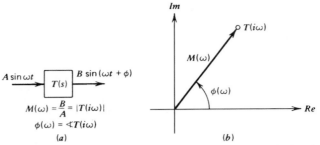

FIGURE 4.1-2   Vector representation of the frequency transfer function $T(i\omega)$.

The magnitude ratio is not dimensionless but will have the dimensions of the response variable x divided by the dimensions of the forcing function amplitude F_o. For example, if the input is a force in newtons and the output is a displacement in meters, the dimensions of $M(\omega)$ will be m/N.

Recalling that we can represent a complex number in terms of its absolute value and its phase angle, we see that the magnitude ratio $M(\omega)$ and the angle $\phi(\omega)$ can be thought of as specifying the complex number $M(\omega)e^{i\phi(\omega)}$, which we will denote by $T(i\omega)$:

$$T(i\omega) = M(\omega)e^{i\phi(\omega)} \tag{4.1-18}$$

This complex number is sometimes called the *frequency transfer function* because it expresses how the effect of the input is transferred to the output. It can be represented as a vector of length $M(\omega)$ and angle $\phi(\omega)$, as shown in Figure 4.1-2.

The Transfer Function

Given the amplitude and frequency of a sinusoidal or cosinusoidal forcing function, we can use the frequency transfer function to determine the amplitude and phase shift of the steady-state forced response. For other types of forcing functions, however, we will need a generalization of the frequency transfer function. This generalized function is called simply the *transfer function*. Some software programs, such as MATLAB, do not recognize the frequency transfer function. Instead, they are designed to work with the transfer function.

The transfer function is formally derived from the application of the Laplace transform to the solution of differential equations. We will see this application in Chapter 5. For now, however, we will show how easy it is to obtain the transfer function from a given differential equation. This will enable us to use MATLAB to analyze harmonic response later in this chapter.

Consider the model we have been using in this section:

$$m\ddot{x} + c\dot{x} + kx = f(t) \tag{4.1-19}$$

To obtain the transfer function of this model, simply replace $f(t)$ with $F(s)e^{st}$ and $x(t)$ with $X(s)e^{st}$. We use the notation $X(s)$ to distinguish it from $X(\omega)$. Since $X(s)$ is not a function of t, then $\dot{x}$ becomes $sX(s)e^{st}$ and $\ddot{x}$ becomes $s^2X(s)e^{st}$. So Equation 4.1-19 gives

$$ms^2X(s)e^{st} + csX(s)e^{st} + kX(s)e^{st} = F(s)e^{st}$$

Factor out e^{st} and solve for the ratio:

$$\frac{X(s)}{F(s)} = \frac{1}{ms^2 + cs + k} \tag{4.1-20}$$

The term $1/(ms^2 + cs + k)$ is the transfer function of Equation 4.1-19. In some sense it expresses the ratio of the response $x(t)$ to the forcing function $f(t)$. By analogy to the frequency transfer function, we define the transfer function $T(s)$ for Equation 4.1-19 as

$$T(s) = \frac{X(s)}{F(s)} = \frac{1}{ms^2 + cs + k} \qquad (4.1\text{-}21)$$

In Chapter 5 we will use this transfer function to obtain the response to other types of forcing functions.

As an example of the usefulness of the transfer function, consider the specific model $5\ddot{x} + 3\dot{x} + 9x = f(t)$, which has the transfer function $1/(5s^2 + 3s + 9)$. This consists of the ratio of two polynomials, 1 and $5s^2 + 3s + 9$, which are represented in MATLAB as the arrays 1 and [5, 3, 9]. We will see that one way of specifying a differential equation model in MATLAB is to enter its transfer function as the ratio of the two polynomials, as follows:

```
≫model_1 = tf(1,[5,3,9]);
```

The function `tf` creates a transfer function model, here named `model_1`, from the numerator and denominator polynomials of the transfer function. This model can be used to solve the differential equation and to obtain the magnitude ratio and phase angle, for example. This is discussed in Section 4.9.

Relation between the Two Transfer Functions

We have made a distinction between the *frequency* transfer function $T(i\omega)$ and the general transfer function $T(s)$. As you may expect, they are related.

When we make the substitution $s = i\omega$ into Equation 4.1-21, we obtain

$$T(s)|_{s=i\omega} = T(i\omega) = \frac{1}{k - m\omega^2 + c\omega i}$$

The absolute value of this complex number is

$$|T(i\omega)| = \frac{1}{\sqrt{(k - m\omega^2)^2 + (c\omega)^2}}$$

which is identical to $M(\omega)$, given by Equation 4.1-17. Its angle is

$$\phi(\omega) = \angle T(i\omega) = \angle 1 - \angle(k - m\omega^2 + c\omega i)$$
$$= 0 - \tan^{-1}\frac{c\omega}{k - m\omega^2} = \tan^{-1}\frac{c\omega}{m\omega^2 - k}$$

which is identical to ϕ, given by Equation 4.1-16. Thus the frequency transfer function can be obtained from the general transfer function $T(s)$ by replacing s with $i\omega$. So the frequency transfer function is a special case of the general transfer function for sinusoidal and cosinusoidal forcing functions.

Although we have demonstrated this relation only for the model given by Equation 4.1-1, in general, we can show that for any *stable*, constant-coefficient, *linear* differential equation, the frequency transfer function $T(i\omega)$ can be obtained by replacing s with $i\omega$ in the general transfer function $T(s)$. We will use this technique throughout this chapter.

In summary:

- The steady-state response of such a system to a sinusoidal forcing function $A(\omega) \sin \omega t$ has the form $B(\omega) \sin(\omega t + \phi)$.
- The frequency transfer function $T(i\omega)$ represents the effect of a harmonic forcing function on the steady-state portion of the forced response. It does not give information about the transient part of the forced response.
- The magnitude $M(\omega)$ of the frequency transfer function $T(i\omega)$ is the ratio $B(\omega)/A(\omega)$ of the harmonic steady-state response amplitude to the harmonic forcing function amplitude. Thus

$$M(\omega) = \frac{B(\omega)}{A(\omega)} = |T(i\omega)| \qquad (4.1\text{-}22)$$

- The phase shift of the response relative to the forcing function is $\phi(\omega)$, the angle of $T(i\omega)$. Thus

$$\phi(\omega) = \angle T(i\omega) \qquad (4.1\text{-}23)$$

Use of Dimensionless Parameters

We may express the results so far in terms of two dimensionless parameters, the damping ratio ζ and frequency ratio $r = \omega/\omega_n$, as follows. Recall that

$$\zeta = \frac{c}{2\sqrt{mk}}$$

$$\omega_n = \frac{\sqrt{k}}{m}$$

and

$$\frac{2\zeta}{\omega_n} = \frac{c}{k}$$

$$r = \frac{\omega}{\omega_n} \qquad (4.1\text{-}24)$$

Consider the amplitude given by Equation 4.1-15. Divide its numerator and denominator by k to obtain

$$X(\omega) = \frac{F_o}{k} \frac{1}{\sqrt{(1 - m\omega^2/k)^2 + (c\omega/k)^2}}$$

$$= \frac{F_o}{k} \frac{1}{\sqrt{\left(1 - \omega^2/\omega_n^2\right)^2 + (2\zeta\omega/\omega_n)^2}} = \frac{F_o}{k} \frac{1}{\sqrt{(1 - r^2)^2 + (2\zeta r)^2}}$$

The phase angle expression, Equation 4.1-16, becomes

$$\phi(\omega) = \tan^{-1} \frac{c\omega/k}{m\omega^2/k - 1} = \tan^{-1} \frac{2\zeta\omega/\omega_n}{\omega^2/\omega_n^2 - 1}$$

$$= \tan^{-1} \frac{2\zeta r}{r^2 - 1} \qquad (4.1\text{-}25)$$

where ϕ will be in the third quadrant if $r^2 - 1 > 0$, in the fourth quadrant if $r^2 - 1 < 0$, and $\phi = -90°$ if $r^2 = 1$ and $c \neq 0$. The case where $c = 0$ will be treated later.

Thus the steady-state sine response may be written as

$$x_{ss}(t) = \frac{F_o}{k} \frac{1}{\sqrt{(1 - r^2)^2 + (2\zeta r)^2}} \sin(\omega t + \phi) \tag{4.1-26}$$

and the steady-state magnitude ratio M for the equation $m\ddot{x} + c\dot{x} + kx = f(t)$ is

$$M = \frac{|x_{ss}(t)|}{F_o} = \frac{1}{k} \frac{1}{\sqrt{(1 - r^2)^2 + (2\zeta r)^2}} \tag{4.1-27}$$

For a low forcing frequency ($r \approx 0$), we obtain $M \approx 1/k$, which means that the response amplitude is close to the steady-state displacement $x_{ss} = F_o/k$ caused by a *constant* force F_o. This displacement is called the *static deflection* and is denoted δ_{st}:

$$\delta_{st} = \frac{F_o}{k} \tag{4.1-28}$$

Thus we may express kM as X/δ_{st}, and Equation 4.1-27 becomes

$$\frac{X}{\delta_{st}} = kM = \frac{1}{\sqrt{(1 - r^2)^2 + (2\zeta r)^2}} \tag{4.1-29}$$

when $\omega = \omega_n$, $r = 1$, we obtain

$$\left(\frac{X}{\delta_{st}} \right)_{\omega = \omega_n} = \frac{1}{2\zeta} \tag{4.1-30}$$

Figure 4.1-3 shows plots of X/δ_{st} and ϕ versus the frequency ratio $r = \omega/\omega_n$, for several values of the damping ratio ζ. At high forcing frequencies the response amplitude approaches zero because the system's inertia prevents it from following a rapidly varying forcing function. This effect is due primarily to the system's inertia — not its damping, as one might think. This statement can be proved by noting that the X/δ_{st} curve approaches zero for high frequencies for any non-negative value of ζ.

The phase shift ϕ is exactly $-90°$ when $r = 1$, and thus the velocity is in phase with the forcing function. This condition (when $\omega = \omega_n$) is defined as *resonance*. At high frequencies the phase angle ϕ is almost $-180°$, and the response is a negative sine. For $0 < r < 1$, the plot shows that $0 < \phi < -90°$, and thus the response lags behind the forcing function. For $r > 1$, $-90° < \phi < -180°$, and thus the forcing function lags behind the response.

The maximum value of X/δ_{st} does not occur when $r = 1$ unless $c = 0$. It occurs when the denominator of Equation 4.1-29 has a minimum. Setting the derivative of the denominator with respect to r equal to zero shows that the maximum of X/δ_{st} occurs at $r = \sqrt{1 - 2\zeta^2}$, which corresponds to $\omega = \omega_n \sqrt{1 - 2\zeta^2}$. This frequency is called the *peak frequency* ω_p. The peak of X/δ_{st} exists only when the term under the radical is positive; that is, when $\zeta \leq 1/\sqrt{2}$. Thus,

$$\omega_p = \omega_n \sqrt{1 - 2\zeta^2} \qquad 0 \leq \zeta \leq \frac{1}{\sqrt{2}} \tag{4.1-31}$$

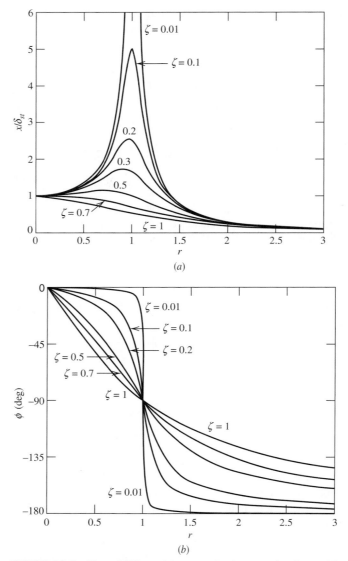

FIGURE 4.1-3 Plots of X/δ_{st} and ϕ versus r for the second-order model $m\ddot{x} + c\dot{x} + kx = F_0 \sin \omega t$.

The value of the peak of X/δ_{st} is found by substituting ω_p into Equation (4.1-29). This gives

$$\left.\frac{X}{\delta_{st}}\right|_p = \frac{1}{2\zeta\sqrt{1-\zeta^2}} \qquad 0 \le \zeta \le \frac{1}{\sqrt{2}} \tag{4.1-32}$$

Note that this peak is larger than the value at resonance, given by Equation (4.1-30). So the peak response is given by

$$X_p = \delta_{st}\frac{1}{2\zeta\sqrt{1-\zeta^2}} \qquad 0 \le \zeta \le \frac{1}{\sqrt{2}} \tag{4.1-33}$$

If ζ is small, the two values are close. If $\zeta > 1/\sqrt{2}$, no peak exists, and the maximum value of X/δ_{st} occurs at $\omega = 0$ and is $X/\delta_{st} = 1$. Note that as $\zeta \to 0$, $\omega_p \to \omega_n$ and $X/\delta_{st} \to \infty$. For an undamped system, the peak frequency is the natural frequency ω_n.

TABLE 4.1-2   Steady-State Sine Response of a Second-Order Model

Model: $m\ddot{x} + c\dot{x} + kx = F_o \sin \omega t$

$$\zeta = \frac{c}{2\sqrt{mk}} \qquad \omega_n = \sqrt{\frac{k}{m}} \qquad r = \frac{\omega}{\omega_n}$$

Steady-state response:

$$x_{ss}(t) = \frac{F_o}{k} \frac{1}{\sqrt{(1-r^2)^2 + (2\zeta r)^2}} \sin(\omega t + \phi)$$

$$\phi = -\tan^{-1}\frac{2\zeta r}{1-r^2}$$

Third quadrant if $1 - r^2 < 0$
Fourth quadrant if $1 - r^2 > 0$
$\phi = -90°$ if $r^2 = 1$

Magnitude ratio: $M = \dfrac{|x_{ss}|}{F_o} = \dfrac{1}{\sqrt{(k-m\omega^2)^2 + (c\omega)^2}} = \dfrac{1}{k}\dfrac{1}{\sqrt{(1-r^2)^2 + (2\zeta r)^2}}$

Dimensionless magnitude ratio: $\dfrac{X}{\delta_{st}} = kM \qquad \delta_{st} = \dfrac{F_o}{k}$

Peak frequency: $\omega_p = \omega_n\sqrt{1-2\zeta^2} \qquad 0 \le \zeta \le \dfrac{1}{\sqrt{2}}$

Peak response: $X_p = \delta_{st}\dfrac{1}{2\zeta\sqrt{1-\zeta^2}} \qquad 0 \le \zeta \le \dfrac{1}{\sqrt{2}}$

These results are summarized in Table 4.1-2.

We may use Equation 4.1-33 with measurements to estimate the damping value. Knowing δ_{st}, varying the forcing frequency, and measuring the peak response amplitude X_p, we can compute ζ from Equation 4.1-33.

Logarithmic Plots

Some instrumentation and some software packages, such as MATLAB, give plots of the magnitude ratio and the phase angle as logarithmic plots with log ω as the independent variable. Instead of plotting the magnitude ratio M, the logarithmic unit known as the *decibel* is often used. It is abbreviated dB and is defined as

$$m = 10 \log M^2 = 20 \log M \quad \text{dB} \tag{4.1-34}$$

where the logarithm is to the base 10. For example, $M = 10$ corresponds to 20 dB and $M = 1$ corresponds to 0 dB. Values of M less than 1 have negative decibel values; for example, $M = 0.1$ corresponds to -20 dB and $M = 1/\sqrt{2}$ corresponds to $m = -3.01$ dB. Note that

$$M = 10^{m/20} \tag{4.1-35}$$

An advantage of using logarithmic plots is that for some models the range of values of M and ω can be quite large.

Consider the first-order model $m\ddot{x} + c\dot{x} + kx = F_o \sin \omega t$, for which

$$\frac{X}{F_o} = \frac{1}{k} \frac{1}{\sqrt{(1-r^2)^2 + (2\zeta r)^2}}$$

If we define the dimensionless ratio $M = kX/F_o$, then in decibel units

$$m = 20 \log \frac{1}{\sqrt{(1-r^2)^2 + (2\zeta r)^2}} \qquad (4.1\text{-}36)$$

When plotting $m(\omega)$, it is customary to use a logarithmic scale for the frequency axis. Figure 4.1-4 shows the plots for m and ϕ versus $\log \omega/\omega_n = \log r$. Note that $M \to 0$ as $\omega \to \infty$, but $m \to -\infty$. When plotted in decibel units, peaks appear less pronounced than on plots of M.

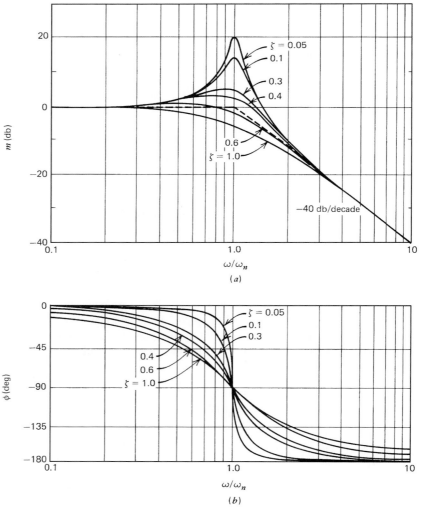

FIGURE 4.1-4   Logarithmic frequency response plots for the underdamped model $T(s) = \omega_n^2/(s^2 + 2\zeta\omega_n s + \omega_n^2)$.

4.2   RESONANCE AND BANDWIDTH

While all real systems will have some damping, it is instructive and useful to obtain some results for the undamped case, where $c = \zeta = 0$. This is because the mathematical results for the undamped case are more easily derived and analyzed, and they give insight into the behavior of many real systems that have a small amount of damping.

We noted that if the model is stable, the free response term disappears in time. The results derived in Section 4.1 therefore must be reexamined for the case where there is no damping because this is not a stable case (it is neutrally stable). In this case the magnitude and phase angle of the transfer function do not give the entire steady-state response. The free response for the undamped equation

$$m\ddot{x} + kx = F_o \sin \omega t$$

has the form $A_1 \sin \omega_n t + A_2 \cos \omega_n t$ and thus the trial solution form

$$x(t) = A_1 \sin \omega_n t + A_2 \cos \omega_n t + B_1 \cos \omega t + B_2 \sin \omega t$$

Substituting this into the differential equation, we obtain $B_1 = 0$ and

$$B_2 = \frac{F_o}{k - m\omega^2} = \frac{F_o}{k} \frac{1}{1 - \omega^2/\omega_n^2} = \frac{F_o}{k} \frac{1}{1 - r^2}$$

Thus the form of the total response is

$$x(t) = A_1 \sin \omega_n t + A_2 \cos \omega_n t + \frac{F_o}{k} \frac{1}{1 - r^2} \sin \omega t$$

Evaluating this solution for the initial conditions gives

$$A_1 = \frac{\dot{x}(0)}{\omega_n} - \frac{F_o}{k} \frac{r}{1 - r^2} \qquad A_2 = x(0)$$

Thus the total response is

$$x(t) = \left(\frac{\dot{x}(0)}{\omega_n} - \frac{F_o}{k} \frac{r}{1 - r^2} \right) \sin \omega_n t + x(0) \cos \omega_n t + \frac{F_o}{k} \frac{1}{1 - r^2} \sin \omega t \qquad (4.2\text{-}1)$$

There is no transient response here; the entire solution is the steady-state response. We may examine the effects of the forcing function independently of the effects of the initial conditions by setting $x(0) = \dot{x}(0) = 0$. The result is the forced response:

$$x(t) = \frac{F_o}{k} \frac{1}{1 - r^2} (\sin \omega t - r \sin \omega_n t) \qquad (4.2\text{-}2)$$

When the forcing frequency ω is substantially different from the natural frequency ω_n, the forced response looks somewhat like Figure 4.2-1 and consists of a higher-frequency oscillation superimposed on a lower-frequency oscillation.

Beating

If the forcing frequency ω is close to the natural frequency ω_n, then $r \approx 1$ and the forced response, Equation 4.2-2, may be expressed as follows:

$$x(t) = \frac{F_o}{k} \frac{1}{1 - r^2} (\sin \omega t - 1 \sin \omega_n t) = \frac{F_o}{m} \frac{1}{\omega_n^2 - \omega^2} (\sin \omega t - \sin \omega_n t)$$

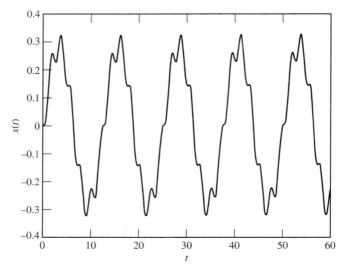

FIGURE 4.2-1   Forced response for the case where the forcing frequency ω differs greatly from the natural frequency ω_n.

Using the sine and cosine identities for the sum and difference of two angles, and the identity $\sin 2\theta = 2\sin\theta\,\cos\theta$, it is straightforward to show that

$$2\sin\frac{\omega - \omega_n}{2}t\,\cos\frac{\omega + \omega_n}{2}t = \sin\frac{\omega}{2}t\,\cos\frac{\omega}{2}t - \cos\frac{\omega_n}{2}t\,\sin\frac{\omega_n}{2}t = \sin\omega t - \sin\omega_n t$$

and thus the forced response is given approximately by

$$x(t) \approx \left(\frac{2F_o}{m}\frac{1}{\omega_n^2 - \omega^2}\sin\frac{\omega - \omega_n}{2}t\right)\cos\frac{\omega + \omega_n}{2}t \qquad (4.2\text{-}3)$$

This may be interpreted as a cosine with a frequency $(\omega + \omega_n)/2$ and a time-varying amplitude of

$$\frac{2F_o}{m}\frac{1}{\omega_n^2 - \omega^2}\sin\frac{\omega - \omega_n}{2}t$$

The amplitude varies sinusoidally with the frequency $(\omega - \omega_n)/2$, which is lower than the frequency of the cosine. This response thus looks like Figure 4.2-2. This behavior, in which the amplitude increases and decreases periodically, is called *beating*. The *beat period* is the time between the occurrence of zeros in $x(t)$ and thus is given by the half-period of the sine wave, which is $2\pi/|\omega - \omega_n|$. The *vibration period* is the period of the cosine wave, $4\pi/(\omega + \omega_n)$.

$$\text{Beat Period } = \frac{2\pi}{|\omega - \omega_n|} \qquad (4.2\text{-}4)$$

$$\text{Vibration Period } = \frac{4\pi}{|\omega + \omega_n|} \qquad (4.2\text{-}5)$$

Resonance

Equation 4.2-2 shows that when $\zeta = 0$, the amplitude of the response becomes infinite when $r = 1$; that is, when the forcing frequency ω equals the natural frequency ω_n.

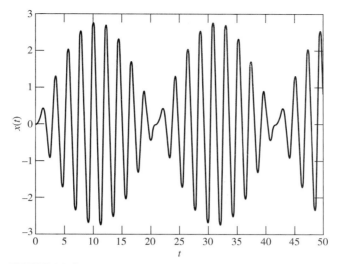

FIGURE 4.2-2 Example of beating, when the forcing frequency ω is close to the natural frequency ω_n.

This phenomenon is called *resonance*, and the frequency at which this occurs is called the *resonant* frequency. The phase shift ϕ is exactly $-90°$ at this frequency. When a system is "resonating," the velocity is in phase with the forcing function. This causes the response $x(t)$ to increase indefinitely if there is no damping.

  To obtain the expression for $x(t)$ at resonance, we compute the limit of $x(t)$ as $r \to 1$ after replacing ω in $x(t)$ using the relation $\omega = \omega_n r$. We must use L'Hôpital's rule to compute the limit:

$$
\begin{aligned}
x(t) &= \lim_{r \to 1} \frac{F_o}{k} \frac{1}{1 - r^2} (\sin \omega_n rt - r \sin \omega_n t) \\[2mm]
&= \frac{F_o}{k} \lim_{r \to 1} \frac{\dfrac{d}{dr}(\sin \omega_n rt - r \sin \omega_n t)}{\dfrac{d}{dr}(1 - r^2)} \\[2mm]
&= \frac{F_o}{k} \lim_{r \to 1} \frac{\omega_n t \cos \omega_n rt - \sin \omega_n t}{1 - 2r} \\[2mm]
&= \frac{F_o \omega_n}{2k} \left(\frac{\sin \omega_n t}{\omega_n} - t \cos \omega_n t \right)
\end{aligned}
\tag{4.2-6}
$$

The plot is shown in Figure 4.2-3 for the case $m = 4$, $c = 0$, $k = 36$, and $F_o = 10$. The amplitude increases linearly with time. The behavior for damped systems is similar, except that the amplitude does not become infinite. Figure 4.2-4 shows the damped case: $m = 4$, $c = 4$, $k = 36$, and $F_o = 10$.

  For large amplitudes, the linear model on which this analysis is based will no longer be accurate. In addition, all physical systems have some damping, so ζ will never be exactly zero, and the response amplitude will never be infinite. The important point, however, is that the amplitude might be large enough to damage the system or cause some other undesirable result. At resonance the output amplitude will continue to increase until either the linear model is no longer accurate or the system fails. Designers of structural systems and suspensions try to avoid resonance because of the damage or discomfort that large motions can produce.

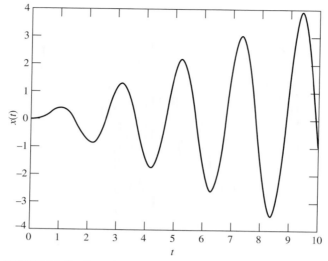

FIGURE 4.2-3 Response of an undamped system at resonance.

Resonance and Startup

Although the peak frequency expression, Equation 4.1-31, was derived from steady-state response formulas, resonance can still occur in transient processes if the input varies slowly enough to allow the steady-state response to begin to appear. For example, resonance can be a problem even if a machine's operating speed is well above its resonant frequency, because the machine's speed must pass through the resonant frequency at startup. If the machine's speed does not pass through the resonant frequency quickly enough, high-amplitude oscillations will result. Figure 4.2-5 shows the transient response of the model

$$2\ddot{x} + \dot{x} + 50x = 150 \sin \left[\omega(t)t\right]$$

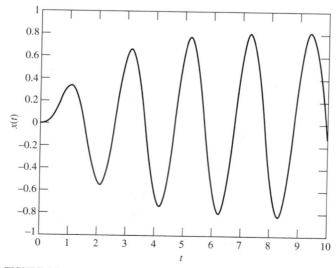

FIGURE 4.2-4 Response of a damped system at resonance.

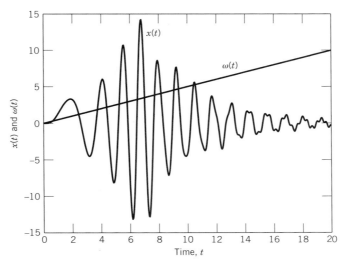

FIGURE 4.2-5 Transient response to an input having an increasing frequency.

where the frequency $\omega(t)$ increases linearly with time from 0 to 10, as $\omega(t) = 0.5t$. Thus the frequency passes through the peak frequency of this model, which is $\omega_p = 4.99$. The plot shows the large oscillations that occur when the input frequency is close to the resonant frequency.

Bandwidth

It is useful to have a specific measure of the range of forcing frequencies to which a system is especially responsive. This single measure may be used as a design specification. The most common such measure is the *bandwidth*. It is defined as the range of frequencies over which the power or energy transmitted or dissipated by the system is no less than one-half of the peak power. As we have seen, the energy dissipated by a damper is proportional to the square of amplitude of its velocity difference. For many systems the power transmitted or dissipated is proportional to M^2, where M is the magnitude of the frequency transfer function. Then the bandwidth is that range of frequencies (ω_1, ω_2) over which

$$M^2(\omega_1) \leq \frac{M_p^2}{2} \geq M^2(\omega_2)$$

where M_p is the peak value of M. This gives

$$M(\omega_1) \leq \frac{M_p}{\sqrt{2}} \geq M(\omega_2) \tag{4.2-7}$$

For this reason, the lower and upper bandwidth points are called the half-power points. For the second-order model $m\ddot{x} + c\dot{x} + kx = F_o \sin \omega t$,

$$M = \frac{X}{F_o} = \frac{1}{k}\frac{1}{\sqrt{(1 - r^2)^2 + (2\zeta r)^2}} \tag{4.2-8}$$

The value of r that makes $M = M_p/\sqrt{2}$ is found from

$$\frac{1}{k}\frac{1}{\sqrt{(1 - r^2)^2 + (2\zeta r)^2}} = \frac{1}{\sqrt{2}}\left(\frac{1}{k}\frac{1}{2\zeta\sqrt{1 - \zeta^2}}\right)$$

This can be solved for r by squaring both sides and rearranging to obtain

$$r^4 + (4\zeta^2 - 2)r^2 + 1 - 8\zeta^2 + 8\zeta^4 = 0$$

The solution is

$$r = \sqrt{1 - 2\zeta^2 \pm 2\zeta\sqrt{1 - \zeta^2}} \tag{4.2-9}$$

If two positive, real solutions of this equation exist, they are r_1 and r_2, from which we obtain the lower and upper bandwidth frequencies ω_1 and ω_2. For low to moderate damping, the bandwidth is approximately $(0, \omega_n)$.

Figure 4.2-6 shows several cases that can occur with various other models. In Figure 4.2-6a, the peak M_p is large enough so that $M_p/\sqrt{2} > M(0)$, and thus the lower bandwidth frequency ω_1 exists and is greater than zero. Such a system is called a *band-pass* system. Figure 4.2-6b shows the plot of a *low-pass* system, for which $M(0) > M_p/\sqrt{2}$, so that $\omega_1 = 0$. This occurs for the magnitude ratio, Equation 4.2-8, when $\zeta > 0.382$. Figure 4.2-6c shows a case where the upper bandwidth frequency is infinite. Such a system is called a *high-pass* system because it responds more to high-frequency forcing functions. We will see an examples of such a system later on. Figure 4.2-6d shows a magnitude ratio plot with two peaks. This can occur only with a model of fourth-order or higher, as we will see later. Depending on the exact shape of the plot, such a system can have two bandwidths, one for each peak.

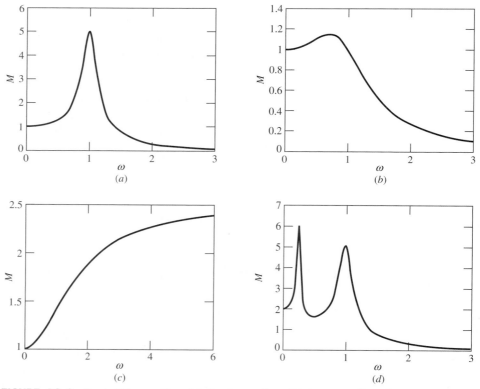

FIGURE 4.2-6   Bandwidth examples. (*a*) Band-pass filter. (*b*) Low-pass filter. (*c*) High-pass filter. (*d*) System having two bandwidth regions.

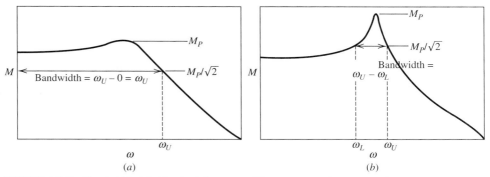

FIGURE 4.2-7 Bandwidth definition for (*a*) one cutoff frequency and (*b*) two cutoff frequencies.

In our definition of bandwidth, the power transmitted by a forcing function having a frequency outside the bandwidth is less than one-half the power transmitted by a forcing function whose frequency corresponds to M_p, *assuming* that both forcing functions have the *same* amplitude. However, often the amplitudes of low-frequency forcing functions are larger than those of high-frequency forcing functions, so the low-frequency forcing functions may account for more power. Thus forcing functions whose frequencies lie below the lower bandwidth frequency ω_1 may contribute significantly to the response and cannot be neglected. Because of this, a modified definition of bandwidth is sometimes used.

With this alternative definition of bandwidth, the lower bandwidth frequency ω_1 is always zero, and the upper bandwidth frequency ω_2 is defined to be that frequency at which the power is one-half of the power at *zero* frequency (Figure 4.2-7).[1] The two definitions give the same bandwidth for those systems whose peak value of M is $M(0)$. For Equation 4.2-8, note that $M(0) = 1/k$. Using Equation 4.2-8, we see that the value of r_2 for this alternative bandwidth definition can be found from

$$\frac{1}{k}\frac{1}{\sqrt{\left(1 - r_2^2\right)^2 + (2\zeta r_2)^2}} = \frac{1}{\sqrt{2}}\frac{1}{k}$$

This can be solved for r by squaring both sides and rearranging to obtain

$$r_2^4 + (4\zeta^2 - 2)r_2^2 - 1 = 0$$

The solution is

$$r_2 = \sqrt{1 - 2\zeta^2 + \sqrt{4\zeta^4 - 4\zeta^2 + 2}} \qquad (4.2\text{-}10)$$

Note that these results apply only to the model form $m\ddot{x} + c\dot{x} + kx = f(t)$. However, the definitions of bandwidth apply to any model form.

Instrument Design

When no *s* term occurs in the numerator of a transfer function, its magnitude ratio is small at high frequencies. However, the presence of an *s* term in the numerator can produce a large magnitude ratio at high frequencies. This effect can be used to advantage in

---

[1] The MATLAB function `bandwidth` uses this alternative definition and thus can give meaningless results for systems having a plot like Figure 4.2-6c.

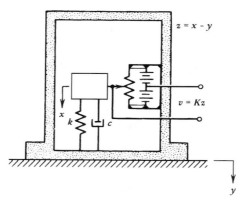

y **FIGURE 4.2-8** Vibration instrument.

instrument design, for example. The instrument shown in Figure 4.2-8 illustrates this point. With proper selection of the natural frequency of the device, it can be used either as a *vibrometer* to measure the amplitude of a sinusoidal displacement $y = A \sin \omega t$ or an *accelerometer* to measure the amplitude of the acceleration $\ddot{y} = -A \omega^2 \sin \omega t$. When used to measure ground motion from an earthquake, for example, the instrument is commonly referred to as a *seismograph*.

The mass displacement x and the support displacement y are defined relative to an inertial reference, with $x = 0$ corresponding to the equilibrium position of m when $y = 0$. With the potentiometer arrangement shown, the voltage v is proportional to the relative displacement z between the support and the mass m, where $z = x - y$. Newton's law gives

$$m\ddot{x} = -c(\dot{x} - \dot{y}) - k(x - y)$$

or

$$m\ddot{z} + c\dot{z} + kz = -m\ddot{y} \tag{4.2-11}$$

The transfer function between the forcing function y and the response z is obtained by substituting $y(t) = Y(s)e^{st}$ and $z(t) = Z(s)e^{st}$ and canceling the e^{st} terms. The result is

$$T(s) = \frac{Z(s)}{Y(s)} = \frac{-ms^2}{ms^2 + cs + k} = \frac{-\dfrac{s^2}{\omega_n^2}}{\dfrac{s^2}{\omega_n^2} + \dfrac{2\zeta s}{\omega_n} + 1} \tag{4.2-12}$$

The frequency transfer function is

$$T(i\omega) = \frac{\omega^2/\omega_n^2}{1 - \omega^2/\omega_n^2 + 2\zeta(\omega/\omega_n)i} = \frac{r^2}{1 - r^2 + 2\zeta ri}$$

The magnitude ratio is

$$M = \frac{Z}{Y} = \frac{r^2}{\sqrt{(1 - r^2)^2 + (2\zeta r)^2}}$$

The plot of M for several values of ζ is shown in Figure 4.2-9.

To make a vibrometer, this plot shows that the device's natural frequency ω_n must be selected so that $\omega \gg \omega_n$, where ω is the oscillation frequency of the displacement to be measured. For $\omega \gg \omega_n$, $r \gg 1$ and $M \approx 1$, and thus $|z| \approx |y| = A$ as desired. The voltage v is directly proportional to A in this case. The physical explanation for this result is the fact

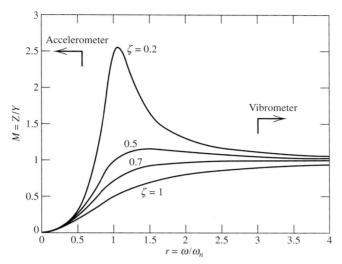

FIGURE 4.2-9 Magnitude ratio curve for the vibration instrument shown in Figure 4.2-8, for several values of ζ.

that the mass m cannot respond to high-frequency forcing function displacements. Its displacement x therefore remains fixed, and the motion z directly indicates the motion y. To design a specific vibrometer, we must know the lower bound of the forcing function displacement frequency ω. The frequency $\omega_n = \sqrt{k/m}$ is then made much smaller than this bound by selecting a large mass and a "soft" spring (small k). However, these choices are governed by constraints on the allowable deflections. For example, a very soft spring will have a large distance between the free length and the equilibrium positions.

An accelerometer can be obtained by using the lower end of the frequency range; that is, selecting ω_n to be very large. This gives

$$T(s) = \frac{Z(s)}{Y(s)} \approx -\frac{s^2}{\omega_n^2}$$

Thus

$$Z(s) \approx -\frac{1}{\omega_n^2} s^2 Y(s)$$

The term $s^2 Y(s)$ represents $\ddot{y}$, so the output of the accelerometer is

$$z \approx \frac{1}{\omega_n^2} |\ddot{y}| = \frac{\omega^2}{\omega_n^2} A$$

With ω_n chosen to be large (using a small mass and a "stiff" spring), the forcing function acceleration amplitude $\omega^2 A$ can be determined by measuring v, which gives z.

4.3 BASE EXCITATION

The most common sources of inputs to vibrating systems are motion of a base support (called base excitation or sometimes *seismic* excitation) and rotating unbalance. We treat base excitation in this section and rotating unbalance in the next section.

A common example of base excitation is caused by a vehicle's motion along a bumpy road surface. This motion produces a displacement input to the suspension system via the wheels. The primary purpose of a vehicle suspension is to maintain tire contact with the road surface, and the secondary purpose is to minimize the motion and force transmitted to the passenger compartment. The suspension is an example of a *vibration isolation system* designed isolate the source of vibration from other parts of the machine or structure. In addition to metal springs and hydraulic dampers, vibration isolators consisting of highly damped materials like rubber provide stiffness and damping between the source of vibration and the object to be protected. For example, the rubber motor mounts of an automobile engine are used to isolate the automobile's frame from the effects of the motor's rotating unbalance. Cork, felt, and pneumatic springs are also used as isolators.

Sometimes we want to reduce the effects of a force transmitted to the supporting structure, and sometimes we want to reduce the output displacement caused by an input displacement. Thus we speak of *force isolation* and *displacement isolation*. We begin our study of isolation system design by analyzing the displacement transmitted through base motion.

Displacement Transmissibility

The motion of the mass shown in Figure 4.3-1 is produced by the motion $y(t)$ of the base. This system is a model of many common displacement isolation systems. Assuming that the mass displacement x is measured from the rest position of the mass when $y = 0$, the weight mg is canceled by the static spring force. The force transmitted to the mass by the spring and damper is denoted f_t and is given by

$$f_t = c(\dot{y} - \dot{x}) + k(y - x) \tag{4.3-1}$$

This gives the following equation of motion:

$$m\ddot{x} = f_t = c(\dot{y} - \dot{x}) + k(y - x)$$

or

$$m\ddot{x} + c\dot{x} + kx = c\dot{y} + ky \tag{4.3-2}$$

We could use the substitution method to obtain the response. For example, if $y(t) = Y \sin \omega t$ and we wanted to obtain the steady-state response, we would use a trial solution of the form $x(t) = A \sin \omega t + B \cos \omega t$. The general transfer function, however, is more convenient to use for the following reasons:

1. Its use is not limited to a specific type of forcing function.

2. It enables us to manipulate the system's model to obtain expressions for several quantities, such as the magnitude ratio X/Y and the magnitude ratio F_t/Y.

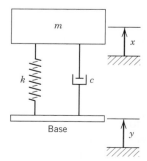

FIGURE 4.3-1 A system driven by base excitation.

3. It can be used with software such as MATLAB.

4. We can easily derive the expressions for the magnitude ratio and phase angle directly from the transfer function by replacing s with $i\omega$.

Substitute $x(t) = X(s)e^{st}$ and $y(t) = Y(s)e^{st}$ into Equation 4.3-2 and cancel the e^{st} terms to obtain the general transfer function between $x(t)$ as the output and $y(t)$ as the input:

$$T_{x/y}(s) = \frac{X(s)}{Y(s)} = \frac{cs + k}{ms^2 + cs + k} \tag{4.3-3}$$

where the subscript notation x/y is used to denote the ratio of the output to the input variables. This ratio can be used to analyze the effects of the base motion $y(t)$ on $x(t)$, the motion of the mass. Notice that this transfer function has an s term in the numerator, so its frequency response plots will be different from those of the transfer function $1/(ms^2 + cs + k)$.

We can express this transfer function in terms of just two parameters as follows. Divide the numerator and denominator by m and use the fact that $\omega_n = \sqrt{k/m}$ and $c/m = 2\zeta\omega_n$ to obtain

$$T_{x/y}(s) = \frac{X(s)}{Y(s)} = \frac{2\zeta\omega_n s + \omega_n^2}{s^2 + 2\zeta\omega_n s + \omega_n^2} \tag{4.3-4}$$

A common application of the transmissibility expressions is to analyze the effect of a sinusoidal input $y(t) = Y \sin \omega t$, having an amplitude Y and frequency ω. In this case we can derive the expression for the *frequency* transfer function by substituting $s = i\omega$ into $T_{x/y}(s)$. The result is

$$T_{x/y}(i\omega) = \frac{2\zeta\omega_n \omega i + \omega_n^2}{-\omega^2 + 2\zeta\omega_n \omega i + \omega_n^2}$$

Dividing the numerator and denominator by ω_n^2 gives

$$T_{x/y}(i\omega) = \frac{2\zeta\dfrac{\omega}{\omega_n}i + 1}{1 - \dfrac{\omega^2}{\omega_n^2} + 2\zeta\dfrac{\omega}{\omega_n}i}$$

Define the *frequency ratio r* to be

$$r = \frac{\omega}{\omega_n} \tag{4.3-5}$$

We can now express the frequency transfer function in terms of r as

$$T_{x/y}(i\omega) = \frac{X(i\omega)}{Y(i\omega)} = \frac{2\zeta ri + 1}{1 - r^2 + 2\zeta ri}$$

The magnitude and angle of this transfer function are

$$\left|T_{x/y}(i\omega)\right| = \frac{X}{Y} = \sqrt{\frac{4\zeta^2 r^2 + 1}{(1 - r^2)^2 + 4\zeta^2 r^2}} \tag{4.3-6}$$

$$\phi(\omega) = \left|T_{x/y}(i\omega)\right| = \angle(2\zeta ri + 1) - \angle(1 - r^2 + 2\zeta ri)$$

$$= \tan^{-1} 2\zeta r - \tan^{-1}\frac{2\zeta r}{1 - r^2}$$

where X denotes the amplitude of the steady-state response, $x_{ss}(t) = X \sin(\omega t + \phi)$, to the forcing function $y(t) = Y \sin \omega t$.

This result can be used to calculate the amplitude X of the steady-state motion caused by a sinusoidal input displacement of amplitude Y. The ratio X/Y is called the *displacement transmissibility*.

Force Transmissibility

Substituting $x(t) = X(s)e^{st}$, $y(t) = Y(s)e^{st}$, and $f_t(t) = F_t(s)e^{st}$ into Equation 4.3-1 and canceling the e^{st} terms gives

$$F_t(s) = (cs + k)[Y(s) - X(s)] \tag{4.3-7}$$

Substituting for $X(s)$ from Equation 4.3-3 gives

$$F_t(s) = (cs + k)\left[Y(s) - \frac{cs + k}{ms^2 + cs + k}Y(s)\right] = (cs + k)\frac{ms^2}{ms^2 + cs + k}Y(s)$$

The desired ratio is

$$T_{f_t/y} = \frac{F_t(s)}{Y(s)} = (cs + k)\frac{ms^2}{ms^2 + cs + k} \tag{4.3-8}$$

This is the transfer function for the transmitted force with the base motion as the input. It can be used with the Laplace transform (Chapter 5) or MATLAB to study the effects of any type of base motion $y(t)$, but for now we limit our discussion to sinusoidal inputs.

The transfer function can be used to compute the transmitted force $f_t(t)$ that results from a specified base motion $y(t)$. Multiply and divide Equation 4.3-8 by k/m.

$$T_{f_t/y}(s) = k\frac{\dfrac{c}{m}s + \dfrac{k}{m}}{k}\frac{ms^2}{s^2 + \dfrac{c}{m}s + \dfrac{k}{m}} = k\frac{2\zeta\omega_n s + \omega_n^2}{\omega_n^2}\frac{s^2}{s^2 + 2\zeta\omega_n s + \omega_n^2} \tag{4.3-9}$$

The frequency transfer function is found from Equation 4.3-9 with $s = i\omega$.

$$T_{f_t/y}(i\omega) = -k\frac{2\zeta\omega_n\omega i + \omega_n^2}{\omega_n^2}\frac{\omega^2}{-\omega^2 + 2\zeta\omega_n\omega i + \omega_n^2}$$

$$= -k\left(2\zeta\frac{\omega}{\omega_n}i + 1\right)\frac{\dfrac{\omega^2}{\omega_n^2}}{1 - \dfrac{\omega^2}{\omega_n^2} + 2\zeta\dfrac{\omega}{\omega_n}i}$$

In terms of r, this becomes

$$T_{f_t/y}(i\omega) = -k(2\zeta ri + 1)\frac{r^2}{1 - r^2 + 2\zeta ri} \tag{4.3-10}$$

It is customary to use instead the ratio $F_t(i\omega)/kY(i\omega)$, which is a dimensionless quantity. This gives

$$\left|\frac{F_t(i\omega)}{kY(i\omega)}\right| = \frac{|T_{f_t/y}(i\omega)|}{k} = \frac{F_t}{kY} = r^2\sqrt{\frac{4\zeta^2 r^2 + 1}{(1 - r^2)^2 + 4\zeta^2 r^2}} \tag{4.3-11}$$

The expression ratio F_t/kY is called the dimensionless *force transmissibility*. It can be used to calculate the steady-state amplitude F_t of the transmitted forced caused

TABLE 4.3-1 Transmissibility Formulas for Base Excitation

Model:

$$m\ddot{x} + c\dot{x} + kx = c\dot{y} + ky \qquad y(t) = Y \sin \omega t$$

$$\zeta = \frac{c}{2\sqrt{mk}} \qquad \omega_n = \sqrt{\frac{k}{m}} \qquad r = \frac{\omega}{\omega_n}$$

Steady-state response: $x(t) = X \sin(\omega t + \phi)$

$$f_t(t) = F_t \sin(\omega t + \theta)$$

Displacement transmissibility:

$$\frac{X}{Y} = \sqrt{\frac{4\zeta^2 r^2 + 1}{(1 - r^2)^2 + 4\zeta^2 r^2}}$$

Force transmissibility:

$$\frac{F_t}{kY} = r^2 \frac{X}{Y} = r^2 \sqrt{\frac{4\zeta^2 r^2 + 1}{(1 - r^2)^2 + 4\zeta^2 r^2}}$$

$$F_t = r^2 k X$$

by a sinusoidal input displacement of amplitude Y. Thus, at steady state, $f_t(t) = F_t \sin(\omega t + \theta)$, where the phase shift θ of f_t is the angle of the complex expression given by Equation 4.3-10.

Comparing Equations 4.3-6 and 4.3-11, we see that

$$\frac{F_t}{kY} = r^2 \frac{X}{Y} \tag{4.3-12}$$

and

$$F_t = r^2 k X \tag{4.3-13}$$

These relations enable us to calculate F_t/kY and F_t easily if we have previously computed X/Y and X.

Table 4.3-1 summarizes these results.

EXAMPLE 4.3-1
Base Excitation and Vehicle Suspension

An example of base excitation occurs when a car drives over a rough road. Figure 4.3-2 shows a quarter-car representation, where the stiffness k is the series combination of the tire and suspension stiffnesses. The equation of motion is given by Equation 4.3-2. Although road surfaces are not truly sinusoidal in shape, we can nevertheless use a sinusoidal profile to obtain an approximate evaluation of the performance of the suspension at various speeds.

Suppose the road profile is given (in feet) by

$$y(t) = 0.03 \sin \omega t$$

where the amplitude of variation of the road surface is 0.03 ft and the frequency ω depends on the vehicle's speed and the road profile's period. Suppose the period of the road surface is 20 ft. Compute the steady-state motion amplitude and the force transmitted to the chassis for two cars traveling at speeds of 20 and 50 mi/hr. The first car has a mass of 900 kg; the second has a mass of 1450 kg. Both cars have the same suspension, whose values are $k = 2.9 \times 10^4$ N/m and $c = 5300$ N·s/m.

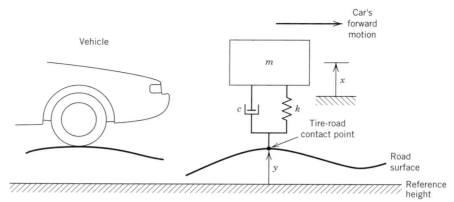

FIGURE 4.3-2 A quarter-car model of a vehicle suspension.

Solution

For a period of 20 feet and a vehicle speed of v (mi/hr), the frequency ω is

$$\omega = \left(\frac{5280}{20}\right)\left(\frac{1}{3600}\right)(2\pi)v = 0.4608v \quad \text{rad/sec}$$

Thus $\omega = 0.4608(20) = 9.216$ rad/sec for $v = 20$ mi/hr, and $\omega = 0.4608(50) = 23.04$ rad/sec for $v = 50$ mi/hr.

For the 900-kg car, the quarter-car mass is $m = 225$ kg. Its natural frequency is $\omega_n = \sqrt{k/m} = \sqrt{2.9 \times 10^4/225} = 11.35$ rad/s. Its frequency ratio at 20 mi/hr is $r = \omega/\omega_n = 9.216/11.35 = 0.812$, and at 50 mi/hr it is $r = 23.04/11.35 = 2.03$. Its damping ratio is

$$\zeta = \frac{5300}{2\sqrt{2.9 \times 10^4(225)}} = 1.04$$

Similarly, for the 1450-kg car, the quarter-car mass is $m = 362.5$ kg. Its natural frequency is $\omega_n = \sqrt{k/m} = \sqrt{2.9 \times 10^4/362.5} = 8.94$ rad/sec. Its frequency ratio at 20 mi/hr is $r = \omega/\omega_n = 9.216/8.94 = 1.03$, and at 50 mi/hr it is $r = 23.04/8.94 = 2.57$. Its damping ratio is

$$\zeta = \frac{5300}{2\sqrt{2.9 \times 10^4(362.5)}} = 0.817$$

Now substitute these values of r, ζ and $Y = 0.03(304.8)$ mm into the following expressions, obtained from Equations 4.3-6 and 4.3-13.

$$X = Y\sqrt{\frac{1 + 4\zeta^2 r^2}{(1 - r^2)^2 + 4\zeta^2 r^2}}$$

$$F_t = r^2 kX$$

This gives the table shown below.

v (mi/hr)	900-kg Car ($\zeta = 1.04$)			1450-kg Car ($\zeta = 0.817$)		
	r	X (mm)	F_t (N)	r	X (mm)	F_t (N)
20	0.812	10	201	1.03	11	331
50	2.03	8	905	2.57	3	652

Both cars have the same suspension, so the table shows the effect of the car's weight on the vibrational motion felt by the occupants and on the force transmitted from the road to the chassis. The motion amplitude of both cars is about the same at the lower speed, but the heavier car has less motion at higher speeds than the lighter car. The heavier car weighs 1.6 times as much as the smaller car, and about 1.6 times as much force is transmitted to its chassis at the lower speed. However, it is interesting to note that at the higher speed, the heavier car has *less* force transmitted, because the heavier car has a smaller value of ζ. ∎

4.4 ROTATING UNBALANCE

A common cause of sinusoidal forcing in machines is the unbalance that exists to some extent in every rotating machine. The unbalance is caused by the fact that the center of mass of the rotating part does not coincide with the center of rotation. Let M be the total mass of the machine and m the rotating mass causing the unbalance. Consider the entire unbalanced mass m to be lumped at its center of mass, a distance R from the center of rotation. This distance is the *eccentricity*. Figure 4.4-1a shows this situation. The main mass is thus $(M - m)$ and is assumed to be constrained to allow only vertical motion. The motion of the unbalanced mass m will consist of the vector combination of its motion relative to the main mass $(M - m)$ and the motion of the main mass. For a constant speed of rotation ω_R, the rotation produces a radial acceleration of m equal to $R\omega_R^2$. This causes a force to be exerted on the bearings at the center of rotation. This force has a magnitude $mR\omega_R^2$ and is directed radially outward. The vertical component of this unbalance force is, from Figure 4.4-1b,

$$f = mR\omega_R^2 \sin \omega_R t \tag{4.4-1}$$

Vibration Isolation and Rotating Unbalance

In many situations involving an unbalanced machine, we are interested in the force that is transmitted to the base or foundation. The equation of motion of a mass-spring-damper system, like that shown in Figure 4.4-1a, with an applied force $f(t)$ is

$$M\ddot{x} + c\dot{x} + kx = f(t) \tag{4.4-2}$$

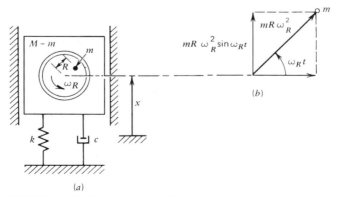

(a)

(b)

FIGURE 4.4-1 Machine with rotating unbalance. (*a*) System components. (*b*) Force diagram.

where x is the displacement of the mass from its rest position. The force transmitted to the foundation is the sum of the spring and damper forces; it is given by

$$f_t = kx + c\dot{x} \tag{4.4-3}$$

The *force transmissibility* of this system is the ratio $F_t(s)/F(s)$, which represents the ratio of the force f_t transmitted to the foundation by the applied force f. The most common case of such an applied force is the rotating unbalance force. From Equation 4.4-1, we see that the amplitude F of the unbalance force is

$$F = mR\omega_R^2 \tag{4.4-4}$$

We can use Equations 4.4-2 and 4.4-3 with the transfer function concept to obtain an expression for the force transmissibility. Substituting $x(t) = X(s)e^{st}$ and $f(t) = F(s)e^{st}$ into Equation 4.4-2 and canceling the e^{st} terms, we obtain

$$T_{x/f}(s) = \frac{X(s)}{F(s)} = \frac{1}{Ms^2 + cs + k} \tag{4.4-5}$$

Thus, with s replaced by $i\omega_R$, we obtain the magnitude of the frequency transfer function:

$$\left| T_{x/f}(i\omega) \right| = \frac{X}{F} = \frac{1}{\sqrt{(k - M\omega_R^2)^2 + (c\omega_R)^2}}$$

Thus, Equation from 4.4-4,

$$X = \frac{mR\omega_R^2}{\sqrt{(k - M\omega_R^2)^2 + (c\omega_R)^2}}$$

Divide the numerator and denominator by the mass M, use the fact that $\omega_n = \sqrt{k/M}$ and $c/M = 2\zeta\omega_n$, and define $r = \omega_R/\omega_n$ to obtain

$$X = |X(i\omega_R)| = \frac{1}{M}\frac{mRr^2}{\sqrt{(1 - r^2)^2 + (2\zeta r)^2}} \tag{4.4-6}$$

and

$$X = |X(i\omega_R)| = \frac{mR}{M}\frac{r^2}{\sqrt{(1 - r^2)^2 + (2\zeta r)^2}} \tag{4.4-7}$$

From Equation 4.4-3,

$$F_t(s) = (k + cs)X(s) \tag{4.4-8}$$

Substituting $X(s)$ from Equation 4.4-5 into Equation 4.4-8 gives

$$F_t(s) = \frac{k + cs}{Ms^2 + cs + k}F(s)$$

Thus the force transmissibility is

$$T_{f_t/f}(s) = \frac{F_t(s)}{F(s)} = \frac{k + cs}{Ms^2 + cs + k} \tag{4.4-9}$$

The frequency transfer function is

$$T_{f_t/f}(i\omega_R) = \frac{k + c\omega_R i}{-M\omega_R^2 + c\omega_R i + k}$$

Divide the numerator and denominator by the mass M and use the fact that $\omega_n = \sqrt{k/M}$ and $c/M = 2\zeta\omega_n$ to obtain

$$T_{f_t/f}(i\omega_R) = \frac{\dfrac{k}{M} + \dfrac{c}{M}\omega_R i}{-\omega_R^2 + \dfrac{c}{M}\omega_R i + \dfrac{k}{M}} = \frac{\omega_n^2 + 2\zeta\omega_n\omega_R i}{-\omega_R^2 + 2\zeta\omega_n\omega_R i + \omega_n^2}$$

Now divide the numerator and denominator by ω_n^2 and use the definition $r = \omega_R/\omega_n$ to obtain

$$T_{f_t/f}(i\omega_R) = \frac{1 + 2\zeta\dfrac{\omega_R}{\omega_n}i}{1 - \dfrac{\omega_R^2}{\omega_n^2} + 2\zeta\dfrac{\omega_R}{\omega_n}i} = \frac{1 + 2\zeta ri}{1 - r^2 + 2\zeta ri}$$

The magnitude of this expression is

$$\left|T_{f_t/f}(i\omega_R)\right| = \frac{F_t}{F} = \sqrt{\frac{1 + 4\zeta^2 r^2}{(1 - r^2)^2 + 4\zeta^2 r^2}} \tag{4.4-10}$$

This is the force transmissibility expression for a sinusoidal applied force. It is the ratio of the amplitude F_t of the transmitted force to the steady-state amplitude F of the applied force. In what follows, we will use the symbol T_f to represent the force transmissibility ratio F_t/F. Thus

$$T_f = \frac{F_t}{F} = \sqrt{\frac{1 + 4\zeta^2 r^2}{(1 - r^2)^2 + 4\zeta^2 r^2}} \tag{4.4-11}$$

Note that although the expression for force transmissibility given by Equation 4.4-11 is identical to that of *displacement* transmissibility $T_d = X/Y$ for base motion, the expressions arise from different physical applications.

If the applied force is due to rotating unbalance and has an amplitude $F = mR\omega_R^2$, we can solve Equation 4.4-11 for the amplitude of the transmitted force as follows:

$$F_t = mR\omega_R^2 \sqrt{\frac{1 + 4\zeta^2 r^2}{(1 - r^2)^2 + 4\zeta^2 r^2}} \tag{4.4-12}$$

Table 4.4-1 summarizes these results.

TABLE 4.4-1   Formulas for Rotating Unbalance

Model:

$$M\ddot{x} + c\dot{x} + kx = mR\omega_R^2 \sin \omega_R t$$

$m =$ unbalanced mass $R =$ eccentricity

$$\zeta = \frac{c}{2\sqrt{Mk}} \qquad \omega_n = \sqrt{\frac{k}{M}} \qquad r = \frac{\omega_R}{\omega_n}$$

Displacement:

$$X = \frac{\dfrac{mR}{M}r^2}{\sqrt{(1 - r^2)^2 + (2\zeta r)^2}}$$

Force transmissibility:

$$T_f = \frac{F_t}{F} = \sqrt{\frac{1 + 4\zeta^2 r^2}{(1 - r^2)^2 + 4\zeta^2 r^2}}$$

EXAMPLE 4.4-1
Foundation Force
Due to Rotating
Unbalance

A system having a rotating unbalance, like that shown in Figure 4.4-1, has a total mass of $M = 10$ kg, an unbalanced mass of $m = 0.01$ kg, and an eccentricity of $R = 0.01$ m. The machine rotates at 3000 rpm. Its vibration isolator has a stiffness of $k = 50\ 000$ N/m. Compute the force transmitted to the foundation if the isolator's damping ratio is **(a)** $\zeta = 0.01$ and **(b)** $\zeta = 0.5$.

Solution

First convert the machine's speed to radians per second:

$$\omega_R = 3000 \text{ rpm} = \frac{3000(2\pi)}{60} = 314.16 \text{ rad/sec}$$

Then

$$mR\omega_R^2 = 0.01(0.01)(314.16)^2 = 9.8699$$

and the frequency ratio is

$$r = \frac{\omega_R}{\omega_n} = \frac{\omega_R}{\sqrt{k/M}} = \frac{314.16}{\sqrt{50\ 000/10}} = 4.4429$$

We can calculate the transmissibility ratio from Equation 4.4-11:

$$T_f = \sqrt{\frac{1 + 78.9572\zeta^2}{351.1636 + 78.9572\zeta^2}}$$

The force transmitted to the foundation is

$$F_t = mR\omega_R^2 T_f = 9.867 T_f$$

(a) If $\zeta = 0.01$, $T_f = 0.0536$ and $F_t = 9.867(0.0536) = 0.5286$ N.

(b) If $\zeta = 0.5$, $T_f = 0.2365$ and $F_t = 9.867(0.2365) = 2.3332$ N. The more highly damped isolator in this case transmits more force to the foundation because $r > \sqrt{2}$. ■

4.5   CRITICAL SPEEDS OF ROTATING SHAFTS

A major application of vibration theory lies in the analysis and design of rotating shafts. Examples of such systems include electrical generators and jet engines in which a number of rotors are attached to a common shaft. Each circular rotor has turbine blades attached to its circumference. It is possible that the rotating shaft can bend enough to fail, thus destroying the device. The rotational speeds at which this resonance occurs are called the *critical speeds* of the rotor-shaft system

Figure 4.5-1 shows a shaft supported by two bearings and carrying a rotor of mass m in the middle. If the rotor is somewhat unsymmetric (due to manufacturing variations or blade loss, for example), then its geometric center P and its center of mass G will be a distance R apart, and the rotor will experience an excitation due to the offset center of mass. The resulting motion is called *whirling*.

The forces acting on the rotor are that due to the bending of the shaft, which we will model as a spring force, any external damping forces (such as those due to the bearings), and any internal damping forces (such as those due to internal friction within the shaft material). The spring constant due to the bending can be calculated from the stiffness formula for a fixed-fixed beam. Here we will model the external damping only. Inclusion of internal damping leads to a model that cannot be analyzed as a system having a single degree of freedom.

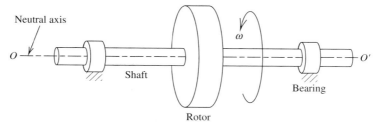

FIGURE 4.5-1 A shaft-rotor system exhibiting whirl.

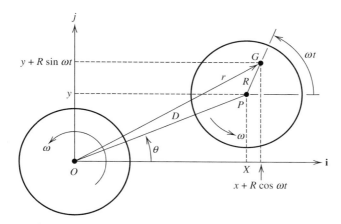

FIGURE 4.5-2 Definitions of variables used in the analysis of whirl.

From Figure 4.5-2, Newton's law in vector form gives

$$m\ddot{\mathbf{r}} = -kx\mathbf{i} - ky\mathbf{j} - c\dot{x}\mathbf{i} - c\dot{y}\mathbf{j} \tag{4.5-1}$$

where m is the mass of the rotor and $\mathbf{r}$ is the position vector from the point O to the mass center at point G. The point O is defined by the shaft *neutral axis*, OO' in Figure 4.5-1, which is the centerline of the shaft when it is at rest. The coefficients c and k represent the external damping and the bending stiffness of the shaft.

The position vector $\mathbf{r}$ can be expressed in component form as

$$\mathbf{r} = (x + R\cos\omega t)\mathbf{i} + (y + R\sin\omega t)\mathbf{j} \tag{4.5-2}$$

where ω is the rotational speed of the shaft. Differentiating this twice gives

$$\ddot{\mathbf{r}} = (\ddot{x} - R\omega^2\cos\omega t)\mathbf{i} + (\ddot{y} - R\omega^2\sin\omega t)\mathbf{j} \tag{4.5-3}$$

Substituting this into Equation 4.5-2 gives

$$(m\ddot{x} - mR\omega^2\cos\omega t + c\dot{x} + kx)\mathbf{i} + (m\ddot{y} - R\omega^2\sin\omega t + c\dot{y} + ky)\mathbf{j} = 0 \tag{4.5-4}$$

Equating each compoment to zero and bringing the forcing terms to the right-hand side, we obtain

$$m\ddot{x} + c\dot{x} + kx = mR\omega^2\cos\omega t \tag{4.5-5}$$

$$m\ddot{y} + c\dot{y} + ky = mR\omega^2\sin\omega t \tag{4.5-6}$$

Each equation has the form of a one-degree-of-freedom system undergoing harmonic forcing due to a rotating unbalanced force of magnitude $mR\omega^2$. Noting that here the unbalanced mass is the entire mass m, the steady-state solutions can therefore be expressed as

$$x(t) = X\cos(\omega t + \phi) \tag{4.5-7}$$

$$y(t) = Y\sin(\omega t + \phi) \tag{4.5-8}$$

where

$$X = Y = \frac{Rr^2}{\sqrt{(1-r^2)^2 + (2\zeta r)^2}} \tag{4.5-9}$$

$$\phi = -\tan^{-1}\frac{2\zeta r}{1-r^2} \tag{4.5-10}$$

$$r = \frac{\omega}{\omega_n} \tag{4.5-11}$$

$$\omega_n = \sqrt{\frac{k}{m}} \tag{4.5-12}$$

From Figure 4.5-2,

$$\theta = \tan^{-1}\frac{y(t)}{x(t)} = \tan^{-1}\frac{\sin(\omega t + \phi)}{\cos(\omega t + \phi)} = \omega t + \phi$$

Therefore $\dot{\theta} = \omega$. Thus ω is the rotational speed of the rotor's geometric center P as it revolves about the neutral axis. This is called the whirling speed and is seen to be identical to the rotational speed of the shaft. Because the two speeds are identical, this particular motion is called *synchronous whirl*.

In Figure 4.5-2, the distance D is the distance from the neutral axis to the shaft center as it rotates. Because $X = Y$ and $\sin^2\theta + \cos^2\theta = 1$, it is easy to show that

$$D = \sqrt{x^2 + y^2} = X = \frac{Rr^2}{\sqrt{(1-r^2)^2 + (2\zeta r)^2}} \tag{4.5-13}$$

Figure 4.5-3 is a plot of D/R versus $r = \omega/\omega_n$. For no damping, the peak of the curve occurs at $r = 1$, and the peak is near $r = 1$ for underdamped systems. The rotational speed ω corresponding to $r = 1$ is $\omega = \sqrt{k/m}$ and is called the *critical speed*. Underdamped systems operating near the critical speed are at risk of failure due to bearing failure (because of the high reaction forces at the bearings) or shaft fracture.

Figure 4.5-3 shows that the precise value of the damping ratio need not be known if r can be made large enough, because the curves approach one another as r increases. For large r values, $D \to R$, so the maximum dynamic deflection at steady state will equal the distance from the neutral axis to the mass center. Of course, the transient deflection will be larger as the rotational speed increases from zero at startup and passes through the region where $r = 1$. Increased damping will limit this deflection. In addition, if the steady-state speed is reached quickly, the deflection will not have time to build up.

Common sense tells us that low-speed systems, low-mass systems, and stiff systems will not have excessive deflection. Figure 4.5-3 confirms this observation because $D/R \to 0$ as $r \to 0$. This occurs if the speed ω is slow or if ω_n is large. The latter case corresponds to a large value of k or a small value of m.

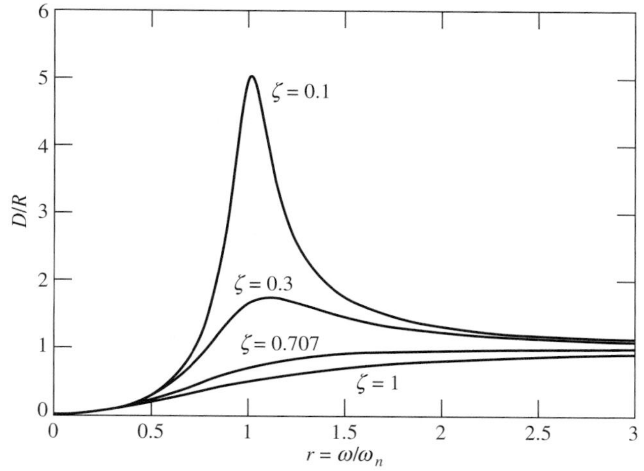

FIGURE 4.5-3   Plot of D/R versus r.

EXAMPLE 4.5-1
Design to Minimize Whirl

A common theme throughout vibration analysis is that damping is very difficult to quantify, and so for systems that are known to be lightly damped, the engineer often ignores damping altogether.

(a) The design value of D/R for large r is $D/R = 1$. Analyze the uncertainty in the ratio D/R due to uncertainty in the value of ζ if we design the rotor-shaft system so that $r = 3$.

(b) How much must the shaft stiffness k be changed so that D/R differs no more than 10% from the design value?

Solution

(a) Setting $r = 3$ in Equation 4.5-13, we obtain

$$\frac{D}{R} = \frac{9}{\sqrt{64 + 36\zeta^2}}$$

Assuming ζ to be in the range $0 \le \zeta \le 1$, we see that $\zeta = 1$ gives the minimum value for D/R, which is $D/R = 0.9$. Setting $\zeta = 0$ gives the maximum value for D/R, which is $D/R = 1.125$. Thus the uncertainty in D/R is from 90% to 112.5% of its design value.

(b) The uncertainty in D/R is greater than 10% if $\zeta = 0$, so we set $\zeta = 0$ in Equation 4.5-13 to obtain

$$\frac{D}{R} = \frac{r^2}{\sqrt{(1 - r^2)^2}} = \frac{r^2}{|1 - r^2|}$$

Setting $D/R = 1.1$ and solving for r gives $r = \sqrt{11}$. Noting that $k = m\omega^2/r^2$, we see that with $r = 3$, $k_1 = m\omega^2/9$. With $r = \sqrt{11}$, $k_2 = m\omega^2/11$. Thus the percent change required in the shaft stiffness k is $(k_1 - k_2)/k_1 = 2/11$, or 18%.   ∎

4.6   EQUIVALENT DAMPING

Damping may be due to many causes besides viscous friction. During oscillation, energy loss may occur due to viscous friction, Coulomb friction, or *internal damping*. Internal damping is also called *solid damping*, *structural damping*, or *hysteretic damping*. The mechanisms that cause internal damping have to do with the internal motion of the material as it deforms. In some materials such as metals, it may be due to the slipping between grain boundaries, while in some materials such as rubber other mechanisms come into play. These mechanisms are quite varied and complex, and they will not be treated here. The results, therefore, are usually empirical, and no general theory is available to explain internal damping.

It is helpful, however, to be able to describe such effects in terms of an equivalent viscous damping coefficient, because such a description leads to a linear equation of motion. To this end we will need an expression for the energy dissipated in a viscous damper during one cycle of oscillation.

Energy Dissipation in Viscous Damping

When a harmonic forcing function with frequency ω is applied to a system, the relative displacement $x(t)$ of the damper at steady state is $x(t) = X \sin \omega t$ and the force exerted by the damper is thus $f(t) = c\dot{x} = \omega X \cos \omega t$. The power dissipated in the damper is the product of the force and the velocity $v = \dot{x}$, or

$$\frac{dE}{dt} = f(t)v(t) = [cv(t)]v(t) = c\dot{x}^2(t) = c\omega^2 X^2 \cos^2 \omega t$$

Over one cycle of period $2\pi/\omega$, the energy dissipated in the damper is

$$E = \int_0^{2\pi/\omega} \frac{dE}{dt}\, dt = \int_0^{2\pi/\omega} c\omega^2 X^2 \cos^2(\omega t)\, dt = \pi c\omega X^2$$

So we see that the energy dissipated by the damper in one cycle is proportional to the square of the amplitude and that it is also dependent on the frequency. So one definition of equivalent viscous damping may be stated as

$$c_e = \frac{E}{\pi \omega X^2} \tag{4.6-1}$$

The relation between the damping force f and the damper displacement x at steady state is

$$f = c\dot{x} = c\omega X \cos \omega t$$

or

$$f^2 = c^2\omega^2 X^2 \cos^2 \omega t = c^2\omega^2 X^2\left(1 - \sin^2 \omega t\right)$$

This may be arranged as

$$\left(\frac{f}{c\omega X}\right)^2 + \left(\frac{x}{X}\right)^2 = 1 \tag{4.6-2}$$

Equation 4.6-2 describes the ellipse shown in Figure 4.6-1. Since the area of an ellipse is π times the product of the semi-major and semi-minor axes, the area of this ellipse is $\pi c\omega X^2$, which is identical to the energy dissipated per cycle.

Equivalent Damping from Resonance Response

We may also define the equivalent viscous damping to be the damping required to give the same amplitude ratio found by measuring the response at resonance. When $\omega = \omega_n$, $r = 1$, and from Equation 4.2-12,

$$\frac{X}{\delta_{st}} = \frac{1}{2\zeta}$$

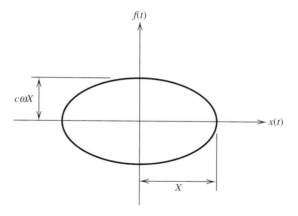

FIGURE 4.6-1 Viscous damping force versus displacement at steady state under harmonic forcing.

Therefore, the equivalent damping ratio is found from

$$\zeta_e = \frac{1}{\left(\dfrac{X}{\delta_{st}}\right)_{\omega=\omega_n}} \tag{4.6-3}$$

and the equivalent damping constant is

$$c_e = 2\sqrt{km}\,\zeta_e \tag{4.6-4}$$

Hysteretic Damping

The effects of internal damping appear in an experimental stress-strain plot such as shown in Figure 4.6-2, which shows one cycle of loading and unloading of a specimen. The loop is known as a *hysteresis loop*. If the plot is force versus displacement, the area of the loop equals the energy loss E during one cycle. If the plot is stress versus strain, the area equals the energy loss *per unit volume* during one cycle. For many materials, including steel and aluminum, it has been found that the area is independent of the frequency of the cycle but

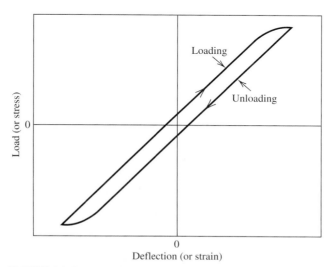

FIGURE 4.6-2 A hysteresis loop.

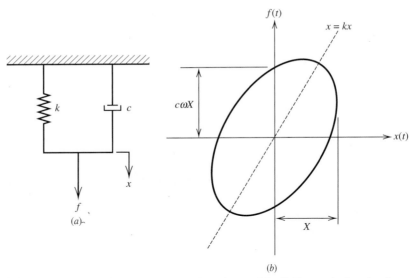

FIGURE 4.6-3   Hysteretic damping. (a) Damping model. (b) Hysteretic damping force versus displacement at steady state under harmonic forcing.

proportional to the square of the amplitude X of the displacement. Thus, by analogy to the formula $E = \pi c \omega X^2$ for viscous damping, we may include a factor of π and write for the hysteresis loop:

$$E = \pi h X^2 \tag{4.6-5}$$

where h is the *hysteresis damping constant*.

A common model of hysteretic damping is a spring element in parallel with a viscous damper. The force across the two elements is $f = kx + c\dot{x}$ (Figure 4.6-3a). When a harmonic forcing function with frequency ω is applied to the system, the displacement $x(t)$ at steady state is $x(t) = X \sin \omega t$ and the force exerted by the spring and damper is thus $f = kx + c\dot{x} = kX \sin \omega t + \omega X \cos \omega t$. This equation may be arranged as

$$f = kx \pm c\omega \sqrt{X^2 - x^2}$$

or

$$\left(\frac{x}{X}\right)^2 + \left(\frac{f - kx}{cX\omega}\right)^2 = 1 \tag{4.6-6}$$

This is the equation of the ellipse shown in Figure 4.6-3b. Because its major axis forms an angle with the x axis, this ellipse more closely resembles a typical hysteresis loop than does the ellipse for damping only (see Figure 4.6-1).

The power dissipated is the product of the force and the velocity $\dot{x}$, or

$$\frac{dE}{dt} = f\dot{x} = (kx + c\dot{x})\dot{x} = kx\dot{x} + c\dot{x}^2 = k\omega X^2 \sin \omega t \cos \omega t + c\omega^2 X^2 \cos^2 \omega t$$

Over one cycle of period $2\pi/\omega$, the energy dissipated is

$$E = \int_0^{2\pi/\omega} \frac{dE}{dt}\, dt = \int_0^{2\pi/\omega} \left[k\omega X^2 \sin(\omega t) \cos(\omega t) + c\omega^2 X^2 \cos^2(\omega t)\right] dt$$

or

$$E = \int_0^{2\pi/\omega} kX^2 \sin(\omega t) \cos(\omega t)\, d(\omega t) + cX^2\omega \int_0^{2\pi/\omega} \cos^2(\omega t)\, d(\omega t)$$

$$= \pi c\omega X^2$$

which is identical to the energy lost with damping only. This is not unexpected because the spring does not dissipate energy but merely stores and releases potential energy over the cycle. Comparing this expression with that of Equation 4.6-5, we see that

$$h = c\omega \tag{4.6-7}$$

Thus the model for hysteretic damping is

$$f = kx + \frac{h}{\omega}\dot{x} \tag{4.6-8}$$

and we note that in this model the equivalent damping coefficient $c_e = h/\omega$ is a function of the driving frequency and cannot be used to predict response at other frequencies.

A value of h can be estimated from an experimentally obtained hysteresis loop either by measuring the area of the loop and using Equation 4.6-5 or by comparing the ordinate intercept of the loop with that of the ellipse shown in Figure 4.6-3b, which is $c\omega X = hX$. Since in practice the hysteresis loop is often narrow, these two methods should give similar answers. The value of X can be taken to be the maximum value of the deflection shown on the hysteresis loop. The value of k is the slope of the major axis of the ellipse and can be estimated from the hysteresis loop either from the slope of a straight line drawn between the two tips of the loop or from the slope of the loop if its sides are relatively straight.

The form of free vibration with hysteretic damping is the same as that with viscous damping. Thus we define a hysteretic damping ratio and hysteretic logarithmic decrement in a similar manner, as follows:

$$\zeta_e = \frac{c_e}{2\sqrt{km}}$$

$$\delta_e = \frac{2\pi\zeta_e}{\sqrt{1 - \zeta_e^2}}$$

The equation of motion for sinusoidal forcing of a mass m with hysteretic damping is thus

$$m\ddot{x} + \frac{h}{\omega}\dot{x} + kx = F \sin \omega t$$

The magnitude and phase angle of the steady-state response $x(t) = X \sin(\omega t + \phi)$ can be found from Equations 4.2-8 and 4.2-9 by substituting ζ_e for ζ to obtain

$$X = \frac{F/k}{\sqrt{(1 - r^2)^2 + (2\zeta_e r)^2}} = \frac{F/k}{\sqrt{(1 - r^2)^2 + (h/k)^2}}$$

$$\phi = -\tan^{-1} \frac{h/k}{1 - r^2}$$

where $r = \omega/\omega_n$. Recall that for viscous damping the peak frequency is $\omega_p = \omega_n\sqrt{1 - 2\zeta^2}$, which is less than the natural frequency ω_n. For hysteretic damping, however, the maximum response occurs at $r = 1$, which corresponds to $\omega = \omega_n$. From the phase angle equation, we see that the response of a system with hysteretic damping is never in phase with the driving force, whereas the two are in phase when there is viscous damping.

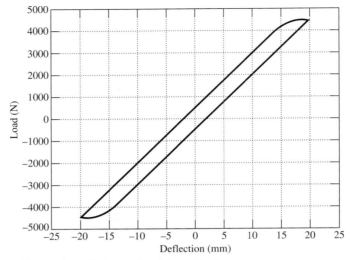

FIGURE 4.6-4 Estimating h and k graphically.

EXAMPLE 4.6-1
Estimating h and k

Figure 4.6-4 is a graph obtained by loading and unloading a structure. Estimate the hysteretic damping coefficient h and the stiffness k.

Solution

The area enclosed by the loop is approximately 31 N·m. This area can be estimated either graphically (by counting squares), numerically (with trapezoidal integration, for example), or analytically (by curve fitting a polynomial and integrating). The maximum amplitude of the displacement is $X = 20$ mm, and so from $E = \pi h X^2$ we have

$$31 = \pi h (0.020)^2$$

which gives $h = 24.67$ kN/m. The maximum force is $F_{max} = 4500$ N, so $k = 4500/0.020 = 225$ kN/m.

Using the alternate method, we note that the loop intercepts the ordinate at 500 N, and we may thus estimate h from $hX = 500$. This gives $h = 25$ kN/m, which is very close to the other estimate. We may also estimate k from the slope of the straight side of the loop as follows: $k = 1200/5 = 240$ kN/m, which is similar to the previous estimate. ∎

Velocity-Squared Damping

The resisting force due to fluid drag is proportional to the square of the velocity. This observation leads to the *velocity-squared* damping model, which is also known as *quadratic damping*. The model of drag force for such a case is

$$f = \frac{1}{2} \rho C_d A \, \text{sign}(v) v^2 \tag{4.6-9}$$

where ρ is the fluid mass density, C_d is the *drag coefficient*, and A is the object's surface area normal to the flow. Letting $b = \rho C_d A / 2$ and following the same procedure as used with viscous damping, we find that the energy dissipated per cycle is

$$E = \frac{8}{3} b \omega^2 X^3 \tag{4.6-10}$$

For viscous damping, $E = \pi c \omega X^2$. Equating these two expressions gives the equivalent damping constant for velocity-squared damping:

$$c_e = \frac{8}{3\pi} b \omega X \tag{4.6-11}$$

4.7 FORCED VIBRATION WITH COULOMB DAMPING

In forced vibration with Coulomb friction, the equation of motion may be expressed compactly as

$$m\ddot{x} + \mu N \text{sign}(\dot{x}) + kx = F \sin \omega t \tag{4.7-1}$$

The steady-state displacement is given by $x = X \sin(\omega t + \phi)$; the work done by the friction force μN is the product of the force and the displacement over four half-cycles. Thus the energy dissipated per cycle, which equals the work done, is $E = 4X\mu N$. Equating this to the energy loss per cycle due to viscous friction, we have

$$E = \pi c \omega X^2 = 4X\mu N$$

This gives the equivalent viscous damping constant for Coulomb damping:

$$c_e = \frac{4\mu N}{\pi \omega X}$$

Thus we may obtain the equivalent damping ratio as follows:

$$\zeta_e = \frac{c_e}{2\sqrt{mk}} = \frac{c_e}{2m\omega_n} = \frac{2\mu N}{\pi m X \omega \omega_n}$$

By this means we have replaced the nonlinear equation, Equation 4.7-1, with the following approximate, but linear, equation

$$\ddot{x} + 2\zeta_e \omega_n \dot{x} + \omega_n^2 x = \frac{F}{m} \sin \omega t \tag{4.7-2}$$

which describes a system that dissipates as much energy per cycle as the original nonlinear system. We may now use Equation 4.1-26 for the amplitude of vibration, with ζ_e replacing ζ:

$$X = \frac{F/k}{\sqrt{(1 - r^2)^2 + (2\zeta_e r)^2}} = \frac{F/k}{\sqrt{(1 - r^2)^2 + (4\mu N/\pi k X)^2}}$$

We solve this equation for X as follows:

$$X = \text{sign}(1 - r^2) \frac{F}{k} \frac{\sqrt{1 - (4\mu N/\pi F)^2}}{1 - r^2} \tag{4.7-3}$$

The phase shift is given by Equation 4.1-25:

$$\phi = -\tan^{-1} \frac{2\zeta_e r}{1 - r^2} = -\tan^{-1} \frac{4\mu N}{\pi k X(1 - r^2)}$$

Substitute X from Equation 4.7-3 to obtain

$$\phi = \text{sign}(r^2 - 1) \tan^{-1} \left[\frac{4\mu N}{\pi F \sqrt{1 - (4\mu N/\pi F)^2}} \right] \tag{4.7-4}$$

Thus ϕ is negative if $r^2 - 1 < 0$ ($r < 1$), ϕ is positive if $r^2 - 1 > 0$ ($r > 1$), and ϕ is discontinuous at $r = 1$. This equation shows that the phase angle is constant for a given value of $F/\mu N$ and is independent of the forcing frequency ω. This solution is valid only if the term under the square root is non-negative; that is, if

$$1 - (4\mu N/\pi F)^2 \geq 0$$

The condition is equivalent to $F \geq 4\mu N/\pi$, which means that the applied force amplitude must be somewhat larger than the friction force.

Comparing the forced response involving Coulomb damping with that involving viscous damping, we see that the following are true:

1. When viscous damping is present, the amplitude of vibration is finite even at resonance. For Coulomb damping, however, Equation 4.7-3 shows that the amplitude is infinite at resonance (when $r = 1$).

2. With viscous damping, the phase shift is continuous, whereas with Coulomb damping, it is discontinuous at resonance.

EXAMPLE 4.7-1
A Mass-Spring System with Coulomb Friction

A sinusoidal force $500 \sin 150t$ N acts on a mass-spring system. The mass is 11 kg and moves on a horizontal surface. The coefficient of dynamic friction between the mass and the surface is $\mu = 0.3$. The spring stiffness is $k = 1.75 \times 10^4$ N/m. Calculate the amplitude and phase angle of the steady-state displacement of the mass.

Solution

First check to see if the inequality $4\mu N < \pi F$ is satisfied. Because the surface is horizontal, the normal force is $N = mg$. Thus

$$4\mu N = 4\mu mg = 4(0.3)11(9.81) = 129$$

$$\pi F = \pi(500) = 1570$$

so the inequality is satisfied.

The natural frequency of the system is

$$\omega_n = \sqrt{\frac{k}{m}} = \sqrt{\frac{1.75 \times 10^4}{11}} = 39.9 \text{ rad/s}$$

The frequency ratio is

$$r = \frac{\omega}{\omega_n} = \frac{150}{39.9} = 3.817$$

The amplitude is calculated from Equation 4.7-3:

$$X = \text{sign}\left[1 - (3.817)^2\right] \frac{500}{1.75 \times 10^4} \frac{\sqrt{1 - [4(0.3)11(9.81)/500\pi]^2}}{1 - (3.817)^2} = 2.1 \text{ mm}$$

The phase angle is calculated from Equation 4.7-4:

$$\phi = \text{sign}\left[(3.817)^2 - 1\right] \tan^{-1} \frac{4(0.3)11(9.81)}{500\pi\sqrt{1 - [4(0.3)11(9.81)/500\pi]^2}} = 0.0825 \text{ rad}$$

Thus the steady-state response is

$$x(t) = 2.1 \sin(150t + 0.0825) \text{ mm} \qquad \blacksquare$$

4.8 NONLINEAR RESPONSE

The response of nonlinear models differs significantly from that of linear models in several ways. In this section we give a brief overview of these differences. Many mathematical results are available for approximate solutions to nonlinear equations of motion, but coverage of the methods used to obtain these results is beyond the scope of our study. Here we take the view that we can obtain the information we need by using computer solution methods. The following examples are a guide to what to look for in nonlinear response.

First consider the simple, static model

$$x(t) = f^2(t) \tag{4.8-1}$$

where $x(t)$ is the response and $f(t)$ is the forcing function. If $f(t) = A \sin \omega t$, then

$$x(t) = A^2 \sin^2 \omega t = \frac{1}{2} A^2 (1 - \cos 2\omega t)$$

Thus the frequency of the response is twice that of the input, and the response is shifted relative to the input. Note also that the response amplitude is a nonlinear function of the input amplitude. For example, if we double the input amplitude, the response amplitude is not doubled. An example is shown in the top graph of Figure 4.8-1, for which $A = 2$ and $\omega = 2\pi$.

Now consider an input that is a sum of two terms:

$$f(t) = \sin \omega_1 t + \sin \omega_2 t$$

The response is

$$x(t) = \sin^2 \omega_1 t + 2 \sin \omega_1 t \sin \omega_2 t + \sin^2 \omega_2 t$$

$$= \frac{1}{2}(1 - \cos 2\omega_1 t) + \cos(\omega_1 - \omega_2)t - \cos(\omega_1 + \omega_2)t + \frac{1}{2}(1 - \cos 2\omega_2 t)$$

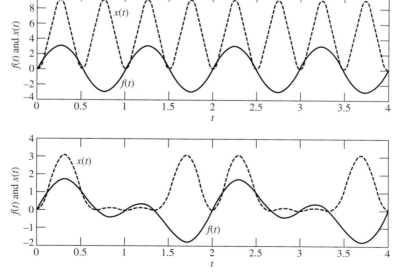

FIGURE 4.8-1 Response of $x(t) = f^2(t)$ for $f(t) = 2 \sin 2\pi t$ (top graph) and $f(t) = \sin \pi t + \sin 2\pi t$ (bottom graph).

Whereas the input contains only two frequencies, the response contains four frequencies, all different from the input frequencies. Also, the response is not a linear combination of the two input terms. In general, the principle of superposition does not apply to nonlinear models. An example is shown in the bottom graph of Figure 4.8-1, for which $A = 1$, $\omega_1 = \pi$, and $\omega_2 = 2\pi$. The response contains the four frequencies π, 2π, 3π, and 4π.

Properties of Nonlinear Dynamic Models

Linear dynamic models have several useful properties not found with nonlinear models, and thus we should try to model a system with a linear model whenever possible. One consequence of the superposition property of linear dynamic models is that their solutions consist of the sum of two terms: the *free response*, which is that part attributable to the initial conditions, and the *forced response*, which is due to the forcing function or input. The entire solution is called the *complete* or *total response*.

This property enables us to find the solution for the free response separately from the solution for the forced response. The free response is independent of the forced response, and vice versa. If we need to find the total response, we simply add the free and the forced responses.

Another consequence of the superposition property is that the responses due to more than one input can be separated.

Few nonlinear models have closed-form solutions, but we can illustrate some of their properties by considering the following first-order model.

EXAMPLE 4.8-1
Solution of a
Nonlinear
Dynamic Model

The equation of motion of a mass m subjected to a square-law damping force and a constant force f is

$$m\frac{dv}{dt} = f - qv^2 \tag{1}$$

where v is the velocity of the mass and q is a constant.

Let the initial speed $v(0)$ at time $t = 0$ be denoted v_0. Obtain the solution for $v(t)$ and discuss the solution for the two cases: **(a)** $f \neq 0$ and **(b)** $f = 0$.

Solution

(a) Using separation of variables, we can write Equation 1 as follows:

$$m\frac{dv}{f - qv^2} = dt$$

Integrate both sides to obtain

$$\int_{v_0}^{v(t)} \frac{dv}{v^2 - \alpha^2} = -\frac{q}{m}\int_0^t dt \tag{2}$$

where $\alpha = \sqrt{f/q}$. The integral tables give the following result, provided that $v^2 \neq \alpha^2$ and $f \neq 0$:

$$\frac{1}{2\alpha}\ln\left(\frac{v - \alpha}{v + \alpha}\right)\Bigg|_{v_0}^{v(t)} = -\frac{q}{m}t \tag{3}$$

Rearrange the above solution as follows:

$$\ln\left(\frac{v - \alpha}{v + \alpha}\right) = \ln\left(\frac{v_0 - \alpha}{v_0 + \alpha}\right) - \frac{2\alpha q}{m}t = \ln\left[\left(\frac{v_0 - \alpha}{v_0 + \alpha}\right)e^{-2\alpha qt/m}\right]$$

Take the inverse logarithm of both sides and solve for $v(t)$ to obtain

$$v(t) = \alpha\frac{a + be^{-ct}}{a - be^{-ct}} \tag{4}$$

where

$$a = v_0\sqrt{q} + \sqrt{f} \tag{5}$$

$$b = v_0\sqrt{q} - \sqrt{f} \tag{6}$$

$$c = \frac{2\alpha q}{m} \tag{7}$$

This solution is valid if $v_0^2 \neq \alpha^2$ and if $f \neq 0$.

If $v_0^2 = \alpha^2$, then $v_0^2 = f/q$, and Equation 1 shows that

$$m\frac{dv}{dt}\bigg|_{t=0} = f - qv_0^2 = f - q\frac{f}{q} = 0$$

Thus $\dot{v}(0) = 0$ and $v(t)$ remains constant at v_0.

If $v_0 = 0$, Equation 4 reduces to

$$v(t) = \alpha\frac{1 - e^{-ct}}{1 + e^{-ct}} \tag{8}$$

(b) If $f = 0$, the integral formula, Equation 3, is not valid, and the solution can be found using separation of variables as follows. From Equation 1 with $f = 0$,

$$\int_{v_0}^{v(t)} \frac{dv}{v^2} = -\frac{q}{m}\int_0^t dt$$

or

$$-v^{-1}\bigg|_{v_0}^{v(t)} = \frac{1}{v(t)} - \frac{1}{v_0} = -\frac{q}{m}t$$

Solve this for $v(t)$ to obtain

$$v(t) = \frac{mv_0}{m + qv_0t} \tag{9}$$

Note that this solution's form is different from that of the general case in Equation 4, where $f \neq 0$. Note also that we *cannot* obtain the total response of Equation 4 for $f \neq 0$, $v_0 \neq 0$ by adding the forced response of Equation 8 (where $f \neq 0$ and $v_0 = 0$) to the free response of Equation 9 (where $f = 0$ and $v_0 \neq 0$). We usually cannot obtain the total response of a nonlinear differential equation by superimposing the free and the forced responses.

Not only can we not use superposition to solve nonlinear equations, but we must also be careful to spot special cases that arise in the mathematics. For example, we had to be careful in evaluating the integral in Equation 2 because of the singularity at $v^2 = f/q = \alpha^2$ (that is, the integrand is undefined when $v^2 = f/q$ because its denominator is zero). This forced us to solve the equation for three cases:

1. $v^2 \neq \alpha^2$, $f \neq 0$

2. $v^2 = \alpha^2$, $f \neq 0$

3. $f = 0$ ∎

Linearized Stability Analysis of Duffing's Equation

A common model of a nonlinear spring is

$$f_s(y) = k_1y + k_3y^3 \tag{4.8-2}$$

which represents a hardening spring if $k_3 > 0$ and a softening spring if $k_3 < 0$. For a mass-spring-damper system with such a spring and an applied force $f(t)$, the equation of motion is

$$m\ddot{y} + c\dot{y} + k_1 y + k_3 y^3 = f(t) \qquad (4.8\text{-}3)$$

When $f(t)$ is a cosine function, this equation is called *Duffing's equation*. Its characteristics have been extensively studied, but a closed-form solution is not available (for details, consult [Harris, 2002] or [Greenberg, 1998]).

Suppose that f is a constant, f_o. Then there can possibly be three equilibrium positions for the mass m, which are the solutions of the equation

$$k_1 y_e + k_3 y_e^3 = f_o \qquad (4.8\text{-}4)$$

where y_e denotes an equilibrium solution of this equation. We can linearize the equation of motion about such a equilibrium by linearizing the cubic term y^3 using a truncated Taylor series of y^3 about the point y_e, as follows:

$$m\ddot{y} + c\dot{y} + k_1 y + k_3 \left[y_e^3 + \frac{\partial y^3}{\partial y}\bigg|_{y=y_e} (y - y_e) \right] = f(t) = f_o + u(t)$$

where $u(t) = f(t) - f_o$ is the deviation in $f(t)$ from its reference value f_o. The equation becomes

$$m\ddot{y} + c\dot{y} + k_1 y + k_3 \left[y_e^3 + 3y_e^2(y - y_e) \right] = f_o + u(t)$$

If we define x to the deviation of y from equilibrium, $x = y - y_e$, then $\dot{x} = \dot{y}$, $\ddot{x} = \ddot{y}$, and

$$m\ddot{x} + c\dot{x} + k_1(x + y_e) + k_3 \left[y_e^3 + 3y_e^2 x \right] = f_o + u(t)$$

In light of Equation 4.8-4, this becomes

$$m\ddot{x} + c\dot{x} + \left(k_1 + 3k_3 y_e^2 \right) x = u(t)$$

The characteristic equation is

$$ms^2 + cs + k_1 + 3k_3 y_e^2 = 0 \qquad (4.8\text{-}5)$$

The equilibrium will be stable in a linearized sense (that is, *locally stable*) if both roots have negative real parts. This occurs if m, c, and $k_1 + 3k_3 y_e^2$ all have the same sign. Assuming, of course, that $m > 0$ and $c > 0$, local stability therefore requires that $k_1 + 3k_3 y_e^2 > 0$. Because $y_e^2 > 0$, we see that the equilibrium of a hardening spring is locally stable, while that of a softening spring can be either locally unstable or locally stable, depending of the values of k_1 and k_3.

Since the stability analysis is based on a linearized model, which assumes that y is "close" to y_e, its conclusions are valid only for local stability characteristics. It is possible that a system with either a hardening or a softening spring can be globally stable or globally unstable.

For example, consider the case where $f_o = 0$. Then the equilibria are found from Equation 4.8-4 to be $y_e = 0$ and $y_e = \pm\sqrt{-k_1/k_3}$. Assume that $k_1 > 0$.

- The only real-valued equilibrium is $y_e = 0$ if $k_3 > 0$.

- If $k_3 < 0$, there are three real-valued equilibria.

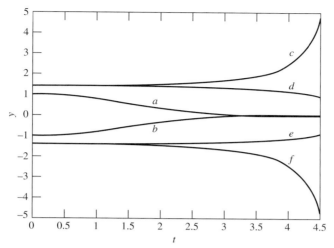

FIGURE 4.8-2   Free response of Duffing's equation for $k_3 = -1$.

- For $y_e = 0$, Equation 4.8-5 becomes $ms^2 + cs + k_1 = 0$, which represents a locally stable equilibrium regardless of the sign of k_3.

- For $y_e^2 = -k_1/k_3$, Equation 4.8-5 becomes $ms^2 + cs - 2k_1 = 0$, which represents a locally unstable equilibrium regardless of the sign of k_3.

Consider the case where $f_o = 0$, $m = 1$, $c = 2$, and $k_1 = 2$. We choose this overdamped case so that the resulting plots will be easier to understand compared to plots showing many oscillations.

1. Suppose that $k_3 = -1$. The equilibria are $y_e = 0$, $y_e = +\sqrt{2}$, and $y_e = -\sqrt{2}$. Figure 4.8-2 shows the response of the nonlinear model, Equation 4.8-3, obtained numerically using MATLAB. For all six cases, $\dot{y}(0) = 0$. Curves a and b result from the initial conditions $y(0) = 1$ and $y(0) = -1$, respectively. They both approach the equilibrium at $y_e = 0$, thus verifying the stability prediction based on the linearized equation. Curves c and d result from the initial conditions $y(0) = 1.42$ and $y(0) = 1.41$, respectively, which are slightly above and slightly below the equilibrium at $\sqrt{2} = 1.414$. Curve d approaches the equilibrium at $y_e = 0$, but curve c approaches ∞ and thus displays unstable behavior. These results show how an equilibrium of a nonlinear model can be stable for some initial conditions but unstable for others. Similar results are observed with the equilibrium at $-\sqrt{2}$.

2. Figure 4.8-3 shows the responses of the model linearized about $y_e = \sqrt{2}$ for the initial conditions $y(0) = 1.42$ and $y(0) = 1.41$, respectively. Both responses display unstable behavior, as predicted by the linearized model. The nonlinear model, however, is stable for $y(0) = 1.41$, as seen with curve d in Figure 4.8-2. This result shows how the linearized model can give erroneous results if y is not close to y_e.

Free Response of a Nonlinear System

Suppose that $f(t) = 0$ in Equation 4.8-3. This gives the model

$$m\ddot{y} + c\dot{y} + k_1 y + k_3 y^3 = 0$$

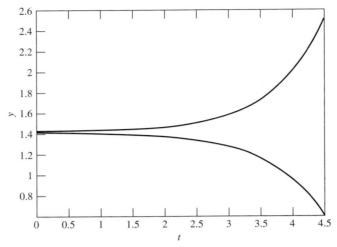

FIGURE 4.8-3 Response of linearized Duffing's equation.

Suppose that $m = 1$, $c = 0$, $k_1 = 2$, and $k_3 = 0.1$. Figure 4.8-4 shows the responses for $\dot{y}(0) = 0$ and two values of $y(0)$: $y(0) = 1$ and $y(0) = 30$. These plots were obtained using the `ode45` function in MATLAB. Although it is difficult to tell from the plots, neither response is a pure harmonic. Each response, however, contains a single harmonic that dominates the response. Using methods to be discussed in Chapter 9, we can determine the frequency of the dominant harmonic. For $y(0) = 1$, the dominant radian frequency is approximately 1.8. For $y(0) = 30$, the dominant radian frequency is approximately 8.

We thus conclude that it is possible for the oscillation frequency of the free response of a nonlinear system to be dependent on the specific initial conditions. This result stands in great contrast to the free response of linear systems, which depends only on the values of the system parameters, not on the initial conditions.

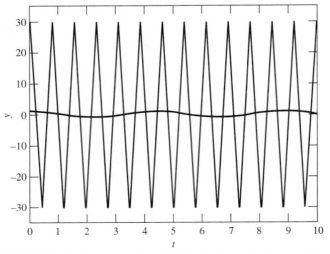

FIGURE 4.8-4 Response of linearized Duffing's equation for $k_3 = 0.1$.

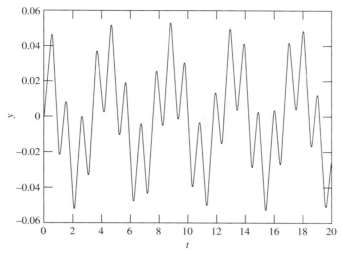

FIGURE 4.8-5 Forced response of Duffing's equation.

Harmonic Response of Duffing's Equation

Suppose that $f(t) = F_o \cos \omega t$ in Equation 4.8-3. This gives the model

$$m\ddot{y} + c\dot{y} + k_1 y + k_3 y^3 = F_o \cos \omega t$$

Suppose that $m = 1$, $c = 0$, $k_1 = 2$, and $k_3 = 0.1$. Figure 4.8-5 shows the response for $y(0) = \dot{y}(0) = 0$ and $\omega = 2$. Although it is difficult to tell from the plot, the response contains two dominant harmonics, with radian frequencies of approximately 1.6 and 6.4. These estimates can be obtained with the methods to be discussed in Chapter 9.

We thus conclude that it is possible for the oscillation frequency of the forced harmonic response of a nonlinear system to contain multiple frequencies that are different from the forcing frequency. We should have expected this after studying the response of the model for Equation 4.8-1. Recall that the harmonic response of a linear system has only one frequency, which is the same as that of the forcing function.

Duffing's equation has other unusual response characteristics, including something called the *jump phenomenon*, in which the response frequency suddenly jumps from one value to another if the forcing frequency gradually changes. This is discussed in the specialized literature; for example, see [Harris, 2002] or [Greenberg, 1998]).

Phase Plane Plots and Limit Cycles

The *phase plane plot* is a plot of one state variable, usually velocity, versus another variable, usually displacement. Such a plot provides a concise description of the response. Figure 4.8-6 shows a phase plane plots of the free response of two second-order models of the form $m\ddot{x} + c\dot{x} + kx = 0$, one with medium damping and one with light damping, using the initial conditions $x(0) = 1$, $\dot{x}(0) = 0$. The greater the damping, the faster the response spirals into the equilibrium point at $x = \dot{x} = 0$. The phase plane plot of the free response of an undamped system never reachs the point $(0, 0)$ but is a single ellipse passing through the initial condition point. The phase plane plot of the free response of an overdamped system does not spiral but heads directly for the point $(0, 0)$.

The phase plane plot of some nonlinear models displays a *limit cycle*, which is a closed curve that represents a dynamic equilibrium. Limit cycles can be stable or

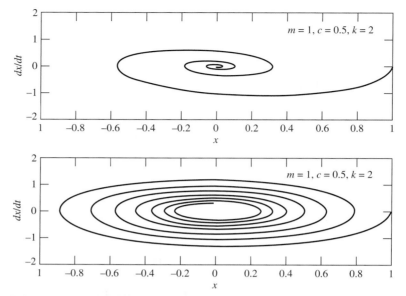

FIGURE 4.8-6 Phase plane plots of two underdamped systems.

unstable, and they can be stable for initial conditions within the closed curve and unstable for initial conditions outside the curve, or vice versa. Van der Pol's equation displays limit cycles. It has been used to model some kinds of oscillations, including some found in electrical circuits and heart dynamics. It is

$$\ddot{y} - \mu(1 - y^2)\dot{y} + y = 0 \qquad \mu > 0 \tag{4.8-6}$$

Figure 4.8-7 shows a limit cycle for Van der Pol's equation with $\mu = 1$. The response is shown for two sets of initial conditions, one inside and one outside the limit cycle. These are marked A and B in the figure. In each case the response approaches the limit cycle, which indicates, but does not prove, that the limit cycle is *bilaterally* stable.

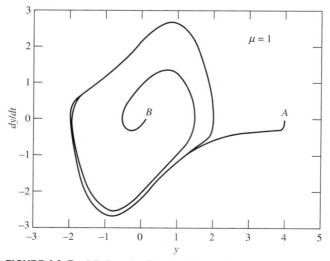

FIGURE 4.8-7 A limit cycle of Van der Pol's equation.

Analysis of limit cycles is an advanced topic, but you should be aware of their existence. One of the classic works on the subject of limit cycles and nonlinear vibration is [Minorsky, 1962].

Summary of Nonlinear Response Characteristics

In general, nonlinear models display the following phenomena:

1. The principle of superposition does not apply to nonlinear models. Thus we cannot separate the effects of the initial conditions from the forced response, and we cannot decompose the forced response into a sum of terms, each corresponding to a term in the forcing function.

2. In a linear model, response characteristics such as stability, natural frequency, and logarithmic decrement depend only on the numerical values of the mass, damping, and stiffness parameters. This is not true with nonlinear models, for which the response characteristics also depend on the initial conditions.

3. Nonlinear models can resonate at frequencies that are different from the forcing frequency, and a periodic forcing function can produce a nonperiodic response.

4. Linear models can have only one equilibrium point, and there is no distinction between local and global stability for such a point. Nonlinear models can have multiple equilibria, each of which can be (1) globally stable, (2) globally unstable, (3) locally stable but globally unstable, or (4) locally unstable but globally stable.

5. Linear models cannot have limit cycles.

4.9 FREQUENCY RESPONSE WITH MATLAB

MATLAB can be used to compute and plot the formulas derived in this chapter. It also contains some built-in functions for solving linear differential equations and for obtaining frequency response plots.

For example, Equations 4.1-25 and 4.1-27 for the the magnitude ratio M and phase angle ϕ are

$$M = \frac{1}{k} \frac{1}{\sqrt{(1 - r^2)^2 + (2\zeta r)^2}}$$

$$\phi = -\tan^{-1} \frac{2\zeta r}{1 - r^2}$$

The following MATLAB file will plot M and ϕ versus r for the case $m = 4$, $c = 4$, and $k = 9$.

```
m=4;c=4;k=9;
zeta=c/(2*sqrt(m*k));
r=[0:0.01:2];
M=(1/k)./sqrt((1-r.^2).^2+(2*zeta*r).^2);
phi=-atan2(2*zeta*r,1-r.^2)*(180/pi);
subplot(2,1,1)
plot(r,M),xlabel('r'),ylabel('M')
subplot(2,1,2)
plot(r,phi),xlabel('r'),ylabel('\phi(degrees)')
```

Transfer Function Methods

MATLAB provides several functions in the Control System Toolbox for solving linear, time-invariant (constant-coefficient) differential equations. They are sometimes more convenient to use and more powerful than the ode solvers discussed thus far because general solutions can be found for linear, time-invariant equations, and therefore less programming is required.

MATLAB uses the concept of an *LTI object*, which describes a linear, time-invariant differential equation or sets of such equations, which we refer to here as the *system*. An LTI object can be created from different descriptions of the system, it can be analyzed with several functions, and it can be accessed to provide alternate descriptions of the system. For example, from the equation

$$2\ddot{x} + 3\dot{x} + 5x = f(t) \tag{4.9-1}$$

we can immediately obtain the transfer function description of the model. It is

$$\frac{X(s)}{F(s)} = \frac{1}{2s^2 + 3s + 5} \tag{4.9-2}$$

To create an LTI object from the transfer function, you use the `tf(num,den)` function, where the vector `num` is the vector of coefficients of the numerator of the transfer function, arranged in order of descending powers of s, and `den` is the vector of coefficients of the denominator of the transfer function, also arranged in descending order. The result is the LTI object that describes the system in the transfer function form. For Equation 4.9-2, you type:

```
≫sys1=tf(1, [2,3,5]);
```

where `sys1` is an arbitrary name given to the system model created by the `tf` function.

Here is another example. The following equation has numerator dynamics:

$$2\ddot{x} + 3\dot{x} + 5x = 7\dot{f}(t) + 4f(t) \tag{4.9-3}$$

Its transfer function is

$$\frac{X(s)}{F(s)} = \frac{7s + 4}{2s^2 + 3s + 5} \tag{4.9-4}$$

To create an LTI object for this model, you type

```
≫sys2=tf([7,4], [2,3,5]);
```

The `lsim` Function

The `lsim` function solves a differential equation specified as an LTI object and plots the forced response of the system to an input defined by the user. The basic syntax is `lsim(sys,f,t)`, where `sys` is the LTI object, `t` is a time vector having regular spacing, as `t = [0:dt:tf]`, and `f` is an array describing the input.

When called with left-hand arguments, as `[x, t] = lsim(sys,f)`, the function returns the output response `y` and the time vector `t` used for the simulation. No plot is drawn, but you can use the output with the MATLAB plotting functions to generate a plot formatted to your specifications.

If $m = c = 4$ and $k = 36$, we can plot the forced response of the equation $m\ddot{x} + c\dot{x} + kx = f(t)$ to the input, $f(t) = 10 \sin 3t$, at 500 points over the time interval $0 \le t \le 10$ by typing

```
m=4;c=4;k=36;
sys3=tf(1,[m,c,k]);
t=linspace(0,10,500);
f=10*sin(3*t);
[x,t]=lsim(sys3,f,t);
plot(t,x),xlabel('t'),ylabel('x(t)')
```

This code was used to create Figure 4.2-4.

The LTI Viewer assists in the analysis of LTI systems. It provides an interactive user interface that allows you to switch between different types of response plots and between the analyses of different systems. The viewer is invoked by typing `ltiview`. See the MATLAB help for more information.

The `bode` Function

MATLAB provides an easier way to obtain the frequency response plots. The `bode` and `evalfr` functions compute and plot frequency response. These functions do not require that we first derive the formulas for M and ϕ. Thus they can be used to obtain plots for transfer functions not covered so far. In addition, when using these functions, less programming is required and you need not guess the range of frequency values, because MATLAB will do this for you.

Recall the definition of the logarithmic magnitude ratio:

$$m = 10 \log M^2 = 20 \log M \quad \text{dB}$$

The `bode` and `evalfr` functions are not part of the core MATLAB program but are available in the Control System Toolbox. The `bode` function generates frequency response plots, using decibel units and logarithmic frequency axes. It is named after H. W. Bode, who applied frequency response methods to electronic circuit design. The logarithmic frequency response plots are sometimes called Bode plots. The `evalfr` function stands for "evaluation of frequency response." It computes the value of a transfer function at a specified frequency.

The basic syntax of the `bode` function is `bode(sys)`, where `sys` is an LTI system model created with the `tf` function or the `ss` function (discussed in Chapter 8). In this basic form the `bode` function produces a screen plot of the logarithmic magnitude ratio m and the phase angle ϕ versus ω. Note that the phase angle is plotted in *degrees*, not radians. MATLAB uses the corner frequencies of the numerator and denominator of the transfer function to automatically select an appropriate frequency range for the plots. For example, to obtain the plots for the transfer function

$$T(s) = \frac{100}{s^2 + 2s + 25} \tag{4.9-5}$$

you type

```
≫sys=tf(100,[1,2,25]);
≫bode(sys)
```

Figure 4.9-1 shows the plots you will see on the screen.

Note that plots are in *decibels* and *degrees*, while the frequencies are given in *rad/sec*. If the time units of your problem are not seconds, you will need to use the label

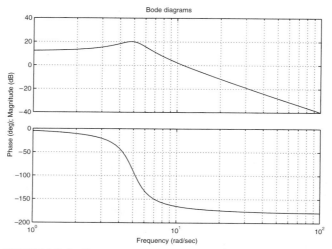

FIGURE 4.9-1 Frequency response plots generated with the bode function.

functions discussed below or edit the plot. To edit the axis label, click on the northwest-facing arrow at the top of the plot window. This enables the editing functions. Then double-click on the axis label, delete the unwanted text, and add your own. Then click outside the label.

Interactive Plot Screen While the plot is on the screen, you can click the right-hand mouse button to display a menu. Click on **Characteristics**, then on **Peak Response**. MATLAB will display dashed lines showing the peak value of m and its corresponding frequency, and it will display these values in a box. In this example, the results are 20.2 dB at 4.8 rad/sec. To obtain the bandwidth, position the cursor on the plotted line and click the left-hand mouse button. A small filled square will appear on the line, along with a box indicating the coordinates. When you move the cursor over the square, the cursor will change to a hand. While depressing the button, you can then slide the square along the line. You can use this method to compute the bandwidth by selecting the two points 3 dB below the peak. In this example, the bandwidth frequencies are 3.6 to 5.7 rad/sec.

You can also use this technique on the phase plot.

Extended Syntax Sometimes you might want to specify a set of frequencies rather than let MATLAB select them. In this case you can use the syntax `bode(sys,w)`, where `w` is a vector containing the desired frequencies. The frequencies must be specified in *radians* per unit time (for example, radians/second). For example, you might want to examine more closely the peak in the m curve near the resonant frequency of Equation 4.8-1, $\omega_p = 4.8$. To see 401 regularly spaced points on the curve for $4 \le \omega \le 6$, you type

```
≫sys=tf(100,[1,2,25]);
≫w=[4:0.005:6];
≫bode(sys,w)
```

To obtain 401 points in the range $4 \le \omega \le 6$ that are logarithmically spaced, instead type `w = logspace(log10(4),log10(6),401)`.

If you want to specify just the lower and upper frequencies `wmin` and `wmax`, and let MATLAB select the spacing, generate the vector `w` by typing

```
≫w={wmin,wmax};
```

You can compare the responses of several systems on the same plot by using the syntax `bode(sys1, sys2, ..., sysN)` or the syntax `bode(sys1, sys2, ..., sysN, w)`. This will display the curves for every model on the same plot. To use this syntax, all the models must have the same number of inputs and the same number of outputs. You can distinguish each curve with a different line style by using the syntax `bode(sys1, 'PlotStyle1', ..., sysN, 'PlotStyleN')` or the syntax `bode(sys1, 'PlotStyle1', ..., sysN, 'PlotStyleN', w)`. The string `'PlotStyle1'` specifies the color, line style, and/or marker to be used. For example, assuming that `sys1` and `sys2` are LTI models that have been previously defined and that the frequency vector `w` has been specified, the following session plots the response for model `sys1` as a green dashed line and the response for model `sys2` with red x markers.

```
≫bode(sys1,'g − −',sys2,'rx',w)
```

A very important syntax form is the following

```
[mag,phase,w]=bode(sys);
```

This form returns the magnitude ratio M (*not* in decibels!), the phase angle ϕ in degrees, and the frequencies in `mag`, `phase`, and `w` but does not display a plot. It is very important to note that `mag` contains the ratio M, *not* the logarithmic ratio m. Note also that `phase` contains the phase angle in *degrees*, not radians. However, `w` contains the frequencies in *radians* per unit time. The frequencies in `w` are the frequencies automatically selected by MATLAB. If you want to use your own set of frequencies, use the syntax

```
[mag, phase]=bode(sys,w)
```

It is important to realize that the returned variables `mag` and `phase` are three-dimensional arrays whose third dimension contains the needed values. The easiest way to access these values is to reassign them as follows, where `M` and `phi` are arbitrarily chosen names.

```
≫M=mag(:);
≫phi=phase(:);
```

The extended syntax forms are useful when you want to save the frequency response calculations generated by the `bode` function. You can use `mag`, `phase`, and `w` for further analysis or to generate plots that you can format as you want. For example, to plot M versus ω for Equation 4.8-3 using 401 regularly spaced points over the range $4 \leq \omega \leq 6$, you type

```
≫sys = tf(100,[1,2,25]);
≫w = [4:0.005:6];
≫[mag] = bode(sys,w);
≫M = mag(:);
≫plot(w,M),xlabel('Frequency(rad/sec)'),...
      ylabel('Magnitude Ratio M')
```

Note that the `phase` array need not be generated if you are not going to use it. Similarly, the `mag` array is optional if you need phase information only.

To plot the logarithmic magnitude ratio m versus log ω using the frequencies generated by MATLAB, you type

```
>>sys=tf(100,[1,2,25]);
>>[mag,phase,w]=bode(sys);
>>M=mag(:);
>>semilogx(w,20*log10(M)),grid,...
       xlabel('Frequency(rad/sec)'),...
       ylabel('Log Magnitude Ratio m(dB)')
```

Note that you must use the decibel conversion $m = 20 \log M$.

The related function `bodemag(sys)` plots the magnitude in decibels but does not generate a phase plot. Its extended syntax is identical to that of the function `bode`.

The `evalfr` Function

The `evalfr` function computes the complex value of a transfer function at a specified value of s. For example, to compute the value of $T(s)$ given by Equation 4.8-1 at $s = 0.5i$ and find its magnitude and phase angle, you type

```
>>fr=evalfr(sys,0.5i);
>>mag=abs(fr);
>>ang=angle(fr);
```

The results are `fr` $= 4.0338 - 0.1630i$, `mag` $= 4.0371$, and `phase` $= -0.0404$ rad. Note that MATLAB uses i, not j, to display imaginary parts.

The `freqresp` Function

The `evalfr` function can be used to evaluate `sys` at one frequency only. For a vector of frequencies, use the `freqresp` function, whose syntax is

```
fr=freqresp(sys,w)
```

where `w` is a vector of frequencies, which must be in radians/unit time.

4.10   SIMULINK APPLICATIONS

Now that we have introduced the transfer function $T(s)$, you can understand why Simulink uses the symbol $1/s$ for an integrator block. Consider the equation $\dot{x} = y$. If we integrate both sides with respect to t, we obtain

$$x = \int y \, dt$$

Thus x is the integral of y. However, the transfer function of $\dot{x} = y$ is

$$T_{x/y} = \frac{X(s)}{Y(s)} = \frac{1}{s}$$

and thus we see that $1/s$ represents an integration.

The MATLAB `lsim` solver is very convenient to use. It is limited, however, to solving *linear* equations. On the other hand, the `ode45` function and its related solvers are excellent for solving nonlinear equations, but only those containing what we may describe as "continuous" nonlinear functions, such as $\sin\theta$. These solvers are more difficult to apply when an equation contains more complicated nonlinear functions, such as Coulomb friction. Such a function is described as "piecewise linear," although in fact it is nonlinear. In Chapter 3 we used Simulink to obtain the free response of a mass-spring system having Coulomb friction.

Simulink is well suited to handle nonlinearities that are difficult to program. Of course, it can also handle linear equations with ease if you prefer to use a graphical interface instead of the MATLAB Command window or M-files. In this section we show how to use Simulink to obtain the forced response of systems having nonlinearities.

Simulating Transfer Function Models

The equation of motion for the mass and spring with viscous surface friction is repeated here:

$$m\ddot{x} + c\dot{x} + kx = f(t) \tag{4.10-1}$$

As with the Control System Toolbox, Simulink can accept a system description in transfer function form. This form was discussed in Section 4.9. In MATLAB the transfer function form of Equation 4.10-1 is entered by typing `tf([1],[m,c,k])`, assuming m, c, and k have already been assigned numerical values. Simulink represents this form graphically using a ratio of polynomials, as follows:

$$\frac{1}{ms^2 + cs + k}$$

If the mass-spring system is subjected to a sinusoidal forcing function $f(t)$, it is easy to use the MATLAB commands presented in Section 4.9 to solve and plot the response $x(t)$. However, suppose that the force $f(t)$ is created by applying a sinusoidal input voltage to a hydraulic piston that has a *saturation nonlinearity*. This means that the piston cannot generate more than a certain amount of force in both the positive and negative directions. All actuators have limits on the amount of force or torque they can generate, and the saturation nonlinearity can be used to model this limitation. Simulink provides the Saturation block in the Discontinuities library for this purpose.

The graph of the saturation nonlinearity is shown in Figure 4.10-1. When the input (the independent variable on the graph) is between the lower-limit and the upper-limit values, the output of the block equals its input. When the input is greater than or equal to the upper limit, the output is set equal to the upper limit. When the input is less than or equal to the lower limit, the output is set equal to the lower limit. For example, if the limits are set to -5 and 7, and if the input is $f(t) = 10\sin 4t$, the output $g(t)$ will be

$$g(t) = \begin{cases} 7 & \text{if } 10\sin 4t \geq 7 \\ 10\sin 4t & \text{if } -5 < 10\sin 4t < 7 \\ -5 & \text{if } 10\sin 4t \leq -5 \end{cases}$$

As you can see from this example, the saturation nonlinearity need not be symmetric about 0.

Simulations with saturation nonlinearities are somewhat tedious to program in MATLAB but are easily done in Simulink. The following example illustrates how it is done.

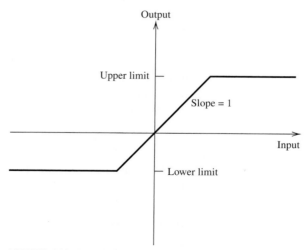

FIGURE 4.10-1 The saturation nonlinearity.

EXAMPLE 4.10-1
A Simulink Model of Saturation Response

Create and run a Simulink simulation of a mass-spring system with viscous friction using the parameter values $m = 1$, $c = 2$, and $k = 4$. The forcing function is the function $f(t) = \sin 1.4t$. The system has a saturation nonlinearity with a lower limit of -0.3 and an upper limit of 0.6.

Solution

To construct the simulation, perform the following steps.

1. In the Command window, type m = 1; c = 2; k = 4;.
2. Start Simulink and open a new model window as described previously.
3. Select and place in the new window the Sine block from the Sources category. Double-click on it, and set the Amplitude to 1, the Frequency to 1.4, the Phase to 0, and the Sample time to 0. Click **OK**.
4. Select and place the Saturation block from the Discontinuities category, double-click on it, and set the Lower limit to -0.3 and the Upper limit to 0.6. Click **OK**.
5. Select and place the Transfer Fcn block from the Continuous category, double-click on it, and set the Numerator to [1] and the Denominator to [m, c, k]. Click **OK**. When the simulation runs, it will use the values for m, c, and k you entered in step 1.
6. Select and place the Scope block from the Sinks category.
7. Once the blocks have been placed, connect the input port on each block to the outport port on the preceding block. To do this, move the cursor to an input port or an output port; the cursor will change to a cross. Hold the mouse button down, and drag the cursor to a port on another block (always connect an output port to an input port). When you release the mouse button, Simulink will connect them with an arrow pointing at the input port. Your model should now look like that shown in Figure 4.10-2.
8. Click on the **Simulation** menu, and click the **Configuration Parameters** item. Click on the Solver tab, and enter 15 for the Stop time. Make sure the Start time is 0. Then click **OK**.
9. Run the simulation by clicking on the **Simulation** menu and then clicking the **Start** item. You can also start the simulation by clicking on the Start icon on the toolbar (this is the black triangle).

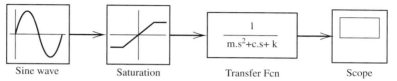

FIGURE 4.10-2   Simulink model of saturation response.

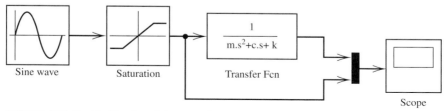

FIGURE 4.10-3   Modification of the saturation model to include a Mux block.

10. You will hear a bell sound when the simulation is finished. Then double-click on the Scope block and then click on the binoculars icon in the Scope display to enable autoscaling. You should see an oscillating curve. The independent variable in the Scope block is time; the input to the block is the dependent variable. This completes the simulation.

It is informative to plot both the input and the output of the Transfer Fcn block versus time on the same graph. To do this, perform the following steps:

1. Delete the arrow connecting the Scope block to the Transfer Fcn block. Do this by clicking on the arrow line and then pressing the **Delete** key.

2. Select and place the Mux block from the Signal Routing Library, double-click on it, and make sure the Number of inputs is set to 2. Click **OK**.

3. Connect the top input port of the Mux block to the output port of the Transfer Fcn block. Then use the same technique to connect the bottom input port of the Mux block to the arrow from the outport port of the Dead Zone block. Just remember to

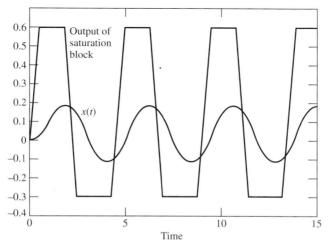

FIGURE 4.10-4   Plot of the response of the saturation model.

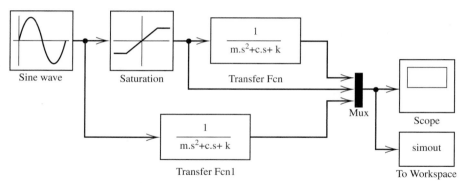

FIGURE 4.10-5 Simulink model to compare response with and without saturation.

start with the input port. Simulink will sense the arrow automatically and make the connection. Your model should now look like that shown in Figure 4.10-3.

4. Set the Stop time to 15, run the simulation as before, and bring up the Scope display. You should see a plot similar to the edited plot shown in Figure 4.10-4. This plot shows the effect of the saturation nonlinearity on the sine wave.

You can bring the simulation results into the MATLAB workspace by using the To Workspace block. For example, suppose we want to examine the effects of the saturation nonlinearity by comparing the response of the system with and without saturation. We can do this with the model shown in Figure 4.10-5. To create this model, perform the following steps:

1. Copy the Transfer Fcn block by right-clicking on it, holding down the mouse button, and dragging the block copy to a new location. Then release the button. Copy the Mux block in the same way.

2. Double-click on the first Mux block and change the number of its inputs to 3.

3. In the usual way, select and place the To Workspace block from the Sinks category. Double-click on the To Workspace block. Specify the Save format as Array. You can specify any variable name you want as the output; the default is `simout`. Change its name to y. The output variable y will have as many rows as there are simulation time steps and as many columns as there are inputs to the block. The first column in the array y will be the output of the upper Transfer Fcn block, because we have connected that output to the top port of the Mux. The second and third columns of y will contain the output of the Saturation block and the output of the lower Transfer Fcn block, respectively. Use the default values for the other parameters (these should be `inf`, `1`, and `-1` for Maximum number of rows, Decimation, and Sample Time, respectively. Click on **OK**.

4. Connect the blocks as shown, and run the simulation.

5. When the simulation runs, it puts the time values in the variable `tout` if that option has been set in the Data I/O tab under the **Simulation Parameters** menu.

6. You can use the MATLAB plotting commands from the Command window to plot the columns of y; for example, to plot the response of the two systems and the output of the Saturation block versus time, type

```
≫plot(tout,y(:,1),tout,y(:,2),tout,y(:,3))
```
∎

Forced Response with Coulomb Friction

The equation of motion for a mass-spring system having Coulomb friction and an applied force $f(t)$ is

$$m\ddot{x} + kx = f(t) - \mu mg \, \text{sign}(\dot{x}) \quad \text{where} \quad \text{sign}(\dot{x}) = \begin{cases} +1 & \text{if } \dot{x} > 0 \\ 0 & \text{if } \dot{x} = 0 \\ -1 & \text{if } \dot{x} < 0 \end{cases} \quad (4.10\text{-}2)$$

We may modify the Simulink program given in Chapter 3 to investigate the effects of the force $f(t)$. The following example shows how to do this for a sinusoidal input.

EXAMPLE 4.10-2
Forced Response with Coulomb Friction

Consider a 1-kg mass moving on a horizontal surface with $\mu = 0.4$ and attached to a spring of stiffness $k = 5$ N/m. A force $f(t) = 10 \sin 2.5t$ acts on the mass. Suppose the mass is initially at rest and it is suddenly given a velocity of 5 m/s. Use Simulink to obtain the total response.

Solution

Here $\mu mg = 3.92$, and the equation of motion becomes

$$\ddot{x} + 5x = 10 \sin 2.5t - 3.92 \, \text{sign}(\dot{x})$$

Modify the model shown in Figure 3.10-9 to obtain that shown in Figure 4.10-6. The output of the Integrator 1 block is $\dot{x}$. Double-click on it and make sure the initial condition is set to 5. The output of the Integrator 2 block is x. Double-click on it and set the initial condition to 0.

The Stop time is unknown here, so we must experiment with the model. Increase the Stop time and run the simulation until you see a constant amplitude develop. Figure 4.10-7 shows a plot for a Stop time of 80 s.

The theory of Equation 4.7-3 predicts a steady-state amplitude of $X = 6.9$, which agrees with the simulation. ∎

The real usefulness of Simulink, of course, is for those problems for which we do not have theory to make predictions. Such problems include nonlinear systems for which no analytical solution has been obtained. One such case is given by the following example.

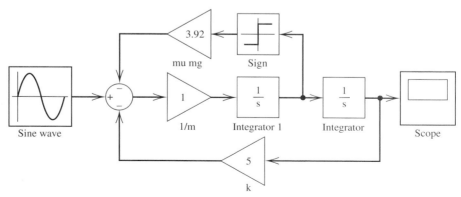

FIGURE 4.10-6 Simulink model of harmonic response with Coulomb friction.

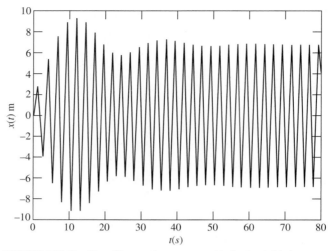

FIGURE 4.10-7   Plot of harmonic response with Coulomb friction.

EXAMPLE 4.10-3
A Nonlinear
Pendulum with
Coulomb Friction

In Example 2.3-4 we derived the equation of motion for a crane with a movable base (Figure 4.10-8). We now include a Coulomb friction moment M_F in the pivot. Thus the equation of motion is

$$L\ddot{\theta} + g\sin\theta = a\cos\theta - M_F$$

where a is the acceleration of the base and

$$M_F = 0.03\,\text{sign}(\dot{\theta})\ \ \text{N}\cdot\text{m}$$

We are given $L = 10$ m and $a = A\sin\omega t$. Create a Simulink model for this problem.

(a) Check the model with the linearized, small-angle approximation, using a small amplitude $A = 0.5$ m/s² and a forcing frequency $\omega = 0.5$ rad/s that is not close to resonance.

(b) Use the Simulink model to plot the response for a larger amplitude $A = 1.5$ m/s² and a forcing frequency $\omega = 0.8$ rad/s closer to resonance.

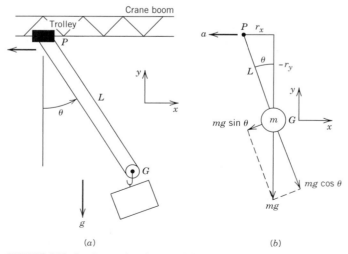

FIGURE 4.10-8   An overhead crane with an accelerating base.

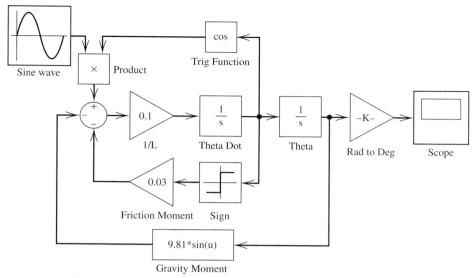

FIGURE 4.10-9   Simulink model of an overhead crane with an accelerating base.

Solution

The model is shown in Figure 4.10-9. We have used two new blocks. The Cos block implements cos θ. Select it from the Trigonometric Function category and place it. Double-click on it and select **cos** from the menu of functions. The Product block is in the Math category. After placing it, double-click on it and specify it to have two inputs. The Gain block before the Scope is used to convert from radians to degrees. Its gain value is $180/\pi$.

(a) The linearized model for small angles is

$$L\ddot{\theta} + g\theta = a - M_F \operatorname{sign}(\dot{\theta})$$

From the theory developed in Section 4.7 with $A = 0.5$ and $\omega = 0.5$, the predicted steady-state amplitude is $\Theta = 3.9°$.

   Double-click on the Sine wave block and set the Amplitude to 0.5, the Frequency to 0.5, and the Phase to 0. Make sure the initial value for each integrator is 0. Run the simulation and check the results. They will agree with those predicted from the linearized equation.

(b) Now double-click on the Sine wave block and set the Amplitude to 1.5, the Frequency to 0.8, and the Phase to 0. Run the simulation and examine the results. The plot will show beating and an amplitude of about 50° at steady state (Figure 4.10-10). Note that the resonant frequency predicted by the linearized model is $\omega_n = \sqrt{g/L} = \sqrt{0.98} = 0.99$ rad/s. So the forcing frequency of 0.8 is close to the resonant frequency. This causes the large amplitude. ∎

The Dead Zone Block

Consider the system shown in Figure 4.10-11. The block of mass m slides on a frictionless surface under the influence of the applied force $f(t)$ and the spring force when the mass is in contact with the spring. The width of the block is W. When the block is centered at $x = 0$, each side of the block is a distance D from a spring.

   The distance D is an example of a *dead zone*, in which no spring force acts on the mass. A graph of a particular dead zone nonlinearity is shown in Figure 4.10-12. When

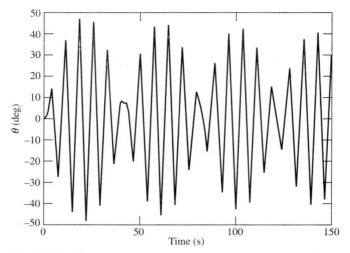

FIGURE 4.10-10 Plot of the response of an overhead crane to harmonic base acceleration.

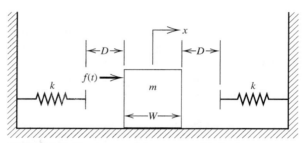

FIGURE 4.10-11 A mass-spring system with dead zone.

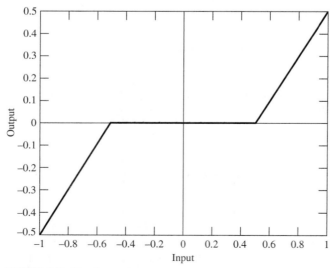

FIGURE 4.10-12 Graph of a particular dead zone nonlinearity.

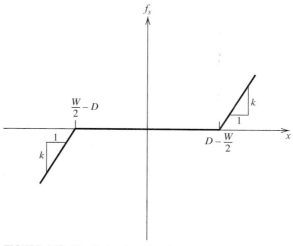

FIGURE 4.10-13   Spring force as a function of mass displacement for the system shown in Figure 4.10-11.

the input (the independent variable on the graph) is between -0.5 and 0.5, the output is zero. When the input is greater than or equal to the upper limit of 0.5, the output is the input minus the upper limit. When the input is less than or equal to the lower limit of -0.5, the output is the input minus the lower limit. In this example, the dead zone is symmetric about 0, but it need not be in general. The spring force as a function of x is graphed in Figure 4.10-13.

Simulations with dead zone nonlinearities are somewhat tedious to program in MATLAB but are easily done in Simulink. The following example illustrates how it is done.

EXAMPLE 4.10-4
A Simulink Model of Dead Zone Response

Create and run a Simulink simulation of the mass-spring system shown in Figure 4.10-11 using the parameter values $m = 2$, $k = 4$, $D = 4$, and $W = 2$. The forcing function is $f(t) = 5 \sin 0.5t$.

Solution

Construct the model shown in Figure 4.10-14. Set the two gains to $1/m$ and k respectively. Set the Start of dead zone to `-(D-W/2)`, and set the End of dead zone to `D-W/2`. In the Command window, type `m = 2; k = 4; D = 4; W = 2;`. Set the initial conditions on the integrators to 0. Set the Amplitude of the Sine Wave block to 5 and the Frequency to 0.5. Finally, set the Stop time to 30 and run the simulation.

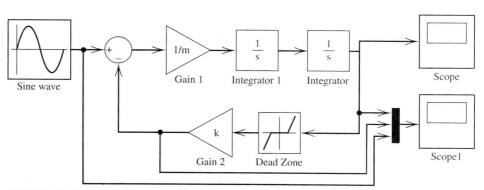

FIGURE 4.10-14   Simulink model for Example 4.10-4.

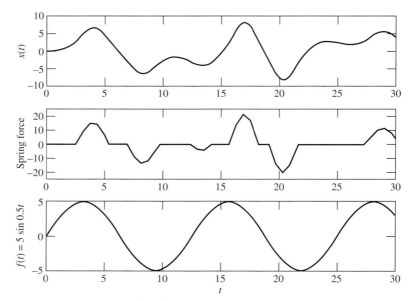

FIGURE 4.10-15 Plots of the displacement, spring force, and forcing function for Example 4.10-4.

Use the `subplot` function to obtain a plot like that shown in Figure 4.10-15, which has been edited. Note the irregular shape of the plot of $x(t)$. This is due to the nonlinear characteristics of the system. ∎

4.11 CHAPTER REVIEW

This chapter treated the harmonic response of systems having one degree of freedom. We introduced the transfer function and demonstrated its usefulness for analyzing a system's frequency response. A sinusoidal input applied to a stable linear system produces a steady-state sinusoidal output of the same frequency, but with a different amplitude and a phase shift. The frequency transfer function is the transfer function with the variable s replaced by $i\omega$, where ω is the input frequency. The magnitude M of the frequency transfer function is the amplitude ratio between the input and output, and its angle is the phase shift ϕ.

Both M and ϕ are functions of ω, and a system's frequency response plots consist of plots of M and ϕ versus ω or log ω. When log ω is used, M is expressed in decibels, where $m = 20 \log M$. The plots are useful for determining the bandwidth, which is the frequency range for which $M \geq M_p$, where M_p is the peak value. The bandwidth indicates the filtering property of the system.

With base excitation the input is a displacement. Applications include ground motion caused by earthquakes and displacement inputs caused by vehicle motion. Two quantities of interest are the displacement of the mass and the force transmitted to the mass as a result of the base motion.

Rotating unbalance in machinery produces a harmonic force. The resulting motion of the mass and the force transmitted to the supporting structure are important in many applications. An important application is the design of vibration isolators to minimize the transmission of vibration between a support and a machine. Another example of rotating unbalance is shaft whirl.

Sources of damping other than linear viscous damping include hysteresis, drag, and Coulomb friction. These types of damping introduce nonlinear effects into the equation of motion. We used energy concepts to model the effects of such damping as an equivalent linear viscous damping term, which simplifies the required mathematics. Nonlinear damping and nonlinear stiffness functions introduce response properties not found in linear systems.

The chapter concluded with coverage of computer methods. We introduced several MATLAB functions that are especially intended for harmonic response analysis. Simulink contains several features useful for simulating systems containing nonlinearities that are difficult to program in MATLAB or other languages. These include the Sign block, the Dead Zone block, and the Saturation block.

Now that you have finished this chapter, you should be able to do the following:

1. Obtain the harmonic response of systems having a single degree of freedom.

2. Obtain the transfer function from the equation of motion.

3. Obtain the frequency transfer function.

4. Use the plots or the frequency transfer function to determine the steady-state output amplitude and phase that result from a sinusoidal input.

5. Determine the resonance frequency, peak response, and bandwidth, and identify the system as a low-pass, high-pass, or band-pass filter.

6. Analyze the displacement and transmitted force of systems having base excitation, rotating unbalance, or rotor-shaft vibration.

7. Determine the critical speed of a rotor-shaft system.

8. Design simple vibration isolation systems.

9. Model and analyze nonlinear sources of damping, such as hysteresis, drag, and Coulomb friction.

10. Describe the differences between the response characteristics of linear and nonlinear systems.

11. Apply MATLAB and Simulink to analyze harmonic response.

PROBLEMS

SECTION 4.1 SOLUTION FOR THE HARMONIC RESPONSE

4.1 Use the transfer functions given below to find the steady-state response $x_{ss}(t)$ to the given input function $f(t)$, where $T(s) = X(s)/F(s)$.

(a) $T(s) = \dfrac{5}{40s^2 + 14s + 1}$ $f(t) = 20\sin 0.01t$

(b) $T(s) = \dfrac{4}{s^2 + 10s + 100}$ $f(t) = 4\sin 5t$

4.2 The equation of motion of a certain mass-spring-damper system is
$$5\ddot{x} + c\dot{x} + 10x = f(t)$$
Suppose that $f(t) = F\sin\omega t$. Define the magnitude ratio as $M = X/F$. Determine the natural frequency ω_n, the peak frequency ω_p, and the peak magnitude ratio M_p for (a) $\zeta = 0.1$ and (b) $\zeta = 0.3$.

4.3 The equation of motion of a certain mass-spring-damper system is
$$5\ddot{x} + c\dot{x} + 10x = f(t)$$

How large must the damping constant c be so that the maximum steady-state amplitude of x is no greater than 3, if the input is $f(t) = 22\sin\omega t$, for an arbitrary value of ω?

SECTION 4.2 RESONANCE AND BANDWIDTH

4.4 Determine the beat period and the vibration period of a mass-spring system whose natural frequency is 500 rad/s and whose forcing frequency is 503 rad/s.

4.5 A certain mass-spring system oscillates with an amplitude of 5 mm when the forcing frequency is 20 Hz, and with an amplitude of 1 mm when the forcing frequency is 40 Hz. Estimate the natural frequency of the system.

4.6 When the forcing frequency is 3500 rpm, a certain mass-spring system oscillates with an amplitude equal to 60% of the amplitude if the forcing frequency is 1750 rpm. Estimate the natural frequency of the system.

4.7 A certain metal frame supports a platform and motor. The combined mass of the platform and motor is 500 kg. At a motor speed of 800 rpm, the platform oscillates with an amplitude of 5 mm. When the speed is 1000 rpm, the amplitude is 14 mm. Determine the natural frequency of the system and the spring constant of the frame.

4.8 Find the bandwidth for each of the transfer functions listed in Problem 4.1.

4.9 A certain system has the following transfer function:

$$T(s) = \frac{10^{-6}s}{3 \times 10^{-6}s^2 + 10^{-2}s + 1}$$

Find the bandwidth.

4.10 A certain vibration instrument like that shown in Figure 4.2-8 has the displacement input $y(t) = A \sin 236t$, where t is in seconds. The instrument mass is 0.1 kg.

(a) For what range of spring constants can the instrument be used to measure the displacement amplitude A?

(b) For what range of spring constants can the instrument be used to measure the acceleration amplitude?

SECTION 4.3 BASE EXCITATION

4.11 The 0.5-kg mass shown in Figure P4.11 is attached to the frame with a spring of stiffness $k = 500$ N/m. Neglect the spring weight and any damping. The frame oscillates vertically with an amplitude of 4 mm at a frequency of 3 Hz. Compute the steady-state amplitude of motion of the mass.

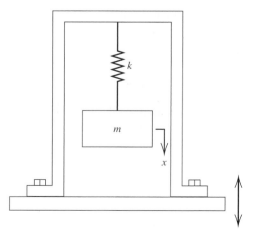

FIGURE P4.11

4.12 A machine mounted on an elastic support with negligible damping is observed to vibrate with an amplitude of 6 mm at 3 Hz and 9 mm at 6 Hz. Determine the resonant frequency.

4.13 A certain mass is driven by base excitation through a spring (Figure P4.13). Its parameter values are $m = 100$ kg, $c = 400$ N·s/m, and $k = 10\ 000$ N/m. Determine its peak frequency ω_p, its peak M_p, and the lower and upper bandwidth frequencies.

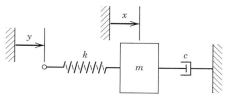

FIGURE P4.13

4.14 A certain mass is driven by base excitation through a spring (Figure P4.13). Its parameter values are $m = 100$ kg, $c = 1000$ N·s/m, and $k = 10\ 000$ N/m. Determine its peak frequency ω_p, its peak M_p, and its bandwidth.

4.15 A quarter-car representation of a certain car has a stiffness $k = 4 \times 10^4$ N/m, which is the series combination of the tire stiffness and suspension stiffness, and a damping constant of $c = 4000$ N·s/m. The car mass is 1500 kg. Suppose the road profile is given (in meters) by

$$y(t) = 0.015 \sin \omega t$$

where the amplitude of variation of the road surface is 0.015 m and the frequency ω depends on the vehicle's speed and the road profile's period. Suppose the period of the road surface is 10 m. Compute the steady-state motion amplitude and the force transmitted to the chassis if the car is traveling at speeds of 13 and 26 m/s.

4.16 A certain factory contains a heavy rotating machine that causes the factory floor to vibrate. We want to operate another piece of equipment nearby, and we measure the amplitude of the floor's motion at that point to be 0.01 m. The mass of the equipment is 1500 kg and its support has a stiffness of $k = 20\ 000$ N/m and a damping ratio of $\zeta = 0.04$. Calculate the maximum force that will be transmitted to the equipment at resonance.

4.17 Figure P4.17 shows a system being driven by base excitation through a damping element. Assume that the base displacement is sinusoidal: $y(t) = Y \sin \omega t$.

(**a**) Derive the expression for X, the steady-state amplitude of motion of the mass m.

(**b**) Derive the expression for F_t, the steady-state amplitude of the force transmitted to the support.

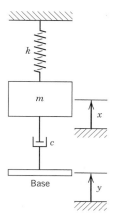

FIGURE P4.17

4.18 An electronics module inside an aircraft must be mounted on an elastic pad to protect it from vibration of the airframe. The largest-amplitude vibration produced by the airframe's motion has a frequency of 40 cycles per second. The module weighs 200 N, and its amplitude of motion is limited to 0.003 m to save space. Neglect damping and calculate the percent of the airframe's motion transmitted to the module.

4.19 An electronics module used to control a large crane must be isolated from the crane's motion. The module has a mass of 1 kg.

(a) Design an isolator so that no more than 10% of the crane's motion amplitude is transmitted to the module. The crane's vibration frequency is 3000 rpm.

(b) What percentage of the crane's motion will be transmitted to the module if the crane's frequency can be anywhere between 2500 and 3500 rpm?

4.20 An instrument of mass 2 kg is mounted on the housing of a pump that rotates at 30 rpm. The amplitude of motion of the housing is 0.9 mm. We want no more than 10% of the housing's motion to be transmitted to the instrument. Design a suitable isolator having negligible damping. Compute the force transmitted to the instrument.

4.21 Consider the vehicle suspension problem in Example 4.3-1. Investigate the choice of the damping constant c for each car so that the displacement transmissibility X/Y is as small as possible for the case where the cars' speed gives a frequency ratio of $r = 2$. Do this by trying different values for the damping ratio ζ. Investigate your design's sensitivity to a $\pm 20\%$ variation in r.

4.22 Consider the vehicle suspension problem in Example 4.3-1. Suppose the amplitude of variation of the road surface is 16 mm. Determine a set of values for the suspension's stiffness k and damping c so that the force transmitted to the chassis of the lighter car will be as small as possible at a speed of 18 m/s. For your design, calculate the transmitted force at this speed.

SECTION 4.4   ROTATING UNBALANCE

4.23 When a certain motor is started, it is noticed that its supporting frame begins to resonate when the motor speed passes through 900 rpm. At the operating speed of 1750 rpm, the support oscillates with an amplitude of 8 mm. Determine the amplitude that would result at 1750 rpm if the support were replaced with one having one-half the stiffness.

4.24 A 225-kg motor is supported by an elastic pad that deflects 6 mm when the motor is placed on it. When the motor operates at 1750 rpm, it oscillates with an amplitude of 2.5 mm. Suppose a 680-kg platform is placed between the motor and the pad. Compute the oscillation amplitude that would result at 1750 rpm.

4.25 A certain pump has a mass of 23 kg and has a rotating unbalance. The unbalanced weight is 0.2 N and has an eccentricity of 2.5 mm. The pump rotates at 1000 rpm. Its vibration isolator has a stiffness of $k = 7300$ N/m. Compute the force transmitted to the foundation if the isolator's damping ratio is (a) $\zeta = 0.05$ and (b) $\zeta = 0.7$.

4.26 In order to calculate the effects of rotating unbalance, we need to know the value of the product mR, where m is the unbalanced mass and R is the eccentricity. These two quantities are sometimes difficult to calculate separately, but sometimes an experiment can be performed to estimate the product mR. An experiment was performed on a particular rotating machine whose mass is 75 kg. The machine's support has negligible damping and a stiffness of $k = 2500$ N/m. When the machine operates at 200 rpm, the measured force transmitted to the foundation was 15 N. Estimate the value of mR.

4.27 A computer disk drive is mounted to the computer's chassis with an isolator consisting of an elastic pad. The disk drive motor weighs 3 kg and runs at 3000 rpm. Calculate the pad stiffness required to provide a 90% reduction in the force transmitted from the motor to the chassis.

4.28 Figure P4.28 shows a motor mounted on four springs (the second pair of springs is behind the front pair and is not visible). Each spring has a stiffness $k = 2000$ N/m. The inertia of the motor is

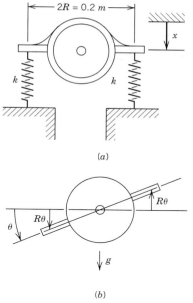

(a)

(b)

FIGURE P4.28

$I = 0.2$ kg·m^2; its mass is $m = 25$ kg, and its speed is 1750 rpm. Because the motor mounts allow the motor to vibrate both vertically and rotationally, the system has a vertical force transmissibility F_t/F and a torque transmissibility T_t/T, where F is the vertical unbalance force and T is the torque due to the unbalance.

Neglect damping in the system and compute the vertical force transmissibility F_t/F and the torque transmissibility T_t/T.

4.29 A particular system has a rotating unbalanced mass m with eccentricity R. The machine's total mass is M and it rotates at a speed ω_R. Derive the expression for the ratio MX/mR as a function of the frequency ratio r, where $r = \omega_R/\sqrt{k/M}$.

4.30 The following data were compiled by driving a machine on its support with a rotating unbalance force at various frequencies. The machine's mass is 100 kg, but the stiffness and damping in the support are unknown. The frequency of the driving force is f Hz. The measured steady-state displacement of the machine is X mm.

(a) Estimate the stiffness and damping in the support.

(b) Estimate what percentage of the rotating unbalance force is transmitted to the foundation at 6 Hz.

f (Hz)	X (mm)	f (Hz)	X (mm)
0.2	1	3.8	13
1	2	4	11
2	4	5	8
2.6	12	6	7
2.8	18	7	6
3	25	8	6
3.4	18	9	6
3.6	15	10	5

SECTION 4.5   CRITICAL SPEEDS OF ROTATING SHAFTS

4.31 A 100-kg rotor rotates on a shaft whose bending stiffness is 2×10^7 N/m. Assuming a damping ratio of $\zeta = 0.05$ due to external damping and an eccentricity of $R = 1$ mm, calculate the critical speed, the steady-state shaft deflection D at the operating speed of 5000 rpm, and the steady-state shaft deflection D at the critical speed.

4.32 For a rotor-shaft system, the design value of D/R for large r is $D/R = 1$.

    (a) Analyze the uncertainty in the ratio D/R due to uncertainty in the value of ζ if we design the rotor-shaft system so that $r = 2.5$.

    (b) How much must the shaft stiffness k be changed so that D/R differs no more than 10% from the design value?

SECTION 4.6   EQUIVALENT DAMPING

4.33 When an electric motor is mounted on rubber isolators, the measured peak of the normalized amplitude ratio kX/F is 4. What must the frequency ratio r be in order to reduce the normalized amplitude ratio to 1/3 if we model the damping in the isolators as viscous damping?

4.34 When an electric motor is mounted on rubber isolators, the measured peak of the normalized amplitude ratio kX/F is 4. What must the frequency ratio r be in order to reduce the normalized amplitude ratio to 1/3 if we model the damping in the isolators as hysteretic damping?

4.35 Suppose that for a certain elastic support, we estimate the spring constant to be $k = 2.2 \times 10^5$ N/m and the hysteretic constant to be $h = 2.6 \times 10^4$ N/m. Compute the maximum value of the normalized amplitude ratio kX/F.

4.36 A certain mass-spring system has hysteresis damping and a spring constant $k = 5 \times 10^4$ N/m. When a sinusoidal force of amplitude 2000 N is applied, the resonant amplitude is 50 mm. Estimate the value of the hysteretic damping constant h.

4.37 Derive the equation of motion of a pendulum having velocity-squared damping. Assume that the damping force is concentrated on the mass m, located a distance L from the pivot.

SECTION 4.7   FORCED VIBRATION WITH COULOMB DAMPING

4.38 A sinusoidal force of $150 \sin 20t$ acts on a mass-spring system. The mass is 10 kg and moves on a horizontal surface. The coefficient of dynamic friction between the mass and the surface is $\mu = 0.4$. The spring stiffness is $k = 20\ 000$ N/m. Calculate the amplitude and phase angle of the steady-state displacement of the mass.

4.39 A sinusoidal force of $200 \sin 16t$ acts on a mass-spring system. The mass is 10 kg and moves on a horizontal surface. The spring stiffness is $k = 10\ 000$ N/m. The observed amplitude of motion is $X = 25$ mm. Calculate the coefficient of dynamic friction between the mass and the surface.

SECTION 4.8   NONLINEAR RESPONSE

4.40 The following is a model of the velocity of an object subjected to cubic damping.

$$m\frac{dv}{dt} = -cv^3$$

Suppose that $m = 1$ and $c = 4$. Obtain the solution in terms of the initial condition $v(0)$.

4.41 Find the equilibria of Equation 4.8-3 for $m = 1$, $c = 12$, $k_1 = 16$, and $k_3 = -4$, and use a numerical method to solve and plot the solution for $\dot{y}(0) = 0$ and four values of $y(0)$: ± 1, ± 1.9, and ± 2.1. Compare the results with the stability properies predicted from the linearized model.

4.42 Use a numerical method to compute and plot the free response of Equation 4.8-3 for $m = 1$, $c = 0$, $k_1 = 2$, and $k_3 = 0.1$ for $\dot{y}(0) = 0$ and two initial conditions: $y(0) = 10$ and $y(0) = 40$. Compare the results with those shown in Figure 4.8-4. How does the initial condition affect the frequency of the response?

4.43 Plot the phase plane plots for the following equations and the initial conditions: $x(0) = 1$, $\dot{x}(0) = 0$:

(a) $\ddot{x} + 0.1\dot{x} + 2x = 0$

(b) $\ddot{x} + 2\dot{x} + 2x = 0$

(c) $\ddot{x} + 4\dot{x} + 2x = 0$

4.44 Plot the phase plane plot for the following equation with the initial conditions $y(0) = 1$ and $\dot{y}(0) = 0$.

$$\ddot{y} + 2\dot{y} + 2y + 3y^3 = 0$$

4.45 Plot the phase plane plot and identify the limit cycle for van der Pol's equation, Equation 4.8-6, with $\mu = 5$ and the initial conditions $y(0) = 1$ and $\dot{y}(0) = 0$.

SECTION 4.9 FREQUENCY RESPONSE WITH MATLAB

Note: In addition to the problems listed under this section, other problems in the chapter may be easily solved with MATLAB. These include Problems 4.1, 4.2, 4.8, and 4.9.

4.46 Use MATLAB to obtain the frequency response plots for the following transfer functions:

(a) $T(s) = \dfrac{5}{(10s + 1)(4s + 1)}$

(b) $T(s) = \dfrac{4}{s^2 + 10s + 100}$

4.47 Use MATLAB to obtain the frequency response plots for the following transfer functions:

(a) $T(s) = \dfrac{s}{(2s + 1)(5s + 1)}$

(b) $T(s) = \dfrac{s^2}{(2s + 1)(5s + 1)}$

4.48 Use the `lsim` function to compute and plot the forced response of a mass-spring-damper system having the values $m = 10$ kg, $c = 100$ N $\cdot$ s/m, and $k = 2500$ N/m. The forcing function is $f(t) = 100 \sin 13t$ N.

4.49 Consider the following model of a vehicle suspension:

$$m\ddot{x} + c\dot{x} + kx = k_1(y - x)$$

where m is the mass of the wheel-tire assembly, c and k are the damping and stiffness of the shock absorber, and k_1 is the tire stiffness. The displacement of the wheel-tire assembly is x and the displacement of the road surface is y. The wheel-tire assembly has a mass of 14 kg and the tire stiffness is $k_1 = 1.75 \times 10^5$ N/m. The damping in the shock absorber is $c = 5250$ N $\cdot$ s/m.

The nominal value for shock absorber stiffness $k = 3.5 \times 10^4$ N/m. Investigate the effects of changing this stiffness on the bandwidth of the suspension and on the force transmitted to the car body.

4.50 Use MATLAB to compute and plot the forced response of a mass-spring-damper system having the nonlinear equation of motion

$$5\ddot{y} + 100\dot{y} + 900y + 1700y^3 = 5g + f(t)$$

where $g = 9.8$ m/s^2. The forcing function is $f(t) = 100 \sin 9t$ N.

SECTION 4.10 SIMULINK APPLICATIONS

4.51 For each transfer functions and input $f(t)$ given in parts (a) and (b) of Problem 4.1, create and run a Simulink model and compare its output with that found analytically in that problem.

4.52 Create and run a Simulink model of a mass-spring system with viscous friction using the parameter values $m = 10$ kg, $c = 30$ N·s/m, and $k = 250$ N/m. The forcing function is the function $f(t) = 4 \sin 4.5t$ N. The forcing function is limited by a saturation nonlinearity that limits the value of the forcing function to the range $-3 \leq f(t) \leq 3$ N. Choose an appropriate stop time, export the response to MATLAB, and plot it.

4.53 The following is the equation of motion of a mass-spring-damper system with base excitation: $m\ddot{x} + c\dot{x} + kx = c\dot{y} + ky$. Use the values $m = 10$ kg, $c = 100$ N·s/m, and $k = 2500$ N/m. The base excitation is $y(t) = 0.05 \sin 13t$ m. Create and run a Simulink model of this system. The model should have two outputs: the displacement $x(t)$ and the transmitted force $f_t(t)$. Use the model to obtain a plot of both outputs.

4.54 Consider a 10-kg mass moving on a horizontal surface with $\mu = 0.3$ and attached to a spring of stiffness $k = 2500$ N/m. A force $f(t) = 50 \sin 14t$ acts on the mass. Suppose the mass is initially at rest and is given an initial velocity of 2 m/s. Use Simulink to obtain and plot the total response.

4.55 The equation of motion for the payload of an overhead crane whose base is moving horizontally with an acceleration a is

$$L\ddot{\theta} + g \sin \theta = a \cos \theta$$

Suppose that $L = 9$ m. Create a Simulink model and use it to find the maximum allowable acceleration a required to keep the angle θ less than 20° assuming that $\theta(0) = \dot{\theta}(0) = 0$.

4.56 Use Simulink to compute and plot the forced response of a mass-spring-damper system having the nonlinear equation of motion

$$5\ddot{y} + 100\dot{y} + 900y + 1700y^3 = f(t)$$

The forcing function is $f(t) = 100 \sin 9t$ N.

4.57 A certain mass-spring system is subjected to velocity-squared damping. Its equation of motion is

$$15\ddot{x} + 200 \, \text{sign}(\dot{x})\dot{x}^2 + 3000x = 140 \sin 9t$$

Use Simulink to compute and plot the forced response.

4.58 Run the Simulink model shown in Figure 4.10-14 using the same parameter values except that $D = 2$. Compare the results with those of Example 4.10-4.

4.59 Create and run a Simulink model for the system shown in Figure 4.10-11 having the dead zone shown in Figure 4.10-13, except that now there is viscous damping between the block and the surface, having the value $c = 3$. Compare the results with those of Example 4.10-4.

GENERAL FORCED RESPONSE

UP TO now we have investigated the free response and the harmonic response of systems having one degree of freedom. There are, however, many other types of forcing functions, other than pure sine and cosine functions, that occur in vibration applications. These include a suddenly applied constant force, called a *step* input; a constantly increasing force, called a *ramp* input; exponentially changing forces; and a number of other types. In this chapter we will see how to obtain the response to such forcing functions.

We start, in Section 5.1, with inputs that are periodic but not purely harmonic. The Fourier series representation of a periodic function as a series of sines and cosines enables us to use the results of Chapter 4 to obtain the response. Then, in Sections 5.2 through 5.4, we develop and apply the method based on the Laplace transform, which provides a means of obtaining the response of a linear system, of *any* order, for most of the commonly found forcing functions, including the step, ramp, and exponential functions.

Section 5.5 deals with two common inputs, the *impulse* and the *pulse* functions. Obtaining the response to an impulse, which is a limiting case of a pulse input and models a suddenly applied and suddenly removed input, requires special care, and so a separate section is devoted to it.

Section 5.6 shows how MATLAB can be used to perform some of the algebra required to use the Laplace transform method, and Section 5.7 covers the application of MATLAB to the other chapter topics. Section 5.8 shows how Simulink can be used to obtain the general forced response.

Although the examples in this chapter are restricted to systems having one degree of freedom, the methods are applicable to systems with more degrees of freedom. We will see this in Chapters 6 and 7.

LEARNING OBJECTIVES

After you have finished this chapter, you should be able to do the following:

- Apply the Fourier series method to obtain the response of a linear system to a periodic forcing function and apply the amplitude spectrum plot.
- Use the Laplace transform method to obtain the response of a linear system to a variety of forcing functions, such as step, ramp, exponential, pulse, and impulse functions.
- Apply MATLAB and Simulink to implement the methods of the chapter.

5.1 RESPONSE TO GENERAL PERIODIC INPUTS

The application of the sinusoidal input response is not limited to cases involving a single sinusoidal input. A basic theorem of analysis states that under some assumptions, which are generally satisfied in most practical applications, any *periodic* function can be expressed by a constant term plus an infinite series of sines and cosines with increasing frequencies. This theorem is the *Fourier theorem*, and its associated series is the *Fourier series*.

Using the Fourier series representation of the forcing function and the superposition property of linear systems, we can obtain the response of a linear system to any periodic forcing function by adding the responses due to each term in the series. We will also see that the frequency transfer function greatly facilitates this process.

EXAMPLE 5.1-1
Superposition and Forced Response

The model of the vibration isolator shown in Figure 5.1-1 is

$$c\dot{y}_1 + ky_1 = ky_2$$

Consider the case where $c = 0.5$ and $k = 1$. Suppose the displacement $y_2(t)$ is the following sum of a constant, a sine function, and a cosine function:

$$y_2(t) = 10 + 5 \sin t + 3 \cos 6t$$

Obtain the expression for the steady-state response.

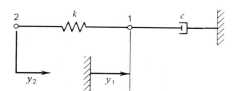

FIGURE 5.1-1 Vibration isolator.

Solution

The total forced response at steady state is the sum of the steady-state forced responses due to each of the three terms in $y_2(t)$. For the constant term 10, the steady-state response is $y_1 = 10$.

The transfer function can be found as in Chapter 4 by replacing $y_1(t)$ with $Y_1 e^{st}$ and $y_2(t)$ with $Y_2 e^{st}$. The transfer function is

$$T(s) = \frac{Y_1}{Y_2} = \frac{k}{cs + k} = \frac{1}{0.5s + 1}$$

The frequency transfer function is found as shown in Chapter 4 by replacing s with $i\omega$ to obtain

$$T(i\omega) = \frac{1}{0.5\omega i + 1}$$

The magnitude ratio M and phase angle ϕ are found from the magnitude and phase angle of $T(i\omega)$ as follows:

$$M(\omega) = \frac{1}{\sqrt{1 + 0.25\omega^2}}$$

$$\phi(\omega) = -\tan^{-1}(0.5\omega)$$

For the input term $5 \sin t$, $\omega = 1$, $M(1) = 1/\sqrt{1.25} = 2/\sqrt{5}$, and $\phi(1) = -\tan^{-1}(0.5) = -0.464$ rad. For the input term $3 \cos 6t$, $\omega = 6$, $M(6) = 1/\sqrt{10}$, and $\phi(6) = -\tan^{-1}(3) = -1.25$ rad. Thus the steady-state response is

$$y_1(t) = 10 + 5M(1) \sin[t + \phi(1)] + 3M(6) \cos[6t + \phi(2)]$$

$$= 10 + \frac{10}{\sqrt{5}} \sin(t - 0.464) + \frac{3}{\sqrt{10}} \cos(6t - 1.25)$$

Figure 5.1-2 shows the input and the response. The isolator does not filter the constant term at all; it filters the lower-frequency term somewhat and heavily filters the higher-frequency term. ∎

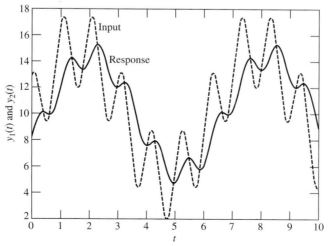

FIGURE 5.1-2 The input and the response for Example 5.1-1.

Fourier Series

We now show how to obtain a Fourier series representation of a given periodic function. We do not discuss some of the more theoretical concepts such as convergence of the series. These are covered in texts on engineering mathematics. See, for example, [Kreyszig, 1997].

If a function $f(t)$ is periodic with period P, then $f(t + P) = f(t)$. The Fourier series for this function defined on the interval $t_1 \leq t \leq t_1 + P$, where t_1 and P are constants and $P > 0$, is

$$f(t) = \frac{a_0}{2} + \sum_{n=1}^{\infty} \left(a_n \cos \frac{2n\pi t}{P} + b_n \sin \frac{2n\pi t}{P} \right) \tag{5.1-1}$$

where

$$a_n = \frac{2}{P} \int_{t_1}^{t_1+P} f(t) \cos \frac{2n\pi t}{P} \, dt \tag{5.1-2}$$

$$b_n = \frac{2}{P} \int_{t_1}^{t_1+P} f(t) \sin \frac{2n\pi t}{P} \, dt \tag{5.1-3}$$

If $f(t)$ is defined outside the specified interval $[t_1, t_1 + P]$ by a periodic extension of period P, and if $f(t)$ and df/dt are piecewise continuous, then the Fourier series converges to $f(t)$ if t is a point of continuity and otherwise to the average value $[f(t_+) + f(t_-)]/2$.

EXAMPLE 5.1-2
Fourier Series of a
Train of Pulses

Obtain the Fourier series for the train of unit pulses of width π and alternating in sign, as shown in Figure 5.1-3. The function is described by

$$f(t) = \begin{cases} 1 & 0 < t < \pi \\ -1 & \pi < t < 2\pi \end{cases} \tag{1}$$

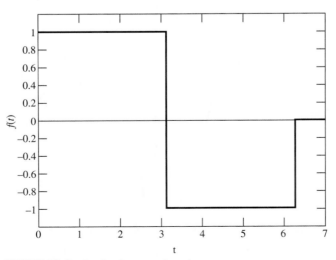

FIGURE 5.1-3   A train of rectangular pulses.

Solution | The period is $P = 2\pi$, and we may take the constant t_1 to be 0. Using a table of integrals, we find that

$$a_n = 0 \qquad \text{for all } n$$

$$b_n = \frac{4}{n\pi} \qquad \text{for } n \text{ odd}$$

$$b_n = 0 \qquad \text{for } n \text{ even}$$

The Fourier series is

$$f(t) = \frac{4}{\pi}\left(\frac{\sin t}{1} + \frac{\sin 3t}{3} + \frac{\sin 5t}{5} + \cdots\right) \tag{2}$$

The *fundamental or apparent frequency* of a periodic function is $2\pi/P$. For the function given by Equation 1, the apparent frequency is 1. Note that this frequency appears in the Fourier series, Equation 2, in the term $\sin t$. ∎

In general, the constant term a_0 and the cosine terms will not appear in the series if the function is odd; that is, if $f(-t) = -f(t)$. If the function is even, then $f(-t) = f(t)$, and no sine terms will appear in the series.

An alternative form of the Fourier series contains only sine terms. It is

$$f(t) = A_0 + \sum_{n=1}^{\infty} A_n \sin\left(\frac{2n\pi t}{P} + \phi_n\right) \tag{5.1-4}$$

where

$$A_0 = \frac{a_0}{2} \tag{5.1-5}$$

$$A_n = +\sqrt{a_n^2 + b_n^2} \tag{5.1-6}$$

$$\phi_n = \tan^{-1}\frac{a_n}{b_n} \tag{5.1-7}$$

This form is useful for plotting the *spectrum* of the function $f(t)$. The spectrum is a plot of the amplitudes A_n versus the frequencies $2n\pi/P$. We will soon see how the spectrum is used.

Steady-State Response to a Periodic Forcing Function

When the forcing function $f(t)$ is periodic and expressed in the form of Equation 5.1-1, the superposition principle states that the complete steady-state response is the sum of the steady-state responses due to each term in Equation 5.1-1. Although this is an infinite series, in practice we have to deal with only a few of its terms, because those terms whose frequencies lie outside the system's bandwidth have less influence on the response as a result of the filtering property of the system.

The constant term $a_0/2$ in the Fourier series produces a constant *steady-state* response when applied to a *stable* system. This response can be found by setting the derivatives to zero in the differential equation. For the first-order model

$$a\dot{x} + bx = f(t)$$

the steady-state response to $f = a_0/2$ can be found by setting $\dot{x} = 0$ to obtain $x = a_0/2b$. This is true if the model is stable; that is, if $b/a > 0$.

For the second-order model

$$m\ddot{x} + c\dot{x} + kx = f(t)$$

the steady-state response to $f = a_0/2$ can be found by setting $\ddot{x} = \dot{x} = 0$ to obtain $x = a_0/2k$. This is true if the model is stable; that is, if m, c, and k all have the same sign.

The remaining terms in the Fourier series are sine and cosine terms. Because $\cos \omega t = \sin(\omega t + \pi/2)$, the response to the cosine terms can be obtained from the response to a sine function with the phase shift increased by $\pi/2$ rad. Using the steady-state response to a sine function, which was obtained in Chapter 4, we can compute the steady-state response to a general periodic forcing function by adding the steady-state responses of the constant, the sine terms, and the cosine terms.

EXAMPLE 5.1-3
Response of a Vibration Isolator

For the vibration isolator shown in Figure 5.1-1, suppose that $c/k = \tau = 0.1$ sec and the displacement $y_2(t)$ consists of a sine for 0.5 sec followed by zero displacement for the second half of the period (Figure 5.1-4). Such a motion might be produced by a rotating cam with a dwell (the zero displacement portion). Determine the steady-state response of y_1.

Solution

The Fourier series' representation of $y_2(t)$ is

$$y_2(t) = \frac{1}{\pi} + \frac{1}{2}\sin 2\pi t - \frac{2}{\pi}\left(\frac{\cos 4\pi t}{1(3)} + \frac{\cos 8\pi t}{3(5)} + \frac{\cos 12\pi t}{5(7)} + \cdots\right) \qquad (1)$$

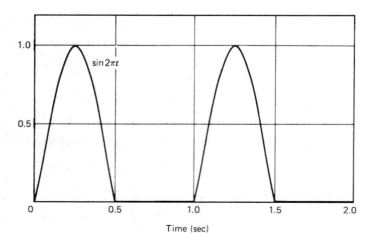

FIGURE 5.1-4   Half-sine function.

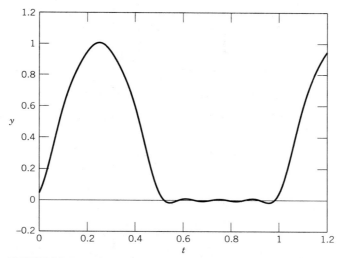

FIGURE 5.1-5 Fourier series representation of the half-sine function, using only the first five terms.

Figure 5.1-5 is a plot of Equation 1 including only those series terms shown. The plot illustrates how well the Fourier series represents the input function. The isolator's differential equation model is

$$c\dot{y}_1 + ky_1 = ky_2(t)$$

Its time constant is $\tau = c/k$, and its transfer function is

$$T(s) = \frac{Y_1}{Y_2} = \frac{1}{\frac{c}{k}s + 1} = \frac{1}{0.1s + 1}$$

The magnitude ratio M and phase angle ϕ are

$$M(\omega) = \frac{1}{\sqrt{1 + 0.01\omega^2}} \tag{2}$$

The phase angle is

$$\phi(\omega) = \tan^{-1}(-0.1\omega)$$

The isolator's bandwidth is $0 \le \omega \le 1/\tau$, which is $0 \le \omega \le 10$ rad/sec. Thus the only Fourier series terms lying within the bandwidth are the constant term $1/\pi$ (whose frequency is 0) and the term $\sin 2\pi t$.

The first term in the series expansion for y_2 is a constant: $1/\pi = 0.318$. This corresponds to an input with zero frequency; at steady state, it produces a response of $y_1 = 0.318$ because $M(0) = 1$.

The second term in the series is a sine function with an amplitude of 0.5 and a frequency of 2π rad/sec. Therefore, we evaluate $M(\omega)$ at $\omega = 2\pi$:

$$M(2\pi) = \frac{1}{\sqrt{1 + 0.01(2\pi)^2}} = 0.847$$

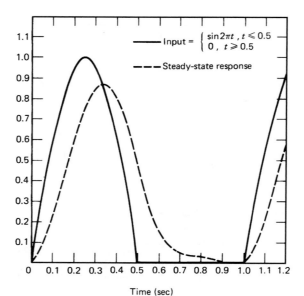

FIGURE 5.1-6 Steady-state response of the isolator to a half-sine input.

The second term produces a sinusoidal response of the same frequency with an amplitude of $0.847(0.5) = 0.424$, with a phase shift of $\phi(2\pi) = \tan^{-1}(-0.2\pi) = -0.56$ rad.

These are the only terms within the bandwidth of the isolator. However, to demonstrate the filtering property, we will treat the third term in the series. Here,

$$M(4\pi) = \frac{1}{\sqrt{1 + 0.01(4\pi)^2}} = 0.623$$

The amplitude of the response is $0.623(2/3\pi) = 0.132$, and the phase shift is $\phi(4\pi) = -0.9$ rad.

From superposition, the total steady-state response resulting from the three series terms retained is

$$y_1(t) = 0.318 + 0.423 \sin(2\pi t - 0.56) - 0.132 \cos(4\pi t - 0.9) \qquad (3)$$

The input and the response are plotted in Figure 5.1-6. The difference between the input and the response results from the resistive, or lag, effect of the system, not from the omission of the higher-order terms in the series. To see this, we compute that part of the response amplitude resulting from the fourth term in the series ($\cos 8\pi t$). The amplitude contribution is $2M(8\pi)/15\pi = 0.016$, which is about 10% of the amplitude of the third term. The decreasing amplitude of the higher-order terms in the series for $y_2(t)$, when combined with the filtering property of the system, allows us to truncate the series beyond some finite number of terms. ∎

The Spectrum

Using Equations 5.1-5 through 5.1-7, we obtain the following table for the first five terms in the series in Equation 1 of Example 5.1-3.

i	Frequency ω_i	Amplitude A_i	M_i
0	0	$1/\pi = 0.318$	1
1	2π	$1/2 = 0.5$	0.847
3	4π	$2/3\pi = 0.212$	0.623
5	8π	$2/15\pi = 0.042$	0.37
7	12π	$2/35\pi = 0.018$	0.256

The plot of A_i versus ω_i shown in Figure 5.1-7 is the spectrum of Equation 1. It graphically shows how the forcing function is composed of harmonic functions of different frequencies. The spectrum alone, however, does not tell the entire story about the system response. To do that, we need to examine the values of M given by Equation 2 in Example 5.1-3. These are shown in the last column of the table and plotted versus ω_i in the top graph of Figure 5.1-8. The bottom graph shows the response amplitudes $B = MA$. These are the amplitudes of the response series in Equation 3 of Example 5.1-3; they are $0.318, 0.413, 0.132, \cdots$. The three graphs in Figures 5.1-7 and 5.1-8 show how the spectrum A_i of the forcing function interacts with the magnitude ratio M_i of the system to produce the steady-state response.

Obtaining the closed-form solution for the total response of a system subjected to a periodic forcing function is very tedious. Before the widespread availability of computers, solving for the response numerically was also very tedious, and the Fourier series was a very useful tool for vibration engineers to use to obtain the closed-form solution for at least the steady-state response. Now, with programs such as MATLAB and Simulink, we can easily obtain a plot of the total response due to a periodic forcing function without expressing it as a Fourier series.

The Fourier series, however, is still a useful tool, primarily because we can use it to obtain the spectrum of the forcing function. Comparing the spectrum with the plot of the magnitude ratio M, we can estimate the effect of the forcing function on the steady-state response. If we are designing the system by picking values for its stiffness k and damping c, and thus its natural frequency ω_n, we can tell whether or not our choice for ω_n will lead

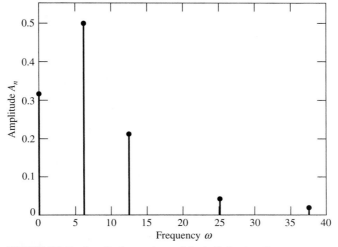

FIGURE 5.1-7 Amplitude spectrum of the half-sine function.

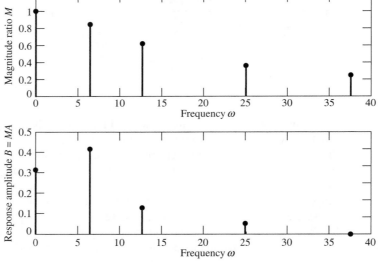

FIGURE 5.1-8   Plots of *M* and *B* versus frequency for Example 5.1-2.

to a resonant condition. This will occur if ω_n is near one of the spectrum frequencies that has a significant amplitude.

   In general, if the forcing function is periodic, not a pure sine or cosine, its Fourier series will contain terms with frequencies above the apparent frequency of the forcing function. If one of these terms has a frequency close to the system's resonant frequency, that term could dominate the response. Thus even if we design the system to have a resonant frequency far from the apparent frequency of the forcing function, we could still get resonance.

EXAMPLE 5.1-4

Apparent Frequency versus Resonant Frequency

The forcing function shown in Figure 5.1-9 acts on a system whose model is

$$3\ddot{x} + 6\dot{x} + 1200x = f(t)$$

Evaluate the steady-state response.

Solution

The Fourier series of the forcing function is

$$f(t) = \frac{80}{\pi}\left(\sin 2\pi t + \frac{\sin 6\pi t}{3} + \frac{\sin 10\pi t}{5} + \frac{\sin 14\pi t}{7} + \cdots\right)$$

The apparent period is 1 sec. The apparent frequency is 2π rad/sec and is the same as the frequency of the first term in the series. The natural frequency is $\omega_n = \sqrt{k/m} = 20$ rad/sec, which is close to the frequency $6\pi = 18.85$ rad/sec of the second term in the series. Thus we can expect the response to contain a large amplitude term corresponding to the frequency 18.85 rad/sec.

   The transfer function is

$$T(s) = \frac{X(s)}{F(s)} = \frac{1}{3s^2 + 6s + 1200}$$

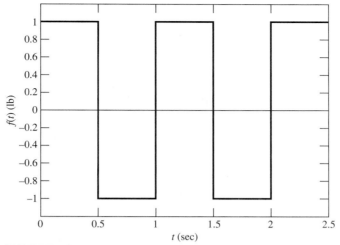

FIGURE 5.1-9 Forcing function for Example 5.1-4.

This gives

$$T(i\omega) = \frac{1}{-3\omega^2 + 6\omega i + 1200}$$

and

$$M(\omega) = \frac{1}{\sqrt{(1200 - 3\omega^2)^2 + (6\omega)^2}} \tag{1}$$

$$\phi(\omega) = -\tan^{-1}\left(\frac{6\omega}{1200 - 3\omega^2}\right) \tag{2}$$

The following table was computed using these equations and the fact that $B_i = M_i A_i$. A MATLAB program for computing this table is given in Section 5.7.

i	ω_i	A_i	M_i	B_i	ϕ_i (rad)
1	2π	$80/\pi$	9×10^{-4}	2.35×10^{-2}	-0.0348
3	6π	$80/3\pi$	5.7×10^{-3}	4.84×10^{-2}	-0.7007
5	10π	$80/5\pi$	6×10^{-4}	2.9×10^{-3}	-3.035
7	14π	$80/7\pi$	2×10^{-4}	8×10^{-4}	-3.0843

Using this table, we can express the steady-state response as follows:

$$x(t) = B_1 \sin(2\pi t + \phi_1) + B_3 \sin(6\pi t + \phi_3)$$
$$+ B_5 \sin(10\pi t + \phi_5) + B_7 \sin(14\pi t + \phi_7)$$

The term whose frequency is 6π dominates the response because its amplitude, B_3, is more than twice that of the next largest amplitude, B_1, the one whose frequency equals 2π, the apparent frequency. ∎

These calculations are easily done in MATLAB and Simulink. The total response, including the transient response, is easily calculated in MATLAB, using the `lsim` function, and in Simulink, which has several blocks that are useful for creating periodic forcing functions. These are discussed in Sections 5.7 and 5.8.

5.2   THE LAPLACE TRANSFORM

The Fourier series method enables us to obtain the response to a periodic forcing function, and the primary response of interest is the steady-state response. There are, however, many other types of forcing functions that are not periodic, and thus the Fourier series method cannot be used. Because these functions are not periodic, the primary response of interest is often the transient response. For such applications, the Laplace transform is very useful because it provides a systematic method for obtaining the total response. If the Laplace transform of the forcing function is available, and if one is patient in working through the required algebra, the method automatically produces a closed-form expression for the total response. Some of the required algebra can be done in MATLAB; this is illustrated in Section 5.7.

The Laplace transform can be used to convert linear differential equations into algebraic relations. With proper algebraic manipulation of the resulting quantities, the solution of the differential equation can be recovered in an orderly fashion by reversing the transformation process to obtain a function of time. Thus the transform provides a systematic method for obtaining the response to a variety of forcing functions without the need to guess the form of the response. The transform also provides a more direct way to obtain the response of models containing derivatives of the forcing function.

Definition

The Laplace transform $\mathcal{L}[y(t)]$ of a function $y(t)$ is defined to be

$$\mathcal{L}[y(t)] = \int_0^\infty y(t)e^{-st}dt \tag{5.2-1}$$

The integration removes t as a variable, and the transform is thus a function of only the Laplace variable s, which may be a complex number. The integral exists for most of the commonly encountered functions if suitable restrictions are placed on s. An alternative notation is the use of the uppercase symbol to represent the transform of the corresponding lowercase symbol; that is,

$$Y(s) = \mathcal{L}[y(t)] \tag{5.2-2}$$

We will use the *one-sided* transform, which uses the integral from $t = 0$ to $t = \infty$ and assumes that the variable $y(t)$ is zero for $t < 0$. For example, the unit-step function $u_s(t)$ is such a function. It is defined as

$$u_s(t) = \begin{cases} 0 & t < 0 \\ 1 & t \geq 0 \end{cases} \tag{5.2-3}$$

Suppose that $y(t) = Mu_s(t)$, where M is a constant. Its plot looks like Figure 5.2-1. It is called a step function because its shape resembles a stair step. The value of M is the height, or *magnitude*, of the step function; if $M = 1$, the function is called the *unit*-step function.

The step function is an approximate description of an input that can be switched on in a time interval that is very short compared to the smallest time constant of the system.

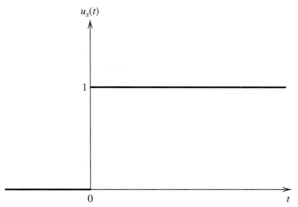

$u_s(t)$

1

0

t

FIGURE 5.2-1 The unit-step function $u_s(t)$.

Figure 5.2-2 illustrates how thrust varies with time for two types of solid-fuel rocket motors. The arrows indicate the direction in which the fuel burns. For a propellant grain having a tubular cross section, the fuel burns from the inside to the outside. The thrust increases with time because the propellant's burning area increases. A step function usually would not be used to model the thrust unless we needed a quick, very approximate answer, in which case we would take the average thrust as the step magnitude. The rod-and-tube shape was designed to give a constant thrust by keeping the total burning area approximately constant. The step function is a good model for this case if the time to reach constant thrust is short compared to the system time constant.

If $y(t) = Mu_s(t)$, then its transform is

$$\mathcal{L}[y(t)] = \int_0^\infty Mu_s(t)e^{-st}dt = M\int_0^\infty e^{-st}dt = -M\frac{e^{-st}}{s}\Big|_0^\infty = \frac{M}{s}$$

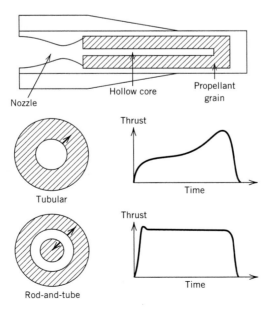

Nozzle

Hollow core

Propellant grain

Thrust

Time

Tubular

Thrust

Time

Rod-and-tube

FIGURE 5.2-2 Solid-fuel rocket thrust as a function of time.

where we have assumed that the real part of s is greater than zero, so that the limit of e^{-st} exists as $t \to \infty$. Similar considerations of the region of convergence of the integral apply for other functions of time. However, we need not concern ourselves with this here, because the transforms of all the common functions have been derived and tabulated. Table 5.2-1 is a table of transforms of common functions. We will discuss entry 1, the unit impulse, in Section 5.5. Entries 3, 4, 5, 11, and 12 can be derived in a manner similar to that used to derive entry 2. We will soon see how to derive the other entries in the table.

We have assumed that the process under study starts at time $t = 0$. Thus the given initial conditions—say $x(0), \dot{x}(0), \cdots$—represent the situation at the start of the process and are the result of any inputs applied prior to $t = 0$. That is, we need not know what the inputs were before $t = 0$ because their effects are contained in the initial conditions. The effects of any inputs starting at $t = 0$ are not felt by the system until an infinitesimal time later, at $t = 0+$. This distinction becomes important when we analyze the response of models containing an impulsive input or derivatives of discontinuous inputs such as the step function. This topic will be discussed in detail in Sections 5.4 and 5.5.

TABLE 5.2-1 Laplace Transform Pairs

$F(s)$	$f(t), t \geq 0$
1. 1	$\delta(t)$, unit impulse at $t = 0$
2. $\dfrac{1}{s}$	$u_s(t)$, unit step
3. $\dfrac{n!}{s^{n+1}}$	t^n
4. $\dfrac{1}{s+a}$	e^{-at}
5. $\dfrac{1}{(s+a)^n}$	$\dfrac{1}{(n-1)!}t^{n-1}e^{-at}$
6. $\dfrac{a}{s(s+a)}$	$1 - e^{-at}$
7. $\dfrac{1}{(s+a)(s+b)}$	$\dfrac{1}{b-a}(e^{-at} - e^{-bt})$
8. $\dfrac{s+p}{(s+a)(s+b)}$	$\dfrac{1}{b-a}[(p-a)e^{-at} - (p-b)e^{-bt}]$
9. $\dfrac{1}{(s+a)(s+b)(s+c)}$	$\dfrac{e^{-at}}{(b-a)(c-a)} + \dfrac{e^{-bt}}{(c-b)(a-b)} + \dfrac{e^{-ct}}{(a-c)(b-c)}$
10. $\dfrac{s+p}{(s+a)(s+b)(s+c)}$	$\dfrac{(p-a)e^{-at}}{(b-a)(c-a)} + \dfrac{(p-b)e^{-bt}}{(c-b)(a-b)} + \dfrac{(p-c)e^{-ct}}{(a-c)(b-c)}$
11. $\dfrac{b}{s^2+b^2}$	$\sin bt$
12. $\dfrac{s}{s^2+b^2}$	$\cos bt$
13. $\dfrac{b}{(s+a)^2+b^2}$	$e^{-at}\sin bt$
14. $\dfrac{s+a}{(s+a)^2+b^2}$	$e^{-at}\cos bt$
15. $\dfrac{\omega_n^2}{s^2+2\zeta\omega_n s+\omega_n^2}$	$\dfrac{\omega_n}{\sqrt{1-\zeta^2}}e^{-\zeta\omega_n t}\sin\omega_n\sqrt{1-\zeta^2}\,t \qquad \zeta < 1$
16. $\dfrac{\omega_n^2}{s(s^2+2\zeta\omega_n s+\omega_n^2)}$	$1 + \dfrac{\omega_n}{\sqrt{1-\zeta^2}}e^{-\zeta\omega_n t}\sin\left(\omega_n\sqrt{1-\zeta^2}\,t+\phi\right) \qquad \zeta < 1$ $\phi = \tan^{-1}\dfrac{\sqrt{1-\zeta^2}}{\zeta} + \pi \quad \text{(third quadrant)}$

Transform Properties

The transform of a derivative will be useful to us. Applying integration by parts to the definition of the transform, we obtain

$$\mathcal{L}\left(\frac{dy}{dt}\right) = \int_0^\infty \frac{dy}{dt} e^{-st} dt = y(t)e^{-st}\Big|_0^\infty + s\int_0^\infty y(t)e^{-st} dt$$

$$= s\mathcal{L}[y(t)] - y(0) = sY(s) - y(0) \tag{5.2-4}$$

This procedure can be extended to higher derivatives. For example, the result for the second derivative is

$$\mathcal{L}\left(\frac{d^2y}{dt^2}\right) = s^2Y(s) - sy(0) - \dot{y}(0) \tag{5.2-5}$$

The general result for a derivative of any order is given in Table 5.2-2, which lists the properties of the transform. These properties are due to the properties of integrals in general. For example, we have just seen how a multiplicative constant can be factored out of the integral. In addition, the integral of a sum equals the sum of the integrals. These facts point out the linearity property of the transform; namely, that

$$\mathcal{L}[af_1(t) + bf_2(t)] = a\mathcal{L}[f_1(t)] + b\mathcal{L}[f_2(t)] \tag{5.2-6}$$

TABLE 5.2-2 Properties of the Laplace Transform

$f(t)$	$F(s) = \int_0^\infty f(t)e^{-st} dt$
1. $af_1(t) + bf_2(t)$	$aF_1(s) + bF_2(s)$
2. $\dfrac{df}{dt}$	$sF(s) - f(0)$
3. $\dfrac{d^2f}{dt^2}$	$s^2F(s) - sf(0) - \dfrac{df}{dt}\Big|_{t=0}$
4. $\dfrac{d^nf}{dt^n}$	$s^nF(s) - \sum_{\kappa=1}^n s^{n-\kappa}g_{\kappa-1}$
	$g_{\kappa-1} = \dfrac{d^{\kappa-1}f}{dt^{\kappa-1}}\Big|_{t=0}$
5. $\int_0^t f(t)\,dt$	$\dfrac{F(s)}{s} + \dfrac{h(0)}{s}$
	$h(0) = \int f(t)\,dt\big|_{t=0}$
6. $g(t) = \begin{cases} 0 & t < D \\ f(t-D) & t \geq D \end{cases}$	$G(s) = e^{-sD}F(s)$
7. $e^{-at}f(t)$	$F(s+a)$
8. $tf(t)$	$-\dfrac{dF(s)}{ds}$
9. $f(t) = \int_0^t x(t-\tau)y(\tau)\,d\tau = \int_0^t y(t-\tau)x(\tau)\,d\tau$	$F(s) = X(s)Y(s)$
10. $f(\infty) = \lim_{s\to 0} sF(s)$	
11. $f(0+) = \lim_{s\to\infty} sF(s)$	

Partial-Fraction Expansion

Entries 6 through 10 in Table 5.2-1 can be derived from entry 4 by expressing the transform as a sum of terms of the form of entry 4; namely, $1/(s+a)$. Each term corresponds to a factor of the denominator. This sum of terms is called a *partial-fraction expansion*.

For example, we can expand entry 6 as follows:

$$F(s) = \frac{a}{s(s+a)} = \frac{C_1}{s} + \frac{C_2}{s+a} \tag{5.2-7}$$

Multiplying by the least common denominator $s(s+a)$, we have

$$\frac{a}{s(s+a)} = \frac{C_1(s+a) + C_2 s}{s(s+a)}$$

Comparing the numerators, we see that

$$a = C_1(s+a) + C_2 s = (C_1 + C_2)s + aC_1$$

This is true only if the coefficient of s on the right side is zero and if the constants on the left and right sides are equal to each other. That is, $C_1 + C_2 = 0$ and $a = aC_1$. Thus $C_1 = 1$ and $C_2 = -1$. Equation 5.2-7 therefore becomes

$$F(s) = \frac{a}{s(s+a)} = \frac{1}{s} - \frac{1}{s+a}$$

Using entries 2 and 4, we see that

$$f(t) = u_s(t) - e^{-at} = 1 - e^{-at} \qquad \text{for } t \geq 0$$

This is entry 6.

Entries 7 through 10 can also be derived using a partial-fraction expansion. In Section 5.3 we will present a systematic procedure for finding the coefficients in the partial-fraction expansion.

Several of the properties listed in Table 5.2-2 can be used to derive transforms given in Table 5.2-1. For example, property 7 can be used to derive entry 4 from entry 2. Property 7 can also be used to derive entry 5 from entry 3. Property 2 can be used to derive entry 12 from entry 11. Entries 13 and 14 can be derived from entries 11 and 12 using property 7. The derivations are covered in the end-of-chapter problems.

Sine and Cosine Transforms

A partial-fraction expansion is not always needed to use Table 5.2-1. For example, consider the transform

$$F(s) = \frac{6s + 11}{s^2 + 6s + 34}$$

The factors of the denominator are complex: $s = -3 \pm 5i$, so we can express $F(s)$ as a sum of terms similar to entries 13 and 14 as follows (note that $a = 3$ and $b = 5$):

$$F(s) = \frac{6s + 11}{(s+3)^2 + 25} = C_1 \frac{s+3}{(s+3)^2 + 25} + C_2 \frac{5}{(s+3)^2 + 25} = \frac{C_1(s+3) + 5C_2}{(s+3)^2 + 25}$$

Comparing numerators, we see that

$$6s + 11 = C_1(s + 3) + 5C_2 = C_1s + 3C_1 + 5C_2$$

This is true only if $C_1 = 6$ and $C_2 = -7/5$. Thus

$$f(t) = C_1 e^{-3t} \sin 5t + C_2 e^{-3t} \cos 5t = 6e^{-3t} \cos 5t - \frac{7}{5} e^{-3t} \sin 5t$$

Response in Terms of ζ and ω_n

The model

$$m\ddot{x} + c\dot{x} + kx = ky(t) \tag{5.2-8}$$

can be expressed in terms of the damping ratio ζ and the natural frequency ω_n as

$$\ddot{x} + 2\zeta\omega_n \dot{x} + \omega_n^2 x = \omega_n^2 y(t) \tag{5.2-9}$$

where

$$\zeta = \frac{c}{2\sqrt{mk}} \tag{5.2-10}$$

$$\omega_n = \sqrt{\frac{k}{m}} \tag{5.2-11}$$

Assuming that $\zeta < 1$, the characteristic roots can be expressed as

$$s = -\zeta\omega_n \pm \omega_n \sqrt{1 - \zeta^2} \tag{5.2-12}$$

If $x(0) = 0$ and $\dot{x}(0) = \omega_n^2$, the transform of the free response is

$$X(s) = \frac{\omega_n^2}{s^2 + 2\zeta\omega_n s + \omega_n^2} = \frac{\omega_n^2}{(s + \zeta\omega_n)^2 + \omega_n^2(1 - \zeta^2)}$$

Comparing this with entry 13 in Table 5.2-1, we see that the free response is

$$x(t) = \frac{\omega_n}{\sqrt{1 - \zeta^2}} e^{-\zeta\omega_n t} \sin \omega_n \sqrt{1 - \zeta^2} t \tag{5.2-13}$$

This is entry 15 in Table 5.2-1.

Entry 16 is the response to a unit-step input, for zero initial conditions. Its form will be derived in Section 5.4.

Other Transform Properties

The transform properties listed in Table 5.2-2 can be derived from the properties of integrals and can be used to simplify the algebra and calculus required to apply the transform. Property 5 is useful for solving equations containing integrals of the dependent variable. Property 6 is the *shifting theorem*. We will illustrate its use in Section 5.5. Property 7 can be used to obtain the transform of te^{-at}, which is entry 5 in Table 5.2-1 with $n = 2$. Let $f(t) = t$. Then $F(s) = 1/s^2$ from entry 3, and thus

$$\mathcal{L}[e^{-at}f(t)] = \mathcal{L}(te^{-at}) = \frac{1}{s^2}\bigg|_{s \to s+a} = \frac{1}{(s + a)^2}$$

Property 8 can be used with entry 12 to derive the transform of $t \cos bt$, as follows. Let $f(t) = \cos bt$. Then $F(s) = s/(s^2 + b^2)$ and thus

$$\mathcal{L}[tf(t)] = \mathcal{L}(t \cos bt) = -\frac{d}{ds}\left(\frac{s}{s^2 + b^2}\right) = \frac{s^2 - b^2}{(s^2 + b^2)^2}$$

Property 9 is the *convolution integral*. It has been used for numerical computation of the forced response, and it useful for expressing the response to an arbitrary input in general form, as we will see in Section 5.4.

Property 10 is the *final value theorem* and is useful for quickly calculating the steady-state response. Property 11 is the *initial value theorem*. We will apply it in Sections 5.4 and 5.5.

We will now use these results to solve differential equations.

5.3   FORCED RESPONSE FROM THE LAPLACE TRANSFORM

We now show how to apply the Laplace transform to solve differential equations to obtain the forced response. A linear model of a damped single-DOF system contains most of the possible types of behavior that can occur in linear systems. This is because a second-order characteristic polynomial with real coefficients must have roots that fall into one of the following three categories:

1. Real and distinct
2. Real and repeated
3. Distinct but complex conjugate pairs

Once we understand the behavior generated by each of the three root cases, we can apply the results to a linear system of any order. The only exception is the case where the roots are repeated complex conjugate pairs. This case requires a model of at least a two-DOF system, whose characteristic polynomial is fourth order, or a forcing function—like a damped sinusoid—that introduces a complex root pair.

Application to a Single-DOF Model

Consider the following equation of motion of a linear single-DOF system having the forcing function $f(t)$:

$$m\ddot{x} + c\dot{x} + kx = f(t) \qquad (5.3\text{-}1)$$

Apply the Laplace transform to both sides of this equation, as follows:

$$\int_0^\infty (m\ddot{x} + c\dot{x} + kx)e^{-st}dt = \int_0^\infty f(t)e^{-st}dt$$

or, from the linearity property of the integral,

$$m\int_0^\infty \ddot{x}e^{-st}dt + c\int_0^\infty \dot{x}e^{-st}dt + k\int_0^\infty xe^{-st}dt = \int_0^\infty f(t)e^{-st}dt$$

Applying the derivative properties (properties 2 and 3 in Table 5.2-2), we obtain

$$m\left[s^2X(s) - sx(0) - \dot{x}(0)\right] + c[sX(s) - x(0)] + kX(s) = F(s)$$

Collect the $X(s)$ terms

$$(ms^2 + cs + k)X(s) = mx(0)s + m\dot{x}(0) + cx(0) + F(s) \tag{5.3-2}$$

and solve for $X(s)$:

$$X(s) = \frac{mx(0)s + m\dot{x}(0) + cx(0)}{ms^2 + cs + k} + \frac{F(s)}{ms^2 + cs + k} \tag{5.3-3}$$

The first term on the right depends on the initial conditions and is the transform of the free response. Thus

$$X_{\text{free}}(s) = \frac{mx(0)s + m\dot{x}(0) + cx(0)}{ms^2 + cs + k} \tag{5.3-4}$$

The Transfer Function

The last term on the right of Equation 5.3-3 depends on the forcing function and is the transform of the forced response:

$$X_{\text{forced}}(s) = \frac{F(s)}{ms^2 + cs + k} \tag{5.3-5}$$

This can be expressed as

$$\frac{X_{\text{forced}}(s)}{F(s)} = \frac{1}{ms^2 + cs + k}$$

This is the transfer function of Equation 5.3-1 and is the same transfer function found by substituting $x(t) = X(s)e^{st}$ and $f(t) = F(s)e^{st}$ into Equation 5.3-1, as was done in Chapter 4. Now, however, we see that the transfer function has a more general meaning than simply the ratio of two constants, X and F, involving the exponential response Xe^{st} to an exponential input Fe^{st}. The general meaning of the transfer function is now seen to be the ratio of the forced response transform $X(s)$ to the transform $F(s)$ of the input. Thus the application of the transfer function is not limited to exponential functions. It may be used to obtain the forced response to any input that has a Laplace transform.

Evaluating the Response

The characteristic equation is found from the coefficient of $X(s)$ in Equation 5.3-2. It is

$$ms^2 + cs + k = 0 \tag{5.3-6}$$

The characteristic roots are given by

$$s = \frac{-c \pm \sqrt{c^2 - 4mk}}{2m} \tag{5.3-7}$$

The following examples illustrate how to find the response when the roots are distinct, repeated, and complex.

The free response is formally expressed as the inverse Laplace transform

$$x(t) = \mathcal{L}^{-1}\left[\frac{mx(0)s + m\dot{x}(0) + cx(0)}{ms^2 + cs + k}\right] \tag{5.3-8}$$

and the forced response as

$$x(t) = \mathcal{L}^{-1} \left[\frac{F(s)}{ms^2 + cs + k} \right] \tag{5.3-9}$$

In order to evaluate either response, we must expand the transforms within the square brackets into a series of transforms that appear in Table 5.2-1. To perform this expansion, we need to know the values of m, c, and k to determine whether the roots of $ms^2 + cs + k$ are (1) real and distinct, (2) real and repeated, or (3) complex, because the form of the expansion is different for each case. In addition, we also need to know the forcing function $f(t)$ so that we can find $F(s)$.

Consider the transform, Equation 5.3-9, of the forced response for several examples of different forcing functions.

1. A step function: $f(t) = Mu_s(t)$. Then $F(s) = M/s$ and Equation 5.3-9 becomes

$$X_{\text{forced}}(s) = \frac{M}{s(ms^2 + cs + k)}$$

The roots of the denominator are $s = 0$ and the roots of $ms^2 + cs + k = 0$, which may be real and distinct, real and repeated, or complex conjugates.

2. A ramp function: $f(t) = Bt$ for $t \geq 0$. Then $F(s) = B/s^2$ and Equation 5.3-9 becomes

$$X_{\text{forced}}(s) = \frac{M}{s^2(ms^2 + cs + k)}$$

The roots of the denominator are $s = 0$, $s = 0$, and the roots of $ms^2 + cs + k = 0$.

3. An exponential function: $f(t) = Be^{-at}$ for $t \geq 0$. Then

$$F(s) = \frac{B}{s + a}$$

and Equation 5.3-9 becomes

$$X_{\text{forced}}(s) = \frac{B}{(s + a)(ms^2 + cs + k)}$$

The roots of the denominator are $s = -a$ and the roots of $ms^2 + cs + k = 0$.

4. A cosine function: $f(t) = B \cos \omega t$ for $t \geq 0$. Then

$$F(s) = \frac{Bs}{s^2 + \omega^2}$$

and Equation 5.3-9 becomes

$$X_{\text{forced}}(s) = \frac{Bs}{(s^2 + \omega^2)(ms^2 + cs + k)}$$

The roots of the denominator are $s = \pm i\omega$ and the roots of $ms^2 + cs + k = 0$.

From these examples we see that the factors that appear in the partial-fraction expansion of $X_{\text{forced}}(s)$ are the factors introduced by the characteristic roots (the roots of $ms^2 + cs + k = 0$) and the factors introduced into the *denominator* of $X_{\text{forced}}(s)$ by the factors of the *denominator* of $F(s)$. Thus the factor s in the numerator of cosine transform $Bs/(s^2 + \omega^2)$ does *not* introduce a factor into the partial-fraction expansion.

The General Case

Most transforms occur in the form of a ratio of two polynomials, such as

$$X(s) = \frac{N(s)}{D(s)} = \frac{b_m s^m + b_{m-1} s^{m-1} + \cdots + b_1 s + b_0}{s^n + a_{n-1} s^{n-1} + \cdots + a_1 s + a_0} \tag{5.3-10}$$

If $X(s)$ is of the form of Equation 5.3-10, the method of partial-fraction expansion can be used. Note that we assume that the coefficient $a_n = 1$. If not, divide the numerator and denominator by a_n. We do this so that we can easily express the denominator in factored form.

Distinct, Real Roots Case

The first step is to solve for the n roots of the denominator. If all the roots are real and distinct, we can express $X(s)$ in Equation 5.3-10 in factored form as

$$X(s) = \frac{N(s)}{(s + r_1)(s + r_2) \ldots (s + r_n)} \tag{5.3-11}$$

where the roots are $s = -r_1, -r_2, \ldots, -r_n$. This form can be expanded as

$$X(s) = \frac{C_1}{s + r_1} + \frac{C_2}{s + r_2} + \cdots + \frac{C_n}{s + r_n} \tag{5.3-12}$$

where

$$C_i = \lim_{s \to -r_i} \left[X(s)(s + r_i) \right] \tag{5.3-13}$$

Multiplying by the factor $(s + r_i)$ cancels that term in the denominator before the limit is taken. This is a good way of remembering Equation 5.3-13. Each factor corresponds to an exponential function of time, and the solution is

$$x(t) = C_1 e^{-r_1 t} + C_2 e^{-r_2 t} + \ldots + C_n e^{-r_n t} \tag{5.3-14}$$

EXAMPLE 5.3-1
Step Response
with Distinct, Real
Roots

Obtain the response of the model

$$2\ddot{x} + 10\dot{x} + 8x = 15u_s(t)$$

for the initial conditions $x(0) = 2$, $\dot{x}(0) = 5$.

Solution

Applying the Laplace transform, we obtain

$$2\left[s^2 X(s) - sx(0) - \dot{x}(0)\right] + 10[sX(s) - x(0)] + 8X(s) = \frac{15}{s}$$

Multiply both sides by s and collect the $X(s)$ terms:

$$s\left(2s^2 + 10s + 8\right)X(s) = 15 + 2s[sx(0) + \dot{x}(0)] + 10sx(0) = 4s^2 + 30s + 15$$

Divide by the coefficient of the highest power of s on the left side and solve for $X(s)$:

$$X(s) = \frac{2s^2 + 15s + 7.5}{s(s^2 + 5s + 4)} = \frac{2s^2 + 15s + 7.5}{s(s + 1)(s + 4)}$$

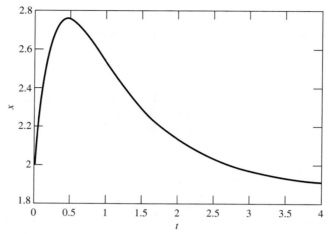

FIGURE 5.3-1   Response for Example 5.3-1.

Because the roots are $s = 0$, $s = -1$, and $s = -4$, which are real and distinct, we can express $X(s)$ as the following partial-fraction expansion:

$$X(s) = \frac{C_1}{s} + \frac{C_2}{s+1} + \frac{C_3}{s+4}$$

Using Equation 5.3-13, we have

$$C_1 = \lim_{s \to 0} X(s)s = \lim_{s \to 0} \frac{2s^2 + 15s + 7.5}{(s+1)(s+4)} = \frac{15}{8}$$

$$C_2 = \lim_{s \to -1} X(s)(s+1) = \lim_{s \to -1} \frac{2s^2 + 15s + 7.5}{s(s+4)} = \frac{11}{6}$$

$$C_3 = \lim_{s \to -4} X(s)(s+4) = \lim_{s \to -4} \frac{2s^2 + 15s + 7.5}{s(s+1)} = -\frac{41}{24}$$

The response is

$$x(t) = C_1 + C_2 e^{-t} + C_3 e^{-4t} = \frac{15}{8} + \frac{11}{6} e^{-t} - \frac{41}{24} e^{-4t}$$

The plot is shown in Figure 5.3-1.   ■

The Laplace transform can be used in this manner to obtain the general solution for the free and the step responses, for the distinct, real roots case. The free response is given in Table 3.5-1 in Chapter 3, and the step response is given in Table 5.3-1.

Repeated-Roots Case

Suppose that p of the roots have the same value $s = -r_1$, and the remaining $(n - p)$ roots are real and distinct. Then $X(s)$ is of the form

$$X(s) = \frac{N(s)}{(s + r_1)^p (s + r_{p+1})(s + r_{p+2}) \ldots (s + r_n)} \tag{5.3-15}$$

TABLE 5.3-1 Unit-Step Response of a Second-Order Model

Model: $m\ddot{x} + c\dot{x} + kx = f(t)$

$f(t) = 1 \qquad t \geq 0$

$x(0) = \dot{x}(0) = 0$

Characteristic roots: $s = \dfrac{-c \pm \sqrt{c^2 - 4mk}}{2m} = s_1, s_2$

Case 1. Real, distinct roots: $s_1 \neq s_2$

$$x(t) = A_1 e^{s_1 t} + A_2 e^{s_2 t} + \frac{1}{k} = \frac{1}{k}\left(\frac{s_2}{s_1 - s_2} e^{s_1 t} - \frac{s_1}{s_1 - s_2} e^{s_2 t} + 1\right)$$

Case 2. Repeated roots: $s_1 = s_2$

$$x(t) = (A_1 + A_2 t)e^{s_1 t} + \frac{1}{k} = \frac{1}{k}\left[(s_1 t - 1)e^{s_1 t} + 1\right]$$

Case 3. Complex conjugate roots: $s = -a \pm ib, b > 0$

$$x(t) = \frac{1}{k}\left[\frac{1}{b}\sqrt{\frac{k}{m}}e^{-at}\sin(bt + \phi) + 1\right]$$
$$\phi = \angle(-a - ib)$$

Alternative form:

$$x(t) = \frac{1}{k}\left[\frac{1}{\sqrt{1 - \zeta^2}}e^{-\zeta\omega_n t}\sin(\omega_d t + \phi) + 1\right]$$

$$\phi = \tan^{-1}\frac{\sqrt{1 - \zeta^2}}{\zeta} + \pi \qquad \omega_d = \omega_n\sqrt{1 - \zeta^2}$$

The expansion is

$$X(s) = \frac{C_1}{(s + r_1)^p} + \frac{C_2}{(s + r_1)^{p-1}} + \cdots + \frac{C_p}{s + r_1} + \cdots + \frac{C_{p+1}}{s + r_{p+1}} + \cdots + \frac{C_n}{s + r_n} \qquad (5.3\text{-}16)$$

The coefficients for the repeated roots are found from

$$C_1 = \lim_{s \to -r_1}\left[X(s)(s + r_1)^p\right] \qquad (5.3\text{-}17)$$

$$C_2 = \lim_{s \to -r_1}\left\{\frac{d}{ds}[X(s)(s + r_1)^p]\right\} \qquad (5.3\text{-}18)$$

$$\vdots$$

$$C_i = \lim_{s \to -r_1}\left\{\frac{1}{(i - 1)!}\frac{d^{i-1}}{ds^{i-1}}[X(s)(s + r_1)^p]\right\} \qquad i = 1, 2, \ldots, p \qquad (5.3\text{-}19)$$

The coefficients for the distinct roots are found from Equation 5.3-3. The solution for the time function is

$$x(t) = C_1\frac{t^{p-1}}{(p - 1)!}e^{-r_1 t} + C_2\frac{t^{p-2}}{(p - 2)!}e^{-r_1 t} + \cdots + C_p e^{-r_1 t}$$
$$+ \cdots + C_{p+1}e^{-r_{p+1} t} + \cdots + C_n e^{-r_n t} \qquad (5.3\text{-}20)$$

EXAMPLE 5.3-2
Step Response with Repeated Roots

Obtain the response of the model

$$5\ddot{x} + 20\dot{x} + 20x = 12u_s(t)$$

for zero initial conditions.

Solution

We have

$$(5s^2 + 20s + 20)X(s) = \frac{12}{s}$$

Divide by the coefficient of the highest power of s on the left side and solve for $X(s)$:

$$X(s) = \frac{2.4}{s(s^2 + 4s + 4)} = \frac{2.4}{s(s+2)^2}$$

Because two of the roots are repeated, we can express $X(s)$ as the following partial-fraction expansion:

$$X(s) = \frac{2.4}{s(s+2)^2} = \frac{C_1}{s} + \frac{C_2}{(s+2)^2} + \frac{C_3}{s+2}$$

Using Equations 5.3-13 through 5.3-19, we have

$$C_1 = \lim_{s \to 0} X(s)s = \lim_{s \to 0} \frac{2.4}{(s+2)^2} = 0.6$$

$$C_2 = \lim_{s \to -2} X(s)(s+2)^2 = \lim_{s \to -2} \frac{2.4}{s} = -1.2$$

$$C_3 = \lim_{s \to -2} \frac{d}{ds}\left[X(s)(s+2)^2\right] = \lim_{s \to -2} \frac{d}{ds}\left(\frac{2.4}{s}\right) = \lim_{s \to -2.4} \frac{-2}{s^2} = -0.6$$

The response is

$$x(t) = C_1 + C_2 te^{-2t} + C_3 e^{-2t} = 0.6 - 1.2te^{-2t} - 0.6e^{-2t}$$

The response is shown in Figure 5.3-2. ∎

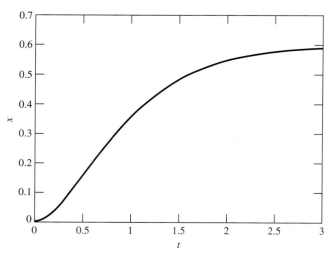

FIGURE 5.3-2 Response for Example 5.3-2.

The Laplace transform can be used in this manner to obtain the general solution for the step response for the repeated-roots case. This solution is given in Table 5.3-1.

Response Calculations with Complex Roots

When the roots are a complex conjugate pair, the expansion has the same form as Equation 5.3-14 because the roots are in fact distinct. The coefficients C_i can be found from Equation 5.3-13. However, these will be complex numbers, and the solution form given by Equation 5.3-14 will not be convenient to use. The following example shows how complex roots can be handled.

EXAMPLE 5.3-3
Step Response with Complex Characteristic Roots

Obtain the response of the following model:

$$10\ddot{x} + 80\dot{x} + 200x = 23u_s(t) \qquad x(0) = 0 \qquad \dot{x}(0) = 0$$

Solution

Transform both sides of the differential equation:

$$(10s^2 + 80s + 200)X(s) = \frac{23}{s}$$

Divide by 10, note that $s^2 + 8s + 20 = (s+4)^2 + 4$, and solve for $X(s)$:

$$X(s) = \frac{2.3}{s[(s+4)^2 + 4]} \tag{1}$$

The denominator factors are s and $(s+4)^2 + 4$, so we expand $X(s)$ as

$$X(s) = \frac{C_1}{s} + C_2 \frac{s+4}{(s+4)^2 + 4} + C_3 \frac{2}{(s+4)^2 + 4} \tag{2}$$

where

$$C_1 = \lim_{s \to 0} \frac{2.3}{(s+4)^2 + 4} = \frac{23}{200}$$

Now express Equation 2 as a single fraction:

$$X(s) = \frac{C_1[(s+4)^2 + 4] + C_2 s(s+4) + 2C_3 s}{s[(s+4)^2 + 4]}$$

Collect terms:

$$X(s) = \frac{(C_1 + C_2)s^2 + (8C_1 + 4C_2 + 2C_3)s + 20C_1}{s[(s+4)^2 + 4]} \tag{3}$$

Compare the numerator of Equation 3 with the numerator of Equation 1:

$$(C_1 + C_2)s^2 + (8C_1 + 4C_2 + 2C_3)s + 20C_1 = 2.3$$

Thus

$$C_1 + C_2 = 0 \quad \text{or} \quad C_2 = -C_1 = -\frac{23}{200}$$

$$8C_1 + 4C_2 + 2C_3 = 0 \quad \text{or} \quad C_3 = -\frac{23}{100}$$

Thus

$$x(t) = C_1 + C_2 e^{-4t} \cos 2t + C_3 e^{-4t} \sin 2t = \frac{23}{200}\left(1 - e^{-4t} \cos 2t - 2e^{-4t} \sin 2t\right) \tag{4}$$

The response is shown in Figure 5.3-3. ∎

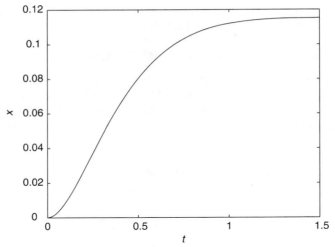

FIGURE 5.3-3 Response for Example 5.3-3.

Alternative Forms of the Step Response

Using the identity

$$\sin(bt + \phi) = \sin bt \cos \phi + \cos bt \sin \phi$$

we can express the response due to complex roots in a more compact form, as a single sine function with a phase shift. Using this approach with Equation 4 in Example 5.3-3, we see that

$$\cos 2t + 2 \sin 2t = B \sin 2t \cos \phi + B \cos 2t \sin \phi$$

Thus

$$B \cos \phi = 2 \qquad B \sin \phi = 1$$

This gives $B = \sqrt{5}$ and $\phi = 0.4636$ rad. Thus Equation 4 in Example 5.3-3 becomes

$$x(t) = \frac{23}{200}[1 - \sqrt{5}e^{-4t} \sin(2t + 0.4636)]$$

Using this approach, we can express the step response for complex roots as shown for case 3 in Table 5.3-1. This general form is derived in Example 5.4-7 in the next section.

The step response can also be expressed in terms of ζ and ω_n using the fact that the roots $s = -a \pm ib$ can be expressed as

$$s = -\zeta\omega_n \pm i\omega_n\sqrt{1 - \zeta^2} \tag{5.3-21}$$

Therefore $a = \zeta\omega_n$ and $b = \omega_n\sqrt{1 - \zeta^2}$ and the response becomes

$$x(t) = \frac{1}{k}\left[\frac{1}{\sqrt{1 - \zeta^2}}e^{-\zeta\omega_n t} \sin(\omega_d t + \phi) + 1\right] \tag{5.3-22}$$

where

$$\phi = \tan^{-1} \frac{\sqrt{1 - \zeta^2}}{\zeta} + \pi \qquad (5.3\text{-}23)$$

and

$$\omega_d = \omega_n \sqrt{1 - \zeta^2} \qquad (5.3\text{-}24)$$

Because $\zeta > 0$, ϕ lies in the third quadrant. This form is also given in Table 5.3-1.

Analysis of the Step Response

The step response is illustrated in Figure 5.3-4, with a plot of the normalized response variable kx as a function of the normalized time variable $\omega_n t$. The plot gives the response for several values of the damping ratio, with ω_n held constant. When $\zeta > 1$, the response is sluggish and does not overshoot the steady-state value. As ζ is decreased, the speed of response increases. The critically damped case $\zeta = 1$ is the case in which the steady-state value is reached most quickly but without oscillation.

As ζ is decreased below 1, the response overshoots and oscillates about the final value. The smaller ζ is, the larger the overshoot and the longer it takes for the oscillations to die out. There are design applications in which we wish the response to approach its final value as quickly as possible, with some oscillation tolerated. As ζ is decreased to zero (no damping), the oscillations never die out.

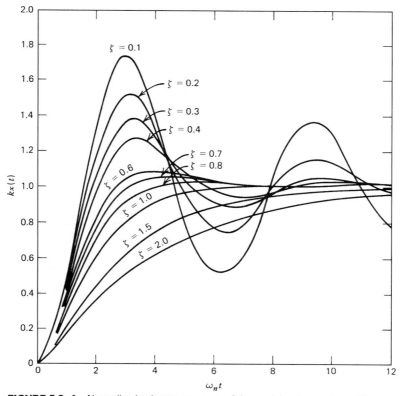

FIGURE 5.3-4 Normalized unit-step response of the model $m\ddot{x} + c\dot{x} + kx = f(t)$.

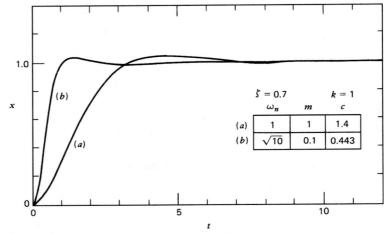

FIGURE 5.3-5 Unit-step response for $\zeta = 0.7$, $k = 1$, and variable ω_n.

Because the axes of Figure 5.3-4 have been normalized by k and ω_n, respectively, the plot shows only the variation in the response as ζ is varied, with k and ω_n held constant. Let us see how each of the parameters affects the step response.

Figure 5.3-5 shows what happens when ζ and k are held constant while ω_n is changed. The effect of increasing ω_n is to speed up the response and make the overshoot occur earlier. Figure 5.3-6 shows two cases in which ζ and ω_n are held constant while k is changed. The effect of increasing k is to decrease the magnitude of the response.

Finally, Figure 5.3-7 shows three cases for which c and k are fixed while ζ is varied by changing m. The plot shows that as ζ is decreased, the response becomes more oscillatory, the overshoot becomes larger and occurs later. This occurs even though the damping constant c is held fixed. Notice that the overshoot in Figure 5.3-7 occurs *earlier* as ζ is decreased.

The discrepancy between Figures 5.3-4 and 5.3-7 is explained by the fact that if $k = 1$, then ω_n and m must be fixed in order to interpret Figure 5.3-4. That is, the patterns shown in Figure 5.3-4 for variable ζ are valid only if k and ω_n (and thus m) are fixed values.

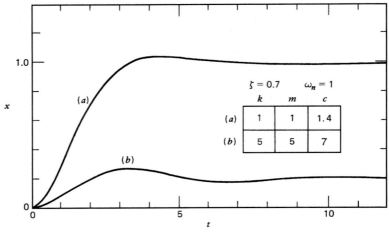

FIGURE 5.3-6 Unit-step response for $\zeta = 0.7$, $\omega_n = 1$, and variable k.

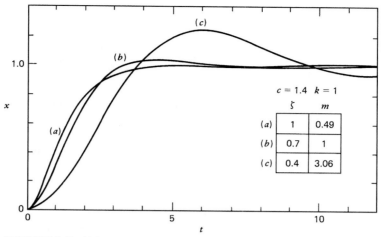

FIGURE 5.3-7   Unit-step response for $c = 1.4$, $k = 1$, and variable ζ.

Transient-Response Specifications

Description of and performance criteria for the transient response of a system are frequently stated in terms of the parameters shown in Figure 5.3-8 for a typical step response of an underdamped system (not necessarily a second-order system). The *maximum overshoot* is the maximum deviation of the output x above its steady-state value x_{ss}. It is sometimes expressed as a percentage of the final value. Because the maximum overshoot increases with decreasing ζ, it is sometimes used as an indicator of the relative stability of the system. The *peak time* t_p is the time at which the maximum overshoot occurs. The *settling time* t_s is the time required for the oscillations to stay within

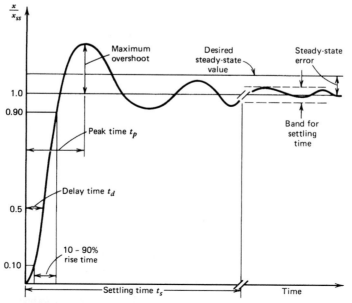

FIGURE 5.3-8   Transient performance specifications based on step response.

some specified small percentage of the final value. The most common values used are 2% and 5%; the latter choice is shown in the figure. If the final value of the response differs from some desired value, a steady-state error exists.

The *rise time* t_r can be defined as the time required for the output to rise from 10% to 90% of its final value. However, no agreement exists on this definition. Sometimes the rise time is taken to be the time required for the response to reach the final value for the first time. Other definitions are also in use. Finally, the *delay time* t_d is the time required for the response to reach 50% of its final value.

These parameters are relatively easy to obtain from an experimentally determined step-response plot. However, if they are to be determined in analytical form from a differential equation model, the task is difficult for models of order greater than two. Here, we obtain expressions for these quantities from the expression for the step response given by Equation 5.3-22.

Setting the derivative dx/dt of Equation 5.3-22 equal to zero gives expressions for both the maximum overshoot and the peak time t_p. After some trigonometric manipulation, the result is

$$\frac{dx}{dt} = \frac{1}{k}\left(\frac{\omega_n}{\sqrt{1-\zeta^2}}e^{-\zeta\omega_n t}\sin \omega_n\sqrt{1-\zeta^2}t\right) = 0$$

For $t < \infty$, this gives

$$\omega_n\sqrt{1-\zeta^2}t = n\pi \qquad n = 0, 1, 2, \ldots$$

The times at which extreme values of the oscillations occur are thus

$$t = \frac{n\pi}{\omega_n\sqrt{1-\zeta^2}} \tag{5.3-25}$$

The odd values of n give the times of overshoots, and the even values correspond to the times of undershoots. The maximum overshoot occurs when $n = 1$. Thus

$$t_p = \frac{\pi}{\omega_n\sqrt{1-\zeta^2}} \tag{5.3-26}$$

The magnitudes of the overshoots and undershoots are found by substituting Equation 5.3-25 into Equation 5.3-22. After some manipulation, the result is

$$x_{\text{extremum}} = \frac{1}{k}\left[1 + (-1)^{n-1}e^{-n\pi\zeta/\sqrt{1-\zeta^2}}\right] \tag{5.3-27}$$

The maximum overshoot is found when $n = 1$:

$$\text{Maximum Overshoot} = x_{\max} - x_{ss} = \frac{1}{k}e^{-\pi\zeta/\sqrt{1-\zeta^2}} \tag{5.3-28}$$

The preceding expressions show that the maximum overshoot and the peak time are functions of only the damping ratio ζ for a second-order system. The percent overshoot is

$$\text{Maximum Percent Overshoot} = \frac{x_{\max} - x_{ss}}{x_{ss}}100 = 100e^{-\pi\zeta/\sqrt{1-\zeta^2}} \tag{5.3-29}$$

This is shown graphically in Figure 5.3-9a. The normalized peak time $\omega_n t_p$ is plotted versus ζ in Figure 5.3-9b.

Analytical expressions for the delay time, the rise time, and the settling time are difficult to obtain. For the delay time, set $x = 0.5x_{ss} = 0.5/k$ in Equation 5.3-22 to obtain

$$e^{-\zeta\omega_n t}\sin\left(\omega_n\sqrt{1-\zeta^2}t + \phi\right) = -0.5\sqrt{1-\zeta^2} \tag{5.3-30}$$

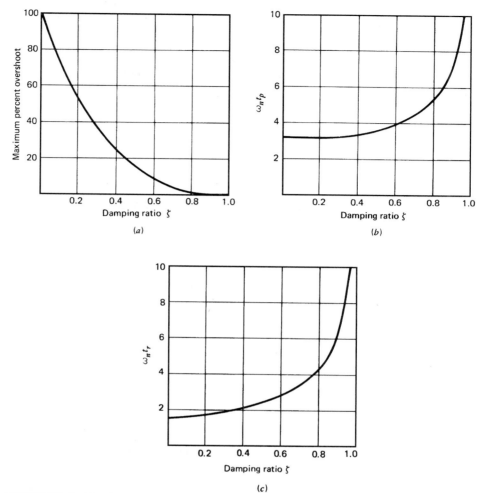

FIGURE 5.3-9 Transient–response specifications as a function of the damping ratio ζ. (a) Maximum percent overshoot. (b) Normalized peak time $\omega_n t_p$. (c) Normalized 100% rise time $\omega_n t_r$.

where ϕ is given by Equation 5.3-23. For a given ζ and ω_n, t_d can be obtained by a numerical procedure, such as Newton's method. This is easily done on a calculator, especially if the following straight-line approximation is used as a starting guess:

$$t_d \approx \frac{1 + 0.7\zeta}{\omega_n} \qquad 0 \leq \zeta \leq 1 \tag{5.3-31}$$

A similar procedure can be applied to find the rise time t_r. In this case, two equations must be solved, one for the 10% time and one for the 90% time. The difference between these times is the rise time. These calculations are made easier by using the following straight-line approximation:

$$10\% - 90\% \text{ Rise Time } t_r = \frac{0.8 + 2.5\zeta}{\omega_n} \qquad 0 \leq \zeta \leq 1 \tag{5.3-32}$$

This approximation was obtained by plotting the results of many such computer solutions.

If we choose the alternative definition of rise time—that is, the first time at which the final value is crossed—then the solution for t_r is easier to obtain. Set $x = x_{ss} = 1/k$ in Equation 5.3-22 to obtain

$$e^{-\zeta\omega_n t}\sin\left(\omega_n\sqrt{1-\zeta^2}\,t+\phi\right)=0$$

This implies that for $t < \infty$,

$$\omega_n\sqrt{1-\zeta^2}\,t+\phi=n\pi \qquad n=0,1,2,\ldots \tag{5.3-33}$$

For $t_r > 0$, $n = 2$ because ϕ is in the third quadrant. Thus

$$100\%\ \text{Rise Time}\ t_r = \frac{2\pi-\phi}{\omega_n\sqrt{1-\zeta^2}} \tag{5.3-34}$$

where ϕ is given by Equation 5.3-23. The rise time is inversely proportional to the natural frequency ω_n for a given value of ζ. A plot of the normalized rise time $\omega_n t_r$ versus ζ is given in Figure 5.3-9c.

In order to express the settling time in terms of the parameters ζ and ω_n, we can use the fact that the exponential term in Equation 5.3-22 provides the envelopes of the oscillations. These envelopes are found by setting the sine term to $+1$ in Equation 5.3-22. The magnitude of the difference between each envelope and the final value $1/k$ is

$$\frac{1}{k}\frac{e^{-\zeta\omega_n t}}{\sqrt{1-\zeta^2}}$$

Both envelopes are within 2% of the final value when

$$\frac{e^{-\zeta\omega_n t}}{\sqrt{1-\zeta^2}}\le 0.02$$

The 2% settling time can be found from the preceding expression. For ζ less than about 0.7, t_s can also be approximated by noting that $e^{-4}\approx 0.02$ and using the formula

$$2\%\ \text{Settling Time}\ t_s \approx \frac{4}{\zeta\omega_n} \tag{5.3-35}$$

Thus t_s is approximately four time constants for ζ less than approximately 0.7. However, in practice, we usually can take the 2% settling time to be 4τ as long as $\zeta < 1$.

Table 5.3-2 summarizes these formulas.

Table 5.3-2 Step-Response Specifications for the Linear Second-Order Model

Maximum percent overshoot:	$M_p = 100e^{-\pi\zeta/\sqrt{1-\zeta^2}}$
	$\zeta = \dfrac{A}{\sqrt{\pi^2+A^2}} \qquad A = \ln\dfrac{100}{M_p}$
Peak time:	$t_p = \dfrac{\pi}{\omega_n\sqrt{1-\zeta^2}}$
Delay time:	$t_d \approx \dfrac{1+0.7\zeta}{\omega_n} \qquad 0 \le \zeta \le 1$
100% rise time:	$t_r = \dfrac{2\pi-\phi}{\omega_n\sqrt{1-\zeta^2}}$
	$\phi = \tan^{-1}\dfrac{\sqrt{1-\zeta^2}}{\zeta}+\pi$

Figure 5.3-10 shows the effect of root location on decay rate, peak time, and overshoot. Roots lying on the same vertical line have the same decay rate because they have the same time constant. Roots lying on the same horizontal line have the same frequency and peak time. This can be shown to be true by using the formula

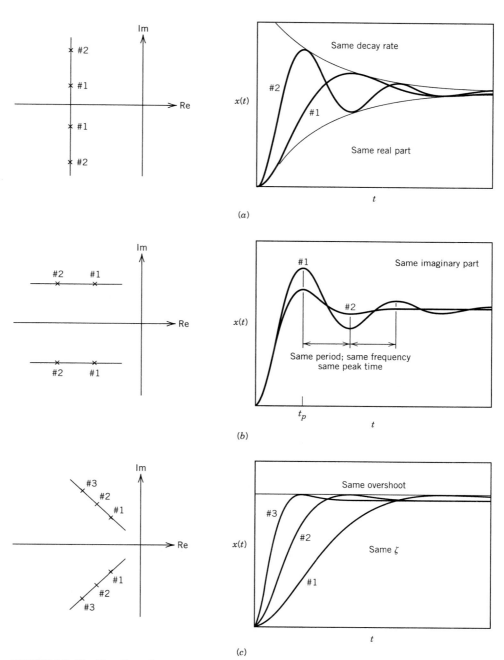

FIGURE 5.3-10 The effect of root location on decay rate, peak time, and overshoot.

$t_p = \pi/\omega_n \sqrt{1 - \zeta^2}$ and the formulas for ω_n and ζ in terms of root location. Figure 5.3-10c shows that roots lying on the same radial line have the same maximum percent overshoot M_p because M_p depends only on ζ, which is the same for all roots lying on the same radial line.

5.4 ADDITIONAL RESPONSE EXAMPLES

The ramp function $f(t) = at + b$ is shown in Figure 5.4-1a. Although no real forcing function lasts forever, the ramp function nevertheless is useful for determining the response to a forcing function that is a linear function of time and that lasts long enough for the transient response to disappear. This occurs after about four time constants. Such ramp-like forcing functions are shown in Figures 5.4-1b–e.

Applications of the ramp response occur when a structure, or a weighing scale, or a vehicle has a weight dumped on it at a constant rate. An example of this is illustrated in Figure 5.4-2, which shows a liquid being dumped into a railcar. If the liquid flow rate is a constant, say r, the mass added to the vehicle is $m_a(t) = rt$. If the vehicle's empty

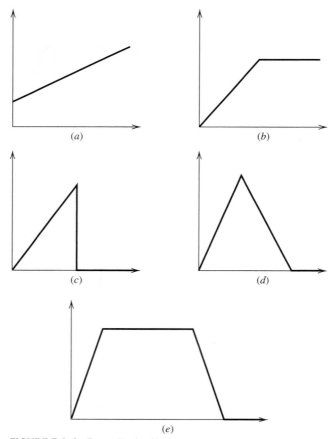

FIGURE 5.4-1 Ramp-like forcing functions.

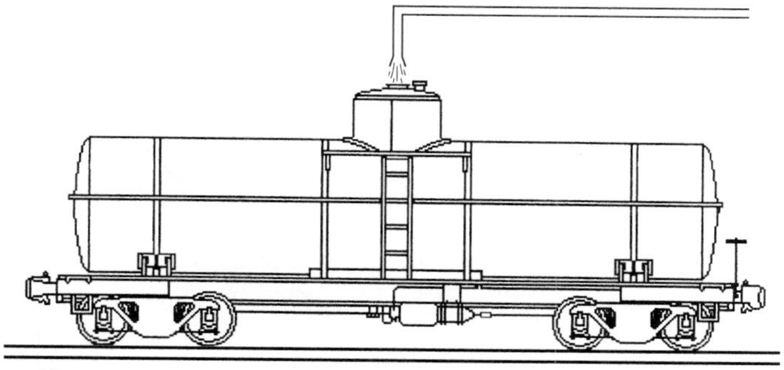

FIGURE 5.4-2   Filling a railcar is an example of a ramp input.

weight is m_e, then the equation of motion of the vehicle, when modeled as a single-DOF system, is

$$[m_e + m_a(t)]\ddot{x} + c\dot{x} + kx = rt$$

where c and k represent the properties of the vehicle suspension. If the total added mass is a significant fraction of the total mass $m_e + m_a(t)$, then the model is one having a variable coefficient. If, however, total added mass is a small fraction of the total mass $m_e + m_a(t)$, then the system can be approximated by the constant-coefficient = model $m_e\ddot{x} + c\dot{x} + kx = rt$.

EXAMPLE 5.4-1
Ramp Response

Use the Laplace transform to solve the following problem:

$$\ddot{x} + 8\dot{x} + 15x = 0.05t \qquad x(0) = 0 \qquad \dot{x}(0) = 0$$

Solution

Taking the transform of both sides of the equation and noting that both initial conditions are zero, we obtain

$$s^2X(s) + 8sX(s) + 15X(s) = \frac{0.05}{s^2}$$

Solve for $X(s)$:

$$X(s) = \frac{0.05}{s^2(s^2 + 8s + 15)} = \frac{0.05}{s^2(s + 3)(s + 5)}$$

The inverse transform can be obtained as follows. The denominator roots are $s = 0$, $s = 0$, $s = -3$, and $s = -5$. Thus we can express $X(s)$ as

$$X(s) = \frac{C_1}{s^2} + \frac{C_2}{s} + \frac{C_3}{s + 3} + \frac{C_4}{s + 5}$$

where

$$C_1 = \lim_{s \to 0} \frac{0.05}{(s + 3)(s + 5)} = \frac{1}{300}$$

$$C_2 = \lim_{s \to 0} \frac{d}{ds}\left[\frac{0.05}{(s + 3)(s + 5)}\right] = \lim_{s \to 0} \frac{-0.1s - 0.4}{(s^2 + 8s + 15)^2} = -\frac{2}{1125}$$

$$C_3 = \lim_{s \to -3} \frac{0.05}{s^2(s + 5)} = \frac{1}{360}$$

$$C_4 = \lim_{s \to -5} \frac{0.05}{s^2(s + 3)} = -\frac{1}{1000}$$

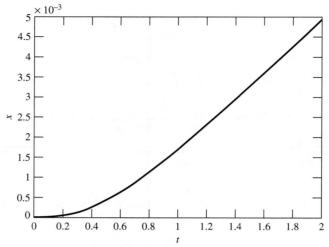

FIGURE 5.4-3 The response for Example 5.4-1.

Thus

$$x(t) = C_1 t + C_2 + C_3 e^{-3t} + C_4 e^{-5t} = \frac{1}{300} t - \frac{2}{1125} + \frac{1}{360} e^{-3t} - \frac{1}{1000} e^{-5t}$$

The response is shown in Figure 5.4-3. ∎

Consider the forcing function $(t/\tau)e^{-t/\tau}$ shown in Figure 5.4-4. It can be used to represent a force that rapidly increases and then decays slowly. After four time constants (at $t = 4\tau$), the function is still about 22% of its peak value, as compared to 2% for the pure exponential $(1/\tau)e^{-t/\tau}$. The function $(t/\tau)e^{-t/\tau}$ does not reach 2% of its peak value

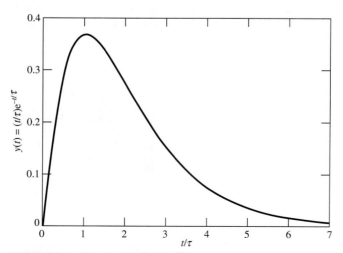

FIGURE 5.4-4 The function $(t/\tau)e^{-t/\tau}$.

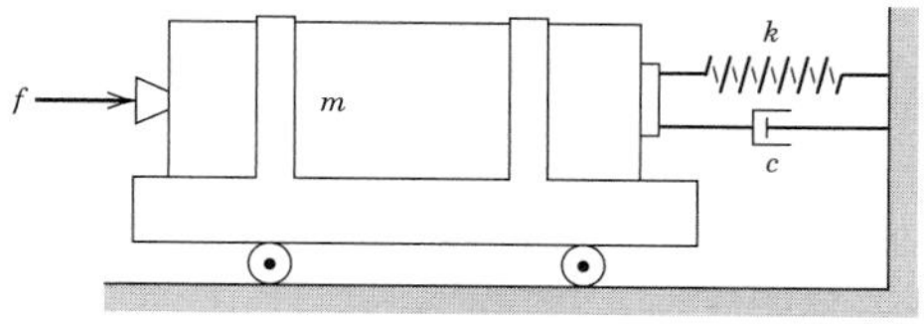

FIGURE 5.4-5 A rocket-engine test stand.

until about seven time constants ($t = 7\tau$). So its effects last longer than those due to the pure exponential.

The system shown in Figure 5.4-5 is a model of certain force isolation systems. In such systems, the mass and spring are used to reduce the force $f(t)$ felt by an object represented by the vertical support. An example is a rocket-engine test stand in which the force $f(t)$ is the rocket's thrust. Depending on the rocket characteristics, its thrust may be modeled with the function $f(t) = bte^{-at}$.

EXAMPLE 5.4-2
Response to te^{-at}

Obtain the response of the following model:

$$\ddot{x} + 4\dot{x} + 3x = 4te^{-10t} \qquad x(0) = 0 \qquad \dot{x}(0) = 0$$

Solution

Taking the transform of both sides of the equation and noting that both initial conditions are zero, we obtain

$$s^2 X(s) + 4sX(s) + 3X(s) = \frac{4}{(s+10)^2}$$

Solve for $X(s)$:

$$X(s) = \frac{4}{(s+10)^2(s^2+4s+3)} = \frac{4}{(s+10)^2(s+1)(s+3)}$$

The inverse transform can be obtained as follows. The denominator roots are $s = -10$, $s = -10$, $s = -1$, and $s = -3$. Thus we can express $X(s)$ as

$$X(s) = \frac{C_1}{(s+10)^2} + \frac{C_2}{s+10} + \frac{C_3}{s+1} + \frac{C_4}{s+3}$$

where

$$C_1 = \lim_{s \to -10} \frac{4}{(s+1)(s+3)} = \frac{2}{81}\frac{18}{7}$$

$$C_2 = \lim_{s \to -10} \frac{d}{ds}\left[\frac{4}{(s+1)(s+3)}\right] = \lim_{s \to -10} \frac{-4(2s+4)}{(s^2+4s+3)^2} = \frac{2}{81}\frac{32}{49}$$

$$C_3 = \lim_{s \to -1} \frac{4}{(s+10)^2(s+3)} = \frac{2}{81}$$

$$C_4 = \lim_{s \to -3} \frac{4}{(s+10)^2(s+1)} = -\frac{2}{81}\frac{81}{49}$$

Thus

$$x(t) = C_1 te^{-10t} + C_2 e^{-10t} + C_3 e^{-t} + C_4 e^{-3t}$$

$$= \frac{2}{81}\left(\frac{18}{7}te^{-10t} + \frac{32}{49}e^{-10t} + e^{-t} - \frac{81}{49}e^{-3t}\right)$$

The response is shown in Figure 5.4-6. ∎

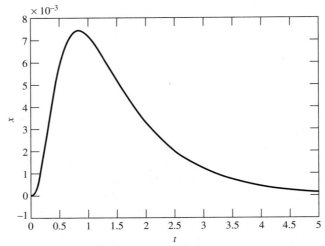

FIGURE 5.4-6   The response for Example 5.4-2.

EXAMPLE 5.4-3
Sine Response of a
Second-Order
Model

Use the Laplace transform to solve the following problem:

$$\ddot{x} + 6\dot{x} + 34x = 5 \sin 6t \qquad x(0) = 0 \qquad \dot{x}(0) = 0$$

Solution

Taking the transform of both sides of the equation and noting that both initial conditions are zero, we obtain

$$s^2 X(s) + 6sX(s) + 34X(s) = 5\frac{6}{s^2 + 6^2}$$

Solve for $X(s)$:

$$X(s) = \frac{30}{(s^2 + 6s + 34)(s^2 + 6^2)}$$

The inverse transform can be obtained as follows. The denominator roots are $s = -3 \pm 5i$ and $s = \pm 6i$. Thus we can express $X(s)$ as

$$X(s) = \frac{30}{\left[(s+3)^2 + 5^2\right](s^2 + 6^2)} \tag{1}$$

which can be expressed as the sum of terms that are proportional to entries 8 through 11 in Table 5.2-1.

$$X(s) = C_1 \frac{5}{(s+3)^2 + 5^2} + C_2 \frac{s+3}{(s+3)^2 + 5^2} + C_3 \frac{6}{s^2 + 6^2} + C_4 \frac{s}{s^2 + 6^2} \tag{2}$$

We can obtain the coefficients by noting that $X(s)$ can be written as

$$X(s) = \frac{N(s)}{\left[(s+3)^2 + 5^2\right](s^2 + 6^2)} \tag{3}$$

where

$$N(s) = 5C_1(s^2 + 6^2) + C_2(s+3)(s^2 + 6^2) + 6C_3\left[(s+3)^2 + 5^2\right] + C_4 s\left[(s+3)^2 + 5^2\right]$$

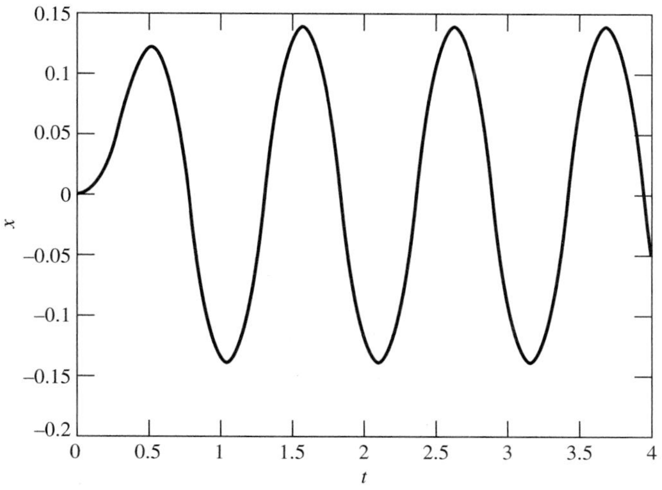

FIGURE 5.4-7 The response for Example 5.4-3.

Comparing the numerators of Equations 1 and 3, and collecting powers of s, we see that

$$(C_2 + C_4)s^3 + (5C_1 + 3C_2 + 6C_3 + 6C_4)s^2 + (36C_2 + 36C_3 + 34C_4)s$$
$$+ 180C_1 + 108C_2 + 204C_3 = 30$$

or

$$C_2 + C_4 = 0 \qquad 5C_1 + 3C_2 + 6C_3 + 6C_4 = 0$$

$$36C_2 + 36C_3 + 34C_4 = 0 \qquad 180C_1 + 108C_2 + 204C_3 = 30$$

These are four equations in four unknowns. Note that the first equation gives $C_4 = -C_2$. Thus we can easily eliminate C_4 from the equations and obtain a set of three equations in three unknowns. The solution is $C_1 = 6/65$, $C_2 = 9/65$, $C_3 = -1/130$, and $C_4 = -9/65$. The inverse transform is

$$x(t) = C_1 e^{-3t} \sin 5t + C_2 e^{-3t} \cos 5t + C_3 \sin 6t + C_2 \cos 6t$$

$$= \frac{6}{65} e^{-3t} \sin 5t + \frac{9}{65} e^{-3t} \cos 5t - \frac{1}{130} \sin 6t - \frac{9}{65} \cos 6t$$

The response is shown in Figure 5.4-7. ∎

EXAMPLE 5.4-4
Response to a Decaying Sinusoid

Obtain the response of the following model:

$$\ddot{x} + 12\dot{x} + 20x = 4e^{-5t} \sin 40t \qquad x(0) = 0 \qquad \dot{x}(0) = 0$$

Solution

Taking the transform of both sides of the equation and noting that both initial conditions are zero, we obtain

$$s^2 X(s) + 12sX(s) + 20X(s) = 4 \frac{40}{(s+5)^2 + 40^2}$$

Solve for $X(s)$:

$$X(s) = \frac{160}{(s^2 + 12s + 20)(s^2 + 10s + 1625)}$$

The denominator roots are $s = -2$, $s = -10$, and $s = -5 \pm 40i$. Thus we can express $X(s)$ as

$$X(s) = \frac{160}{(s+2)(s+10)[(s+5)^2 + 40^2]} \tag{1}$$

which can be expressed as the sum of terms that are proportional to entries 10 and 11 in Table 5.2-1.

$$X(s) = \frac{C_1}{s+2} + \frac{C_2}{s+10} + C_3 \frac{s+5}{(s+5)^2 + 40^2} + C_4 \frac{40}{(s+5)^2 + 40^2} \tag{2}$$

where

$$C_1 = \lim_{s \to -2} \frac{160}{(s+10)[(s+5)^2 + 40^2]} = \frac{20}{1609} = 0.01243$$

$$C_2 = \lim_{s \to -10} \frac{160}{(s+2)[(s+5)^2 + 40^2]} = -\frac{4}{325} = -0.012308$$

We can obtain the remaining coefficients by noting that Equation 2 can be written as a single fraction,

$$X(s) = \frac{N(s)}{(s+2)(s+10)[(s+5)^2 + 40^2]}$$

where

$$N(s) = C_1(s+10)(s^2 + 10s + 1625) + C_2(s+2)(s^2 + 10s + 1625)$$
$$+ C_3(s+5)(s+2)(s+10) + 40C_4(s+2)(s+10)$$

Comparing $N(s)$ with the numerator of Equation 1, and collecting powers of s, we see that

$$(C_1 + C_2 + C_3)s^3 + (20C_1 + 12C_2 + 17C_3 + 40C_4)s^2$$
$$+ (1725C_1 + 1645C_2 + 80C_3 + 480C_4)s + 16250C_1$$
$$+ 3250C_2 + 204C_3 + 800C_4 = 160$$

This gives

$$C_3 = -1.22 \times 10^{-4}$$

$$C_4 = -2.471 \times 10^{-3}$$

Thus

$$x(t) = C_1 e^{-2t} + C_2 e^{-10t} + C_3 e^{-5t} \cos 40t + C_4 e^{-5t} \sin 40t$$
$$= 0.01243 e^{-2t} - 0.012308 e^{-10t} - 1.22 \times 10^{-4} e^{-5t} \cos 40t - 2.471 \times 10^{-3} e^{-5t} \sin 40t$$

The response is shown in Figure 5.4-8. ∎

EXAMPLE 5.4-5
A Model with an
Input Derivative

Obtain the transfer function and investigate the response of the following model in terms of the parameter a. The input $g(t)$ is a unit-step function.

$$3\ddot{x} + 18\dot{x} + 24x = a\dot{g}(t) + 6g(t) \qquad x(0) = 0 \qquad \dot{x}(0) = 0$$

Solution

Transforming the equation with zero initial conditions and solving for the ratio $X(s)/G(s)$ gives the transfer function

$$\frac{X(s)}{G(s)} = \frac{as + 6}{3s^2 + 18s + 24}$$

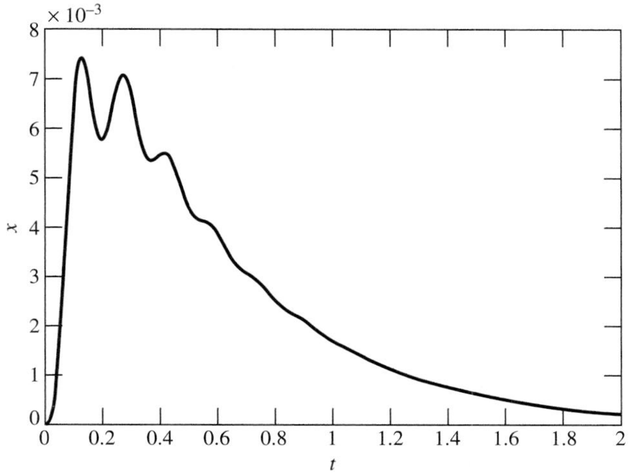

FIGURE 5.4-8   The response for Example 5.4-4.

The s in the numerator indicates the presence of a derivative of the input function and is called *numerator dynamics*.

For a unit-step input, $G(s) = 1/s$ and

$$X(s) = \frac{as + 6}{s(3s^2 + 18s + 24)} = \frac{1}{4}\frac{1}{s} + \frac{a - 3}{6}\frac{1}{s + 2} + \frac{3 - 2a}{12}\frac{1}{s + 4}$$

Thus the response is

$$x(t) = \frac{1}{4} + \frac{a - 3}{6}e^{-2t} + \frac{3 - 2a}{12}e^{-4t}$$

From this solution or from the initial value theorem (see property 11 in Table 5.2-2), we find that $x(0+) = 0$, which is equal to $x(0)$, but that $\dot{x}(0+) = a/3$, which is not equal to $\dot{x}(0)$ unless $a = 0$ (which corresponds to the absence of numerator dynamics). The plot of the response is given in Figure 5.4-9 for several values of a. Notice that a "hump" in the response (called an "overshoot") does not occur for smaller values of a and that the height of the hump increases as a increases. However, the value of a does not affect the steady-state response. ∎

EXAMPLE 5.4-6
Damper Location and Numerator Dynamics

(a) By obtaining the equations of motion and the transfer functions of the two systems shown in Figure 5.4-10, investigate the effect of the location of the damper on the step response of the system. The displacement $u(t)$ is a step function.

(b) Obtain the unit-step response for each system for the specific case $m = 1$, $c = 6$, and $k = 8$, with zero initial conditions.

(c) Use the initial value theorem to investigate the response.

Solution

(a) For the system in Figure 5.4-10a,

$$m\ddot{x} + c\dot{x} + kx = c\dot{u} + ku$$

and thus

$$\frac{X(s)}{U(s)} = \frac{cs + k}{ms^2 + cs + k} \tag{1}$$

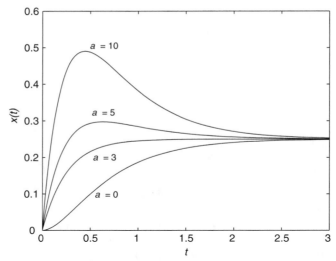

FIGURE 5.4-9 The response for Example 5.4-5.

Therefore this system has numerator dynamics.

For the system in Figure 5.4-10b,

$$m\ddot{x} + c\dot{x} + kx = ku$$

and thus

$$\frac{X(s)}{U(s)} = \frac{k}{ms^2 + cs + k} \tag{2}$$

Therefore this system does not have numerator dynamics. Both systems have the same characteristic equation and thus have the same time constant, damping ratio, and natural frequency, but response predictions based on the time constant and damping ratio (such as the settling time, maximum overshoot, and rise time) may not be accurate for the system with numerator dynamics.

(b) Substituting the given values into Equation 1, using $U(s) = 1/s$, and performing a partial-fraction expansion, we obtain

$$X(s) = \frac{6s + 8}{s(s^2 + 6s + 8)} = \frac{1}{s} + \frac{1}{s + 2} - \frac{2}{s + 4} \tag{3}$$

The response is

$$x(t) = 1 + e^{-2t} - 2e^{-4t} \qquad [\textit{numerator dynamics}] \tag{4}$$

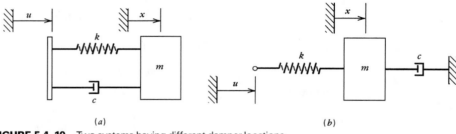

(a)   (b)

FIGURE 5.4-10 Two systems having different damper locations.

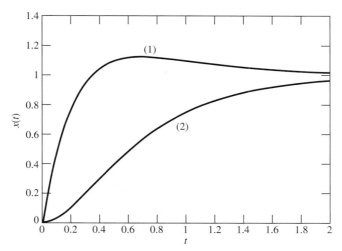

FIGURE 5.4-11   The response for Example 5.4-6.

For Equation 2,

$$X(s) = \frac{8}{s(s^2 + 6s + 8)} = \frac{1}{s} - \frac{2}{s+2} + \frac{1}{s+4} \tag{5}$$

The response is

$$x(t) = 1 - 2e^{-2t} + e^{-4t} \qquad [no\ numerator\ dynamics] \tag{6}$$

The responses are shown in Figure 5.4-11. Curve (1) corresponds to Equation 1, which has numerator dynamics. Curve (2) corresponds to the system without numerator dynamics, Equation 2. The numerator dynamics causes an overshoot in the response but does not affect the steady-state response. Thus the damper location can affect the response.

(c) The given initial conditions are $x(0) = \dot{x}(0) = 0$. Evaluating Equation 6 at $t = 0$ gives $x(0) = 0$ and $\dot{x}(0) = 0$. However, evaluating Equation 4 at $t = 0$ gives $x(0) = 0$ but $\dot{x}(0) = 6$. Thus the numerator dynamics has caused the velocity $\dot{x}$ to change instantaneously from 0 at $t = 0$ to 6 at $t = 0+$.

These results can be confirmed with the initial value theorem (property 11 in Table 5.2-2), as follows. Using Equation 1 with the given values of m, c, and k, with a unit-step input, we have

$$x(0+) = \lim_{s \to \infty} s \frac{6s + 8}{s^2 + 6s + 8} \frac{1}{s} = 0$$

$$\dot{x}(0+) = \lim_{s \to \infty} s \frac{s(6s + 8)}{s^2 + 6s + 8} \frac{1}{s} = 6$$

Because velocity cannot change instantaneously in the real world, this example shows that we must be careful when applying the step response of a model that has numerator dynamics. This issue is discussed in more detail in Section 5.5.   ■

Response in the Form of a Sine Function with a Phase Shift

For the complex-roots case, the following method can be used to give the response in the form of a sine function with a phase shift.

Assume that the Laplace transformation of the system model results in the following expression for the transform of the response $X(s)$:

$$X(s) = \frac{P(s)}{Q(s)} \tag{5.4-1}$$

where $Q(s)$ is a quadratic polynomial of the form

$$Q(s) = (s + a)^2 + b^2$$

Thus the roots are $s = -a + ib$. The term $P(s)$ represents the remaining terms in $X(s)$. A partial-fraction expansion of $X(s)$ for the quadratic roots gives

$$X(s) = \frac{C_1}{s + a - ib} + \frac{C_2}{s + a + ib} \tag{5.4-2}$$

where

$$C_1 = \lim_{s \to -a+ib} \left[\frac{P(s)}{Y(s)} (s + a - ib) \right] \tag{5.4-3}$$

$$C_2 = \lim_{s \to -a-ib} \left[\frac{P(s)}{Q(s)} (s + a + ib) \right] \tag{5.4-4}$$

After the factor $(s + a + ib)$ is canceled, C_1 has the form

$$C_1 = \lim_{s \to -a+ib} \left[\frac{P(s)}{(s + a + ib)} \right] = \frac{1}{2ib} R(-a + ib) \tag{5.4-5}$$

where R is a function of the complex number $-a + ib$ and is defined as

$$R(-a + ib) = \lim_{s \to -a+ib} [P(s)] \tag{5.4-6}$$

Because $R(-a + ib)$ can be complex in general, it can be expressed as a magnitude and a phase angle, as shown in Figure 5.4-12:

$$R(-a + ib) = |R(-a + ib)|e^{i\phi}$$

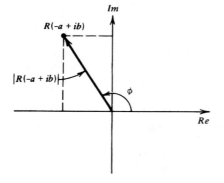

FIGURE 5.4-12 Polar representation of the complex factor $R(-a + ib)$.

Thus

$$C_1 = \frac{1}{2ib}|R(-a+ib)|e^{i\phi}$$

and because C_2 is the conjugate of C_1,

$$C_2 = -\frac{1}{2ib}|R(-a+ib)|e^{-i\phi}$$

Reverting to the time domain, we obtain

$$
\begin{aligned}
x(t) &= C_1 e^{-at} e^{ibt} + C_2 e^{-at} e^{-ibt} \\
&= \frac{1}{b}|R(-a+ib)|e^{-at} \frac{e^{ibt}e^{i\phi} - e^{-ibt}e^{-i\phi}}{2i} \qquad (5.4\text{-}7) \\
&= \frac{1}{b}|R(-a+ib)|e^{-at} \sin(bt+\phi)
\end{aligned}
$$

where we have used the identity

$$\sin\theta = \frac{e^{i\theta} - e^{-i\theta}}{2i}$$

with $\theta = bt + \phi$. The phase angle is given by

$$\phi = \angle R(-a+ib) \qquad (5.4\text{-}8)$$

The preceding expressions can be used to determine that part of the free or forced response resulting from a pair of complex roots. If the system is of an order greater than two, or if the input function introduces more roots, then Equation 5.4-7 gives only part of the response. The rest of the response is found by doing a complete partial-fraction expansion for all the roots and adding all of the resulting response terms. The method is summarized in Table 5.4-1 and is illustrated by Example 5.4-7. The method does not cover the case where the complex roots are repeated. In such a case you need to use the repeated-roots formula, Equation 5.3-20.

TABLE 5.4-1 Response Calculation for Complex Roots

$X(s) =$ response transform

$$X(s) = \frac{P(s)}{Q(s)}$$
$$Q(s) = (s+a)^2 + b^2$$

$P(s)$ absorbs all other terms

Solution form:

$$x(t) = \frac{1}{b}|R(-a+ib)|e^{-at}\sin(bt+\phi)$$

+terms due to the roots of $P(s)$.

$$R(-a+ib) = \lim_{s \to -a+ib} P(s)$$

$\phi = \angle R(-a+ib)$ (see Figure 5.4-12)

EXAMPLE 5.4-7
Sine Form of the
Step Response

Obtain the response in the form of a sinusoidal function of the model

$$m\ddot{x} + c\dot{x} + kx = u_s(t)$$

for zero initial conditions, assuming that the characteristic roots are complex.

Solution

Transforming the differential equation using the linearity and derivative properties, we obtain

$$ms^2 X(s) + csX(s) + kX(s) = \frac{1}{s}$$

Collecting terms and dividing by m gives

$$\left(s^2 + \frac{c}{m}s + \frac{k}{m}\right)X(s) = \frac{1}{ms} \tag{1}$$

The characteristic roots are assumed to be of the form $s = -a \pm ib$, so the solution for $X(s)$ can be expressed as

$$X(s) = \frac{1/m}{\left[(s+a)^2 + b^2\right]s} = \frac{A(s)}{(s+a)^2 + b^2} + \frac{C_3}{s}$$

where we need not find $A(s)$.

From Equation 5.3-13,

$$C_3 = \lim_{s \to 0} \frac{(1/m)s}{\left[(s+a)^2 + b^2\right]s} = \frac{1}{m}\frac{1}{a^2 + b^2}$$

In the notation of Table 5.4-1,

$$P(s) = \frac{1}{ms}$$

$$Q(s) = (s+a)^2 + b^2$$

From Table 5.4-1, the form of the step response is

$$x(t) = \frac{1}{b}|R(-a+ib)|e^{-at}\sin(bt + \phi) + C_3 u_s(t)$$

where

$$R(-a+ib) = \lim_{s \to -a+ib} P(s) = \frac{1}{m(-a+ib)}$$

Multiplying numerator and denominator by the complex conjugate of the denominator gives

$$R(-a+ib) = \frac{-a-ib}{m(-a+ib)(-a-ib)} = \frac{1}{m}\frac{-a-ib}{a^2+b^2}$$

Thus

$$|R(-a+ib)| = \frac{1}{m}\frac{\sqrt{a^2+b^2}}{a^2+b^2} = \frac{1}{m}\frac{1}{\sqrt{a^2+b^2}}$$

and

$$\phi = \angle R(-a+ib) = \angle(-a-ib)$$

Thus the step response is

$$x(t) = \frac{1}{b}|R(-a+ib)|e^{-at}\sin(bt+\phi) + C_3 u_s(t)$$

$$= \frac{1}{mb\sqrt{a^2+b^2}}e^{-at}\sin(bt+\phi) + \frac{1}{m(a^2+b^2)}u_s(t)$$

Note that

$$(s+a)^2 + b^2 = s^2 + 2as + a^2 + b^2 = s^2 + \frac{c}{m}s + \frac{k}{m}$$

Comparing the constant terms, we see that $k/m = a^2 + b^2$. Substituting $k = m(a^2 + b^2)$ and $\sqrt{a^2+b^2} = \sqrt{k/m}$ into the expression for $x(t)$ gives

$$x(t) = \frac{1}{k}\left[\frac{\sqrt{k}}{b\sqrt{m}}e^{-at}\sin(bt+\phi) + u_s(t)\right]$$

These results are summarized as case 3 in Table 5.3-1.   ■

Application of the Shifting Theorem

A *rectangular pulse* is shown in Figure 5.4-13. The pulse response of the model $\tau\dot{y} + y = f(t)$ can be obtained by using the step response to find $y(T)$, which is then used as the initial condition for a zero input solution. Alternatively, the *shifting property* of Laplace transforms can be applied (Table 5.2-2, property 9). The shifted function is defined as

$$g(t) = \begin{cases} 0 & t < D \\ f(t-D) & t \geq D \end{cases} \tag{5.4-9}$$

where $f(t)$ is a transformable function, as shown in Figure 5.4-14. The transform of $g(t)$ is

$$\mathcal{L}[g(t)] = \int_0^\infty f(t-D)e^{-st}\,dt = \int_D^\infty f(t-D)e^{-st}\,dt$$

$$= \int_0^\infty f(q)e^{-s(q+D)}\,dq = e^{-sD}\int_0^\infty f(q)e^{-sq}\,dq \tag{5.4-10}$$

where the integration variable was changed to $q = t - D$. Thus

$$\mathcal{L}[g(t)] = e^{-sD}F(s) \tag{5.4-11}$$

If $u_s(t)$ is the unit-step function, the shifted function $g(t)$ can be expressed as $f(t-D)u_s(t-D)$.

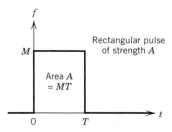

FIGURE 5.4-13   A rectangular pulse.

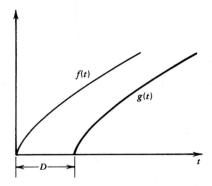

FIGURE 5.4-14   A translated function.

The pulse in Figure 5.4-15 is taken to be composed of a step input of magnitude M starting at $t = 0$, followed at $t = T$ by a negative step input of magnitude M. The pulse input can now be expressed as follows:

$$f(t) = Mu_s(t) - Mu_s(t - T) \tag{5.4-12}$$

Its transform is

$$
\begin{aligned}
F(s) &= M\mathcal{L}[u_s(t)] - M\mathcal{L}[u_s(t - T)] \\
&= M\frac{1}{s} - Me^{-sT}\frac{1}{s} \\
&= \frac{M}{s}(1 - e^{-sT})
\end{aligned}
\tag{5.4-13}
$$

Assuming that $y(0) = 0$ the pulse response of $\tau\dot{y} + y = f(t)$ is found from a partial-fraction expansion:

$$
\begin{aligned}
Y(s) &= \frac{1}{\tau s + 1}F(s) = \frac{1}{\tau}\frac{M}{s + \dfrac{1}{\tau}}\left(\frac{1 - e^{-sT}}{s}\right) \\[2ex]
&= \left(\frac{C_1}{s + \dfrac{1}{\tau}} + \frac{C_2}{s}\right)(1 - e^{-sT}) \\[2ex]
&= \left[\frac{C_1 s + C_2(s + 1/\tau)}{s(s + 1/\tau)}\right](1 - e^{-sT})
\end{aligned}
$$

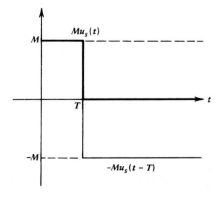

FIGURE 5.4-15   Rectangular pulse constructed by super-imposing two step functions.

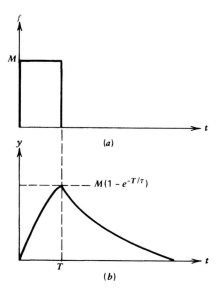

FIGURE 5.4-16 Pulse response of the linear first-order model $\tau \dot{y} + y = f(t)$. (*a*) Pulse input. (*b*) Pulse response.

using the least common denominator $s(s + 1/\tau)$. This is true if $C_1 + C_2 = 0$ and if $C_2 = M$. Thus

$$Y(s) = \left(\frac{-M}{s + \dfrac{1}{\tau}} + \frac{M}{s} \right) \left(1 - e^{-sT} \right)$$

In the time domain, we obtain

$$y(t) = M[u_s(t) - u_s(t - T)] - M\left[e^{-t/\tau} u_s(t) - e^{-(t-T)/\tau} u_s(t - T) \right]$$

For $0 < t < T$, $u_s(t) = 1$, $u_s(t - T) = 0$, and

$$y(t) = M\left(1 - e^{-t/\tau} \right) \qquad 0 < t < T \tag{5.4-14}$$

For $t \geq T$, $u_s(t) = u_s(t - T) = 1$, which gives

$$y(t) = Me^{-(t-T)/\tau} - Me^{-t/\tau} = Me^{-t/\tau}\left(e^{T/\tau} - 1 \right) \qquad t \geq T \tag{5.4-15}$$

This response is shown in Figure 5.4-16. When written in terms of the pulse strength $A = MT$, the response is

$$y(t) = \frac{A}{T} e^{-t/\tau}\left(e^{T/\tau} - 1 \right) \tag{5.4-16}$$

5.5 PULSE AND IMPULSE RESPONSE

The step function models an input that rapidly reaches a constant value. Two other commonly used input functions are the *rectangular pulse function*, which models a constant input that is suddenly removed, and the *impulse function*, which models an input that is applied for a very short time.

In our development of the Laplace transform and its associated methods, we have assumed that the process under study starts at time $t = 0$. Thus the given initial conditions, say $x(0)$, $\dot{x}(0)$, $\cdots$, represent the situation at the start of the process and are the result of any inputs applied prior to $t = 0$. That is, we need not know what the inputs were before $t = 0$ because their effects are contained in the initial conditions.

The effects of any inputs starting at $t = 0$ are not felt by the system until an infinitesimal time later, at $t = 0+$. For some models the dependent variable $x(t)$ and its derivatives do not change between $t = 0$ and $t = 0+$, and thus the solution $x(t)$ obtained from the differential equation will match the given initial conditions when $x(t)$ and its derivatives are evaluated at $t = 0$. The results obtained from the initial value theorem will also match the given initial conditions.

However, we will now investigate the behavior of some systems for which $x(0) \neq x(0+)$, or $\dot{x}(0) \neq \dot{x}(0+)$, and so forth for higher derivatives. The initial value theorem gives the value at $t = 0+$, which for some models is not necessarily equal to the value at $t = 0$. In these cases the solution of the differential equation is correct only for $t > 0$. This phenomenon occurs in models having impulse inputs and in models containing derivatives of a discontinuous input.

The Pulse Function

The rectangular pulse function shown in Figure 5.5-1a is defined for $t \geq 0$ as

$$f(t) = \begin{cases} M & 0 \leq t \leq T \\ 0 & t > T \end{cases} \tag{5.5-1}$$

Its transform is

$$F(s) = \int_0^\infty Me^{-st}\,dt = \int_0^T Me^{-st}\,dt = -\frac{Me^{-st}}{s}\bigg|_{t=0}^{t=T}$$

$$= M\frac{1 - e^{-sT}}{s} \tag{5.5-2}$$

The area A under the pulse is MT.

The Impulse Function

If we let the area $A = MT$ under the pulse remain constant at the value A and let the pulse width T approach zero, we obtain the impulse function, represented in Figure 5.5-1b. Because $M = A/T$, the transform $F(s)$ is

$$F(s) = \lim_{T \to 0} \frac{A}{T}\frac{1 - e^{-sT}}{s} = \lim_{T \to 0} \frac{Ase^{-sT}}{s} = A$$

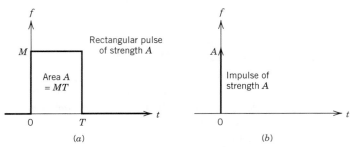

FIGURE 5.5-1   Rectangular pulse and impulse functions.

using L'Hôpital's limit rule. The area A is the *strength* of the impulse. If $A = 1$, the function is a *unit impulse*.

Besides the step function, the pulse function and its approximation, the impulse, appear quite often in vibration applications. The impulse, called the *Dirac delta function* $\delta(t)$ in the mathematics literature, is often used in vibration analysis. The impulse is an analytically convenient approximation of an input applied for only a very short time, such as when a high-speed object strikes another object. The impulse is also useful for estimating a system's parameters experimentally.

In keeping with our interpretation of the initial conditions, we consider the impulse $\delta(t)$ to start at time $t = 0$ and finish at $t = 0+$, with its effects first felt at $t = 0+$.

Impulse Response of a First-Order Model

Consider the first-order model $\tau \dot{y} + y = f(t)$. The impulse response of this model can be obtained by the Laplace transform method. Transforming the differential equation and using $F(s) = A$ gives

$$Y(s) = \frac{\tau y(0)}{\tau s + 1} + \frac{1}{\tau s + 1} A = \frac{\tau y(0) + A}{\tau s + 1} = \frac{y(0) + \dfrac{A}{\tau}}{s + \dfrac{1}{\tau}} \tag{5.5-3}$$

In the time domain, this becomes

$$y(t) = \left[y(0) + \frac{A}{\tau} \right] e^{-t/\tau} \tag{5.5-4}$$

Thus the impulse can be thought of as being equivalent to an additional initial condition of magnitude A/τ.

We can also see this with the initial value theorem (property 11 in Table 5.2-2), which gives

$$y(0+) = \lim_{s \to \infty} sY(s) = \lim_{s \to \infty} s \frac{\tau y(0) + A}{\tau s + 1} = \frac{\tau y(0) + A}{\tau} = y(0) + \frac{A}{\tau}$$

The value of y after the impulse has been applied is $y(0+)$. The effect of the impulse is to change y from $y(0)$ at $t = 0$ to $y(0) + A/\tau$ at $t = 0+$, which is an infinitesimal time later than $t = 0$.

EXAMPLE 5.5-1 *Impulse Response* *of a Simple* *Second-Order* *Model*	Obtain the unit-impulse response of the following model in two ways: **(a)** by separation of variables and **(b)** with the Laplace transform. The initial conditions are $x(0) = 5$ and $\dot{x}(0) = 10$. What are the values of $x(0+)$ and $\dot{x}(0+)$? $$\ddot{x} = \delta(t) \tag{1}$$
Solution	**(a)** Let $v(t) = \dot{x}(t)$. Then Equation 1 becomes $\dot{v} = \delta(t)$, which can be integrated to obtain $v(t) = v(0) + 1 = 10 + 1 = 11$. Thus $\dot{x}(0+) = 11$ and is not equal to $\dot{x}(0)$. Now integrate $\dot{x} = v = 11$ to obtain $x(t) = x(0) + 11t = 5 + 11t$. Thus $x(0+) = 5$, which is the same as $x(0)$. So for this model the unit-impulse input changes $\dot{x}$ from $t = 0$ to $t = 0+$ but does not change x. **(b)** The transformed equation is $$s^2 X(s) - sx(0) - \dot{x}(0) = 1 \tag{2}$$

or

$$X(s) = \frac{sx(0) + \dot{x}(0) + 1}{s^2} = \frac{5s + 11}{s^2} = \frac{5}{s} + \frac{11}{s^2}$$

which gives the solution $x(t) = 5 + 11t$ and $\dot{x}(t) = 11$. Note that the initial values used with the derivative property in Equation 2 are the values at $t = 0$.

The initial value theorem gives

$$x(0+) = \lim_{s \to \infty} sX(s) = \lim_{s \to \infty} s\frac{5s + 11}{s^2} = 5$$

and because $\mathcal{L}[\dot{x}] = sX(s) - x(0)$,

$$\dot{x}(0+) = \lim_{s \to \infty} s[sX(s) - x(0)] = \lim_{s \to \infty} s\left(\frac{5s + 11}{s} - 5\right) = 11$$

as we found in part (a).   ∎

Impulse Response of a Second-Order Model

Consider the second-order model

$$m\ddot{x} + c\dot{x} + kx = f(t) \tag{5.5-5}$$

If the input $f(t)$ is zero, the transform gives

$$(ms^2 + cs + k)X(s) = (ms + c)x(0) + m\dot{x}(0)$$

However, if the initial conditions are zero and the input is a unit impulse $[F(s) = 1]$, then

$$(ms^2 + cs + k)X(s) = 1$$

Comparing the last two expressions shows that the response to a unit impulse is equivalent to the free response for the special set of initial conditions such that

$$1 = (ms + c)x(0) + m\dot{x}(0)$$

or

$$x(0) = 0$$

$$m\dot{x}(0) = 1$$

For a mass-spring system with viscous friction, this means that if the mass is started at the equilibrium $x(0) = 0$ with a velocity $\dot{x}(0) = 1/m$, the resulting free response is equivalent to the unit-impulse response for zero initial conditions. This corresponds to the impulse-momentum principle of Newtonian mechanics, which says that the change in momentum, here $m\dot{x}(0)$, must equal the impulse, which is the time integral of the applied force.

These results can also be obtained with the initial value theorem:

$$x(0+) = \lim_{s \to \infty} sX(s) = \lim_{s \to \infty} \frac{s}{ms^2 + cs + k} = 0$$

Because $\mathcal{L}(\dot{x}) = sX(s) - x(0+) = sX(s)$,

$$\dot{x}(0+) = \lim_{s \to \infty} s[sX(s)] = \lim_{s \to \infty} \frac{s^2}{ms^2 + cs + k} = \frac{1}{m}$$

In summary, be aware that the solution $x(t)$ and its derivatives $\dot{x}(t)$, $\ddot{x}(t)$, ... will match the given initial conditions at $t = 0$ only if there are no impulse inputs and no derivatives of inputs that are discontinuous at $t = 0$.

Applicability of the Initial Value Theorem

If $X(s)$ is a rational function and if the degree of the numerator of $X(s)$ is less than the degree of the denominator, then the initial value theorem will give a finite value for $x(0+)$. If the degrees are equal, then initial value is undefined and the initial value theorem is invalid. The latter situation corresponds to an impulse in $x(t)$ at $t = 0$ and therefore $x(0+)$ is undefined. When the degrees are equal, the transform can be expressed as a constant plus a partial-fraction expansion. For example, consider the transform

$$X(s) = \frac{9s + 4}{s + 3} = 9 - \frac{23}{s + 3}$$

The inverse transform is $x(t) = 9\delta(t) - 23e^{-3t}$, and therefore $x(0+)$ is undefined.

Collision Applications

An example of a force often modeled as impulsive is the force generated when two objects collide. Recall that when Newton's law of motion $m\dot{v} = f(t)$ is integrated over time, we obtain the *impulse-momentum principle* for a system having constant mass:

$$mv(t) - mv(0) = \int_0^t f(t)\, dt \tag{5.5-6}$$

This states that the change in momentum mv equals the time integral of the applied force $f(t)$. In the terminology of mechanics, the force integral — the area under the force-time curve—is called the *linear impulse*. The linear impulse is the strength of an impulsive force, but a force need not be impulsive to produce a linear impulse.

If $f(t)$ is an impulsive input of strength A, that is, if $f(t) = A\delta(t)$, then

$$mv(t) - mv(0) = \int_0^t A\delta(t)\, dt = A \int_0^t \delta(t)\, dt = A \tag{5.5-7}$$

since the area under the $\delta(t)$ curve is 1. So the change in momentum equals the strength of the impulsive force.

In practice, the strength is often all we can determine about an input modeled as impulsive. Sometimes we need not determine the input characteristics at all, as the next example illustrates.

EXAMPLE 5.5-2
Inelastic Collision

Suppose a mass m_1 moving with a speed v_1 becomes embedded in mass m_2 after striking it (Figure 5.5-2). Suppose $m_2 = 8m_1 = 8m$. Determine the expression for the displacement $x(t)$ after the collision.

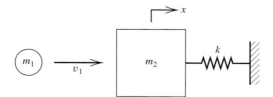

FIGURE 5.5-2 Inelastic collision.

Solution   If we take the entire system to consist of both masses, then the force of collision is *internal* to the system. Because the displacement of m_2 immediately after the collision will be small, we may neglect the spring force initially. Thus the *external* force $f(t)$ in Equation 5.5-6 is zero, and we have

$$(m + 8m)v(0+) - [mv_1 + 8m(0)] = 0$$

or

$$v(0+) = \frac{mv_1}{9m} = \frac{1}{9}v_1$$

The equation of motion for the combined mass is $9m\ddot{x} + kx = 0$. We can solve it for $t \geq 0+$ by using the initial conditions at $t = 0+$; namely, $x(0+) = 0$ and $\dot{x}(0+) = v(0+)$. The solution is

$$x(t) = \frac{v(0+)}{\omega_n} \sin \omega_n t = \frac{v_1}{9}\sqrt{\frac{9m}{k}} \sin \sqrt{\frac{k}{9m}}t = \frac{v_1}{3}\sqrt{\frac{m}{k}} \sin \frac{1}{3}\sqrt{\frac{k}{m}}t$$

Note that it was unnecessary to determine the impulsive collision force. Note also that this force did not change x initially; it changed only $\dot{x}$.   ∎

Consider the two colliding masses shown in Figure 5.5-3a, which shows the situation before collision, and Figure 5.5-3b, which shows the situation after collision. When the two masses are treated as a single system, no external force is applied to the system, and Equation 5.5-6 shows that the momentum is conserved, so that

$$m_1v_1 + m_2v_2 = m_1v_3 + m_2v_4$$

or

$$m_1(v_1 - v_3) = -m_2(v_2 - v_4) \tag{5.5-8}$$

If the collision is *perfectly elastic*, kinetic energy is conserved, so that

$$\frac{1}{2}m_1v_1^2 + \frac{1}{2}m_2v_2^2 = \frac{1}{2}m_1v_3^2 + \frac{1}{2}m_2v_4^2$$

or

$$\frac{1}{2}m_1\left(v_1^2 - v_3^2\right) = -\frac{1}{2}m_2\left(v_2^2 - v_4^2\right) \tag{5.5-9}$$

Using the algebraic identities:

$$v_1^2 - v_3^2 = (v_1 - v_3)(v_1 + v_3)$$

$$v_2^2 - v_4^2 = (v_2 - v_4)(v_2 + v_4)$$

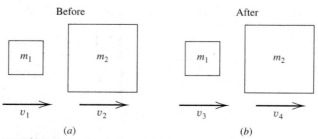

FIGURE 5.5-3   Two colliding masses. (*a*) Before collision. (*b*) After collision.

we can write Equation 5.5-9 as

$$\frac{1}{2}m_1(v_1 - v_3)(v_1 + v_3) = -\frac{1}{2}m_2(v_2 - v_4)(v_2 + v_4) \tag{5.5-10}$$

Divide Equation 5.5-10 by Equation 5.5-8 to obtain

$$v_1 + v_3 = v_2 + v_4 \qquad \text{or} \qquad v_1 - v_2 = v_4 - v_3 \tag{5.5-11}$$

This relation says that in a perfectly elastic collision the *relative* velocity of the masses changes sign but its magnitude remains the same.

The most common application is where we know v_1 and mass m_2 is initially stationary, so that $v_2 = 0$. In this case we may solve Equation 5.5-8 and 5.5-11 for the velocities after collision as follows:

$$v_3 = \frac{m_1 - m_2}{m_1 + m_2}v_1 \qquad v_4 = \frac{2m_1}{m_1 + m_2}v_1 \tag{5.5-12}$$

EXAMPLE 5.5-3
Perfectly Elastic
Collision

Consider again the system treated in Example 5.5-2 and shown again in Figure 5.5-4a. Suppose now that the mass $m_1 = m$ moving with a speed v_1 rebounds from the mass $m_2 = 8m$ after striking it. Assume that the collision is perfectly elastic. Determine the expression for the displacement $x(t)$ after the collision.

Solution

For a perfectly elastic collision, the velocity v_3 of the mass m after the collision is given by Equation 5.5-12:

$$v_3 = \frac{m_1 - m_2}{m_1 + m_2}v_1 = \frac{m - 8m}{m + 8m}v_1 = -\frac{7}{9}v_1$$

Thus the change in the momentum of m is

$$m\left(-\frac{7}{9}v_1\right) - mv_1 = \int_0^t f(t)\, dt$$

Thus the linear impulse applied to the mass m during the collision is

$$\int_0^t f(t)\, dt = -\frac{16}{9}mv_1$$

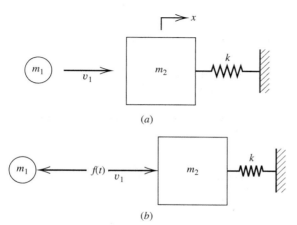

(a)

(b)

FIGURE 5.5-4 Perfectly elastic collision.

From Newton's law of action and reaction (Figure 5.5-4b), we see that the linear impulse applied to the $8m$ mass is $+16mv_1/9$, and thus its equation of motion is

$$8m\ddot{x} + kx = \frac{16}{9}mv_1\delta(t)$$

We may solve this equation for $t \geq 0$ using the initial conditions $x(0) = 0$ and $\dot{x}(0) = 0$. The Laplace transform gives

$$(8ms^2 + k)X(s) = \frac{16}{9}mv_1$$

$$X(s) = \frac{16mv_1/9}{8ms^2 + k} = \frac{2v_1}{9}\frac{1}{s^2 + \frac{k}{8m}}$$

Thus

$$x(t) = \frac{2v_1}{9}\sqrt{\frac{8m}{k}}\sin\sqrt{\frac{k}{8m}}t = \frac{4v_1}{9}\sqrt{\frac{2m}{k}}\sin\frac{1}{2}\sqrt{\frac{k}{2m}}t$$

This gives $\dot{x}(0) = 2v_1/9$, which is identical to the solution for v_4 from Equation 5.5-12, as it should be. ∎

Effect of Numerator Dynamics

Consider the following transfer function, which has numerator dynamics:

$$\frac{X(s)}{G(s)} = \frac{2s + 10}{5s + 10} \tag{5.5-13}$$

With numerator dynamics models we must proceed carefully if the input is discontinuous, as is the case with the step function, because the input derivative produces an impulse when acting on a discontinuous input. To help you understand this, we state without rigorous proof that the unit impulse $\delta(t)$ is the time derivative of the unit-step function $u_s(t)$; that is,

$$\delta(t) = \frac{d}{dt}u_s(t) \tag{5.5-14}$$

This result does not contradict common sense, because the step function changes from 0 at $t = 0-$ to 1 at $t = 0+$ in an infinitesimal amount of time. Therefore its derivative should be infinite at $t = 0$. To further indicate the correctness of this relation, we integrate both sides and note that the area under the unit impulse is unity. Thus

$$\int_{0-}^{0+}\delta(t)\,dt = \int_{0-}^{0+}\frac{d}{dt}u_s(t)\,dt = u_s(0+) - u_s(0-) = 1 - 0 = 1$$

which gives $1 = 1$.

Thus an input derivative will create an impulse in response to a step input. For example, consider the differential equation that corresponds to Equation 5.5-13:

$$5\dot{x} + 10x = 2\dot{g}(t) + 10g(t)$$

If the input $g(t) = u_s(t)$, the model is equivalent to

$$5\dot{x} + 10x = 2\delta(t) + 10u_s(t)$$

which has an impulse input.

Approximation of Pulse Response

A rule of thumb for determining when a pulse can be approximated by an impulse is obtained as follows. The Taylor series expansion for $e^{T/\tau}$ is

$$e^{T/\tau} = 1 + \frac{T}{\tau} + \frac{1}{2}\left(\frac{T}{\tau}\right)^2 + \cdots + \frac{1}{n!}\left(\frac{T}{\tau}\right)^n + \cdots$$

The response of the model $\tau\dot{y} + y = f(t)$ to a rectangular pulse of area A and duration T was obtained in Section 5.4:

$$y(t) = \frac{A}{T}e^{-t/\tau}\left(e^{T/\tau} - 1\right)$$

If the first two terms in the series expansion are retained and substituted into the expression for $y(t)$, the result is

$$y(t) = -\frac{A}{T}\left(1 - 1 - \frac{T}{\tau} - \cdots\right)e^{-t/\tau} \approx \frac{A}{\tau}e^{-t/\tau}$$

This is identical to the impulse response for zero initial conditions.

To see when it is justifiable to truncate the series in this way, consider the case where $T/\tau = 0.1$. Thus

$$e^{T/\tau} = 1 + 0.1 + 0.005 + 0.00017 + \cdots$$

For accuracy to the second decimal place, only the first two terms need be kept. This leads to the following guide, which can also be shown true for an arbitrarily shaped pulse.

> *If the pulse duration is of the order of $\tau/10$ or less, the first-order system response is nearly the same as the impulse response.*

If we are willing to accept the dominant-root approximation, then the preceding guide can be extended to higher-order systems, where τ is the dominant-time constant.

Shock

A *shock* is a sudden input, such as an impulse or a pulse input of short duration. The *shock response* is a system's response to a particular shock input. The *shock spectrum* is a plot of the maximum response versus some parameter that describes the input.

The response of interest is sometimes the maximum absolute value of the time response $|x|_{\text{max}}$, and the shock spectrum is often a plot of $|x|_{\text{max}}$ versus the duration of the input. Another response characteristic sometimes used is the maximum absolute value acceleration $|\ddot{x}|_{\text{max}}$.

Shock spectra are usually obtained only for undamped systems, because damping tends to limit the response to a shock input. The unit-impulse response of the undamped system $m\ddot{x} + kx = f(t)$ is denoted $h(t)$ and is found from

$$(ms^2 + k)X(s) = 1$$

or

$$X(s) = \frac{1}{m}\frac{1}{s^2 + k/m} = \frac{1}{m\omega_n}\frac{\omega_n}{s^2 + \omega_n^2}$$

Thus the unit-impulse response is

$$h(t) = \frac{1}{m\omega_n}\sin \omega_n t$$

From the convolution theorem (see property 9 of Table 5.2-2), the response to an arbitrary input $f(t)$ can be expressed in terms of the unit-impulse response as

$$x(t) = \int_0^t f(\tau)h(t - \tau)\,d\tau \tag{5.5-15}$$

The evaluation of $|x|_{max}$ or $|\ddot{x}|_{max}$ from Equation 5.5-15 in terms of m, k, and other parameters is usually very tedious if done for other than very simple input functions. Some results are available in standard references such as [Harris, 2002]. Here we will illustrate the process for the shock due to a falling object striking the ground.

EXAMPLE 5.5-4
Shock from a Fall

Figure 5.5-5 shows a mass m with an attached stiffness, such as that due to protective packaging. The mass drops a distance h, at which time the stiffness element contacts the ground. Let x denote the displacement of m after contact with the ground. (**a**) Determine the shock response associated with the maximum acceleration magnitude felt by the mass. (**b**) What is the maximum required stiffness of the packaging if a 10-kg package cannot experience a deceleration greater than $8g$ when dropped from a height of 2 m?

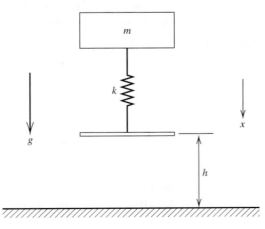

FIGURE 5.5-5 A model of a falling object with protective packaging.

Solution

(a) At the time of contact, the velocity can be determined from conservation of energy

$$mgh = \frac{1}{2}mv^2$$

which gives $v = \sqrt{2gh}$. Thus the initial conditions for the process that starts at the time of contact are $x(0) = 0$ and $\dot{x}(0) = v = \sqrt{2gh}$.

The equation of motion after contact is

$$m\ddot{x} + kx = mg$$

This has the solution

$$x(t) = \frac{g}{\omega_n^2}(1 - \cos \omega_n t) + \frac{v}{\omega_n} \sin \omega_n t$$

where $\omega_n = \sqrt{k/m}$. The acceleration is

$$\ddot{x}(t) = g \cos \omega_n t - v\omega_n \sin \omega_n t = B \sin(\omega_n t + \phi)$$

where $B = \sqrt{g^2 + (v\omega_n)^2}$ and $\tan \phi = -g/v\omega_n$, $\pi/2 \le \phi \le \pi$.

The mass is decelerating, and the maximum acceleration magnitude is $|\ddot{x}|_{max} = |a|_{max} = B$, which in relative terms is

$$\frac{|a|_{max}}{g} = \sqrt{1 + \left(\frac{v\omega_n}{g}\right)^2} = \sqrt{1 + \frac{2h\omega_n^2}{g}} \tag{1}$$

A plot of this function versus the dimensionless quantity $h\omega_n^2/g$ is shown in Figure 5.5-6. It shows the obvious result that increasing the drop height h will increase $|a|_{max}$, but it also shows that increasing ω_n also increases $|a|_{max}$. So if m is constant, increasing the stiffness k will increase $|a|_{max}$.

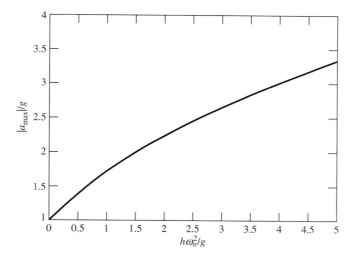

FIGURE 5.5-6 The shock response curve for a falling object.

(b) For $|a|_{\max} = 8g$, Equation 1 gives

$$\frac{8g}{g} = \sqrt{1 + \frac{2(2)\omega_n^2}{g}}$$

This gives

$$\omega_n^2 = \frac{63g}{4}$$

and thus $k \leq 10(63g)/4 = 1545$ N/m. ∎

5.6 MATLAB AND THE LAPLACE TRANSFORM

MATLAB provides the `residue` function to compute the coefficients in the partial-fraction expansion. When the transform of the solution to a differential equation has been obtained, it is necessary to invert the transform to obtain the solution as a function of time. Unless the transform is a simple one appearing in the transform table, it will have to be represented as a combination of simple transforms. Most transforms occur in the form of a ratio of two polynomials, such as

$$X(s) = \frac{N(s)}{D(s)} = \frac{b_m s^m + b_{m-1}s^{m-1} + \cdots + b_1 s + b_0}{s^n + a_{n-1}s^{n-1} + \cdots + a_1 s + a_0} \tag{5.6-1}$$

Note that we assume that the coefficient $a_n = 1$. If not, divide the numerator and denominator by a_n.

If all the roots are real and distinct, we can express $X(s)$ in factored form as

$$X(s) = \frac{N(s)}{(s + r_1)(s + r_2)\ldots(s + r_n)} \tag{5.6-2}$$

where the roots are $s = -r_1, -r_2, \ldots, -r_n$. This form can be expanded as

$$X(s) = \frac{C_1}{s + r_1} + \frac{C_2}{s + r_2} + \cdots + \frac{C_n}{s + r_n} \tag{5.6-3}$$

Each factor corresponds to an exponential function of time, and the solution is

$$x(t) = C_1 e^{-r_1 t} + C_2 e^{-r_2 t} + \ldots + C_n e^{-r_n t} \tag{5.6-4}$$

Suppose that p of the roots have the same value $s = -r_1$ and the remaining $(n - p)$ roots are real and distinct. Then $X(s)$ is of the form

$$X(s) = \frac{N(s)}{(s + r_1)^p (s + r_{p+1})(s + r_{p+2})\ldots(s + r_n)} \tag{5.6-5}$$

The expansion is

$$X(s) = \frac{C_1}{(s + r_1)^p} + \frac{C_2}{(s + r_1)^{p-1}} + \cdots + \frac{C_p}{s + r_1} + \cdots + \frac{C_{p+1}}{s + r_{p+1}} + \cdots + \frac{C_n}{s + r_n} \tag{5.6-6}$$

The solution for the time function is

$$x(t) = C_1 \frac{t^{p-1}}{(p-1)!}e^{-r_1 t} + C_2 \frac{t^{p-2}}{(p-2)!}e^{-r_1 t} + \cdots + C_p e^{-r_1 t}$$
$$+ \cdots + C_{p+1}e^{-r_{p+1}t} + \cdots + C_n e^{-r_n t} \tag{5.6-7}$$

The expansion coefficients C_i are called the *residues*, and the `residue` function can be used to compute them. The set of roots of the denominator of $X(s)$ are called the *poles*, and they consist of the characteristic roots plus any roots introduced by the transform of the input function.

In the unusual case where the order m of the numerator is greater than the order n of the denominator, the transform can be represented by a polynomial $K(s)$, called the *direct term*, plus a ratio of two polynomials. For example,

$$X(s) = \frac{5s^3 + 39s^2 + 95s + 71}{s^2 + 7s + 12} = 5s + 4 + \frac{7s + 23}{s^2 + 7s + 12} \tag{5.6-8}$$

Although we will not encounter this situation, MATLAB nevertheless provides for its occurrence. The syntax of the `residue` function is as follows:

```
[r,p,K]=residue(num,den)
```

where `num` and `den` are vectors containing the coefficients of the numerator and denominator of $X(s)$. The output of the function consists of the vector `r`, which contains the residues, the vector `p`, which contains the poles, and the vector `K`, which contains the coefficients of the direct term $K(s)$ in polynomial form.

For example, using Equation 5.6-8 as an example, you would type

```
[r,p,K]=residue([5,39,95,71],[1,7,12])
```

The answer given by MATLAB is `r = [5, 2]`, `p = [-4, -3]`, and `K = [5, 4]`. This corresponds to the expression

$$X(s) = 5s + 4 + \frac{5}{s + 4} + \frac{2}{s + 3}$$

Note that the order in which the residues are displayed corresponds to the order in which the poles are displayed.

The following example shows how repeated poles are handled. To solve the model $\ddot{x} + 5\dot{x} + 6x = 2\dot{f} + f$ where $f(t) = 5e^{-2t}$, for zero initial conditions, the transform is

$$X(s) = \frac{5(2s + 1)}{(s^2 + 5s + 6)(s + 2)} = \frac{10s + 5}{s^3 + 7s^2 + 16s + 12}$$

In this case, the repeated poles are $s = -2, -2$. One of the repeated poles is a characteristic root; the other is due to the transform of the input. To obtain the expansion, you type

```
[r,p,K]=residue([10,5],[1,7,16,12])
```

The answer given by MATLAB is `r = [-25, 25, -15]`, `p = [-3, -2, -2]`, and `K = [0]`. This corresponds to the expression

$$X(s) = \frac{-25}{s + 3} + \frac{25}{s + 2} + \frac{-15}{(s + 2)^2}$$

and the solution

$$x(t) = -25e^{-3t} + 25e^{-2t} - 15te^{-2t}$$

Note that the *last* residue returned by the `residue` function corresponds to the highest power in the expansion.

The following example shows how complex poles are handled. To solve the model $\ddot{x} + 4\dot{x} + 20x = 5\dot{f} + f$ where $f(t)$ is a unit-step function, for zero initial conditions, the transform is

$$X(s) = \frac{5s + 1}{(s^2 + 4s + 20)s} = \frac{5s + 1}{s^3 + 4s^2 + 20s}$$

To obtain the expansion, you type

```
[r,p,K]=residue([5,1],[1,4,20,0])
```

Note that the last coefficient in the denominator is 0. The answer given by MATLAB is
`r = [-0.025-0.6125i, -0.025+0.6125i, 0.05]`, `p = [-2+4i, -2-4i, 0]`, and
`K = [0]`. This corresponds to the expression

$$X(s) = \frac{-0.025 - 0.6125i}{s + 2 - 4i} + \frac{-0.025 + 0.6125i}{s + 2 + 4i} + \frac{0.05}{s}$$

and the solution

$$x(t) = (-0.025 - 0.6125i)e^{(-2+4i)t} + (-0.025 + 0.6125i)e^{(-2-4i)t} + 0.05 \qquad (5.6\text{-}9)$$

This form is not very useful, but we can convert to a more useful form by noting that the first two terms in the expansion have the form

$$\frac{C + iD}{s + a - ib} + \frac{C - iD}{s + a + ib} \qquad (5.6\text{-}10)$$

which corresponds to the time function

$$(C + iD)e^{(-a+ib)t} + (C - iD)e^{(-a-ib)t}$$

Using Euler's identities, $e^{\pm ibt} = \cos bt \pm \sin bt$, the previous form can be written as

$$2e^{-at}(C \cos bt - D \sin bt) \qquad (5.6\text{-}11)$$

Using the identity

$$B \sin(bt + \phi) = B \sin bt \cos \phi + B \cos bt \sin \phi$$

we can write the expression as

$$Be^{-at} \sin(bt + \phi) \qquad (5.6\text{-}12)$$

where

$$B = 2\sqrt{C^2 + D^2} \qquad (5.6\text{-}13)$$

$$\phi = \tan^{-1}\left(\frac{-C}{D}\right) \qquad (5.6\text{-}14)$$

The quadrant of ϕ is determined from the fact that

$$\sin \phi = \frac{2C}{B} \qquad \cos \phi = \frac{-2D}{B} \qquad B > 0 \qquad (5.6\text{-}15)$$

In summary, if the residue corresponding to the root $s = -a + ib$ is denoted as $C + iD$, then the response corresponding to the root pair $s = -a \pm ib$ can be expressed either as Equation 5.6-11 or as Equation 5.6-12, using Equations 5.6-13 through 5.6-15. For example, we can write the solution, Equation 5.6-9, as

$$x(t) = 2e^{-2t}(-0.025 \cos 4t + 0.6125 \sin 4t) + 0.05$$

or as

$$x(t) = 1.226e^{-2t} \sin(4t - 0.04) + 0.05$$

5.7 MATLAB APPLICATIONS

There are several MATLAB functions that are useful for implementing the methods of this chapter.

The `freqresp` Function

As we saw in Chapter 4, the `evalfr` function can be used to evaluate `sys` at one frequency only. For a vector of frequencies, use the `freqresp` function, whose syntax is

```
fr = freqresp(sys,w)
```

where `w` is a vector of frequencies, which must be in radians/unit time.

The following MATLAB script file calculates the coefficients and phase angles of the steady-state response described in Example 5.1-4:

```
sys1=tf(1,[3,6,1200]);
w=[2*pi,6*pi,10*pi,14*pi];
A=(80/pi)*[1,1/3,1/5,1/7];
fr=freqresp(sys1,w);
M=abs(fr);
B=A.*M(:)'
ph=angle(fr)
```

The results are shown in the table in Example 5.1-4.

Computing Fourier Series Coefficients

Equations 5.1-1 through 5.1-3 for the Fourier series can be evaluated in MATLAB as follows. These are repeated here:

$$f(t) = \frac{a_0}{2} + \sum_{n=1}^{\infty} \left(a_n \cos \frac{2n\pi t}{P} + b_n \sin \frac{2n\pi t}{P} \right)$$

where

$$a_n = \frac{2}{P} \int_{t_1}^{t_1+P} f(t) \cos \frac{2n\pi t}{P} \, dt$$

$$b_n = \frac{2}{P} \int_{t_1}^{t_1+P} f(t) \sin \frac{2n\pi t}{P} \, dt$$

We distinguish between two cases. In the first case, the periodic input is described as a set of data rather than as a set of functions. In this case we must use the MATLAB function `trapz` to compute the integrals. This function uses the trapezoidal rule to compute the integral. In the second case, where the periodic function is given in closed form, we can use the `quad` function, which is more accurate than `trapz` but cannot be used when the input is given as data.

The following program can be used for the first case. It computes the coefficients, evaluates and plots the series along with the given data, and then plots the spectrum of the series.

```
function[a0,a,b]=Fouriercoeffs(t,f,L,N)
%Computes the Fourier coefficients from data in the arrays t and f,
and plots the series and the original data, and the spectrum.
% f=dependent variable
% t=independent variable
% P=period
% 2N+1=number of series terms to be computed
a0=(2/P)*trapz(t,f);
for n=1:N
    a(n)=(2/P)*trapz(t,f.*cos(2*n*pi*t/P));
    b(n)=(2/P)*trapz(t,f.*sin(2*n*pi*t/P));
end
% Define the limits and increment for plotting.
mint=min(t);
maxt=max(t);
trange=maxt-mint;
% Plot 1001 points
dt=trange/1000;
tp=[mint:dt:maxt];
fp=0;
for n=1:N
    fp=fp+a(n)*cos(2*pi*n*tp/P)+b(n)*sin(2*pi*n*tp/P);
    om(n)=2*pi*n/P;
    A(n)=sqrt(a(n)^2+b(n)^2);
end
A0=a0/2;
fp=a0/2+fp;
subplot(2,1,1)
plot(t,f,'o',tp,fp),xlabel('t'),ylabel('f')
subplot(2,1,2)
stem([0,om],[A0,A]),xlabel('Radian Frequency'),...
    ylabel('Amplitude')
```

This program can be modified to use the quad function. The syntax of this function is quad('name', a, b), where name is the name of the function defining the integrand, and a and b are the lower and upper limits of integration. For example, to compute the integral

$$A = \int_0^\pi \sin t \, dt$$

you type A = quad('sin', 0, pi). The returned answer is 2.000, which is correct to four decimal places.

For some applications the integrand cannot be expressed as a simple combination of built-in MATLAB functions, so in these cases you will need to create a user-defined function that describes the integrand.

The step Function

The Control System toolbox provides several solvers for linear models. These solvers are categorized by the type of input function they can accept: some of these are a step input, an impulse input, and a general input function.

In their basic form, each of the following functions automatically puts a title and axis labels on the plot. You can change these by activating the Plot Editor or by right-clicking on the plot. This brings up a menu that includes the Properties as a choice. Selecting Properties enables you to change the labels as well as other features such as limits, units, and style.

The menu obtained by right-clicking on the plot also contains Characteristics as a choice. The contents of the subsequent menu depend on the particular function. When the `step` function is used, the Characteristics menu includes

1. Peak Response

2. Rise Time

3. Settling Time

4. Steady State

When you select Peak Response, for example, MATLAB identifies the peak value of the response curve and marks its location with a dot and dashed lines. Moving the cursor over the dot displays the numerical values of the peak response and the time at which it occurs. The Rise Time is the time required for the response to go from 10% to 90% of its steady-state value. The Settling Time is the time required for the response to settle within 2% of its steady-state value. You can change these percents by selecting the Characteristics tab under the Properties menu.

The `step` function plots the unit-step response, assuming that the initial conditions are zero. The basic syntax is `step(sys)`, where `sys` is the LTI object. The time span and number of solution points are chosen automatically. To specify the final time `ft`, use the syntax `step(sys,ft)`. To specify a vector of times of the form `t = [0:dt:ft]`, at which to obtain the solution, use the syntax `step(sys,t)`. When called with left-hand arguments, as `[y, t] = step(sys,...)`, the function returns the output response `y` and the time vector `t` used for the simulation. No plot is drawn. The array `y` is $p \times q \times m$, where p is `length(t)`, q is the number of outputs, and m is the number of inputs.

The syntax `step(sys1, sys2,...,t)` plots the step response of multiple LTI systems on a single plot. The time vector `t` is optional. You can specify line color, line style, and marker for each system; for example, `step(sys1,'r', sys2,'y', sys3,'gx')`. The steady-state response and the time to reach that state are automatically determined. The steady-state response is indicated by a horizontal dotted line.

EXAMPLE 5.7-1

Step Response of a Second-Order Model

Consider the following model:

$$\frac{X(s)}{F(s)} = \frac{cs + 5}{10s^2 + cs + 5}$$

(a) Plot the unit-step response for $c = 3$ using the time span selected by MATLAB.

(b) Plot the unit-step response for $c = 3$ over the range $0 \le t \le 15$.

(c) Plot the unit-step responses for $c = 3$ and $c = 8$ over the range $0 \le t \le 15$. Put the plots on the same graph.

(d) Plot the step response for $c = 3$, where the magnitude of the step input is 20. Use the time span selected by MATLAB.

Solution

The MATLAB programs are shown below for each case.

(a) This illustrates the use of the `step` function in its most basic form:

```
sys1=tf([3,5],[10,3,5])
step(sys1)
```

The plot is shown in Figure 5.7-1.

(b) This illustrates how to use a user-selected time span and spacing:

```
sys1=tf([3,5],[10,3,5])
t=[0:0.01:15];
step(sys1,t)
```

The plot is shown in Figure 5.7-2.

(c) This illustrates how to plot two responses on the same graph:

```
sys1=tf([3,5],[10,3,5]);
sys2=tf([8,5],[10,8,5]);
t=[0:0.01:15];
step(sys1,' ',sys2,'--',t)
```

The plot is shown in Figure 5.7-3.

(d) This illustrates how to obtain the response when the magnitude of the step input is not unity. The output of the `step(sys1)` function is for a unit-step input, and so it must be multiplied by 20. This multiplication can be performed within the `plot` function:

```
sys1=tf([3,5],[10,3,5])
[y,t]=step(sys1);
plot(t,20*y),xlabel('t'),ylabel('x(t)')
```

The plot is shown in Figure 5.7-4. ∎

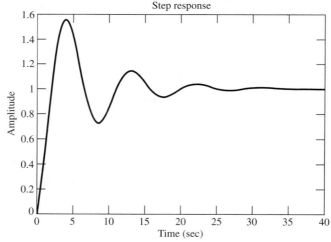

FIGURE 5.7-1 Response for part (a) of Example 5.7-1.

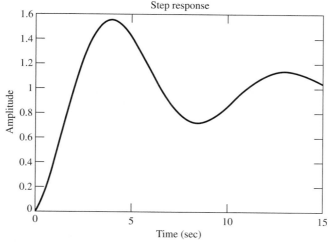

FIGURE 5.7-2 Response for part (b) of Example 5.7-1.

Note that when the `step` function is used without an assignment on the left-hand side of the equal sign, it automatically computes and plots the steady-state response and puts a title and axis labels on the plot, with the assumption that the unit of time is seconds. When the form `[y, t] = step(sys)` is used, however, the steady-state response is not computed, and you must put the labels on the plot.

The `impulse` Function

The `impulse` function plots the unit-impulse response, assuming that the initial conditions are zero. The basic syntax is `impulse(sys)`, where `sys` is the LTI object.

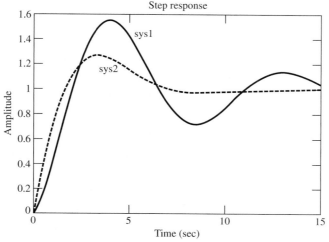

FIGURE 5.7-3 Response for part (c) of Example 5.7-1.

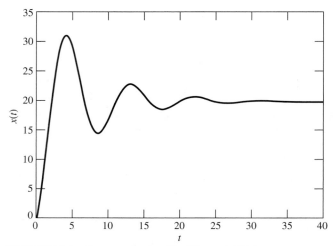

FIGURE 5.7-4 Response for part (d) of Example 5.7-1.

The time span and number of solution points are chosen automatically. For example, the impulse response of the transfer function

$$\frac{X(s)}{F(s)} = \frac{1}{5s^2 + 9s + 4}$$

is found as follows:

```
≫sys1=tf(1,[5,9,4]);
≫impulse(sys1)
```

To specify the final time ft, use the syntax impulse(sys,ft). To specify a vector of times of the form t = [0:dt:ft], at which to obtain the solution, use the syntax impulse(sys,t). When called with left-hand arguments, as [y, t] = impulse(sys,...), the function returns the output response y and the time vector t used for the simulation. No plot is drawn. The array y is $p \times q \times m$, where p is length(t), q is the number of outputs, and m is the number of inputs.

The syntax impulse(sys1,sys2,...,t) plots the impulse response of multiple LTI systems on a single plot. The time vector t is optional. You can specify line color, line style, and marker for each system; for example, impulse(sys1,'r',sys2, 'y--',sys3,'gx').

The characteristics available with the impulse function by right-clicking on the plot are the Peak Response and the Settling Time.

EXAMPLE 5.7-2
Impulse Response of Second-Order Models

Using the Laplace transform and the initial value theorem, an analysis of the unit-impulse response of the following model would show that if $x(0) = \dot{x}(0) = 0$, then $x(0+) = 0$ and $\dot{x}(0+) = 1/2$.

$$\frac{X(s)}{F(s)} = \frac{1}{2s^2 + 14s + 20}$$

Use the impulse function to verify these results.

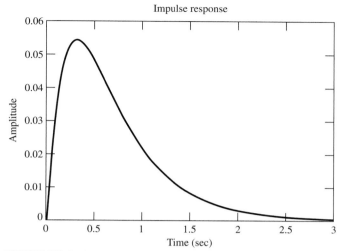

FIGURE 5.7-5 Impulse response of $x(t)$ for Example 5.7-2.

Solution

The program is shown below:

```
≫sys1=tf(1,[2,14,20]);
≫impulse(sys1)
```

The plot is shown in Figure 5.7-5. From it we see that $x(0+) = 0$ and that $\dot{x}(0+)$ is positive as predicted. We are unable to determine the exact value of $\dot{x}(0+)$ from this plot, so we multiply the transfer function by s to obtain the transfer function for $v = \dot{x}$:

$$\frac{V(s)}{F(s)} = \frac{s}{2s^2 + 14s + 20}$$

We now use the impulse function on this transfer function.

```
≫ sys2=tf([1,0],[2,14,20]);
≫ impulse(sys2)
```

The plot is shown in Figure 5.7-6. From it we see that $\dot{x}(0+) = 0.5$ as predicted. ∎

The lsim Function

The lsim function was covered in Section 4.8 and was used to compute and plot the sine response. This function can also be used to compute the response to any forcing function. When used without the left-hand arguments, the Peak Response is available from the Characteristics menu.

EXAMPLE 5.7-3
Ramp Response
with the lsim
Command

Plot the forced response of

$$\ddot{x} + 3\dot{x} + 5x = 10f(t)$$

to a *ramp* input, $f(t) = 1.5t$, over the time interval $0 \le t \le 2$.

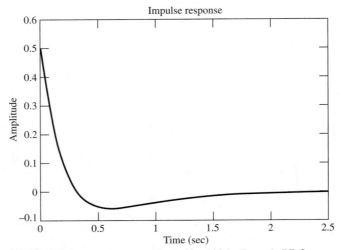

FIGURE 5.7-6   Impulse response of $v(t) = \dot{x}(t)$ for Example 5.7-2.

Solution   We choose to generate the plot with 300 points. The MATLAB session is the following:

```
≫ t=linspace(0,2,300);
≫ f=1.5*t;
≫ sys =tf(10,[1,3,5]);
≫ [y,t]=lsim(sys,f,t)
≫ plot(t,y),xlabel('t'),ylabel('x(t) and f(t)')
```

The plot is shown in Figure 5.7-7.   ■

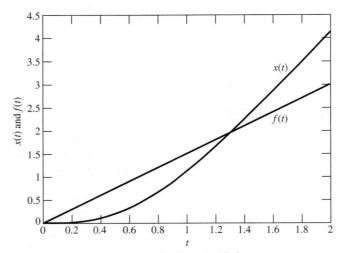

FIGURE 5.7-7   Ramp response for Example 5.7-3.

5.8 SIMULINK APPLICATIONS

Simulink is especially useful for computing the response to periodic and nonperiodic forcing functions.

Periodic Forcing Functions

Simulink is well equipped to handle periodic forcing functions, such as a train of pulses, that must be defined with different functions over different intervals of its period. We now demonstrate how to use Simulink to compute the response to a repeating series of (1) rectangular pulses, (2) half-sine pulses, and (3) trapezoidal pulses.

Figure 5.8-1*a* shows a series of rectangular pulses. Figure 5.8-1*b* is the Simulink diagram to compute the response of the system $m\ddot{x} + c\dot{x} + kx = f(t)$ to such an input. As we have seen before, in the Property window of the Transfer Function block, you set the numerator to [1] and the denominator to [m c k]. Then in the MATLAB Command window you enter the values of *m*, *c*, and *k* before running the Simulink program. In the Property window of the Pulse Generator block, enter 2 for the Amplitude, 2*pi for the period, and 50 for the Pulse Width (which means that the pulse is "on" 50% of the time and "off" 50% of the time). The amplitude value is the height of the pulse above 0, and the Property window does not provide for shifting the pulse up or down. To shift the pulse down by 1 unit, we include the Constant block of value 1 and use the Sum block to subtract 1 from the output of the Pulse Generator.

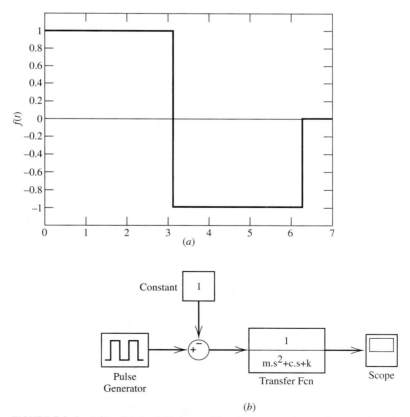

FIGURE 5.8-1 A Simulink model for computing the response to a series of rectangular pulses.

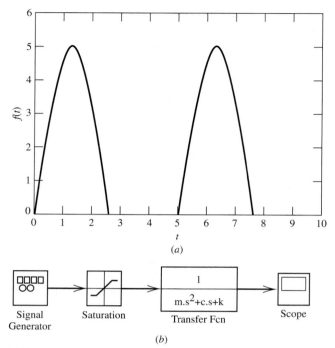

FIGURE 5.8-2 A Simulink model for computing the response to a series of half-sine functions.

Figure 5.8-2*a* shows a series of half-sine functions. Figure 5.8-2*b* is the Simulink diagram to compute the response of the system $m\ddot{x} + c\dot{x} + kx = f(t)$ to such an input. In the Property window of the Signal Generator block, select `sine` for the Wave form, enter 5 for the Amplitude, enter `0.2` for the Frequency, and select `Hertz` for the Units. The Saturation block serves to cut off the negative part of the sine wave. To do this, in the Property window of the block, set the upper limit to some value greater than the amplitude of 5 and set the Lower limit to 0.

Figure 5.8-3*a* shows a series of trapezoidal pulses. Figure 5.8-3*b* is the Simulink diagram to compute the response of the system $m\ddot{x} + c\dot{x} + kx = f(t)$ to such an input. In the Property window of the Repeating Sequence block, enter `[0 2 4 6]` for the Time values and `[0 5 5 0]` for the Output values. These values define the four "corner" points of the trapezoid.

Nonperiodic Forcing Functions

Simulink provides blocks for simulating step, ramp, and pulse inputs:

- The Step block implements a step function. The Step time in the Block properties window is the time at which the step function switches from one constant value (the Initial value) to another constant value (the Final value). For most of our examples, the Step time and the Initial value have been 0. The Sample time should be left at 0.

- The Ramp block implements a ramp function. The Slope must be specified in the Block properties window, along with the Start time (usually 0) and the Initial output (which is the intercept, the value the ramp has at the Start time). Usually the Initial output is 0.

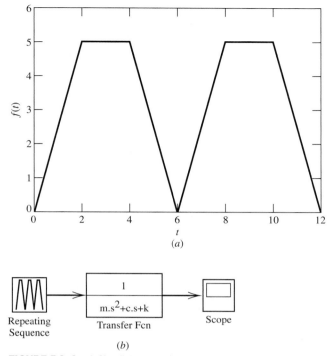

FIGURE 5.8-3 A Simulink model for computing the response to a series of trapezoidal pulses.

- The Pulse Generator block can be used to produce a single rectangular pulse by setting the Period to a large value and setting the Pulse width percentage to a value required to produce the desired pulse duration. For example, if the simulation is to run for less than 20 seconds, to create a pulse duration of 2 seconds, we set the Period to 20 seconds and the Pulse width to 10%.

5.9 CHAPTER REVIEW

In this chapter we saw many types of forcing functions, other than pure sine and cosine functions, that occur in vibration applications. These include forcing functions that are periodic but not harmonic. The chapter showed how to use the Fourier series to represent a periodic forcing function. The amplitude spectrum plot based on this series is a useful tool for determining whether or not the forcing function will cause excessive vibration such as resonance.

Other types of forcing functions often found include a step input, which is a suddenly applied constant force; a ramp input, which is a constantly increasing force; exponentially changing forces; the pulse; the impulse; and a number of other types. The Laplace transform is a useful tool for obtaining the total response due to such forcing functions. The advantage of this method is that it is systematic and does not require a guess as to the form of the solution.

The response to an impulse, which is a limiting case of a pulse input and models a suddenly applied and suddenly removed input, requires special care, and this was treated in Section 5.5.

MATLAB can be used to perform some of the algebra required to use the Laplace transform method, as shown in Section 5.6. Section 5.7 covered the application of MATLAB to the other chapter topics. Section 5.8 showed how Simulink can be used to obtain the forced response to a series of pulses of different shapes, something that is very difficult to do analytically and somewhat difficult to program in MATLAB.

Although the examples in this chapter were restricted to systems having one degree of freedom, the methods are applicable to systems with more degrees of freedom. We will see this in Chapters 6 and 7.

Now that you have finished this chapter, you should be able to do the following:

1. Apply the Fourier series method to obtain the response of a linear system to a periodic forcing function and apply the amplitude spectrum plot.

2. Use the Laplace transform method to obtain the response of a linear system to a variety of forcing functions, such as step, ramp, exponential, pulse, and impulse functions.

3. Apply MATLAB and Simulink to implement the methods of the chapter.

PROBLEMS

SECTION 5.1   RESPONSE TO GENERAL PERIODIC INPUTS

5.1   Given the model

$$0.1\dot{y} + y = f(t)$$

with the following Fourier series representation of the input

$$f(t) = 0.5 \sin 4t + 2 \sin 8t + 0.02 \sin 12t + 0.03 \sin 16t + \cdots$$

(a) Find the bandwidth of the system.

(b) Plot the amplitude spectrum of the input.

(c) Find the steady-state response $y(t)$ by considering only those components of the $f(t)$ expansion that lie within the bandwidth of the system.

5.2   A mass-spring-damper system is described by the model

$$m\ddot{x} + c\dot{x} + kx = f(t)$$

where $m = 0.05$ slug, $c = 0.4$ lb-sec/ft, $k = 5$ lb/ft, and $f(t)$ (lb) is the externally applied force shown in Figure P5.2. The forcing function can be expanded in a Fourier series as follows:

$$f(t) = -0.2\left(\sin 3t + \frac{1}{3}\sin 9t + \frac{1}{5}\sin 15t + \frac{1}{7}\sin 21t + \cdots + \frac{1}{n}\sin 3nt \pm \cdots\right) \quad n \text{ odd}$$

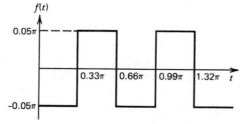

FIGURE P5.2

(a) What is the bandwidth of the model?

(b) Plot the amplitude spectrum of the input.

(c) Find an approximate description of the output $x(t)$ at steady state, using only those input components that lie within the bandwidth.

5.3 The Fourier series approximation to the function shown in Figure P5.3 is

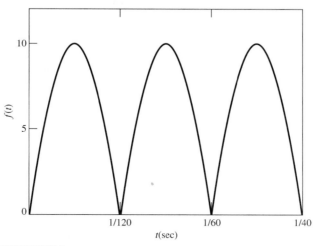

FIGURE P5.3

$$f(t) = \frac{20}{\pi} - \frac{40}{\pi}\left[\frac{\cos 240\pi t}{1(3)} + \frac{\cos 480\pi t}{3(5)} + \frac{\cos 720\pi t}{5(7)} + \cdots\right]$$

Suppose this force is applied to an isolator whose transfer function is

$$\frac{X(s)}{F(s)} = \frac{1}{6 \times 10^{-4}s + 1}$$

(a) Plot the amplitude spectrum of the input.

(b) Keeping only those terms in the Fourier series whose frequencies lie within the system's bandwidth, obtain the expression for the steady-state voltage response $x(t)$.

5.4 The rectangular wave shown in Figure P5.4 has the Fourier series

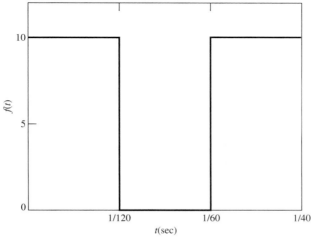

FIGURE P5.4

$$f(t) = 5\left[1 + \frac{4}{\pi}\left(\frac{\sin 120\pi t}{1} + \frac{\sin 360\pi t}{3} + \frac{\sin 600\pi t}{5} + \cdots\right)\right]$$

(a) Plot the amplitude spectrum.

(b) Suppose this force is applied to an isolator whose transfer function is

$$\frac{X(s)}{F(s)} = \frac{1000}{s + 1000}$$

Keeping only those terms in the Fourier series whose frequencies lie within the system's bandwidth, obtain the expression for the steady-state response $x(t)$.

5.5 The displacement shown in Figure P5.5a is produced by the cam shown in Figure P5.5b. The Fourier series approximation to this function is

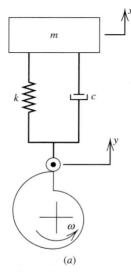

(a)

FIGURE P5.5a

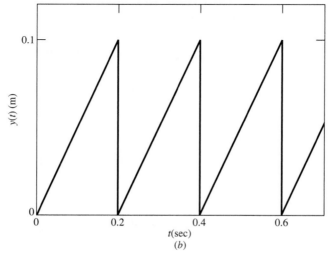

(b)

FIGURE P5.5b

$$y(t) = \frac{1}{20\pi} \left[\pi - 2 \left(\frac{\sin 10\pi t}{1} + \frac{\sin 20\pi t}{2} + \frac{\sin 30\pi t}{3} + \cdots \right) \right]$$

(a) Plot the amplitude spectrum.

(b) For the values $m = 1$ kg, $c = 98$ N·s/m, and $k = 4900$ N/m, keeping only those terms in the Fourier series whose frequencies lie within the system's bandwidth, obtain the expression for the steady-state displacement $x(t)$.

5.6 Given the model

$$0.5\dot{y} + 5y = f(t)$$

with the following Fourier series representation of the input:

$$f(t) = \sin 4t + 4 \sin 8t + 0.04 \sin 12t + 0.06 \sin 16t + \cdots$$

Find the steady-state response $y_{ss}(t)$ by considering only those components of the $f(t)$ expansion that lie within the bandwidth of the system.

5.7 Refer to Figure P5.7, which shows a cylindrical hydraulic valve subjected to a pressure $p(t)$. The damper c models the fluid resistance to valve motion, and the spring acts to return the valve to its neutral position when there is no pressure applied. The pressure varies with time as shown in the figure. Obtain the expression for the steady-state displacement $x(t)$ of the valve for the case where $m = 0.5$ kg, $c = 20$ N·s/m, $k = 5000$ N/m, $D = 60$ mm, $L = 1$, and $P_0 = 22$ kPa.

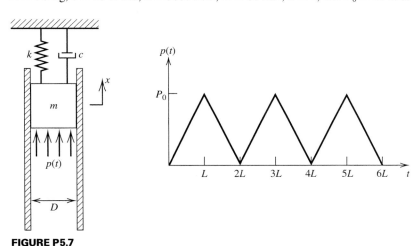

FIGURE P5.7

5.8 Refer to Figure P5.8, which shows a slider-crank mechanism driving a mass-spring-damper system. Note that $y(t)$ is determined entirely by $\theta(t)$ and the values of D and L.

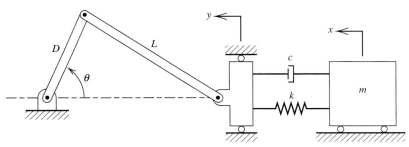

FIGURE P5.8

(a) Derive the expression for $y(t)$ as a function of $\theta(t)$.

(b) Assuming that the crank rotates at 900 rpm, express $y(t)$ as a Fourier series.

(c) Use the results from part (b) to obtain the expression for the steady-state response $x(t)$ for the case where $D = 0.1$ m, $L = 1$ m, $m = 1.5$ kg, $k = 1750$ N/m, and $c = 15$ N·s/m.

SECTION 5.2   THE LAPLACE TRANSFORM

5.9   Refer to the Laplace transform Tables 5.2-1 and 5.2-2 in Section 5.2. Table 5.2-1 contains the transforms; Table 5.2-2 contains the properties of the transforms.

(a) Use property 5 in Table 5.2-2 to derive entry 6 in Table 5.2-1.

(b) Use property 7 in Table 5.2-2 to derive entry 4 in Table 5.2-1.

(c) Use property 2 in Table 5.2-2 to derive entry 12 in Table 5.2-1.

5.10   Use entries 7 and 8 in Table 5.2-1 to find $f(t)$, where

$$F(s) = \frac{3s + 5}{s^2 + 6s + 8}$$

5.11   Use entries 13 and 14 in Table 5.2-1 to find $f(t)$, where

$$F(s) = \frac{10s + 3}{s^2 + 4s + 29}$$

SECTION 5.3   FORCED RESPONSE FROM THE LAPLACE TRANSFORM

5.12   Find the forced response of the model $\dot{v} = f(t) - 2v$ for each of the following cases. The initial condition is $v(0) = 0$.

(a) $f(t) = t$

(b) $f(t) = t^2$

(c) $f(t) = e^{-t}$

(d) $f(t) = te^{-t}$

5.13   Find the forced response of the model $\dot{v} = f(t) - 5v$ for $f(t) = 6e^{-2t}$. The initial condition is $v(0) = 0$.

5.14   Use the principle of superposition to find the forced response of the model $\dot{v} = f(t) - 4v$ for $f(t) = 6(1 - e^{-5t})$. The initial condition is $v(0) = 0$.

5.15   Find the unit-step response of the following model with $x(0) = 2$, $\dot{x}(0) = -4$, and $\ddot{x}(0) = 1$.

$$\frac{d^3x}{dt^3} + 6\frac{d^2x}{dt^2} + 11\frac{dx}{dt} + 6x = u_s(t)$$

5.16   The equation of motion of a certain system is $\dot{v} = f(t)$, where the forcing function is $3e^{-2t} + 5\sin 2t$. The initial condition is $v(0) = 10$. Find the total response.

5.17   Find the step response for the following models. The initial conditions are zero.

(a) $3\ddot{x} + 21\dot{x} + 30x = 40u_s(t)$

(b) $5\ddot{x} + 20\dot{x} + 20x = 72u_s(t)$

(c) $2\ddot{x} + 8\dot{x} + 58x = 95u_s(t)$

5.18   Find the total response for the following models. In all cases the initial conditions are $x(0) = 0$ and $\dot{x}(0) = 1$.

(a) $\ddot{x} + 4\dot{x} + 8x = 2$

(b) $\ddot{x} + 8\dot{x} + 12x = 2$

(c) $\ddot{x} + 4\dot{x} + 4x = 2$

SECTION 5.4   ADDITIONAL RESPONSE EXAMPLES

5.19 Refer to Figure P5.19. A piston of mass 10 kg and area 2.4×10^{-2} m^2 slides with negligible friction inside a cylinder. At $t = 0$ the air valve is opened and the pressure $p(t)$ decays as follows:

$$p(t) = p_a + p_0 e^{-t/\tau}$$

The initial pressure $p(0)$ is 30 kPa above atmospheric pressure p_a, and at time $t = 0.2$ s it is 15 kPa above atmospheric. The stiffness is $k = 1000$ N/m. Obtain the expression for the piston displacement $x(t)$.

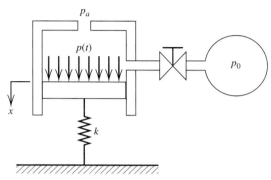

FIGURE P5.19

5.20 Figure P5.20 is a representation of an instrument package of mass m in a space capsule supported by a suspension of stiffness k. When the rocket fires, the acceleration $\ddot{y}$ increases as $\ddot{y} = bt$ where b is a constant. Let $z = x - y$ and assume that $z(0) = \dot{z}(0) = 0$. Obtain the expression for the relative displacement $z(t)$ and the acceleration $\ddot{x}(t)$ felt by the package.

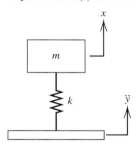

FIGURE P5.20

5.21 Figure P5.21 is a representation of an astronaut and seat in a space capsule supported by a suspension of stiffness k. When the final stage fires, the acceleration $\ddot{y}$ increases as $\ddot{y} = ae^{bt}$ where a and b are constants. Let $z = x - y$ and assume that $z(0) = \dot{z}(0) = 0$. Obtain the expression for the acceleration $\ddot{x}(t)$ felt by the astronaut.

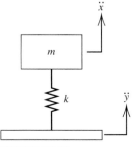

FIGURE P5.21

5.22 Refer to Figure P5.22. At time $t = 0$ a freight elevator is moving at velocity v_0 and then decelerates to zero velocity at time T with a constant deceleration so that

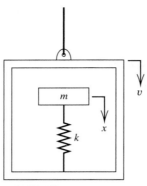

FIGURE P5.22

$$v(t) = v_0\left(1 - \frac{t}{T}\right) \quad 0 \leq t \leq T$$

$$v(t) = 0 \quad t > T$$

The elevator contains a package of mass m with cushioning of stiffness k. Obtain the expression for the displacement $x(t)$ for $0 \leq t \leq T$.

5.23 Refer to Figure P5.23, which shows a device for compacting loose material such as powder. A hydraulic pressure $p(t)$ is suddenly applied at the constant value P_0, and thus is modeled as a step forcing function. The masses of the piston, the material to be compacted, and the supporting structure have been included in the mass m. The area of the piston is A. Assuming that the system is underdamped with zero initial conditions, obtain the expression for $x(t)$.

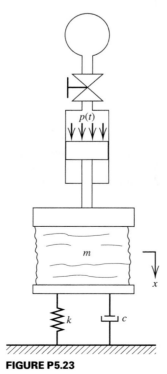

FIGURE P5.23

5.24 Refer to Figure P5.24a, which shows a water tank subjected to a blast force $f(t)$. We will model the tank and its supporting columns as the mass-spring system shown in Figure P5.24b. The blast force as a function of time is shown in Figure P5.24c. Assuming zero initial conditions, obtain the expression for $x(t)$.

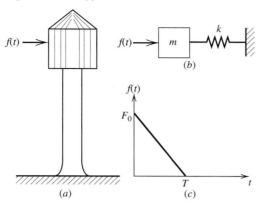

FIGURE P5.24

5.25 Refer to Figure P5.25, which is a simplified representation of a vehicle striking a bump. The vertical displacement x is 0 when the tire first meets the bump. Assuming that the vehicle's horizontal speed v remains constant and that the system is critically damped, obtain the expression for $x(t)$.

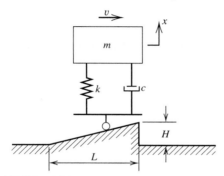

FIGURE P5.25

5.26 Refer to Figure P5.26, which is a simplified representation of a vehicle striking a bump. The vertical displacement x is 0 when the tire first meets the bump. Assuming that the vehicle's horizontal speed v remains constant at 13 m/s and that the system is undamped with a natural frequency of 1 Hz, obtain the expression for $x(t)$, the vertical displacement of the vehicle. The vehicle mass associated with this wheel is 225 kg.

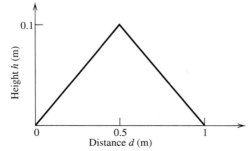

FIGURE P5.26

5.27 Refer to Figure P5.27, in which the displacement $y(t)$ of the lower end of the left-hand spring is given to be $y(t) = 0.013e^{-t}$ m. Assuming that the rod is rigid, has a mass of 7.3 kg, and has an inertia $I_O = 2.44$ kg·m^2, obtain the expression for the angle $\theta(t)$. Use the values $L = 0.23$ m, $D = 0.69$ m, and $k = 5250$ N/m.

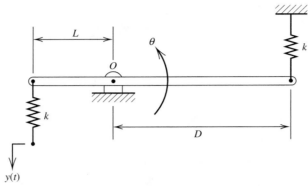

FIGURE P5.27

5.28 Refer to Figure P5.28. The applied force is $f(t) = 100e^{-t}$ N. The pulley inertia about point O is $I_O = 0.5$ kg·m^2. The remaining values are $m_1 = 20$ kg, $k_1 = 2000$ N/m, $k_2 = 1000$ N/m, $R_1 = 0.2$ m, and $R_2 = 0.05$ m. Obtain the expression for $x(t)$ assuming zero initial conditions.

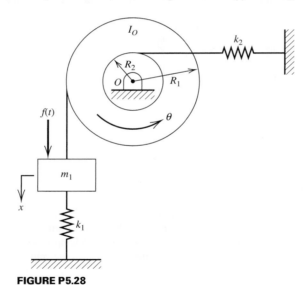

FIGURE P5.28

5.29 Find the forced response of the model $\dot{v} = f(t) - 5v$ for the following input function. The initial condition is $v(0) = 0$.

$$f(t) = \begin{cases} 4t, & 0 \le t \le \\ -4t + 16, & 2 < t \le 4 \\ 0, & t > 4 \end{cases}$$

(*Hint*: Use the ramp response and the shifting theorem.)

5.30 Find the response for the following models. The initial conditions are zero.

(**a**) $3\ddot{x} + 21\dot{x} + 30x = 4t$

(b) $5\ddot{x} + 20\dot{x} + 20x = 7t$

(c) $2\ddot{x} + 8\dot{x} + 58x = 5t$

SECTION 5.5   PULSE AND IMPULSE RESPONSES

5.31 Compare the pulse and impulse response of the model $\dot{y} = f(t) - 0.2y$ for two values of the pulse duration T: **(a)** $T = 1$ **(b)** $T = 10$.

5.32 Find the response of the following model to a rectangular pulse of height 50 and duration 1. The initial conditions are zero.

$$3\ddot{x} + 21\dot{x} + 30x = f(t)$$

5.33 **(a)** Find the response of the following model to a rectangular pulse of height 50 and duration 0.03. The initial conditions are zero.

$$3\ddot{x} + 21\dot{x} + 30x = f(t)$$

(b) Compare your answer in part (a) with the estimate obtained by approximating the pulse response with the impulse response.

5.34 Refer to Figure P5.34. A mass m drops from a height h and hits and sticks to a simply supported beam of equal mass. Obtain an expression for the maximum deflection of the center of the beam. Your answer should be a function of h, g, m, and the beam stiffness k.

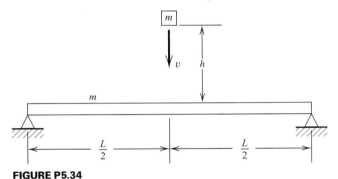

FIGURE P5.34

SECTION 5.6   MATLAB AND THE LAPLACE TRANSFORM

5.35 Use the MATLAB `residue` function to find the forced response of the model $\dot{v} = f(t) - 2v$ for each of the following cases. The initial condition is $v(0) = 0$.

(a) $f(t) = t$

(b) $f(t) = t^2$

(c) $f(t) = e^{-t}$

(d) $f(t) = te^{-t}$

5.36 Use the MATLAB `residue` function to find the total response of the following model. The initial condition is $y(0) = 1$.

$$6\dot{y} + 5y = 20u_s(t)$$

5.37 Use the MATLAB `residue` function to find the total response of the model $\dot{v} = f(t) - 5v$ for $f(t) = 6e^{-2t}$. The initial condition is $v(0) = 10$.

5.38 Use the MATLAB `residue` function to find the response of the model $\dot{v} = f(t) - 4v$ for $f(t) = 6(1 - e^{-5t})$. The initial condition is $v(0) = 0$.

5.39 Use the MATLAB `residue` function to find the unit-step response of the following models with zero initial conditions.

(a) $\ddot{x} + 4\dot{x} + 8x = 2u_s(t)$
(b) $\ddot{x} + 8\dot{x} + 12x = 2u_s(t)$
(c) $\ddot{x} + 4\dot{x} + 4x = 2u_s(t)$

5.40 Use the MATLAB `residue` function to find the total response of the following model with $x(0) = 2$, $\dot{x}(0) = -4$, and $\ddot{x}(0) = 1$.

$$\frac{d^3x}{dt^3} + 6\frac{d^2x}{dt^2} + 11\frac{dx}{dt} + 6x = u_s(t)$$

5.41 Use the MATLAB `residue` function to find the response of the following model with zero initial conditions:

$$\frac{d^3x}{dt^3} + 22\frac{d^2x}{dt^2} + 131\frac{dx}{dt} + 110x = u_s(t)$$

5.42 The equation of motion of a certain system is $\dot{v} = f(t)$, where the forcing function is $3e^{-2t} + 5\sin 2t$. The initial condition is $v(0) = 10$. Use the MATLAB `residue` function to find the total response.

5.43 Use the MATLAB `residue` function to find the ramp response for the following models. The initial conditions are zero.

(a) $3\ddot{x} + 21\dot{x} + 30x = 4t$
(b) $5\ddot{x} + 20\dot{x} + 20x = 7t$
(c) $2\ddot{x} + 8\dot{x} + 58x = 5t$

SECTION 5.7 MATLAB APPLICATIONS

5.44 The following M-file contains the data on the torque output of a six-cylinder, four-cycle gasoline engine. The time is in seconds; the torque is in Newton-meters.

```
time=[0.0015:0.0015:0.036];
torque=[820,840,880,960,1080,1280,1480,1920,1960,1840,1720,1600,...
        1500,1400,1300,1200,1120,1060,1000,940,880,840,820,800];
```

Use MATLAB to compute the Fourier coefficients and plot the amplitude spectrum.

5.45 Use MATLAB to compute the Fourier coefficients of the function given in Figure 5.1-3.

5.46 Use MATLAB to compute the Fourier coefficients of the function given in Figure 5.1-4.

5.47 Use MATLAB to compute the Fourier coefficients of the function given in Example 5.1-4.

5.48 Use MATLAB to solve for and plot the unit-step response of the following models:

(a) $3\ddot{x} + 21\dot{x} + 30x = f(t)$
(b) $5\ddot{x} + 20\dot{x} + 65x = f(t)$
(c) $4\ddot{x} + 32\dot{x} + 60x = 3\dot{f}(t) + 2f(t)$

5.49 Use MATLAB to solve for and plot the unit-impulse response of the following models:

(a) $3\ddot{x} + 21\dot{x} + 30x = f(t)$
(b) $5\ddot{x} + 20\dot{x} + 65x = f(t)$

5.50 Use MATLAB to solve for and plot the impulse response of the following model, where the strength of the impulse is 5:

$$3\ddot{x} + 21\dot{x} + 30x = f(t)$$

5.51 Use MATLAB to solve for and plot the step response of the following model, where the magnitude of the step input is 5:

$$3\ddot{x} + 21\dot{x} + 30x = f(t)$$

5.52 Use MATLAB to solve for and plot the response of the following models for $0 \leq t \leq 1.5$, where the input is $f(t) = 5t$ and the initial conditions are zero:

(a) $3\ddot{x} + 21\dot{x} + 30x = f(t)$

(b) $5\ddot{x} + 20\dot{x} + 65x = f(t)$

(c) $4\ddot{x} + 32\dot{x} + 60x = 3\dot{f}(t) + 2f(t)$

5.53 Use MATLAB to solve for and plot the response of the following models for $0 \leq t \leq 6$, where the input is $f(t) = 6 \cos 3t$ and the initial conditions are zero:

(a) $3\ddot{x} + 21\dot{x} + 30x = f(t)$

(b) $5\ddot{x} + 20\dot{x} + 65x = f(t)$

(c) $4\ddot{x} + 32\dot{x} + 60x = 3\dot{f}(t) + 2f(t)$

SECTION 5.8 SIMULINK APPLICATIONS

5.54 Use Simulink to compute and plot the forced response of the system described in Example 5.1-3.

5.55 Use Simulink to compute and plot the forced response of the system described in Example 5.1-4.

5.56 Use Simulink to compute and plot the forced response of the system $160\ddot{x} + 560\dot{x} + 10x = f(t)$, where $f(t)$ is a series of trapezoidal pulses given in Figure 5.8-3a. Discuss the role of the apparent frequency and the resonant frequency in the response.

5.57 Use Simulink to compute and plot the unit-step response of the following models:

(a) $3\ddot{x} + 21\dot{x} + 30x = f(t)$

(b) $4\ddot{x} + 32\dot{x} + 60x = 3\dot{f}(t) + 2f(t)$

5.58 Use Simulink to solve for and plot the forced response of the following models for $0 \leq t \leq 1.5$, where the input is $f(t) = 5t$.

(a) $3\ddot{x} + 21\dot{x} + 30x = f(t)$

(b) $4\ddot{x} + 32\dot{x} + 60x = 3\dot{f}(t) + 2f(t)$

CHAPTER 6

TWO-DEGREES-OF-FREEDOM SYSTEMS

CHAPTER OUTLINE

UP TO now we have considered only models that have one degree of freedom. There are, however, many practical vibration applications that require a model having more degrees of freedom in order to describe the important features of the system response. In this chapter we consider models that have two degrees of freedom. Then in Chapter 7 we consider three or more degrees of freedom.

We will see that two new phenomena occur in systems having two or more degrees of freedom. Such systems have two or more natural frequencies instead of only one, and their motion is composed of *modes*. These modes are critical to understanding the response of such systems. The mathematics now becomes more complicated than for single-degree-of-freedom models, and we restrict our attention in this chapter to two degrees of freedom only so that the important concepts and methods can be clearly presented and understood more easily. Then in Chapter 8 we will use matrix methods to generalize the results of this chapter to models having more than two degrees of freedom.

Newton's laws of motion can be used to develop models of multi-degree-of-freedom systems, and we give several examples in Section 6.1. There is, however, another method available for developing such models. This method, which is based on energy concepts rather than forces, is due to Lagrange, and the corresponding equations are called *Lagrange's equations*. These are treated in Section 6.2.

The concept of modes is introduced in Section 6.3 by considering the free response. In Section 6.4 we consider the harmonic response of systems with two degrees of freedom. The transient response under harmonic forcing and the total response to a nonharmonic forcing function can be found with the help of the Laplace transform. This is the topic of Section 6.5.

A more detailed analysis of modal response, including the phenomenon of *nodes*, is explored in Section 6.6 by analyzing the pitch response of a suspension system.

Section 6.7, 6.8, and 6.9 treat applications of MATLAB and Simulink to the chapter's topics. As we will see, the mathematics of models having two degrees of freedom can be quite formidable, and MATLAB has powerful capabilities for handling such analysis. One of these, the root locus plot, was developed for control system design and has useful applications in vibrations, but it has been ignored in the vibration literature. In addition, MATLAB can be used not only to solve the differential equations numerically but also to perform some of the algebra required to obtain closed-form solutions. Simulink is useful in applications where the stiffness and/or the damping elements are nonlinear and where the input is not a simple function of time. We will analyze a two-mass suspension model to show how Simulink can be used for such problems.

LEARNING OBJECTIVES

After you have finished this chapter, when dealing with systems having two degrees of freedom, you should be able to do the following:

- Apply Newton's laws and Lagrange's equations to develop equations of motion for such systems.
- Identify the modes of a system and compute its natural frequencies.
- Analyze the harmonic response.
- Obtain the response to a general forcing function.
- Analyze the role of nodes in the modal response.
- Use MATLAB and Simulink to support the methods of the chapter.

6.1   MODELS OF TWO-DEGREES-OF-FREEDOM SYSTEMS

The following examples show how to use Newton's laws of motion to obtain the equations of motion for systems having two degrees of freedom.

EXAMPLE 6.1-1
A Vibration Absorber

A *vibration absorber* consists of a mass and elastic element that is attached to another mass in order to reduce its vibration. In Chapter 7 we will show how to select the absorber's mass and elasticity. Figure 6.1-1a shows a vibration absorber (with mass m_2 and stiffness k_2) attached to a cantilever beam that supports a motor. In Chapter 7 we will show that if m_2 and k_2 are selected properly, the beam vibration due to rotating unbalance in the motor can be minimized. Figure 6.1-1b is a representation showing the beam stiffness k and effective mass m.

(a) Obtain the equation of motion for the system. The force f is a specified force acting on the mass m and is due to the rotating unbalance of the motor. The displacements x and x_2 are measured from the rest positions of the masses when $f = 0$.

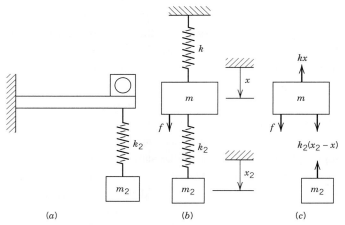

FIGURE 6.1-1 (*a*) A vibration absorber. (*b*) The equivalent representation. (*c*) Free-body diagrams assuming that $x_2 > x$.

(b) Obtain the transfer functions $X(s)/F(s)$ and $X_2(s)/F(s)$ as well as the reduced-form differential equation models.

Solution

(a) In multi-mass problems such as this, you must make *consistent* assumptions about the relative displacements of the masses, draw the free-body diagrams accordingly, and write the corresponding equations of motion. For example, if we assume that $x_2 > x$, then we obtain the free-body diagram in Figure 6.1-1*c*. Note that the spring force $k_2(x_2 - x)$ acts in opposite directions on the masses because of Newton's law of action and reaction. From the free-body diagrams we obtain the following equations of motion:

$$m\ddot{x} = f + k_2(x_2 - x) - kx \tag{1}$$

$$m_2\ddot{x}_2 = -k_2(x_2 - x) \tag{2}$$

By a *consistent* assumption, we mean that the same assumptions concerning relative motion are used throughout the entire analysis. A common mistake of beginners is to draw the free-body diagram for mass m_1 assuming that $x_2 > x$, and then draw the free-body diagram for m_2 assuming that $x > x_2$. This leads to incorrect equations. Applying Newton's law of action and reaction is a good way to catch such mistakes.

(b) To obtain the transfer functions, apply the Laplace transform to each equation using zero initial conditions. After collecting terms, this gives

$$(ms^2 + k + k_2)X(s) - k_2X_2(s) = F(s)$$

$$-k_2X(s) + (m_2s^2 + k_2)X_2(s) = 0$$

These are two equations in two unknowns, $X(s)$ and $X_2(s)$. They can be solved by substitution or by using Cramer's method. Because we want both transfer functions, Cramer's method is preferred. This gives

$$X(s) = \frac{D_1(s)}{D(s)}$$

$$X_2(s) = \frac{D_2(s)}{D(s)}$$

where

$$D(s) = \begin{vmatrix} (ms^2 + k + k_2) & -k_2 \\ -k_2 & (m_2 s^2 + k_2) \end{vmatrix}$$

$$= (ms^2 + k + k_2)(m_2 s^2 + k_2) - k_2^2$$

$$= mm_2 s^4 + [m_2(k + k_2) + mk_2]s^2 + kk_2$$

$$D_1(s) = \begin{vmatrix} F(s) & -k_2 \\ 0 & (m_2 s^2 + k_2) \end{vmatrix} = (m_2 s^2 + k_2)F(s)$$

$$D_2(s) = \begin{vmatrix} (ms^2 + k + k_2) & 0 \\ -k_2 & F(s) \end{vmatrix} = (ms^2 + k + k_2)F(s)$$

Thus the transfer functions are

$$\frac{X(s)}{F(s)} = \frac{m_2 s^2 + k_2}{mm_2 s^4 + [m_2(k + k_2) + mk_2]s^2 + kk_2}$$

$$\frac{X_2(s)}{F(s)} = \frac{ms^2 + k + k_2}{mm_2 s^4 + [m_2(k + k_2) + mk_2]s^2 + kk_2}$$

The denominator of the transfer functions is the characteristic polynomial. Because it is a fourth-order polynomial, we see that the model is fourth order. Note that both transfer functions have numerator dynamics.

The reduced-form differential equation models can be obtained from the transfer functions. From the first transfer function, we have

$$\{mm_2 s^4 + [m_2(k + k_2) + mk_2]s^2 + kk_2\}X(s) = (m_2 s^2 + k_2)F(s)$$

Using the fact that $sX(s)$ corresponds to $\dot{x}$ and $s^2 X(s)$ corresponds to $\ddot{x}$, and so forth, we see that

$$mm_2 \frac{d^4 x}{dt^4} + (m_2 k + m_2 k_2 + mk_2)\frac{d^2 x}{dt^2} + kk_2 x = m_2 \frac{d^2 f}{dt^2} + k_2 f$$

Similarly, the second transfer function gives

$$mm_2 \frac{d^4 x_2}{dt^4} + (m_2 k + m_2 k_2 + mk_2)\frac{d^2 x_2}{dt^2} + kk_2 x_2 = m \frac{d^2 f}{dt^2} + (k + k_2)f$$ ∎

When damping acts between two masses, you must make an assumption about the relative velocities of the masses so that you can draw the free-body diagrams. Because of Newton's law of action and reaction, the damping force on one mass has the same magnitude but opposite direction on the other mass.

EXAMPLE 6.1-2
A Two-Mass Model of a Suspension

Consider a suspension model that includes the mass of the wheel-tire-axle assembly (Figure 6.1-2). The mass m_1 is one-fourth the mass of the car body, and m_2 is the mass of the wheel-tire-axle assembly. The spring constant k_1 represents the suspension's elasticity, and k_2 represents the tire's elasticity. Derive the equations of motion for m_1 and m_2 in terms of the displacements from equilibrium, x_1 and x_2.

Solution

Assuming that $x_2 > x_1$, $\dot{x}_2 > \dot{x}_1$, and $y > x_2$, we obtain the free-body diagram shown. For mass m_1, Newton's law gives

$$m_1 \ddot{x}_1 = c(\dot{x}_2 - \dot{x}_1) + k_1(x_2 - x_1)$$

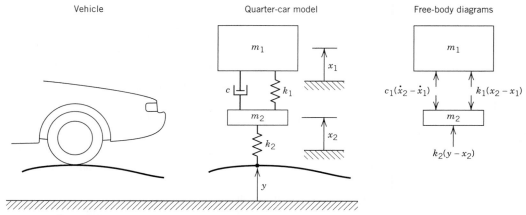

FIGURE 6.1-2 Two-mass model of a suspension and its free-body diagram.

or

$$m_1\ddot{x}_1 + c\dot{x}_1 + k_1 x_1 - c\dot{x}_2 - k_1 x_2 = 0 \tag{1}$$

For mass m_2,

$$m_2\ddot{x}_2 = -c(\dot{x}_2 - \dot{x}_1) - k_1(x_2 - x_1) + k_2(y - x_2)$$

or

$$m_2\ddot{x}_2 + c\dot{x}_2 + (k_1 + k_2)x_2 - c\dot{x}_1 - k_1 x_1 = k_2 y \tag{2}$$

∎

EXAMPLE 6.1-3
Shaft Elasticity in a Drive Train

Figure 6.1-3a shows a drive train with spur gears. The first shaft turns N times faster than the second shaft. Develop a model of the system, including the stiffness k of the second shaft. Assume the first shaft is rigid, and neglect the gear and shaft masses.

Solution

When the inertia I_1 and torque T_1 on the first shaft are reflected to the second shaft, the result is the model shown in Figure 6.1-3b, where $I_{1e} = N^2 I_1$. The free-body diagrams are shown in Figure 6.1-3c for the case $\theta_2 > \theta_3$. Summing moments on I_{1e} and I_2 gives

$$I_{1e}\ddot{\theta}_2 = NT_1 - k(\theta_2 - \theta_3) \tag{1}$$

$$I_2\ddot{\theta}_3 = k(\theta_2 - \theta_3) - c\dot{\theta}_3 \tag{2}$$

If desired, this model can be expressed in terms of θ_1 by substituting $\theta_2 = \theta_1/N$. ∎

EXAMPLE 6.1-4
Vibration with Rigid-Body Motion

Figure 6.1-4a shows two identical cylinders of mass m and radius R, whose axles are connected with a spring. Obtain the equations of motion, and solve them. Assume cylinders roll without slipping. The spring is at its free length when $x_1 = x_2$.

Solution

The effective mass of each cylinder is $m_e = m + I^2/R^2 = 1.5m$, since $I = mR^2$. So the system is equivalent to the one shown in Figure 6.1-4b. Summing forces on each mass gives

$$1.5m\ddot{x}_1 = -k(x_1 - x_2) \tag{1}$$

$$1.5m\ddot{x}_2 = k(x_1 - x_2) \tag{2}$$

Note that if we subtract Equation 2 from Equation 1, we obtain

$$1.5m(\ddot{x}_1 - \ddot{x}_2) = -2k(x_1 - x_2)$$

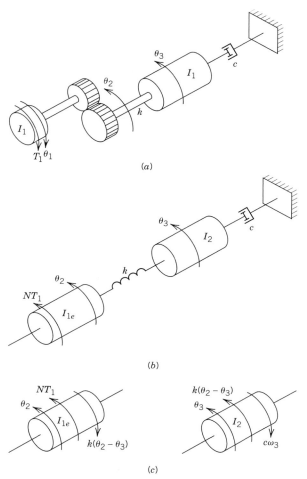

FIGURE 6.1-3 A model of shaft elasticity in a drive train. (*a*) The original system. (*b*) Equivalent gearless system. (*c*) Free-body diagrams.

If we define $y_1 = x_1 - x_2$, then this equation can be written as

$$1.5m\ddot{y}_1 = -2ky_1 \tag{3}$$

which has the solution

$$y_1 = \frac{\dot{y}_1(0)}{\omega_n} \sin \omega_n t + y_1(0) \cos \omega_n t$$

where

$$\omega_n = \sqrt{\frac{4k}{3m}}$$

Thus

$$x_2 = x_1 - y_1 = x_1 - \frac{\dot{y}_1(0)}{\omega_n} \sin \omega_n t + y_1(0) \cos \omega_n t \tag{4}$$

This says that the motion of mass 2 relative to mass 1 is oscillatory with a frequency ω_n.

Now add Equations 1 and 2 to obtain

$$1.5m(\ddot{x}_1 + \ddot{x}_2) = 0$$

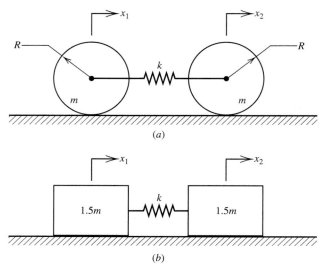

FIGURE 6.1-4 (a) A system capable of rigid-body motion. (b) The equivalent nonrotating system.

If we define $y_2 = x_1 + x_2$, we obtain

$$\ddot{y}_2 = 0$$

whose solution is $\dot{y}_2 = \dot{y}_2(0)$ and $y_2 = \dot{y}_2(0)t + y_2(0)$.

The particular motion displayed by the system depends on the initial conditions. Suppose that $\dot{x}_1(0) = \dot{x}_2(0) = 0$ and $x_1(0) = -x_2(0)$. That is, the masses are pulled apart an equal distance from the free length position and released from rest. Then $y_2(0) = \dot{y}_2(0) = 0$, so that $y_2(t) = 0$. This implies that $x_1(t) = -x_2(t)$, which says that the masses are always moving in the opposite directions. Also,

$$x_1(t) = x_2(t) + 2x_1(0) \cos \omega_n t$$

which says that the masses oscillate with the frequency ω_n.

If the masses are displaced equally and in the same direction, and released with the same velocity, then $x_1(0) = x_2(0)$ and $\dot{x}_1(0) = \dot{x}_2(0)$, and thus $y_1(0) = \dot{y}_1(0) = 0$, and so $y_1(t) = 0$. Thus $x_1(t) = x_2(t)$. This motion is called the *rigid-body mode* because the entire system translates without oscillation, as if it were a rigid body. ■

EXAMPLE 6.1-5
A Pendulum
Coupled to a
Translating Mass

Consider the system shown in Figure 6.1-5a. The main body of mass M is propelled along a horizontal track by a traction force f. The main body contains an actuator for rotating the pendulum. The actuator applies a torque T to the arm. The pendulum has a total mass m and moment of inertia I relative to its mass center at point C. Obtain the equations of motion with f and T as the given inputs.

Solution

One choice of coordinates is x_1 (the location of the mass M) and (x_2, y_2), which describe the location of the mass center of the pendulum. However, x_2 and y_2 are not independent, so a better choice is x_1 and θ. The free-body diagrams are shown in Figures 6.1-5b,c. F_x and F_y are the reaction forces in the x and y directions due to the physical connection between the pendulum and the mass M. R_1 and R_2 are the vertical reaction forces due to the wheels.

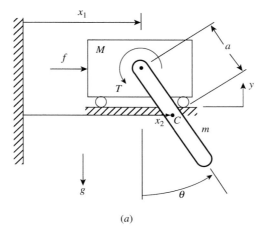

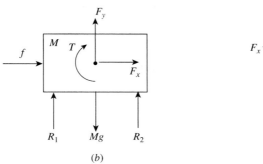

FIGURE 6.1-5   (a) A cart-pendulum system. (b) Free-body diagrams.

Summing forces on M in the horizontal and vertical directions gives

$$M\ddot{x}_1 = F_x + f \tag{1}$$

$$M\ddot{y}_1 = F_y + R_1 + R_2 - Mg \tag{2}$$

Equation 2 is of no interest unless we need to find R_1 and R_2.

For mass m, summing forces in the horizontal and vertical directions gives

$$m\ddot{x}_2 = -F_x \tag{3}$$

$$m\ddot{y}_2 = -F_y - mg \tag{4}$$

Summing moments about the mass center gives

$$I\ddot{\theta} = aF_x \cos\theta + aF_y \sin\theta + T \tag{5}$$

To eliminate F_x, use Equations 1 and 3 to obtain

$$M\ddot{x}_1 + m\ddot{x}_2 = f \tag{6}$$

Use Equations 1 and 4 to eliminate F_x and F_y in Equation 5. This gives

$$I\ddot{\theta} + m\ddot{x}_2 a \cos\theta + ma(\ddot{y}_2 + g)\sin\theta = T \tag{7}$$

The displacements are related as follows:

$$x_2 = x_1 + a \sin\theta$$

$$y_2 = -a \cos\theta$$

Thus

$$\ddot{x}_2 = \ddot{x}_1 - a\dot{\theta}^2 \sin\theta + a\ddot{\theta}\cos\theta$$

$$\ddot{y}_2 = a\dot{\theta}^2 \cos\theta + a\ddot{\theta}\sin\theta$$

Substituting $\ddot{x}_2$ and $\ddot{y}_2$ into Equations 6 and 7 gives the desired result.

$$(I + ma^2)\ddot{\theta} + ma\ddot{x}_1 \cos\theta + mga \sin\theta = T \tag{8}$$

$$(M + m)\ddot{x}_1 - ma\dot{\theta}^2 \sin\theta + ma\ddot{\theta}\cos\theta = f \tag{9}$$

∎

EXAMPLE 6.1-6
Two Degrees of Freedom with Damping

Obtain the equations of motion for the system shown in Figure 6.1-6a. The forces f_1 and f_2 are given functions of time. The displacements x_1 and x_2 are measured from the equilibrium positions of the masses when $f_1 = f_2 = 0$.

Solution

Assuming that $x_1 > x_2$ and $\dot{x}_1 > \dot{x}_2$, we obtain the free-body diagrams shown in Figure 6.1-6b. From these we obtain

$$m_1\ddot{x}_1 = f_1 - k_1x_1 - c_1\dot{x}_1 - k_2(x_1 - x_2) - c_2(\dot{x}_1 - \dot{x}_2)$$

$$m_2\ddot{x}_2 = f_2 + k_2(x_1 - x_2) + c_2(\dot{x}_1 - \dot{x}_2)$$

Bringing all the dependent variables to the left-hand sides gives

$$m_1\ddot{x}_1 + (c_1 + c_2)\dot{x}_1 + (k_1 + k_2)x_1 - k_2x_2 = f_1 \tag{1}$$

$$m_2\ddot{x}_2 + c_2\dot{x}_2 + k_2x_2 - k_2x_1 - c_2\dot{x}_1 = f_2 \tag{2}$$

∎

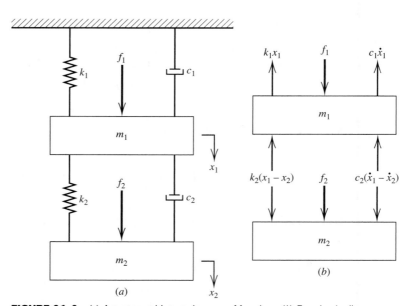

FIGURE 6.1-6   (a) A system with two degrees of freedom. (b) Free-body diagrams.

Free Response

The algebra required to obtain the closed-form expression for the free response of two-degrees-of-freedom systems is more tedious than for systems with one degree. This fact means that you should carefully consider whether or not you need a closed-form expression. In many applications only a plot of the response is needed. If so, a numerical method can be used to obtain the plot with less effort than required to obtain the closed-form solution. Equations 1 and 2 of Example 6.1-6 are not in a form that is useful for numerical solution. Instead, the model must be put into either transfer function form or state-variable form, which consists of coupled first-order equations.

A linear model of a system with two degrees of freedom has four characteristic roots. Thus the free response consists of the linear combination of four terms, each term corresponding to one root. If a root pair is complex, its solution form may be expressed as a sine and cosine or as a sine function with a phase shift.

There will be four undetermined constants in the free response form. Thus we require four conditions to evaluate these constants. Usually these conditions will be given as initial conditions on the displacements and the velocities of the masses.

EXAMPLE 6.1-7
Free Response with Two Degrees of Freedom

Obtain the closed-form expression for the free response of the system shown in Figure 6.1-6 using the values $m_1 = 4$, $m_2 = 1$, $k_1 = 16$, $k_2 = 12$, and $c_1 = 0$. The initial conditions are $x_1(0) = 1$, $x_2(0) = \dot{x}_1(0) = \dot{x}_2(0) = 0$. Consider two cases: **(a)** no damping present $(c_2 = 0)$ and **(b)** damping present, with $c_2 = 1$.

Solution

(a) Substituting the given values into Equations 1 and 2 of Example 6.1-6, and setting $f_1 = f_2 = 0$, since only the free response is required, we obtain

$$4\ddot{x}_1 + 28x_1 - 12x_2 = 0 \tag{1}$$

$$\ddot{x}_2 + 12x_2 - 12x_1 = 0 \tag{2}$$

The Laplace transform provides a systematic way of obtaining the response. Recall the derivative property: $\mathcal{L}(\ddot{x}) = s^2 X(s) - sx(0) - \dot{x}(0)$. Applying the transform to Equations 1 and 2 and using the given initial conditions, we obtain

$$4\left[s^2 X_1(s) - s\right] + 28X_1(s) - 12X_2(s) = 0$$

$$s^2 X_2(s) + 12X_2(s) - 12X_1(s) = 0$$

These can be solved with Cramer's rule:

$$X_1(s) = \frac{4s^3 + 48s}{4s^4 + 76s^2 + 192} = \frac{s^3 + 12s}{s^4 + 19s^2 + 48}$$

$$X_2(s) = \frac{48s}{4s^4 + 76s^2 + 192} = \frac{12s}{s^4 + 19s^2 + 48}$$

Note that the denominators, which are identical, are quadratic in s^2. This occurs in undamped systems. The roots of the denominators are $s^2 = -3$ and $s^2 = -16$. Thus the characteristic roots are $s = \pm i\sqrt{3}$ and $s = \pm 4i$.

We can thus express $X_1(s)$ and $X_2(s)$ in factored form as

$$X_1(s) = \frac{s^3 + 12s}{(s^2 + 3)(s^2 + 16)}$$

$$X_2(s) = \frac{12s}{(s^2 + 3)(s^2 + 16)}$$

Let us concentrate on finding $x_1(t)$ first. The imaginary roots imply a partial-fraction expansion of the form

$$X_1(s) = \frac{\sqrt{3}C_1}{s^2 + 3} + \frac{C_2 s}{s^2 + 3} + \frac{4C_3}{s^2 + 16} + \frac{C_4 s}{s^2 + 16}$$

Using the least common denominator and comparing the numerator with $s^3 + 12s$, we obtain

$$C_2 + C_4 = 1 \qquad \sqrt{3}C_1 + 4C_3 = 0$$

$$16C_2 + 3C_4 = 12 \qquad 16\sqrt{3}C_1 + 12C_3 = 0$$

These have the solution $C_1 = C_3 = 0$, $C_2 = 9/13$, and $C_4 = 4/13$. Thus the free response is

$$x_1(t) = \frac{9}{13}\cos \sqrt{3}t + \frac{4}{13}\cos 4t \tag{3}$$

Note that this satisfies the two initial conditions on x_1 and $\dot{x}_1$.

To find $x_2(t)$, we can repeat this process with $X_2(s)$ or we can solve for $x_2(t)$ from Equation 1. Using the latter approach we obtain

$$x_2(t) = \frac{1}{3}\ddot{x}_1 + \frac{7}{3}x_1 \tag{4}$$

where, from Equation 3,

$$\dot{x}_1 = -\frac{9\sqrt{3}}{13}\sin \sqrt{3}t - \frac{16}{13}\sin 4t$$

$$\ddot{x}_1 = -\frac{27}{13}\cos \sqrt{3}t - \frac{64}{13}\cos 4t$$

Thus, from Equation 4,

$$x_2(t) = \frac{12}{13}\cos \sqrt{3}t - \frac{12}{13}\cos 4t$$

The plots of the responses are shown in Figure 6.1-7.

(b) With $c_2 = 1$, we have

$$4\ddot{x}_1 + \dot{x}_1 + 28x_1 - 12x_2 = 0 \tag{5}$$

$$\ddot{x}_2 + \dot{x}_2 + 12x_2 - 12x_1 = 0 \tag{6}$$

Applying the transform to Equations 5 and 6 and using the given initial conditions, we obtain

$$4\left[s^2 X_1(s) - s\right] + sX_1(s) - 1 + 28X_1(s) - 12X_2(s) = 0$$

$$s^2 X_2(s) + sX_2(s) + 12X_2(s) - 12X_1(s) = 0$$

These can be solved with Cramer's rule:

$$X_1(s) = \frac{(4s + 1)(s^2 + s + 12)}{4s^4 + 5s^3 + 77s^2 + 40s + 192}$$

$$X_2(s) = \frac{12(4s + 1)}{4s^4 + 5s^3 + 77s^2 + 40s + 192}$$

Note that the denominators are no longer quadratic in s^2. This occurs in damped systems. The characteristic roots are now complex rather than imaginary. They are

$$s = -0.3827 \pm 3.9629i \qquad \text{and} \qquad s = -0.2423 \pm 1.7232i$$

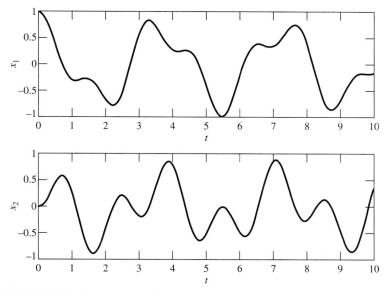

FIGURE 6.1-7   Free response of an undamped system.

Because $X_2(s)$ has the simplest numerator, we start with it. Expanding $X_2(s)$ gives the form

$$X_2(s) = \frac{12s + 3}{\left[(s+a)^2 + b^2\right]\left[(s+c)^2 + d^2\right]} = \frac{C_1 b + C_2(s+a)}{\left[(s+a)^2 + b^2\right]} + \frac{C_3 d + C_4(s+c)}{\left[(s+c)^2 + d^2\right]}$$

where

$$a = 0.3827 \qquad b = 3.9629$$

and

$$c = 0.2423 \qquad d = 1.7232$$

Using the least common denominator and comparing the numerator with $12s + 3$, we obtain $C_1 = 0.0506$, $C_2 = -0.9393$, $C_3 = -0.0398$, and $C_4 = 0.9393$.[1] Thus the free response is

$$x_2(t) = e^{-0.3827t}(0.0506 \sin 3.9629t - 0.9393 \cos 3.9629t)$$
$$+ e^{-0.2423t}(-0.0398 \sin 1.7232t + 0.9393 \cos 1.7232t)$$

Note that this satisfies the two initial conditions on x_2 and $\dot{x}_2$.

The response of $x_1(t)$ can be found in a similar manner. The responses are shown in Figure 6.1-8. The dominant time constant is $\tau = 1/0.2423 = 4.127$, and the damping ratio of the dominant root pair is $\zeta = \cos\left[\tan^{-1}(1.7232/0.2423)\right] = 0.139$.   ∎

Neglecting Damping

We can draw some important insights from this example. When damping is present, the characteristic roots are complex, and the algebra required to obtain the closed-form response is more complicated than for undamped systems. This is why damping is often

---

[1] These coefficients are easily calculated with the MATLAB `residue` function. See Section 5.6.

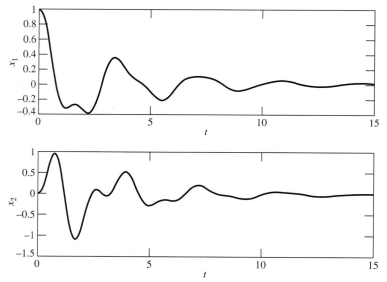

FIGURE 6.1-8 Free response of a damped system.

assumed to be negligible in many vibration analyses. This assumption may not be justified for systems containing hydraulic dampers, but it is often justified in many other practical applications, especially those involving structural vibration.

Example 6.1-7 illustrates that an undamped model can give reasonably accurate results, especially with predictions based on the characteristic roots. For example, the radian oscillation frequencies predicted by the undamped model, 1.732 and 4, are very close to those of the damped model, 1.7232 and 3.9629.

6.2 LAGRANGE'S EQUATIONS

Direct application of Newton's law to each mass in a multi-mass system requires you to account for the reaction forces explicitly. Then, once all the governing equations have been obtained, the reaction forces must be eliminated algebraically in order to obtain the equations of motion in terms of only the given input and output variables, as was done in Example 6.1-5. This is not too difficult to accomplish in many applications, but as the number of masses increases or the kinematic constraints become more complex, it becomes tedious to eliminate the reaction forces.

There is, however, another method available that avoids the need to deal with the reaction forces. This method, due to Lagrange, is briefly presented here so that the student will be aware that there is more than one way to solve a dynamics problem. The derivation of Lagrange's equations is beyond the scope of this book, and the reader is referred to a text on advanced dynamics for more discussion.

The Lagrangian method is energy based rather than force based, and it requires you to derive expressions for the system's potential and kinetic energies. Denote these energies by P and K, respectively, and define the *Lagrangian L* as $L = K - P$. Then Lagrange's equations are

$$\frac{d}{dt}\left(\frac{\partial L}{\partial \dot{q}_j}\right) + \frac{\partial D}{\partial \dot{q}_j} - \frac{\partial L}{\partial q_j} = Q_j \quad j = 1, 2, \ldots, n \tag{6.2-1}$$

where the variables q_j are a set of generalized coordinates that completely describe the system's motion and n is the number of such coordinates. The corresponding velocities are $\dot{q}_j$, and the Q_j terms represent externally applied forces or moments at the coordinates q_j. The term D represents the power dissipation, such as that due to damping, from either a force or moment at that coordinate.

Generalized Coordinates

Identification of the coordinates q_j might seem somewhat vague at this point, but it is usually easily done in practice. Figure 6.2-1 shows some examples. The motion of a simple pendulum can be described with the rectangular coordinates x and y or with just

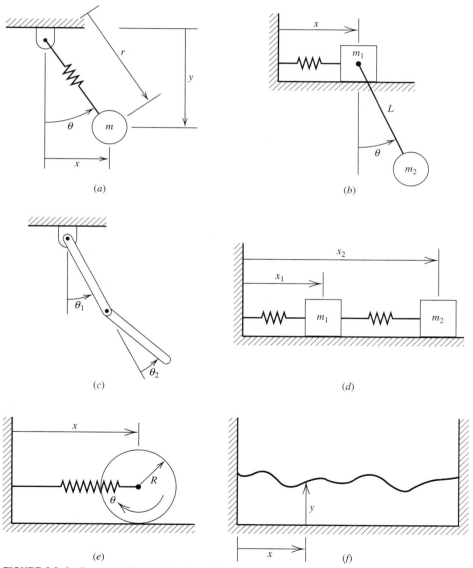

FIGURE 6.2-1 Examples of generalized coordinates.

the single coordinate θ, since the length L is constant. So the constant length L acts as a *constraint* on the motion of the mass at the end of the pendulum. Not only do the rectangular coordinates result in more complicated expressions for velocity and acceleration, but the two coordinates are unnecessary because the pendulum rotating in a single plane has only one degree of freedom. The coordinates x and y are constrained by the fact that $x^2 + y^2 = L^2$. Thus only one of the two coordinates is independent.

Generalized coordinates are those that reflect the existence of constraints and yet still provide a complete description of the motion of the object. Referring again to the pendulum, we see that the angle θ is a generalized coordinate but the coordinates x and y are not. However, if a spring replaces the rigid rod, as in Figure 6.2-1b, then the system has two degrees of freedom, and either (r, θ) or (x, y) may be used as generalized coordinates.

If the base of the pendulum is free to slide, as in Figure 6.2-1c, the system has two degrees of freedom. The coordinates (x, θ) may be used as generalized coordinates.

Two pendula coupled as shown in Figure 6.2-1d have two degrees of freedom. The coordinates (θ_1, θ_2) may be used as generalized coordinates. Similarly, the system shown in Figure 6.2-1e has two degrees of freedom. The coordinates (x_1, x_2) may be used as generalized coordinates.

In Figure 6.2-1e, if the wheel does not slip on the surface, we may describe its motion with either the displacement x or the angle θ. The object has one degree of freedom because $\Delta x = R\Delta\theta$. Either x or θ may be used as a generalized coordinate.

Generalized Force

The generalized force (or moment) Q_j is defined from

$$\delta W_j = Q_j \delta q_j$$

where δq_j is a *virtual displacement* and δW_j is the *virtual work* done by Q_j in causing the virtual displacement. A virtual displacement is an infinitesimal displacement that does not violate the constraints. If the coordinate q_j is a translational coordinate, Q_j is a force; if q_j is a rotational coordinate, Q_j is a moment.

The following example shows how to apply Lagrange's equations. It illustrates how to handle a translating and rotating system having an external force and torque, with damping, a stiffness element, and a gravitational restoring force.

EXAMPLE 6.2-1 *Motion of a* *Pendulum and a* *Translating Mass*	Consider the system shown in Figure 6.2-2. The main body of mass M is propelled along a horizontal track by a traction force f. The main body contains an actuator for rotating the pendulum. The actuator applies a torque T to the arm. The pendulum has a total mass m and moment of inertia I relative to its mass center at point C. Use Lagrange's equations to obtain the equations of motion with f and T as the given inputs.
Solution	The first step is to choose the generalized coordinates q_1 and q_2. It is easily seen that x_1 and θ are convenient choices for q_1 and q_2 because it is then possible to choose $Q_1 = f$ and $Q_2 = T$.

The next step is to develop the expressions for the kinetic and potential energies. The kinetic energy of the mass M is due solely to horizontal translation, while that for m is due to translation of the mass center C in the horizontal and vertical directions, plus the rotation about the mass center. Thus the system's kinetic energy is

$$K = \frac{1}{2}M\dot{x}_1^2 + \frac{1}{2}mv_2^2 + \frac{1}{2}I\dot{\theta}^2$$

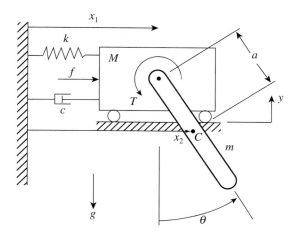

FIGURE 6.2-2 A cart-pendulum system with elasticity and damping between the cart and the support.

where, from the Pythagorean theorem,

$$v_2^2 = \dot{x}_2^2 + \dot{y}_2^2 = \left(\dot{x}_1 + a\dot{\theta}\cos\theta\right)^2 + \left(a\dot{\theta}\sin\theta\right)^2$$
$$= \dot{x}_1^2 + 2a\dot{x}_1\dot{\theta}\cos\theta + a^2\dot{\theta}^2$$

since $\sin^2\theta + \cos^2\theta = 1$. Thus

$$K = \frac{1}{2}(M+m)\dot{x}_1^2 + ma\dot{x}_1\dot{\theta}\cos\theta + \frac{1}{2}\left(ma^2 + I\right)\dot{\theta}^2$$

The system's potential energy is due to the spring k and the vertical displacement of the center of mass of the pendulum. Taking the potential energy to be zero when $y_2 = -a$, we obtain

$$P = mga(1 - \cos\theta) + \frac{1}{2}kx_1^2$$

Thus

$$L = K - P$$
$$= \frac{1}{2}(M+m)\dot{x}_1^2 + ma\dot{x}_1\dot{\theta}\cos\theta + \frac{1}{2}\left(ma^2 + I\right)\dot{\theta}^2 - mga(1 - \cos\theta) - \frac{1}{2}kx_1^2$$

Only the coordinate x_1 has damping associated with it, so the dissipation function D is

$$D = \frac{1}{2}c\dot{q}_1^2 = \frac{1}{2}c\dot{x}_1^2$$

We can now form Lagrange's equations. For $j = 1$,

$$\frac{d}{dt}\left[(M+m)\dot{x}_1 + ma\dot{\theta}\cos\theta\right] + c\dot{x}_1 + kx_1 = f$$

This gives

$$(M+m)\ddot{x}_1 + ma\ddot{\theta}\cos\theta - ma\dot{\theta}^2\sin\theta + c\dot{x}_1 + kx_1 = f \qquad (1)$$

For $j = 2$,

$$\frac{d}{dt}\left[ma\dot{x}_1\cos\theta + (ma^2 + I)\dot{\theta}\right] + mga\sin\theta = T$$

After completing the derivative, some terms will cancel, leaving

$$ma\ddot{x}_1\cos\theta + (ma^2 + I)\ddot{\theta} + mga\sin\theta = T \qquad (2)$$

Equations 1 and 2 are the equations of motion. It is left as an exercise to show that these equations, with $k = c = 0$, are the same as those found with the Newtonian method used in Example 6.1-5. ∎

In some applications the reaction forces must be found for mechanical design purposes. In such cases the Lagrangian approach does not directly give all the necessary information. In addition, although not the case in the previous example, Lagrange's method sometimes produces a set of equations that is not as computationally efficient as those derived with the Newtonian method.

6.3 MODES AND SYSTEM RESPONSE

Consider an undamped two-mass system like that shown in Figure 6.3-1, where x_1 and x_2 are the displacements of the masses. The equations of motion are

$$m_1\ddot{x}_1 + (k_1 + k_2)x_1 - k_2x_2 = 0 \tag{6.3-1}$$

$$m_2\ddot{x}_2 + (k_2 + k_3)x_2 - k_2x_1 = 0 \tag{6.3-2}$$

The characteristic roots will be imaginary; we denote them by $s = \pm i\omega_1$ and $s = \pm i\omega_2$. The free response for x_1 has the form

$$x_1(t) = A_1 \sin \omega_1 t + B_1 \cos \omega_1 t + C_1 \sin \omega_2 t + D_1 \cos \omega_2 t \tag{6.3-3}$$

Similarly, the response of x_2 has the form

$$x_2(t) = A_2 \sin \omega_1 t + B_2 \cos \omega_1 t + C_2 \sin \omega_2 t + D_2 \cos \omega_2 t \tag{6.3-4}$$

If we substitute these expressions into the equations of motion, we find that

$$\frac{A_2}{A_1} = \frac{B_2}{B_1} = \frac{k_1 + k_2 - m_1\omega_1^2}{k_2} = r_1$$

$$\frac{C_2}{C_1} = \frac{D_2}{D_1} = \frac{k_1 + k_2 - m_1\omega_2^2}{k_2} = r_2$$

Thus $x_2(t)$ can be expressed as

$$x_2(t) = r_1A_1 \sin \omega_1 t + r_1B_1 \cos \omega_1 t + r_2C_1 \sin \omega_2 t + r_2D_1 \cos \omega_2 t \tag{6.3-5}$$

If the initial conditions are such that $C_1 = D_1 = 0$, then both x_1 and x_2 will oscillate at the same frequency, ω_1. Similarly, if the initial conditions are such that $A_1 = B_1 = 0$, then both x_1 and x_2 will oscillate at the same frequency, ω_2. So a particular set of initial condition can generate motion at only one frequency, either ω_1 or ω_2.

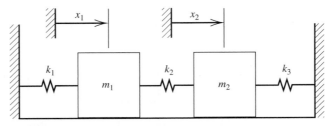

FIGURE 6.3-1 An undamped two-mass system.

Suppose that the initial velocities are zero. Then differentiating Equations 6.3-3 and 6.3-5 and applying the zero initial velocities gives $A_1 = C_1 = 0$. Applying the initial displacement to Equations 6.3-3 and 6.3-5 gives

$$x_1(0) = B_1 + D_1$$

$$x_2(0) = r_1 B_1 + r_2 D_1$$

Solve for B_1 and D_1:

$$B_1 = \frac{r_2 x_1(0) - x_2(0)}{r_2 - r_1}$$

$$D_1 = \frac{x_2(0) - r_1 x_1(0)}{r_2 - r_1}$$

For the special initial conditions where $x_2(0) = r_1 x_1(0)$, we obtain $B_1 = x_1(0)$ and $D_1 = 0$, as long as $r_1 \neq r_2$. Thus

$$x_1(t) = x_1(0) \sin \omega_1 t$$

$$x_2(t) = r_1 x_1(0) \sin \omega_1 t$$

Both masses oscillate in phase at the frequency ω_1.

Similarly, if $x_2(0) = r_2 x_1(0)$, then $B_1 = 0$, $D_1 = x_1(0)$, and both masses oscillate in phase at the frequency ω_2.

Equations 6.3-3 and 6.3-5 show that the motion of each mass is the sum of two basic motions, each of which is associated with a particular natural frequency. A *mode* is a fundamental motion of the system associated with one pair of its characteristic roots. The constants r_1 and r_2 are called the *mode ratios*. Thus the displacement in each mode is harmonic with a fixed amplitude. Therefore the velocity in each mode is harmonic with the same frequency as the displacements.

The mode concept enables us to understand the behavior of complex systems. One application will be treated in Section 6.6, where modes are used to understand the dynamics of a vehicle suspension. In Chapter 8 we show how modes help us to analyze systems having more than two degrees of freedom.

The previous analysis was needed to develop the mode concept. Computing the mode ratios in practice, however, can be done more easily, as shown in the following example.

EXAMPLE 6.3-1
Modes of Two Masses in Translation

Find and interpret the mode ratios for the system shown in Figure 6.3-1, with $k_1 = k_2 = k_3 = k$, for two cases: **(a)** $m_1 = m_2 = m$ and **(b)** $m_1 = m$, $m_2 = 2m$.

Solution

The equations of motion for the system are

$$m_1 \ddot{x}_1 = -k x_1 - k(x_1 - x_2)$$

$$m_2 \ddot{x}_2 = k(x_1 - x_2) - k x_2$$

Substitute $x_1(t) = A_1 e^{st}$ and $x_2(t) = A_2 e^{st}$ into the above differential equations, cancel the e^{st} terms, and collect the coefficients of A_1 and A_2 to obtain

$$(m_1 s^2 + 2k)A_1 - kA_2 = 0 \tag{1}$$

$$-kA_1 + (m_2 s^2 + 2k)A_2 = 0 \tag{2}$$

In order to have nonzero solutions for A_1 and A_2, the determinant of the above equations must be zero. Thus

$$\begin{vmatrix} m_1 s^2 + 2k & -k \\ -k & m_2 s^2 + 2k \end{vmatrix} = 0$$

Expanding this determinant gives

$$(m_1 s^2 + 2k)(m_2 s^2 + 2k) - k^2 = 0 \tag{3}$$

or

$$m_1 m_2 s^4 + (2m_1 k + 2m_2 k)s^2 + 3k^2 = 0$$

This polynomial has four roots because it is fourth order. We can solve it for s^2 using the quadratic formula because it is quadratic in s^2 (there is no s term or s^3 term). To see why this is true, let $u = s^2$. Then the above equation becomes

$$m_1 m_2 u^2 + (2m_1 k + 2m_2 k)u + 3k^2 = 0 \tag{4}$$

The mode ratio can be found from either Equation 1 or Equation 2. Choosing the former, we obtain

$$\frac{A_2}{A_1} = \frac{m_1 s^2 + 2k}{k} \tag{5}$$

The mode ratio A_2/A_1 can be thought of as the ratio of the amplitudes of x_2 and x_1 in that mode.

(a) For this case, $m_1 = m_2 = m$, and Equation 4 becomes

$$m^2 u^2 + 4mku + 3k^2 = 0$$

From the quadratic formula,

$$u = \frac{-4mk \pm \sqrt{16m^2 k^2 - 12m^2 k^2}}{2m^2} = -\frac{k}{m}, -3\frac{k}{m}$$

Because $s = \sqrt{u}$, we have $s = \pm i\sqrt{k/m}$ and $s = \pm i\sqrt{3k/m}$. Thus the two modal frequencies are $\omega_1 = \sqrt{k/m}$ and $\omega_2 = \sqrt{3k/m}$. From Equation 5, the mode ratios are

$$\frac{A_2}{A_1} = \frac{ms^2 + 2k}{k} = \frac{-m\dfrac{k}{m} + 2k}{k} = 1$$

for mode 1 and

$$\frac{A_2}{A_1} = \frac{ms^2 + 2k}{k} = \frac{-3m\dfrac{k}{m} + 2k}{k} = -1$$

for mode 2.

Thus in mode 1 the masses move in the same direction with the same amplitude. This oscillation has a frequency of $\omega_1 = \sqrt{k/m}$. In mode 2 the masses move in the opposite direction but with the same amplitude. This oscillation has a higher frequency of $\omega_2 = \sqrt{3k/m}$.

Figure 6.3-2 is a modal diagram that illustrates the mode shapes. Figure 6.3-2a is the first mode, for which $A_2 = A_1$. Figure 6.3-2b illustrates the second mode, for which $A_2 = -A_1$ and for which a *node* exists. A node is a point at which no motion occurs. The node for the second mode is located halfway between the two masses. At this point the spring does not move.

The specific motion depends on the initial conditions and, in general, is a combination of both modes. If the masses are initially displaced an equal distance in the same direction and then released from rest, only the first mode will be stimulated. Only the second mode will be stimulated if the masses are initially displaced an equal distance but in opposite directions.

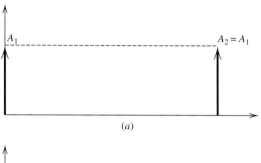

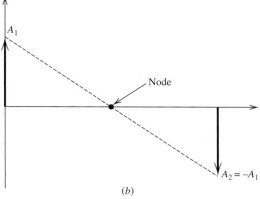

FIGURE 6.3-2 Modal diagrams for part (a) of Example 6.3-1 for the case where $m_1 = m_2$. (a) First mode. (b) Second mode.

(b) For this case, $m_1 = m$ and $m_2 = 2m$, and Equation 4 becomes

$$2m^2u^2 + 6mku + 3k^2 = 0$$

From the quadratic formula, we obtain $u = -0.63397k/m$ and $u = -2.366k/m$. Because $s = \sqrt{u}$, we have $s = \pm i\sqrt{0.63397k/m}$ and $s = \pm i\sqrt{2.366k/m}$. Thus the two modal frequencies are $\omega_1 = \sqrt{0.63397k/m} = 0.796\sqrt{k/m}$ and $\omega_2 = \sqrt{2.366k/m} = 1.538\sqrt{k/m}$. From Equation 5, the mode ratios are

$$\frac{A_2}{A_1} = \frac{ms^2 + 2k}{k} = \frac{-0.63397m\dfrac{k}{m} + 2k}{k} = 1.366$$

for mode 1, and

$$\frac{A_2}{A_1} = \frac{ms^2 + 2k}{k} = \frac{-2.366m\dfrac{k}{m} + 2k}{k} = -0.366$$

for mode 2.

Thus in mode 1 the masses move in the same direction with the amplitude of mass m_2 equal to 1.366 times the amplitude of mass m_1. This oscillation has a frequency of $\omega_1 = 0.796\sqrt{k/m}$. In mode 2 the masses move in the opposite direction with amplitude of mass m_2 equal to 0.366 times the amplitude of mass m_1. This oscillation has a higher frequency of $\omega_2 = 1.538\sqrt{k/m}$.

To stimulate the first mode, displace mass m_2 1.366 times the initial displacement of mass m_1, in the same direction. To stimulate the second mode, displace mass m_2 0.366 times the initial displacement of mass m_1, but in the opposite direction.

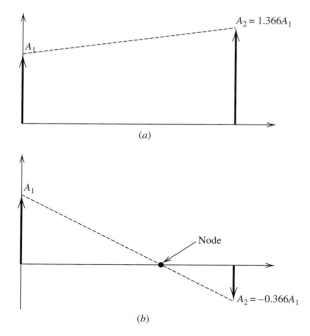

FIGURE 6.3-3 Modal diagrams for part (b) of Example 6.3-1 for the case where $m_2 = 2m_1$. (a) First mode. (b) Second mode.

Figure 6.3-3 is the modal diagram for the case where $m_1 = m$ and $m_2 = 2m$. Figure 6.3-3a is the first mode, for which $A_2 = 1.366A_1$. There is no node point in this mode because the masses are always moving in the same direction. Figure 6.3-3b illustrates the second mode, for which $A_2 = -0.366A_1$ and for which a node exists. The node for the second mode is located 73.2% of the distance from m_1 to m_2. At this point the spring does not move. ∎

In practice, we usually need not solve for the mode amplitudes in terms of the initial conditions because sufficient insight for design purposes can be gained from the mode ratios alone. The vehicle suspension example in Section 6.6 will illustrate how the mode ratios are used.

Modes and Damping

When there is no damping in a vibratory system, the model can be arranged so that the mode ratios are real and are ratios of displacements only and thus are easy to interpret.

However, when damping is present in the model, the characteristic roots are complex numbers and so are the mode ratios. This makes the analysis more difficult and the results harder to interpret. This is one reason damping is often neglected when making a modal analysis of a vibratory system. If the damping is slight, the characteristic roots and modes will be almost the same as those of the undamped model. Even if the damping is not small, the insight gained from the undamped analysis is often quite useful for design purposes.

Modes, State-Variable Models, and Stability

Modes can be used to explain some unusual results seen earlier. If an equilibrium of a linear model is stable, then it not possible to find a set of initial conditions for which

$x(t) \to \infty$. However, if the equilibrium is unstable, there might still be certain initial conditions that result in the system returning to equilibrium $[x(t) \to 0]$.

For example, the equation $\ddot{x} - x = 0$ has the roots $s = \pm 1$, and thus it represents an unstable system. However, if the initial conditions are $x(0) = 1$, $\dot{x}(0) = -1$, then the free response is $x(t) = e^{-t}$, which certainly approaches zero as $t \to \infty$. Note that the exponential e^t corresponding to the root at $s = +1$ does not appear in the response because of the special nature of the initial conditions.

We will now explore this situation further to improve our understanding of dynamic system behavior. An alternative form of the model $\ddot{x} - x = 0$ is

$$\dot{x}_1 = x_2 \tag{6.3-6}$$

$$\dot{x}_2 = x_1 \tag{6.3-7}$$

where $x_1 = x$ and $x_2 = \dot{x}$. To find the free response, substitute $x_1(t) = A_1 e^{st}$ and $x_2(t) = A_2 e^{st}$ into the differential equations. This substitution gives, after dividing through by e^{st} and collecting all terms on the left,

$$sA_1 - A_2 = 0 \tag{6.3-8}$$

$$sA_2 - A_1 = 0 \tag{6.3-9}$$

Because the right-hand sides of these equations are zero, the only way they can give a nonzero solution for A_1 and A_2 is if their determinant is zero. Thus we require that

$$\begin{vmatrix} s & -1 \\ -1 & s \end{vmatrix} = s^2 - 1 = 0$$

which is the model's characteristic equation.

The next step is to solve the characteristic equation for the roots, which are $s = \pm 1$. To distinguish the coefficients belonging to each root, we number them as A_{1i} and A_{2i}, where $i = 1, 2$ denotes the first and the second root. If we substitute the first root, $s = +1$, into either Equation 6.3-8 or Equation 6.3-9, we obtain $A_{21} = A_{11}$.

Substitution of the second root, $s = -1$, gives $A_{22} = -A_{12}$. This process gives four coefficients, two of which are related to the other two.

The superposition principle states that the free response is

$$x_1(t) = A_{11}e^t + A_{12}e^{-t} \tag{6.3-10}$$

$$x_2(t) = A_{21}e^t + A_{22}e^{-t} \tag{6.3-11}$$

But $A_{21} = A_{11}$ and $A_{22} = -A_{12}$. Thus Equation 6.3-8 can be written as

$$x_2(t) = A_{11}e^t - A_{12}e^{-t} \tag{6.3-12}$$

The solutions provided by Equations 6.3-10 and 6.3-11 can be expressed in vector form as

$$\begin{bmatrix} x_1(t) \\ x_2(t) \end{bmatrix} = \begin{bmatrix} A_{11}e^t \\ A_{11}e^t \end{bmatrix} + \begin{bmatrix} A_{12}e^{-t} \\ -A_{12}e^{-t} \end{bmatrix}$$

or

$$\mathbf{x}(t) = \begin{bmatrix} x_1(t) \\ x_2(t) \end{bmatrix} = A_{11}e^t \begin{bmatrix} 1 \\ 1 \end{bmatrix} + A_{12}e^{-t} \begin{bmatrix} 1 \\ -1 \end{bmatrix} \tag{6.3-13}$$

The response consists of two modes. Each mode has a magnitude, which is a function of time, and a vector that shows how the state variables are related to one another for that mode. For example, the magnitude of the first mode is $A_{11}e^t$ and that of the second mode is $A_{12}e^{-t}$. Thus the second mode eventually disappears, while the first mode grows with time. The vector $\begin{bmatrix} 1 & 1 \end{bmatrix}^T$ corresponding to the first mode shows that $x_1 = x_2$ in that mode. In the second mode, the modal vector is $\begin{bmatrix} 1 & -1 \end{bmatrix}^T$, and thus $x_1 = -x_2$ in that mode. The free response is the sum of both modes.

Effect of the Initial Conditions

Evaluating Equation 6.3-13 at $t = 0$, we obtain

$$\mathbf{x}(0) = \begin{bmatrix} x_1(0) \\ x_2(0) \end{bmatrix} = A_{11} \begin{bmatrix} 1 \\ 1 \end{bmatrix} + A_{12} \begin{bmatrix} 1 \\ -1 \end{bmatrix} \tag{6.3-14}$$

The initial conditions determine the values of A_{11} and A_{12}. If the initial conditions are such that $A_{11} = 0$, then the first mode does not appear in the free response at all. The initial conditions that cause this mode to be absent from the response can be found by setting $A_{11} = 0$ in Equation 6.3-14. This gives

$$\mathbf{x}(0) = A_{12} \begin{bmatrix} 1 \\ -1 \end{bmatrix}$$

which can only be satisfied if $\mathbf{x}(0)$ is proportional to the modal vector $\begin{bmatrix} 1 & -1 \end{bmatrix}^T$; that is, if $x_1(0) = -x_2(0)$.

Similarly, the initial conditions that cause the second mode to be absent can be found by setting $A_{12} = 0$ in Equation 6.3-14. This gives

$$\mathbf{x}(0) = A_{11} \begin{bmatrix} 1 \\ 1 \end{bmatrix}$$

which can be satisfied only if $\mathbf{x}(0)$ is proportional to the modal vector $\begin{bmatrix} 1 & 1 \end{bmatrix}^T$; that is, if $x_1(0) = x_2(0)$.

Thus any initial conditions satisfying $x_1(0) = -x_2(0)$ will suppress the unstable first mode, and only the second mode will appear. Because the second mode is stable, $\mathbf{x}(t)$ will approach zero as time increases.

This method and its associated concepts apply to a state-variable model of any order. The free response is the sum of the modes, and a mode will be absent from the response if the initial condition vector is proportional to the modal vector for that mode.

The following example demonstrates another physical application of the mode concept.

EXAMPLE 6.3-2

Free Response of a Fluid Coupling

The fluid coupling shown in Figure 6.3-4a can be modeled as two inertia elements connected by a rotational damper, as in Figure 6.3-4b.

$$I_1\dot{\omega}_1 = T_1 - c(\omega_1 - \omega_2) \tag{1}$$

$$I_2\dot{\omega}_2 = T_2 + c(\omega_1 - \omega_2) \tag{2}$$

where c represents the rotational damping due to the fluid forces acting on the inertias. Determine its modes and the form of its free response.

Solution

The free response does not depend on the inputs, so we can set the inputs T_1 and T_2 to zero. Make the following substitutions into Equations 1 and 2: $\omega_1 = A_1e^{st}$ and $\omega_2 = A_2e^{st}$. After collecting all terms on the left and factoring out the exponential terms, we obtain

$$(I_1s + c)A_1 - cA_2 = 0 \tag{3}$$

$$-cA_1 + (I_2s + c)A_2 = 0 \tag{4}$$

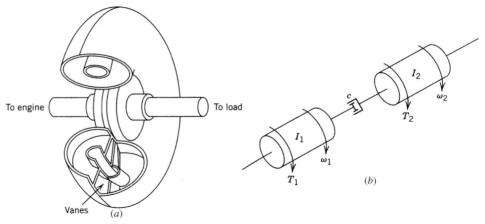

FIGURE 6.3-4 (*a*) A fluid coupling. (*b*) Fluid coupling representation with two inertia elements.

Set the determinant equal to zero to obtain the model's characteristic equation:

$$\begin{vmatrix} (I_1 s + c) & -c \\ -c & (I_2 s + c) \end{vmatrix} = (I_1 s + c)(I_2 s + c) - c^2 = 0$$

This equation has the two roots:

$$s_1 = 0 \tag{5}$$

$$s_2 = -\frac{c(I_1 + I_2)}{I_1 I_2} \tag{6}$$

The system is neutrally stable because $s_1 = 0$. Let the second root be denoted by $s_2 = -b$, where

$$b = \frac{c(I_1 + I_2)}{I_1 I_2} \tag{7}$$

The modes are found in the usual way.

For the root $s_1 = 0$, the first mode ratio is

$$\frac{A_2}{A_1} = 1 \tag{8}$$

For $s_2 = -b$, the second mode ratio is

$$\frac{A_2}{A_1} = -\frac{I_1}{I_2} \tag{9}$$

In the first mode the inertias rotate at the same speed (because $A_2 = A_1$), and the speeds do not change (because $e^{s_1 t} = e^0 = 1$). This is a rigid-body mode because the entire system rotates as if it were a rigid body. In the second mode the inertias rotate in the opposite direction at different speeds (because A_1 and A_2 have opposite signs and different magnitudes). The speed ratio is $\omega_2 / \omega_1 = -I_1 / I_2$. In this mode, the speeds will decay to zero with a time constant equal to $1/b$ (because $s = -b$).

For general initial conditions, the free response of the system consists of the sum of a mode 1 response (in which the inertias rotate together at the same speed) and a mode 2 response (in which the inertias rotate in opposite directions and at different speeds). The proportion of each mode's response is determined by the initial conditions. If $\omega_1(0) = \omega_2(0)$, only the first mode will be present, and the inertias will rotate at the same constant speed. If $\omega_2(0) = -(I_1/I_2)\omega_1(0)$, then only the second mode will appear in the free response. ∎

6.4 HARMONIC RESPONSE

As with single-degree-of-freedom systems, systems with two degrees of freedom can also display resonance when the forcing frequency is close to the resonant frequency. One difference is that there can be two resonant frequencies when there are two degrees of freedom. Another difference is that for some forcing frequencies, one of the masses can be motionless at steady state. This observation forms the basis for the design of vibration absorbers, which are discussed in Chapter 7.

Frequency Transfer Functions

Consider the system shown in Figure 6.4-1, in which a forcing function $f_1(t)$ acts on the mass m_1. The equations of motion are

$$m_1\ddot{x}_1 = f_1(t) - k_1 x_1 - k_2(x_1 - x_2) \tag{6.4-1}$$

$$m_2\ddot{x}_1 = -k_3 x_2 + k_2(x_1 - x_2) \tag{6.4-2}$$

The transfer functions are found by applying the Laplace transform to both equations, using zero initial conditions:

$$\left(m_1 s^2 + k_1 + k_2\right)X_1(s) - k_2 X_2(s) = F_1(s) \tag{6.4-3}$$

$$-k_2 X_1(s) + \left(m_2 s^2 + k_2 + k_3\right)X_2(s) = 0 \tag{6.4-4}$$

These can be solved for $X_1(s)$ and $X_2(s)$, using Cramer's method, as follows:

$$\frac{X_1(s)}{F_1(s)} = \frac{m_2 s^2 + k_2 + k_3}{D(s)} \tag{6.4-5}$$

$$\frac{X_2(s)}{F_1(s)} = \frac{k_2}{D(s)} \tag{6.4-6}$$

$$D(s) = \left(m_1 s^2 + k_1 + k_2\right)\left(m_2 s^2 + k_2 + k_3\right) - k_2^2 \tag{6.4-7}$$

Suppose the forcing function is harmonic. Then $f_1(t) = F_1 \sin \omega t$ and the steady-state displacement response will also be harmonic with the same frequency ω. We can obtain the frequency transfer functions by substituting $s = i\omega$. This gives

$$\frac{X_1(i\omega)}{F_1(i\omega)} = \frac{k_2 + k_3 - m_2\omega^2}{D(i\omega)} \tag{6.4-8}$$

$$\frac{X_2(i\omega)}{F_1(i\omega)} = \frac{k_2}{D(i\omega)} \tag{6.4-9}$$

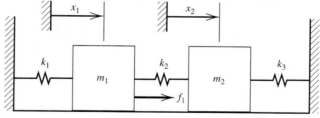

FIGURE 6.4-1 A two-mass system.

$$D(i\omega) = \left(k_1 + k_2 - m_1\omega^2\right)\left(k_2 + k_3 - m_2\omega^2\right) - k_2^2 \tag{6.4-10}$$

Let the two roots of $D(i\omega) = 0$ be denoted ω_1 and ω_2. Then $D(i\omega)$ can be expressed as

$$D(i\omega) = m_1 m_2 \left(\omega^2 - \omega_1^2\right)\left(\omega^2 - \omega_2^2\right) \tag{6.4-11}$$

Thus Equations 6.4-8 and 6.4-9 become

$$\frac{X_1(i\omega)}{F_1(i\omega)} = \frac{k_2 + k_3 - m_2\omega^2}{m_1 m_2 \left(\omega^2 - \omega_1^2\right)\left(\omega^2 - \omega_2^2\right)} \tag{6.4-12}$$

$$\frac{X_2(i\omega)}{F_1(i\omega)} = \frac{k_2}{m_1 m_2 \left(\omega^2 - \omega_1^2\right)\left(\omega^2 - \omega_2^2\right)} \tag{6.4-13}$$

Both frequency transfer functions become infinite when either $\omega = \omega_1$ or $\omega = \omega_2$. These two frequencies are the resonant frequencies.

The numerator of Equation 6.4-12 is zero if $k_2 + k_3 - m_2\omega^2 = 0$; that is, when the forcing frequency is

$$\omega = \sqrt{\frac{k_2 + k_3}{m_2}} \tag{6.4-14}$$

When the forcing function has this frequency, the steady-state motion of mass m_1 will be zero. The insight from this remarkable and nonintuitive result can be used to design systems to reduce unwanted motion in a vibrating system. This is the basis for the design of a vibration absorber, which is discussed in Chapter 7.

Figure 6.4-2 shows the normalized frequency response plots for the case where $m_1 = m_2 = m$, $k_1 = k_2 = k_3 = k$, and $f_1 = f$. For this case, Equations 6.4-12 and 6.4-13 can be expressed as

$$\frac{kX_1(i\omega)}{F(i\omega)} = \frac{2 - r^2}{(r^2 - 1)(r^2 - 3)}$$

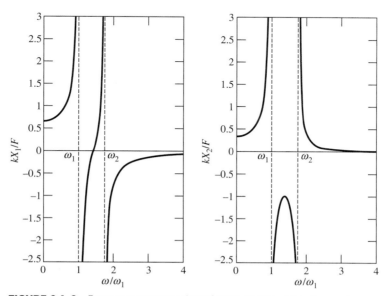

FIGURE 6.4-2   Frequency response plots of a two-degree-of-freedom system.

$$\frac{kX_2(i\omega)}{F(i\omega)} = \frac{1}{(r^2 - 1)(r^2 - 3)}$$

where $r = \omega/\omega_1$, $\omega_1^2 = k/m$, and $\omega_2^2 = 3k/m$. These plots show that resonance occurs at $\omega = \omega_1$ and at $\omega = \omega_2$. For this system, the mass m_1 is motionless at steady state if the forcing frequency ω equals $\sqrt{2}\omega_1$.

Interpreting the Phase Plot

Frequency response plots of two-degrees-of-freedom systems are readily obtained from the transfer functions in the same manner as with one-degree-of-freedom systems. We have seen in this section how the magnitude plots aid in interpreting the system's response. The phase angle plot can also give useful information about the motion of the system relative to the input or the motion of the constituent masses relative to each other.

Consider the following two-mass suspension model whose equations of motion were derived in Section 6.1. The equations of motion are

$$m_1\ddot{x}_1 = c(\dot{x}_2 - \dot{x}_1) + k_1(x_2 - x_1)$$

$$m_2\ddot{x}_2 = -c(\dot{x}_2 - \dot{x}_1) - k_1(x_2 - x_1) + k_2(y - x_2)$$

We will use the following numerical values: $m_1 = 250$ kg, $m_2 = 40$ kg, $k_1 = 1.5 \times 10^4$ N/m, $k_2 = 1.5 \times 10^5$ N/m, and $c = 1917$ N·s/m. The transfer functions can be derived in the usual way with Cramer's rule. They are

$$\frac{X_1(s)}{Y(s)} = \frac{(0.2876s + 2.25)10^5}{s^4 + 55.6s^3 + 4185s^2 + 43.14 \times 10^4 s + 2.75 \times 10^5}$$

$$\frac{X_2(s)}{Y(s)} = \frac{(0.0375s^2 + 0.2876s + 2.25)10^5}{s^4 + 55.6s^3 + 4185s^2 + 4.314 \times 10^4 s + 2.75 \times 10^5}$$

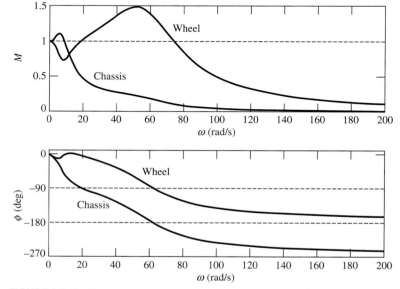

FIGURE 6.4-3 Frequency response plots of a two-mass suspension model.

The frequency response plots (obtained with the MATLAB `bode` function, as described in Section 4.9) are shown in Figure 6.4-3. The input frequency ω in rad/s depends on the spatial period P (in meters) of the road surface and on the vehicle speed v in m/s, as $\omega = 2\pi v/P$. The amplitude plot shows that the chassis amplitude is at most 10% greater than the road motion at low frequencies, where $M \approx 1.1$. At higher frequencies the chassis motion is well isolated from the road motion because $M < 0.5$ at those frequencies. On the other hand, the wheel amplitude is 50% greater than the road motion when $\omega \approx 50$ rad/s.

The relative motion of the chassis, wheel, and road can be determined from the phase plot. For example, at $\omega = 60$, the wheel motion and the chassis motion lag behind the road's sinusoidal motion by 90° and 180°, respectively. Thus at this frequency the chassis motion is in the opposite direction of the ground motion, and it lags the wheel motion by 90°. Such information is useful to designers for understanding the suspension behavior.

6.5 GENERAL FORCED RESPONSE

In the previous section we saw how to use the transfer function to obtain the steady-state response to a harmonic forcing function. We now consider the problem of obtaining the total response to other types of forcing functions, such as step inputs and ramps. For such applications, the transfer function and the Laplace transform method provide the most systematic way of analyzing the response.

The mathematics of obtaining the response of systems with two degrees of freedom is much more tedious than for systems with one degree of freedom. Because of this, we need to carefully plan our analysis, keeping in mind the ultimate purpose of the analysis. If, for example, all we need is a plot of the response, then it makes no sense to spend the time and effort to obtain a closed-form solution; a numerical solution, such as that obtained with MATLAB, is the best choice. We will see that it can be difficult to obtain formulas that we can use to assess the transient or the total response in terms of a parameter, such as a stiffness or a damping constant. Often we must be satisfied with a numerical solution.

Frequently in engineering we first analyze the behavior of a system using a model with only one degree of freedom to keep the mathematics at a manageable level. This permits a quick preliminary analysis to determine the general response characteristics. Then, if these characteristics are acceptable, we develop and analyze a more accurate model, one having more degrees of freedom. This model can provide more detailed information about the response. We will now illustrate this approach.

Suspension System Transfer Functions

For example, consider the quarter-car model of a suspension system shown in Figure 6.5-1. This single-mass model ignores the dynamics of the tire-wheel assembly and focuses on the dynamics of the main body. Figure 6.5-2 shows a more detailed model that describes the tire-wheel dynamics. The stiffness k in Figure 6.5-1 is the series combination of the suspension stiffness k_1 and the tire stiffness k_2, so $k = k_1 k_2/(k_1 + k_2)$. The transfer function is

$$\frac{X(s)}{Y(s)} = \frac{k}{ms^2 + cs + k} \tag{6.5-1}$$

Although this does not model the wheel dynamics, nevertheless, because it has only one degree of freedom, it is easier to analyze than the model to follow.

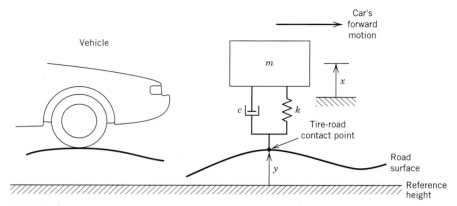

FIGURE 6.5-1 A one-degree of freedom model of a suspension.

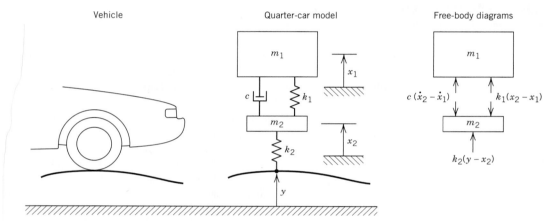

FIGURE 6.5-2 A two-degrees-of-freedom model of a suspension and its free-body diagrams.

From Example 6.1-2, the two-mass model of the suspension system shown in Figure 6.5-2 is

$$m_1\ddot{x}_1 + c\dot{x}_1 + k_1 x_1 - c\dot{x}_2 - k_1 x_2 = 0 \tag{6.5-2}$$

$$m_2\ddot{x}_2 + c\dot{x}_2 + (k_1 + k_2)x_2 - c\dot{x}_1 - k_1 x_1 = k_2 y \tag{6.5-3}$$

Applying the Laplace transform to these equations with zero initial conditions gives

$$(m_1 s^2 + cs + k_1)X_1(s) - (cs + k_1)X_2(s) = 0$$

$$-(cs + k_1)X_2(s) + (m_2 s^2 + cs + k_1 + k_2)X_1(s) = k_2 Y(s)$$

The transfer functions can be found with Cramer's rule. They are

$$\frac{X_1(s)}{Y(s)} = \frac{k_2(cs + k_1)}{m_1 m_2 s^4 + (m_1 + m_2)cs^3 + (k_3 m_1 + k_1 m_2)s^2 + k_2 cs + k_1 k_2} \tag{6.5-4}$$

$$\frac{X_2(s)}{Y(s)} = \frac{k_2(m_1 s^2 + cs + k_1)}{m_1 m_2 s^4 + (m_1 + m_2)cs^3 + (k_3 m_1 + k_1 m_2)s^2 + k_2 cs + k_1 k_2} \tag{6.5-5}$$

Step Response

To analyze this system, we must next select an appropriate input function $y(t)$ to model the effects of road surface variation. In doing this, we must balance the desire for an accurate description of the actual road surface variation versus the need to keep the required mathematics from becoming too cumbersome. For this reason, the step function is often used to model the effects of a sudden change in the road surface, although, in reality, even the most severe bump does not change height instantaneously.

Let us consider some realistic values. Suppose that $k_1 = 15\ 000$ N/m, $k_2 = 150\ 000$ N/m, $m_1 = 250$ kg, and $m_2 = 30$ kg; also suppose that we need to select a value for the damping constant c so that the maximum percent overshoot of mass m_1 due to a step input is no more than 10%.

For the single-mass model, $m = 250$ kg and $k = 13\ 636$ N/m, and its transfer function is

$$\frac{X(s)}{Y(s)} = \frac{13\ 636}{250s^2 + cs + 13\ 636} \tag{6.5-6}$$

Using a *unit*-step function for $y(t)$ (the response can be scaled later by the actual change in the height of the road surface), we can compute the maximum percent overshoot M_p from Table 5.3-2 as follows:

$$M_p = 100e^{-\pi\zeta/\sqrt{1-\zeta^2}}$$

For the single-mass model,

$$\zeta = \frac{c}{2\sqrt{250(13\ 636)}} = \frac{c}{3693}$$

For $M_p = 10\%$, from Table 5.3-2 we obtain

$$A = \ln\frac{100}{10} = 2.3026$$

$$\zeta = \frac{A}{\sqrt{\pi^2 + A^2}} = 0.5912$$

This gives $c = 3693(0.5912) = 2183$ N $\cdot$ s/m. This is a preliminary estimate of the required value of c.

We now use this estimate in the two-mass model. Substituting the given values into the transfer function for $X_1(s)$, we obtain, after simplifying,

$$\frac{X_1(s)}{Y(s)} = \frac{1500(cs + 1.5 \times 10^4)}{75s^4 + 2.8cs^3 + 4.17 \times 10^5 s^2 + 1500cs + 2.25 \times 10^7} \tag{6.5-7}$$

Because the damping c is the design parameter here, we would like to be able to analyze the response for an arbitrary value of c. Unfortunately, there is no clear or easy answer because we have no simple formulas like the formula for M_p to use for this model, which is fourth order. From the transfer function, we can see that increasing c can increase the overshoot because c appears in the numerator dynamics term cs. The characteristic roots are also affected by c.

We now have to decide how to obtain the response $x_1(t)$ for a step input. The choices are the following:

1. Obtain the closed-form expression for the response by using, for example, the Laplace transform and evaluating the coefficients of the partial-fraction expansion by hand using algebra.

2. Obtain the closed-form expression for the response by using the MATLAB `residue` function to evaluate the coefficients in the partial-fraction expansion.

3. Obtain a plot of the response directly by using a numerical solution method such as the MATLAB `step` function.

The third method is the easiest if all we need to do is to evaluate the overshoot, but it has the disadvantage of not giving us a formula to use to gain insight into how c affects the response. We can examine a number of plots for various values of c, but this might not give us the needed insight. The first method requires some tedious algebra to find the coefficients and is impractical if c is left as a parameter. The second method requires that a specific value of c be used, but it will show the components of the response due to each of the characteristic roots.

A plot of the step response of the fourth-order model using $c = 2183$ can be found with MATLAB, as shown in Section 6.7. For this value of c, the overshoot is $M_p = 33\%$, which is much larger than desired. If we make c smaller, say $c = 500$, the response of the fourth-order model is almost identical to that of the second-order model, but the overshoot is 65%. Making c larger than 2183 results in a smaller overshoot until $c > 2(2183) = 4366$. Then the overshoot increases for increasing c. The table below shows the results, along with the 2% settling time t_s.

c	M_p	t_s
2183	33%	0.969
1.5(2183)	24%	0.62
2(2183)	23%	0.691
2.5(2183)	27%	0.712
3(2183)	32%	0.686

We can continue this process until we determine the best value for c. Using only the values shown in the table, the value that gives the smallest overshoot is $c = 2(2183) = 4366$. We conclude that we cannot obtain an overshoot of less than 23% by selecting only c.

We can obtain the closed-form expression for the response from the partial-fraction expansion. We can use the MATLAB `residue` function to evaluate the roots and the coefficients in the expansion, as shown in Section 6.7. The response for $c = 4366$ is

$$x_1(t) = -0.0621e^{-123.56t} + 2.4712e^{-17.41t}\sin(14.94t - 2.5286) + 0.4838e^{-4.61t} + 1$$

From the amplitudes and the time constants, we see that the roots $s = -17.41 \pm 14.94i$ dominate the response. The response is shown in Figure 6.5-3.

Alternatives to the Step Input

Consider alternatives to the step input for a model of a bump. Three such alternatives are shown in Figure 6.5-4. All three have Laplace transforms much more complicated than $1/s$. For example, the function in Figure 6.5-4a has the transform

$$Y(s) = \frac{1}{s^2}\left(1 - 2e^{-s} + 2e^{-2s}\right)$$

If we were to use this in the solution for $X_1(s)$ from Equation 6.5-7, the algebra required to obtain the partial-fraction expansion would be very cumbersome, and the `residue` function cannot be used with this transform because it contains the transcendental (nonpolynomial) expressions e^{-s} and e^{-2s}. The transforms of the functions in Figures 6.5-4a, b also contain transcendental terms.

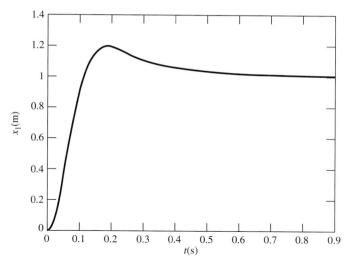

FIGURE 6.5-3   Suspension system response.

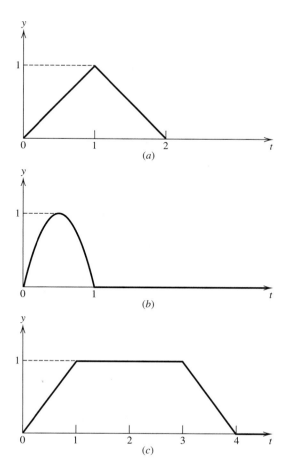

FIGURE 6.5-4   Bump models.

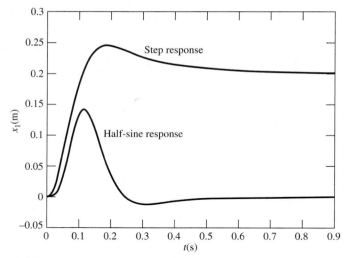

FIGURE 6.5-5 Response to a half-sine pulse.

MATLAB and Simulink can be used to compute the response to inputs such as those shown in Figure 6.5-4. Figure 6.5-5 compares the step response with the response due to a half-sine input that represents a bump 0.2 m high and 1.4 m long, for a vehicle traveling at 50 km/hr, for $c = 4366$. Although overshoot is not defined for a nonstep input, we can see from the plot that the peak response of the mass m_1 is approximately 0.14 m, which is 70% of the height of the bump (0.2 m). The plot was obtained using the MATLAB `lsim` function which was discussed in Section 5.7. In Section 6.8 we show how the trapezoidal input shown in Figure 6.5-4c can be handled in Simulink.

6.6 NODES AND MODAL RESPONSE

Sometimes we need not solve for the mode amplitudes in terms of the initial conditions because sufficient insight for design purposes can be gained from the mode ratios alone. The following vehicle suspension example illustrates how the mode ratios can be used.

EXAMPLE 6.6-1
Modes of a
Bounce-Pitch
Suspension Model

A representation of a car's suspension suitable for modeling the bounce and pitch motions is shown in Figure 6.6-1a, which is a side view of the vehicle's body showing the front and rear suspensions. Assume that the car's motion is constrained to a vertical translation x of the mass center and rotation θ about a single axis that is perpendicular to the page. The body's mass is m and its moment of inertia about the mass center is I_G. As usual, x and θ are the displacements from the equilibrium position corresponding to $y_1 = y_2 = 0$. The displacements $y_1(t)$ and $y_2(t)$ can be found if you know the vehicle's speed and the road surface profile.

(**a**) Assume that x and θ are small, and derive the equations of motion for the bounce motion x and pitch motion θ.

(**b**) Neglect the damping in the system, and determine its modes.

(**c**) Evaluate the results for the case: $k_1 = 1.6 \times 10^4$ N/m, $k_2 = 2.23 \times 10^4$ N/m, $L_1 = 1.46$ m, $L_2 = 1.1$ m, $m = 730$ kg, and $I_G = 1356$ kg · m².

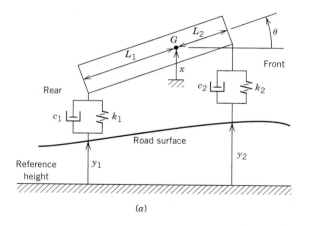

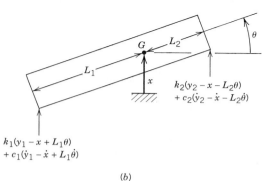

FIGURE 6.6-1 (*a*) A model of bounce-pitch motion in a suspension. (*b*) Free-body diagram.

Solution

(**a**) The small displacement assumption implies that the suspension forces are nearly perpendicular to the centerline of the mass m and thus are nearly vertical (see Figure 6.6-1*b*). To draw the free body-diagrams, assume arbitrarily that

$$y_1 > x - L_1\theta$$
$$\dot{y}_1 > \dot{x} - L_1\dot{\theta}$$
$$y_2 > x + L_2\theta$$
$$\dot{y}_2 > \dot{x} + L_2\dot{\theta}$$

We obtain the following moment equation from the free-body diagram:

$$I_G\ddot{\theta} = -c_1(\dot{y}_1 - \dot{x} + L_1\dot{\theta})L_1 - k_1(y_1 - x + L_1\theta)L_1 + c_2(\dot{y}_2 - \dot{x} - L_2\dot{\theta})L_2 + k_2(y_2 - x - L_2\theta)L_2$$

Rearranging gives

$$I_G\ddot{\theta} + (c_2L_2^2 + c_1L_1^2)\dot{\theta} + (k_1L_1^2 + k_2L_2^2)\theta - (c_1L_1 - c_2L_2)\dot{x} - (k_1L_1 - k_2L_2)x$$
$$= -c_1L_1\dot{y}_1 + c_2L_2\dot{y}_2 - k_1L_1y_1 + k_2L_2y_2 \tag{1}$$

The force equation in the vertical direction gives

$$m\ddot{x} = c_1(\dot{y}_1 - \dot{x} + L_1\dot{\theta}) + k_1(y_1 - x + L_1\theta) + c_2(\dot{y}_2 - \dot{x} - L_2\dot{\theta}) + k_2(y_2 - x - L_2\theta)$$

Rearranging gives

$$m\ddot{x} + (c_1 + c_2)\dot{x} + (k_1 + k_2)x - (c_1L_1 - c_2L_2)\dot{\theta} - (k_1L_1 - k_2L_2)\theta$$
$$= c_1\dot{y}_1 + c_2\dot{y}_2 + k_1y_1 + k_2y_2 \tag{2}$$

This is the desired model.

(**b**) Setting the damping coefficients to zero in the model gives

$$m\ddot{x} + (k_1 + k_2)x = (k_1L_1 - k_2L_2)\theta \tag{3}$$

$$I_G\ddot{\theta} + (k_1L_1^2 + k_2L_2^2)\theta = (k_1L_1 - k_2L_2)x \tag{4}$$

where the input $y(t)$ has been set to zero because we need only the free response to find the modes. Make the substitutions $x(t) = A_1e^{st}$ and $\theta(t) = A_2e^{st}$ into the above differential equations, cancel the e^{st} terms, and collect the coefficients of A_1 and A_2 to obtain

$$(ms^2 + k_1 + k_2)A_1 - (k_1L_1 - k_2L_2)A_2 = 0 \tag{5}$$

$$-(k_1L_1 - k_2L_2)A_1 + (I_Gs^2 + k_1L_1^2 + k_2L_2^2)A_2 = 0 \tag{6}$$

In order to have nonzero solutions for A_1 and A_2, the determinant of the above equations must be zero. Thus

$$\begin{vmatrix} ms^2 + k_1 + k_2 & -(k_1L_1 - k_2L_2) \\ -(k_1L_1 - k_2L_2) & I_Gs^2 + k_1L_1^2 + k_2L_2^2 \end{vmatrix} = 0$$

Expanding this determinant gives

$$(ms^2 + k_1 + k_2)(I_Gs^2 + k_1L_1^2 + k_2L_2^2) - (k_1L_1 - k_2L_2)^2 = 0 \tag{7}$$

or

$$mI_Gs^4 + [m(k_1L_1^2 + k_2L_2^2) + I_G(k_1 + k_2)]s^2 + k_1k_2(L_1 + L_2)^2 = 0 \tag{8}$$

This polynomial has four roots because it is fourth order. We can solve it for s^2 using the quadratic formula because it is quadratic in s^2 (there is no s term or s^3 term).

The mode ratio can be found from either Equation 5 or Equation 6. Choosing the former, we obtain

$$\frac{A_1}{A_2} = \frac{k_1L_1 - k_2L_2}{ms^2 + k_1 + k_2} \tag{9}$$

The mode ratio A_1/A_2 can be thought of as the ratio of the amplitudes of x and θ in that mode. From Figure 6.6-2a, we find that

$$\tan\theta = \frac{x}{D}$$

and for small angles θ,

$$D \approx \frac{x}{\theta} = \frac{A_1}{A_2} \tag{10}$$

The distance D locates a point called a node or "motion center" at which no motion occurs (that is, a passenger located at a node would not move if the vehicle were moving in the corresponding modal motion). Thus there are two nodes, one for each mode.

Equation 9 shows that A_1, the amplitude of x, will be zero if

$$k_1L_1 - k_2L_2 = 0 \tag{11}$$

In this case, Equation 10 shows that $D = 0$. Thus no coupling exists between the bounce motion x and the pitch motion θ, and the node for each mode is at the mass center. As we will discuss shortly, this condition will result in poor ride quality. Note also, that if Equation 11 is satisfied, the characteristic equation, Equation 7, can be factored as follows:

$$(ms^2 + k_1 + k_2)(I_Gs^2 + k_1L_1^2 + k_2L_2^2) = 0$$

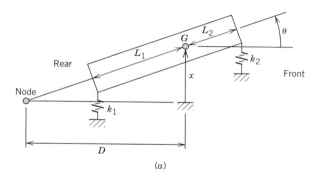

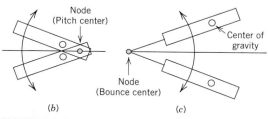

FIGURE 6.6-2 (*a*) Locations of nodes for the bounce-pitch model. (*b*) Pitch center. (*c*) Bounce center.

or

$$ms^2 + k_1 + k_2 = 0 \tag{12}$$

$$I_G s^2 + k_1 L_1^2 + k_2 L_2^2 = 0 \tag{13}$$

Each of these equations has a pair of imaginary roots. Thus both modes are oscillatory.

(**c**) Dividing the characteristic equation, Equation 7, by mI_G and using the given values for the constants, we obtain

$$s^4 + 97.5159s^2 + 2361.9 = 0$$

The quadratic formula gives the roots $s^2 = -52.687, -44.829$. Thus the four characteristic roots are

$$s = \pm 7.259i, \quad \pm 6.695i$$

These correspond to frequencies of 1.16 Hz and 1.07 Hz.

The mode ratio (Equation 9) becomes

$$\frac{A_1}{A_2} = \frac{x}{\theta} = \frac{-1.603}{s^2 + 52.5} \tag{14}$$

For mode 1 ($s^2 = -52.6174$),

$$\frac{x}{\theta} = 8.57 \, \text{m}$$

Thus the node is located 8.57 m *behind* the mass center. Because this node is so far from the mass center, the motion in this mode is predominantly a bounce motion, and this node is called the "bounce center" (see Figure 6.6-2*c*).

For mode 2 ($s^2 = -44.829$),

$$\frac{x}{\theta} = -0.209 \, \text{m}$$

Thus the node is located 0.209 m *ahead* of the mass center (because $x/\theta < 0$). Because this node is close to the mass center, the motion in this mode is predominantly a pitching motion, and this node is called the "pitch center" (see Figure 6.6-2*b*). ∎

The calculation of the node locations can be very sensitive to slight variations in the parameter values. Rearrange the mode ratio expression, Equation 9 in Example 6.6-1, to obtain

$$\frac{A_1}{A_2} = \frac{x}{\theta} = \frac{k_1 L_1 - k_2 L_2}{m} \frac{1}{\omega_c^2 - \omega^2}$$

where we have defined $s^2 = -\omega^2$, and ω_c as the natural frequency of a concentrated mass m attached to two parallel springs k_1 and k_2:

$$\omega_c = \sqrt{\frac{k_1 + k_2}{m}}$$

The value of $s^2 = \omega^2$ is obtained from the characteristic equation. If ω^2 is close to ω_c^2, then the denominator of x/θ will be very small and x/θ will be very large. In addition, any slight change in the values of ω_c^2 or ω^2 due to changes in the parameters k_1, k_2, L_1, L_2, m, or I_G will cause large changes in the value of x/θ and thus large changes in the node location. In Example 6.6-1, $\omega_c^2 = 52.5$. One of the roots gives $s^2 = -52.6$, which is close to -52.5. Thus we can expect large changes in the node location if slightly different parameter values are used. The other root gives $s^2 = -45$, which is not as close to -52.5; thus the corresponding node location will not be as sensitive.

If it is important to have a reliable prediction for the node location, the parameters must be selected—usually by trial and error—so that ω^2 is not close to ω_c^2.

Ride Isolation and Suspension Design

The primary purpose of a road vehicle's suspension is to keep the tires in contact with the road. The secondary purpose is to give the passengers a smooth ride. Proper design of the wheel suspension provides the primary source of isolation from the road roughness. The quarter-car model is the basis for the suspension system's design. Referring to Figure 6.6-3, the *ride rate* k_e is the effective stiffness of the suspension spring and the tire stiffness, neglecting the wheel mass m_2. Thus $k_e = k_1 k_2 / (k_1 + k_2)$. If $m_2 << m_1$, the suspension system's undamped natural frequency is approximately $\omega_n = \sqrt{k_e/m_1}$ and its damping ratio is $\zeta = c_1 / 2\sqrt{k_e m_1}$. Most modern cars have a suspension damping ratio ζ between 0.2 and 0.4, and thus their damped natural frequency ω_d is very close to ω_n.

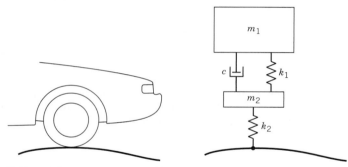

FIGURE 6.6-3 Suspension model illustrating ride rate.

Bounce motion is produced by road surfaces having short wavelengths.

Pitch motion is produced by longer wavelengths, and usually appears only at lower speeds.

FIGURE 6.6-4   Bounce and pitch motions.

The suspension's *static deflection* due to the weight $m_1 g$ is $\Delta = m_1 g/k_e$, and thus $\Delta = m_1 g/m_1 \omega_n^2 = g/\omega_n^2$. A natural frequency of 1 Hz ($\omega_n = 2\pi$ rad/sec) is considered to be a design optimum for highway vehicles, and this corresponds to a static deflection of $\Delta = 0.25$ m. For a vehicle weighing 1.4×10^4 N, this deflection requires an effective stiffness of $k_e = m_1 g/\Delta = 0.25(1.4 \times 10^4)/0.25 = 1.4 \times 10^4$ N/m. With a tire stiffness of $k_2 = 2.1 \times 10^5$ N/m, the suspension stiffness must be $k_1 = k_e k_2/(k_2 - k_e) = 1.5 \times 10^4$ N/m ([Gillespie, 1992]).

Ride Quality

In the 1930s Maurice Olley discovered that the best road vehicle ride is obtained by placing the bounce center behind the rear axle and the pitch center near the front axle (because bounce motion is less annoying than pitch motion). This is accomplished by designing the front suspension to have a lower natural frequency than the rear suspension. Since road excitation affects the front wheels first, a lower front suspension natural frequency will tend to induce bounce rather than pitch. See Figure 6.6-4.

Olley developed other rules of thumb, which are still used today ([Gillespie, 1992]):

1. If the front-rear weight distribution is approximately equal, the front suspension should have a 30% lower ride rate than the rear, or the spring center should be at least 6.5% of the wheelbase behind the center of gravity. (The *spring center* is the point at which an applied vertical force will produce only vertical displacement.) When the spring center is at the center of gravity, there is no coupling between the bounce and pitch motion, and poor ride quality results because the total motion is irregular.

2. The bounce frequency should be within 20% of the pitch frequency. Otherwise, the superposition of the two modes will result in irregular, annoying motions. The roll frequency should be close to the bounce and pitch frequencies for the same reason.

3. The bounce and pitch frequencies should be kept as low as possible (≤ 1.3 Hz if possible), consistent with any upper limit on static deflection. This is because the most severe road acceleration inputs occur at higher frequencies.

6.7 ROOT LOCUS ANALYSIS WITH MATLAB

The MATLAB Control System toolbox provides several useful commands for producing root locus plots and for extracting information from them. The basic function is `rlocus(sys)`, which displays the root locus. Additional commands are available for enhancing the plot and for obtaining root locations and gain values from the plot.

The function `rlocus(sys)` computes and displays the root locus plot for the equation

$$D(s) + KN(s) = 0 \qquad (6.7\text{-}1)$$

where `sys` corresponds to the transfer function $N(s)/D(s)$ and the parameter K varies from 0 through a large positive value determined by MATLAB. The order of $N(s)$ must be no greater than the order of $D(s)$. It is recommended that the highest coefficients in $N(s)$ and $D(s)$ both be *unity* so that the values of K displayed by MATLAB will be interpreted correctly.

MATLAB refers to the parameter K as the "gain." The starting points of the roots corresponding to $K = 0$ are called the "poles" and are marked with a cross ($\times$). These are the roots of $D(s) = 0$. The finite termination points that the roots approach as $K \rightarrow \infty$ are called the "zeros" and are marked with a small circle ($\bigcirc$). These are the roots of $N(s) = 0$.

You should use equal scaling on the real and imaginary axes so that circular root loci will display as circles and so that the angle θ associated with the damping ratio, $\zeta = \cos\theta$, may be properly interpreted. This may be done by including the command `axis equal` after the `rlocus` function.

The plot shown in Figure 3.5-7 corresponds to the characteristic equation $2s^2 + cs + 8 = 0$. This may be put into the standard form (Equation 6.7-1) by rewriting it as

$$s^2 + 4 + \frac{c}{2}s = 0$$

where $K = c/2$, $D(s) = s^2 + 4$, and $N(s) = s$. The poles are $s = \pm i2$. The zero is $s = 0$. To obtain this root locus plot, the session is

```
≫sys1 = tf([1,0],[1,0,4]);
≫rlocus(sys1), axis equal
```

Note that the numerator must be expressed as `[1, 0]` because the numerator polynomial is actually $s + 0$. The result is shown in Figure 6.7-1.

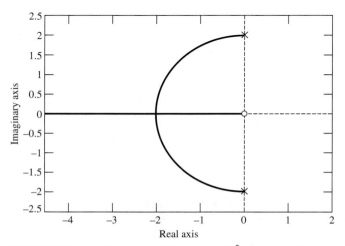

FIGURE 6.7-1 Root locus plot of the equation $2s^2 + cs + 8 = 0$ as c varies through positive values.

Placing a (ζ, ω_n) Grid

The `sgrid` command superimposes an s-plane grid of constant ζ and constant ω_n lines on the root locus plot. This grid is useful for locating roots that satisfy performance specifications stated in terms of ζ and ω_n. The alternate syntax `sgrid(zeta,omega)` superimposes an s-plane grid of constant ζ and constant ω_n lines on the root locus plot, using the values contained in the vectors `zeta` and `omega`.

Suppose we want to see if any dominant roots of the following equation have damping ratios in the range $0.5 \le \zeta \le 0.707$ and undamped natural frequencies in the range $0.5 \le \omega_n \le 0.75$:

$$s^3 + 3s^2 + 2s + K = 0$$

The required session is

```
≫sys2 = tf(1,[1,3,2,0])
≫rlocus(sys2),sgrid([0.5,0.707],[0.5,0.75]),axis equal
```

You may use empty brackets if you want to omit either lines of constant ζ or lines of constant ω_n. For example, to see just the damping ratio lines corresponding to $\zeta = 0.5$ and $\zeta = 0.707$, you would use the form `sgrid([0.5,0.707],[])`.

Obtaining Information from the Plot

Once the plot is displayed, you can pick a point from this plot to find the gain K required to achieve ζ and ω_n values in the desired range. To do this, you can use the `rlocfind` function discussed later in this section, or you can place the cursor on the plot at the desired point. Left-click to display the root value (although labeled "pole," this is not the same as the starting points marked by $\times$). The gain value, the damping ratio (labeled "damping"), the percent overshoot, and the undamped natural frequency ω_n (labeled "frequency") are also displayed. You can move the cursor along the plot and view the updated values. Right-clicking anywhere within the plot area brings up a menu that enables the plot to be edited.

Note that the percent overshoot value displayed on the screen with this method is computed from the formula for M_p given in Table 5.3-2, and thus it does *not* include the effects of numerator dynamics on the overshoot.

You can use this method to see how sensitive the root location is to a change in the value of the gain K. Sometimes it is inadvisable to design a critically damped system, which corresponds to two or more equal, real roots, because such designs can be very sensitive to parameter variation and parameter uncertainty. Near such roots, a slight change in the gain K can cause a large change in the root location.

An alternate syntax to obtain the root values for specified values of K is `r = rlocus(sys,K)`. This returns the root locations in the vector `r` corresponding to the user-specified vector of gain values K. The matrix `r` has `length(K)` columns, and its *j*th column contains the roots for the gain value `K(j)`. No plot is displayed.

MATLAB provides the `rlocfind` function to obtain information about roots and gain values from the plot on the screen. The syntax `[K,r] = rlocfind(sys)` enables you to use a cursor to obtain the value of the gain K corresponding to a specified point on a root locus plot. The vector `r` contains the roots corresponding to this gain value. The advantage of using `rlocfind` is that it returns *all* the roots for a given value of K, whereas left-clicking on the curve shows only some of the roots. The `rlocfind` function, which must follow the `rlocus` command, generates a cursor on the screen and waits for the user to press the mouse button after positioning the cursor over the desired point on the locus. Once the button is pressed, you will see on the screen the coordinates of the selected point, the gain value at that point, the roots closest to that point, and the other roots that correspond to the gain value.

For example, the following session displays the root locus with a grid line corresponding to $\zeta = 0.707$. This enables us to use the cursor to find the dominant roots that have a damping ratio of $\zeta = 0.707$ and the corresponding gain value K.

```
≫sys3 = tf(1,[1,3,2,0])
≫rlocus(sys3),axis equal,sgrid(0.707,[]),...
    [K,r]=rlocfind(sys3)
```

The dominant roots selected this way will not have a damping ratio of $\zeta = 0.707$ exactly, because you cannot position the cursor exactly at the intersection of the locus and the $\zeta = 0.707$ line. To reduce this inaccuracy, you may enlarge the plot by enabling the Edit Plot button and clicking the magnifying glass icon on the menu bar of the figure window.

Another syntax is $[K,r] = rlocfind(sys,p)$, which computes a root locus gain K for each desired root location specified in the vector p (or a gain for which one of the closed-loop roots is near the desired location). The jth entry in the vector K is the computed gain for the root location r(j). The jth column of the matrix r contains the resulting closed-loop roots.

The real power of the root locus method only becomes apparent when it is applied to a model higher than second order, as in the following example.

EXAMPLE 6.7-1
Adjusting Damping in a Suspension

Consider the two-mass suspension model developed in Example 6.1-2 and shown again in Figure 6.7-2. The equations of motion are

$$m_1\ddot{x}_1 = c(\dot{x}_2 - \dot{x}_1) + k_1(x_2 - x_1)$$

$$m_2\ddot{x}_2 = -c(\dot{x}_2 - \dot{x}_1) - k_1(x_2 - x_1) + k_2(y - x_2)$$

We will use the following numerical values: $m_1 = 250$ kg, $m_2 = 40$ kg, $k_1 = 15\ 000$ N/m, and $k_2 = 150\ 000$ N/m.

(a) Use the root locus plot to determine the value of the damping c required to give a dominant-root pair having a damping ratio of $\zeta = 0.707$.

(b) Using the value of c found in part (a), obtain a plot of the unit-step response.

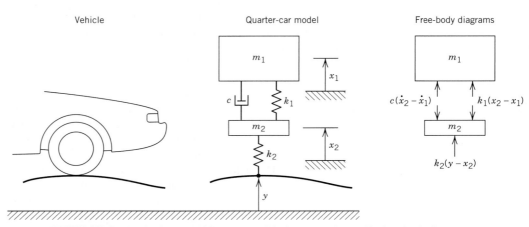

FIGURE 6.7-2 A two-degrees-of-freedom model of a suspension and its free-body diagrams.

Solution (a) The transfer functions for $X_1(s)$ and $X_2(s)$ with $Y(s)$ as the input were obtained in Section 6.5 and are

$$\frac{X_1(s)}{Y(s)} = \frac{k_2(cs + k_1)}{D(s)}$$

$$\frac{X_2(s)}{Y(s)} = \frac{k_2(m_1 s^2 + cs + k_1)}{D_3(s)}$$

where

$$D(s) = (m_1 s^2 + cs + k_1)(m_2 s^2 + cs + k_1 + k_2) - (cs + k_1)^2$$

Now substitute the given values of m_1, m_2, k_1, and k_2 into $D(s)$ to obtain the root locus equation:

$$D(s) = (250s^2 + cs + 15\ 000)(40s^2 + cs + 165\ 000) - (cs + 15\ 000)^2$$

Factoring out c and rearranging into the form of Equation 6.7-1 so that the highest coefficients of $N(s)$ and $D(s)$ are unity, we obtain

$$s^4 + 4185s^2 + 2.25 \times 10^5 + K(s^3 + 775.86s) = 0$$

where the root locus parameter K is related to c as $K = 0.029c$. The MATLAB session for obtaining the root locus plot is as follows.

```
≫sys4 = tf([1, 0, 775.86,0],[1,0,4185,0,2.25*10^5])
≫rlocus(sys4),axis equal,sgrid(0.707,[])
```

The root locus plot is shown in Figure 6.7-3.

Using the magnifying glass to expand the plot, and clicking on the intersection of the dominant-root path with the $\zeta = 0.707$ line, we find that the dominant-root pair is $s = -5.53 \pm 5.53j$, with a gain of $K = 55.6$. Using the command r = rlocus(sys4,K), we find that the other roots are at $s = -22.2 \pm 56j$, and thus the first root pair is dominant. So to achieve a dominant-root pair having a damping ratio of $\zeta = 0.707$, we must set c to be $c = K/0.029 = 1917\ \text{N}\cdot\text{s/m}$.

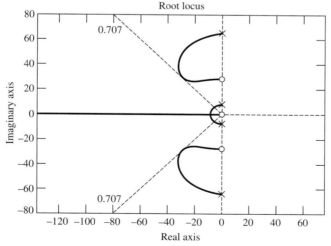

FIGURE 6.7-3 Root locus plot for Example 6.7-1.

(b) Using $c = 1917$ and the other parameter values, we find that

$$D_3(s) = 250(40)\left[s^4 + 4185s^2 + 2.25 \times 10^5 + K(s^3 + 775.86s)\right]$$

$$\frac{X_1(s)}{Y(s)} = \frac{150\ 000(1917s + 15\ 000)}{D_3(s)}$$

$$\frac{X_2(s)}{Y(s)} = \frac{150\ 000(250s^2 + 1917s + 15\ 000)}{D_3(s)}$$

We continue the previous session as follows.

```
≫[num,den] = tfdata(sys4,'v');
≫D3 = 250*40*(den+55.6*num);
≫sysx1 = tf(150000*[1917,15000],D3);
≫sysx2 = tf(150000*[250,1917,15000],D3);
≫step(sysx1,'-',sysx2,'--')
```

The plot is shown in Figure 6.7-4. By right-clicking on the plot and selecting the Characteristics menu, we find that for $x_1(t)$, the peak response is 11% at $t = 0.357$ s and the settling time is 0.655 s. For $x_2(t)$, the peak response is 21% at $t = 0.051$ s and the settling time is 0.494 s.

The response characteristics computed from the dominant-root pair $s = -5.53 \pm 5.53j$ using the formulas given in Table 5.3-2 for the second-order model without numerator dynamics are a peak response of 4.3% at $t = 0.568$ s and a settling time of 0.723 s. The difference between these results is due to the fact that the present model is fourth order and has numerator dynamics. ∎

The MATLAB Control System toolbox also provides two graphical interfaces, rltool, which enables you to do more detailed analysis of root locus plots for system design, and ltiview, which simplifies the application of the LTI solvers. For example, with rltool you can obtain the step response of frequency response plots once you have used the root locus plot to set the parameter values.

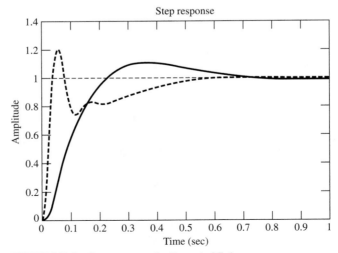

FIGURE 6.7-4 Step response for Example 6.7-1.

6.8 RESPONSE ANALYSIS WITH MATLAB

The Laplace transform and the transfer function play a central role in the methods of this chapter. They provide a systematic way of obtaining the free response, the characteristic roots, stability properties, system modes, and the forced response. In this section we show how you can use MATLAB to obtain the transfer function from the equations of motion and how to implement the methods of this chapter. We will use MATLAB to do some of the tedious algebra and calculations.

As our example we will use the system shown in Figure 6.8-1, which represents many types of systems having two degrees of freedom. It is possible also to include a damper between masses m_1 and m_2 and an external force acting on mass m_1, but we omit these to show the method more clearly. The equations of motion are

$$m_1\ddot{x}_1 + c\dot{x}_1 + (k_1 + k_2)x_1 - k_2x_2 = 0$$

$$m_2\ddot{x}_2 + k_2x_2 - k_2x_1 = f(t)$$

Applying the Laplace transform to these equations and collecting terms gives:

$$(m_1s^2 + cs + k_1 + k_2)X_1(s) - k_2X_2(s) = m_1x_1(0)s + m_1\dot{x}_1(0) + cx_1(0)$$

$$-k_2X_1(s) + (m_2s^2 + k_2)X_2(s) = F(s) + m_2x_2(0)s + m_2\dot{x}_2(0)$$

Define the following terms, which are due to the initial conditions:

$$I_1(s) = m_1x_1(0)s + m_1\dot{x}_1(0) + cx_1(0)$$

$$I_2(s) = m_2x_2(0)s + m_2\dot{x}_2(0)$$

Then we have

$$(m_1s^2 + cs + k_1 + k_2)X_1(s) - k_2X_2(s) = I_1(s)$$

$$-k_2X_1(s) + (m_2s^2 + k_2)X_2(s) = F(s) + I_2(s)$$

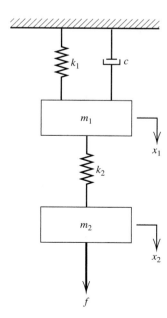

FIGURE 6.8-1 A system having two degrees of freedom.

These equations can be expressed in matrix form as

$$\begin{bmatrix} (m_1 s^2 + cs + k_1 + k_2) & -k_2 \\ -k_2 & (m_2 s^2 + k_2) \end{bmatrix} \begin{bmatrix} X_1(s) \\ X_2(s) \end{bmatrix} = \begin{bmatrix} I_1(s) \\ I_2(s) \end{bmatrix} + \begin{bmatrix} 0 \\ F(s) \end{bmatrix} \qquad (6.8\text{-}1)$$

Obtaining the Free Response

To find the free response, first set $F(s) = 0$ and solve for $X_1(s)$ and $X_2(s)$ with Cramer's method. The results are

$$X_1(s) = \frac{D_1(s)}{D(s)} \qquad (6.8\text{-}2)$$

$$X_2(s) = \frac{D_2(s)}{D(s)} \qquad (6.8\text{-}3)$$

where

$$\begin{aligned} D(s) &= \begin{vmatrix} (m_1 s^2 + cs + k_1 + k_2) & -k_2 \\ -k_2 & (m_2 s^2 + k_2) \end{vmatrix} \\ &= (m_1 s^2 + cs + k_1 + k_2)(m_2 s^2 + k_2) - k_2^2 \end{aligned} \qquad (6.8\text{-}4)$$

$$D_1(s) = \begin{vmatrix} I_1(s) & -k_2 \\ I_2(s) & (m_2 s^2 + k_2) \end{vmatrix} = I_1(s)(m_2 s^2 + k_2) + k_2 I_2(s) \qquad (6.8\text{-}5)$$

$$\begin{aligned} D_2(s) &= \begin{vmatrix} (m_1 s^2 + cs + k_1 + k_2) & I_1(s) \\ -k_2 & I_2(s) \end{vmatrix} \\ &= (m_1 s^2 + cs + k_1 + k_2) I_2(s) + k_2 I_1(s) \end{aligned} \qquad (6.8\text{-}6)$$

Because the transform of a unit-step input is $1/s$, we can use the `step` function to obtain a plot of the free response by multiplying the numerators of $X_1(s)$ and $X_2(s)$ by s. We denote the artificial transfer function thus obtained by $G_1(s)$ and $G_2(s)$. Thus

$$G_1(s) = \frac{sD_1(s)}{D(s)} \qquad G_2(s) = \frac{sD_2(s)}{D(s)} \qquad (6.8\text{-}7)$$

A MATLAB program file to perform the rest of the algebra is shown below. Recall that, in MATLAB, to add or subtract polynomials whose coefficients are represented by arrays, we must create arrays of the same size (see Appendix A for a discussion of polynomial algebra in MATLAB).

```
% Program TwoDOF.m
% The following coefficients must be given values
% before using this program.
% m1, m2 = masses
% c, k1, k1 = damping and spring constants
% x10, x20 = initial displacements
% x1d0, x2d0 = initial velocities
% Form the determinants.
D = conv([m1, c, k1+k2],[m2, 0, k2]) - [0, 0, 0, 0, k2^2];
I1 = [m1*x10, m1*x1d0 + c*x10];
I2 = [m2*x20, m2*x2d0];
D1 = conv(I1, [m2, 0, k2]) + k2*[0, 0, I2];
D2 = conv(I2, [m1, c, k1+k2]) + k2*[0, 0, I1];
chroots = roots(D)
```

```
G1 = tf(conv([1, 0], D1), D);
G2 = tf(conv([1, 0], D2), D);
[x1free, t1free] = step(G1);
[x2free, t2free] = step(G2);
plot(t1free, x1free, t2free, x2free)
```

This program solves for the characteristic roots (in the array `chroots`) and produces plots of the free responses of x_1 and x_2. The closed-form expression for the free response may be found with the `residue` function. This method is discussed later in this section.

Obtaining the Transfer Functions and Forced Response

The transfer functions $X_1(s)/F(s)$ and $X_2(s)/F(s)$ can be found from Equation 6.8-1 by setting $I_1(s) = I_2(s) = 0$ and solving for the ratios as follows:

$$X_1(s) = \frac{D_3(s)}{D(s)} \tag{6.8-8}$$

$$X_2(s) = \frac{D_4(s)}{D(s)} \tag{6.8-9}$$

where

$$D_3(s) = \begin{vmatrix} 0 & -k_2 \\ F(s) & (m_2 s^2 + k_2) \end{vmatrix} = k_2 F(s) \tag{6.8-10}$$

$$D_4(s) = \begin{vmatrix} (m_1 s^2 + cs + k_1 + k_2) & 0 \\ -k_2 & F(s) \end{vmatrix}$$
$$= (m_1 s^2 + cs + k_1 + k_2) F(s) \tag{6.8-11}$$

So the transfer functions are

$$T_1(s) = \frac{X_1(s)}{F(s)} = \frac{k_2}{(m_1 s^2 + cs + k_1 + k_2)(m_2 s^2 + k_2) - k_2^2} \tag{6.8-12}$$

$$T_2(s) = \frac{X_2(s)}{F(s)} = \frac{m_1 s^2 + cs + k_1 + k_2}{(m_1 s^2 + cs + k_1 + k_2)(m_2 s^2 + k_2) - k_2^2} \tag{6.8-13}$$

These can be constructed in MATLAB and the unit-step response plotted by appending the following lines to the program `TwoDOF.m`.

```
T1 = tf(k2, D);
T2 = tf([m1, c, k1+k2], D);
[x1step, t1step] = step(G1);
[x2step, t2step] = step(G2);
plot(t1step, x1step, t2step, x2step)
```

The closed-form expression for the step response may be found with the `residue` function. This method is discussed later in this section.

The response to other forcing functions may be computed with the `lsim` function. For example, to compute the response of x_1 to $f(t) = 10 \sin 2t$ for $0 \le t \le 15$, you would replace the last three lines with

```
tsine = [0:0.001:15];
f = 10*sin(2*tsine);
x1sine = lsim(T1, f, tsine];
plot(tsine, x1sine)
```

The frequency response plots can be obtained with the `bode(sys)` function and its alternate form `[mag, phase, w] = bode(sys)`, where `sys` may be either `T1` or `T2`. Remember, the phase angle `phase` is returned in degrees while the frequency w is returned as a radian frequency.

Obtaining the Mode Ratios

The mode ratios can be obtained as follows. Referring to Equation 6.8-1, with $I_1(s)$, $I_2(s)$, and $F(s)$ set to 0, the ratio $X_2(s)/X_1(s)$ is found to be

$$\frac{X_2(s)}{X_1(s)} = \frac{m_1 s^2 + cs + k_1 + k_2}{k_2} \qquad (6.8\text{-}14)$$

Therefore the mode ratio $R = X_2/X_1$ can be found with MATLAB by appending the following line to the program `TwoDOF.m`.

```
R = polyval([m1, c, k1+k2]/k2, chroots)
```

If $c = 0$, you will obtain real values for R; otherwise, R will be complex.

EXAMPLE 6.8-1
Test Case

For the system shown in Figure 6.8-1, suppose that $m_1 = 4$, $m_2 = 1$, $k_1 = 18$, $k_2 = 6$.
(a) Suppose there is no damping ($c = 0$) and that the initial conditions are $x_1(0) = 1$, $x_2(0) = \dot{x}_1(0) = \dot{x}_2(0) = 0$. Obtain the free response; the transfer functions $X_1(s)/F(s)$, $X_1(s)/F(s)$; the characteristic roots; and the mode ratios.

(b) Suppose that $c = 5$ and that the initial conditions are zero. Obtain the transfer functions $X_1(s)/F(s)$, $X_1(s)/F(s)$; the characteristic roots; and the unit-step response.

Solution

(a) With $c = 0$, the program `TwoDOF.m` gives the characteristic roots $s = \pm 3i$ and $s = \pm 1.7321i$. The transfer functions are

$$T_1(s) = \frac{X_1(s)}{F(s)} = \frac{6}{4s^4 + 48s^2 + 108}$$

$$T_2(s) = \frac{X_2(s)}{F(s)} = \frac{4s^2 + 24}{4s^4 + 48s^2 + 108}$$

The program gives one mode ratio for each of the four roots. The results are $R = -2, -2$, corresponding to the first root pair listed, namely, $s = \pm 3i$, and $R = 2, 2$, which corresponds to $s = \pm 1.7321i$. Thus the mode ratios are

$$R = \frac{X_2}{X_1} = -2 \qquad \text{for} \quad s = \pm 3i$$

and

$$R = \frac{X_2}{X_1} = 2 \qquad \text{for} \quad s = \pm 1.732i$$

The free response is shown in Figure 6.8-2.

(b) With $c = 5$, the program `TwoDOF.m` gives the characteristic roots $s = -0.2891 \pm 2.8831i$ and $s = -0.2891 \pm 2.8831i$. The transfer functions are

$$T_1(s) = \frac{X_1(s)}{F(s)} = \frac{6}{4s^4 + 5s^3 + 48s^2 + 30s + 108}$$

$$T_2(s) = \frac{X_2(s)}{F(s)} = \frac{4s^2 + 24}{4s^4 + 5s^3 + 48s^2 + 30s + 108}$$

The unit-step response is shown in Figure 6.8-3.   ∎

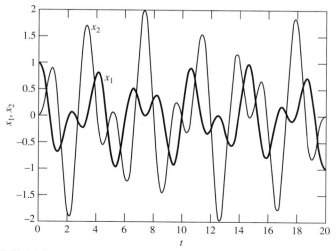

FIGURE 6.8-2 Free response for Example 6.8-1.

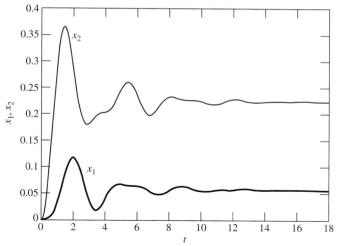

FIGURE 6.8-3 Unit-step response for Example 6.8-1.

Obtaining Expressions for the Response

The closed-form expression for the response may be found with the `residue` function. As an example, consider the step response of the suspension system application discussed in Section 6.5. The transfer function $X_1(s)/Y(s)$ from Equation 6.5-7 is

$$\frac{X_1(s)}{Y(s)} = \frac{1500(cs + 1.5 \times 10^4)}{75s^4 + 2.8cs^3 + 4.17 \times 10^5 s^2 + 1500cs + 2.25 \times 10^7}$$

We can obtain the closed-form expression for the response by using the MATLAB `residue` function to evaluate the coefficients in the partial-fraction expansion.

To do this, use the following program (see Equations 5.6-12 through 5.6-15 in Section 5.6).

```
c = 2183;
num = 1500*[c,1.5e+4];
den = [75,2.8*c,4.17e+5,1500*c,2.25e+7];
[r,p,K] = residue(num,[den,0]);
disp('The poles are')
p
disp('The amplitudes are')
B = 2*abs(r)
C = real(r);
D = imag(r);
phi = atan2(C,-D)
```

where `phi` represents the phase angle ϕ. The array `p` contains the factors (the poles) in the partial-fraction expansion. These factors are the characteristic roots $s = -123.5592$, $-17.413 \pm 14.9406i$, and $s = -4.6121$, and the factor $s = 0$, which is due to the step input.

The results for the amplitudes are $B = 0.1242, 2.4712, 2.4712, 0.9676$, and 2. The amplitudes and phase angles corresponding to the complex poles are $B = 2.4712$, $\phi = -2.5286$ rad, and $B = 2.4712$, $\phi = -0.613$, which is equivalent to -2.5286. Realizing that complex conjugate pairs are involved, we need only one of the two B, ϕ pairs.

The values of B corresponding to the real poles must be divided by 2 because of the line `B = 2*abs(r)`. Of course, we ignore the values of ϕ computed for the real poles. Thus the response is

$$x_1(t) = \frac{0.1242}{2}e^{-123.56t} + 2.4712e^{-17.41t}\sin(14.94t - 2.5286) + \frac{0.9676}{2}e^{-46.1t} + \frac{2}{2}$$
$$= 0.0621e^{-123.56t} + 2.4712e^{-17.41t}\sin(14.94t - 2.5286) + 0.4838e^{-4.61t} + 1$$

6.9 SIMULINK APPLICATIONS

As we have seen throughout our study, linear or linearized models are useful for predicting the behavior of dynamic systems because powerful analytical techniques are available for such models, especially when the inputs are relatively simple functions such as the impulse, step, ramp, and sine. Often in the design of an engineering system, however, we must eventually deal with nonlinearities in the system and with more complicated inputs such as trapezoidal functions, and this must often be done with simulation. In this section we introduce four additional Simulink elements that enable us to model a wide range of nonlinearities and input functions.

As our example, we will use the two-mass suspension model shown in Figure 6.9-1, where the spring and damper elements have the nonlinear models shown in Figures 6.9-2 and 6.9-3. These models represent a hardening spring and a degressive damper. In addition, the damper model is asymmetric. It represents a damper whose force during rebound is higher than during jounce (in order to minimize the force transmitted to the passenger compartment when the vehicle strikes a bump). The bump is represented by the trapezoidal function $y(t)$ shown in Figure 6.9-4. This function corresponds approximately to a vehicle traveling at 30 mi/hr over a 0.2-m-high road surface elevation 48 m long.

Vehicle   Quarter-car model   Free-body diagrams

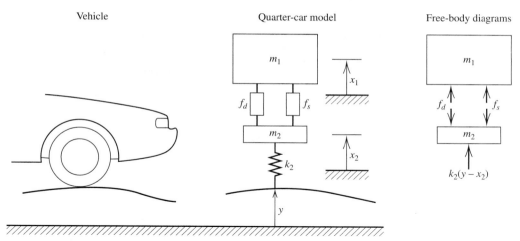

FIGURE 6.9-1 A two-degrees-of-freedom model of a nonlinear suspension and its free-body diagrams.

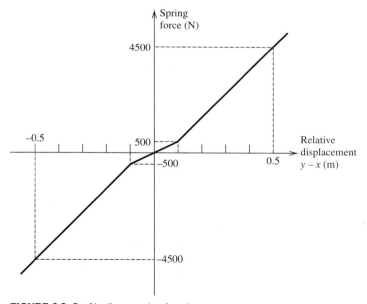

FIGURE 6.9-2 Nonlinear spring function.

The system model from Newton's law is

$$m_1\ddot{x}_1 = f_s + f_d \tag{6.9-1}$$

$$m_2\ddot{x}_2 = k_2(y - x_2) - f_s - f_d \tag{6.9-2}$$

where $m_1 = 350$ kg, $m_2 = 30$ kg, $k_2 = 20\,000$ N/m, f_s is the nonlinear spring function shown in Figure 6.9-2, and f_d is the nonlinear damper function shown in Figure 6.9-3. The corresponding simulation diagram is shown in Figure 6.9-5.

Place the Signal Builder block, then double-click on it. A plot window appears that enables you to place points to define the input function. Follow the directions in the

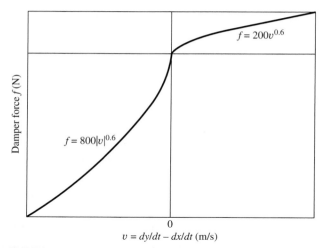

FIGURE 6.9-3 Nonlinear damping function.

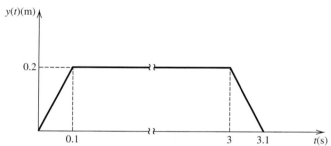

FIGURE 6.9-4 Trapezoidal model of a bump.

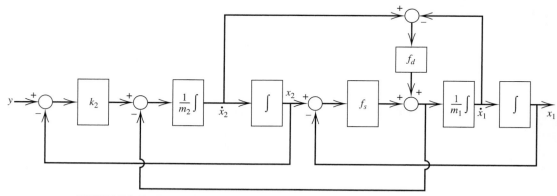

FIGURE 6.9-5 Simulation diagram for the suspension model.

window to create the function shown in Figure 6.9-4. The spring function f_s is created with the Look-Up Table block. After placing it, double-click on it and enter [-0.5,-0.1,0, 0.1,0.5] for the Vector of input values and [-4500,-500,0,500,4500] for the Vector of output values. Use the default settings for the remaining parameters.

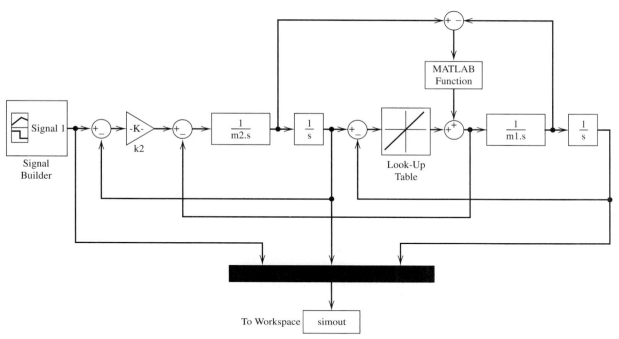

FIGURE 6.9-6   Simulink model of the suspension.

Place the two integrators as shown, and make sure the initial values are set to 0. Then place the Gain block and set its gain to 1/400. The To Workspace block will enable us to plot $x_1(t)$, $x_2(t)$ and $y(t)$ versus t in the MATLAB Command window.

We must write a user-defined function to describe the damping function shown in Figure 6.9-3. This function is as follows:

```
function f = damper(v)
if v <= 0
  f = -800*(abs(v)).^(0.6);
else
  f = 200*v.^(0.6);
end
```

Create and save this function file. After placing the MATLAB Function block, double-click on it and enter the function name `damper`. Make sure `Output dimensions` is set to `-1` and the `Output signal type` is set to `auto`.

The Fcn, MATLAB Function, Math Function, and S-Function blocks can be used to implement functions, but each has its advantages and limitations. The Fcn block can contain an expression, but its output must be a scalar and it cannot call a function file. The MATLAB Function block is slower than the Fcn block, but its output can be an array and it can call a function file. The Math Function block can produce an array output but it is limited to a single MATLAB function and cannot use an expression or call a file. The S-Function block provides more advanced features, such as the ability to use C language code.

The Simulink model when completed should look like Figure 6.9-6. You can plot the input $y(t)$, along with the responses $x_1(t)$ and $x_2(t)$, in the Command window as follows. Note how the inputs to the Mux block are numbered.

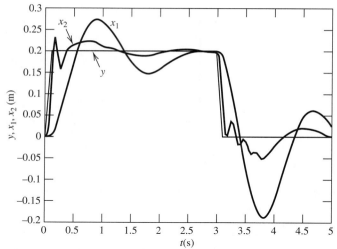

FIGURE 6.9-7 Response of the suspension model.

```
≫y = simout(:,1);
≫x2 = simout(:,2);
≫x1 = simout(:,3);
≫t = simout(:,3);
≫plot(t, y, t, x1, t, x2)
```

The result is shown in Figure 6.9-7. The maximum response of the chassis m_2 is found by typing `max(simout(:,3))` to be 0.274, and thus the maximum overshoot is $(0.274 - 0.2) = 0.074$ m, but the maximum *under*shoot is seen to be much greater, -0.187 m. It is found by typing `min(simout(:,3))`.

6.10 CHAPTER REVIEW

Many practical vibration applications require a model having more than one degree of freedom in order to describe the important features of the system response. In this chapter we considered models that have two degrees of freedom. In Chapter 8 we will consider three or more degrees of freedom.

Newton's laws and Lagrange's equations can be used to develop models of multi-degree of freedom systems. These methods were discussed in Sections 6.1 and 6.2. Two phenomena occur in systems having two or more degrees of freedom that do not occur in system with only one degree of freedom. Such systems have two natural frequencies instead of only one, and their motion is composed of modes. These modes are critical to understanding the response of such systems and were discussed in Section 6.3.

In Section 6.4 we considered the harmonic response of systems with two degrees of freedom. The transient response under harmonic forcing and the total response to a nonharmonic forcing function can be found with the help of the transfer function and the Laplace transform. Several examples were given in Section 6.5. The role of nodes in understanding modal response, with applications to suspension system design, was covered in Section 6.6.

Sections 6.7, 6.8, and 6.9 treated applications of MATLAB and Simulink to the chapter's topics. As we will see, the mathematics of models having two degrees of freedom can be quite formidable, and MATLAB can be used not only to solve the differential equations numerically but also to perform some of the algebra required to obtain closed-form solutions. Simulink is useful in applications where the stiffness and/or the damping elements are nonlinear and where the input is not a simple function of time. We analyzed a two-mass suspension model to show how Simulink can be used for such problems.

Now that you have finished this chapter, when dealing with systems having two degrees of freedom, you should be able to do the following:

1. Apply Newton's laws and Lagrange's equations to develop equations of motion for such systems.

2. Identify the modes of a system and compute its natural frequencies.

3. Analyze the harmonic response.

4. Obtain the response to a general forcing function.

5. Analyze the role of nodes in the modal response.

6. Use MATLAB and Simulink to support the methods of the chapter.

PROBLEMS

SECTION 6.1   MODELS OF TWO-DEGREES-OF-FREEDOM SYSTEMS

6.1 Derive the equations of motion for the system shown in Figure P6.1.

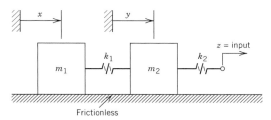

FIGURE P6.1

6.2 Derive the equations of motion for the system shown in Figure P6.2.

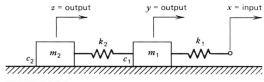

FIGURE P6.2

6.3 Derive the equations of motion for the system shown in Figure P6.3.

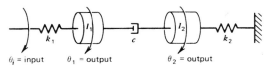

FIGURE P6.3

6.4 Refer to Figure P6.4. Obtain the equations of motion.

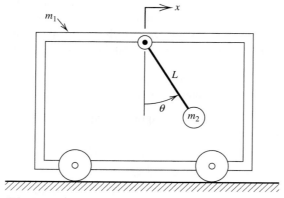

FIGURE P6.4

6.5 Refer to Figure P6.5. A cylinder rolls without slipping on a translating platform. Obtain the equations of motion for the case where $m_1 = m_2 = m$.

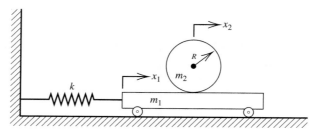

FIGURE P6.5

6.6 Refer to Figure P6.6. Suppose that $R_2 = 2R_1$ and that $m_1 = m$, $m_2 = 2m$. Treat the pulley as a solid cylinder. Obtain the equations of motion.

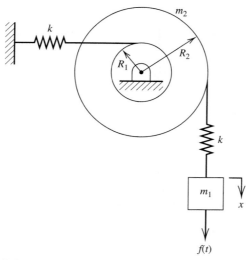

FIGURE P6.6

6.7 Refer to Figure P6.7. Suppose that $k_1 = k$, $k_2 = k_3 = 2k$, and that $m_1 = m_2 = m$. Obtain the equations of motion.

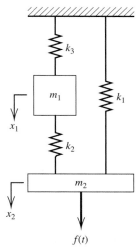

FIGURE P6.7

6.8 Refer to Figure P6.8. The mass m_1 translates without friction along the platform. The pendulum is attached to m_1 and is free to rotate. Suppose that $m_1 = m_2 = 5$ kg, $k = 400$ N/m, and $L = 0.25$ m. Assume small-angle oscillations and obtain the equations of motion.

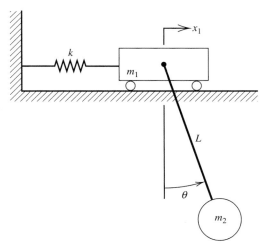

FIGURE P6.8

6.9 Refer to Figure P6.9. Obtain the equations of motion for the case where $m_1 = m$ and $m_2 = 2m$. The cylinder is solid and rolls without slipping.

6.10 Refer to Figure P6.10. The solid cylinder rolls without slipping and the platform translates without friction on the horizontal surface. Obtain the equations of motion for the case where $m_1 = m_2 = m$.

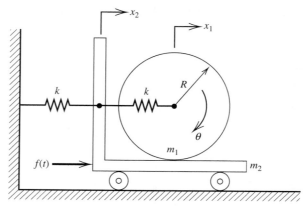

FIGURE P6.9

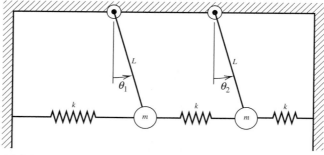

FIGURE P6.10

6.11 Refer to Figure P6.11. Obtain the equations of motion, assuming small angles.

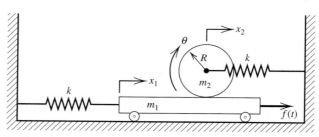

FIGURE P6.11

SECTION 6.2 LAGRANGE'S EQUATIONS

6.12 Refer to Problem 6.4 and Figure P6.4. Use Lagrange's equation to obtain the equations of motion.

6.13 Refer to Problem 6.6 and Figure P6.6. Use Lagrange's equation to obtain the equations of motion.

6.14 Refer to Problem 6.7 and Figure P6.7. Use Lagrange's equation to obtain the equations of motion.

6.15 Refer to Problem 6.8 and Figure P6.8. Use Lagrange's equation to obtain the equations of motion.

6.16 Refer to Problem 6.9 and Figure P6.9. Use Lagrange's equation to obtain the equations of motion.

6.17 Refer to Problem 6.10 and Figure P6.10. Use Lagrange's equation to obtain the equations of motion.

6.18 Refer to Problem 6.11 and Figure P6.11. Use Lagrange's equation to obtain the equations of motion.

SECTION 6.3 MODES AND SYSTEM RESPONSE

6.19 Find and interpret the mode ratios for the system shown in Figure P6.19. The masses are $m_1 = 10$ kg, $m_2 = 30$ kg. The spring constants are $k_1 = 10\ 000$ N/m and $k_2 = 20\ 000$ N/m.

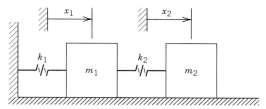

FIGURE P6.19

6.20 Find and interpret the mode ratios for the coupled pendulum system shown in Figure P6.20. Use the values $m_1 = 1$, $m_2 = 4$, $L = 5$, $d = 2$, and $k = 2$.

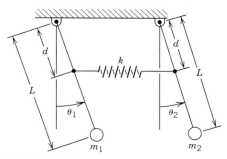

FIGURE P6.20

6.21 Find and interpret the mode ratios for the torsional system shown in Figure P6.21. Use the values $I_1 = 1$, $I_2 = 4$, $k_1 = 1$, and $k_2 = 3$.

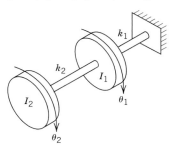

FIGURE P6.21

6.22 Find and interpret the mode ratios for the system shown in Figure P6.22.

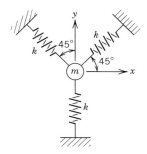

FIGURE P6.22

6.23 Refer to Problem 6.4 and Figure P6.4. Determine the expressions for the natural frequencies of the frame and pendulum system.

6.24 Refer to Problem 6.5 and Figure P6.5. Obtain the expressions for the natural frequencies and mode shapes.

6.25 Refer to Figure P6.25. The masses translate without friction. Obtain the expressions for the natural frequencies and mode shapes for the case where $m_1 = m_2 = m$.

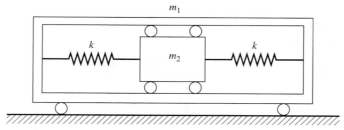

FIGURE P6.25

6.26 Refer to Problem 6.6 and Figure P6.6. Determine the expressions for the natural frequencies of the system.

6.27 Refer to Problem 6.7 and Figure P6.7. Obtain the expressions for the natural frequencies and mode shapes.

6.28 Refer to Problem 6.8 and Figure P6.8. Determine the expressions for the natural frequencies of the system.

6.29 Refer to Problem 6.9 and Figure P6.9. Obtain the expressions for the natural frequencies and mode shapes.

6.30 Refer to Problem 6.10 and Figure P6.10. Determine the expressions for the natural frequencies.

6.31 Refer to Figure P6.31. Obtain the expressions for the natural frequencies and mode shapes for the case where $m_1 = m_2 = m$.

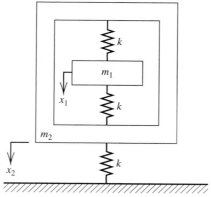

FIGURE P6.31

6.32 Refer to Problem 6.11 and Figure P6.11. Obtain the expressions for the natural frequencies and mode shapes.

6.33 Put the model $\ddot{x} + 3\dot{x} + 2x = 0$ into state-variable form, and find its modes.

6.34 Find the modes of the model

$$\dot{x}_1 = x_2$$

$$\dot{x}_2 = -x_2$$

Identify the rigid-body mode, if any. Find the free response for the following initial conditions:
   (a) $x_1(0) = 10$, $x_2(0) = 0$
   (b) $x_1(0) = 10$, $x_2(0) = -10$
   (c) $x_1(0) = 10$, $x_2(0) = 10$

6.35 Refer to Figure P6.35a, which shows a ship's propeller, drive train, engine, and flywheel. The diameter ratio of the gears is $D_1/D_2 = 1.5$. The inertias in kg·m² of gear 1 and gear 2 are 500 and 100, respectively. The flywheel, engine, and propeller inertias are 10^4, 10^3, and 2500, respectively. The torsional stiffness of shaft 1 is 5×10^6 N·m/rad, and that of shaft 2 is 10^6 N·m/rad. Since the flywheel inertia is so much larger than the other inertias, a simpler model of the shaft vibrations can be obtained by assuming the flywheel does not rotate. In addition, since the shaft between the engine and gears is short, we will assume that it is very stiff compared to the other shafts. If we also neglect the shaft inertias, the resulting model consists of two inertias, one obtained by lumping the engine and gear inertias, and one for the propeller (Figure P6.35b). Using these assumptions, obtain the natural frequencies and mode shapes of the system.

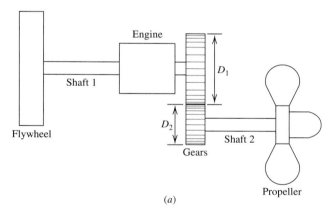

(a)

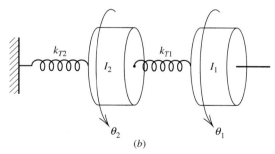

(b)

FIGURE P6.35

SECTION 6.4 HARMONIC RESPONSE

6.36 Figure P6.36 shows a type of vibration absorber that uses only mass and damping, not stiffness, to reduce vibration. The main mass is m_1 and the absorber mass is m_2. Suppose the applied force $f(t)$ is sinusoidal. Derive the expressions for $X_1(i\omega)/F(i\omega)$ and $X_2(i\omega)/F(i\omega)$.

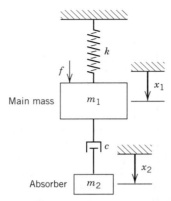

FIGURE P6.36

6.37 Figure P6.37 shows a type of vibration absorber that uses mass, stiffness, and damping to reduce vibration. The damping can be used to reduce the amplitude of vibration near resonance. The main mass is m_1 and the absorber mass is m_2. Suppose the applied force $f(t)$ is sinusoidal. Derive the expressions for $X_1(i\omega)/F(i\omega)$ and $X_2(i\omega)/F(i\omega)$.

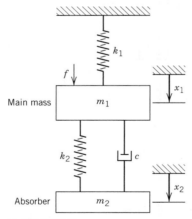

FIGURE P6.37

SECTION 6.5 GENERAL FORCED RESPONSE

6.38 Consider the suspension model shown in Figure 6.5-2. The masses and stiffness are $m_1 = 350$ kg, $m_2 = 20$ kg, $k_1 = 20\ 000$ N/m, and $k_2 = 100\ 000$ N/m. For a step input, is it possible to select a value of the damping c to obtain a maximum overshoot of no more than 15%? If so, determine the required value of c. If not, determine the value of c that will give the smallest possible overshoot.

SECTION 6.6   NODES AND MODAL RESPONSE

6.39 For the roll-pitch vehicle model described in Example 6.6-1, the suspension stiffnesses are to be changed to $k_1 = 1.8 \times 10^4$ N/m and $k_2 = 2.16 \times 10^4$ N/m. Find the natural frequencies, the mode ratios, and the node locations.

6.40 A particular road vehicle weighs 1.8×10^4 N. Using the quarter-car model, determine a suitable value for the suspension stiffness, assuming that the tire stiffness is 2.3×10^5 N/m.

6.41 The car model shown in Figure 6.6-1a has the following parameter values: weight = 2.1×10^4 N, $I = 2440$ kg $\cdot$ m^2, $L_1 = 1.07$ m, and $L_2 = 0.76$ m. Neglect damping and design the front and rear suspension stiffnesses to achieve good ride quality.

SECTION 6.7   ROOT LOCUS ANALYSIS WITH MATLAB

6.42 Use MATLAB to obtain the root locus plot of $5s^2 + cs + 45 = 0$ for $c \geq 0$.

6.43 Use MATLAB to obtain the root locus plot of the equation $ms^2 + cs + k = 0$ in terms of the variable $k \geq 0$. Use the values $m = 4$, $c = 8$. What is the smallest possible dominant time constant and the associated value of k?

6.44 Use MATLAB to obtain the root locus plot of the equation $ms^2 + cs + k = 0$ in terms of the variable $c \geq 0$. Use the values $m = 4$, $k = 64$. What is the smallest possible dominant time constant and the associated value of c?

6.45 Use MATLAB to obtain the root locus plot of the model $m\ddot{x} + c\dot{x} + (k_1 + k_2)x = k_2 y$ in terms of the variable $k_2 \geq 0$. Use the values $m = 2$, $c = 8$, $k_1 = 26$. What is the value of k_2 required to give $\zeta = 0.707$?

6.46 Use MATLAB to obtain the root locus plot of the model $m\ddot{x} + (c_1 + c_2)\dot{x} + kx = c_2 \dot{y}$ in terms of the variable $c_2 \geq 0$. Use the values $m = 2$, $c_1 = 8$, $k = 26$. What is the smallest possible dominant time constant and the associated value of c_2?

6.47 Consider the two-mass model treated in Figure 6.8-1. Its characteristic equation is given by Equation 6.8-4. Use the following numerical values: $m_1 = m_2 = 1$, $k_1 = 1$, and $k_2 = 4$.

(**a**) Use MATLAB to obtain the root locus plot in terms of the parameter c.

(**b**) Use the root locus plot to determine the value of c required to give a dominant-root pair having a damping ratio of $\zeta = 0.707$.

(**c**) Use the root locus plot to determine the value of c required to give a dominant root that is real and has a time constant equal to 4.

(**d**) Using the value of c found in part (c), obtain a plot of the unit-step response.

SECTION 6.8   RESPONSE ANALYSIS WITH MATLAB

6.48 Refer to Figure 6.8-1, where $m_1 = 4$, $m_2 = 1$, $k_1 = 16$, $k_2 = 12$, and $c = 0$. Obtain the characteristic roots, the transfer functions, and the mode ratios.

6.49 Refer to Figure 6.8-1, where $m_1 = 4$, $m_2 = 1$, $k_1 = 16$, $k_2 = 12$, and $c = 5$. Obtain the characteristic roots, the transfer functions, and a plot of the unit-step response.

6.50 Refer to Figure 6.8-1, where $m_1 = 4$, $m_2 = 1$, $k_1 = 12$, $k_2 = 4$, and $c = 1$. Obtain the characteristic roots, the transfer functions, and the closed-form expression for the unit-step response.

6.51 The equations of motion of the vibration absorber shown in Figure 6.1-1 are

$$m\ddot{x} = -k_1 x_1 - k_2(x_1 - x_2) + f$$

$$m_2\ddot{x}_2 = k_2(x_1 - x_2)$$

(**a**) Obtain the transfer functions $X(s)/F(s)$ and $X_2(s)/F(s)$.

(**b**) Suppose that $m = m_2 = 10$, $k_1 = 5$, and $k_2 = 10,000$. Plot the unit step response of $x(t)$ and $x_2(t)$, assuming zero initial conditions.

6.52 The quarter-car suspension model given in Example 6.1-2 is repeated below:

$$m_1\ddot{x}_1 = c(\dot{x}_2 - \dot{x}_1) + k_1(x_2 - x_1)$$

$$m_2\ddot{x}_2 = -c(\dot{x}_2 - \dot{x}_1) - k_1(x_2 - x_1) + k_2(y - x_2)$$

Suppose that $m_1 = 1$, $m_2 = 2$, $c = 1$, $k_1 = 10$, and $k_2 = 20$. The initial conditions are $x_1(0) = 10$, $x_2(0) = 3$, $\dot{x}_1(0) = \dot{x}_2(0) = 0$. Use MATLAB to plot the free response and to find the transfer functions $X_1(s)/Y(s)$ and $X_2(s)/Y(s)$.

6.53 The following model describes a certain two-degrees-of-freedom system. Use MATLAB to plot the step response for three cases: $a = 0.2$, $a = 1$, and $a = 10$. The step input has a magnitude of 2500.

$$\frac{d^4y}{dt^4} + 24\frac{d^3y}{dt^3} + 225\frac{d^2y}{dt^2} + 900\frac{dy}{dt} + 2500y = f + a\frac{df}{dt}$$

Compare the response to that estimated by the maximum overshoot, peak time, 100% rise time, and 2% settling time calculated from the dominant root. Discuss the effect of the numerator dynamics on your estimates.

SECTION 6.9 SIMULINK APPLICATIONS

6.54 Run the Simulink model developed in Section 6.9 for the case where $m_1 = 400$ kg and $m_2 = 40$ kg. Obtain a plot of the response, and evaluate the overshoot and undershoot.

6.55 Redo the Simulink suspension model developed in Section 6.9, using the spring relation and input function shown in Figure P6.55 and the following damper relation:

$$f_d(v) = \begin{cases} -500|v|^{1.2} & v \le 0 \\ 50v^{1.2} & v > 0 \end{cases}$$

Use the simulation to plot the response. Evaluate the overshoot and undershoot.

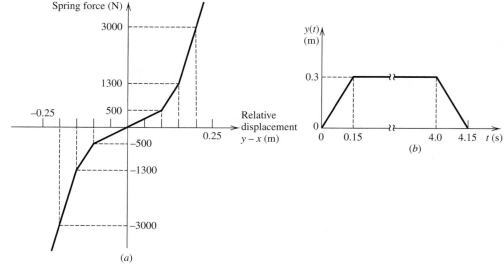

FIGURE P6.55

VIBRATION SUPPRESSION AND CONTROL

Now that we have seen how to model and analyze vibrating systems, we will now consider how to design systems to eliminate or at least reduce the effects of unwanted vibration. In order to determine how much the vibration should be reduced, we need to know what levels of vibration are harmful, or at least disagreeable. This topic is discussed in Section 7.1. To reduce vibration, it is often important to understand the vibration source, and this is discussed in Section 7.2.

Unwanted vibration can be in the form of displacement, velocity, acceleration, transmitted force, or all of these. There are several ways to eliminate or reduce vibration:

1. Reduction of the source causing the vibration. For example,

   (a) Balancing translating or rotating masses. Examples include balancing of fans and reciprocating machines such as internal combustion engines in which the translating pistons and rotating crank arms can cause serious vibration.

   (b) Minimizing clearances. Clearances in bearings and pin joints, for example, can cause impact forces that produce vibration.

   (c) Streamlining objects exposed to wind or currents.

Such topics more properly belong to the disciplines of machine design and fluid mechanics, rather than vibrations, and so we will not treat them further.

2. Redesign of the system. For example,

(**a**) Changing the natural frequency. If the forcing frequency is known and is constant, it may be possible to increase or decrease the mass of the vibrating object to move its natural frequency away from that of the forcing function to avoid near-resonance conditions. It may also be possible, for example, to add stiffeners to increase the stiffness and thus increase the natural frequency.

(**b**) Dissipating the energy of vibration by adding damping. Methods for doing this include friction dampers and *damping treatments*, which consist of coatings of damping material. Such treatments include polymeric materials in the form of mastic materials, pressure-sensitive adhesives, damping tapes, laminates, and elastomeric materials. See Chapter 37 of [Harris, 2002].

(**c**) Isolating the source. For example, drop forges and punch presses produce forces on floors that can cause nearby equipment to vibrate. In such cases an *isolator*, consisting of a stiffness element and perhaps a damping element, often can be placed between the source and the surrounding environment. Isolators can be used to reduce displacement, transmitted force, or both. Proper design of isolators is treated in Sections 7.3 and 7.4.

(**d**) Using a *vibration absorber*. With this method, an auxiliary mass and in some cases a stiffness element are attached to the main mass. If properly sized, the auxiliary mass "absorbs" the energy of motion and the main mass does not vibrate. Such a device is a *vibration absorber*. Design of vibration absorbers is treated in Section 7.4.

(**e**) Using an active control system. An active vibration control system uses a power source such as a hydraulic cylinder or an electric motor to provide forces needed to counteract the forces producing the unwanted vibration. An advantage of such systems, which usually contain a computer, is that they can change the effective stiffness and damping to respond to changing conditions. Section 7.5 treats active vibration control.

LEARNING OBJECTIVES

After you have finished this chapter, you should be able to do the following:

- Design isolators for fixed-base and movable-base systems.
- Design vibration absorbers for single-degree-of-freedom systems.
- Design active vibration control systems for single-degree-of-freedom systems.
- Apply MATLAB to the relevant topics of the chapter.

7.1 ACCEPTABLE VIBRATION LEVELS

The vibration engineer cannot design a proper vibration suppression system without knowing what levels of vibration are harmful, or at least disagreeable. If the vibration is affecting people, the engineer needs to know what levels affect health and comfort; if it affects buildings or some other structure, the vibration levels at which damage may occur need to be known.

We begin by discussing how vibration levels are specified.

Vibration Nomographs

We have seen that for harmonic motion the displacement x, velocity $\dot{x}$, and acceleration $\ddot{x}$ are the following:

$$x(t) = A \sin(\omega t + \phi)$$

$$\dot{x}(t) = A\omega \cos(\omega t + \phi)$$

$$\ddot{x}(t) = -A\omega^2 \sin(\omega t + \phi)$$

Denoting $|x|$, $|\dot{x}|$, and $|\ddot{x}|$ as the amplitudes, we see that $|x| = A$, $|\dot{x}| = \omega A = \omega|x|$, and $|\ddot{x}| = \omega^2 A = \omega^2|x| = \omega|\dot{x}|$. Thus

$$\log |\dot{x}| = \log \omega + \log |x| \qquad (7.1\text{-}1)$$

and

$$\log |\ddot{x}| = \log \omega + \log |\dot{x}| \qquad (7.1\text{-}2)$$

which gives

$$\log |\dot{x}| = \log |\ddot{x}| - \log \omega \qquad (7.1\text{-}3)$$

Equations 7.1-1 and 7.1-3 describe straight lines when plotted versus $\log \omega$. Such a plot is often used to specify the acceptable vibration limits for a given application.

First consider Equation 7.1-1. It describes the velocity amplitude as a function of frequency ω for a given displacement amplitude, which determines the intercept. This describes a family of lines for different values of displacement amplitude, all having a slope of $+1$ (Figure 7.1-1). Note the direction of increasing displacement amplitude.

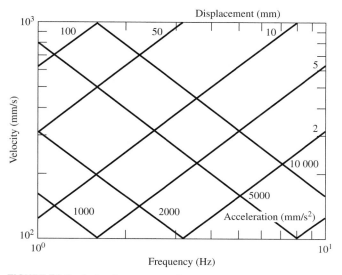

FIGURE 7.1-1 A vibration nomograph.

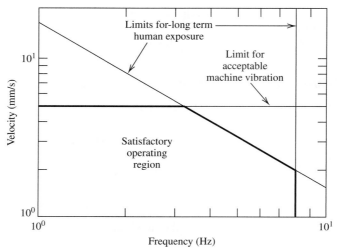

FIGURE 7.1-2   Specification of vibration levels on a nomograph.

Equation (7.1-3) describes the velocity amplitude as a function of frequency for a given acceleration amplitude. Its family of lines has a slope of -1 (Figure 7.1-1). Note the direction of increasing acceleration amplitude.

The third family of lines corresponds to $\log |\dot{x}| = $ a constant. When these three families of lines are plotted on the same graph, they form a *vibration nomograph* that is useful for specifying vibration limits. Figure 7.1-2 shows a typical case for which the maximum allowable amplitudes of displacement, velocity, and acceleration have been specified. The boundary formed by the lines corresponding to these maximum values defines the allowable operating region for the system.

Acceleration values are often quoted as *rms* values, which stands for *root mean square*. For harmonic acceleration a, the rms value is

$$a_{\mathrm{rms}} = \frac{a}{\sqrt{2}} \tag{7.1-4}$$

Effects of Vibration on People

Many studies have been done on the effects of vibration on people, but it is difficult to quantify the effects precisely, partly because of individual variablity and subjective responses in some cases. So the numbers given here are not to be treated as definitive or exact.

There are several aspects to this problem: immediate mechanical damage to the body, longer-term health effects, and discomfort. Maximum acceleration amplitude is the limit most often specified for comfort and health, and it is often specified in terms of the gravitational acceleration constant g. Maximum displacement amplitude is often a function of available space and is not usually related to discomfort. Vibrations with frequencies above approximately 9 Hz are normally beyond the threshold of perception by humans.

Tolerance of vibration has been found to depend on frequency as well as on acceleration. For example, in one study of the effects of vertical vibration of approximately $2g$, the subjects indicated difficulty breathing when the vibration was in the range from 1 to 4 Hz, and they reported chest and/or abdominal pain for vibration in the range from 3 to 9 Hz.

Experimental and computer simulation studies have been performed to identify the resonant frequencies of various parts of the body. For example, for a person seated in a rigid seat, the seat-to-head transmissibility has a maximum near 4.5 Hz, while the thorax/abdomen system has a resonance frequency in the range from 3 to 4 Hz.

In a study of the probable subjective reactions to vibration by passengers in public transport vehicles, the following comfort levels were identified:

Acceleration	Comfort level
0.03g	Not uncomfortable
0.03g to 0.08g	Somewhat uncomfortable
0.08g to 0.13g	Uncomfortable
0.13g to 0.2g	Very uncomfortable
> 0.2g	Extremely uncomfortable

The *duration* of the vibration also affects both comfort and health. Figure 7.1-3 shows the frequency dependence of the fatigue limits for vertical rms acceleration of a person seated in a vehicle, for different exposure durations. The plot shows that people fatigue most quickly when the frequency is between 4 and 8 Hz.

Figure 7.1-4 shows the safe exposure limit, which is the recommended vertical acceleration limit to avoid health problems, when the exposure time is 8 hours. Vibration below the reduced comfort level allows activities such as reading, writing, and eating to take place comfortably.

The International Standards Organization (ISO) has developed detailed recommendations concerning acceptable vibration limits for both people and structures; for example, see their publication Standard ISO 2631. For example, structural effects such as cracks and wall damage can occur when the velocity of vibration exceeds approximately 30 mm/s (ISO DP4866).

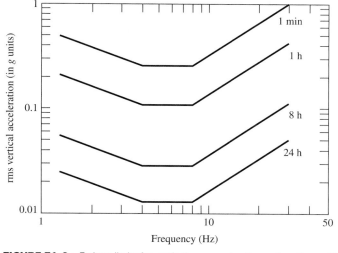

FIGURE 7.1-3 Fatigue limits for vertical rms acceleration as functions of frequency.

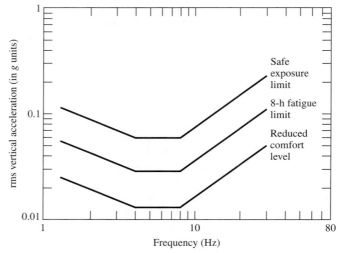

FIGURE 7.1-4 Safe exposure limits and reduced comfort level as functions of frequency.

7.2 SOURCES OF VIBRATION

Vibration can be caused by many types of excitation. These include:

1. Fluid flow
2. Reciprocating machinery
3. Rotating unbalanced machinery
4. Motion induced in vehicles traveling over uneven surfaces
5. Ground motion caused by earthquakes

To reduce the effects of unwanted vibration, the engineer should understand the source.

Vibration Induced by Fluid Flow

Vibration can be generated by the forces exerted on an object by fluid motion. Such situations can be complicated by the fact that the motion of the vibrating object can alter the fluid flow conditions, thus changing the fluid forces. Another complicating factor is the mass of the fluid, which increases the effective mass of the system.

Examples of vibration caused by fluid motion include:

1. Wave action on structures
2. Vortex-induced vibration
3. Vibration caused by internal flows, such as flow through pipes and hoses having bends
4. Structural vibration caused by fluctuating aerodynamic forces such as turbulence

Vortex-Induced Vibration

Fluid flowing over an object can sometimes separate from the downstream side of the object. In such cases the wake can contain vortices like those shown in Figure 7.2-1. The vortices shed alternately from the top and the bottom of the object, and this produces an

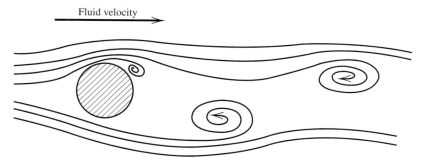

FIGURE 7.2-1 Vortices shedding from a cylinder.

oscillating lift on the object. Taking a cylindrical object as an example, the resulting vibration of the cylinder can do the following:

1. Increase the lift generated by the shedding vortices
2. Cause the shedding frequency to shift from that occuring with a stationary cylinder to the natural frequency of the cylinder
3. Increase the drag on the cylinder
4. Change the vortex pattern

Common examples of vortex-induced vibration are vibration of electrical transmission cables, bridge and tower cables, underwater cables used for towing and structural support, and cooling towers and chimneys.

In certain situations, the steady-state excitation force due to vortex shedding is sinusoidal with an amplitude proportional to the square of the forcing frequency. The model of a single-degree-of-freedom system undergoing such excitation is

$$m\ddot{x} + c\dot{x} + kx = F_o\omega^2 \sin\omega t \tag{7.2-1}$$

Further analysis of such phenomena is quite complicated and requires detailed consideration of the fluid mechanics of the situation. See Chapter 29 of [Harris, 2002] for more details.

Vibration from Reciprocating Engines

The dynamics of reciprocating engines is a complex subject. Vibration is transmitted to the chassis or foundation by the engine, and the crankshaft undergoes torsional vibration. These are the result of the unbalanced motions of the pistons, connecting rods, and cranks as well as of the fluctuating steam or gas pressure in the cylinders. Designers try to balance the engine as much as possible, but it is not possible to eliminate all vibration. Balancing methods are treated in [Den Hartog, 1985].

We now present a simplified analysis of the piston-rod-crank system in order to estimate the vibration frequencies that are generated by their motions. Referring to Figure 7.2-2, let x_p be the displacement of the piston from top dead center, ωt the crank angle from top dead center, R the crank radius, and L the length of the connecting rod. We assume that the rotational speed ω is constant, which is a simplification.

From the figure, note that $L\sin\phi = R\sin\omega t$, and thus

$$\cos\phi = \sqrt{1 - \frac{R^2}{L^2}\sin^2\omega t}$$

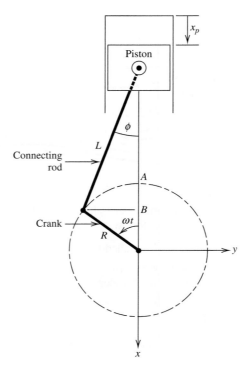

FIGURE 7.2-2 A piston, connecting rod, and crank of a reciprocating engine.

The distance $\overline{AB}$ is given by $R(1 - \cos \omega t)$, and thus the piston displacement is

$$x_p = \overline{AB} + L(1 - \cos \phi) = R(1 - \cos \omega t) + L\left(1 - \sqrt{1 - \frac{R^2}{L^2} \sin^2 \omega t}\right)$$

In practice, $R/L < 1$ (in fact, usually $R/L < 1/4$), and since $\sin^2 \omega t \leq 1$, we can use the approximation

$$\sqrt{1 - \epsilon} \approx 1 - \frac{\epsilon}{2} \quad \text{if } \epsilon \ll 1$$

Therefore we can express x_p approximately as

$$x_p \approx R(1 - \cos \omega t) + \frac{R^2}{2L} \sin^2 \omega t$$

To identify the frequency components, we use the trigonometric identity

$$\sin^2 \omega t = \frac{1 - \cos 2\omega t}{2}$$

to obtain

$$x_p \approx R(1 - \cos \omega t) + \frac{R^2}{2L} \frac{1 - \cos 2\omega t}{2}$$

which can be rearranged as

$$x_p \approx \left(R + \frac{R^2}{4L}\right) - R\left(\cos \omega t + \frac{R}{4L} \cos 2\omega t\right) \tag{7.2-2}$$

Differentiating twice with respect to time gives the piston acceleration:

$$\ddot{x}_p \approx R\omega^2 \left(\cos \omega t + \frac{R}{L} \cos 2\omega t \right) \tag{7.2-3}$$

The vertical force generated by the piston motion and causing vibration is obtained by multiplying the piston acceleration by the piston mass. Thus we can see that this force contains two frequencies, the frequency of crank rotation ω and twice this frequency, 2ω. Thus the excitation is not harmonic. It is close to being harmonic if the ratio $R/L \ll 1$. For shorter connecting rods, say where $R/L = 1/4$, the force is not close to being harmonic.

This analysis, which is for a single-cylinder engine, ignores the dynamics of the crank and connecting rod, the lateral force, and the effects of cylinder pressure variation. A detailed analysis of the torque acting on the crankshaft as a result of the inertial forces indicates that the torque has a frequency component at 3ω, in addition to the frequencies ω and 2ω. The effects of cylinder pressure variation depend on whether the engine is a two-cycle or a four-cycle engine. This, and the effects due to multiple cylinders, can produce vibration at many frequencies [Den Hartog, 1985].

Motion Induced by the Ground

The most obvious source of ground motion is an earthquake, but this category also includes the base motion that occurs when a vehicle travels over an uneven surface. These types of motion cannot be easily represented by simple functions because they are essentially somewhat random.

It is possible, however, to characterize them somewhat by means of statistical averages and spectrum plots, in which Fourier analysis is used to identify the major frequency components in the excitation. Earthquake motion has been studied extensively, while some studies have been done to characterize the road profiles produced by standard road construction techniques.

The methods of Chapter 9 can be used to analyze such excitations.

7.3 ISOLATOR DESIGN FOR FIXED-BASE SYSTEMS

In many applications an excitation acts on a mass whose resulting motion will produce vibration in adjacent objects unless it is isolated. For example, during forging the hammer strikes the object to be formed. The impact can damage the supporting structure and floor if the system is not properly designed. Another example is a rotating machine that has an unbalance.

In such applications we can insert an *isolator* between the supporting structure and the mass being excited. In this section we consider the design of such isolators and treat the supporting structure as being fixed. In such applications the excitation is a *force*. In Section 7.4 we consider the design of isolators where it is the *motion* of the supporting structure that causes the vibration. In both types of applications we are usually interested in both the motion and the force transmitted to the mass by the vibration. Thus we may speak of *displacement isolation* and *force isolation*.

We will see that the design of an effective isolator can be highly dependent on the nature of the excitation. For example, an isolator designed to handle a harmonic excitation of a certain frequency may not be effective for excitations at other frequencies or for other types of excitations such as step inputs.

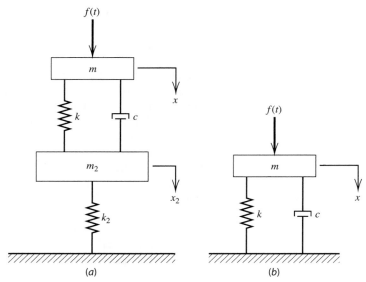

FIGURE 7.3-1 (*a*) A model of an isolation system that includes the mass of the base. (*b*) A model of an isolation system that treats the base as fixed.

In our pictorial representations the stiffness and damping are usually represented by a helical-coil spring and a damper. There are, however, many other types of stiffness and damping elements that can be used for isolation. Some of these are discussed near the end of this section.

Fixed-Base Model

Figure 7.3-1*a* shows a machine of mass m subjected to a forcing function $f(t)$ that can be due to, for example, a drop forge, a punch press, or rotating unbalance. The isolator consists of a stiffness element k and perhaps a damping element c. The isolator is intended to reduce the motion of the mass m or the force transmitted to the base mass m_2, which can be a floor or another machine, for example. The stiffness k_2 represents the stiffness of the floor or other supporting structure. It is important to realize that without the isolator, the force $f(t)$ would be transmitted completely to the base mass m_2.

Now if the base mass m_2 is very large or the stiffness k_2 is very stiff, as would be the case with a concrete foundation, then we can approximate the two-DOF system by the single-DOF system shown in Figure 7.3-1*b*, in which the base is taken to be fixed.

Isolation for Harmonic Excitation

The equation of motion for the system in Figure 7.3-2 is

$$m\ddot{x} + c\dot{x} + kx = f(t)$$

and the force transmitted to the base is

$$f_t = kx + c\dot{x}$$

So the transfer functions are

$$\frac{X(s)}{F(s)} = \frac{1}{ms^2 + cs + k} \tag{7.3-1}$$

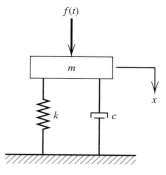

FIGURE 7.3-2   A fixed-base isolation system.

$$\frac{F_t(s)}{F(s)} = \frac{F_t(s)}{X(s)}\frac{X(s)}{F(s)} = \frac{k+cs}{ms^2+cs+k} \tag{7.3-2}$$

The frequency transfer functions are

$$\frac{X(i\omega)}{F(i\omega)} = \frac{1}{k-m\omega^2+c\omega i} \tag{7.3-3}$$

$$\frac{F_t(i\omega)}{F(i\omega)} = \frac{k+c\omega i}{k-m\omega^2+c\omega i} \tag{7.3-4}$$

Thus the amplitude ratios are

$$\frac{X}{F} = \frac{1}{\sqrt{(k-m\omega^2)^2+(c\omega)^2}} = \frac{1}{k}\frac{1}{\sqrt{(1-r^2)^2+(2\zeta r)^2}} \tag{7.3-5}$$

which is called the *displacement transmissibility*, and

$$\frac{F_t}{F} = \frac{\sqrt{k^2+(c\omega)^2}}{\sqrt{(k-m\omega^2)^2+(c\omega)^2}} = \sqrt{\frac{1+(2\zeta r)^2}{(1-r^2)^2+(2\zeta r)^2}} \tag{7.3-6}$$

which is called the *force transmissibility*. The parameters ζ and r are defined as usual.

$$\zeta = \frac{c}{2\sqrt{mk}} \tag{7.3-7}$$

$$r = \frac{\omega}{\omega_n} = \frac{\omega}{\sqrt{k/m}} \tag{7.3-8}$$

These transmissibility relations apply to the case where the forcing function $f(t)$ is harmonic with a radian frequency of ω and an amplitude F. For example, $f(t) = F\sin\omega t$ or $f(t) = F\cos\omega t$.

Defining $\delta_{st} = F/k$ as the static deflection caused by a *constant* force F, we may express Equation 7.3-5 in nondimensional form as

$$\frac{kX}{F} = \frac{X}{\delta_{st}} = \frac{1}{\sqrt{(1-r^2)^2+(2\zeta r)^2}} \tag{7.3-9}$$

This is plotted in Figure 7.3-3 for several values of the damping ratio ζ. At high forcing frequencies, the response amplitude approaches zero because the system's inertia prevents it from following a rapidly varying forcing function. This effect is due primarily to the system's inertia, not its damping as one might think.

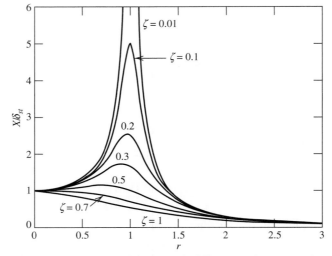

FIGURE 7.3-3 Displacement transmissibility versus frequency ratio.

Assuming that m is a given value, to obtain good *displacement* isolation you must choose k so that r is not close to 1. In some cases, however, we may also be able to increase or decrease the mass m to improve the isolation. Usually, however, we cannot decrease the mass because well-designed machines will probably have their mass minimized already. Whether or not we can increase the mass depends on the application. For systems carried aboard aircraft or satellites, for example, additional mass may not be allowed.

Let $T_r = F_t/F$. The force transmissibility is plotted versus the frequency ratio r in Figure 7.3-4, for several values of ζ. Note that all the curves pass through the point $\zeta = 1$, $T_r = \sqrt{2}$.

To obtain good *force* isolation, that is, to decrease the force transmitted to the foundation, we need to make F_t/F small. Near resonance, when r is near 1, F_t/F is highly dependent on the value of ζ. When $r < \sqrt{2}$, *increasing* ζ will decrease F_t/F and

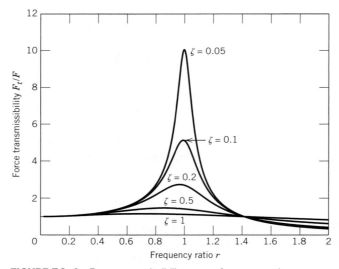

FIGURE 7.3-4 Force transmissibility versus frequency ratio.

TABLE 7.3-1 Formulas for Fixed-Base Harmonic Excitation

Model:
$$m\ddot{x} + c\dot{x} + kx = F \sin \omega t$$

$$\zeta = \frac{c}{2\sqrt{mk}} \qquad \omega_n = \sqrt{\frac{k}{m}} \qquad r = \frac{\omega}{\omega_n}$$

Displacement transmissibility:

$$\frac{X}{F} = \frac{1}{k} \frac{1}{\sqrt{(1 - r^2)^2 + (2\zeta r)^2}}$$

Force transmissibility:

$$\frac{F_t}{F} = \sqrt{\frac{1 + (2\zeta r)^2}{(1 - r^2)^2 + (2\zeta r)^2}}$$

If $\zeta = 0$, then

$$\frac{F_t}{F} = T_r = \frac{1}{r^2 - 1}$$

$$r^2 = \frac{1 + T_r}{T_r}$$

thus improve the force isolation. For $r \geq \sqrt{2}$, F_t/F is not so highly dependent on ζ, and F_t/F decreases as ζ *decreases*.

Thus, for $r \geq \sqrt{2}$, we can improve the force isolation by *decreasing* ζ. We want to have some damping, however, because when the machine is started, its speed increases from zero and eventually passes through the resonance region near $r = 1$, and large forces can result if the machine's speed does not pass through the resonance region quickly enough. Some damping helps to limit the force buildup near resonance.

If $r > 1$ and if ζ is small, then the formula for F_t/F can be replaced with the approximate formula

$$\frac{F_t}{F} = \frac{1}{r^2 - 1} \tag{7.3-10}$$

This formula can be easily solved for r as a function of T_r, as follows:

$$r^2 = \frac{1 + T_r}{T_r} \tag{7.3-11}$$

Table 7.3-1 summarizes these results.

EXAMPLE 7.3-1
Undamped
Isolator Design

Design an undamped isolator for a 20-kg mass subjected to a harmonic forcing function whose amplitude and frequency are 600 N and 17 Hz. The isolator should transmit to the base no more than 10% of the applied force. Determine the resulting displacement amplitude.

Solution

Here $T_r = 0.1$, and from Equation 7.3-11 we obtain

$$r^2 = \frac{1 + 0.1}{0.1} = 11$$

Since $r^2 = \omega^2/\omega_n^2 = m\omega^2/k$, the isolator stiffness should be

$$k = \frac{m\omega^2}{r^2} = \frac{20[(2\pi)(17)]^2}{11} = 2.0744 \times 10^4 \text{ N/m}$$

The resulting displacement amplitude is found from

$$X = \frac{600}{k}\frac{1}{r^2 - 1} = \frac{600}{2.0744 \times 10^4}\frac{1}{11 - 1} = 0.0029 \text{ m} \qquad \blacksquare$$

Isolation from Rotating Unbalance

A common cause of harmonic forcing in machines is the unbalance that exists to some extent in every rotating machine. The unbalance is caused by the fact that the center of mass of the rotating part does not coincide with the center of rotation. Let M be the total mass of the machine and m the rotating mass causing the unbalance. Consider the entire unbalanced mass m to be lumped at its center of mass, a distance R from the center of rotation. This distance is the *eccentricity*. Figure 7.3-5a shows this situation. The main mass is thus $(M - m)$ and is assumed to be constrained to allow only vertical motion. The motion of the unbalanced mass m will consist of the vector combination of its motion relative to the main mass $(M - m)$ and the motion of the main mass. For a constant speed of rotation ω_R, the rotation produces a radial acceleration of m equal to $R\omega_R^2$. This causes a force to be exerted on the bearings at the center of rotation. This force has a magnitude $mR\omega_R^2$ and is directed radially outward. The vertical component of this unbalance force is, from Figure 7.3-5b,

$$f = mR\omega_R^2 \sin \omega_R t \qquad (7.3\text{-}12)$$

Thus the amplitude of the excitation is $mR\omega_R^2$.

The equation of motion is

$$M\ddot{x} + c\dot{x} + kx = mR\omega_R^2 \sin \omega_R t \qquad (7.3\text{-}13)$$

To apply the transmissibility relations, Equations 7.3-5 and 7.3-6, we must substitute M for m, $mR\omega_R^2$ for F, and ω_R for ω. The results are

$$\frac{X}{mR\omega_R^2} = \frac{1}{\sqrt{(k - M\omega_R^2)^2 + (c\omega_R)^2}} = \frac{1}{k}\frac{1}{\sqrt{(1 - r^2)^2 + (2\zeta r)^2}} \qquad (7.3\text{-}14)$$

$$\frac{F_t}{mR\omega_R^2} = T_r = \frac{\sqrt{k^2 + (c\omega_R)^2}}{\sqrt{(k - M\omega_R^2)^2 + (c\omega_R)^2}} = \sqrt{\frac{1 + (2\zeta r)^2}{(1 - r^2)^2 + (2\zeta r)^2}} \qquad (7.3\text{-}15)$$

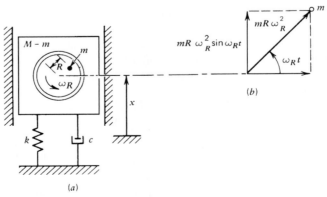

(a)

(b)

FIGURE 7.3-5   A system having rotating unbalance.

TABLE 7.3-2   Formulas for Rotating Unbalance

Model:

$$M\ddot{x} + c\dot{x} + kx = mR\omega_R^2 \sin \omega_R t$$

m = unbalanced mass R = eccentricity

$$\zeta = \frac{c}{2\sqrt{Mk}} \qquad \omega_n = \sqrt{\frac{k}{M}} \qquad r = \frac{\omega_R}{\omega_n}$$

Displacement:

$$X = \frac{mR}{M} \frac{r^2}{\sqrt{(1 - r^2)^2 + (2\zeta r)^2}}$$

Force transmissibility:

$$T_r = \frac{F_t}{mR\omega_R^2} = \sqrt{\frac{1 + 4\zeta^2 r^2}{(1 - r^2)^2 + 4\zeta^2 r^2}}$$

where

$$\zeta = \frac{c}{2\sqrt{Mk}} \tag{7.3-16}$$

$$r = \frac{\omega_R}{\omega_n} = \frac{\omega_R}{\sqrt{k/M}} \tag{7.3-17}$$

Table 7.3-2 summarizes these relations.

EXAMPLE 7.3-2
Support Vibration Due to Rotating Unbalance

Alternating-current motors are often designed to run at a constant speed, typically either 1750 or 3500 rpm. One such motor for a power tool has a mass of 8 kg and is to be mounted on a steel cantilever beam as shown in Figure 7.3-6a. Static-force calculations and space considerations suggest that a beam 15 cm long, 10 cm wide, and 1 cm thick would be suitable. The rotating part of the motor has a mass of 4 kg and has an eccentricity of 0.3 mm. (This distance can be determined by standard balancing methods or applying the analysis in this example in reverse, using a vibration test.) The damping ratio for such beams is difficult to determine but is usually very small, say, $\zeta \leq 0.1$. Estimate the amplitude of vibration of the beam at steady state.

Solution

For the equivalent system shown in Figure 7.3-6b, we have the following spring constant and equivalent mass:

$$k = \frac{Ewh^3}{4L^3} = \frac{(2 \times 10^{11})(0.1)(0.01)^3}{4(0.15)^3}$$

$$= 1.4815 \times 10^6 \text{ N/m}$$

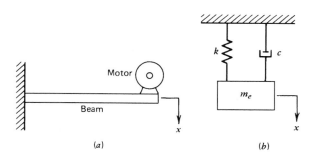

FIGURE 7.3-6 A cantilever-supported motor and its equivalent mass-spring model.

We compute the system mass M as follows. Using 7.8×10^3 kg/m³ for the density of steel, and including 23% of the beam mass, we obtain

$$M = 8 + 0.23(7.8 \times 10^3)(0.1)(0.15)(0.01) = 8.269 \text{ kg}$$

The unbalanced mass is $m = 4$ kg. The model for the system is

$$M\ddot{x} + c\dot{x} + kx = f(t) = mR\omega_R^2 \sin \omega_R t$$

Note that $\omega_n = \sqrt{k/M} = 423$ rad/s. The motor speed of 1750 rpm gives $\omega_R = 183$ rad/s. Thus $r = \omega_R/\omega_n = 183/423 = 0.433$. The amplitude of the forcing function is $mR\omega_R^2 = 4(0.0003)(183)^2 = 40.4$ N. From Equation 7.3-14, using $\zeta = 0.1$, we see that the steady-state amplitude is $X = 3.3 \times 10^{-5}$ m. On the downward oscillation, the total amplitude as measured from horizontal is the preceding value plus the static deflection, or $3.3 \times 10^{-5} + 8(9.8)/1.4815 \times 10^6 = 8.59 \times 10^{-5}$ m.

If this deflection would cause too much stress in the beam, one or more of the following changes could be made in the design:

1. Add a damper to the system.

2. Reduce the unbalance in the motor.

3. Increase the separation between the forcing frequency and the natural frequency either by selecting a beam with more stiffness (a larger k value), by reducing the mass of the system, or by adding mass to shift ω_n far below ω_R.

Because ω_R is not close to ω_n in this example, the preceding results are not very sensitive to the assumed value of $\zeta = 0.1$. For example, the calculated amplitudes of vibration for $\zeta = 0.05$ and 0.2 are very close to the amplitude for $\zeta = 0.1$.

However, if we had used a motor with a speed of 3500 rpm = 366 rad/s, this choice would put the forcing frequency very close to the natural frequency. In this region, the assumed value of ζ would be critical in the amplitude calculation. In practice, such a design would be avoided. That is, in vibration analysis, the most important quantity to know is the natural frequency ω_n. If the damping is slight, the resonant frequency is near ω_n. If ω_n is designed so that it is not close to the forcing frequency, it is not usually necessary to know the precise amount of damping. ∎

Static Deflection versus Rattle Space

The previous example illustrates that the maximum deflection is the sum of the static deflection and the dynamic deflection from the equilibrium. The space available for the total deflection is sometimes called the *rattle space*. The total deflection must always be considered when designing a vibration isolation system, because the rattle space might not be large enough to accommodate the motion. This is frequently a serious limitation in the design of an isolation system, because sometimes to minimize the maximum acceleration we must accept a large deflection.

EXAMPLE 7.3-3
Isolation of a Motor

Often motors are mounted to a base with an isolator consisting of an elastic pad. The pad serves to reduce the motor's rotating unbalance force transmitted to the base. A particular motor has a mass of 2 kg and runs at 5000 rpm. Neglect damping in the pad and calculate the pad stiffness required to provide a 95% reduction in the force transmitted from the motor to the base.

Solution | A 95% force reduction corresponds to a transmissibility ratio of $T_r = 0.05$. Using the approximate formula of Equation 7.3-11, we obtain

$$r^2 = \frac{1 + T_r}{T_r} = \frac{1.05}{0.05} = 21$$

From the definition of r and the fact that $\omega_n = \sqrt{k/M}$, we have

$$r^2 = \frac{\omega_R^2}{\omega_n^2} = \omega_R^2 \frac{M}{k}$$

Thus

$$k = \frac{\omega_R^2}{r^2} M = \frac{[5000(2\pi)/60]^2}{21} 2 = 2.61 \times 10^4 \text{ N/m}$$

If the pad's damping is slight, say $\zeta = 0.1$, the exact expression of Equation 7.3-15 gives $T_r = 0.07$, a 93% reduction that is close to the desired value of 95%. ∎

Sometimes we are not given the values of the unbalanced mass m and the eccentricity R, but instead we are given the measured value of the force due to rotating unbalance.

EXAMPLE 7.3-4
Calculation of
Transmitted Force

In the machine shown in Figure 7.3-5a, when the machine is rotating at 3600 rpm, the effect of the rotating unbalance is to exert a force of 350 N on the machine, whose total mass is 150 kg. The isolator values are $k = 1.6 \times 10^7$ N/m and $\zeta = 0.3$.

(a) Compute the steady-state amplitude of the displacement.

(b) Compute the magnitude of the force transmitted to the foundation at steady state.

Solution | **(a)** We are given that $mR\omega_R^2 = 350$ N and that $\omega_R = 3600(2\pi)/60 = 377$ rad/s. Thus

$$mR = \frac{350}{(377)^2} = 2.463 \times 10^{-3} \text{ kg} \cdot \text{m}$$

and

$$r^2 = \left(\frac{\omega_R}{\omega_n}\right) = \frac{150(377)^2}{1.6 \times 10^5} = 1.333$$

From Equation 7.3-14,

$$X = \frac{mR\omega_R^2}{k} \frac{1}{\sqrt{(1 - r^2)^2 + (2\zeta r)^2}} = 2.8 \times 10^{-5} \text{ m}$$

(b) From Equation 7.3-15,

$$F_t = mR\omega_R^2 \sqrt{\frac{1 + (2\zeta r)^2}{(1 - r^2)^2 + (2\zeta r)^2}} = 554 \text{ N}$$ ∎

EXAMPLE 7.3-5
Shaft Design
with Rotating
Unbalance

Rotating machines such as pumps and fans must have their shafts supported by bearings and often are enclosed in a housing. Suppose that the rotor (the rotating element) of a specific machine has a mass of 500 kg and a measured unbalance of $mR = 1$ kg·m. The machine will be run at a speed of 1750 rpm, and there is a clearance of 5 mm between the shaft and the housing. The shaft length from the bearings is $L = 0.1$ m. Assuming the shaft is steel, compute the minimum required shaft diameter. Model the shaft as a cantilever beam supported by the bearings, and neglect any damping in the system.

Solution First convert the speed into rad/s: 1750 rpm = $1750(2\pi)/60$ = 183.26 rad/s. From Equation 7.3-14 with $\zeta = 0$, since $\omega_R^2 = r^2\omega_n^2 = r^2 k/M$,

$$X = \frac{mR\omega_R^2}{k}\frac{1}{|1 - r^2|} = \frac{mR}{M}\frac{r^2}{|1 - r^2|}$$

Because all the mass is rotating here, $M = m = 500\,\text{kg}$. With $mR = 1$ and $X = 5\,\text{mm} = 0.005\,\text{m}$, we have

$$0.005 = \frac{1}{500}\frac{r^2}{|1 - r^2|}$$

Solve this for r^2 assuming that $r^2 > 1$:

$$0.005 = \frac{1}{500}\frac{r^2}{r^2 - 1}$$

which gives $r^2 = 1.66$ and thus

$$\omega_n^2 = \frac{\omega^2}{r^2} = \frac{(183.26)^2}{1.66} = 20\,231$$

But $\omega_n^2 = k/m = k/500$. Thus $k = 500(20\,231) = 1.012 \times 10^7$ N/m.
  From the formula for a cantilever spring,

$$k = \frac{3EI_A}{L^3}$$

where the area moment of inertia for a cylinder is $I_A = \pi d^4/64$. Solving for the diameter d in terms of k, we obtain

$$d^4 = \frac{64kL^3}{3\pi E} = \frac{64(10.12 \times 10^6)(0.1)^3}{3\pi(2 \times 10^{11})} = 34 \times 10^{-8}$$

which gives a minimum shaft diameter of $d = 0.024\,\text{m} = 24\,\text{mm}$. ∎

Shock Inputs

Vibrational motion is characterized by oscillation, which can be caused by the initial conditions or by a forcing function. The vibration literature uses the term *shock* to describe certain types of forcing functions, but there is no strict definition of what constitutes a shock input. Most vibration engineers consider a shock to be a sudden and severe nonperiodic excitation. Examples include an impulsive input and pulses of various shapes—rectangular, triangular, and half-sine. According to this definition, a step input is a shock, but some do not consider an input to be a shock unless it has a short duration. We will, however, not dwell on such semantics.

Response Spectra

Consider the typical system shown in Figure 7.3-2 subjected to the forcing function shown in Figure 7.3-7. With no damping, the equation of motion for the system in Figure 7.3-2 is $m\ddot{x} + kx = f(t)$. The closed-form expression for the response can be obtained with the Laplace transform by treating the forcing function as composed of the sum of two ramp functions, one with a slope F_o/t_1 starting at $t = 0$ and the other with

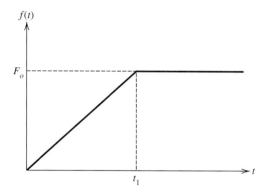

FIGURE 7.3-7 A modified ramp input.

slope of $-F_o/t_1$ starting at $t = t_1$. Using the expression for the response $x(t)$ to obtain the expression for maximum response x_{max}, we obtain the following result.

$$\frac{kx_{max}}{F_o} = 1 + \frac{T_n}{2\pi t_1} \sqrt{2(1 - \cos 2\pi t_1/T_n)} \tag{7.3-18}$$

Here T_n is the natural period corresponding to the natural frequency ω_n: $T_n = 2\pi/\omega_n$. Figure 7.3-8 is a plot of this relation in terms of the dimensionless variables kx_{max}/F_o and t_1/T_n. Such a plot of the maximum response is called a *response spectrum*. If the input is a shock, the plot is sometimes called a *shock spectrum*.

Instead of the force shown in Figure 7.3-7, suppose that the force is a constant F_o. The static deflection caused by such a force would be $\delta = F_o/k$. In the dynamic case where the force is suddenly applied, limiting the deflection x to no more than the static deflection is the best we can do in isolating the motion of the mass from the input. Since the dimensionless variable kx_{max}/F_o is the same as the ratio x_{max}/δ, then the value $kx_{max}/F_o = 1$ represents the best result we can expect to achieve.

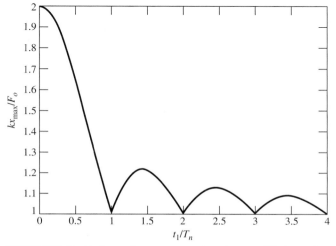

FIGURE 7.3-8 Undamped response spectrum for a modified ramp input.

Now consider the plot in Figure 7.3-8. It shows that the best design is achieved when $t_1/T_n = 1, 2, 3, \ldots$ because $kx_{max}/F_o = 1$ at these points. Using the first design point gives

$$t_1 = T_n = \frac{2\pi}{\omega_n} = 2\pi\sqrt{\frac{m}{k}}$$

or

$$k = \frac{4m\pi^2}{t_1^2} \qquad (7.3\text{-}19)$$

If k is selected to be this value, then the maximum response x_{max} is no greater than the static deflection kx_{max}/F_o caused by a constant force F_o.

The transfer function for the force f_t transmitted to the base is

$$\frac{F_t(s)}{F(s)} = \frac{cs + k}{ms^2 + cs + k}$$

Thus, with no damping, $c = 0$, and the transmitted force is seen to be simply proportional to the displacement: $f_t = kx$, and thus the maximum transmitted force is $kx_{max} = F_o$.

EXAMPLE 7.3-6

Selecting Stiffness for Optimum Response

For the system shown in Figure 7.3-2 and the force shown in Figure 7.3-7, $m = 50$ kg, $F_o = 500$ N, and $t_1 = 0.1$ s. Determine the optimum value of k to minimize the response, and plot $x(t)$ for this value of k. First assume that $c = 0$; then investigate the effects of damping.

Solution

From Equation 7.3-19,

$$k = \frac{4(50)\pi^2}{(0.1)^2} = 1.9739 \times 10^5 \text{ N/m}$$

Realizing that this precise value is probably not available in a stock component, we round off the value to $k = 2 \times 10^5$. The resulting static deflection is $F_o/k = 0.0025$ m. The response is shown in Figure 7.3-9 and behaves as predicted; that is, the maximum displacement is no greater than the static deflection, and the maximum transmitted force will be 500 N.

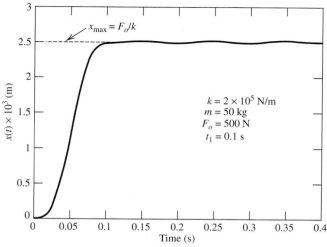

FIGURE 7.3-9 Response of the undamped isolator of Example 7.3-6.

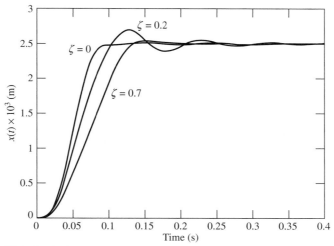

FIGURE 7.3-10   Response of the isolator of Example 7.3-6 for $\zeta = 0, 0.2$, and 0.7.

All isolators have some inherent damping, and Figure 7.3-10 shows the response of this system for three damping values. For $\zeta = 0.7$, the maximum response is essentially the same as in the undamped case but is somewhat more sluggish. For $\zeta = 0.2$, the maximum response is about 10% greater than with the undamped case.

We mentioned earlier that the design of an effective isolator can be highly dependent on the nature of the excitation. To illustrate this, suppose that the excitation acting on the mass in this example is instead the harmonic function $f(t) = 500 \sin 20\pi t$. The natural frequency is $\omega_n = 63.2$ rad/s and the frequency ratio is $r = 20\pi/63.2 = 0.99$. From the displacement transmissibility formula, Equation 7.3-5, we find that the steady-state displacement amplitude is $X = 0.126$ m, which is approximately 50 times the static deflection!

On the other hand, if the forcing function is a pure step of magnitude 500, then the steady-state response will be

$$x(t) = 0.0025(1 - \cos 63.2t)$$

which has an amplitude of 0.005, twice the static deflection.

Isolating against different types of inputs is a great challenge for the designer. For example, suppose the input is harmonic with a period of 0.2 s and an amplitude of 500 N. For good force isolation, Figure 7.3-4 shows that we should select $r > \sqrt{2}$. For the present system, this means that the stiffness should be $k < 2.4674 \times 10^4$ N/m, which is an order of magnitude below the stiffness required to isolate against the truncated ramp input shown in Figure 7.3-7. ∎

The advantage of obtaining a closed-form solution for the response is that it enables us to develop a response plot in terms of dimensionless variables such as kx_{max}/F_o and t_1/T_n. Unfortunately this requires detailed mathematical analysis that may not be feasible for more complicated forcing functions or for systems with damping. In practice, however, we usually know the system's mass value and the characteristics of the forcing function, such as the values of F_o and t_1. All we would need to determine is the optimum value of the stiffness k and, if damping is present, the optimum value of c. In such cases we can obtain satisfactory response plots of x_{max} versus k by numerical simulation. This topic is treated in Section 7.7.

For the case where the isolator has damping, we can reduce the number of parameters to be used in the simulation as follows. The equation of motion is

$$m\ddot{x} + c\dot{x} + kx = f(t)$$

or

$$\ddot{x} + 2\zeta\omega_n\dot{x} + \omega_n^2 x = \frac{1}{m}f(t)$$

Define a new time scale T such that $T = t/t_1$, where t_1 is a characteristic time associated with the input, such as the ramp time shown in Figure 7.3-7. Thus

$$\frac{dx}{dT} = x' = \frac{dx}{dt}\frac{dt}{dT} = t_1\frac{dx}{dt}$$

and

$$\frac{d^2x}{dT^2} = x'' = \frac{d}{dT}\left(\frac{dx}{dT}\right) = \frac{d}{dT}\left[t_1\frac{dx}{dt}\right] = t_1^2\frac{d^2x}{dt^2}$$

Then the equation of motion becomes

$$\frac{1}{t_1^2}x'' + \frac{2\zeta\omega_n}{t_1}x' + \omega_n^2 x = \frac{1}{m}f(T)$$

or

$$x'' + 2\zeta\omega_n t_1 x' + \omega_n^2 t_1^2 x = \frac{t_1^2}{m}f(T)$$

Noting that the natural period is

$$T_n = \frac{2\pi}{\omega_n} \tag{7.3-20}$$

we obtain

$$x'' + \frac{4\zeta\pi t_1}{T_n}x' + \frac{4\pi^2 t_1^2}{T_n^2}x = \frac{4\pi^2 t_1^2}{kT_n^2}f(T)$$

If we define

$$b = \frac{t_1}{T_n} \tag{7.3-21}$$

then

$$x'' + 4\zeta\pi b x' + 4\pi^2 b^2 x = \frac{4\pi^2 b^2}{k}f(T) \tag{7.3-22}$$

In terms of the new time scale, the forcing function shown in Figure 7.3-7 can be expressed as

$$f(T) = \begin{cases} F_o T & 0 \leq T \leq 1 \\ F_o & T > 1 \end{cases}$$

Since F_o/k is the static deflection δ, we can express the right-hand side of Equation 7.3-22 as

$$\frac{4\pi^2 b^2}{k}f(T) = 4\pi^2 b^2 \begin{cases} \delta T & 0 \leq T \leq 1 \\ \delta & T > 1 \end{cases}$$

Thus we have reduced the problem from one having five parameters to one having only three parameters: b, ζ, and $\delta = F_o/k$. Furthermore, for the input shown in Figure 7.3-7, we can reduce the number of parameters to just two, b and ζ, by defining a new displacement variable:

$$z = \frac{x}{\delta} = \frac{kx}{F_o} \qquad (7.3\text{-}23)$$

Then the equation of motion becomes

$$z'' + 4\zeta\pi b z' + 4\pi^2 b^2 z = 4\pi^2 b^2 \begin{cases} T & 0 \le T \le 1 \\ 1 & T > 1 \end{cases} \qquad (7.3\text{-}24)$$

Damping Effects

Section 7.7 contains a MATLAB program based on Equation 7.3-24. The program was used to obtain the plot shown in Figure 7.3-11, which shows the relative maximum response kx_{max}/F_o for four values of damping. Recall that for no damping, unless we know the value of t_1 accurately, the response will be much greater than desired. With damping, however, the plot shows that if you can use an isolator that is critically damped ($\zeta = 1$), the input in Figure 7.3-7 will never produce a displacement greater than the static deflection. With metal and elastomer isolators, however, damping values are usually low, and something like a friction damper would be needed. The situation is improved even with slight damping, say $\zeta = 0.2$. So damping can be used to protect against uncertainty in the value of t_1.

Damping can also reduce the transmitted force. Figure 7.3-12 shows the maximum relative force transmitted to the base, F_t/F_o, as a function of t_1/T_n. Again we see that the critically damped isolator provides the greatest isolation. For small values of t_1/T_n, the isolation is essentially the same for all values of damping. This plot was obtained by numerical simulation using a dimensionless parameter analysis similar to that used to obtain Equation 7.3-22.

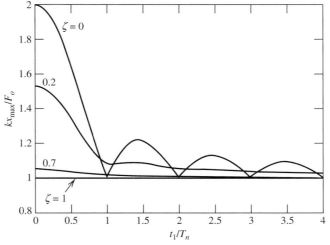

FIGURE 7.3-11   Damped response spectrum for the modified ramp input.

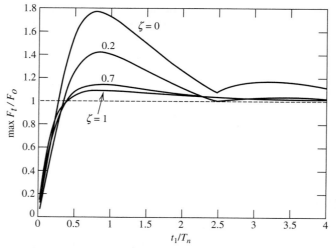

FIGURE 7.3-12 Spectrum of transmitted force for the modified ramp input.

Practical Isolator Design

Commercially available isolators consist of a mount and an elastic material. Vibration isolator design consists of using the above formulas to compute the required values for the material's damping and stiffness. Designing the isolator also must take into account any requirements or constraints on size, shape, and weight imposed on the mount by the particular application. The designer then must look at vendor catalogs for existing mounts and materials that have values of damping and stiffness near the required values.

If none can be found, there is often enough latitude to recompute another set of damping and stiffness values (that is, usually there will be more than one isolator design that will meet the specifications). Of course, other factors normally considered in engineering design, such as cost, ease of installation, reliability, and availability, must also be considered.

The transmissibility formulas in this section assume that the input is harmonic and that the motion has reached steady state. In many applications, such as with reciprocating engines, vehicle suspensions, and earthquakes, the input does not have a constant frequency, may not be harmonic, and may even be somewhat random. In light of this, the forcing frequency used for the analysis can be chosen to represent a bound on the range of actual frequencies to be encountered, or it might represent the frequency of the input that has the greatest effect on the system (for example, the one that has the largest amplitude). Often the engineer must design an isolator to provide isolation over a range of frequencies; in such cases, the analysis must consider the isolator's performance over a range of frequency ratios r.

Applications where the input is not random but not harmonic must be handled with the transient-response methods developed in Chapter 5. Sometimes the transient requirements conflict with the steady-state requirements. For example, when the excitation is a step function, the isolator must provide protection over a wide range of frequencies, whereas to protect against sinusoidal inputs the isolator's stiffness and damping are chosen to give a small ζ and $r > \sqrt{2}$. Thus we often find that an isolator that provides good protection against harmonic inputs will often provide poor protection against shock inputs, and vice versa.

Usually the mass is specified by the machine or object we are trying to isolate, and we choose the isolator's stiffness so that $r > \sqrt{2}$. If that is difficult to do, noting that $r = \omega\sqrt{m/k}$, we see that adding mass to the machine increases r and thus can enable an acceptable design to be achieved.

Commercially Available Isolators

A large number of companies offer vibration isolators, and there is a wide range of designs available. In addition to the helical-coil springs and leaf springs we have seen in earlier chapters, a common type of isolator is made of rubber or another elastomer. Two designs are shown in Figure 7.3-13, and Figure 7.3-14 shows a typical load-deflection plot supplied by the vendor for the type of isolator shown in Figure 7.3-13a. The short dashes denote the maximum safe static deflection, although the curve above the dash can be used for dynamic situations. Such isolators typically have damping ratios from 0.02 to 0.1, with the damping increasing with stiffness. While the graph shown is for compressive loading, note that these isolators can also be used in shear or torsion, with a different load-deflection curve for each direction.

Rubber isolators can be molded in different shapes to provide desired nonlinear load-deflection characteristics. Air springs, often seen in the suspensions of large moving vans, consist of a rubber bladder whose stiffness can be changed by varying the air pressure.

A *Belleville spring* is shown in Figure 7.3-15a. These can be stacked in series or parallel arrangements, as shown in parts b, c, and d to achieve different equivalent stiffness values.

Nonlinear load-deflection curves can be achieved in a variety of ways. The variable pitch and tapered springs shown in Figure 7.3-16 are one method. Figure 7.3-17 shows the use of mechanical stops. Figure 7.3-18 shows two ways of creating a hardening spring with a cantilever beam. The screws in Figure 7.3-18b enable the deflection characteristics to be adjusted.

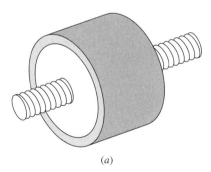

(a)

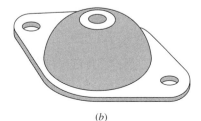

(b)

FIGURE 7.3-13 Two types of rubber isolators.

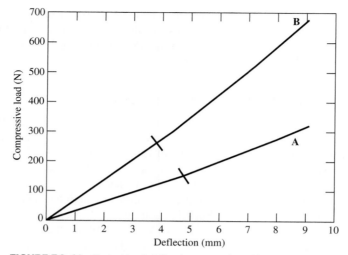

FIGURE 7.3-14 Typical load-deflection curves for rubber isolators.

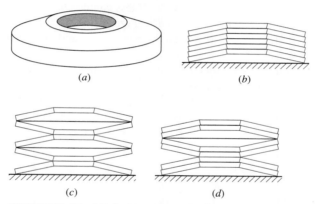

FIGURE 7.3-15 A Belleville spring with series and parallel arrangements.

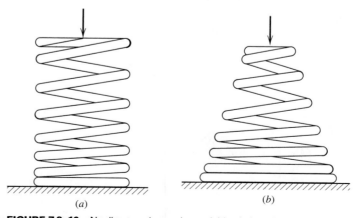

FIGURE 7.3-16 Nonlinear springs using variable pitch and tapering.

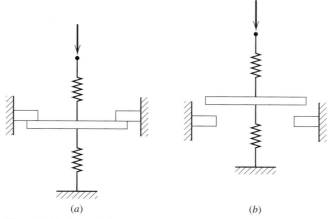

FIGURE 7.3-17 Nonlinear springs created with mechanical stops.

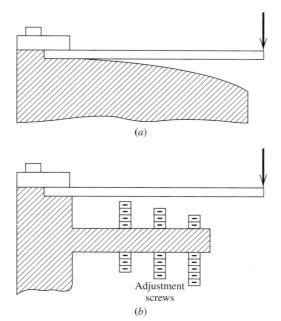

(a)

Adjustment
screws
(b)

FIGURE 7.3-18 Hardening springs created
by varying the effective beam length.

7.4 ISOLATION WITH BASE MOTION

A common input is the motion of a base support (called base excitation or sometimes *seismic* excitation). Sometimes we must reduce the effects of a force transmitted to the supporting structure, and sometimes we must reduce the output displacement caused by an input displacement. So both force isolation and displacement isolation are important factors in the design of the isolator.

The motion of the mass shown in Figure 7.4-1 is produced by the motion $y(t)$ of the base. This system is a model of many common displacement isolation systems. Assuming

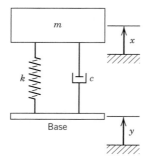

FIGURE 7.4-1 A mass having base motion excitation.

that the mass displacement x is measured from the rest position of the mass when $y = 0$, the weight mg is canceled by the static spring force. The force transmitted to the mass by the spring and damper is denoted f_t and is given by

$$f_t = c(\dot{y} - \dot{x}) + k(y - x) \tag{7.4-1}$$

This gives the following equation of motion:

$$m\ddot{x} = f_t = c(\dot{y} - \dot{x}) + k(y - x)$$

or

$$m\ddot{x} + c\dot{x} + kx = c\dot{y} + ky \tag{7.4-2}$$

Base motion was studied in Chapter 4, and here we summarize the results derived in that chapter. The pertinent transfer functions are

$$\frac{X(s)}{Y(s)} = \frac{cs + k}{ms^2 + cs + k} = \frac{2\zeta\omega_n s + \omega_n^2}{s^2 + 2\zeta\omega_n s + \omega_n^2} \tag{7.4-3}$$

where $\omega_n^2 = k/m$ and $\zeta = c/2\sqrt{km}$. This ratio can be used to analyze the effects of the base motion $y(t)$ on $x(t)$, the motion of the mass. Notice that this transfer function has numerator dynamics, so its frequency response plots will be different from those of the transfer function $1/(ms^2 + cs + k)$.

The second transfer function is

$$\frac{F_t(s)}{Y(s)} = (cs + k)\frac{ms^2}{ms^2 + cs + k} \tag{7.4-4}$$

This is the ratio of the transmitted force to the base motion. It is customary to use instead the ratio $F_t(s)/kY(s)$, which is a dimensionless quantity representing how the base displacement y affects the force transmitted to the mass. Thus

$$\frac{F_t(s)}{kY(s)} = \frac{cs + k}{k}\frac{ms^2}{ms^2 + cs + k} = \frac{2\zeta\omega_n s + \omega_n^2}{\omega_n^2}\frac{s^2}{s^2 + 2\zeta\omega_n s + \omega_n^2} \tag{7.4-5}$$

The ratio F_t/kY is called the *force transmissibility*. It can be used to compute the transmitted force $f_t(t)$ that results from a specified base motion $y(t)$.

The magnitude of the displacement transmissibility is

$$\frac{X}{Y} = \sqrt{\frac{1 + (2\zeta r)^2}{(1 - r^2)^2 + (2\zeta r)^2}} \tag{7.4-6}$$

where the *frequency ratio* is

$$r = \frac{\omega}{\omega_n} \tag{7.4-7}$$

This expression can be used to calculate the amplitude X of the steady-state motion caused by a sinusoidal input displacement of amplitude Y.

Because ωX and $\omega^2 X$ are the magnitudes of velocity and acceleration, the ratio X/Y is also the ratio of the velocity magnitudes $|\dot{x}/\dot{y}|$ and the acceleration magnitudes $|\ddot{x}/\ddot{y}|$. Thus, for example, if we want the isolator to transmit only 20% of the base *acceleration*, then we can use Equation 7.4-6 and set $X/Y = 0.2$.

Note that although the expression for displacement transmissibility given by Equation 7.4-6 is identical to that of *force* transmissibility given by Equation 7.3-8 for a fixed base, the expressions arise from different physical applications.

We can express the magnitude of the force transmissibility as follows:

$$\frac{F_t}{kY} = r^2 \sqrt{\frac{1 + (2\zeta r)^2}{(1 - r^2)^2 + (2\zeta r)^2}} \tag{7.4-8}$$

This expression can be used to calculate the steady-state amplitude F_t of the transmitted forced caused by a sinusoidal input displacement of amplitude Y. Comparing Equations 7.4-6 and 7.4-8 we see that

$$\frac{F_t}{kY} = r^2 \frac{X}{Y} \tag{7.4-9}$$

and

$$F_t = r^2 k X \tag{7.4-10}$$

These relations enable us to calculate F_t/kY and F_t easily if we have previously computed X/Y and X.

It is instructive to plot X/Y and F_t/kY versus the frequency ratio r for various values of the damping ratio ζ. These plots are shown in Figures 7.4-2 and 7.4-3. Consider

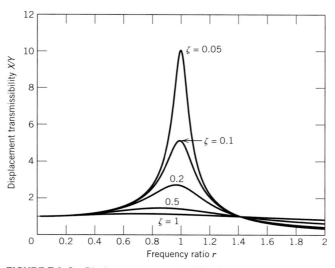

FIGURE 7.4-2 Displacement transmissibility for base excitation.

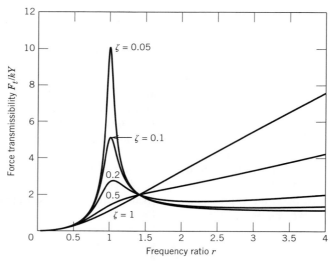

FIGURE 7.4-3 Force transmissibility for base excitation.

first the displacement transmissibility plot in Figure 7.4-2. When r is near 1, the forcing frequency ω is near the system's resonance frequency, and the curve is at a maximum. This means that the maximum base motion is transferred to the mass when r is near 1. Note also that the transmissibility can be greater than 1, which indicates that the base motion can be amplified.

In fact, the displacement transmissibility is greater than 1 if $r < \sqrt{2}$ and is less than 1 when $r \geq \sqrt{2}$. If $r \geq \sqrt{2}$, the displacement transmissibility decreases as r is increased. Note also that for a specific value of r, the displacement transmissibility decreases as ζ is increased.

Figure 7.4-3 shows the plots of F_t/kY versus r, for several values of ζ. When $r \geq \sqrt{2}$, the force transmissibility does not necessarily decrease as r is increased. For example, if $\zeta = 1$, the force transmissibility increases with r. If ζ is small, say $\zeta = 0.05$, the force transmissibility decreases with r. This is in contrast with the behavior of the displacement transmissibility, which always decreases with r for any value of ζ, as long as $r \geq \sqrt{2}$.

The formulas and plots for force and displacement transmissibility can be used to design isolators to protect objects from unwanted vibration. The plots of X/Y and F_t/Y versus r show that the values of X/Y and F_t/Y are very sensitive to the value of ζ when r is near 1. Therefore, because the value of c is usually difficult to estimate from data, and thus ζ is difficult to compute accurately, you should be careful in using the formulas for X/Y and F_t/Y to design isolators when r is near 1.

Table 7.4-1 summarizes these results.

Frequency Ratio as a Function of Displacement Transmissibility

Usually the displacement transmissibility $T_r = X/Y$ is specified and we must determine the required frequency ratio r. Expressing Equation 7.4-6 as

$$T_r^2 = \frac{1 + (2\zeta r)^2}{(1 - r^2)^2 + (2\zeta r)^2}$$

TABLE 7.4-1 **Transmissibility Formulas for Base Excitation**

Model:

$$m\ddot{x} + c\dot{x} + kx = c\dot{y} + ky \qquad y(t) = Y\sin\omega t$$

$$\zeta = \frac{c}{2\sqrt{mk}} \qquad \omega_n = \sqrt{\frac{k}{m}} \qquad r = \frac{\omega}{\omega_n}$$

Displacement transmissibility:

$$\frac{X}{Y} = T_r = \sqrt{\frac{1 + (2\zeta r)^2}{(1 - r^2)^2 + (2\zeta r)^2}}$$

Force transmissibility:

$$\frac{F_t}{kY} = r^2\frac{X}{Y} = r^2 T_r$$

and rearranging, we obtain

$$T_r^2 r^4 + (4\zeta^2 T_r^2 - 2T_r^2 - 4\zeta^2)r^2 + T_r^2 - 1 = 0 \qquad (7.4\text{-}11)$$

This is quadratic in r^2 and can be solved easily if ζ and T_r are specified.

The following example illustrates that we sometimes must add weight to achieve the desired transmissibility. Note also that we do not always need the value of the isolator stiffness, but rather we can sometimes work directly from its load-deflection curve.

Figure 7.4-4 shows the load-deflection curve of two typical cylindrical rubber isolators, like that shown in Figure 7.3-13a. The damping ratio for such isolators is typically between 0.02 and 0.1, with the stiffer isolator having the higher damping ratio. The safe static load for each isolator is marked with a dash on the graph. Data on the curve above the safe static load can be used for calculations involving dynamic loads.

EXAMPLE 7.4-1
Design of an
Isolator

A 40-kg instrument is used on a table that vibrates due to nearby machinery. The principal frequency component is 1100 rpm. Determine whether or not the isolator whose load-deflection curve is given in Figure 7.4-4 can be used at each of the four corners of the instrument so that no more than 20% of the table motion is transmitted to the instrument.

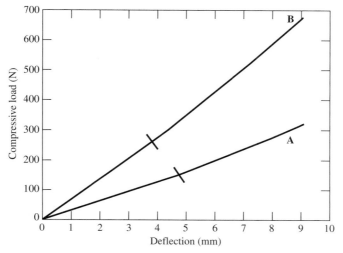

FIGURE 7.4-4 Load-deflection curves of two rubber isolators.

Solution

The specified transmissibility is the displacement transmissibility, so that we require $X/Y = 0.2$. Choosing isolator A and assuming its damping ratio to be 0.02, we find that Equation 7.4-6 gives

$$\frac{X}{Y} = \sqrt{\frac{1 + (0.04r)^2}{(1 - r^2)^2 + (0.04r)^2}} = 0.2$$

Squaring both sides and collecting terms, we obtain

$$r^4 - 2.0384r^2 - 24 = 0$$

The solution for r^2 is $r^2 = 6.023$, which gives $r = 2.454$.

Therefore, the natural frequency of the isolated system must be

$$\omega_n = \frac{2\pi(1100)/60}{2.454} = 47 \text{ rad/s}$$

The static deflection will be

$$\delta = \frac{g}{\omega_n^2} = 4.43 \times 10^{-3} \text{ m}$$

Because we are using four isolators, the static load on each isolator (due to the weight of the instrument) will be $40(9.8)/4 = 98$ N. The graph for isolator A shows that the isolator deflection for a load of 98 N will be 3.1 mm. The safe static load for this isolator, however, is 150 N. So we should mount the instrument on a block weighing $4(150 - 98) = 208$ N, and place the isolator between the block and the table. ∎

EXAMPLE 7.4-2
Transmissibility of a Printed Circuit Board

A printed circuit board (PCB) may be sensitive to vibration because vibration can loosen the soldered joints attaching the components (resistors and capacitors, for example). Figure 7.4-5 shows a PCB supported by a chassis that is attached to a vibrating motor. The board is 1.6 mm thick, 178 mm wide, 200 mm long, and has a mass of 0.45 kg. Neglect damping, model the board as a fixed-fixed beam, determine its stiffness and natural frequency, and compute the displacement transmissibility if the chassis vibrates at 60 Hz due to motor unbalance. The board is made from epoxy fiberglass for which $E = 1.38 \times 10^{10}$ N/m².

Solution

The board stiffness is found from

$$k = \frac{16Ewh^3}{L^3} = \frac{16(1.38 \times 10^{10})(0.178)(0.0016)^3}{(0.2)^3} = 2.012 \times 10^4 \text{ N/m}$$

The equivalent mass m_e is one-half the beam mass, so $m_e = 0.45/2 = 0.225$ kg. Thus the natural frequency is $\omega_n = \sqrt{(2.012 \times 10^4)/0.225} = 299$ rad/s.

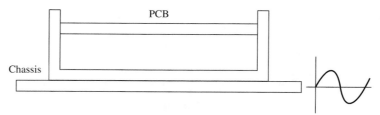

FIGURE 7.4-5 A PCB supported by a chassis undergoing base excitation.

The frequency ratio is

$$r = \frac{60(2\pi)}{299} = \frac{377}{299} = 1.261$$

Using the displacement transmissibility formula for base motion, we obtain

$$T_r = \frac{1}{|1 - r^2|} = 1.69$$

which is 169%. ∎

EXAMPLE 7.4-3 *Isolation of a* *Printed Circuit* *Board*	Consider the printed circuit board of Example 7.4-2. We want to reduce the displacement transmissibility from 169% to 20% by using an isolator, as shown in Figure 7.4-6. We are told that the chassis mass is one-tenth the mass of the PCB. Obtain the required values for the isolator stiffness k_2 and the damping c.
Solution	If we account for the flexibility of the PCB, we will have a system with two degrees of freedom. Since our design formulas apply only for one degree of freedom, we will initially treat the board as rigid and examine the effects of this assumption later. So we model the PCB-chassis system as a rigid body having a mass m equal to the board mass plus the chassis mass. Thus $m = 0.45 + 0.045 = 0.495$ kg.

Standard rubber isolators typically have damping ratios between 0.02 and 0.1. Choosing $\zeta = 0.02$ and using $T_r = 0.2$, we obtain from Equation 7.4-6 $r = 2.4634$. The required stiffness is found from

$$k_2 = m\frac{\omega^2}{r^2} = 0.495\frac{(377)^2}{(2.4634)^2} = 1.159 \times 10^4 \text{ N/m}$$

The damping is found from

$$c = 2\zeta\sqrt{mk_2} = 3.03 \text{ N} \cdot \text{s/m}$$

Other solutions are possible. For example, choosing $\zeta = 0.1$, we obtain $r = 2.5686$ and $k_2 = 1.066 \times 10^4$ N/m.

Usually the precise value required for the isolator stiffness is not available, and in practice we must choose an isolator whose stiffness is close to the required value. ∎

EXAMPLE 7.4-4 *Simulation of an* *Isolator Design*	In Example 7.4-3 we designed an isolator by treating the PCB-chassis system as a rigid body. Incorporating the board flexibility in the model gives the two-degrees-of-freedom system shown in Figure 7.4-7. Determine the displacement transmissibility of this system using the isolator values from Example 7.4-3, $k_2 = 1.159 \times 10^4$ N/m and $c = 3.03$ N·s/m.

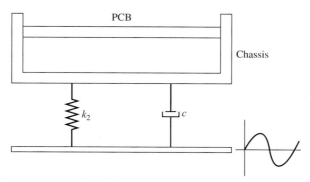

FIGURE 7.4-6 Use of an isolator with a PCB.

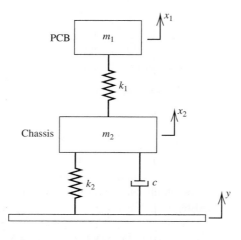

FIGURE 7.4-7 Isolation model of a PCB that includes board flexibility.

Solution | The equations of motion are

$$m_1\ddot{x}_1 = -k_1(x_1 - x_2)$$

$$m_2\ddot{x}_2 = k_1(x_1 - x_2) + k_2(y - x_2) + c(\dot{y} - \dot{x}_2)$$

where $k_1 = 2.012 \times 10^4$ N/m is the PCB stiffness and $m_1 = m_e = 0.225$ kg, the equivalent mass of the PCB. The transfer function is

$$\frac{X_1(s)}{Y(s)} = \frac{k_1(cs + k_2)}{m_1m_2s^4 + m_1cs^3 + (m_1k_1 + m_1k_2 + m_2k_1)s^2 + ck_1s + k_1k_2}$$

Since the forcing frequency is 377 rad/s, we substitute $s = 377i$ and use the given parameter values to obtain

$$\left|\frac{X_1(377i)}{Y(377i)}\right| = 0.2882$$

(This calculation is easily done with the MATLAB `freqresp` function.) So the displacement transmissibility is 29%, as compared to the desired value of 20%.

The difference is due to the fact that we computed k_2 by ignoring the flexibility of the PCB. The difference could be greater if we were forced to use the stiffness provided by a commercially available isolator.

If instead the chassis mass were 40% of the board mass, the required stiffness would be $k_2 = 1.476 \times 10^4$ N/m and the transmissibility would be 54%. For a chassis mass equal to 70% of the board mass, transmissibility would be 120%. However, for a chassis mass equal to twice the board mass, the transmissibility would be 77%, and it would be 34% if the chassis mass is three times the board mass.

To understand what is happening, let the frequency ω_2 denote the design frequency for the isolator:

$$\omega_2 = \sqrt{\frac{k_2}{m_{\text{board}} + m_{\text{chassis}}}} = \frac{377}{r}$$

Thus ω_2 is constant for a given r value. In our case, since $r = 2.4634$, $\omega_2 = 153$ rad/s regardless of the ratio $m_{\text{chassis}}/m_{\text{board}}$.

Note also that in treating the PCB-chassis system as a rigid body, we neglected the mode in which the board and the chassis are oscillating in opposite directions. The frequency of this mode, denoted ω_3, is approximately given by

$$\omega_3 = \sqrt{\frac{k_1}{m_e + m_{\text{chassis}}}}$$

The following table shows these frequencies.

Case	$m_{\text{chassis}}/m_{\text{board}}$	$T_r - 20\%$	ω_2 (rad/s)	ω_3 (rad/s)
1	0.1	$28\% - 20\% = 8\%$	153	273
2	0.4	$54\% - 20\% = 34\%$	153	223
3	0.7	$120\% - 20\% = 100\%$	153	193
4	1.4	$166\% - 20\% = 146\%$	153	153
5	2	$77\% - 20\% = 57\%$	153	134
6	3	$34\% - 20\% = 14\%$	153	113

Notice that the system having the largest discrepancy between the actual transmissibility and the desired transmissibility is case 4, the system for which the approximate mode frequency ω_3 is closest to the isolator design frequency ω_2. That is, for this case the isolator was designed to have a frequency near the frequency of the mode that was neglected in the model.

Base Motion as a Shock Input

A function commonly used to model a pulse input is the half-sine function shown in Figure 7.4-8. The duration of the pulse is t_1. If this pulse represents the base motion $y(t)$ of our standard base motion model (Figure 7.4-1), then its shock spectrum is that shown in Figure 7.4-9, where T_n is the natural undamped period of the system: $T_n = 2\pi/\omega_n$. This plot was obtained by numerical simulation of the following model, which was derived using an approach similar to that used in Section 7.3 (see Section 7.7).

$$x'' + 4\zeta\pi b x' + 4\pi^2 b^2 x = 4\pi\zeta b y' + 4\pi^2 b^2 y \tag{7.4-12}$$

where

$$b = \frac{t_1}{T_n} \tag{7.4-13}$$

and the time scale is $T = t/t_1$.

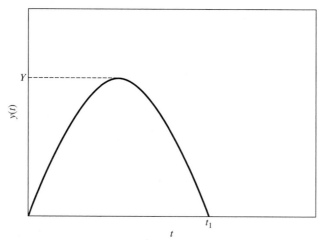

FIGURE 7.4-8 The half-sine function.

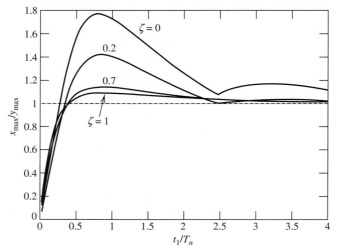

FIGURE 7.4-9 Shock spectrum for base excitation with a half-sine displacement.

If we are interested by isolating the displacement of the mass from the base motion, then the isolator is considered to be effective if $x_{max} \leq y_{max}$; that is, if $x_{max}/y_{max} \leq 1$. This effective range is indicated in Figure 7.4-9. We see that increasing the damping from $\zeta = 0$ to $\zeta = 1$ slightly increases the effective range from $t_1/T_n = 0.25$ to $t_1/T_n = 0.38$ approximately, but increasing the damping also greatly decreases the maximum response x_{max} outside the effective range.

7.5 DYNAMIC VIBRATION ABSORBERS

A *vibration absorber* is useful for situations in which the disturbance has a constant frequency. As opposed to a vibration *isolator*, which contains stiffness and damping elements, a vibration *absorber* is a device consisting of another mass and a stiffness element that are attached to the main mass to be protected from vibration. The new system consisting of the main mass and the absorber mass has two degrees of freedom, and thus the new system has two natural frequencies.

If we know the frequency of the disturbing input and the natural frequency of the original system, we can select values for the absorber's mass and stiffness so that the motion of the original mass is very small, which means that its kinetic and potential energies will be small. In order to achieve this small motion, the energy delivered to the system by the disturbing input must be "absorbed" by the absorber's mass and stiffness. Thus the resulting absorber motion will be large.

Because the principle of the absorber depends on the absorber's motion, such devices are sometimes called *dynamic* vibration absorbers. Their purpose is different from that of vibration *isolators*, which are intended to isolate one part of the structure from an excitation or from another part that is vibrating. Another term used for vibration absorber is a *tuned mass damper*.

Examples of Vibration Absorbers

Vibration absorbers are often found on devices that run at constant speed. These include saws, sanders, shavers, and devices powered by ac motors, because such motors are usually designed to operate at constant speed.

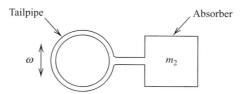

Tailpipe

Absorber

ω

m_2

FIGURE 7.5-1   A vibration absorber for an exhaust pipe.

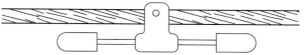

FIGURE 7.5-2   A Stockbridge damper.

Figure 7.5-1 illustrates a vibration absorber that is used on some passenger cars to reduce the vibration of the exhaust pipe. It consists of a cantilever beam about 3 in. long and clamped to the pipe. The block at the beam end is the absorber mass.

Power lines often have vibration absorbers, called Stockbridge dampers, that are shaped somewhat like elongated dumbbells (Figure 7.5-2). These protect the lines from excessive vibration caused by the wind.

Many modern buildings and bridges have vibration absorbers to counteract the effects of wind and earthquakes. These are quite large; for example, the one in the Citicorp building in New York City uses a 370-ton concrete block that is allowed to slide horizontally like that shown in Figure 7.5-3. Some bridges and buildings use a pendulum damper. The designer must properly select the pendulum inertia and the pendulum length.

Analysis of a Vibration Absorber

Figure 7.5-4 illustrates a simple vibration absorber. The absorber consists of a mass m_2 and stiffness element k_2 connected to the main mass m_1. The disturbing input is the

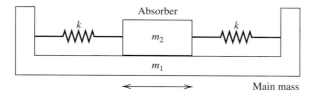

Absorber

k

m_2

k

m_1

Main mass

FIGURE 7.5-3   Vibration absorber for a tall building.

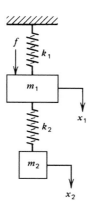

f

k_1

m_1

x_1

k_2

m_2

x_2 **FIGURE 7.5-4**   Model of a vibration absorber.

applied force $f(t)$, which might be due to a rotating unbalance or ground motion, for example.

The equations of motion for the system are

$$m_1\ddot{x}_1 = -k_1 x_1 - k_2(x_1 - x_2) + f$$

$$m_2\ddot{x}_2 = k_2(x_1 - x_2)$$

We find the transfer functions in the usual way:

$$(m_1 s^2 + k_1 + k_2)X_1(s) - k_2 X_2(s) = F(s)$$

$$-k_2 X_1(s) + (m_2 s^2 + k_2)X_2(s) = 0$$

Solve the second equation for $X_2(s)$ and substitute the result into the first equation to eliminate $X_2(s)$. Then solve for the transfer function $T_1(s) = X_1(s)/F(s)$ to obtain

$$T_1(s) = \frac{X_1(s)}{F(s)} = \frac{m_2 s^2 + k_2}{(m_1 s^2 + k_1 + k_2)(m_2 s^2 + k_2) - k_2^2} \tag{7.5-1}$$

Similarly, solve for the transfer function $T_2(s) = X_2(s)/F(s)$ to obtain

$$T_2(s) = \frac{X_2(s)}{F(s)} = \frac{k_2}{(m_1 s^2 + k_1 + k_2)(m_2 s^2 + k_2) - k_2^2} \tag{7.5-2}$$

If the applied force $f(t)$ is sinusoidal with a frequency ω, we can apply the frequency transfer functions to design a vibration absorber. These transfer functions are obtained by substituting $s = i\omega$ into the transfer functions $T_1(s)$ and $T_2(s)$ to obtain

$$T_1(i\omega) = \frac{k_2 - m_2\omega^2}{(k_1 + k_2 - m_1\omega^2)(k_2 - m_2\omega^2) - k_2^2}$$

$$= \frac{1}{k_1} \frac{1 - \dfrac{m_2}{k_2}\omega^2}{\left(1 + \dfrac{k_2}{k_1} - \dfrac{m_1}{k_1}\omega^2\right)\left(1 - \dfrac{m_2}{k_2}\omega^2\right) - \dfrac{k_2}{k_1}}$$

Define the frequency ratios r_1 and r_2 to be

$$r_1 = \frac{\omega}{\omega_{n1}} = \frac{\omega}{\sqrt{k_1/m_1}} \tag{7.5-3}$$

$$r_2 = \frac{\omega}{\omega_{n2}} = \frac{\omega}{\sqrt{k_2/m_2}} \tag{7.5-4}$$

Then we can write $T_1(i\omega)$ as follows:

$$T_1(i\omega) = \frac{1}{k_1} \frac{1 - r_2^2}{\left(1 + \dfrac{k_2}{k_1} - r_1^2\right)\left(1 - r_2^2\right) - \dfrac{k_2}{k_1}}$$

Similarly we obtain

$$T_2(i\omega) = \frac{k_2}{(k_1 + k_2 - m_1\omega^2)(k_2 - m_2\omega^2) - k_2^2} = \frac{1}{k_1} \frac{1}{\left(1 + \dfrac{k_2}{k_1} - r_1^2\right)\left(1 - r_2^2\right) - \dfrac{k_2}{k_1}}$$

Define

$$b = \frac{\omega_{n2}}{\omega_{n1}}$$

$$\mu = \frac{m_2}{m_1}$$

and note that

$$\frac{k_2}{k_1} = \frac{m_2}{m_1} \frac{m_1}{k_1} \frac{k_2}{m_2} = \mu \left(\frac{\omega_{n2}}{\omega_{n1}} \right)^2 = \mu b^2$$

and

$$r_1^2 = \left(\frac{\omega}{\omega_{n1}} \right)^2 = \left(\frac{\omega_{n2}}{\omega_{n1}} \frac{\omega}{\omega_{n2}} \right)^2 = b^2 r_2^2$$

Thus $T_1(i\omega)$ and $T_2(i\omega)$ become

$$T_1(i\omega) = \frac{X_1(i\omega)}{F(i\omega)} = \frac{1}{k_1} \frac{1 - r_2^2}{(1 + \mu b^2 - b^2 r_2^2)(1 - r_2^2) - \mu b^2}$$

$$= \frac{1}{k_1} \frac{1 - r_2^2}{b^2 r_2^4 - [1 + (1 + \mu)b^2]r_2^2 + 1} \qquad (7.5\text{-}5)$$

$$T_2(i\omega) = \frac{X_2(i\omega)}{F(i\omega)} = \frac{1}{k_1} \frac{1}{b^2 r_2^4 - [1 + (1 + \mu)b^2]r_2^2 + 1} \qquad (7.5\text{-}6)$$

Note that there are no imaginary terms in these expressions. Thus both frequency transfer functions are real numbers.

Because $X_1(s) = T_1(s)F(s)$, we have $X_1(i\omega) = T_1(i\omega)F(i\omega)$. Therefore, if the applied force is $f(t) = F \sin \omega t$, the steady-state motion of mass m_1 will be given by $x_1(t) = X_1 \sin(\omega t + \phi_1)$, where

$$X_1 = |T_1(i\omega)|F$$

For m_1 to be motionless requires that $X_1 = 0$, which can be achieved if $|T_1(i\omega)| = 0$. From Equation 7.5-6, we see that $|T_1(i\omega)| = 0$ if $1 - r_2^2 = 0$, which implies that $r_2 = \pm 1$. Because r_2 cannot be negative by definition, we see that the absorber design equation is given by $r_2 = 1$; that is,

$$r_2 = \frac{\omega}{\omega_{n2}} = 1 \qquad (7.5\text{-}7)$$

or

$$\omega_{n2} = \sqrt{\frac{k_2}{m_2}} = \omega \qquad (7.5\text{-}8)$$

Thus the mass m_1 will be motionless if we select an absorber having the same natural frequency ω_{n2} as the frequency ω of the applied force. If this is done, the absorber is said to be "tuned" to the input frequency.

If $r_2 = 1$, the expression for $T_2(i\omega)$ becomes

$$T_2(i\omega) = \frac{X_2(i\omega)}{F(i\omega)} = \frac{1}{k_1} \frac{1}{b^2 - 1 - (1 + \mu)b^2 + 1} = -\frac{1}{k_2}$$

Thus, if the absorber is designed so that $r_2 = 1$, then

$$X_2(i\omega) = -\frac{1}{k_2} F(i\omega) \qquad (7.5\text{-}9)$$

and the amplitude of the absorber's motion will be

$$X_2 = |X_2(i\omega)| = \frac{1}{k_2}F \qquad (7.5\text{-}10)$$

If $r_2 = 1$, then $X_1 = 0$, and from Equation 7.5-10, because the transfer function $T_2(i\omega)$ is real and is negative, we can see that the absorber's spring force acting on the main mass is

$$k_2(x_2 - x_1) = k_2x_2 = k_2X_2 \sin(\omega t + \pi) = -k_2X_2 \sin \omega t$$

Because $X_2 = F/k_2$,

$$k_2x_2 = -k_2\left(\frac{1}{k_2}F\right)\sin \omega t = -F \sin \omega t \qquad (7.5\text{-}11)$$

Thus, if the absorber is tuned to the input frequency and its motion has reached steady state, the force acting on the absorber's mass has the same magnitude F as the applied force but is in the opposite direction. This causes the net force acting on the main mass to be zero; therefore, it does not move.

Equation 7.5-11 shows that in practice, the allowable clearance for the absorber's motion X_2 puts a limit on the allowable range of the absorber's stiffness k_2. Note also that the absorber's stiffness element k_2 must be able to support the force F and the resulting compression or extension X_2.

The frequency range over which the absorber is effective—that is, when the amplitude X_1 is less than it would be if no absorber were attached—is the frequency range such that $|k_1X_1/F| < 1$. This range is found by setting $k_1T_1(i\omega) = \pm 1$ in Equation 7.5-5 to obtain

$$k_1T_1(i\omega) = \frac{1 - r_2^2}{(1 + \mu b^2 - b^2r_2^2)(1 - r_2^2) - \mu b^2} = \pm 1$$

This results in two equations:

$$r_2 = \sqrt{1 + \mu} \qquad (7.5\text{-}12)$$

and

$$b^2r_2^4 - (2 + b^2 + \mu b^2)r_2^2 + 2 = 0 \qquad (7.5\text{-}13)$$

This is quadratic in r_2^2 and yields two solutions for r_2.

When the exciting force is due to rotating unbalance, the force amplitude is given by

$$F = mR\omega^2 \qquad (7.5\text{-}14)$$

where m is the unbalance mass, R is the eccentricity, and ω is the rotation speed.

EXAMPLE 7.5-1
Design of a Vibration Absorber

A certain machine with supports has a measured natural frequency of 3.43 Hz. The machine will be subjected to a rotating unbalance force having an amplitude of 13 N and a frequency of 3 Hz. Design a dynamic vibration absorber for this machine. The available clearance for the absorber's motion is 25 mm.

Solution

The natural frequency of the machine with its supports, in radians per second, is

$$\omega_{n1} = 2\pi(3.43) = 6.86\pi \text{ rad/s}$$

The frequency of the applied force in radians per second is

$$\omega = 2\pi(3) = 6\pi \text{ rad/s}$$

The absorber's design requires that $\omega_{n2} = \omega = 6\pi$. Thus

$$\omega_{n2} = \sqrt{\frac{k_2}{m_2}} = 6\pi$$

Solve for the mass:

$$m_2 = \frac{k_2}{36\pi^2}$$

The maximum allowable clearance is 25 mm. Using Equation 7.5-11, we obtain

$$0.025 = \frac{1}{k_2}F = \frac{13}{k_2}$$

or

$$k_2 = 520 \text{ N/m}$$

Thus the absorber's mass must be

$$m_2 = \frac{k_2}{36\pi^2} = \frac{520}{36\pi^2} = 1.46 \text{ kg} \qquad \blacksquare$$

Successful application of a vibration absorber depends on the disturbing frequency being known and constant. If the absorber's natural frequency is not exactly equal to the input frequency, the main mass will oscillate, and the amplitude of oscillation depends on the difference between the input frequency and the absorber's frequency.

Another difficulty with absorber design is that the combined system now has two natural frequencies. If the input frequency is close to one of the natural frequencies, resonance will occur.

Finally, the model used here does not include damping. If appreciable damping is present, motion of the main mass will not be small. It is advisable to perform a sensitivity analysis of the combined system to determine the effects of uncertainty in the input frequency.

EXAMPLE 7.5-2
Sensitivity Analysis in Absorber Design

Suppose the main mass in the system in Example 7.5-1 has the value $m_1 = 7.3$ kg. Evaluate the sensitivity of the absorber design to variations in the input frequency.

Solution

The absorber values from the previous example are $k_2 = 520$ N/m and $m_2 = 1.46$ kg. The original system's natural frequency was given as 6.86π rad/s. Thus the stiffness is

$$k_1 = (6.86\pi)^2 7.3 = 3391 \text{ N/m}$$

Figure 7.5-5 shows a plot of $k_1 T_1(i\omega)$ versus ω. Resonance occurs when the input frequency ω is near one of the two natural frequencies, which can be found from the roots of the denominator of either transfer function. These frequencies are 16.22 and 25.075 rad/s.

Here the mass ratio is $\mu = 1.46/7.3 = 0.2$ and the frequency ratio is $b = \omega_{n2}/\omega_{n1} = 6\pi/6.86\pi = 0.8746$. The solutions of Equations 7.5-12 and 7.5-13 are

$$r_2 = 0.946, \qquad 1.095, \qquad 1.708$$

Since $r_2 = \omega/6\pi$, these correspond to input frequencies of

$$\omega = 17.8, \qquad 20.7, \qquad 32.2 \text{ rad/s}$$

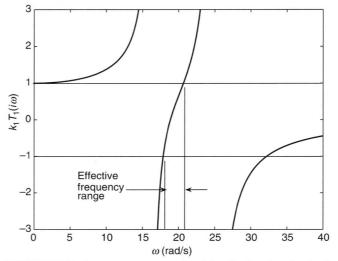

FIGURE 7.5-5 Frequency response plot of the vibration absorber for Example 7.5-2.

The first two frequencies define the effective frequency range of the absorber, which is $17.8 \leq \omega \leq 20.7$ rad/s. However, since we are discussing the sensitivity of the design to changes in the forcing frequency from its nominal value, the third solution, $\omega = 32.2$, is not needed for this analysis. ■

Damping in Vibration Absorbers

The analysis is this section is based on a model with no damping. This has two implications. First of all, the equations

$$X_1 = |T_1(i\omega)|F$$

$$X_2 = |T_2(i\omega)|F$$

are strictly true only if the system is *stable* and at steady state. Because our absorber model has no damping, it is not stable but is *neutrally stable*. Nevertheless, these equations are widely used because the inclusion of damping complicates the mathematics. The assumption made is that in practice every real system will have some damping. In addition, the *transient* response has not been accounted for, and this could affect the acceptability of absorber design. These aspects are explored in what follows.

The second implication is that the existence of damping in the real system could also affect the acceptability of absorber design. There has been much analysis of the design of absorbers with damping intentionally added, but this is a very complicated topic, and the results are not easily presented in a concise form.

Transient Response and Damping in a Vibration Absorber

The design equations for a vibration absorber are based on the *steady-state* response. Before accepting the design as final, however, you should examine the *transient* response

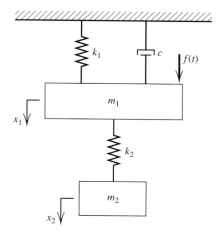

FIGURE 7.5-6   A vibration absorber model that includes damping between the main mass and the support.

of the absorber. To show the effects of damping, we now include a damper between the mass m_1 and the support (Figure 7.5-6). The transfer functions are

$$\frac{X_1(s)}{F(s)} = \frac{m_2 s^2 + k_2}{(m_1 s^2 + cs + k_1 + k_2)(m_2 s^2 + k_2) - k_2^2} \tag{7.5-15}$$

$$\frac{X_2(s)}{F(s)} = \frac{k_2}{(m_1 s^2 + cs + k_1 + k_2)(m_2 s^2 + k_2) - k_2^2} \tag{7.5-16}$$

Let us use the values calculated in Examples 7.5-1 and 7.5-2. These are $m_1 = 7.3$ kg, $m_2 = 1.46$ kg, $k_1 = 3391$ N/m, and $k_2 = 520$ N/m. The forcing function is $f(t) = 13 \sin 6\pi t$. The response without damping ($c = 0$) is shown in Figure 7.5-7. Note that beating occurs and so the masses do not oscillate at a single frequency, the frequency of the forcing function. This is due to the two natural frequencies, which are 6.16 and 20.88 rad/s. Note that the amplitude of the main mass m_1 is not zero, and the amplitude of the absorber mass m_2 is greater than the allowable clearance of 0.025 m, so we have obtained results different from those predicted by the steady-state analysis. The reason is that the steady-state analysis is based on an equation such as

$$X_1 = |T_1(i\omega)|F$$

which is true only for a *stable* system and predicts that the output x_1 will oscillate at the same frequency as the forcing function. Because the system on which the design equations are based has no damping, it is *neutrally stable*.

If we add damping, say $c = 70$ N·s/m, the total response is like that shown in Figure 7.5-8. Now we see that the absorber is oscillating at the forcing frequency, and its amplitude at steady state is 25 mm as predicted. The main mass is motionless at steady state, also as predicted.

From Figure 7.5-8, you might erroneously conclude that damping is causing the main mass m_1 to be motionless at steady state, but this is not true. Figure 7.5-9 shows the response for the case with damping, but where the absorber mass is one-fourth of its design value. Now we see that the main mass is not motionless but oscillates with an amplitude of approximately 5 mm.

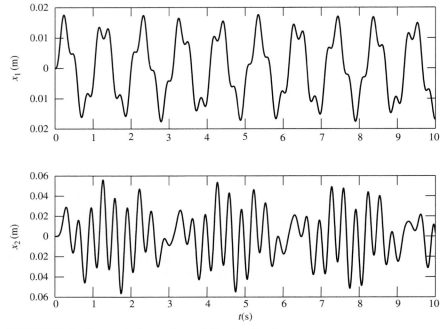

FIGURE 7.5-7 Response of an undamped vibration absorber.

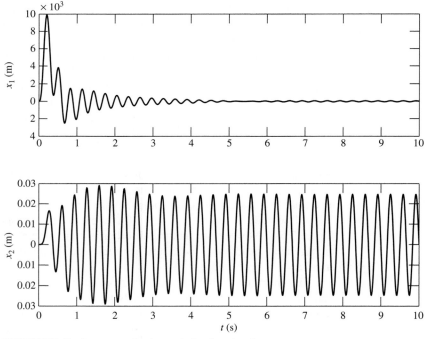

FIGURE 7.5-8 Response of a damped vibration absorber.

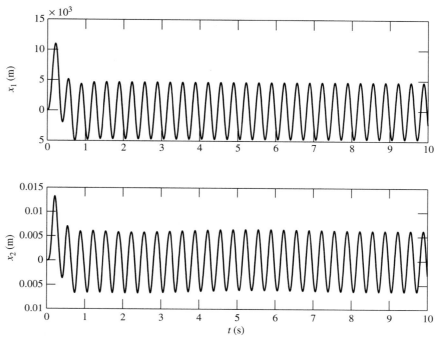

FIGURE 7.5-9   Response of a damped vibration absorber having a nonoptimal mass value.

7.6   ACTIVE VIBRATION CONTROL

Sophisticated actuators and sensors along with miniaturized, low-cost computers have greatly increased the number of applications for active control of vibration. The core of an active vibration controller is typically a microprocessor-based system with analog-to-digital converters to process sensor inputs and digital-to-analog converters to convert the microprocessor's output commands into signals to the actuator (Figure 7.6-1). The term *active* originates from the fact that such systems require an active power source.

   The actuator applies a force to the mass whose vibration is to be reduced. The sensor measures the motion of the mass (either its displacement, velocity, or acceleration,

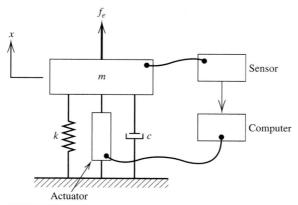

FIGURE 7.6-1   Structure of an active vibration control system.

depending on the application). The control logic, called the *control algorithm*, programmed in the computer uses these measurements to decide how much force should be applied.

The vibration suppression techniques considered thus far in this text have the disadvantage of working only at certain frequencies, because their stiffness and damping—and in the case of vibration absorbers, their mass—are fixed and cannot be changed without replacing the appropriate component. Another advantage of active vibration control systems is that they can change the effective stiffness and damping without adding much weight. Because such systems require a power source, many are designed to be *fail-safe*. For example, a fail-safe damper provides the minimum required damping in the event of a power supply or computer system failure.

Several terms are used to describe active vibration control systems. Structures such as buildings, bridges, aircraft cabins, and aircraft wings containing such systems are called *smart* or *intelligent structures*. Structural materials having the controller embedded in them are called *smart materials*. *Smart dampers* and *active shock absorbers* are dampers that can change their damping properties under computer control.

Applications of Active Vibration Control

Examples of current applications of this technology include the following:

- Reduction of building sway caused by wind and seismic waves using *hybrid mass dampers* that apply a controlled force to a large movable weight. Recall that the vibration absorbers we have thus far seen are limited because they must be designed for a known, fixed frequency.
- Active suspensions in motor vehicles.
- Reduction of aircraft cabin noise by reducing the vibration of the large panels of thin metal that form the cabin walls.
- Piezoelectric devices installed on the trailing edge of helicopter blades to dampen vibrations.
- Reduction of tabletop vibrations to the level necessary for precision manufacturing, such as the manufacture of semiconductors.
- Damping out floor vibrations with an electromagnetic shaker.
- Smart skis that dampen vibrations.
- Active dampers in bicycles.
- Vibration-dampening devices embedded in sporting equipment such as tennis racquets, snowboards, baseball bats, and golf clubs.
- Active stiffness control in running shoes.

Actuators for Active Vibration Control

Piezoelectric materials can convert electrical current into motion, and vice versa. They change shape when an electric current passes through them, and they generate an electric signal when they flex. Thus they can be used as actuators, to create force or motion, and as sensors, to sense motion.

Other materials used for high-precision actuation include electrostrictive and magnetostrictive materials, which are similar to piezoelectric materials. These are ferromagnetic materials that expand or contract when subjected to an electric or a magnetic field.

A magneto-rheological fluid changes its viscosity when subjected to a magnetic field. The viscosity of such fluids can be changed from that of water to that of pudding in milliseconds. Some engineering development has been done using so-called *magneto-rheological fluid dampers* for smart dampers.

Such actuators cannot provide enough force for larger applications such as vehicle suspensions and control of building motions. In building applications, hydraulic cylinders are usually used. With working hydraulic pressures commonly 2000 psi, a cylinder containing a piston whose area is only 1 square inch will generate 1 ton of force! Active vehicle suspensions use hydraulic devices, electric motors, and magneto-rheological fluid dampers.

Elementary Control Theory for Active Vibration Control

Consider the system shown in Figure 7.6-1. We wish to control the vibration of the mass m with the actuator that applies a force f to the mass. The force $f_e(t)$ is disturbance force such as that caused by rotating unbalance. The equation of motion is

$$m\ddot{x} + c\dot{x} + kx = f(t) + f_e(t) \qquad (7.6\text{-}1)$$

Assume we have a sensor that measures the displacement x and the velocity $\dot{x}$ of the mass. A computer takes these measurements in real time, computes the required force f necessary to control the motion, and commands the actuator to supply that amount of force. The stiffness k and damping c are either inherently part of the system structure or are inserted to provide for fail-safe operation.

If we program the computer to control the force f in a manner that is proportional to x and to $\dot{x}$, then the control algorithm is

$$f(t) = -K_P x - K_D \dot{x} \qquad (7.6\text{-}2)$$

where K_P and K_D are constants whose value must be determined by the designer and programmed into the computer. In control system terminology, these constants are called the *control gains*. K_P is the *proportional gain* and K_D is called the *derivative gain* or the *rate gain*. The control algorithm is called *proportional-plus-derivative control*, abbreviated as *PD control*.

Substituting Equation 7.6-2 into Equation 7.6-1 and collecting terms gives

$$m\ddot{x} + (c + K_D)\dot{x} + (k + K_P)x = f_e(t) \qquad (7.6\text{-}3)$$

We immediately see that the gain K_P acts like an artificial stiffness and the gain K_D like an artificial damper. Thus we can change the effectiveness stiffness and damping by properly selecting the gains.

For example, the effective damping ratio is

$$\zeta = \frac{c + K_D}{2\sqrt{m(k + K_P)}} \qquad (7.6\text{-}4)$$

and the effective natural frequency is

$$\omega_n = \sqrt{\frac{k + K_P}{m}} \qquad (7.6\text{-}5)$$

If $\zeta \leq 1$, then the time constant is

$$\tau = \frac{2m}{c + K_P} \qquad (7.6\text{-}6)$$

So if we know m, c, and k, we can compute the values of K_P and K_D to achieve desired values of ζ, ω_n, or τ. Note that the gains need not be positive. If the effective stiffness or damping needs to be smaller than k or c, then the gains will be negative.

EXAMPLE 7.6-1
Calculating
Control Gains

Suppose that $m = 100\,\text{kg}$, $c = 6000\,\text{N·m/s}$, and $k = 9 \times 10^6\,\text{N/m}$. The rotating unbalance force has an amplitude of 50 N with a frequency of 3000 rpm. Thus $f_e(t) = 50\sin(100\pi t)$. The resonant frequency of this system is 300 rad/s and is close to the frequency of the disturbance. In addition, the damping ratio is small ($\zeta = 0.1$). Assuming that c and k cannot be changed, calculate the control gains required to give a damping ratio of $\zeta = 0.5$ and a resonant frequency of 100 rad/s, well below the disturbance frequency. Evaluate the actuator force requirement at steady state.

Solution

From Equation 7.6-5,

$$\omega_n = 300 = \sqrt{\frac{9 \times 10^6 + K_P}{100}}$$

which gives $K_P = -8 \times 10^6$ N/m. This means that the effectiveness stiffness must be reduced to $k + K_P = 10^6$ N/m.

Using this value of K_P in Equation 7.6-4, we obtain

$$0.5 = \frac{6000 + K_D}{2\sqrt{100(10^6)}}$$

or $K_D = 4000$ N·s/m.

Substitute these values into the equation of motion to obtain

$$100\ddot{x} + 10^4\dot{x} + 10^6 x = 50\sin 100\pi t$$

or

$$\ddot{x} + 10^2\dot{x} + 10^4 x = 0.5\sin 100\pi t$$

The steady-state = response is

$$X(100\pi) = \left| \frac{0.5}{10^4 - (100\pi)^2 + 10^2(100\pi)i} \right| = 3.5731 \times 10^{-6}\,\text{m}$$

The actuator response is $f = -K_P x - K_D \dot{x} = 8 \times 10^6 x - 4000\dot{x}$, and at steady state the actuator force required is

$$F(100\pi) = \left| 8 \times 10^6 - 4000(100\pi)i \right| X(100\pi)$$
$$= X(100\pi)\sqrt{6.4 \times 10^{13} + 16\pi^2 \times 10^{10}} = 28.94\,\text{N} \qquad \blacksquare$$

The success of this scheme depends on several factors:

- The ability of the sensors to provide accurate measurements of x and $\dot{x}$ in real time. Noise in the measured signal will degrade the system performance. Some systems do not measure the velocity but rather differentiate the x measurement. This will increase the noise in the signal representing $\dot{x}$.

- The ability of the computer to perform the necessary calculations in real time. Although microprocessor speed is continually increasing, some microprocessors may not be fast enough for certain applications.

• The ability of the actuator to provide the needed force level. All actuators have a limit on the force they can generate. This limitation frequently gets lost in the mathematics of the design and must be considered when setting the desired values of ζ, ω_n, or τ. If the magnitudes of the gains K_P and K_D are too large, the actuator might not be able to provide the expected force.

Adaptive Control

The term *adaptive control* is used quite frequently and often incorrectly. In control system terminology, an adaptive control system is one that can change (or *adapt*) either its gain values or even its control algorithm to accommodate changing conditions. For example, the gains calculated in the previous example always give a damping ratio of 0.5 and a natural frequency of 100 rad/s, so it is not adaptive. A truly adaptive system would be capable of recomputing its gains to adjust to a change in the forcing frequency, for example.

Much work in control theory for adaptive systems has been done since the 1970s. However, the mathematics for such systems is very difficult, and it is often unclear how the control gains or the control algorithm should adapt to changing conditions. Thus relatively few truly adaptive systems are in use, and their widespread application remains a challenging goal for the future.

Control of Multi-Mass Systems

Figure 7.6-2 illustrates the principle of an active suspension for one wheel. An actuator located between the wheel (mass m_1) and the chassis (mass m_2) supplies force f to both the wheel and the chassis. This force supplements the passive spring k_2 and passive damper c. The tire stiffness is k_1.

If we choose the control force f so that it imitates a spring and damper, like the PD control, then the control algorithm is

$$f = K_P(x_1 - x_2) + K_D(\dot{x}_1 - \dot{x}_2)$$

This requires measurements of the relative displacement $x_1 - x_2$ and the relative velocity $\dot{x}_1 - \dot{x}_2$. If this is substituted into the equations of motion, theoretically we can compute

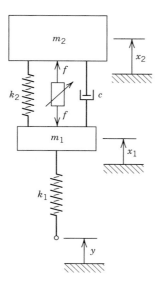

FIGURE 7.6-2   Active suspension system for one wheel of a vehicle.

values for K_P and K_D to achieve a desired set of characteristic roots corresponding to desired damping ratios, natural frequencies, and time constants. The difficulty with this approach is that there will be four roots, because this is a fourth-order system, and so we cannot expect to be able to specify all four roots by choosing the values of only two gains.

Another approach is to use measurements of all four variables — x_1, x_2, $\dot{x}_1$, and $\dot{x}_2$—and the control algorithm

$$f = K_1 x_1 + K_2 x_2 + K_3 \dot{x}_1 + K_4 \dot{x}_2$$

We now have four gains to select in order to place the four roots where desired. It turns out that this is not always possible and that the mathematics required to solve for the gain values is quite difficult. In practice, solution algorithms provided by MATLAB will do this (see the `acker` function). Such treatment is a topic in advanced control theory and will not be covered here. See, for example, Chapter 11 of [Palm, 2000], which contains the theory and a solution to a suspension problem in one of the chapter problems.

Examples of Current Applications

Active Suspensions   Traditional vehicle suspensions with fixed springs and dampers cannot provide both the smooth ride of a luxury car and the precise handling of a sports car, but with an active suspension such a goal is closer to being realized. Versions of active suspensions have been available since the 1990s. As of 2005, however, there is no truly active suspension available commercially. The most sophisticated system currently available, made by Delphi, uses a damper filled with a magneto-rheological fluid. The system uses a controlled magnetic field that can change the viscosity 1000 times a second.

Under development by the Bose Corporation since 1980, another system uses speaker-coil technology to drive linear electrical motors attached to each wheel. Such motors produce translation rather than rotation, and so imitate the forces provided by springs and dampers. One of the challenges is to develop control algorithms that provide effective coordination of all four wheel subsystems. This is necessary for proper vehicle control in sharp turns and sudden stops, for example.

The sports industry has developed a number of applications for active vibration control. These include skis, bike suspensions, and running shoes.

Skis   Skis can vibrate excessively at high speeds, on hard snow, and on rough terrain. This vibration decreases the contact area between the ski edge and the snow surface and results in the skier's reduced ability to control the motion. Smart skis designed to keep the edge of the ski in contact with the snow have been developed by the K2 Corporation and by Head. Because tests have indicated that most vibrations are concentrated near the ski binding and propagate outward to the rest of the ski, a piezoelectric ceramic plate is mounted in front of the ski binding. When current is applied, the element flexes and applies force to the ski to change its shape. As the ski flexes, it generates electricity to power the system.

Bikes   Piezoelectric materials have been used to develop a smart bicycle shock absorber to assist in keeping the wheel in contact with the ground and to minimize the force transmitted to the rider. Some bikes have suspensions in both the front and the rear, but the control of the front suspension is more critical. The front suspension consists of a pair of telescoping cylinders—called fork tubes—that contain springs and dampers.

These can be arranged in parallel, series, or any desired combination. The system monitors the position and velocity of the piston inside the fluid-filled cylinder, using a magnet attached to the piston and a sensor that detects the flux density of the magnetic field and converts it to a voltage.

Compression damping helps to control the rate of shock compression. Too much compression damping will transmit a large force to the rider, but too little compression damping will cause the shock to bottom-out quickly. The smart shock absorber automatically adjusts compression damping with varying trail conditions by commanding a piezoelectric bypass valve to open or close, thus increasing or decreasing the fluid resistance. To do this, some designs use a main hydraulic flow path (for fail-safe operation) and a bypass flow path that is normally closed by a flexible cantilever beam that acts like a flapper valve. An element consisting of layers of piezoelectric ceramic wafers is bonded to the beam. These layers expand when an electric current is applied. When a current is applied to the piezoelectric element by the controller, the beam flexes and opens an additional hydraulic flow path to reduce the resistance. The beam displacement can be controlled in increments to provide fine control of the damping.

Running Shoes   In 2005 Adidas introduced a running shoe that uses active stiffness control with a battery-powered microprocessor and motor. The motor turns a gear train and shaft whose rotation changes the tension of cables in the base of the shoe. The system uses a magnetic sensor to measure the shoe compression. Based on this measurement, the microprocessor commands the motor and adjusts the shoe stiffness before the shoe strikes the ground again.

7.7 MATLAB APPLICATIONS

Suppose the typical system shown in Figure 7.7-1 is subjected to the forcing function shown in Figure 7.7-2. In Section 7.3 we used the closed-form expression for the response to obtain a response plot of kx_{max}/F_o versus t_1/T_n. We then used this plot to select the value of the stiffness k so that the maximum response x_{max} is no greater than the static deflection kx_{max}/F_o due to a constant force F_o.

The advantage of obtaining a closed-form solution for the response is that it enables us to develop a response spectra in terms of dimensionless variables such as kx_{max}/F_o and t_1/T_n. Unfortunately, this requires detailed mathematical analysis that may not be feasible for more complicated forcing functions or for systems with damping. We sometimes can reduce the number of parameters needed by defining a new time scale and a new

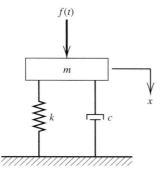

FIGURE 7.7-1 Model of fixed-base isolation.

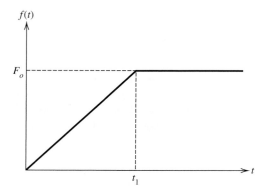

FIGURE 7.7-2 A modified ramp input.

displacement scale. This was the method used in Section 7.3 to obtain the following equation, where the input is shown in Figure 7.7-2.

$$z'' + 4\zeta\pi b z' + 4\pi^2 b^2 z = 4\pi^2 b^2 \begin{cases} T & 0 \le T \le 1 \\ 1 & T > 1 \end{cases} \tag{7.7-1}$$

where

$$z = kx_{max}/F_o, \quad T = t/t_1, \quad \text{and} \quad b = t_1/T_n.$$

Sometimes it is not possible or convenient to reduce the parameters sufficiently, and in some cases we know the system's mass value and the characteristics of the forcing function, such as the values of F_o and t_1. Then all we would need to determine is the optimum value of the stiffness k and, if damping is present, the optimum value of c. In such cases we can obtain satisfactory response plots of x_{max} versus k by numerical simulation.

The MATLAB `lsim` function is very useful for obtaining the response due to a forcing function of arbitrary shape. One form of its syntax is x = lsim(sys, f, t), where `sys` is the transfer function of the system, `t` is a user-supplied vector of time values, and `f` is the forcing function. The calculated response is stored the array x.

The following program is based on Equation 7.7-1 and generates the response spectrum shown in Figure 7.3-11.

```
% Response to a modified ramp input.
t_max = 75;
[f, t] = modified_ramp(t_max)
L = 0;
a = [0.02:0.01:4];
for zeta = [0,0.2,0.7,1]
    L = L+1;
    N = 0;
    for b = a
        N = N+1;
        sys = tf(4*pi^2*b^2,[1,4*pi*zeta*b,4*pi^2*b^2]);
        x = lsim(sys,f,t);
        xmax(L,N) = max(abs(x));
    end
end
plot(a,xmax(1,:),a,xmax(2,:),a,xmax(3,:),a,xmax(4,:)),...
    xlabel('t_1/T_n'), ylabel('kx_{max}/F_o')
```

The value of t_max is selected (by trial and error) to ensure that enough time is given for the maximum responses to appear.

The program calls upon the user-defined function modified_ramp(t_max), which is the following:

```
function [f, t] = modified_ramp(t_max)
ta = [0:0.001:1];
tb = [1.001:0.001:t_max];
fa = ta;
fb = ones(size(tb));
t = [ta, tb];
f = [fa, fb];
```

Note that the program uses the nondimensional time scale $T = t/t_1$, and so the ramp ends at $T = 1$.

The user-defined function modified_ramp(t_max) can be replaced with another function to obtain the response spectra for other types of inputs. For example, to describe the half-sine function with unit amplitude, the program is

```
function [f, t] = half_sine(t_max)
ta = [0:0.001:1];
tb = [1.001:0.001:t_max];
fa = sin(pi*ta);
fb = zeros(size(tb));
t = [ta, tb];
f = [fa, fb];
```

Since the program uses the nondimensional time scale $T = t/t_1$, the half-sine ends at $T = 1$.

7.8 CHAPTER REVIEW

We have studied several ways to reduce unwanted vibration. These include redesign of the system by, for example,

(a) Changing the natural frequency either by increasing or decreasing the mass or by increasing the stiffness

(b) Dissipating the energy of vibration by adding damping

(c) Isolating the source by using an isolator consisting of a stiffness element and perhaps a damping element placed between the source and the surrounding environment

(d) Using a vibration absorber

(e) Using an active control system

Now that you have finished this chapter, you should be able to do the following:

1. Design isolators for fixed-base and movable-base systems.

2. Design vibration aborbers for single-degree-of-freedom systems.

3. Design active vibration control systems for single-degree-of-freedom systems.

4. Apply MATLAB to the relevant topics of the chapter.

PROBLEMS

SECTION 7.1   ACCEPTABLE VIBRATION LEVELS

7.1   An accelerometer measurement of a mass in harmonic motion shows that the frequency is 5 Hz and the maximum acceleration is 4.1 mm/s^2. Compute the maximum velocity and the maximum displacement.

7.2   If a mass in harmonic motion is vibrating at 4 Hz and has maximum displacement of 2 mm, what is its maximum acceleration and maximum velocity?

7.3   An undamped mass-spring system, when displaced from rest by 30 mm, oscillates with a frequency of 3 Hz. Compute its maximum displacement, maximum acceleration, and maximum velocity.

7.4   An undamped mass-spring system, when displaced from rest by 20 mm, oscillates with a maximum acceleration of 3 m/s^2. Compute its natural frequency and maximum velocity.

7.5   Figure 7.1-3 shows that maximum rms acceleration for 24-h endurance is approximately 0.13g over the frequency range from 4 to 8 Hz. What are the corresponding ranges for maximum displacement and maximum acceleration?

7.6   Figure 7.1-4 shows that maximum rms acceleration for 8-h endurance is approximately 0.029g over the frequency range from 4 to 8 Hz. The safe exposure limit and the reduced comfort level over this frequency range are 0.06g and 0.013g, respectively.

   (a) What are the ranges for maximum displacement and maximum acceleration corresponding to the safe exposure limit?

   (b) What are the ranges for maximum displacement and maximum acceleration corresponding to the reduced comfort level?

SECTION 7.2   SOURCES OF VIBRATION

7.7   In certain situations, the steady-state excitation force due to vortex shedding is sinusoidal with an amplitude proportional to the square of the forcing frequency. The model of a particular 20-kg single-degree-of-freedom system undergoing such excitation is

$$20\ddot{x} + 180\dot{x} + 800x = 10\omega^2 \sin \omega t$$

Solve for the steady-state response if the excitation frequency is 1 Hz ($\omega = 2\pi$).

7.8   A single-cylinder engine rotates at 1000 rpm. Plot the piston acceleration $\ddot{x}_p$ versus time for the case where the crank length is 25 mm and the connecting rod is 100 mm long.

SECTION 7.3   ISOLATOR DESIGN FOR FIXED-BASE SYSTEMS

7.9   Design an undamped isolator for a 30-kg mass subjected to a harmonic forcing function whose amplitude and frequency are 1000 N and 20 Hz. The isolator should transmit to the base no more than 10% of the applied force. Determine the resulting displacement amplitude.

7.10   A 40-kg mass is acted upon by a harmonic force of magnitude 30 N and frequency 3 Hz. Design an isolator using $\zeta = 0.1$ so that no more than 10% of the applied force is transmitted to the base.

7.11   A pump of mass 100 kg rotates at 500 rpm. Compute the static deflection of an undamped isolator required for 80% isolation.

7.12   Compute the minimum static deflection for an undamped isolator to provide 80% isolation for a 250-kg washing machine rotating at 900 rpm.

7.13 A floor supporting a machine tool operating at 900 rpm can withstand a harmonic force amplitude of no more than 3000 N. Without an isolator, the measured force at the floor is 21 000 N. Compute the maximum stiffness required for an undamped isolator to provide the required isolation.

7.14 A certain machine has a mass of 400 kg. Its motor rotates at 1500 rpm, and measurements indicate that it has a rotating unbalance of 0.5 kg·m. Design an isolation system to limit the force transmitted to the floor to 5000 N. The available isolator, which can be used in series, has a damping ratio of 0.2 and a stiffness of 6×10^6 N/m. Consider adding mass to the system to achieve the required transmissibility.

7.15 What stiffness is required of an isolator having a damping ratio of 0.02 so that it will provide 80% isolation of a 50-kg machine operating at 1000 rpm?

7.16 It is desired to provide at least 80% isolation for a 200-kg machine operating over the speed range from 1000 rpm to 2000 rpm. An available isolator has a damping ratio of 0.02 and a stiffness of 10^6 N/m. Will this isolator meet the requirements? If not, compute the added mass required.

7.17 A 500-kg machine operates between 600 and 800 rpm and has a rotating unbalance of 0.1 kg·m. Compute the maximum stiffness of an undamped isolator required so that the force transmitted to the floor is no more than 200 N.

7.18 A motor operating at 3500 rpm for a power tool has a mass of 10 kg and is to be mounted on a steel cantilever beam. Static-force calculations and space considerations suggest that a beam 16 cm long, 12 cm wide, and 1 cm thick would be suitable. The rotating part of the motor has a mass of 5 kg and has an eccentricity of 0.4 mm. Neglect damping in the beam and estimate the amplitude of vibration of the beam at steady state.

7.19 A certain 3-kg motor is mounted to a base with an isolator consisting of an elastic pad. The motor operates at 4000 rpm. Neglect damping in the pad and calculate the pad stiffness required to provide a 90% reduction in the force transmitted from the motor to the base.

7.20 When a 200-kg machine is rotating at 2000 rpm, the effect of the rotating unbalance is to exert a force of 250 N on the machine. The isolator values are $k = 3 \times 10^7$ N/m and $\zeta = 0.1$.

(a) Compute the steady-state amplitude of the displacement.

(b) Compute the magnitude of the force transmitted to the foundation at steady state.

7.21 The rotor of a specific machine has mass of 600 kg and a measured unbalance of $mR = 1.3$ kg·m. The machine will be run at a speed of 1750 rpm, and there is a clearance of 6 mm between the shaft and the housing. The shaft length from the bearings is $L = 0.15$ m. Assuming the shaft is steel, compute the minimum required shaft diameter. Model the shaft as a cantilever beam supported by the bearings, and neglect any damping in the system.

7.22 For the system shown in Figure 7.3-2 and the force shown in Figure 7.3-7, $m = 75$ kg, $F_o = 800$ N, and $t_1 = 0.2$ s. Determine the optimum value of k to minimize the response.

7.23 The following data were taken by driving a machine on its support with a rotating unbalance force at various frequencies. The machine's mass is 100 kg, but the stiffness and damping in the support are unknown. The frequency of the driving force is f Hz. The measured steady-state displacement of the machine is X mm.

(a) Estimate the stiffness and damping in the support.

(b) Estimate what percentage of the rotating unbalance force is transmitted to the foundation at 6 Hz.

(c) Calculate what isolator stiffness must be added in order that no more than 10% of the rotating unbalance force be transmitted to the foundation.

f (Hz)	X (mm)	f (Hz)	X (mm)
0.2	1	3.8	13
1	2	4	11
2	4	5	8
2.6	12	6	8
2.8	18	7	7
3	25	8	6
3.4	18	9	6
3.6	15	10	5

SECTION 7.4   ISOLATION WITH BASE MOTION

7.24 A 15-kg instrument is to be used on a factory floor that vibrates with a frequency of 3500 rpm and an amplitude of 3 mm due to nearby equipment. The instrument can withstand an acceleration no greater than $4g$. Assuming a damping ratio of 0.02, compute the required stiffness of an isolator, the maximum deflection of the isolator, and the maximum displacement of the instrument.

7.25 A 50-kg instrument is used on a table that vibrates due to nearby machinery. The principal frequency component is 1500 rpm. Determine whether isolator A or isolator B, whose load-deflection curves are given in Figure 7.4-4, can be used at each of the four corners of the instrument so that no more than 25% of the table motion is transmitted to the instrument. Determine how much added mass is required, if any.

7.26 Consider the problem posed in Example 7.4-1, which used isolator A in Figure 7.4-4. Redo the problem using isolator B.

7.27 A printed circuit board (PCB) is supported by a chassis that is attached to a vibrating motor. The board is 1.6 mm thick, 200 mm wide, 254 mm long, and has a mass of 0.6 kg. Neglect damping, model the board as a fixed-fixed beam, determine its stiffness and natural frequency, and compute the displacement transmissibility if the chassis vibrates at 60 Hz due to motor unbalance. The board is made from epoxy fiberglass for which $E = 1.38 \times 10^{10}$ N/m^2.

7.28 Consider the printed circuit board of Problem 7.27. We want to reduce the displacement transmissibility to 25% by using an isolator, as shown in Figure 7.4-6. We are told that the chassis mass is one-tenth the mass of the PCB. Obtain the required values for the isolator stiffness k_2 and the damping c.

7.29 In Problem 7.27 an isolator was designed by treating the PCB-chassis system as a rigid body. Incorporating the board flexibility in the model gives the two-degrees-of-freedom system shown in Figure 7.4-7. Determine the displacement transmissibility of this system using the isolator values from Problem 7.28.

7.30 The base of a 25-kg machine is subjected to a half-sine displacement of duration 0.05 s. Compute the stiffness of an undamped isolator required so that $x_{\max}/y_{\max} = 1$.

SECTION 7.5   DYNAMIC VIBRATION ABSORBERS

7.31 A certain machine with supports has a measured natural frequency of 4.5 Hz. The machine will be subjected to a rotating unbalance force having an amplitude of 18 N and a frequency of 3.5 Hz. Design a dynamic vibration absorber for this machine. The available clearance for the absorber's motion is 2.5 cm.

7.32 Suppose the main mass in the system in Problem 7.31 has the value $m_1 = 16$ kg. Evaluate the sensitivity of the absorber design to variations in the input frequency.

7.33 Consider the motor mounted on a cantilever beam in Figure P7.33. The motor has a mass of 8 kg and runs at the constant speed of 3500 rpm. The steel beam is 15 cm long, 10 cm wide, and 1 cm thick. The unbalanced part of the motor has a mass of 4 kg and an eccentricity of 0.3 mm. The damping in the beam is very slight. Design a vibration absorber for this system. The available clearance for the absorber's motion is 2.5 cm.

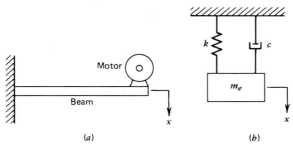

(a)   (b)

FIGURE P7.33

7.34 A motor mounted on a beam vibrates too much when it runs at a speed of 6000 rpm. At that speed the measured force produced on the beam is 240 N. Design a vibration absorber to attach to the beam. Because of space limitations, the absorber's mass cannot have an amplitude of motion greater than 2 mm.

7.35 The supporting table of a radial saw has a mass of 64 kg. When the saw operates at 200 rpm, it transmits a force of 16 N to the table. Design a vibration absorber to be attached underneath the table. The absorber's mass cannot vibrate with an amplitude greater than 2.5 cm.

7.36 A certain machine with supports has an experimentally determined natural frequency of 6 Hz. It will be subjected to a rotating unbalance force with an amplitude of 50 N and a frequency of 4 Hz. The machine mass is 8 kg.

(a) Design a vibration absorber for this machine. The available clearance for the absorber's motion is 0.1 m.

(b) Let x_1 be the vertical displacement of the machine. The amplitude of the rotating unbalance force is $mR\omega_R^2$. Plot X_1/mR versus the frequency ω_R, and use the plot to discuss the sensitivity of the absorber to changes in the frequency ω_R.

7.37 Figure P7.37 illustrates a vibration absorber that is used on some passenger cars to reduce the vibration of the exhaust pipe. Suppose that the forcing frequency is 5 Hz. If we use an absorber mass $m_2 = 0.3$ kg, what must be the value of the stiffness of the cantilever beam that attaches m_2 to the exhaust pipe?

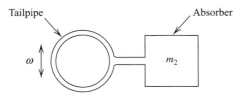

FIGURE P7.37

7.38 Consider the vibration absorber designed in Examples 7.5-1 and 7.5-2.

(a) Investigate its performance by plotting the responses x_1 and x_2 for the case where some damping ($c = 45$ N·s/m) exists between the two masses. Use zero initial conditions.

(b) For the value given in part (a), suppose the forcing function amplitude differs from its design value so that it is 22 N instead of 13 N. Investigate the absorber's performance by plotting the responses x_1 and x_2.

(c) For the value given in part (a), suppose the forcing function frequency differs from its design value of 6π so that it is 5π rad/s. Investigate the absorber's performance by plotting the responses x_1 and x_2.

7.39 Figure P7.39 shows a type of vibration absorber that uses only mass and damping, not stiffness, to reduce vibration. The main mass is m_1 and the absorber mass is m_2. Suppose the applied force $f(t)$ is sinusoidal.

(a) Derive the expressions for $X_1(i\omega)/F(i\omega)$ and $X_2(i\omega)/F(i\omega)$.

(b) Use these expressions to discuss the selection of values for m_2 and c in order to minimize the motion of mass m_1. To aid in your discussion, plot kX_1/F versus $\omega/\sqrt{k/m_1}$ for several values of $\zeta = c/2\sqrt{km_1}$.

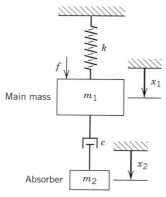

FIGURE P7.39

7.40 Figure P7.40 shows a type of vibration absorber that uses mass, stiffness, and damping to reduce vibration. The damping can be used to reduce the amplitude of vibration near resonance. The main mass is m_1 and the absorber mass is m_2. Suppose the applied force $f(t)$ is sinusoidal.

(a) Derive the expressions for $X_1(i\omega)/F(i\omega)$ and $X_2(i\omega)/F(i\omega)$.

(b) Use these expressions to discuss the selection of values for m_2, k_2, and c in order to minimize the motion of mass m_1. To aid in your discussion, plot k_1X_1/F versus $\omega/\sqrt{k_1/m_1}$ for several values of $\zeta = c/2\sqrt{k_1m_1}$.

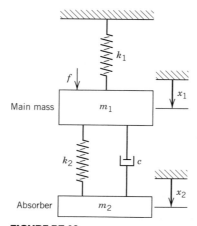

FIGURE P7.40

SECTION 7.6   ACTIVE VIBRATION CONTROL

7.41 A 125-kg machine has a passive isolation system for which $c = 5000$ N·m/s and $k = 7 \times 10^6$ N/m. The rotating unbalance force has an amplitude of 100 N with a frequency of 2500 rpm. The resonant frequency of this system is 216 rad/s and is close the frequency of the disturbance. In addition, the damping ratio is small ($\zeta = 0.08$). Assuming that c and k cannot be changed, calculate the control gains required to give a damping ratio of $\zeta = 0.5$ and a resonant frequency of 100 rad/s, well below the disturbance frequency. Evaluate the actuator force requirement at steady state.

7.42 A 20-kg machine has a passive isolation system whose damping ratio is 0.28 and whose undamped natural frequency is 13.2 rad/s. Assuming that the passive system remains in place, calculate the control gains required to give a damping ratio of $\zeta = 0.707$ and a resonant frequency of 141 rad/s.

7.43 With the increased availability of powered wheelchairs, improved suspension designs are required for safety and comfort. One chair uses an active suspension like the one shown in Figure P7.43 for each driving wheel. An actuator exerts a force f between the tire and the vehicle. Assuming that the control algorithm is such that

$$f = -K_P(x_s - x_w) - K_D(\dot{x}_s - \dot{x}_w)$$

derive the equations of motion of the system, neglecting the tire mass and tire damping.

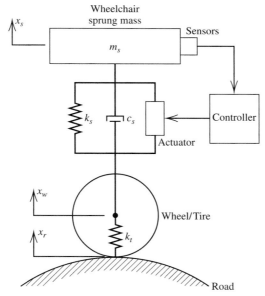

FIGURE P7.43

SECTION 7.7 MATLAB APPLICATIONS

7.44 Use the program given in Section 7.7 to obtain the response spectrum of a fixed-base isolator having the pulse input shown in Figure P7.44. Plot kx_{max}/F_o versus t_1/T_n for $\zeta = 0$, 0.2, 0.7, and 1.

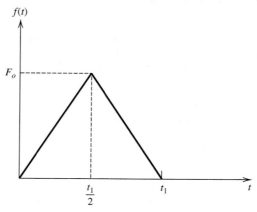

FIGURE P7.44

7.45 Use the program given in Section 7.7 to obtain the response spectrum of a fixed-base isolator having the pulse input shown in Figure P7.45. Plot kx_{max}/F_o versus t_1/T_n for $\zeta = 0$, 0.2, 0.7, and 1.

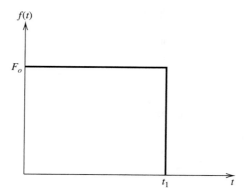

FIGURE P7.45

7.46 Use the program given in Section 7.7 to obtain the response spectrum of a fixed-base isolator having the pulse input shown in Figure P7.46. Plot kx_{max}/F_o versus t_1/T_n for $\zeta = 0$, 0.2, 0.7, and 1.

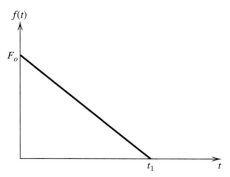

FIGURE P7.46

7.47 Modify the program given in Section 7.7 to obtain the response spectrum of the force transmitted to a fixed-base isolator for the input shown in Figure 7.7-2. Plot $maxF_t/F_o$ versus t_1/T_n for $\zeta = 0$, 0.2, 0.7, and 1.

7.48 Modify the program given in Section 7.7 to obtain the response spectrum of the force transmitted to a fixed-base isolator for the half-sine input whose half-period is t_1 and whose magnitude is F_o. Plot $maxF_t/F_o$ versus t_1/T_n for $\zeta = 0$, 0.2, 0.7, and 1.

7.49 Modify the program given in Section 7.7 to obtain the response spectrum of a moving-base isolator having the input shown in Figure 7.7-2, where instead the magnitude is Y_o. Plot x_{max}/Y_o versus t_1/T_n for $\zeta = 0$, 0.2, 0.7, and 1.

7.50 Modify the program given in Section 7.7 to obtain the response spectrum of a moving-base isolator having the half-sine input whose half-period is t_1 and whose magnitude is Y_o. Plot x_{max}/Y_o versus t_1/T_n for $\zeta = 0$, 0.2, 0.7, and 1.

CHAPTER 8

MATRIX METHODS FOR MULTI-DEGREE OF FREEDOM SYSTEMS

CHAPTER OUTLINE

WE HAVE already studied systems having two degrees of freedom. Although such systems are "multi-degree-of-freedom systems," we will see that the algebra required to analyze systems having more than two degrees of freedom becomes very complicated. For such systems it is more convenient to use matrix representation of the equations of motion and matrix methods to do the analysis. This is the topic of the chapter.

We begin by showing how to represent the equations of motion in compact matrix form. Then we return to the topic of modes and develop systematic procedures for analyzing the modal response. Besides providing a compact form for representing the equations of motion and performing the analysis, matrix methods also form the basis of software packages such as MATLAB and Simulink that provide powerful tools for modal analysis and simulation.

LEARNING OBJECTIVES

After you have finished this chapter, you should be able to do the following:

■ Express the equations of motion of a linear system in matrix form.
■ Obtain the frequencies and mode shapes of a linear model expressed in matrix form.

- Obtain the forced response of a multi-degree-of-freedom system.
- Use approximate methods to include the effects of damping.
- Apply MATLAB and Simulink to the methods of this chapter.

8.1 MATRIX FORM OF THE EQUATIONS OF MOTION

In this section we show how to represent the equations of motion of a linear system in matrix form. This compact way of representing equations is very useful for working with models of multi-degree of freedom systems, as we will see in this chapter. Equations of motion for such systems can be obtained either by applying Newton's laws or by using Lagrange's equations. As discussed in Chapter 6, each method has its own advantages.

EXAMPLE 8.1-1
Application of Newton's Laws

Use Newton's laws to obtain the equations of motion for the system shown in Figure 8.1-1a. The masses slide on a frictionless surface. The forces $f_1(t)$, $f_2(t)$, and $f_3(t)$ are known functions of time. Note that this system has three degrees of freedom.

Solution

The free-body diagrams are shown in Figure 8.1-1b, where we have assumed that $x_1 > x_2$ and $x_2 > x_3$. Recall that you can make other assumptions as long as you make them consistently throughout the derivation. The equations of motion obtained from the free-body diagrams are:

$$m_1\ddot{x}_1 = -k_1x_1 - k_2(x_1 - x_2) + f_1(t)$$
$$m_2\ddot{x}_2 = k_2(x_1 - x_2) - k_3(x_2 - x_3) + f_2(t)$$
$$m_3\ddot{x}_3 = k_3(x_2 - x_3) + f_3(t)$$

These can be arranged as follows by collecting like terms:

$$m_1\ddot{x}_1 + (k_1 + k_2)x_1 - k_2x_2 = f_1(t)$$
$$m_2\ddot{x}_2 - k_2x_1 + (k_2 + k_3)x_2 - k_3x_3 = f_2(t)$$
$$m_3\ddot{x}_3 - k_3x_2 + k_3x_3 = f_3(t)$$

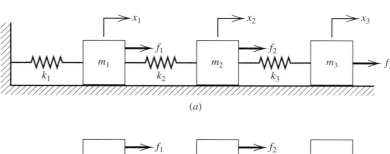

(a)

(b)

FIGURE 8.1-1 A System with three degrees of freedom.

EXAMPLE 8.1-2
Application of Lagrange's Equations

Use Lagrange's equations to obtain the equations of motion for the system shown in Figure 8.1-1a. The masses slide on a frictionless surface.

Solution

The kinetic energy of the system is

$$KE = \frac{1}{2}\left(m_1\dot{x}_1^2 + m_2\dot{x}_2^2 + m_3\dot{x}_3^2\right)$$

The elastic potential energy of the system is

$$PE = \frac{1}{2}[k_1x_1^2 + k_2(x_2 - x_1)^2 + k_3(x_3 - x_2)^2]$$

$$= \frac{1}{2}[k_1x_1^2 + k_2\left(x_2^2 - 2x_1x_2 + x_1^2\right) + k_3\left(x_3^2 - 2x_2x_3 + x_2^2\right)]$$

Note that because of the squared terms, it makes no difference what assumption we make about the relative motions of the masses.

The Lagrangian is $L = KE - PE$, and the generalized forces are $Q_j = f_j(t)$. The generalized coordinates are $q_j = x_j$. Since there is no friction or damping, the dissipation function D is 0, and thus Lagrange's equations have the form

$$\frac{d}{dt}\left(\frac{\partial L}{\partial \dot{x}_j}\right) - \frac{\partial L}{\partial x_j} = f_j(t) \qquad j = 1, 2, 3$$

For $j = 1$, we obtain

$$\frac{d}{dt}(m_1\dot{x}_1) + k_1x_1 + k_2x_1 - k_2x_2 = f_1(t)$$

For $j = 2$,

$$\frac{d}{dt}(m_2\dot{x}_2) + k_2x_2 - k_2x_1 - k_3x_3 + k_3x_2 = f_2(t)$$

For $j = 3$,

$$\frac{d}{dt}(m_3\dot{x}_3) + k_3x_3 - k_3x_2 = f_3(t)$$

Evaluating the time derivatives and collecting terms gives

$$m_1\ddot{x}_1 + (k_1 + k_2)x_1 - k_2x_2 = f_1(t)$$
$$m_2\ddot{x}_2 - k_2x_1 + (k_2 + k_3)x_2 - k_3x_3 = f_2(t)$$
$$m_3\ddot{x}_3 - k_3x_2 + k_3x_3 = f_3(t)$$

These are identical to the equations derived from Newton's laws. ∎

EXAMPLE 8.1-3
A System with Damping

Use Newton's laws to obtain the equations of motion for the system shown in Figure 8.1-2a. The forces $f_1(t)$, $f_2(t)$, and $f_3(t)$ are known functions of time.

Solution

The free-body diagrams are shown in Figure 8.1-2b, where we have assumed that the displacements are such that $x_1 > x_2$, $x_2 > x_3$ and the velocities are such that $\dot{x}_1 > 0$, $\dot{x}_1 > \dot{x}_2$, and $\dot{x}_2 > \dot{x}_3$. Recall that you can make other assumptions as long as you make them consistently throughout the derivation. The equations of motion obtained from the free-body diagrams are

$$m_1\ddot{x}_1 = -k_1x_1 - k_2(x_1 - x_2) - c_1\dot{x}_1 - c_2(\dot{x}_1 - \dot{x}_2) + f_1(t)$$
$$m_2\ddot{x}_2 = k_2(x_1 - x_2) - k_3(x_2 - x_3) + c_2(\dot{x}_1 - \dot{x}_2) - c_3(\dot{x}_2 - \dot{x}_3) + f_2(t)$$
$$m_3\ddot{x}_1 = k_3(x_2 - x_3) + c_3(\dot{x}_2 - \dot{x}_3) + f_3(t)$$

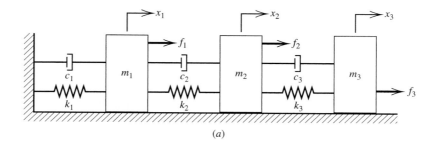

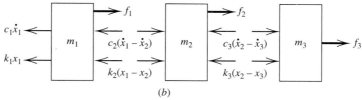

FIGURE 8.1-2 A damped system with three degrees of freedom.

These can be arranged as follows by collecting like terms.

$$m_1\ddot{x}_1 + (c_1 + c_2)\dot{x}_1 + (k_1 + k_2)x_1 - c_2\dot{x}_2 - k_2x_2 = f_1(t)$$

$$m_2\ddot{x}_2 - c_2\dot{x}_1 - k_2x_1 + (c_2 + c_3)\dot{x}_2 + (k_2 + k_3)x_2 - c_3\dot{x}_3 - k_3x_3 = f_2(t)$$

$$m_3\ddot{x}_3 - c_3\dot{x}_2 - k_3x_2 + c_3\dot{x}_3 + k_3x_3 = f_3(t)$$

∎

We now proceed to show how these equations of motion can be represented in compact matrix form. If you are already familiar with matrix algebra, you can skip to the subsection "Matrix Representation."

Matrix Algebra

Matrix notation enables us to represent multiple equations as a single equation. For example, consider the following set of linear algebraic equations:

$$a_{11}x_1 + a_{12}x_2 + \cdots + a_{1n}x_n = b_1$$

$$a_{21}x_1 + a_{22}x_2 + \cdots + a_{2n}x_n = b_2$$

$$\cdots \qquad\qquad (8.1\text{-}1)$$

$$a_{n1}x_1 + a_{n2}x_2 + \cdots + a_{nn}x_n = b_n$$

It can be represented in the following compact form:

$$\mathbf{Ax} = \mathbf{b} \qquad\qquad (8.1\text{-}2)$$

The *matrix* $\mathbf{A}$ is an array of numbers (or expressions) that correspond in an ordered fashion to the coefficients a_{ij}. The matrix is ordered as follows:

$$\mathbf{A} = \begin{bmatrix} a_{11} & a_{12} & \cdots & a_{1n} \\ a_{21} & a_{22} & \cdots & a_{2n} \\ & & \cdots & \\ a_{n1} & a_{n2} & \cdots & a_{nn} \end{bmatrix} \qquad\qquad (8.1\text{-}3)$$

This particular matrix has n rows and n columns, so its *dimension* is expressed as $(n \times n)$. A matrix with equal numbers of rows and columns is a *square* matrix. In general, if a matrix has n rows and m columns, its dimension is $(n \times m)$. In print a matrix symbol, like **A**, is usually identified by boldface type.

A matrix should not be confused with a determinant, which also has rows and columns but which can be reduced to a *single* number. A determinant is usually denoted by a pair of parallel lines; square brackets usually denote matrices. For example, $|\mathbf{A}|$ is the determinant of the matrix **A**. If $|\mathbf{A}| = 0$, the matrix is said to be *singular*.

A *vector* is a special case of a matrix that has either one row or one column. A *row* vector has one row; thus its dimension is $(1 \times n)$. A *column* vector has one row, so its dimension is $(n \times 1)$. In this text a vector is taken to be a column vector unless otherwise specified. Examples of column vectors are the vectors **x** and **b** in Equation 8.1-2, where

$$\mathbf{x} = \begin{bmatrix} x_1 \\ x_2 \\ . \\ . \\ . \\ x_n \end{bmatrix} \qquad \mathbf{b} = \begin{bmatrix} b_1 \\ b_2 \\ . \\ . \\ . \\ b_n \end{bmatrix}$$

Usually lowercase boldface letters are used to denote a vector.

Matrix Addition

Two matrices **A** and **B** are equal to one another if they have the same dimensions and if their corresponding elements are equal; that is, if $a_{ij} = b_{ij}$ for every value of i and j. Two matrices can be added if they have the same dimension. Their sum is obtained by adding all their corresponding elements. Thus $\mathbf{A} + \mathbf{B} = \mathbf{C}$ implies that $c_{ij} = a_{ij} + b_{ij}$. The matrix **C** has the same dimension as **A** and **B**.

For example,

$$\begin{bmatrix} 6 & -2 \\ 10 & 3 \end{bmatrix} + \begin{bmatrix} 9 & 8 \\ -12 & 14 \end{bmatrix} = \begin{bmatrix} 15 & 6 \\ -2 & 17 \end{bmatrix}$$

Matrix subtraction is defined in a similar way.

Matrix addition and subtraction are associative and commutative. This means that

$$(\mathbf{A} + \mathbf{B}) + \mathbf{C} = \mathbf{A} + (\mathbf{B} + \mathbf{C}) \tag{8.1-4}$$

$$\mathbf{A} + \mathbf{B} + \mathbf{C} = \mathbf{B} + \mathbf{C} + \mathbf{A} = \mathbf{A} + \mathbf{C} + \mathbf{B} \tag{8.1-5}$$

Matrix Multiplication

Equation 8.1-2, which is the matrix form of Equation 8.1-1, implies a multiplication rule for matrices. Suppose **A** has dimension $(n \times p)$ and **B** has dimension $(p \times q)$. If **C** is the product **AB**, then **C** has dimension $(n \times q)$ and its elements are given by

$$c_{ij} = \sum_{k=1}^{p} a_{ik} b_{kj} \tag{8.1-6}$$

for all $i = 1, 2, \ldots, n$ and $j = 1, 2, \ldots, q$. In order for the product to be defined, the matrices **A** and **B** must be *conformable*; that is, the number of *rows* in **B** must equal the number of *columns* in **A**.

The algorithm defined by Equation 8.1-6 is easily remembered. Each element in the ith row of $\mathbf{A}$ is multiplied by the corresponding element in the jth column of $\mathbf{B}$. The sum of the products is the element c_{ij}. For example,

$$\begin{bmatrix} 6 & -2 \\ 10 & 3 \\ 4 & 7 \end{bmatrix} \begin{bmatrix} 9 & 8 \\ -5 & 12 \end{bmatrix} = \begin{bmatrix} (6)(9)+(-2)(-5) & (6)(8)+(-2)(12) \\ (10)(9)+(3)(-5) & (10)(8)+(3)(12) \\ (4)(9)+(7)(-5) & (4)(8)+(7)(12) \end{bmatrix} = \begin{bmatrix} 64 & 24 \\ 75 & 116 \\ 1 & 116 \end{bmatrix}$$

Matrix multiplication does not have the commutative property; that is, in general, $\mathbf{AB} \neq \mathbf{BA}$. A simple example will demonstrate this fact:

$$\mathbf{AB} = \begin{bmatrix} 6 & -2 \\ 10 & 3 \end{bmatrix} \begin{bmatrix} 9 & 8 \\ -12 & 14 \end{bmatrix} = \begin{bmatrix} 78 & 20 \\ 54 & 122 \end{bmatrix}$$

whereas

$$\mathbf{BA} = \begin{bmatrix} 9 & 8 \\ -12 & 14 \end{bmatrix} \begin{bmatrix} 6 & -2 \\ 10 & 3 \end{bmatrix} = \begin{bmatrix} 134 & 6 \\ 68 & 66 \end{bmatrix}$$

Reversing the order of matrix multiplication is a common and easily made mistake.

A special case of matrix multiplication occurs when one of the matrices is a *scalar*, which is a single number. In this case the multiplication is defined as

$$w\mathbf{A} = [wa_{ij}] \tag{8.1-7}$$

where w is a scalar. Multiplying $\mathbf{A}$ by a scalar w produces a matrix whose elements are the elements of $\mathbf{A}$ multiplied by w. Thus $w\mathbf{A} = \mathbf{A}w$.

The general exceptions to the noncommutative property are the *null* matrix, denoted by $\mathbf{0}$, and the *identity* or *unity* matrix, denoted $\mathbf{I}$. The null matrix contains all zeros. The identity matrix is a square matrix whose diagonal elements are all equal to 1, with the remaining elements equal to 0. For example, the (2×2) identity matrix is

$$\mathbf{I} = \begin{bmatrix} 1 & 0 \\ 0 & 1 \end{bmatrix}$$

These matrices have the following properties:

$$\mathbf{0A} = \mathbf{A0} = \mathbf{0}$$

$$\mathbf{IA} = \mathbf{AI} = \mathbf{A}$$

Because vectors are matrices, they are multiplied like matrices. To test your understanding of this, convince yourself that Equations 8.1-1 and 8.1-2 are equivalent.

The associative and distributive properties hold for matrix multiplication. The associative property states that

$$\mathbf{A}(\mathbf{B} + \mathbf{C}) = \mathbf{AB} + \mathbf{AC} \tag{8.1-8}$$

The distributive property states that

$$(\mathbf{AB})\mathbf{C} = \mathbf{A}(\mathbf{BC}) \tag{8.1-9}$$

Matrix Transpose and Positive Definite Matrices

A column vector can be converted to a row vector by the *transpose* operation, in which the rows and columns are interchanged. We denote this operation by the superscript T.

The transpose operation can be applied to matrices as well as vectors. For an $(n \times m)$ matrix $\mathbf{A}$ with n rows and m columns, $\mathbf{A}^T$ (read "A transpose") is an $(m \times n)$ matrix. If $\mathbf{A}^T = \mathbf{A}$, the matrix $\mathbf{A}$ is *symmetric*. Only a square matrix can be symmetric.

The matrix $\mathbf{A}$ is said to be *positive definite* if

$$\mathbf{v}^T \mathbf{A} \mathbf{v} > 0 \qquad (8.1\text{-}10)$$

for every nonzero vector $\mathbf{v}$. Note that the product $\mathbf{v}^T \mathbf{A} \mathbf{v}$ is a *scalar*. The matrix $\mathbf{A}$ is said to be *positive semi-definite* if $\mathbf{v}^T \mathbf{A} \mathbf{v} \geq 0$.

The Singular Case

One of our important applications for matrices is in the solution of equations of the form

$$\mathbf{A}\mathbf{x} = \lambda \mathbf{x} \qquad (8.1\text{-}11)$$

where λ is a parameter, $\mathbf{A}$ is a square matrix, and $\mathbf{x}$ is a column vector. Move all terms to the left-hand side and factor out $\mathbf{x}$:

$$\mathbf{A}\mathbf{x} - \lambda \mathbf{x} = \mathbf{A}\mathbf{x} - \lambda \mathbf{I}\mathbf{x} = (\mathbf{A} - \lambda \mathbf{I})\mathbf{x} = \mathbf{0}$$

Note that we had to use the fact that $\lambda \mathbf{x} = \lambda \mathbf{I}\mathbf{x}$ to maintain the proper dimensions in the matrix equation. This equation set will have a nonzero solution for $\mathbf{x}$ if and only if the matrix $(\mathbf{A} - \lambda \mathbf{I})$ is singular; that is, if its determinant is zero. Thus we require that

$$|\mathbf{A} - \lambda \mathbf{I}| = 0 \qquad (8.1\text{-}12)$$

for a nonzero solution $\mathbf{x}$ to exist. We will use this condition many times in this chapter.

For example, consider the set of equations:

$$x_2 = \lambda x_1$$
$$-2x_1 - 4x_2 = \lambda x_2$$

Here

$$\mathbf{A} = \begin{bmatrix} 0 & 1 \\ -2 & -4 \end{bmatrix}$$

and

$$\begin{aligned} |\mathbf{A} - \lambda \mathbf{I}| &= \left| \begin{bmatrix} 0 & 1 \\ -2 & -4 \end{bmatrix} - \begin{bmatrix} \lambda & 0 \\ 0 & \lambda \end{bmatrix} \right| \\ &= \begin{vmatrix} -\lambda & 1 \\ -2 & -4-\lambda \end{vmatrix} = \lambda(4 + \lambda) + 2 = \lambda^2 + 4\lambda + 2 \end{aligned}$$

Thus, if a nonzero solution is to exist for x_1 and x_2, λ must satisfy $\lambda^2 + 4\lambda + 2 = 0$, which gives $\lambda = -3.41$ and $\lambda = -0.586$. For any other value of λ, the solution is $x_1 = x_2 = 0$.

Matrix Representation

The equations of motion of a linear multi-degree-of-freedom system can be expressed in the following compact form:

$$\mathbf{M}\ddot{\mathbf{x}} + \mathbf{C}\dot{\mathbf{x}} + \mathbf{K}\mathbf{x} = \mathbf{F}(t) \qquad (8.1\text{-}13)$$

The matrices $\mathbf{M}$, $\mathbf{C}$, and $\mathbf{K}$ are called the mass, damping, and stiffness matrices. To see how equations can be represented in this form, it is best to consider a specific example.

EXAMPLE 8.1-4
System Matrices
for Three Degrees
of Freedom

Obtain the mass, damping, and stiffness matrices for the system shown in Figure 8.1-2.

Solution

The equations of motion from Example 8.1-2 are

$$m_1\ddot{x}_1 = -k_1 x_1 - k_2(x_1 - x_2) - c_1\dot{x}_1 - c_2(\dot{x}_1 - \dot{x}_2) + f_1(t)$$
$$m_2\ddot{x}_2 = k_2(x_1 - x_2) - k_3(x_2 - x_3) + c_2(\dot{x}_1 - \dot{x}_2) - c_3(\dot{x}_2 - \dot{x}_3) + f_2(t)$$
$$m_3\ddot{x}_1 = k_3(x_2 - x_3) + c_3(\dot{x}_2 - \dot{x}_3) + f_3(t)$$

To put these into matrix form, you must first bring all terms involving the dependent variables to the left-hand side of each equation and collect coefficients. This results in

$$m_1\ddot{x}_1 + (c_1 + c_2)\dot{x}_1 + (k_1 + k_2)x_1 - c_2\dot{x}_2 - k_2 x_2 = f_1(t)$$
$$m_2\ddot{x}_2 - c_2\dot{x}_1 - k_2 x_1 + (c_2 + c_3)\dot{x}_2 + (k_2 + k_3)x_2 - c_3\dot{x}_3 - k_3 x_3 = f_2(t)$$
$$m_3\ddot{x}_3 - c_3\dot{x}_2 - k_3 x_2 + c_3\dot{x}_3 + k_3 x_3 = f_3(t)$$

The matrix form is

$$
\begin{bmatrix} m_1 & 0 & 0 \\ 0 & m_2 & 0 \\ 0 & 0 & m_3 \end{bmatrix}
\begin{bmatrix} \ddot{x}_1 \\ \ddot{x}_2 \\ \ddot{x}_3 \end{bmatrix}
+
\begin{bmatrix} c_1 + c_2 & -c_2 & 0 \\ -c_2 & c_2 + c_3 & -c_3 \\ 0 & -c_3 & c_3 \end{bmatrix}
\begin{bmatrix} \dot{x}_1 \\ \dot{x}_2 \\ \dot{x}_3 \end{bmatrix}
$$

$$
+
\begin{bmatrix} k_1 + k_2 & -k_2 & 0 \\ -k_2 & k_2 + k_3 & -k_3 \\ 0 & -k_3 & k_3 \end{bmatrix}
\begin{bmatrix} x_1 \\ x_2 \\ x_3 \end{bmatrix}
=
\begin{bmatrix} f_1(t) \\ f_2(t) \\ f_3(t) \end{bmatrix}
$$

The mass, damping, and stiffness matrices are

$$
\mathbf{M} = \begin{bmatrix} m_1 & 0 & 0 \\ 0 & m_2 & 0 \\ 0 & 0 & m_3 \end{bmatrix}
\qquad
\mathbf{C} = \begin{bmatrix} c_1 + c_2 & -c_2 & 0 \\ -c_2 & c_2 + c_3 & -c_3 \\ 0 & -c_3 & c_3 \end{bmatrix}
$$

$$
\mathbf{K} = \begin{bmatrix} k_1 + k_2 & -k_2 & 0 \\ -k_2 & k_2 + k_3 & -k_3 \\ 0 & -k_3 & k_3 \end{bmatrix}
$$

The external force vector $\mathbf{F}(t)$ is

$$
\mathbf{F}(t) = \begin{bmatrix} f_1(t) \\ f_2(t) \\ f_3(t) \end{bmatrix}
$$

∎

Matrix Forms of the Energy Expressions

In the scalar case the expressions for kinetic energy *KE*, elastic potential energy *PE*, and power dissipation *PD* are

$$KE = \frac{1}{2}m\dot{x}^2$$

$$PE = \frac{1}{2}kx^2$$

$$PD = c\dot{x}^2$$

From the previous examples, we can conclude that the energy expressions, which are scalars, can be written in matrix form as

$$KE = \frac{1}{2}\dot{\mathbf{x}}^T \mathbf{M}\dot{\mathbf{x}} \tag{8.1-14}$$

$$PE = \frac{1}{2}\mathbf{x}^T \mathbf{K}\mathbf{x} \tag{8.1-15}$$

$$PD = \dot{\mathbf{x}}^T \mathbf{C}\dot{\mathbf{x}} \tag{8.1-16}$$

These are all in the scalar form $b = \mathbf{v}^T \mathbf{A}\mathbf{v}$, from which we see that

$$b^T = \left(\mathbf{v}^T \mathbf{A}\mathbf{v}\right)^T = \mathbf{v}^T \mathbf{A}^T \mathbf{v}$$

Since $b^T = b$, this implies that $\mathbf{A}^T = \mathbf{A}$ and thus that $\mathbf{A}$ is symmetric. Therefore $\mathbf{M}$, $\mathbf{K}$, and $\mathbf{C}$ are symmetric.

Flexibility Influence Coefficients

The *flexibility* is the inverse of the stiffness. In the scalar case, the flexibility a is $1/k$. Thus, because $f = kx$, if $f = 1$, then $x = 1/k = a$, and we see that the flexibility a is the displacement caused by a *unit* force. In the matrix case, the flexibility matrix $\mathbf{A}$ is defined as

$$\mathbf{A} = \mathbf{K}^{-1} \tag{8.1-17}$$

The *flexibility influence coefficient* a_{ij} is the displacement at coordinate i due to a unit force applied at coordinate j, with all other forces equal to zero. The first column of $\mathbf{A}$ consists of the displacements corresponding to $f_1 = 1$, with $f_2 = f_3 = \cdots = f_n = 0$. Similarly, the second column consists of the displacements corresponding to $f_2 = 1$, with $f_1 = f_3 = \cdots = f_n = 0$, and so forth.

The practical importance of the flexibility matrix is that it can be calculated from a static test in which a load f_j is applied at coordinate j and the resulting displacement x_i is measured at coordinate i. The flexibility coefficient can be calculated from $a_{ij} = x_i/f_j$. This test is repeated at the location of every lumped mass in the system. Once the flexibility matrix has been calculated, the stiffness matrix can be calculated by computing the matrix inverse of $\mathbf{A}$. Such a test is easier to perform than one that measures the force resulting from an applied displacement.

Our first example will demonstrate the procedure for a lumped-mass system, even though the stiffness matrix is usually easy to determine from Newton's laws. The usefulness of the flexibility influence coefficients, however, is most apparent with distributed systems. Our second example demonstrates this.

EXAMPLE 8.1-5
Flexibility Influence Coefficients for a Lumped-Mass System

(a) Determine the flexibility matrix for the system shown in Figure 8.1-1.

(b) Compute $\mathbf{K}$ from $\mathbf{A}$ for the case where $k_1 = k_2 = k_3 = k$.

Solution

(a) At equilibrium, $\mathbf{x} = \mathbf{A}\mathbf{f}$. For this system, we have

$$\begin{bmatrix} x_1 \\ x_2 \\ x_3 \end{bmatrix} = \mathbf{A}\begin{bmatrix} f_1 \\ f_2 \\ f_3 \end{bmatrix}$$

First start by imagining a unit force $f_1 = 1$ applied to mass m_1, with $f_2 = f_3 = 0$. From *statics* we know that $x_1 = 1/k_1$. If the forces at m_2 and m_3 are zero, springs k_2 and k_3 are unstretched, and thus $x_2 = x_3 = x_1 = 1/k_1$. So we have

$$\begin{bmatrix} x_1 \\ x_2 \\ x_3 \end{bmatrix} = \begin{bmatrix} 1/k_1 & 0 & 0 \\ 1/k_1 & 0 & 0 \\ 1/k_1 & 0 & 0 \end{bmatrix} \begin{bmatrix} 1 \\ 0 \\ 0 \end{bmatrix}$$

Next imagine a unit force $f_2 = 1$ applied to mass m_2, with $f_1 = f_3 = 0$. From statics we know that $x_1 = 1/k_1$, but since k_1 and k_2 are in series, the displacement of m_2 will be

$$x_2 = \frac{1}{k_1} + \frac{1}{k_2}$$

With no force at x_3, spring k_3 is unstretched and the mass m_3 must have the same displacement as m_2. Thus

$$\begin{bmatrix} x_1 \\ x_2 \\ x_3 \end{bmatrix} = \begin{bmatrix} 0 & \dfrac{1}{k_1} & 0 \\ 0 & \dfrac{1}{k_1} + \dfrac{1}{k_2} & 0 \\ 0 & \dfrac{1}{k_1} + \dfrac{1}{k_2} & 0 \end{bmatrix} \begin{bmatrix} 0 \\ 1 \\ 0 \end{bmatrix}$$

Finally, imagine a unit force $f_3 = 1$ applied to mass m_3, with $f_1 = f_2 = 0$. From statics we know that $x_1 = 1/k_1$. Since k_1 and k_2 are in series, the displacement of m_2 will be

$$x_2 = \frac{1}{k_1} + \frac{1}{k_2}$$

Since k_1, k_2, and k_3 are in series, the displacement of m_3 will be

$$x_3 = \frac{1}{k_1} + \frac{1}{k_2} + \frac{1}{k_3}$$

Thus

$$\begin{bmatrix} x_1 \\ x_2 \\ x_3 \end{bmatrix} = \begin{bmatrix} 0 & 0 & \dfrac{1}{k_1} \\ 0 & 0 & \dfrac{1}{k_1} + \dfrac{1}{k_2} \\ 0 & 0 & \dfrac{1}{k_1} + \dfrac{1}{k_2} + \dfrac{1}{k_3} \end{bmatrix} \begin{bmatrix} 0 \\ 0 \\ 1 \end{bmatrix}$$

The flexibility matrix is obtained by adding the three previous matrices:

$$\mathbf{A} = \begin{bmatrix} \dfrac{1}{k_1} & \dfrac{1}{k_1} & \dfrac{1}{k_1} \\ \dfrac{1}{k_1} & \dfrac{1}{k_1} + \dfrac{1}{k_2} & \dfrac{1}{k_1} + \dfrac{1}{k_2} \\ \dfrac{1}{k_1} & \dfrac{1}{k_1} + \dfrac{1}{k_2} & \dfrac{1}{k_1} + \dfrac{1}{k_2} + \dfrac{1}{k_3} \end{bmatrix}$$

(b) With $k_1 = k_2 = k_3 = k$, we obtain

$$\mathbf{A} = \begin{bmatrix} 1/k & 1/k & 1/k \\ 1/k & 2/k & 2/k \\ 1/k & 2/k & 3/k \end{bmatrix} = 1/k \begin{bmatrix} 1 & 1 & 1 \\ 1 & 2 & 2 \\ 1 & 2 & 3 \end{bmatrix}$$

The inverse can be computed with MATLAB as follows:

```
A = [1,1,1;1,2,2;1,2,3];
K = inv(A)
```

The result is

$$\mathbf{K} = k \begin{bmatrix} 2 & -1 & 0 \\ -1 & 2 & -1 \\ 0 & -1 & 1 \end{bmatrix}$$

∎

EXAMPLE 8.1-6
Flexibility
Influence
Coefficients for a
Beam

Figure 8.1-3 shows a uniform beam with three concentrated masses attached to it. The three masses are much greater than the beam mass, so we will neglect the mass of the beam. An experiment was performed in which a 1000-N vertical weight f_j was attached to the x_1, x_2, and x_3 locations successively, and the resulting static beam deflections x_1, x_2, and x_3 were measured. Use the measurements in the following table to compute the stiffness matrix of the beam:

	$f_1 = 100$ (N)	$f_2 = 1000$ (N)	$f_3 = 1000$ (N)
x_1 (mm)	3.4	4.1	2.6
x_1 (mm)	4.1	6	4.1
x_1 (mm)	2.6	4.1	3.4

Solution

Because a nonunit force of 1000 N was applied, we must divide the displacement data by 1000. This assumes that the system response is linear. We obtain the following table:

	$f_1 = 1$ (N)	$f_2 = 1$ (N)	$f_3 = 1$ (N)
x_1 (10^{-6} m)	3.4	4.1	2.6
x_1 (10^{-6} m)	4.1	6	4.1
x_1 (10^{-6} m)	2.6	4.1	3.4

In this form the flexibility matrix can be read directly from the table. It is

$$\mathbf{A} = \begin{bmatrix} 3.4 & 4.1 & 2.6 \\ 4.1 & 6 & 4.1 \\ 2.6 & 4.1 & 3.4 \end{bmatrix} 10^{-6} \text{ (m/N)}$$

Thus the stiffness matrix is, after rounding to four significant figures,

$$\mathbf{K} = \mathbf{A}^{-1} = \begin{bmatrix} 1.886 & -1.723 & 0.6355 \\ -1.723 & 2.521 & -1.723 \\ 0.6355 & -1.723 & 1.886 \end{bmatrix} 10^6 \text{ (N/m)}$$

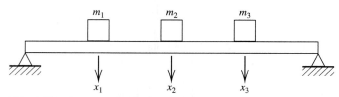

FIGURE 8.1-3 A beam with concentrated masses.

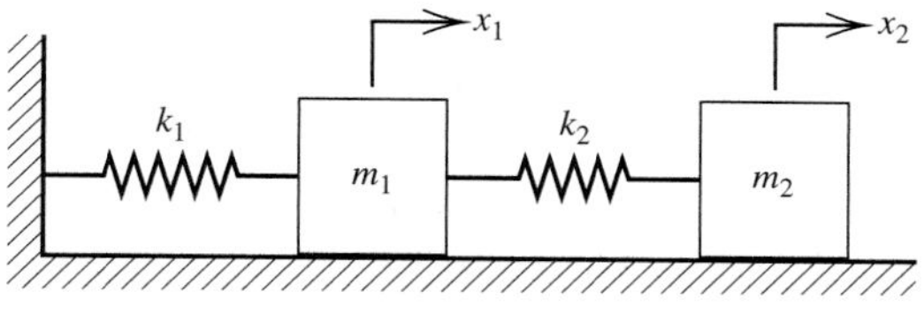

FIGURE 8.1-4 Illustration of different coordinates.

Thus the lumped-parameter model of this system is

$$
\begin{bmatrix} m_1 & 0 & 0 \\ 0 & m_2 & 0 \\ 0 & 0 & m_3 \end{bmatrix} \begin{bmatrix} \ddot{x}_1 \\ \ddot{x}_2 \\ \ddot{x}_3 \end{bmatrix} + 10^6 \begin{bmatrix} 1.886 & -1.723 & 0.6355 \\ -1.723 & 2.521 & -1.723 \\ 0.6355 & -1.723 & 1.886 \end{bmatrix} \begin{bmatrix} x_1 \\ x_2 \\ x_3 \end{bmatrix} = \begin{bmatrix} 0 \\ 0 \\ 0 \end{bmatrix}
$$

■

Nonsymmetric Matrices

The results to be derived in Section 8.3 are based on the assumption that the mass and stiffness matrices are symmetric. Although some models will produce nonsymmetric matrices, it turns that we can always change coordinates to produce a set of symmetric matrices.

Consider the system shown in Figure 8.1-4. Suppose we use the coordinates x_1 and the relative displacement $y = x_2 - x_1$. Note that the absolute acceleration of m_2 is $\ddot{x}_1 + \ddot{y}$. The equations of motion are

$$
m_1 \ddot{x}_1 = -k_1 x_1 + k_2 y
$$

$$
m_2 (\ddot{x}_1 + \ddot{y}) = -k_2 y
$$

and the matrices are

$$
\mathbf{M} = \begin{bmatrix} m_1 & 0 \\ m_2 & m_2 \end{bmatrix} \qquad \mathbf{K} = \begin{bmatrix} k_1 & -k_2 \\ 0 & k_2 \end{bmatrix}
$$

which are not symmetric.

However, if we use the coordinates x_1 and x_2, we obtain symmetric matrices. We will always obtain symmetric mass and stiffness matrices if we derive the equations of motion using Lagrange's equations. This is because Lagrange's equations are obtained from the expressions for the kinetic and potential energies of the system. These expressions are quadratic, because kinetic energy is proportional to the velocity squared and elastic potential energy is proportional to the displacement squared. For example, the energy expressions for the system shown in Figure 8.1-4 are

$$
KE = \frac{1}{2} m_1 \dot{x}_1^2 + \frac{1}{2} m_2 \dot{x}_2^2 = \frac{1}{2} [\dot{x}_1 \quad \dot{x}_2] \begin{bmatrix} m_1 & 0 \\ 0 & m_2 \end{bmatrix} \begin{bmatrix} \dot{x}_1 \\ \dot{x}_2 \end{bmatrix}
$$

$$
PE = \frac{1}{2} k_1 x_1^2 + \frac{1}{2} k_2 (x_2 - x_1)^2 = \frac{1}{2} [x_1 \quad x_2] \begin{bmatrix} k_1 + k_2 & -k_2 \\ -k_2 & k_2 \end{bmatrix} \begin{bmatrix} x_1 \\ x_2 \end{bmatrix}
$$

In general, as we have seen, the energy expressions will be of the form

$$
KE = \frac{1}{2} \dot{\mathbf{x}}^T \mathbf{M} \dot{\mathbf{x}}
$$

$$
PE = \frac{1}{2} \mathbf{x}^T \mathbf{K} \mathbf{x}
$$

and thus $\mathbf{M}$ and $\mathbf{K}$ are symmetric.

Derivation of the equations of motion from Newton's laws can result in either symmetric or nonsymmetric matrices, depending on the coordinates used. If they are nonsymmetric, a new set of coordinates can always be found to yield symmetric matrices.

Inertial and Stiffness Coupling

The equations of motion are usually *coupled*. This means, for example, that x_2 appears in the equation containing $\ddot{x}_1$, and vice versa, so the equations cannot be solved separately one at a time.

Coupling can be due to the spring forces, the accelerations, or the damping forces. When the stiffness matrix $\mathbf{K}$ is not diagonal, then the equations of motion are said to have *stiffness* coupling. The equations have *inertial* or *mass* coupling when the mass matrix $\mathbf{M}$ is not diagonal. The equations can also be coupled through the damping terms. The term *dynamic coupling* is also used for inertial coupling and *static coupling* for stiffness coupling.

The choice of coordinates influences the existence and the type of coupling, as shown in the following example.

EXAMPLE 8.1-7
Coupling in a
Bounce-Pitch
Model

Figure 8.1-5a shows an identical system consisting of a rigid bar, modeled in parts b, c, and d with three different coordinate systems. It is similar to the bounce-pitch model of a vehicle suspension treated in Chapter 6. Investigate the type of coupling that exists for each coordinate system.

Solution

For Figure 8.1-5b, the equations of motion are

$$\begin{bmatrix} m & mL_1 \\ mL_1 & I_1 \end{bmatrix} \begin{bmatrix} \ddot{x} \\ \ddot{\theta} \end{bmatrix} + \begin{bmatrix} (k_1 + k_2) & k_2 L \\ k_2 L & k_2 L^2 \end{bmatrix} \begin{bmatrix} x \\ \theta \end{bmatrix} = \begin{bmatrix} 0 \\ 0 \end{bmatrix}$$

where I_1 is the inertia about the left-hand endpoint. These equations are always both statically and dynamically coupled since $L_1 > 0$ and $L > 0$.

For Figure 8.1-5c, the equations of motion are

$$\begin{bmatrix} m & 0 \\ 0 & I_G \end{bmatrix} \begin{bmatrix} \ddot{x} \\ \ddot{\theta} \end{bmatrix} + \begin{bmatrix} (k_1 + k_2) & (k_2 L_2 - k_1 L_1) \\ (k_2 L_2 - k_1 L_1) & (k_1 L_1^2 + k_2 L_2^2) \end{bmatrix} \begin{bmatrix} x \\ \theta \end{bmatrix} = \begin{bmatrix} 0 \\ 0 \end{bmatrix}$$

where I_G is the inertia about the center of gravity G. These equations are statically coupled as long as $k_2 L_2 - k_1 L_1 \neq 0$.

For Figure 8.1-5d, the point A is a point at which a force applied normal to the bar produces only translation. The condition for this to occur is $k_1 L_3 = k_2 L_4$. The equations of motion are

$$\begin{bmatrix} m & mL_5 \\ mL_5 & I_A \end{bmatrix} \begin{bmatrix} \ddot{x} \\ \ddot{\theta} \end{bmatrix} + \begin{bmatrix} (k_1 + k_2) & 0 \\ 0 & (k_1 L_3^2 + k_2 L_4^2) \end{bmatrix} \begin{bmatrix} x \\ \theta \end{bmatrix} = \begin{bmatrix} 0 \\ 0 \end{bmatrix}$$

where I_A is the inertia about point A. These equations are dynamically coupled. ∎

8.2 MODE SHAPES AND EIGENVALUE PROBLEMS

Consider the two-degrees-of-freedom model shown in Figure 8.2-1. Its equations of motion are

$$m_1 \ddot{x}_1 + (k_1 + k_2) x_1 - k_2 x_2 = f_1 \tag{8.2-1}$$

$$m_2 \ddot{x}_2 - k_2 x_1 + k_2 x_2 = f_2 \tag{8.2-2}$$

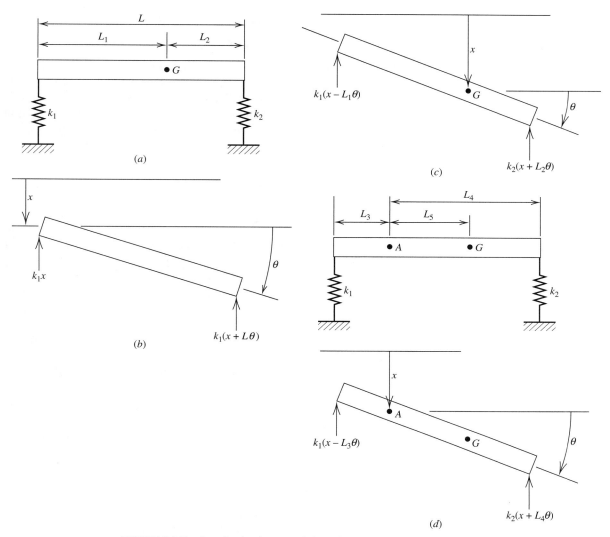

FIGURE 8.1-5 Coupling in a bounce-pitch model.

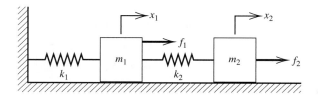

FIGURE 8.2-1 Example with two degrees of freedom.

For now let us concentrate on the free response and set $f_1 = f_2 = 0$. Consider the following specific parameter values to make the discussion easier to follow.

Suppose that $m_1 = 4$, $m_2 = 1$, $k_1 = 12$, and $k_2 = 4$. The equations become

$$4\ddot{x}_1 + 16x_1 - 4x_2 = 0 \tag{8.2-3}$$

$$\ddot{x}_2 - 4x_1 + 4x_2 = 0 \tag{8.2-4}$$

As we did in Chapter 6, we could make the substitutions

$$x_1(t) = X_1 e^{st} \quad x_2(t) = X_2 e^{st}$$

and obtain a determinant equation in terms of s. We found, however, that the roots s of the determinant equation are purely imaginary and of the form $s = \pm i\omega$, where ω is a natural frequency of the system. The s roots will always be purely imaginary if the system is undamped. So, since we are limiting our treatment to undamped systems for now, instead we will make the following substitutions into Equations 8.2-3 and 8.2-4:

$$x_1(t) = X_1 e^{i\omega t} \quad\quad x_2(t) = X_2 e^{i\omega t}$$

Note that

$$\ddot{x}_1 = -\omega^2 X_1 e^{i\omega t} \quad\quad \ddot{x}_2 = -\omega^2 X_2 e^{i\omega t}$$

After canceling the $e^{i\omega t}$ terms, we obtain

$$(16 - 4\omega^2)X_1 - 4X_2 = 0 \tag{8.2-5}$$

$$-4X_1 + (4 - \omega^2)X_2 = 0 \tag{8.2-6}$$

The determinant of these equations must be zero. This gives

$$(16 - 4\omega^2)(4 - \omega^2) - 16 = 0$$

The solutions are $\omega_1^2 = 2$ and $\omega_2^2 = 6$, which correspond to the natural frequencies $\omega_1 = \sqrt{2}$ and $\omega_2 = \sqrt{6}$. For $\omega_1^2 = 2$ the mode ratio is $X_2/X_1 = 2$, and for $\omega_2^2 = 6$ the mode ratio is $X_2/X_1 = -2$. Thus we can express the mode ratios as the following vectors:

$$\mathbf{X}_1 = X_1 \begin{bmatrix} 1 \\ 2 \end{bmatrix} \quad\quad \mathbf{X}_2 = X_1 \begin{bmatrix} 1 \\ -2 \end{bmatrix} \tag{8.2-7}$$

We call these vectors the *mode shape vectors*. Note that we can determine only the *direction*, not the *length*, of the vectors.

Eigenvalue Problems

Equations 8.2-5 and 8.2-6 can be expressed as follows, by moving the terms involving ω^2 to the right-hand side and dividing the first equation by 4:

$$4X_1 - X_2 = \omega^2 X_1$$

$$-4X_1 + 4X_2 = \omega^2 X_2$$

These can be expressed in matrix-vector form as

$$\begin{bmatrix} 4 & -1 \\ -4 & 4 \end{bmatrix} \begin{bmatrix} X_1 \\ X_2 \end{bmatrix} = \omega^2 \begin{bmatrix} X_1 \\ X_2 \end{bmatrix} \tag{8.2-8}$$

If we let

$$\mathbf{X} = \begin{bmatrix} X_1 \\ X_2 \end{bmatrix}$$

and

$$\mathbf{A} = \begin{bmatrix} 4 & -1 \\ -4 & 4 \end{bmatrix} \tag{8.2-9}$$

then we can express the equations in the compact form

$$\mathbf{AX} = \omega^2 \mathbf{X} \qquad (8.2\text{-}10)$$

Equation 8.2-10 represents an *eigenvalue problem*. In such a problem we are given the matrix $\mathbf{A}$, and we must find the values of the scalar ω and the vectors $\mathbf{X}$ that satisfy Equation 8.2-10. For the matrix $\mathbf{A}$ given in Equation 8.2-9, we have already found the solutions, which are $\omega_1^2 = 2$ and $\mathbf{X}_1$, and $\omega_2^2 = 6$ and $\mathbf{X}_2$, where the vectors are given by Equation 8.2-7. The values of ω^2 are called the *eigenvalues* of $\mathbf{A}$, and the vectors $\mathbf{X}_1$ and $\mathbf{X}_2$ are called the *eigenvectors* of $\mathbf{A}$.

We will see that the squared natural frequency ω^2 is not always the eigenvalue; for some problems the eigenvalue is $1/\omega^2$, for example. Because of this, we will represent the eigenvalue in general by the symbol λ and the eigenvectors by $\mathbf{v}$, so the general eigenvalue problem is expressed as

$$\mathbf{Av} = \lambda \mathbf{v} \qquad (8.2\text{-}11)$$

Matrix Inverse

Before proceeding, we need to deal with the subject of the matrix inverse. Consider the equations

$$10x_1 - 6x_2 = 40$$

$$4x_1 + 3x_2 = 43$$

In matrix form these become

$$\begin{bmatrix} 10 & -6 \\ 4 & 3 \end{bmatrix} \begin{bmatrix} x_1 \\ x_2 \end{bmatrix} = \begin{bmatrix} 40 \\ 43 \end{bmatrix}$$

or, in compact form,

$$\mathbf{Ax} = \mathbf{b}$$

We can formally solve this equation as follows:

$$\mathbf{x} = \mathbf{A}^{-1}\mathbf{b}$$

The question arises as to what is $\mathbf{A}^{-1}$? The matrix $\mathbf{A}^{-1}$ is called the *inverse* of $\mathbf{A}$ and is defined only for square matrices (those having the same number of rows as columns). Calculation of the inverse is tedious for matrices having more than three rows and is best done with a calculator or computer. Here we give the results for a (2×2) matrix so that you can follow the calculations in our examples. Let

$$\mathbf{A} = \begin{bmatrix} a & b \\ c & d \end{bmatrix}$$

Then its inverse is

$$\mathbf{A}^{-1} = \frac{1}{|\mathbf{A}|} \begin{bmatrix} d & -b \\ -c & a \end{bmatrix}$$

where $|\mathbf{A}|$ is the determinant formed from the rows and columns of $\mathbf{A}$. Thus

$$|\mathbf{A}| = \begin{vmatrix} a & b \\ c & d \end{vmatrix} = ad - bc$$

If $|\mathbf{A}| = 0$, then the inverse is not defined (this situation corresponds to a set of equations having no unique solution). A matrix that does not have an inverse is called a *singular* matrix.

You can check the correctness of an inverse calculation by checking to see if $\mathbf{A}^{-1}\mathbf{A} = \mathbf{I}$, where $\mathbf{I}$ is the *identity* matrix that consists of 1's on the diagonal and zeros everywhere else. The (2×2) identity matrix is

$$\mathbf{I} = \begin{bmatrix} 1 & 0 \\ 0 & 1 \end{bmatrix}$$

We note that it is easy to compute the inverse of a matrix having zero entries everywhere but on its diagonal. The inverse is a diagonal matrix whose diagonal elements are the reciprocals of the original diagonal elements. For example,

$$\mathbf{A} = \begin{bmatrix} 5 & 0 \\ 0 & 4 \end{bmatrix} \qquad \mathbf{A}^{-1} = \begin{bmatrix} 0.2 & 0 \\ 0 & 0.25 \end{bmatrix}$$

General Matrix Formulation

In general, the equations of motion of a linear, undamped system having n degrees of freedom will have the form

$$\mathbf{M}\ddot{\mathbf{x}} + \mathbf{K}\mathbf{x} = \mathbf{F}(t) \tag{8.2-12}$$

where $\mathbf{M}$ is the $(n \times n)$ *mass matrix*, $\mathbf{K}$ is the $(n \times n)$ *stiffness matrix*, $\mathbf{F}(t)$ is an $(n \times 1)$ vector consisting of the externally applied forces, and $\mathbf{x}$ is the vector consisting of the displacement variables:

$$\mathbf{x} = \begin{bmatrix} x_1 \\ x_2 \\ \cdot \\ \cdot \\ \cdot \\ x_n \end{bmatrix}$$

In rotational models the coefficients of the acceleration terms will be inertia values, and so for this reason the matrix $\mathbf{M}$ is also called the *inertia matrix*.

To find the free response, we set $\mathbf{F}(t)$ equal to zero to obtain

$$\mathbf{M}\ddot{\mathbf{x}} + \mathbf{K}\mathbf{x} = \mathbf{0}$$

Making the substitution $\mathbf{x} = \mathbf{X}e^{i\omega t}$, we obtain

$$-\omega^2 \mathbf{M}\mathbf{X}e^{i\omega t} + \mathbf{K}\mathbf{X}e^{i\omega t} = \mathbf{0}$$

Canceling the scalar $e^{i\omega t}$ gives

$$-\omega^2 \mathbf{M}\mathbf{X} + \mathbf{K}\mathbf{X} = \mathbf{0} \tag{8.2-13}$$

For systems having more than two degrees of freedom, the process of solving for the natural frequencies and modes is very difficult. We will see that eigenvalue problems are easily solved with computer software such as MATLAB. So for this reason we will concentrate on formulating our problems as eigenvalue problems.

Equation 8.2-13 can be arranged as an eigenvalue problem $\mathbf{A}\mathbf{X} = \lambda \mathbf{X}$ in several ways. The first way consists of multiplying both sides from the left by $\mathbf{M}^{-1}$ to obtain

$$-\omega^2 \mathbf{M}^{-1}\mathbf{M}\mathbf{X} + \mathbf{M}^{-1}\mathbf{K}\mathbf{X} = \mathbf{0}$$

or

$$-\omega^2 \mathbf{IX} + \mathbf{M}^{-1}\mathbf{KX} = 0$$

which gives

$$\mathbf{M}^{-1}\mathbf{KX} = \omega^2 \mathbf{X} \tag{8.2-14}$$

In this case $\mathbf{A} = \mathbf{M}^{-1}\mathbf{K}$, and the eigenvalues are $\lambda = \omega^2$.

Another way consists of multiplying both sides from the left by $\mathbf{K}^{-1}$ to obtain

$$-\omega^2 \mathbf{K}^{-1}\mathbf{MX} + \mathbf{K}^{-1}\mathbf{KX} = 0$$

or

$$-\omega^2 \mathbf{K}^{-1}\mathbf{MX} + \mathbf{IX} = 0$$

which gives

$$\mathbf{K}^{-1}\mathbf{MX} = \frac{1}{\omega^2}\mathbf{X} \tag{8.2-15}$$

In this case $\mathbf{A} = \mathbf{K}^{-1}\mathbf{M}$, and the eigenvalues are $\lambda = 1/\omega^2$.

EXAMPLE 8.2-1
Frequencies and Mode Shapes for Two Degrees of Freedom

Determine the frequencies and mode shapes for the model given by Equations 8.2-3 and 8.2-4.

Solution

From Equations 8.2-3 and 8.2-4,

$$\mathbf{M} = \begin{bmatrix} 4 & 0 \\ 0 & 1 \end{bmatrix} \qquad \mathbf{K} = \begin{bmatrix} 16 & -4 \\ -4 & 4 \end{bmatrix}$$

For the first method, Equation 8.2-14,

$$\mathbf{A} = \mathbf{M}^{-1}\mathbf{K} = \begin{bmatrix} 1/4 & 0 \\ 0 & 1 \end{bmatrix}\begin{bmatrix} 16 & -4 \\ -4 & 4 \end{bmatrix} = \begin{bmatrix} 4 & -1 \\ -4 & 4 \end{bmatrix} \tag{1}$$

The matrix equation $\mathbf{M}^{-1}\mathbf{KX} = \omega^2 \mathbf{X}$ yields the following equations:

$$(4 - \omega^2)X_1 - X_2 = 0$$
$$-4X_1 + (4 - \omega^2)X_2 = 0$$

The determinant equation is $(4 - \omega^2)^2 - 4 = 0$, and it has the roots $\omega^2 = 2$ and $\omega^2 = 6$. For $\omega_1^2 = 2$ the mode ratio is $X_2/X_1 = 2$, and for $\omega_2^2 = 6$ the mode ratio is $X_2/X_1 = -2$. These are the same results found earlier. Thus we can express the mode ratios as the following vectors:

$$\mathbf{X}_1 = X_1 \begin{bmatrix} 1 \\ 2 \end{bmatrix} \qquad \mathbf{X}_2 = X_1 \begin{bmatrix} 1 \\ -2 \end{bmatrix} \tag{2}$$

For the second method, Equation 8.2-15,

$$\mathbf{A} = \mathbf{K}^{-1}\mathbf{M} = \begin{bmatrix} 1/12 & 1/12 \\ 1/12 & 1/3 \end{bmatrix}\begin{bmatrix} 4 & 0 \\ 0 & 1 \end{bmatrix} = \begin{bmatrix} 1/3 & 1/12 \\ 1/3 & 1/3 \end{bmatrix} \tag{3}$$

The matrix equation $\mathbf{K}^{-1}\mathbf{M}\mathbf{X} = \lambda\mathbf{X}$ yields the equations

$$\left(\frac{1}{3} - \lambda\right)X_1 + \frac{1}{12}X_2 = 0$$

$$\frac{1}{3}X_1 + \left(\frac{1}{3} - \lambda\right)X_2 = 0$$

where $\lambda = 1/\omega^2$. The determinant equation is

$$\left(\frac{1}{3} - \lambda\right)^2 - \frac{1}{36} = 0$$

and it has the roots $\lambda = 1/2$ and $\lambda = 1/6$, which correspond to $\omega_1^2 = 2$ and $\omega_1^2 = 6$. The mode ratios are $X_2/X_1 = 2$ and $X_2/X_1 = -2$. These are the same results found earlier. This is to be expected because the two eigenvalue problems are based on the same set of differential equations, so they should yield the same natural frequencies. In addition, since both methods are based on the same x_1, x_2 coordinate system, the mode shapes should be the same for both methods. ∎

In practice, we choose the eigenvalue problem, Equation 8.2-14 or 8.2-15, that minimizes the difficulty of computing a matrix inverse. For example, if $\mathbf{M}$ has zero entries everywhere but on its diagonal, its inverse is easily computed and we would use Equation 8.2-14. This is an important issue when doing the computations by hand, and it can be an important issue when using a computer because of the potential for round-off error when computing the inverse of a large matrix.

EXAMPLE 8.2-2
Frequencies and Mode Shapes for Three Degrees of Freedom

Obtain by hand the natural frequencies and mode shapes of the system shown in Figure 8.2-2 for the case where $m_1 = m_2 = m_3 = m$ and $k_1 = k_2 = k_3 = k$.

Solution

The equations of motion are

$$m_1\ddot{x}_1 = -k_1 x_1 - k_2(x_1 - x_2)$$
$$m_2\ddot{x}_2 = k_2(x_1 - x_2) - k_3(x_2 - x_3)$$
$$m_3\ddot{x}_3 = k_3(x_2 - x_3)$$

or

$$m\ddot{x}_1 = -2kx_1 + kx_2$$
$$m\ddot{x}_2 = kx_1 - 2kx_2 + kx_3$$
$$m\ddot{x}_3 = k(x_2 - x_3)$$

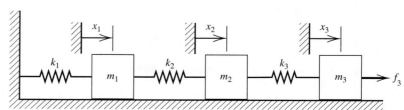

FIGURE 8.2-2   Example with three degrees of freedom.

Substitute

$$x_1(t) = X_1 e^{i\omega t} \qquad x_2(t) = X_2 e^{i\omega t} \qquad x_3(t) = X_3 e^{i\omega t}$$

and cancel the $e^{i\omega t}$ terms to obtain

$$(2k - m\omega^2)X_1 - kX_2 = 0$$
$$-kX_1 + (2k - m\omega^2)X_2 - kX_3 = 0$$
$$-kX_2 + (k - m\omega^2)X_3 = 0$$

Frequently we can simplify such expressions by introducing the parameter $r = m\omega^2/k$. So, dividing these equation by k gives

$$(2 - r)X_1 - X_2 = 0 \tag{1}$$

$$-X_1 + (2 - r)X_2 - X_3 = 0 \tag{2}$$

$$-X_2 + (1 - r)X_3 = 0 \tag{3}$$

Setting the determinant of these equations to zero gives

$$\begin{vmatrix} (2-r) & -1 & 0 \\ -1 & (2-r) & -1 \\ 0 & -1 & (1-r) \end{vmatrix} = (2-r)[(2-r)(1-r) - 1] - (1-r) = 0$$

This reduces to

$$1 - 6r + 5r^2 - r^3 = 0$$

which has the roots $r = 0.19806$, 1.5550, and 3.2470. Since $\omega = \sqrt{kr/m}$, these roots correspond to the natural frequencies $\omega = 0.4450\sqrt{k/m}$, $1.2470\sqrt{k/m}$, and $1.8019\sqrt{k/m}$.

The mode shapes are found as follows. From Equation 1,

$$\frac{X_2}{X_1} = 2 - r \tag{4}$$

From Equations 3 and 4,

$$\frac{X_3}{X_1} = \frac{X_3}{X_2}\frac{X_2}{X_1} = \frac{2 - r}{1 - r}$$

For mode 1, $r = 0.19806$ and

$$\frac{X_2}{X_1} = 2 - r = 1.8019 \qquad \frac{X_3}{X_1} = \frac{2 - r}{1 - r} = 2.2470$$

For mode 2, $r = 1.5550$ and

$$\frac{X_2}{X_1} = 2 - r = 0.4450 \qquad \frac{X_3}{X_1} = \frac{2 - r}{1 - r} = -0.8018$$

For mode 3, $r = 3.2470$ and

$$\frac{X_2}{X_1} = 2 - r = -1.2470 \qquad \frac{X_3}{X_1} = \frac{2 - r}{1 - r} = 0.5550$$

These mode ratios can be represented graphically as shown in Figure 8.2-3, which is drawn for the case $X_1 = 1$. In mode 1 all the masses are moving in the same direction and oscillating with a frequency of $\omega = 0.4450\sqrt{k/m}$.

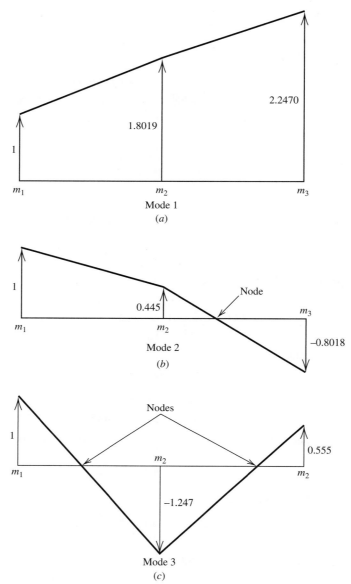

FIGURE 8.2-3 Mode shapes for three degrees of freedom.

In mode 2 the masses m_1 and m_2 are moving in the same direction, while mass m_3 moves in the opposite direction. They oscillate with a frequency of $\omega = 1.2470\sqrt{k/m}$. The diagram shows that in this mode there is a point 36% of the distance from mass m_2 to mass m_3 that does not move. This point is called a *node*.

In mode 3 masses m_1 and m_3 are moving in the same direction, while mass m_2 moves in the opposite direction. They oscillate with a frequency of $\omega = 1.8019\sqrt{k/m}$. The diagram shows that in this mode there are two nodes, one between masses m_1 and m_2, and one between masses m_2 and m_3. For some vibration problems it is important to know the location of any nodes (as, for example, in the bounce-pitch model of a vehicle suspension treated in Chapter 6). ∎

EXAMPLE 8.2-3
Solution as an Eigenvalue Problem

Formulate the problem of Example 8.2-2 as an eigenvalue problem.

Solution

Because **M** is diagonal here, it is easier to use the form $\mathbf{M}^{-1}\mathbf{K}\mathbf{X} = \omega^2\mathbf{X}$ for the eigenvalue problem. Here

$$\mathbf{M} = \begin{bmatrix} m & 0 & 0 \\ 0 & m & 0 \\ 0 & 0 & m \end{bmatrix} = m \begin{bmatrix} 1 & 0 & 0 \\ 0 & 1 & 0 \\ 0 & 0 & 1 \end{bmatrix} = m\mathbf{I}$$

so here

$$\mathbf{M}^{-1} = \frac{1}{m}\mathbf{I}$$

Also,

$$\mathbf{K} = \begin{bmatrix} 2k & -k & 0 \\ -k & 2k & -k \\ 0 & -k & k \end{bmatrix} = k \begin{bmatrix} 2 & -1 & 0 \\ -1 & 2 & -1 \\ 0 & -1 & 1 \end{bmatrix}$$

Thus

$$\mathbf{M}^{-1}\mathbf{K} = \frac{k}{m}\mathbf{I} \begin{bmatrix} 2 & -1 & 0 \\ -1 & 2 & -1 \\ 0 & -1 & 1 \end{bmatrix} = \frac{k}{m} \begin{bmatrix} 2 & -1 & 0 \\ -1 & 2 & -1 \\ 0 & -1 & 1 \end{bmatrix}$$

So the eigenvalue problem can be expressed as

$$\begin{bmatrix} 2 & -1 & 0 \\ -1 & 2 & -1 \\ 0 & -1 & 1 \end{bmatrix} \mathbf{v} = \frac{m}{k}\omega^2\mathbf{v} = r\mathbf{v}$$

The eigenvalues can be found with MATLAB by using the `eig` function. The session is

```
≫A = [2,-1,0;-1,2,-1;0,-1,1]:
≫r = eig(A)
```

The results are $r = 0.19806$, 1.5550, and 3.2470, the same as found in Example 8.2-2.

The syntax `[v, D] = eig(A)` computes the arrays `v` and `D`. The eigenvectors of **A** are stored in the columns of `v`. The eigenvalues are the diagonal elements of `D`. In this example the results displayed on the screen after typing `[v, D] = eig(A)` are

```
v =

  -0.3280  0.7370   -0.5910
  -0.5910  0.3280    0.7370
  -0.7370 -0.5910   -0.3280

D =

  0.1981  0        0
  0       1.5550   0
  0       0        3.2470
```

Thus the eigenvectors corresponding to the three modes are

$$\mathbf{X}_1 = \begin{bmatrix} -0.3280 \\ -0.5910 \\ -0.7370 \end{bmatrix} \quad \mathbf{X}_2 = \begin{bmatrix} 0.7370 \\ 0.3280 \\ -0.5910 \end{bmatrix} \quad \mathbf{X}_3 = \begin{bmatrix} -0.5910 \\ 0.7370 \\ -0.3280 \end{bmatrix}$$

Note that the scalar factor X_1 has been computed so that the length of each vector is 1. That is, $\sqrt{(0.3280)^2 + (0.5910)^2 + (0.7370)^2} = 1$. In the next section we will discuss why this has been done. Note also that we can factor -1 out of each vector without changing its meaning.

The mode ratios in these eigenvectors are the same as those we calculated in Example 8.2-1. For example, for mode 1, $X_2/X_1 = 0.5910/0.3280 = 1.8018$ and $X_3/X_1 = 0.7370/0.3280 = 2.2470$. The slight difference in the value of X_2/X_1 is due to round-off error caused by carrying a smaller number of significant figures in Example 8.2-1. ∎

EXAMPLE 8.2-4
Eigenvalues and Eigenvectors for a Rigid-Body Mode

Obtain the eigenvalues and the mode shape vectors for the system shown in Figure 8.2-4.

Solution

The equations of motion are

$$m_1\ddot{x}_1 = -k(x_1 - x_2)$$
$$m_2\ddot{x}_2 = k(x_1 - x_2)$$

In matrix form these are

$$\begin{bmatrix} m_1 & 0 \\ 0 & m_2 \end{bmatrix}\begin{bmatrix} \ddot{x}_1 \\ \ddot{x}_2 \end{bmatrix} + \begin{bmatrix} k & -k \\ -k & k \end{bmatrix}\begin{bmatrix} x_1 \\ x_2 \end{bmatrix} = \begin{bmatrix} 0 \\ 0 \end{bmatrix}$$

The mass and stiffness matrices are

$$\mathbf{M} = \begin{bmatrix} m_1 & 0 \\ 0 & m_2 \end{bmatrix} \qquad \mathbf{K} = \begin{bmatrix} k & -k \\ -k & k \end{bmatrix}$$

Note that $\mathbf{K}$ does not have an inverse because its determinant is 0:

$$|\mathbf{K}| = k^2 - k^2 = 0$$

Therefore we cannot use the eigenvalue formulation given by Equation 8.2-15. Equation 8.2-14 is easier to use in this example anyway, because $\mathbf{M}$ is a diagonal matrix. Thus $\mathbf{M}^{-1}\mathbf{K}\mathbf{X} = \omega^2\mathbf{X}$ becomes

$$\begin{bmatrix} 1/m_1 & 0 \\ 0 & 1/m_2 \end{bmatrix}\begin{bmatrix} k & -k \\ -k & k \end{bmatrix}\mathbf{X} = \omega^2\mathbf{X}$$

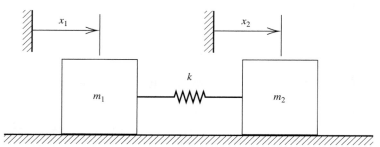

FIGURE 8.2-4 Example of a system having a rigid-body mode.

or

$$\begin{bmatrix} k/m_1 & -k/m_1 \\ -k/m_2 & k/m_2 \end{bmatrix} \mathbf{X} = \omega^2 \mathbf{X}$$

This gives

$$\left(\frac{k}{m_1} - \omega^2 \right) X_1 - \frac{k}{m_1} X_2 = 0$$

$$-\frac{k}{m_2} X_1 + \left(\frac{k}{m_2} - \omega^2 \right) X_2 = 0$$

The determinant equation is

$$\left(\frac{k}{m_1} - \omega^2 \right) \left(\frac{k}{m_2} - \omega^2 \right) - \frac{k^2}{m_1 m_2} = 0$$

This reduces to

$$\omega^4 - \omega^2 \left(\frac{k}{m_1} + \frac{k}{m_2} \right) = 0$$

The roots are

$$\omega^2 = 0 \qquad \omega^2 = \frac{k}{m_1} + \frac{k}{m_2} = k \frac{m_1 + m_2}{m_1 m_2}$$

For the first mode (the rigid-body mode with $\omega^2 = 0$), the mode ratio is

$$\frac{X_2}{X_1} = 1$$

as expected. In this mode the two masses are translating with the same velocity. The corresponding eigenvector is

$$\mathbf{X}_1 = X_1 \begin{bmatrix} 1 \\ 1 \end{bmatrix}$$

For the second mode, the mode ratio is

$$\frac{X_2}{X_1} = -\frac{m_1}{m_2}$$

In this mode the two masses are oscillating in opposite directions with the frequency $\sqrt{k(m_1 + m_2)/m_1 m_2}$. The corresponding eigenvector is

$$\mathbf{X}_2 = X_1 \begin{bmatrix} 1 \\ -m_1/m_2 \end{bmatrix}$$

For systems having a rigid-body mode, the stiffness matrix is singular and one of the natural frequencies is zero. However, the eigenvector corresponding to the zero frequency is *not* zero. It represents the translational mode. The general motion of this system consists of a translation and an oscillation. ∎

8.3   MODAL ANALYSIS

We have indicated that it is important to obtain an eigenvalue formulation from the equation

$$-\omega^2 \mathbf{M} \mathbf{X} + \mathbf{K} \mathbf{X} = \mathbf{0} \qquad (8.3\text{-}1)$$

so that the eigenvalue matrix will be symmetric. We now explain why this is true and show how to formulate an eigenvalue problem to achieve this goal.

Orthogonal Eigenvectors

Two vectors, $\mathbf{v}_1$ and $\mathbf{v}_2$, are said to be *orthogonal* if

$$\mathbf{v}_1^T \mathbf{v}_2 = 0 \tag{8.3-2}$$

or, equivalently, if $\mathbf{v}_2^T \mathbf{v}_1 = 0$. When the vectors have two or three components, this orthogonality condition means that the vectors are mutually perpendicular. This geometric interpretation cannot be made for vectors having four or more components.

A matrix is *symmetric* if it remains the same after its rows and columns are interchanged. The transpose operation interchanges rows and columns; the transpose of $\mathbf{A}$ is denoted $\mathbf{A}^T$. Thus we see that the matrix $\mathbf{A}$ is symmetric if $\mathbf{A}^T = \mathbf{A}$.

We state without proof that the eigenvectors will be orthogonal if the eigenvalue matrix is symmetric. The importance of having orthogonal eigenvectors and the role of a symmetric $\mathbf{A}$ matrix will become apparent later. A hint of the importance of orthogonal vectors comes from statics, in which we use the orthogonal unit vectors $\mathbf{i}$, $\mathbf{j}$, and $\mathbf{k}$ to represent force and displacement vectors.

The mode shape vectors corresponding to Equations 8.2-3 and 8.2-4 are given by Equation 8.2-7 as

$$\mathbf{X}_1 = X_1 \begin{bmatrix} 1 \\ 2 \end{bmatrix} \qquad \mathbf{X}_2 = X_1 \begin{bmatrix} 1 \\ -2 \end{bmatrix}$$

Figure 8.3-1 is a plot of these vectors in terms of the standard unit vectors $\mathbf{i}$ and $\mathbf{j}$ for the case where the vector lengths have been adjusted using $X_1 = 1$. Note that they are not perpendicular.

The two methods we used earlier for obtaining an eigenvalue problem from the matrix equation

$$\mathbf{M}\ddot{\mathbf{x}} + \mathbf{K}\mathbf{x} = \mathbf{0} \tag{8.3-3}$$

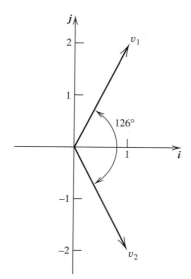

FIGURE 8.3-1   Plot of mode shape vectors.

are

$$\mathbf{M}^{-1}\mathbf{K}\mathbf{X} = \omega^2 \mathbf{X} \qquad \text{for which } \mathbf{A} = \mathbf{M}^{-1}\mathbf{K}$$

and

$$\mathbf{K}^{-1}\mathbf{M}\mathbf{X} = \frac{1}{\omega^2}\mathbf{X} \qquad \text{for which } \mathbf{A} = \mathbf{K}^{-1}\mathbf{M}$$

In our specific example, we saw that neither method produced a symmetric matrix $\mathbf{A}$. See Equations 1 and 3 of Example 8.2-1.

In general, neither method is guaranteed to produce an eigenvalue problem having a symmetric matrix $\mathbf{A}$, and thus neither method is guaranteed to produce orthogonal eigenvectors. Thus we must seek another method of obtaining an eigenvalue problem from Equation 8.3-3.

Suppose we make a change of variables in Equations 8.2-5 and 8.2-6 such that $X_1 = B_1$ and $X_2 = 2B_2$. Then the equations become

$$4B_1 - 2B_2 = \omega^2 B_1 \tag{8.3-4}$$

$$-2B_1 + B_2 = \omega^2 B_2 \tag{8.3-5}$$

or, in matrix form,

$$\begin{bmatrix} 4 & -2 \\ -2 & 4 \end{bmatrix} \begin{bmatrix} B_1 \\ B_2 \end{bmatrix} = \omega^2 \begin{bmatrix} B_1 \\ B_2 \end{bmatrix} \tag{8.3-6}$$

Note that this matrix is symmetric. Let us find its eigenvalues and eigenvectors.

Rearrange Equations 8.3-4 and 8.3-5 as follows:

$$(4 - \omega^2)B_1 - 2B_2 = 0$$

$$-2B_1 + (4 - \omega^2)B_2 = 0$$

Using the same procedure as before, we obtain the determinant equation

$$(4 - \omega^2)(4 - \omega^2) - 4 = 12 - 8\lambda + \lambda^2 = 0$$

This gives the eigenvalues $\lambda_1 = \omega_1^2 = 2$ and $\lambda_2 = \omega_2^2 = 6$, the same as for the equations expressed in terms of the original variables X_1 and X_2.

So we have discovered that a change of variables does not change the eigenvalues. This is to be expected based on physics, because the eigenvalues are related to the natural frequencies, which remain the same regardless of what coordinate system is used to describe the motion.

The eigenvectors of Equations 8.3-4 and 8.3-5 are found as follows. For $\omega_1^2 = 2$ the mode ratio is $B_2/B_1 = 1$, and for $\omega_2^2 = 6$ the mode ratio is $B_2/B_1 = -1$. Thus we can express the mode ratios as the following vectors:

$$\mathbf{v}_1 = B_1 \begin{bmatrix} 1 \\ 1 \end{bmatrix} \qquad \mathbf{v}_2 = B_1 \begin{bmatrix} 1 \\ -1 \end{bmatrix} \tag{8.3-7}$$

These vectors are orthogonal because

$$\mathbf{v}_1^T \mathbf{v}_2 = B_1^2 [1 \quad 1] \begin{bmatrix} 1 \\ -1 \end{bmatrix} = B_1^2 (1 - 1) = 0$$

Figure 8.3-2 is a plot of these vectors for $B_1 = 1$ Note that they are perpendicular.

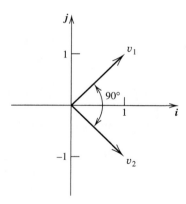

FIGURE 8.3-2 Plot of orthogonal eigenvectors.

This illustrates an important point, which we have not proved. The eigenvectors of the eigenvalue problem $\mathbf{Av} = \lambda\mathbf{v}$ are orthogonal if the matrix $\mathbf{A}$ is symmetric. We will use this property later on.

Orthonormal Eigenvectors

A vector $\mathbf{u}$ is said to be a *unit* vector if it has the property that

$$\mathbf{u}^T\mathbf{u} = 1 \tag{8.3-8}$$

A set of vectors $\mathbf{u}_1, \mathbf{u}_2, \ldots$ is said to be *orthonormal* if they are orthogonal and if they are unit vectors; that is, if

$$\mathbf{u}_i^T\mathbf{u}_j = 0 \qquad i \neq j \tag{8.3-9}$$

and

$$\mathbf{u}_i^T\mathbf{u}_i = 1 \qquad i = 1, 2, \ldots \tag{8.3-10}$$

for every vector in the set. We reserve the symbol $\mathbf{u}$ to represent a set of orthonormal eigenvectors because u stands for "unit" vector.

Recall that the solution of an eigenvalue problem determines only the *direction*, not the *length*, of eigenvectors. For reasons to be seen later, we want the eigenvectors to be orthonormal vectors. We can create a set of orthonormal eigenvectors as follows. From Equation 8.3-7,

$$\mathbf{v}_1^T\mathbf{v}_1 = B_1^2[1 \quad 1]\begin{bmatrix} 1 \\ 1 \end{bmatrix} = 2B_1^2 = 1$$

if $B_1 = 1/\sqrt{2}$. Similarly,

$$\mathbf{v}_2^T\mathbf{v}_2 = B_1^2[1 \quad -1]\begin{bmatrix} 1 \\ -1 \end{bmatrix} = 2B_1^2 = 1$$

if $B_1 = 1/\sqrt{2}$. Thus the set of orthonormal eigenvectors for Equation 8.3-7 is

$$\mathbf{u}_1 = \frac{1}{\sqrt{2}}\begin{bmatrix} 1 \\ 1 \end{bmatrix}$$

$$\mathbf{u}_2 = \frac{1}{\sqrt{2}}\begin{bmatrix} 1 \\ -1 \end{bmatrix}$$

Principal Coordinates

Consider the system shown in Figure 8.2-1 with the parameter values $m_1 = 4$, $m_2 = 1$, $k_1 = 12$, and $k_2 = 4$. In Section 8.1 we saw that this system's natural frequencies are $\sqrt{2}$ and $\sqrt{6}$. Thus the free response of the system can be expressed as

$$x_1(t) = C_1 \sin\left(\sqrt{2}t + \phi_1\right) + C_2 \sin\left(\sqrt{6}t + \phi_2\right)$$

From Equation 8.3-7, the orthogonal eigenvectors for this system are

$$\mathbf{v}_1 = B_1 \begin{bmatrix} 1 \\ 1 \end{bmatrix} \qquad \mathbf{v}_2 = B_1 \begin{bmatrix} 1 \\ -1 \end{bmatrix}$$

Since the mode ratios are $1/1$ and $-1/1$, $x_2(t)$ has the form

$$x_2(t) = C_1 \sin\left(\sqrt{2}t + \phi_1\right) - C_2 \sin\left(\sqrt{6}t + \phi_2\right)$$

where the constants C_1, C_2, ϕ_1, and ϕ_2 depend on the four initial conditions.

Suppose we define the following functions corresponding to the two modes:

$$r_1(t) = C_1 \sin\left(\sqrt{2}t + \phi_1\right)$$

$$r_2(t) = C_2 \sin\left(\sqrt{6}t + \phi_2\right)$$

Note that these functions satisfy the following differential equations:

$$\ddot{r}_1 + 2r_1 = 0$$

$$\ddot{r}_2 + 6r_2 = 0$$

Note that these equations could be easily solved even if we had not already known their solutions. All we would need are the initial conditions $r_1(0)$, $\dot{r}_1(0)$, $r_2(0)$, and $\dot{r}_2(0)$.

Comparing the expressions for r_1 and r_2 with those for x_1 and x_2, we see that

$$x_1 = r_1 + r_2$$

$$x_2 = r_1 - r_2$$

In matrix form these become

$$\begin{bmatrix} x_1 \\ x_2 \end{bmatrix} = \begin{bmatrix} 1 & 1 \\ 1 & -1 \end{bmatrix} \begin{bmatrix} r_1 \\ r_2 \end{bmatrix}$$

Note that the columns of the (2×2) matrix are proportional to the mode shape vectors $\mathbf{v}_1$ and $\mathbf{v}_2$! This is no coincidence. In fact, we will find that if we create a matrix called $\mathbf{P}$ whose columns are the orthonormal eigenvectors, then the coordinate transformation $\mathbf{x} = \mathbf{Pr}$ will result in a set of uncoupled second-order differential equations that are easily solved for the n modes. All we need to know are the initial conditions, which can be obtained from $\mathbf{r}(0) = \mathbf{P}^{-1}\mathbf{x}(0)$ and $\dot{\mathbf{r}}(0) = \mathbf{P}^{-1}\dot{\mathbf{x}}(0)$. The $r_i(t)$ coordinates are called the *principal coordinates*.

Now we must deal with the problem of developing a systematic way of obtaining a symmetric eigenvalue matrix $\mathbf{A}$ from the equation of motion $\mathbf{M}\ddot{\mathbf{x}} + \mathbf{K}\mathbf{x} = \mathbf{F}(t)$.

The Mass Normalized Stiffness Matrix

Assuming that the mass matrix $\mathbf{M}$ is symmetric and positive definite, we can express it as

$$\mathbf{M} = \mathbf{L}^T\mathbf{L} \tag{8.3-11}$$

where the matrix $\mathbf{L}$ is a *lower triangular matrix*, which is one that has all zeros above the main diagonal. Factoring a matrix in this manner is called *Cholesky decomposition*. We can always formulate the equations of motion so that the mass matrix is symmetric (as shown in Section 8.1).

When the mass matrix is *diagonal*, the matrix L reduces to the matrix square root, which is positive definite if all the diagonal elements of $\mathbf{M}$ are positive, real numbers. In this case,

$$\mathbf{L} = \sqrt{\mathbf{M}} = \begin{bmatrix} \sqrt{m_1} & 0 & 0 & \cdots & 0 \\ 0 & \sqrt{m_2} & 0 & \cdots & 0 \\ & & \cdots & & \\ 0 & 0 & \cdots & 0 & \sqrt{m_n} \end{bmatrix} \tag{8.3-12}$$

Therefore,

$$\mathbf{L}^{-1} = \begin{bmatrix} 1/\sqrt{m_1} & 0 & 0 & \cdots & 0 \\ 0 & 1/\sqrt{m_2} & 0 & \cdots & 0 \\ & & \cdots & & \\ 0 & 0 & \cdots & 0 & 1/\sqrt{m_n} \end{bmatrix} \tag{8.3-13}$$

If $\mathbf{M}$ is not diagonal, then L must be determined by Cholesky decomposition, to be illustrated in Example 8.3-5.

Now define the coordinate vector $\mathbf{q}(t)$ such that

$$\mathbf{x}(t) = \mathbf{L}^{-1}\mathbf{q}(t) \tag{8.3-14}$$

Then the equation of motion $\mathbf{M}\ddot{\mathbf{x}} + \mathbf{K}\mathbf{x} = \mathbf{F}(t)$ becomes

$$\mathbf{M}\mathbf{L}^{-1}\ddot{\mathbf{q}} + \mathbf{K}\mathbf{L}^{-1}\mathbf{q} = \mathbf{F}(t)$$

If we multiply from the left by $\mathbf{L}^{-1}$, we obtain

$$\mathbf{L}^{-1}\mathbf{M}\mathbf{L}^{-1}\ddot{\mathbf{q}} + \mathbf{L}^{-1}\mathbf{K}\mathbf{L}^{-1}\mathbf{q} = \mathbf{L}^{-1}\mathbf{F}(t)$$

Note that $\mathbf{L}^{-1}\mathbf{M}\mathbf{L}^{-1} = \mathbf{I}$, so we obtain

$$\ddot{\mathbf{q}} + \mathbf{L}^{-1}\mathbf{K}\mathbf{L}^{-1}\mathbf{q} = \mathbf{L}^{-1}\mathbf{F}(t)$$

If we define $\mathbf{H}$ as

$$\mathbf{H} = \mathbf{L}^{-1}\mathbf{K}\mathbf{L}^{-1} \tag{8.3-15}$$

then the equation of motion becomes

$$\ddot{\mathbf{q}} + \mathbf{H}\mathbf{q} = \mathbf{L}^{-1}\mathbf{F}(t) \tag{8.3-16}$$

The matrix $\mathbf{H}$ is called the *mass normalized stiffness matrix*. In the scalar case it reduces to $H = k/m = \omega_n^2$.

The matrix $\mathbf{H}$ is symmetric, which means that $\mathbf{H}^T = \mathbf{H}$. This is true since $\mathbf{L}$ and $\mathbf{K}$ are symmetric and since the inverse and the products of symmetric matrices are also symmetric.

We will use $\mathbf{H}$ to obtain an eigenvalue problem that yields orthonormal eigenvectors. Set $\mathbf{F}(t) = \mathbf{0}$ in Equation 8.3-16 and make the following substitution:

$$\mathbf{q}(t) = \mathbf{u}e^{i\omega t} \tag{8.3-17}$$

After canceling the scalar $e^{i\omega t}$, we obtain

$$-\omega^2\mathbf{u} + \mathbf{H}\mathbf{u} = \mathbf{0}$$

or

$$\mathbf{H}\mathbf{u} = \omega^2 \mathbf{u} \qquad (8.3\text{-}18)$$

This is our symmetric eigenvalue problem!

Note that since

$$\mathbf{x}(t) = \mathbf{L}^{-1}\mathbf{q}(t)$$

the eigenvectors **u** are related to the mode shape vector **X** by

$$\mathbf{X} = \mathbf{L}^{-1}\mathbf{u} \qquad \mathbf{u} = \mathbf{L}\mathbf{X} \qquad (8.3\text{-}19)$$

EXAMPLE 8.3-1
Orthonormal Eigenvectors for Two Degrees of Freedom

Consider the system whose mass and stiffness matrices are

$$\mathbf{M} = \begin{bmatrix} 4 & 0 \\ 0 & 1 \end{bmatrix} \qquad \mathbf{K} = \begin{bmatrix} 16 & -4 \\ -4 & 4 \end{bmatrix}$$

Use the mass normalized stiffness matrix to obtain a set of orthonormal eigenvectors.

Solution

Here

$$\mathbf{L} = \sqrt{\mathbf{M}} = \begin{bmatrix} 2 & 0 \\ 0 & 1 \end{bmatrix} \qquad \mathbf{L}^{-1} = \begin{bmatrix} 1/2 & 0 \\ 0 & 1 \end{bmatrix}$$

Thus

$$\mathbf{H} = \mathbf{L}^{-1}\mathbf{K}\mathbf{L}^{-1} = \begin{bmatrix} 1/2 & 0 \\ 0 & 1 \end{bmatrix}\begin{bmatrix} 16 & -4 \\ -4 & 4 \end{bmatrix}\begin{bmatrix} 1/2 & 0 \\ 0 & 1 \end{bmatrix} = \begin{bmatrix} 4 & -2 \\ -2 & 4 \end{bmatrix}$$

Note that **H** is symmetric. The eigenvalue problem $\mathbf{H}\mathbf{u} = \omega^2\mathbf{u}$ is

$$\begin{bmatrix} 4 & -2 \\ -2 & 4 \end{bmatrix}\begin{bmatrix} u_1 \\ u_2 \end{bmatrix} = \omega^2\begin{bmatrix} u_1 \\ u_2 \end{bmatrix}$$

This gives

$$(4 - \omega^2)u_1 - 2u_2 = 0$$
$$-2u_1 + (4 - \omega^2)u_2 = 0$$

The determinant equation is $(4 - \omega^2)^2 - 4 = 0$ and has the roots $\omega^2 = 2$ and $\omega^2 = 6$. For the first mode, where $\omega^2 = 2$, the first mode equation gives

$$(4 - 2)u_1 - 2u_2 = 0$$

Thus the first mode ratio is $u_2/u_1 = 1$.

For the second mode, where $\omega^2 = 6$, the first mode equation gives

$$(4 - 6)u_1 - 2u_2 = 0$$

Thus the second mode ratio is $u_2/u_1 = -1$.

Thus the eigenvectors are

$$\mathbf{u}_1 = u_1\begin{bmatrix} 1 \\ 1 \end{bmatrix}$$

$$\mathbf{u}_2 = u_1\begin{bmatrix} 1 \\ -1 \end{bmatrix}$$

We can create orthonormal vectors from these by letting $u_1 = 1/\sqrt{2}$ to obtain

$$\mathbf{u}_1 = \frac{1}{\sqrt{2}} \begin{bmatrix} 1 \\ 1 \end{bmatrix}$$

$$\mathbf{u}_2 = \frac{1}{\sqrt{2}} \begin{bmatrix} 1 \\ -1 \end{bmatrix}$$

Note that the mode shape vector is

$$\mathbf{X} = \mathbf{L}^{-1}\mathbf{u} = \begin{bmatrix} 1/2 & 0 \\ 0 & 1 \end{bmatrix} \begin{bmatrix} u_1 \\ u_2 \end{bmatrix} = \begin{bmatrix} u_1/2 \\ u_2 \end{bmatrix}$$

or

$$\mathbf{X}_1 = \frac{1}{2\sqrt{2}} \begin{bmatrix} 1 \\ 2 \end{bmatrix}$$

for the first mode and

$$\mathbf{X}_2 = \frac{1}{2\sqrt{2}} \begin{bmatrix} 1 \\ -2 \end{bmatrix}$$

for the second mode.   ∎

The Modal Equations

Let $\mathbf{P}$ be a matrix whose columns are the orthonormal eigenvectors; that is, let

$$\mathbf{P} = \begin{bmatrix} \mathbf{u}_1 & \mathbf{u}_2 & \cdots & \mathbf{u}_n \end{bmatrix} \tag{8.3-20}$$

Define the coordinate transformation

$$\mathbf{q} = \mathbf{Pr} \tag{8.3-21}$$

This will result in a set of uncoupled second-order differential equations in terms of principal coordinates $r_i(t)$. These equations are easily solved for the n modes.

The matrix $\mathbf{P}$ has the property that

$$\mathbf{P}^T\mathbf{P} = \mathbf{I} \tag{8.3-22}$$

Any matrix having this property is called an *orthonormal* matrix. This property follows from the fact that the columns of $\mathbf{P}$ are orthonormal vectors, and it implies that

$$\mathbf{P}^T = \mathbf{P}^{-1} \tag{8.3-23}$$

If we substitute $\mathbf{q} = \mathbf{Pr}$ into the equation $\ddot{\mathbf{q}} + \mathbf{Hq} = \mathbf{0}$, we obtain

$$\mathbf{P}\ddot{\mathbf{r}} + \mathbf{HPr} = \mathbf{0}$$

If we then multiply this equation from the left by $\mathbf{P}^T$, we obtain

$$\mathbf{P}^T\mathbf{P}\ddot{\mathbf{r}} + \mathbf{P}^T\mathbf{HPr} = \mathbf{0}$$

With the property $\mathbf{P}^T\mathbf{P} = \mathbf{I}$ and the fact that $\mathbf{I}\ddot{\mathbf{r}} = \ddot{\mathbf{r}}$, this becomes

$$\ddot{\mathbf{r}} + \mathbf{P}^T\mathbf{HPr} = \mathbf{0} \tag{8.3-24}$$

If we define the *spectral* matrix $\mathbf{\Lambda}$ such that

$$\mathbf{\Lambda} = \mathbf{P}^T\mathbf{HP} \tag{8.3-25}$$

then

$$\ddot{\mathbf{r}} + \Lambda \mathbf{r} = \mathbf{0} \tag{8.3-26}$$

These are called the *modal equations*. The r_i terms are the principal coordinates and are also called the *modal coordinates*.

In the scalar case, $\Lambda = k/m = \omega_n^2$, the eigenvalue. It turns out that in the general case, Λ is a diagonal matrix whose diagonal elements are the eigenvalues!

$$\Lambda = \begin{bmatrix} \lambda_1 & 0 & 0 & \cdots & 0 \\ 0 & \lambda_2 & 0 & \cdots & 0 \\ & & \cdots & & \\ 0 & 0 & \cdots & 0 & \lambda_n \end{bmatrix} = \begin{bmatrix} \omega_1^2 & 0 & 0 & \cdots & 0 \\ 0 & \omega_2^2 & 0 & \cdots & 0 \\ & & \cdots & & \\ 0 & 0 & \cdots & 0 & \omega_n^2 \end{bmatrix} \tag{8.3-27}$$

EXAMPLE 8.3-2
Modal Equations for Two Degrees of Freedom

Create a set of principal coordinates for the system treated in Example 8.3-1 and obtain the modal equations.

Solution

Create the matrices $\mathbf{P}$ and Λ:

$$\mathbf{P} = [\mathbf{u}_1 \quad \mathbf{u}_2] = \frac{1}{\sqrt{2}} \begin{bmatrix} 1 & 1 \\ 1 & -1 \end{bmatrix}$$

$$\Lambda = \mathbf{P}^T \mathbf{H} \mathbf{P} = \frac{1}{\sqrt{2}} \begin{bmatrix} 1 & 1 \\ 1 & -1 \end{bmatrix} \begin{bmatrix} 4 & -2 \\ -2 & 4 \end{bmatrix} \begin{bmatrix} 1 & 1 \\ 1 & -1 \end{bmatrix} \frac{1}{\sqrt{2}} = \begin{bmatrix} 2 & 0 \\ 0 & 6 \end{bmatrix}$$

The modal equations are

$$\begin{bmatrix} \ddot{r}_1 \\ \ddot{r}_2 \end{bmatrix} + \begin{bmatrix} 2 & 0 \\ 0 & 6 \end{bmatrix} \begin{bmatrix} r_1 \\ r_2 \end{bmatrix} = \begin{bmatrix} 0 \\ 0 \end{bmatrix}$$

This gives

$$\begin{bmatrix} \ddot{r}_1 \\ \ddot{r}_2 \end{bmatrix} + \begin{bmatrix} 2r_1 \\ 6r_2 \end{bmatrix} = \begin{bmatrix} 0 \\ 0 \end{bmatrix}$$

or

$$\ddot{r}_1 + 2r_1 = 0$$
$$\ddot{r}_2 + 6r_2 = 0$$

These are the modal equations. ∎

Equation 8.3-26 is a set of uncoupled second-order differential equations in terms of principal coordinates $r_i(t)$. These equations are easily solved for the n modes. All we need to know are the initial conditions, which can be obtained from

$$\mathbf{r}(0) = \mathbf{P}^{-1} \mathbf{x}(0)$$

and

$$\dot{\mathbf{r}}(0) = \mathbf{P}^{-1} \dot{\mathbf{x}}(0)$$

We can use $\mathbf{P}$ and $\mathbf{L}^{-1}$ to transform the solution $\mathbf{r}(t)$ back to the original $\mathbf{x}(t)$ coordinates, as follows:

$$\mathbf{x} = \mathbf{L}^{-1} \mathbf{q} = \mathbf{L}^{-1} \mathbf{P} \mathbf{r} = \mathbf{T} \mathbf{r} \tag{8.3-28}$$

The resulting transformation matrix

$$\mathbf{T} = \mathbf{L}^{-1}\mathbf{P} \tag{8.3-29}$$

is called the *modal matrix* or the *mode shape matrix*, since each column of $\mathbf{T}$ is a mode shape vector.

Properties of the matrix inverse and the matrix transpose are

$$(\mathbf{AB})^{-1} = \mathbf{B}^{-1}\mathbf{A}^{-1} \tag{8.3-30}$$

$$(\mathbf{AB})^{T} = \mathbf{B}^{T}\mathbf{A}^{T} \tag{8.3-31}$$

Therefore

$$\mathbf{T}^{-1} = \left(\mathbf{L}^{-1}\mathbf{P}\right)^{-1} = \mathbf{P}^{-1}\mathbf{L} = \mathbf{P}^{T}\mathbf{L} \tag{8.3-32}$$

$$\mathbf{r} = \mathbf{T}^{-1}\mathbf{x} = \mathbf{P}^{T}\mathbf{L}\mathbf{x} \tag{8.3-33}$$

Each modal equation has the form

$$\ddot{r}_j + \omega_j^2 r_j = 0 \tag{8.3-34}$$

and has the solution

$$r_j(t) = \frac{\dot{r}_j(0)}{\omega_j} \sin \omega_1 t + r_j(0) \cos \omega_j t \tag{8.3-35}$$

unless $\omega_j \neq 0$. In that case the modal equation has the form

$$\ddot{r}_j = 0 \tag{8.3-36}$$

and has the solution

$$r_j(t) = \frac{\dot{r}_j(0)}{t} + r_j(0) \tag{8.3-37}$$

The initial conditions are found from the given initial conditions on the original variables $\mathbf{x}$, using the fact that

$$\mathbf{r} = \mathbf{P}^{T}\mathbf{L}\mathbf{x} \tag{8.3-38}$$

Thus

$$\mathbf{r}(0) = \mathbf{P}^{T}\mathbf{L}\mathbf{x}(0) \tag{8.3-39}$$

and

$$\dot{\mathbf{r}}(0) = \mathbf{P}^{T}\mathbf{L}\dot{\mathbf{x}}(0) \tag{8.3-40}$$

EXAMPLE 8.3-3
Solving the Modal Equations

Use the modal equations from Example 8.3-2 to obtain $x_1(t)$ and $x_2(t)$ for the initial conditions $x_1(0) = 3$, $x_2(0) = 5$, and $\dot{x}_1(0) = \dot{x}_2(0) = 0$.

Solution

First we need to obtain the initial conditions for $\mathbf{r}$. From Equation 8.3-39,

$$\mathbf{r}(0) = \mathbf{P}^{T}\mathbf{L}\mathbf{x}(0)$$

where $\mathbf{P}$ was found in Example 8.3-2 and

$$\mathbf{L} = \sqrt{\mathbf{m}} = \begin{bmatrix} 2 & 0 \\ 0 & 1 \end{bmatrix}$$

and

$$\mathbf{L}^{-1} = \begin{bmatrix} 1/2 & 0 \\ 0 & 1 \end{bmatrix}$$

Thus

$$\mathbf{r}(0) = \frac{1}{\sqrt{2}} \begin{bmatrix} 1 & 1 \\ 1 & -1 \end{bmatrix} \begin{bmatrix} 2 & 0 \\ 0 & 1 \end{bmatrix} \begin{bmatrix} 3 \\ 5 \end{bmatrix} = \frac{1}{\sqrt{2}} \begin{bmatrix} 11 \\ 1 \end{bmatrix}$$

From Equation 8.3-40, $\dot{\mathbf{r}}(0) = \mathbf{0}$.

The modal equations are

$$\ddot{r}_1 + 2r_1 = 0$$
$$\ddot{r}_2 + 6r_2 = 0$$

These have the solution

$$r_1(t) = r_1(0) \cos \sqrt{2}t = \frac{11}{\sqrt{2}} \cos \sqrt{2}t$$

$$r_2(t) = r_2(0) \cos \sqrt{6}t = \frac{1}{\sqrt{2}} \cos \sqrt{6}t$$

From $\mathbf{x} = \mathbf{L}^{-1}\mathbf{Pr}$ we obtain

$$\mathbf{x}(t) = \frac{1}{\sqrt{2}} \begin{bmatrix} 1/2 & 0 \\ 0 & 1 \end{bmatrix} \begin{bmatrix} 1 & 1 \\ 1 & -1 \end{bmatrix} \mathbf{r} = \begin{bmatrix} \frac{1}{2}\left(11 \cos \sqrt{2}t + \cos \sqrt{6}t\right) \\ \frac{1}{2}\left(11 \cos \sqrt{2}t - \cos \sqrt{6}t\right) \end{bmatrix}$$ ∎

Keep in mind that, although this method is a long way to obtain the free response of a system having just two degrees of freedom, this example illustrates the details of the method, which is advantageous primarily for systems having more than two degrees of freedom.

The Forced Response

We now consider the forced response. Recall Equation 8.3-16:

$$\ddot{\mathbf{q}} + \mathbf{Hq} = \mathbf{L}^{-1}\mathbf{F}(t) \tag{8.3-41}$$

where

$$\mathbf{H} = \mathbf{L}^{-1}\mathbf{KL}^{-1} \tag{8.3-42}$$

If we substitute $\mathbf{q} = \mathbf{Pr}$, we obtain

$$\mathbf{P}\ddot{\mathbf{r}} + \mathbf{HPr} = \mathbf{L}^{-1}\mathbf{F}(t)$$

If we then multiply this equation from the left by $\mathbf{P}^T$, we obtain

$$\mathbf{P}^T\mathbf{P}\ddot{\mathbf{r}} + \mathbf{P}^T\mathbf{HPr} = \mathbf{P}^T\mathbf{L}^{-1}\mathbf{F}(t)$$

With the property $\mathbf{P}^T\mathbf{P} = \mathbf{I}$ and the fact that $\mathbf{I}\ddot{\mathbf{r}} = \ddot{\mathbf{r}}$, this becomes

$$\ddot{\mathbf{r}} + \mathbf{P}^T\mathbf{HPr} = \mathbf{P}^T\mathbf{L}^{-1}\mathbf{F}(t) \tag{8.3-43}$$

Then, using the spectral matrix $\mathbf{\Lambda}$,

$$\mathbf{\Lambda} = \mathbf{P}^T\mathbf{HP} \tag{8.3-44}$$

we obtain

$$\ddot{\mathbf{r}} + \mathbf{\Lambda r} = \mathbf{P}^T\mathbf{L}^{-1}\mathbf{F}(t) = \mathbf{T}^T\mathbf{F}(t) \tag{8.3-45}$$

These are used to obtain the forced response of each mode. One very important use of these equations is to determine whether or not the forcing function affects all of the modes. This can provide useful design information.

EXAMPLE 8.3-4

Forced Response for Two Degrees of Freedom

Consider the system:

$$4\ddot{x}_1 + 16x_1 - 4x_2 = f_1(t)$$
$$\ddot{x}_2 - 4x_1 + 4x_2 = f_2(t)$$

(a) Obtain the modal equations for the forced response. What conditions must exist for $f_1(t)$ and $f_2(t)$ to influence both modes?

(b) Examine the case where

$$\mathbf{F}(t) = \begin{bmatrix} 8 \\ 4 \end{bmatrix} \sin \sqrt{6}t$$

Solution

(a) This is the same system considered in Example 8.3-3, so we can use the matrices obtained in that example. We have

$$\mathbf{P}^T\mathbf{L}^{-1}\mathbf{F}(t) = \frac{1}{\sqrt{2}}\begin{bmatrix} 1 & 1 \\ 1 & -1 \end{bmatrix}\begin{bmatrix} 1/2 & 0 \\ 0 & 1 \end{bmatrix}\begin{bmatrix} f_1(t) \\ f_2(t) \end{bmatrix}$$

$$= \frac{1}{\sqrt{2}}\begin{bmatrix} 1/2 & 1 \\ 1/2 & -1 \end{bmatrix}\begin{bmatrix} f_1(t) \\ f_2(t) \end{bmatrix} = \begin{bmatrix} \frac{1}{2}f_1(t) + f_2(t) \\ \frac{1}{2}f_1 - f_2(t) \end{bmatrix}$$

The modal equations are

$$\ddot{r}_1 + 2r_1 = \frac{1}{\sqrt{2}}\left[\frac{1}{2}f_1(t) + f_2(t)\right]$$

$$\ddot{r}_2 + 6r_2 = \frac{1}{\sqrt{2}}\left[\frac{1}{2}f_1 - f_2(t)\right]$$

These can be solved once $f_1(t)$ and $f_2(t)$ are specified. The resulting modal response $r_1(t)$ and $r_2(t)$ can be used to obtain $x_1(t)$ and $x_2(t)$ from $\mathbf{x} = \mathbf{L}^{-1}\,\mathbf{Pr} = \mathbf{Tr}$.

From the modal equations we can see that if $f_1(t) = -2f_2(t)$, the first mode will not be affected. If $f_1(t) = 2\,f_2(t)$, the second mode will not be affected.

(b) For the given function $\mathbf{F}(t)$, $f_1(t) = 8 \sin \sqrt{6}t$ and $f_2(t) = 4 \sin \sqrt{6}t$. Thus $f_1(t) = 2\,f_2(t)$, and so the second mode will not be influenced. For this case the modal equations become

$$\ddot{r}_1 + 2r_1 = \frac{8}{\sqrt{2}} \sin \sqrt{6}t$$

$$\ddot{r}_2 + 6r_2 = 0$$

If the initial conditions are such that $r_2(0) = \dot{r}_2(0) = 0$, then $r_2(t) = 0$ and the second mode does not appear in the response.

This situation is interesting because even though the forcing frequency $\sqrt{6}$ equals the natural frequency of the second mode, no resonance occurs. ■

The right-hand side of Equation 8.3-45, the general form of the modal equations, is $\mathbf{T}^T\mathbf{F}(t)$. We can conclude that if the jth *column* of $\mathbf{T}$, denoted $\mathbf{T}_j$, is such that

$$\mathbf{T}_j^T\mathbf{F}(t) = 0 \tag{8.3-46}$$

then the jth mode will not be affected by the forcing function. This observation can be put to practical use if we want to design a system that is isolated from the effects of the forcing function. Conversely, if we are designing a system to control the response, such as an active suspension system, then we must make sure that the condition of Equation 8.3-46 is not satisfied.

Summary of Modal Analysis

The modal analysis procedure for a linear, undamped system can be summarized as follows. Given the equations of motion $\mathbf{M}\ddot{\mathbf{x}} + \mathbf{K}\mathbf{x} = \mathbf{F}(t)$, the initial conditions $\mathbf{x}(0)$ and $\dot{\mathbf{x}}(0)$, and the forcing function vector $\mathbf{F}(t)$, do the following:

1. Calculate $\mathbf{L}$ such that $\mathbf{L}\mathbf{L}^T = \mathbf{M}$. If $\mathbf{M}$ is diagonal, then $\mathbf{L} = \sqrt{\mathbf{M}}$; otherwise, $\mathbf{L}$ can be found from Cholesky decomposition.

2. Form the mass normalized stiffness matrix $\mathbf{H} = \mathbf{L}^{-1}\mathbf{K}\mathbf{L}^{-1}$.

3. Solve the eigenvalue problem $\mathbf{H}\mathbf{v}_i = \omega_i^2 \mathbf{v}_i$ for $i = 1, 2, \cdots, n$.

4. Create the orthonormal eigenvectors $\mathbf{u}_i$ from the orthogonal eigenvectors $\mathbf{v}_i$. Form the matrix

$$\mathbf{P} = \begin{bmatrix} \mathbf{u}_1 & \mathbf{u}_2 & \cdots & \mathbf{u}_n \end{bmatrix}$$

5. Calculate the modal matrix $\mathbf{T} = \mathbf{L}^{-1}\mathbf{P}$ and its inverse $\mathbf{T}^{-1} = \mathbf{P}^T\mathbf{L}$.

6. Calculate the initial values of the modal coordinates from

$$\mathbf{r}(0) = \mathbf{T}^{-1}\mathbf{x}(0) \qquad \dot{\mathbf{r}}(0) = \mathbf{T}^{-1}\dot{\mathbf{x}}(0)$$

7. Calculate the spectral matrix.

$$\mathbf{\Lambda} = \mathbf{P}^T\mathbf{H}\mathbf{P}$$

$$\mathbf{\Lambda} = \begin{bmatrix} \lambda_1 & 0 & 0 & \cdots & 0 \\ 0 & \lambda_2 & 0 & \cdots & 0 \\ & & \cdots & & \\ 0 & 0 & \cdots & 0 & \lambda_n \end{bmatrix} = \begin{bmatrix} \omega_1^2 & 0 & 0 & \cdots & 0 \\ 0 & \omega_2^2 & 0 & \cdots & 0 \\ & & \cdots & & \\ 0 & 0 & \cdots & 0 & \omega_n^2 \end{bmatrix}$$

8. Solve the modal equations

$$\ddot{\mathbf{r}} + \mathbf{\Lambda}\mathbf{r} = \mathbf{P}^T\mathbf{L}^{-1}\mathbf{F}(t) = \mathbf{T}^T\mathbf{F}(t)$$

9. Obtain the solution in terms of the original variables from $\mathbf{x}(t) = \mathbf{T}\mathbf{r}(t)$.

Cholesky Decomposition

When dynamic coupling exists, the mass matrix $\mathbf{M}$ is not diagonal, and we cannot compute $\mathbf{L}$ by simply taking the square roots of the diagonal elements. Cholesky decomposition can be used instead. The following example illustrates the method, which is not difficult to do by hand for three degrees of freedom. MATLAB provides the `chol` function to perform the decomposition for more degrees of freedom. Note that `chol` gives an *upper* triangular matrix, which equals $\mathbf{L}^T$ in our notation.

EXAMPLE 8.3-5
Cholesky Decomposition for Three Degrees of Freedom

Determine $\mathbf{L}$ so that $\mathbf{L}\mathbf{L}^T = \mathbf{M}$ for the following matrix:

$$\mathbf{M} = \begin{bmatrix} 9 & 3 & 0 \\ 3 & 5 & 2 \\ 0 & 2 & 17 \end{bmatrix}$$

Solution

Let

$$\mathbf{L} = \begin{bmatrix} L_{11} & 0 & 0 \\ L_{21} & L_{22} & 0 \\ L_{31} & L_{32} & L_{33} \end{bmatrix}$$

Then, since $\mathbf{L}$ is defined to be symmetric,

$$\mathbf{L}^T = \begin{bmatrix} L_{11} & L_{21} & L_{31} \\ 0 & L_{22} & L_{32} \\ 0 & 0 & L_{33} \end{bmatrix}$$

and

$$\mathbf{LL}^T = \begin{bmatrix} L_{11}^2 & L_{11}L_{21} & L_{11}L_{31} \\ L_{11}L_{21} & (L_{21}^2 + L_{22}^2) & (L_{21}L_{31} + L_{22}L_{32}) \\ L_{11}L_{31} & (L_{21}L_{31} + L_{22}L_{32}) & (L_{31}^2 + L_{32}^2 + L_{33}^2) \end{bmatrix} = \mathbf{M} = \begin{bmatrix} 9 & 3 & 0 \\ 3 & 5 & 2 \\ 0 & 2 & 17 \end{bmatrix}$$

Note that $\mathbf{LL}^T$ is symmetric.

Comparing term-by-term, starting with the upper left and proceeding to the right and then down, we can successively eliminate each unknown one at a time. The results are

$$L_{11} = 3 \qquad L_{21} = 1 \qquad L_{31} = 0$$
$$L_2 = 2 \qquad L_{32} = 1 \qquad L_{33} = 4$$

Thus

$$\mathbf{L} = \begin{bmatrix} 3 & 0 & 0 \\ 1 & 2 & 0 \\ 0 & 1 & 4 \end{bmatrix}$$

■

8.4 EFFECTS OF DAMPING

Introduction of viscous damping into the vibration model leads to the following matrix equation of motion:

$$\mathbf{M\ddot{x}} + \mathbf{C\dot{x}} + \mathbf{Kx} = \mathbf{F}(t) \tag{8.4-1}$$

The damping matrix $\mathbf{C}$, in general, introduces coupling between the derivatives in the equations. This coupling can prevent the equations from being decoupled by the modal transformation method.

Proportional Damping

Decoupling can be achieved, however, if the damping matrix has a special form called *proportional* or *Rayleigh* damping. The damping is said to be *proportional* if two constants a and b can be found such that the damping matrix is

$$\mathbf{C} = a\mathbf{M} + b\mathbf{K} \tag{8.4-2}$$

In this case the equation of motion becomes

$$\mathbf{M\ddot{x}} + (a\mathbf{M} + b\mathbf{K})\mathbf{\dot{x}} + \mathbf{Kx} = \mathbf{F}(t)$$

Making the standard substitution $\mathbf{x} = \mathbf{L}^{-1}\mathbf{q}$, we obtain

$$\mathbf{ML}^{-1}\mathbf{\ddot{q}} + (a\mathbf{ML}^{-1} + b\mathbf{KL}^{-1})\mathbf{\dot{q}} + \mathbf{KL}^{-1}\mathbf{q} = \mathbf{F}(t)$$

Now multiply by $\mathbf{L}^{-1}$ from the left.

$$\mathbf{L}^{-1}\mathbf{ML}^{-1}\mathbf{\ddot{q}} + (a\mathbf{L}^{-1}\mathbf{ML}^{-1} + b\mathbf{L}^{-1}\mathbf{KL}^{-1})\mathbf{\dot{q}} + \mathbf{L}^{-1}\mathbf{KL}^{-1}\mathbf{q} = \mathbf{L}^{-1}\mathbf{F}(t)$$

Using the fact that $\mathbf{L}^{-1}\mathbf{M}\mathbf{L}^{-1} = \mathbf{I}$ and the definition of the mass normalized stiffness matrix, $\mathbf{H} = \mathbf{L}^{-1}\mathbf{K}\mathbf{L}^{-1}$, we obtain

$$\ddot{\mathbf{q}} + (a\mathbf{I} + b\mathbf{H})\dot{\mathbf{q}} + \mathbf{H}\mathbf{q} = \mathbf{L}^{-1}\mathbf{F}(t)$$

Now substitute $\mathbf{q} = \mathbf{P}\mathbf{r}$ and multiply from the left by $\mathbf{P}^T$:

$$\mathbf{P}^T\mathbf{P}\ddot{\mathbf{r}} + (a\mathbf{P}^T + b\mathbf{P}^T\mathbf{H})\mathbf{P}\dot{\mathbf{r}} + \mathbf{P}^T\mathbf{H}\mathbf{P}\mathbf{r} = \mathbf{P}^T\mathbf{L}^{-1}\mathbf{F}(t)$$

Since $\mathbf{P}^T\mathbf{P} = \mathbf{I}$, $\mathbf{P}^T\mathbf{L}^{-1} = \mathbf{T}^T$, and $\mathbf{\Lambda} = \mathbf{P}^T\mathbf{H}\mathbf{P}$, we obtain

$$\dot{\mathbf{r}} + (a\mathbf{I} + b\mathbf{\Lambda})\dot{\mathbf{r}} + \mathbf{\Lambda}\mathbf{r} = \mathbf{T}^T\mathbf{F}(t) \tag{8.4-3}$$

This gives the decoupled modal equations

$$\ddot{r}_i + (a + b\lambda_i)\dot{r}_i + \lambda_i r = \mathbf{T}_i^T\mathbf{F}(t) \quad i = 1, 2, \ldots, n \tag{8.4-4}$$

where $\mathbf{T}_i^T$ is the ith row of $\mathbf{T}^T$.

Note that the λ_i terms are the eigenvalues of the matrix $\mathbf{H} = \mathbf{L}^{-1}\mathbf{K}\mathbf{L}^{-1}$, which does not depend on the damping. Thus, since $\omega_i^2 = \lambda_i$, the frequencies ω_i are the natural frequencies of the *undamped* system $\mathbf{M}\ddot{\mathbf{x}} + \mathbf{K}\mathbf{x} = \mathbf{0}$.

Modal Damping Ratio

The coefficient of $\dot{r}_i$ is the damping coefficient of the ith mode. Since $\lambda_i = \omega_i^2$, we may write this coefficient as

$$(a + b\lambda_i) = (a + b\omega_i^2) = 2\zeta_i\omega_i$$

where the *modal damping ratio* ζ_i is defined as

$$\zeta_i = \frac{a}{2\omega_i} + \frac{b\omega_i}{2} \tag{8.4-5}$$

Thus we may express the modal equations as

$$\ddot{r}_i + 2\zeta_i\omega_i\dot{r}_i + \omega_i^2 r = \mathbf{T}_i^T\mathbf{F}(t) \quad i = 1, 2, \ldots, n \tag{8.4-6}$$

Each of these equations can be solved using the familiar techniques of differential equations. For example, the free response is given by

$$\begin{aligned} r_i(t) &= A_i e^{-\zeta_i\omega_i t}\sin(\omega_i t + \phi_i) \\ &= e^{-\zeta_i\omega_i t}(B_i \cos \omega_i t + C_i \sin \omega_i t) \end{aligned} \tag{8.4-7}$$

where

$$B_i = r_i(0) \qquad C_i = \frac{\dot{r}(0)}{\omega_i} + \zeta_i r_i(0) \tag{8.4-8}$$

The initial values $r_i(0)$ and $\dot{r}_i(0)$ are found from

$$\mathbf{r}(0) = \mathbf{P}^T\mathbf{L}\mathbf{x}(0) \tag{8.4-9}$$

$$\dot{\mathbf{r}}(0) = \mathbf{P}^T\mathbf{L}\dot{\mathbf{x}}(0) \tag{8.4-10}$$

The solution in terms of the original variables can be obtained from

$$\mathbf{x} = \mathbf{T}\mathbf{r} \tag{8.4-11}$$

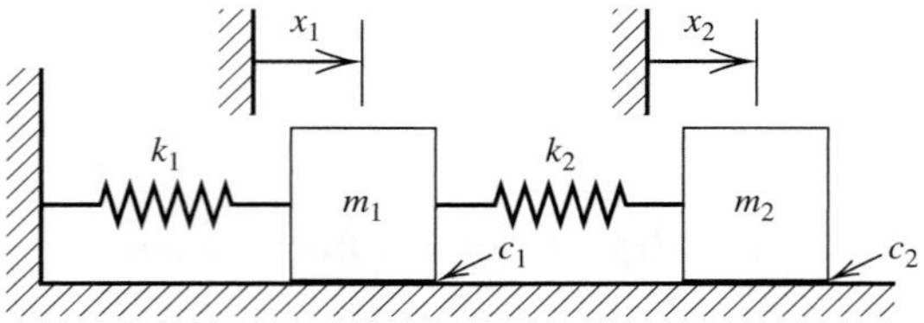

FIGURE 8.4-1   A system with damping.

EXAMPLE 8.4-1
A System with
Proportional
Damping

For the system shown in Figure 8.4-1 there is damping between the surface and the masses. For the case where $m_1 = m_2 = 1$, $c_1 = c_2 = 1$, and $k_1 = k_2 = 1$, determine the modes and frequencies.

Solution

The equations of motion are

$$\ddot{x}_1 + \dot{x}_1 + 2x_1 - x_2 = 0$$
$$\ddot{x}_2 + \dot{x}_2 + x_2 - x_1 = 0$$

In matrix form these are

$$\begin{bmatrix} 1 & 0 \\ 0 & 1 \end{bmatrix}\ddot{\mathbf{x}} + \begin{bmatrix} 1 & 0 \\ 0 & 1 \end{bmatrix}\dot{\mathbf{x}} + \begin{bmatrix} 2 & -1 \\ -1 & 1 \end{bmatrix}\mathbf{x} = \mathbf{0}$$

Therefore

$$\mathbf{M} = \begin{bmatrix} 1 & 0 \\ 0 & 1 \end{bmatrix} \qquad \mathbf{C} = \begin{bmatrix} 1 & 0 \\ 0 & 1 \end{bmatrix} \qquad \mathbf{K} = \begin{bmatrix} 2 & -1 \\ -1 & 1 \end{bmatrix}$$

We can see that $\mathbf{C}$ is proportional to $\mathbf{M}$ and $\mathbf{K}$ as $\mathbf{C} = a\mathbf{M} + b\mathbf{K}$ because

$$\mathbf{C} = a\begin{bmatrix} 1 & 0 \\ 0 & 1 \end{bmatrix} + b\begin{bmatrix} 2 & -1 \\ -1 & 1 \end{bmatrix} = \begin{bmatrix} a & 0 \\ 0 & a \end{bmatrix} + \begin{bmatrix} 2b & -b \\ -b & b \end{bmatrix} = \begin{bmatrix} a+2b & -b \\ -b & a+b \end{bmatrix} = \mathbf{C}$$

if $a = 1$ and $b = 0$. That is, $\mathbf{C} = \mathbf{M}$.

The modal equations are

$$\ddot{r}_i + (a + \omega_i^2 b)\dot{r}_i + \omega_i^2 r_i = 0$$

which for this example becomes

$$\ddot{r}_i + \dot{r}_i + \omega_i^2 r_i = 0 \qquad i = 1, 2$$

where the ω_i terms are the frequencies of the *undamped* system. The equations of motion of the undamped system are

$$\ddot{x}_1 + 2x_1 - x_2 = 0$$
$$\ddot{x}_2 + x_2 - x_1 = 0$$

These have the frequency equation $\left| -\omega^2\mathbf{M} + \mathbf{K} \right| = 0$, which becomes

$$\left| \begin{bmatrix} -\omega^2 & 0 \\ 0 & -\omega^2 \end{bmatrix} + \begin{bmatrix} 2 & -1 \\ -1 & 1 \end{bmatrix} \right| = \left| \begin{bmatrix} 2-\omega^2 & -1 \\ -1 & 1-\omega^2 \end{bmatrix} \right| = 0$$

or $\omega^4 - 3\omega^2 + 1 = 0$. This has the solutions $\omega_1^2 = 0.3820$ and $\omega_2^2 = 2.618$. So the modal equations are

$$\ddot{r}_1 + \dot{r}_1 + 0.382r_1 = 0$$
$$\ddot{r}_2 + \dot{r}_2 + 2.618r_2 = 0$$

From these equations or from Equation 8.4-5, we can compute the modal damping ratios:

$$\zeta_i = \frac{a + b\omega_i^2}{2\omega_i} = \frac{1}{2\omega_i} = 0.809, \qquad 0.309$$

The *damped* modal frequencies are found from $\omega_{di} = \omega_i\sqrt{1 - \zeta_i^2}$. ∎

Effects of Nonproportional Damping

If the damping is proportional, the system mode shapes are those of the undamped system. So, for example, if we displace the system initially in one of the undamped mode shapes, the masses continue to oscillate in that shape, but the motion eventually disappears in time because of the damping. If, however, the damping is not proportional, the masses do not continue to oscillate in that mode shape.

Some Practical Considerations

As we have mentioned throughout this text, damping forces are difficult to quantify. Often the *source* of the damping is unclear (Coulomb friction, viscous damping, or hysteretic damping) and the *form* of the damping relation is uncertain (linear, quadratic, or other). Even when the source and the form are known, obtaining precise values of the parameters in the damping model is usually difficult. Damping is often slight in structural vibration but can be significant in systems where damping is deliberately introduced, such as in suspension systems and in vibration isolation systems.

Because of these considerations and because damping complicates the mathematics quite a bit, vibration engineers often either totally neglect damping in the analysis or assume it is linear and proportional. The latter assumption enables us to decouple the equations, as we have seen.

One approach to simplify the mathematics is to neglect (set to zero) those terms in the damping matrix $\mathbf{C}$ that prevent it from being proportional. The justification for this step, which is not always true, is that the damping model is only approximate at best.

Sometimes a value is chosen for ζ_i based on experience or the results of testing. Some test methods enable the system to be excited in a particular mode. By computing the log decrement, for example, or by other means, we can then estimate the value of ζ_i. If the damping is structural, then a relatively low value, say $\zeta_i = 0.05$, is often assumed.

If the damping values are high but nonproportional, the modal analysis is very complicated because the eigenvalues and the eigenvectors are complex. In such cases, especially where there are more than two degrees of freedom, the best approach is to use computer software. However, most software packages do not support modal analysis based on the second-order model $\mathbf{M\ddot{x} + C\dot{x} + Kx = F}(t)$. Most packages, such as MATLAB, do support modal analysis based on the so-called *state-variable model*

$$\dot{\mathbf{x}} = \mathbf{Ax} + \mathbf{Bf} \tag{8.4-12}$$

where $\mathbf{x}$ is the *state vector* and $\mathbf{f}$ is the vector of forcing functions. The state-variable model represents the equations of motion as a set of first-order differential equations. The eigenvalue problem for a state-variable model is

$$\mathbf{Av} = \lambda\mathbf{v} \tag{8.4-13}$$

The state-variable model of a system with three degrees of freedom will have six differential equations, so the state vector and the eigenvector $\mathbf{v}$ will have six components.

Coverage of these methods is beyond the scope of this text. This approach is sometimes used in the design of feedback control systems for high-order systems, and it is discussed in control system texts. See, for example, Chapter 11 of [Palm, 2000].

The state-variable model is also the form of choice for doing simulation with computer packages. This is discussed in the next section.

8.5 MATLAB APPLICATIONS

This section shows how to use MATLAB to perform modal analysis and to do simulation.

Modal Analysis

MATLAB provides the `eig` and the `chol` functions to support modal analysis. Consider the following eigenvalue problem:

$$\begin{bmatrix} 2 & -1 & 0 \\ -1 & 2 & -1 \\ 0 & -1 & 1 \end{bmatrix} \mathbf{v} = \frac{m}{k}\omega^2 \mathbf{v} = r\mathbf{v}$$

The eigenvalues can be found with MATLAB by using the `eig` function. The session is

```
≫A = [2,-1,0;-1,2,-1;0,-1,1]:
≫r = eig(A)
```

The results are $r = 0.19806$, 1.5550, and 3.2470.

The syntax `[v, D] = eig(A)` computes the arrays v and D. The eigenvectors of **A** are stored in the columns of v. The eigenvalues are the diagonal elements of D. In this example the results displayed on the screen after typing `[v, D] = eig(A)` are

```
v=

    -0.3280  0.7370   -0.5910
    -0.5910  0.3280    0.7370
    -0.7370  -0.5910  -0.3280

D=

    0.1981  0        0
    0       1.5550   0
    0       0        3.2470
```

The eigenvectors are ordered according to the order in which their associated eigenvalue appears in the diagonal of the matrix D. Thus the first eigenvector corresponds to the eigenvalue 0.1981. The three eigenvectors are

$$\mathbf{X}_1 = \begin{bmatrix} -0.3280 \\ -0.5910 \\ -0.7370 \end{bmatrix} \qquad \mathbf{X}_2 = \begin{bmatrix} 0.7370 \\ 0.3280 \\ -0.5910 \end{bmatrix} \qquad \mathbf{X}_3 = \begin{bmatrix} -0.5910 \\ 0.7370 \\ -0.3280 \end{bmatrix}$$

Note that the scalar factors have been computed so that the length of each vector is 1. This makes the vectors orthonormal. For example, for $\mathbf{X}_1$, $\sqrt{(0.3280)^2 + (0.5910)^2 + (0.7370)^2} = 1$. Note also that we can factor -1 out of each vector without changing its meaning.

Cholesky Decomposition

MATLAB provides the `chol` function to perform Cholesky decomposition. Note that `chol` gives an *upper* triangular matrix, which equals $\mathbf{L}^T$ in our notation. For example, determine $\mathbf{L}$ so that $\mathbf{LL}^T = \mathbf{M}$ for the following matrix:

$$\mathbf{M} = \begin{bmatrix} 9 & 3 & 0 \\ 3 & 5 & 2 \\ 0 & 2 & 17 \end{bmatrix}$$

Typing in MATLAB

```
≫ M = [9,3,0;3,5,2;0,2,17]
≫ C = chol(M)
```

returns the result

$$\mathbf{C} = \begin{bmatrix} 3 & 1 & 0 \\ 0 & 2 & 1 \\ 0 & 0 & 4 \end{bmatrix}$$

Thus we obtain

$$\mathbf{L} = \mathbf{C}^T = \begin{bmatrix} 3 & 0 & 0 \\ 1 & 2 & 0 \\ 0 & 1 & 4 \end{bmatrix}$$

Simulation of Linear, Multi-Degree-of-Freedom Systems

In Section 3.9 we saw how to convert a differential equation into a set of first-order equations for use with the MATLAB `ode45` function. This form is called the *state-variable form*. If the mode is nonlinear, then we must use the `ode45` function or its relatives (`ode23`, etc.). However, if the model is linear, then we can take advantage of some powerful MATLAB functions for doing analysis and simulation. These functions include the solvers `initial`, `step`, `impulse`, and `lsim`.

We will now show how to express the state-variable form as a matrix equation for use with the MATLAB functions. This easy process is illustrated with the following two examples.

EXAMPLE 8.5-1
Vector-Matrix Form for One Degree of Freedom

Express the following model as a single vector-matrix equation:

$$m\ddot{x} + c\dot{x} + kx = f(t)$$

Solution

First define the new variables x_1 and x_2 such that

$$x_1 = x \qquad x_2 = \dot{x}$$

Then we can write the model as $m\dot{x}_2 + cx_2 + kx_1 = f$. Next solve for $\dot{x}_2$:

$$\dot{x}_2 = \frac{1}{m}(f - kx_1 - cx_2)$$

The following two equations constitute a state-variable model. The variables x_1 and x_2 are the *state variables*:

$$\dot{x}_1 = x_2$$

$$\dot{x}_2 = \frac{1}{m}f(t) - \frac{k}{m}x_1 - \frac{k}{m}x_2$$

The equations can be written as one equation as follows:

$$\begin{bmatrix} \dot{x}_1 \\ \dot{x}_2 \end{bmatrix} = \begin{bmatrix} 0 & 1 \\ -\dfrac{k}{m} & -\dfrac{c}{m} \end{bmatrix} \begin{bmatrix} x_1 \\ x_2 \end{bmatrix} + \begin{bmatrix} 0 \\ \dfrac{1}{m} \end{bmatrix} f(t)$$

In compact form this becomes

$$\dot{\mathbf{x}} = \mathbf{A}\mathbf{x} + \mathbf{B}f(t)$$

where

$$\mathbf{A} = \begin{bmatrix} 0 & 1 \\ -\dfrac{k}{m} & -\dfrac{c}{m} \end{bmatrix} \qquad \mathbf{B} = \begin{bmatrix} 0 \\ \dfrac{1}{m} \end{bmatrix} \qquad \mathbf{x} = \begin{bmatrix} x_1 \\ x_2 \end{bmatrix}$$

■

EXAMPLE 8.5-2
Vector-Matrix Form for Two Degrees of Freedom

Consider the two-mass system shown in Figure 8.5-1. Suppose the parameter values are $m_1 = 5$, $m_2 = 3$, $c_1 = 4$, $c_2 = 8$, $k_1 = 1$, and $k_2 = 4$. The equations of motion are

$$5\ddot{x}_1 + 12\dot{x}_1 + 5x_1 - 8\dot{x}_2 - 4x_2 = 0 \tag{1}$$

$$3\ddot{x}_2 + 8\dot{x}_2 + 4x_2 - 8\dot{x}_1 - 4x_1 = f(t) \tag{2}$$

Put these equations into state-variable form, and express the state-variable model in vector-matrix form.

Solution

Using the system's potential and kinetic energies as a guide, we see that the displacements x_1 and x_2 describe the system's potential energy and that the velocities $\dot{x}_1$ and $\dot{x}_2$ describe the system's kinetic energy. That is,

$$PE = \frac{1}{2}k_1x_1^2 + \frac{1}{2}k_2(x_1 - x_2)^2$$

and

$$KE = \frac{1}{2}m_1\dot{x}_1^2 + \frac{1}{2}m_2\dot{x}_2^2$$

This indicates that we need four state variables. (Another way to see that we need four variables is to note that the model consists of two coupled second-order equations and thus is effectively a fourth-order model.) Thus we can choose the state variables z_1, z_2, z_3, and z_4 to be

$$z_1 = x_1 \qquad z_2 = \dot{x}_1 \qquad z_3 = x_2 \qquad z_4 = \dot{x}_2 \tag{3}$$

These definitions imply that $\dot{z}_1 = z_2$ and $\dot{z}_3 = z_4$, which are two of the state equations.

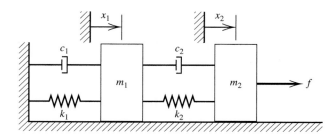

FIGURE 8.5-1 A system with two degrees of freedom.

The remaining two equations can be found by solving Equations 1 and 2 for $\ddot{x}_1$ and $\ddot{x}_2$, noting that $\ddot{x}_1 = \dot{z}_2$ and $\ddot{x}_2 = \dot{z}_4$, and using the substitutions given by Equation 3:

$$\dot{z}_2 = \frac{1}{5}(-12z_2 - 5z_1 + 8z_4 + 4z_3)$$

$$\dot{z}_4 = \frac{1}{3}[-8z_4 - 4z_3 + 8z_2 + 4z_1 + f(t)]$$

Note that the left-hand sides of the state equations must contain only the first-order derivative of each state variable. This is why we divided by 5 and 3, respectively. Note also that the right-hand sides must not contain any derivatives of the state variables. For example, we should not use the substitution $\dot{z}_1$ for $\dot{x}_1$, but rather should substitute z_2 for $\dot{x}_1$. Failure to observe this restriction is a common mistake.

Now list the four state equations in ascending order according to their left-hand sides, after rearranging the right-hand sides so that the state variables appear in ascending order from left to right:

$$\dot{z}_1 = z_2 \tag{4}$$

$$\dot{z}_2 = \frac{1}{5}(-5z_1 - 12z_2 - 4z_3 - 8z_4) \tag{5}$$

$$\dot{z}_3 = z_4 \tag{6}$$

$$\dot{z}_4 = \frac{1}{3}[4z_1 + 8z_2 - 4z_3 - 8z_4 + f(t)] \tag{7}$$

These are the state equations in standard form.

Note that because the potential energy is a function of the difference $x_1 - x_2$, another possible choice of state variables is $z_1 = x_1$, $z_2 = \dot{x}_1$, $z_3 = x_1 - x_2$, and $z_4 = \dot{x}_2$.

In vector-matrix form these equations are

$$\dot{\mathbf{z}} = \mathbf{A}\mathbf{z} + \mathbf{B}f(t)$$

where

$$\mathbf{A} = \begin{bmatrix} 0 & 1 & 0 & 0 \\ -1 & -\dfrac{12}{5} & \dfrac{4}{5} & \dfrac{8}{5} \\ 0 & 0 & 0 & 1 \\ \dfrac{4}{3} & \dfrac{8}{3} & -\dfrac{4}{3} & -\dfrac{8}{3} \end{bmatrix} \qquad \mathbf{B} = \begin{bmatrix} 0 \\ 0 \\ 0 \\ \dfrac{1}{3} \end{bmatrix}$$

and

$$\mathbf{z} = \begin{bmatrix} z_1 \\ z_2 \\ z_3 \\ z_4 \end{bmatrix} = \begin{bmatrix} x_1 \\ \dot{x}_1 \\ x_2 \\ \dot{x}_2 \end{bmatrix}$$

$\blacksquare$

Standard Form of the State Equation

We may use any symbols we choose for the state variables and the input function, although the common choice (and the notation used by MATLAB) is x_i for the state variables and u_i for the input functions. The standard vector-matrix form of the state equations, where the number of state variables is n, is

$$\dot{\mathbf{x}} = \mathbf{A}\mathbf{x} + \mathbf{B}\mathbf{u} \tag{8.5-1}$$

where the vectors $\mathbf{x}$ and $\mathbf{u}$ are column vectors containing the state variables and the inputs, if any. The dimensions are as follows:

- The *state vector* $\mathbf{x}$ is a column vector having n rows.
- The *system matrix* $\mathbf{A}$ is a square matrix having n rows and n columns.
- The *input vector* $\mathbf{u}$ is a column vector having m rows.
- The *control* or *input matrix* $\mathbf{B}$ has n rows and m columns.

In our examples thus far there has been only one input, and for such cases the input vector $\mathbf{u}$ reduces to a scalar u. The standard form, however, allows for more than one input function. Such would be the case in the two-mass model if external forces f_1 and f_2 are applied to each mass.

The Output Equation

Some software packages and some design methods require you to define an *output vector*, usually denoted by $\mathbf{y}$. The output vector contains the variables that are of interest for the particular problem at hand. These variables are not necessarily the state variables; they might be some combination of the state variables and the inputs. For example, in the mass-spring model, we might be interested in the total force $f - kx - c\dot{x}$ acting on the mass and in the momentum $m\dot{x}$. In this case the output vector has two elements. If the state variables are $x_1 = x$ and $x_2 = \dot{x}$, the output vector is

$$\mathbf{y} = \begin{bmatrix} y_1 \\ y_2 \end{bmatrix} = \begin{bmatrix} f - kx - c\dot{x} \\ m\dot{x} \end{bmatrix} = \begin{bmatrix} f - kx_1 - cx_2 \\ mx_2 \end{bmatrix}$$

or

$$\mathbf{y} = \begin{bmatrix} y_1 \\ y_2 \end{bmatrix} = \begin{bmatrix} -k & -c \\ 0 & m \end{bmatrix} \begin{bmatrix} x_1 \\ x_2 \end{bmatrix} + \begin{bmatrix} 1 \\ 0 \end{bmatrix} f = \mathbf{C}\mathbf{x} + \mathbf{D}f$$

where

$$\mathbf{C} = \begin{bmatrix} -k & -c \\ 0 & m \end{bmatrix}$$

and

$$\mathbf{D} = \begin{bmatrix} 1 \\ 0 \end{bmatrix}$$

This is an example of the general form $\mathbf{y} = \mathbf{C}\mathbf{x} + \mathbf{D}\mathbf{u}$.

The standard vector-matrix form of the output equation, where the number of outputs is p, the number of state variables is n, and the number of inputs is m, is

$$\mathbf{y} = \mathbf{C}\mathbf{x} + \mathbf{D}\mathbf{u} \tag{8.5-2}$$

where the vector $\mathbf{y}$ contains the output variables. The dimensions are as follows:

- The *output vector* $\mathbf{y}$ is a column vector having p rows.
- The *state output matrix* $\mathbf{C}$ has p rows and n columns.
- The *control output matrix* $\mathbf{D}$ has p rows and m columns.

The matrices $\mathbf{C}$ and $\mathbf{D}$ can always be found whenever the chosen output vector $\mathbf{y}$ is a linear combination of the state variables and the inputs. However, if the output is a nonlinear function, then the standard form does not apply. This would be the case, for example, if the output is chosen to be the system's kinetic energy: $KE = mx_2^2/2$.

State-Variable Methods with MATLAB

The MATLAB `step`, `impulse`, `lsim`, and `bode` functions can be used with state-variable models as well as with transfer function models. In addition, the `initial` function, which computes the free response, can be used only with a state-variable model. MATLAB also provides functions for converting models between the state-variable and transfer function forms.

Recall that to create an LTI object from the reduced form

$$5\ddot{x} + 7\dot{x} + 4x = f(t) \tag{8.5-3}$$

or the transfer function form

$$\frac{X(s)}{F(s)} = \frac{1}{5s^2 + 7s + 4} \tag{8.5-4}$$

you use the `tf(num,den)` function by typing

≫`sys1=tf(1,[5,7,4]);`

The result, `sys1`, is the LTI object that describes the system in the transfer function form.

The LTI object `sys2` in transfer function form for the equation

$$8\frac{d^3x}{dt^3} - 3\frac{d^2x}{dt^2} + 5\frac{dx}{dt} + 6x = 4\frac{d^2f}{dt^2} + 3\frac{df}{dt} + 5f \tag{8.5-5}$$

is created by typing

≫`sys2 = tf([4,3,5],[8,-3,5,6]);`

LTI Objects and the `ss(A,B,C,D)` Function

To create an LTI object from a state model, you use the `ss(A,B,C,D)` function, where `ss` stands for *state space*. The matrix arguments of the function are the matrices in the following standard form of a state model:

$$\dot{\mathbf{x}} = \mathbf{Ax} + \mathbf{Bu} \tag{8.5-6}$$

$$\mathbf{y} = \mathbf{Cx} + \mathbf{Du} \tag{8.5-7}$$

where **x** is the vector of state variables, **u** is the vector of input functions, and **y** is the vector of output variables. For example, to create an LTI object in state-model form for the system described by

$$\dot{x}_1 = x_2$$

$$\dot{x}_2 = \frac{1}{5}f(t) - \frac{4}{5}x_1 - \frac{7}{5}x_2$$

where x_1 is the desired output, you type

≫`A = [0,1;-4/5,-7/5];`
≫`B = [0;1/5];`
≫`C = [1,0];`
≫`D = 0;`
≫`sys3 = ss(A,B,C,D);`

The ss(sys) and ssdata(sys) Functions

An LTI object defined using the `tf` function can be used to obtain an equivalent state-model description of the system. To create a state model for the system described by the LTI object `sys1` created previously in transfer function form, you type `ss(sys1)`. You will then see the resulting **A**, **B**, **C**, and **D** matrices on the screen. To extract and save the matrices as `A1`, `B1`, `C1`, and `D1` (to avoid overwriting the matrices from the previous example), use the `ssdata` function as follows:

$$\gg \texttt{[A1,B1,C1,D1]=ssdata(sys1);}$$

The results are

$$\mathbf{A1} = \begin{bmatrix} -1.4 & -0.8 \\ 1 & 0 \end{bmatrix}$$

$$\mathbf{B1} = \begin{bmatrix} 0.5 \\ 0 \end{bmatrix}$$

$$\mathbf{C1} = \begin{bmatrix} 0 & 0.4 \end{bmatrix}$$

$$\mathbf{D1} = \begin{bmatrix} 0 \end{bmatrix}$$

which correspond to the state equations

$$\dot{x}_1 = -1.4x_1 - 0.8x_2 + 0.5f(t)$$
$$\dot{x}_2 = x_1$$

and the output equation $y = 0.4x_2$.

Relating State Variables to the Original Variables

When using `ssdata` to convert a transfer function form into a state model, note that the output y will be a scalar that is identical to the solution variable of the reduced form; in this case the solution variable of Equation 8.5-3 is the variable x. To interpret the state model, we need to relate its state variables x_1 and x_2 to x. The values of the matrices **C1** and **D1** tell us that the output variable is $y = 0.4x_2$. Because the output y is the same as x, we then see that $x_2 = x/0.4 = 2.5x$. The other state-variable x_1 is related to x_2 by the second state equation $\dot{x}_2 = x_1$. Thus $x_1 = 2.5\dot{x}$.

The tfdata Function

To create a transfer function description of the system `sys3`, previously created from the state model, you type `tfsys3 = tf(sys3);`. To extract and save the coefficients of the transfer function, use the `tfdata` function as follows:

$$\gg \texttt{[num,den] = tfdata(tfsys3,'v');}$$

The optional parameter `'v'` tells MATLAB to return the coefficients as vectors if there is only one transfer function; otherwise, they are returned as cell arrays.

For this example, the vectors returned are `num = [0, 0, 0.2]` and `den = [1, 1.4, 0.8]`. This corresponds to the transfer function

$$\frac{X(s)}{F(s)} = \frac{0.2}{s^2 + 1.4s + 0.8} = \frac{1}{5s^2 + 7s + 4}$$

which is the correct transfer function.

Because it is difficult to obtain the transfer functions for multi-DOF models directly from the equations of motion, it is often easier to first convert the equations into state-variable form, then use the `ss` function to create an LTI model in MATLAB, and then use the `tf` function to obtain the transfer functions. The following example illustrates this process.

EXAMPLE 8.5-3
Transfer Functions of a Two-Mass System

Obtain the transfer functions $X_1(s)/F(s)$ and $X_2(s)/F(s)$ of the state-variable model obtained in Example 8.5-2. The matrices and state vector of the model are

$$\mathbf{A} = \begin{bmatrix} 0 & 1 & 0 & 0 \\ -1 & -\dfrac{12}{5} & \dfrac{4}{5} & \dfrac{8}{5} \\ 0 & 0 & 0 & 1 \\ \dfrac{4}{3} & \dfrac{8}{3} & -\dfrac{4}{3} & -\dfrac{8}{3} \end{bmatrix} \qquad \mathbf{B} = \begin{bmatrix} 0 \\ 0 \\ 0 \\ \dfrac{1}{3} \end{bmatrix}$$

and

$$\mathbf{z} = \begin{bmatrix} z_1 \\ z_2 \\ z_3 \\ z_4 \end{bmatrix} = \begin{bmatrix} x_1 \\ \dot{x}_1 \\ x_2 \\ \dot{x}_2 \end{bmatrix}$$

Solution

Because we want the transfer functions for x_1 and x_2 (which are the same as z_1 and z_3), we must define the C and D matrices to indicate that z_1 and z_3 are the output variables y_1 and y_2. Thus

$$\mathbf{C} = \begin{bmatrix} 1 & 0 & 0 & 0 \\ 0 & 0 & 1 & 0 \end{bmatrix} \qquad \mathbf{D} = \begin{bmatrix} 0 \\ 0 \end{bmatrix}$$

The MATLAB program is as follows.

```
A = [0,1,0,0;-1,-12/5,4/5,8/5; ...
    0,0,0,1;4/3,8/3,-4/3,-8/3];
B = [0;0;0;1/3];
C = [1,0,0,0;0,0,1,0];D = [0;0]
sys4 = ss(A,B,C,D);
tfsys4 = tf(sys4)
```

The results displayed on the screen are labeled #1 and #2. These correspond to the first and second transfer functions in order. The answers are

$$\frac{X_1(s)}{F(s)} = \frac{0.5333s + 0.2667}{s^4 + 5.067s^3 + 4.467s^2 + 1.6s + 0.2667}$$

$$\frac{X_2(s)}{F(s)} = \frac{0.3333s^2 + 0.8s + 0.3333}{s^4 + 5.067s^3 + 4.467s^2 + 1.6s + 0.2667}$$

■

Table 8.5-1 summarizes these functions.

Linear ode Solvers

The Control System toolbox provides several solvers for linear models. These solvers are categorized by the type of input function they can accept: zero input, impulse input, step input, and a general input function.

TABLE 8.5-1 LTI OBJECT FUNCTIONS

Command	Description
sys = ss(A, B, C, D)	Creates an LTI object in state-space form, where the matrices A, B, C, and D correspond to those in the model $\dot{\mathbf{x}} = \mathbf{Ax} + \mathbf{Bu}, \mathbf{y} = \mathbf{Cx} + \mathbf{Du}$
[A, B, C, D] = ssdata(sys)	Extracts the matrices A, B, C, and D of the LTI object sys corresponding to those in the model $\dot{\mathbf{x}} = \mathbf{Ax} + \mathbf{Bu}, \mathbf{y} = \mathbf{Cx} + \mathbf{Du}$.
sys2 = tf(sys1)	Creates an LTI object sys2 in transfer function form from the LTI object sys1 in state-space form.
sys1 = ss(sys2)	Creates an LTI object sys1 in state-space form from the LTI object sys2 in transfer function form.
sys = tf(num,den)	Creates an LTI object in transfer function form, where the vector num is the vector of coefficients of the transfer function numerator, arranged in descending order, and den is the vector of coefficients of the denominator, also arranged in descending order.
[num,den] = tfdata(sys,'v')	Extracts the coefficients of the numerator and denominator of the transfer function model sys. When the optional parameter 'v' is used, if there is only one transfer function, the coefficients are returned as vectors rather than as cell arrays.

The initial Function

The initial function computes and plots the free response of a state model. This is sometimes called the *initial condition response* or the *undriven response* in the MATLAB documentation. The basic syntax is initial(sys,x0), where sys is the LTI object in state-model form and x0 is the initial condition vector. The time span and number of solution points are chosen automatically.

EXAMPLE 8.5-4
Free Response of the Two-Mass Model

Compute the free response $x_1(t)$ of the state model, derived in Example 8.5-2, for $x_1(0) = 5$, $\dot{x}_1(0) = -3$, $x_2(0) = 4$, and $\dot{x}_2(0) = 2$. The model is

$$\dot{z}_1 = z_2$$

$$\dot{z}_2 = \frac{1}{5}(-5z_1 - 12z_2 + 4z_3 + 8z_4)$$

$$\dot{z}_3 = z_4$$

$$\dot{z}_4 = \frac{1}{3}[4z_1 + 8z_2 - 4z_3 - 8z_4 + f(t)]$$

or

$$\dot{\mathbf{z}} = \mathbf{Az} + \mathbf{B}f(t)$$

where

$$\mathbf{A} = \begin{bmatrix} 0 & 1 & 0 & 0 \\ -1 & -\dfrac{12}{5} & \dfrac{4}{5} & \dfrac{8}{5} \\ 0 & 0 & 0 & 1 \\ \dfrac{4}{3} & \dfrac{8}{3} & -\dfrac{4}{3} & -\dfrac{8}{3} \end{bmatrix} \qquad \mathbf{B} = \begin{bmatrix} 0 \\ 0 \\ 0 \\ \dfrac{1}{3} \end{bmatrix}$$

and

$$\mathbf{z} = \begin{bmatrix} z_1 \\ z_2 \\ z_3 \\ z_4 \end{bmatrix} = \begin{bmatrix} x_1 \\ \dot{x}_1 \\ x_2 \\ \dot{x}_2 \end{bmatrix}$$

Solution

We must first relate the initial conditions given in terms of the original variables to the state variables. From the definition of the state vector $\mathbf{z}$, we see that $z_1(0) = x_1(0) = 5$, $z_2(0) = \dot{x}_1(0) = -3$, $z_3(0) = x_2(0) = 4$, and $z_4(0) = \dot{x}_2(0) = 2$. Next, we must define the model in state-model form. The system `sys4` created in Example 8.5-3 specified two outputs, x_1 and x_2. Because we want to obtain only one output here (x_1), we must create a new state model using the same values for the $\mathbf{A}$, $\mathbf{B}$, and $\mathbf{D}$ matrices, but now using

$$\mathbf{C} = \begin{bmatrix} 1 & 0 & 0 & 0 \end{bmatrix} \qquad \mathbf{D} = [0]$$

The MATLAB program is as follows.

```
A = [0,1,0,0;-1,-12/5,4/5,8/5;...
     0,0,0,1;4/3,8/3,-4/3,-8/3];
B = [0;0;0;1/3];
C = [1,0,0,0];D = [0]
sys5 = ss(A,B,C,D);
initial(sys5,[5,-3,4,2])
```

The plot of $x_1(t)$ will be displayed on the screen. ∎

To specify the final time `tfinal`, use the syntax `initial(sys,x0,tfinal)`. To specify a vector of times of the form `t = [0:dt:tfinal]`, at which to obtain the solution, use the syntax `initial(sys,x0,t)`. When called with left-hand arguments, such as `[y,t,x] = initial(sys,x0,...)`, the function returns the output response `y`, the time vector `t` used for the simulation, and the state vector `x` evaluated at those times. The columns of the matrices `y` and `x` are the outputs and the states, respectively. The number of rows in `y` and `x` equals `length(t)`. No plot is drawn. The syntax `initial(sys1,sys2,...,x0,t)` plots the free response of multiple LTI systems on a single plot. The time vector `t` is optional. You can specify line color, line style, and marker for each system; for example, `initial(sys1,'r',sys2,` `'y--',sys3,'gx',x0)`.

The `impulse`, `step`, `lsim`, and `bode` Functions

You may use the `impulse`, `step`, and `lsim` functions with state-variable models the same way they are used with transfer function models. However, when used with state-variable models, there are some additional features available, which we illustrate with the `step` function. When called with left-hand arguments, such as `[y,t] = step(sys,...)`, the function returns the output response `y` and the time vector `t` used for the simulation. No plot is drawn. The array `y` is $p \times q \times m$, where p is `length(t)`, q is the number of outputs, and m is the number of inputs. To obtain the state vector solution for state-space models, use the syntax `[y,t,x] =` `step(sys,...)`.

To use the `lsim` function for nonzero initial conditions with a state-space model, use the syntax `lsim(sys,u,t,x0)`. The initial condition vector `xo` is needed only if the initial conditions are nonzero.

Obtaining the Characteristic Polynomial

The MATLAB command `poly` can find the characteristic polynomial that corresponds to a specified state matrix **A**. For example, the matrix **A** given in Example 8.5-1 is, for $m = 1$, $c = 5$, $k = 6$,

$$\mathbf{A} = \begin{bmatrix} 0 & 1 \\ -6 & -5 \end{bmatrix} \tag{8.5-8}$$

The coefficients of its characteristic polynomial are found by typing

```
≫ A = [0,1;-6,-5];
≫ poly(A)
```

MATLAB returns the answer [1, 5, 6], which corresponds to the polynomial $s^2 + 5s + 6$. The `roots` function can be used to compute the characteristic roots; for example, `roots(poly(A))`, which gives the roots $[-2, -3]$.

MATLAB provides the `eig` function to compute the characteristic roots without obtaining the characteristic polynomial when the model is given in the state-variable form. Its syntax is `eig(A)`. The function's name is an abbreviation of *eigenvalue*. For example, typing `eig([0, 1; -6, -5])` returns the roots $[-2, -3]$.

The function `bode(sys)` draws the Bode plot of the LTI model `sys`, which can be created with either the `tf` or the `ss` function. Its syntax is described in Section 4.9.

8.6 SIMULINK APPLICATIONS

Simulink provides the State-Space block to perform simulation of linear models expressed in state-variable form and also subsystem blocks to ease the creation of larger models having many blocks. We cover both in this section.

Simulating State-Variable Models

The State-Space block is based on the standard form of the linear state-variable model $\dot{\mathbf{x}} = \mathbf{A}\mathbf{x} + \mathbf{B}\mathbf{u}$, $\mathbf{y} = \mathbf{C}\mathbf{x} + \mathbf{D}\mathbf{u}$. The vector **u** represents the inputs, and the vector **y** represents the outputs. State-variable models, unlike transfer function models, can have more than one input and more than one output. Thus, when connecting inputs to the State-Space block, care must be taken to connect them in the proper order. Similar care must be taken when connecting the block's outputs to another block. The following example illustrates how this is done.

EXAMPLE 8.6-1
Simulink Model of a Two-Mass System

The state-variable model of the two-mass system discussed in Example 8.5-2 is

$$\dot{\mathbf{z}} = \mathbf{A}\mathbf{z} + \mathbf{B}f(t)$$

where

$$\mathbf{A} = \begin{bmatrix} 0 & 1 & 0 & 0 \\ -1 & -\dfrac{12}{5} & \dfrac{4}{5} & \dfrac{8}{5} \\ 0 & 0 & 0 & 1 \\ \dfrac{4}{3} & \dfrac{8}{3} & -\dfrac{4}{3} & -\dfrac{8}{3} \end{bmatrix} \qquad \mathbf{B} = \begin{bmatrix} 0 \\ 0 \\ 0 \\ \dfrac{1}{3} \end{bmatrix}$$

and

$$\mathbf{z} = \begin{bmatrix} z_1 \\ z_2 \\ z_3 \\ z_4 \end{bmatrix} = \begin{bmatrix} x_1 \\ \dot{x}_1 \\ x_2 \\ \dot{x}_2 \end{bmatrix}$$

Develop a Simulink model to plot the unit-step response of the variables x_1 and x_2 with the initial conditions $x_1(0) = 0.2$, $\dot{x}_1(0) = 0$, $x_2(0) = 0.5$, $\dot{x}_2(0) = 0$.

Solution

First select appropriate values for the matrices in the output equation $\mathbf{y} = \mathbf{Cz} + \mathbf{D}f(t)$. Since we want to plot x_1 and x_2, which are z_1 and z_2, we choose $\mathbf{C}$ and $\mathbf{D}$ as follows:

$$\mathbf{C} = \begin{bmatrix} 1 & 0 & 0 & 0 \\ 0 & 0 & 1 & 0 \end{bmatrix} \qquad \mathbf{D} = \begin{bmatrix} 0 \\ 0 \end{bmatrix}$$

To create this simulation, obtain a new model window. Then do the following to create the model shown in Figure 8.6-1

1. Select and place in the new window the Step block. Double-click on it to obtain the Block Parameters window, and set the Step time to 0, the Initial and Final values to 0 and 1, and the Sample time to 0. Click **OK**.

2. Select and place the State-Space block. Double-click on it, and enter [0, 1, 0, 0; -1, -12/5, 4/5, 8/5; 0, 0, 0, 1; 4/3, 8/3, -4/3, -8/3] for **A**, [0; 0; 0; 1/3] for **B**, [1, 0, 0, 0; 0, 0, 1, 0] for **C**, and [0; 0] for **D**. Then enter [0.2; 0; 0.5; 0] for the initial conditions. Click **OK**. Note that the dimension of the matrix **B** tells Simulink that there is one input. The dimensions of the matrices **C** and **D** tell Simulink that there are two outputs.

3. Select and place the Scope block.

4. Once the blocks have been placed, connect the input port on each block to the outport port on the preceding block as shown in Figure 8.6-1.

5. Experiment with different values of the Stop time until the Scope shows that the steady-state response has been reached. For this application, a Stop time of 25 is satisfactory. The plots of both x_1 and x_2 will appear in the Scope. ∎

Subsystem Blocks

One potential disadvantage of a graphical interface such as Simulink is that to simulate a complicated system, such as a system having many masses, the diagram can become rather large and therefore somewhat cumbersome. Simulink, however, provides for the creation of *subsystem blocks*, which play a role analogous to subprograms in a programming language. A subsystem block is actually a Simulink program represented by a single block. A subsystem block, once created, can be used in other Simulink

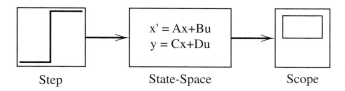

Step   State-Space   Scope

FIGURE 8.6-1 Simulink model of a two-mass system.

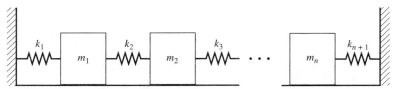

FIGURE 8.6-2 A multi-mass system.

programs. In this section we show how to use subsystem blocks to simulate multi-mass systems.

Creating Subsystem Blocks

You can create a subsystem block in one of two ways: by dragging the Subsystem block from the block library to the model window or by first creating a Simulink model and then "encapsulating" it within a bounding box. We will illustrate the latter method.

EXAMPLE 8.6-2
Subsystem Model for Three Masses

Consider the system shown in Figure 8.6-2. Use the values $m_1 = m_3 = 10$ kg, $m_2 = 30$ kg, $k_1 = k_4 = 10^4$ N/m, and $k_2 = k_3 = 2 \times 10^4$ N/m.

(a) Develop a Subsystem block for one mass-spring combination.

(b) Use the Subsystem block to construct a Simulink model of the entire system of three masses. Plot the displacements of the masses over $0 \le t \le 2$ s if the initial displacement of m_1 is 0.1 m and the initial displacements of m_2 and m_3 are zero.

Solution

(a) For the system shown in Figure 8.6-3, from Newton's law,

$$m\ddot{x} = k_l(x_l - x) - k_r(x - x_r) = -(k_l + k_r)x + k_l x_l + k_r x_r$$

where we have used the subscripts l and r to denote "left" and "right." Using this equation, we can develop the simulation diagram shown in Figure 8.6-4. This diagram can be used to create the Simulink model shown in Figure 8.6-5. The oval blocks are Input and Outport Ports, which are available in the Blocks library. When entering the gains in each of the Gain blocks, note that you can use MATLAB variables and expressions. Before running the program, we will assign values to these variables in the MATLAB Command window. Enter the gains for the Gain blocks using the expressions shown below each block. You may also use a variable as the Initial condition of the Integrator block. Name this variable x10. Save the model and give it a name, such as `mass1`.

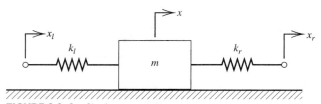

FIGURE 8.6-3 Single-mass system.

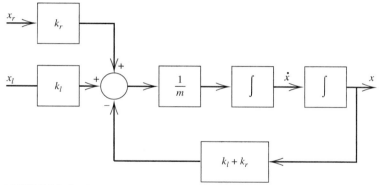

FIGURE 8.6-4 Simulation diagram for a single mass.

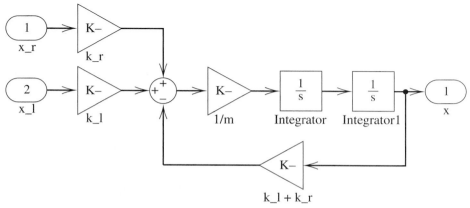

FIGURE 8.6-5 Simulink model based of Figure 8.6-4.

Now create a "bounding box" surrounding the diagram. Do this by placing the mouse cursor in the upper left, holding the mouse button down, and dragging the expanding box to the lower right to enclose the entire diagram. Then choose **Create Subsystem** from the **Edit** menu. Simulink will then replace the diagram with a single block having as many input and output ports as required and will assign default names. You may resize the block to make the labels readable. You may view or edit the subsystem by double-clicking on it. This model can be used to create the subsystem block shown in Figure 8.6-6.

(b) Now consider a system having three masses as shown in Figure 8.6-2 with $n = 3$. The simulation diagram is given in Figure 8.6-7. We can create a Simulink model for this system by connecting the subsystem blocks as shown in Figure 8.6-8. Enter the values of m and k into each gain block by double-clicking on each subsystem. The initial condition on each integrator should be 0 by default, except for 0.1 for the initial condition of the second integrator in the subsystem for mass 1.

In the To Workspace block, set the Save format to array. You can plot the results by typing in MATLAB:

```
» plot(tout,simout)
```

The plot shown in Figure 8.6-9 was obtained by inserting a Max block before the To Workspace block and connecting the outputs x_1, x_2, and x_3 to the Mux. ∎

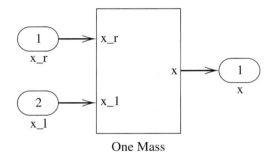

One Mass

FIGURE 8.6-6   Subsystem block.

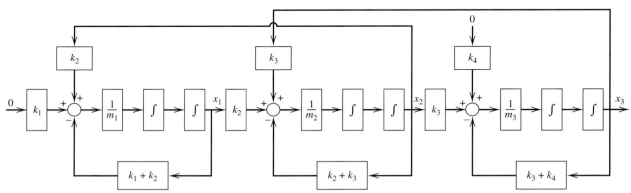

FIGURE 8.6-7   Simulation diagram for the system shown in Figure 8.6-2 for $n = 3$.

Application of Subsystem Blocks

The subsystem pictured in Figure 8.6-3 can be generalized to include the following effects:

- Damping between adjacent masses
- Damping between the mass and the surface
- Nonlinear stiffness or damping elements
- Coulomb friction

In this way a useful library of subsystem blocks can be developed to build models of systems having considerable complexity.

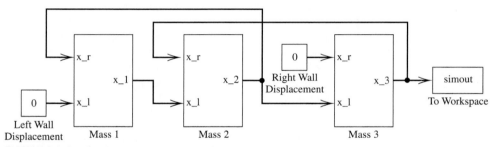

FIGURE 8.6-8   Simulink model composed of subsystem blocks.

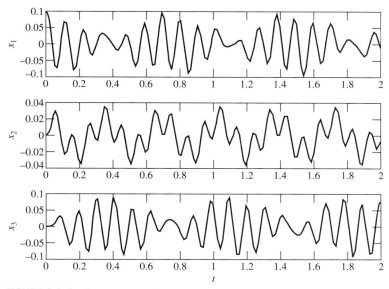

FIGURE 8.6-9   Free response of the system shown in Figure 8.6-2 for $n = 3$.

8.7   CHAPTER REVIEW

Multi-degree-of-freedom systems require complicated and tedious algebra to analyze their response characteristics. For such systems it is more convenient to use matrix representations of the equations of motion and matrix methods to do the analysis.

We began by showing how to represent the equations of motion in compact matrix form. Then we addressed the topic of modes and developed systematic procedures for analyzing the modal response. Besides providing a compact form for representing the equations of motion and performing the analysis, matrix methods also form the basis of software packages such as MATLAB and Simulink that provide powerful tools for modal analysis and simulation.

Now that you have finished this chapter, you should be able to do the following:

1. Express the equations of motion of a linear system in matrix form.
2. Obtain the frequencies and mode shapes of a linear model expressed in matrix form.
3. Obtain the forced response of a multi-degree-of-freedom system.
4. Use approximate methods to include the effects of damping.
5. Apply MATLAB and Simulink to the methods of this chapter.

PROBLEMS

SECTION 8.1   MATRIX FORM OF THE EQUATIONS OF MOTION

8.1   Assuming vertical motion only, determine the mass matrix, and use influence coefficients to determine the stiffness matrix of the system shown in Figure P8.1, for the case where $k_1 = k_2 = k_3 = k$, $m_1 = m_3 = m$, and $m_2 = 2m$.

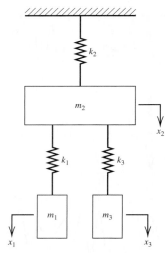

FIGURE P8.1

8.2 In Figure P8.2, a tractor pulling a trailer is used to carry objects, such as large paper rolls or pipes. Assuming the cylindrical load m_3 rolls without slipping, obtain the equations of motion of the system.

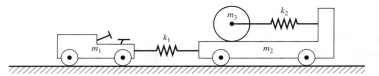

FIGURE P8.2

8.3 Obtain the equations of motion of the system shown in Figure P8.3. Determine the mass and stiffness matrices.

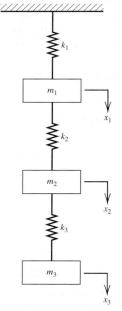

FIGURE P8.3

8.4 A model of a three-story building is shown in Figure P8.4, where each floor is modeled as a lumped mass. The supporting columns are modeled as being capable of deflecting only in shear and have the translational stiffnesses shown. The floor masses are each equal to m kg. Obtain the mass and stiffness matrices.

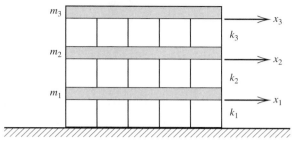

FIGURE P8.4

8.5 Figure P8.5 shows a four-cylinder internal combustion engine driving a generator. The four pistons and connecting rods have been modeled by the effective inertias I_1 through I_4. The torsional stiffnesses of the crankshaft between each cylinder are k_1, k_2, and k_3. The torsional stiffness of the shaft between the engine and the generator is k_4. Obtain the mass and stiffness matrices.

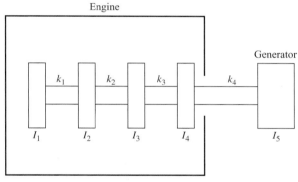

FIGURE P8.5

8.6 Figure P8.6 shows a turbine of inertia I_1 driving a generator through a gear pair. The turbine speed is 5400 rpm, and the speed of the generator is 1800 rpm. The torsional stiffness of the turbine shaft is k_1, and that of the generator shaft is k_2. Neglect the shaft inertias and obtain the mass and stiffness matrices.

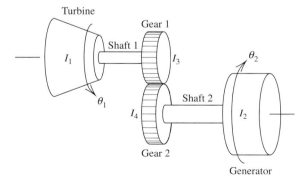

FIGURE P8.6

8.7 Figure 8.1-3 shows a uniform beam with three concentrated masses attached to it. The three masses are much greater than the beam mass, so we will neglect the mass of the beam. An experiment was performed in which a 2000-N vertical weight f_j was attached to the x_1, x_2, and x_3 locations successively, and the resulting static beam deflections x_1, x_2, and x_3 were measured. Use the measurements in the following table to compute the stiffness matrix of the beam.

	$f_1 = 2000$ (N)	$f_2 = 2000$ (N)	$f_3 = 2000$ (N)
x_1 (mm)	1.14	1.37	0.87
x_1 (mm)	1.37	2	1.37
x_1 (mm)	0.87	1.37	1.14

SECTION 8.2   MODE SHAPES AND EIGENVALUE PROBLEMS

8.8 Determine the frequencies and mode shapes for the model whose mass and stiffness matrices are

$$\mathbf{M} = \begin{bmatrix} 5 & 0 \\ 0 & 3 \end{bmatrix} \qquad \mathbf{K} = \begin{bmatrix} 32 & -8 \\ -8 & 8 \end{bmatrix}$$

Use two methods, one based on Equation 8.2-14 and the other based on Equation 8.2-15. Compare the results.

8.9 Obtain the natural frequencies and mode shapes of the system shown in Figure 8.2-2 for the case where $m_1 = m_3 = m$, $m_3 = 4m$, $k_1 = k_2 = k$, and $k_3 = 4k$.

8.10 Obtain the mode shapes and natural frequencies in terms of m and k for the system treated in Problem 8.1.

8.11 Obtain the mode shapes and natural frequencies in terms of m and k for the system treated in Problem 8.2, for the case where $m_1 = m_2 = m$, $m_3 = 2m$, and $k_1 = k_2 = k$.

8.12 Obtain the mode shapes and natural frequencies for the system treated in Example 8.1-6 for $m_1 = m_3 = 100$ kg, $m_2 = 200$ kg.

8.13 Figure P8.13 shows a four-cylinder internal combustion engine driving a generator. The four pistons and connecting rods have been modeled by the effective inertias I_1 through I_4, which are each equal to 1 kg·m². The torsional stiffnesses of the crankshaft between each cylinder are $k_1 = k_2 = k_3 = 1.5 \times 10^5$ N·m/rad. The torsional stiffness of the shaft between the engine and the generator is $k_4 = 2 \times 10^6$ N·m/rad. The generator inertia is $I_5 = 2$ kg · m². Determine the mode shapes and the natural frequencies.

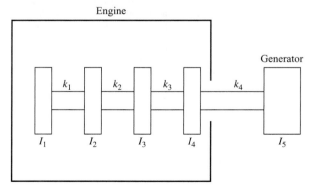

FIGURE P8.13

8.14 Refer to Figure P8.14a, which shows a ship's propeller, drive train, engine, and flywheel. The diameter ratio of the gears is $D_1/D_2 = 1.5$. The inertias in kg·m² of gear 1 and gear 2 are 500 and 100, respectively. The flywheel, engine, and propeller inertias are 10^4, 10^3, and 2500, respectively.

The torsion stiffness of shaft 1 is 5×10^6 N · m/rad, and that of shaft 2 is 10^6 N·m/rad. Since the shaft between the engine and gears is short, we will assume that it is very stiff compared to the other shafts. If we also neglect the shaft inertias, the resulting model, shown in Figure P8.14b, consists of three inertias, the flywheel, one inertia obtained by lumping the engine and gear inertias, and one for the propeller. Using these assumptions, obtain the natural frequencies and mode shapes of the system.

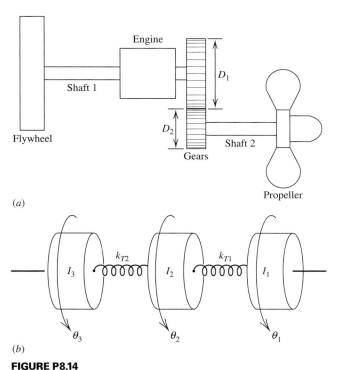

(a)

(b)

FIGURE P8.14

8.15 Refer to Figure P8.15, which shows a turbine driving an electrical generator through a gear pair. The diameter ratio of the gears is $D_2/D_1 = 1.5$. The inertias in kg·m^2 of gear 1 and gear 2 are 100 and 500, respectively. The turbine and generator inertias are 2000 and 1000, respectively. The torsion stiffness of shaft 1 is 3×10^5 N·m/rad, and that of shaft 2 is 8×10^4 N · m/rad. Neglect the shaft inertias and obtain the natural frequencies of the system.

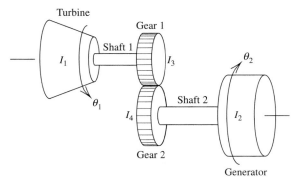

FIGURE P8.15

8.16 Determine the natural frequencies and mode shapes for the system shown in Figure P8.16, for the case where $k_1 = k_2 = k_3 = k$, $m_1 = m_3 = m$, and $m_2 = 2m$.

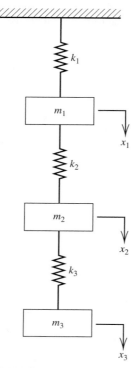

FIGURE P8.16

8.17 Figure P8.17 shows three pendula coupled by spring elements.

(a) Assume small angles and derive the equations of motion.

(b) Determine the natural frequencies and mode shapes for the case where $k_1 = k_2 = k$ and $k_3 = 4k$.

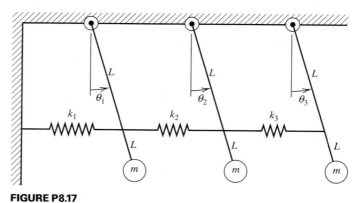

FIGURE P8.17

8.18 Figure P8.18 shows a machine tool support, where $m = 200\,\text{kg}$, $L_1 = 1.2\,\text{m}$, $L_2 = 0.8\,\text{m}$, $k_1 = 1.5 \times 10^4\,\text{N/m}$, and $k_2 = 1.8 \times 10^4\,\text{N/m}$. Determine the natural frequencies and mode shapes in terms of x and θ.

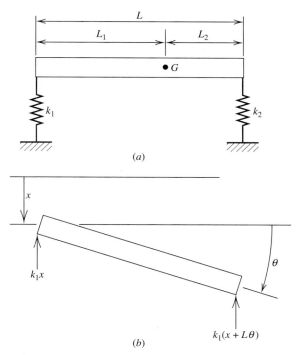

FIGURE P8.18

8.19 Determine the natural frequencies and mode shapes for the system shown in Figure P8.19, for the case where $k_1 = k_2 = k_3 = k$ and $m_1 = m_2 = m_3 = m$.

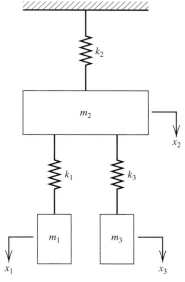

FIGURE P8.19

8.20 The gas turbine in Figure P8.20 drives a generator. Neglect the shaft inertias and the coupling inertias I_4. Determine the natural frequencies and mode shapes using the values $I_1 = 10 \, \text{kg} \cdot \text{m}^2$, $I_2 = I_3 = 5 \, \text{kg} \cdot \text{m}^2$, $k_2 = 10^6 \, \text{N} \cdot \text{m/rad}$, and $k_1 = 2k_2$.

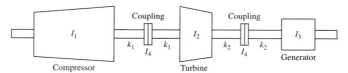

FIGURE P8.20

8.21 Figure P8.21 shows a machine tool support, where $m = 200$ kg, $L_3 = 12/11$ m, $L_4 = 10/11$ m, $L_5 = 12/110$ m, $k_1 = 1.5 \times 10^4$ N/m, and $k_2 = 1.8 \times 10^4$ N/m. Determine the natural frequencies and mode shapes in terms of x and θ.

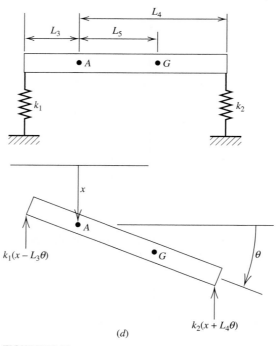

(d)

FIGURE P8.21

8.22 Figure P8.22 shows a six-cylinder diesel engine driving a pump. The six pistons and connecting rods have been modeled by the effective inertias I_1 through I_6, which are each equal to 36 kg·m². Treat the shafts between them as rigid. The flywheel inertia is $I_7 = 80$ kg · m² and the pump inertia reflected to the engine shaft is $I_{10} = 60$ kg · m². Neglect the coupling inertia I_8 and the gear inertia I_9. The torsion stiffnesses in N · m/rad are

$$k_6 = 3 \times 10^6 \quad k_7 = 2.5 \times 10^6$$
$$k_8 = 2.65 \times 10^6 \quad k_9 = 4 \times 10^6$$

Determine the mode shapes and the natural frequencies.

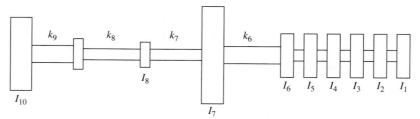

FIGURE P8.22

8.23 Determine the natural frequencies and mode shapes of the geared system shown in Figure P8.23. The shaft connecting I_2 and I_3 runs three times faster than the shaft connecting I_1 with I_4. The inertias and torsional stiffnesses are as follows:

$$I_1 = 1 \text{ kg} \cdot \text{m}^2 \qquad I_2 = 1.5 \text{ kg} \cdot \text{m}^2$$

$$I_3 = 0.5 \text{ kg} \cdot \text{m}^2 \qquad I_4 = 3 \text{ kg} \cdot \text{m}^2 \qquad I_5 = 1 \text{ kg} \cdot \text{m}^2$$

$$k_1 = k_3 = 4000 \text{ N} \cdot \text{m/rad} \qquad k_2 = 10\,000 \text{ N} \cdot \text{m/rad}$$

Neglect the shaft inertias and determine the natural frequencies and mode shapes.

$$k_1 = 4 \times 10^8 \text{ N/m} \qquad k_2 = 3 \times 10^6 \text{ N/m}$$

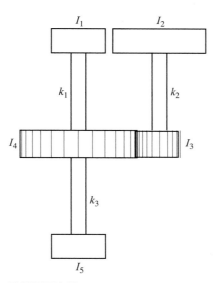

FIGURE P8.23

8.24 A model of a three-story building is shown in Figure P8.24, where each floor is modeled as a lumped mass. The supporting columns are modeled as being capable of deflecting only in shear and have the following stiffnesses in N/m.

$$k_1 = 4 \times 10^8 \qquad k_2 = 3 \times 10^8 \qquad k_3 = 2 \times 10^8$$

The floor masses are each equal to 10^6 kg. Determine the natural frequencies and mode shapes.

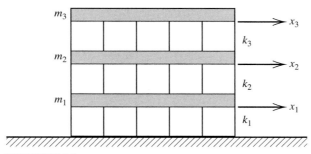

FIGURE P8.24

SECTION 8.3 MODAL ANALYSIS

8.25 Consider the system whose mass and stiffness matrices are

$$\mathbf{M} = \begin{bmatrix} 5 & 0 \\ 0 & 3 \end{bmatrix} \qquad \mathbf{K} = \begin{bmatrix} 32 & -8 \\ -8 & 8 \end{bmatrix}$$

Use the mass normalized stiffness matrix to obtain a set of orthonormal eigenvectors.

8.26 Create a set of principal coordinates for the system treated in Problem 8.25 and obtain the modal equations.

8.27 Use the modal equations from Problem 8.26 to obtain $x_1(t)$ and $x_2(t)$ for the initial conditions $x_1(0) = 4$, $x_2(0) = 3$, and $\dot{x}_1(0) = \dot{x}_2(0) = 0$, with $\mathbf{F}(t) = \mathbf{0}$.

8.28 Consider the system

$$5\ddot{x}_1 + 32x_1 - 8x_2 = f_1(t)$$

$$3\ddot{x}_2 - 8x_1 + 8x_2 = f_2(t)$$

(a) Obtain the modal equations for the forced response.

(b) What conditions must exist for $f_1(t)$ and $f_2(t)$ to influence both modes?

8.29 Each spring in Figure P8.29 has a stiffness of 4000 N/m. Each mass is 0.5 kg.

(a) Obtain the modal differential equations.

(b) Solve the modal equations for the case where the masses are initially at rest and $x_1(0) = x_2(0) = 0$, $x_3(0) = 0.1$ m/s.

(c) Use the results of part (b) to obtain the solutions for $x_1(t)$, $x_2(t)$, and $x_3(t)$.

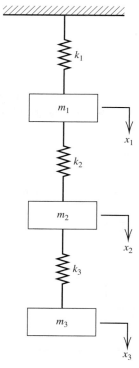

FIGURE P8.29

8.30 Determine **L** so that $\mathbf{LL}^T = \mathbf{M}$ for the following matrix.

$$\mathbf{M} = \begin{bmatrix} 10 & 5 & 0 \\ 5 & 12 & 4 \\ 0 & 4 & 25 \end{bmatrix}$$

SECTION 8.4 EFFECTS OF DAMPING

8.31 For the system shown in Figure 8.4-1 there is damping between the surface and the masses. For the case where $m_1 = 5$, $m_2 = 3$, $c_1 = 0.5$, $c_2 = 0.3$, and $k_1 = k_2 = 1$, determine whether or not the damping is proportional. If it is proportional, determine the modes and frequencies. If it is not proportional, make a suitable approximation that will give a proportional damping matrix, and determine the modes and frequencies for this approximate model. Compute the modal damping ratios.

8.32 For the system shown in Figure 8.4-1 there is damping between the surface and the masses. For the case where $m_1 = 5$, $m_2 = 3$, $c_1 = 0.5$, $c_2 = 0.4$, and $k_1 = k_2 = 1$, determine whether or not the damping is proportional. If it is proportional, determine the modes and frequencies. If it is not proportional, make a suitable approximation that will give a proportional damping matrix, and determine the modes and frequencies for this approximate model. Compute the modal damping ratios.

SECTION 8.5 MATLAB APPLICATIONS

8.33 Figure P8.33 shows a six-cylinder diesel engine driving a pump. The six pistons and connecting rods have been modeled by the effective inertias I_1 through I_6, which are each equal to 36 kg·m². The flywheel inertia is $I_7 = 80$ kg·m² and that of the coupling is $I_8 = 9$ kg·m². The equivalent inertia of the gears is $I_9 = 12$ kg·m², and the equivalent inertia of the pump reflected to the engine shaft is $I_{10} = 60$ kg·m². The torsional stiffnesses of the crankshaft between each cylinder are $k_1 = k_2 = k_3 = k_4 = k_5 = 1.85 \times 10^6$ N·m/rad. The other torsional stiffnesses, in N · m/rad, are as follows.

$$k_6 = 3 \times 10^6 \qquad k_7 = 2.5 \times 10^6$$

$$k_8 = 2.65 \times 10^6 \qquad k_9 = 4 \times 10^6$$

Determine the mode shapes and the natural frequencies.

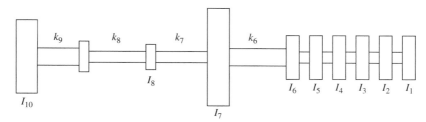

FIGURE P8.33

8.34 The gas turbine in Figure P8.34 drives a generator. Assuming the shaft inertias are negligible, determine the natural frequencies and mode shapes. The inertias and torsional stiffnesses have the following values:

$$I_1 = 10 \text{ kg} \cdot \text{m}^2 \qquad I_2 = 5 \text{ kg} \cdot \text{m}^2$$

$$I_3 = 5 \text{ kg} \cdot \text{m}^2 \qquad I_4 = 1 \text{ kg} \cdot \text{m}^2$$

$$k_1 = 2 \times 10^6 \text{ N} \cdot \text{m/rad} \qquad k_2 = 1 \times 10^6 \text{ N} \cdot \text{m/rad}$$

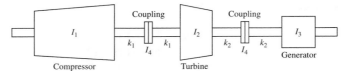

FIGURE P8.34

8.35 A model of a four-story building is shown in Figure P8.35, where each floor is modeled as a lumped mass. The supporting columns are modeled as being capable of deflecting only in shear and have the following stiffnesses in N/m:

$$k_1 = 4 \times 10^8 \qquad k_2 = 3 \times 10^8$$
$$k_3 = 2 \times 10^8 \qquad k_4 = 1 \times 10^8$$

The floor masses are each equal to 10^6 kg. Determine the natural frequencies and mode shapes.

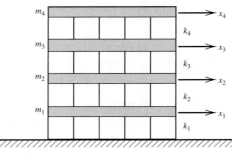

FIGURE P8.35

8.36 Each spring in Figure P8.36 has a stiffness of 4000 N/m. Each mass is 0.5 kg.

(a) Obtain the modal differential equations.

(b) Plot the solutions of the modal equations for the case where the masses are initially at rest and $x_1(0) = x_2(0) = 0$, $x_3(0) = 0.1$ m/s.

(c) Use the results of part (b) to obtain plots of $x_1(t)$, $x_2(t)$, and $x_3(t)$.

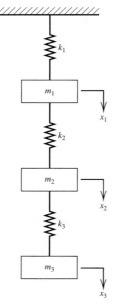

FIGURE P8.36

8.37 Use MATLAB to determine **L** so that $\mathbf{LL}^T = \mathbf{M}$ for the following matrix.

$$\mathbf{M} = \begin{bmatrix} 10 & 5 & 0 \\ 5 & 12 & 4 \\ 0 & 4 & 25 \end{bmatrix}$$

8.38 Consider the two-mass system shown in Figure 8.5-1. Suppose the parameter values are $m_1 = 10$, $m_2 = 4$, $c_1 = 3$, $c_2 = 7$, $k_1 = 4$, and $k_2 = 6$. Put the equations of motion into state-variable form, and express the state-variable model in vector-matrix form.

8.39 Obtain the transfer functions $X_1(s)/F(s)$ and $X_2(s)/F(s)$ of the state-variable model obtained in Problem 8.38.

8.40 Use MATLAB to plot the free response $x_1(t)$ of the state model derived in Example 8.5-3, for $x_1(0) = 7$, $\dot{x}_1(0) = -3$, $x_2(0) = 5$, and $\dot{x}_2(0) = 1$.

8.41 Use MATLAB to plot the free response $x_1(t)$ of the state model derived in Problem 8.38, for $x_1(0) = 3$, $\dot{x}_1(0) = -5$, $x_2(0) = 6$, and $\dot{x}_2(0) = 1$.

8.42 Consider the four-story building model discussed in Problem 8.35. The horizontal ground motion during an earthquake can be decomposed into a spectrum of frequencies, and it is therefore useful to have frequency response plots for the building. Taking the input to the $y(t)$, the horizontal ground displacement, use MATLAB to obtain the magnitude and the phase plots for each floor of the building, using the parameter values given in Problem 8.35.

SECTION 8.6 SIMULINK APPLICATIONS

8.43 Consider the system discussed in Example 8.5-2. Develop a Simulink model to plot the unit-step response of the variables x_1 and x_2 with the initial conditions $x_1(0) = 0.5$, $\dot{x}_1(0) = -0.2$, $x_2(0) = 0.3$, $\dot{x}_2(0) = 0.4$.

8.44. Consider the system shown in Figure P8.44. The equations of motion are

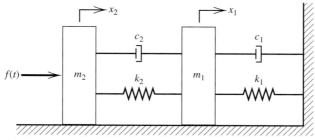

FIGURE P8.44

$$m_1\ddot{x}_1 + (c_1 + c_2)\dot{x}_1 + (k_1 + k_2)x_1 - c_2\dot{x}_2 - k_2x_2 = 0$$

$$m_2\ddot{x}_2 + c_2\dot{x}_2 + k_2x_2 - c_2\dot{x}_1 - k_2x_1 = f(t)$$

Suppose that $m_1 = m_2 = 1$, $c_1 = 3$, $c_2 = 1$, $k_1 = 1$, and $k_2 = 4$.

(a) Develop a Simulink simulation of this system. In doing this, consider whether or not to obtain the transfer functions $X_1(s)/F(s)$ and $X_2(s)/F(s)$.

(b) Use the Simulink program to plot the response $x_1(t)$ for the following input. The initial conditions are zero.

$$f(t) = \begin{cases} t & 0 \le t \le 1 \\ 2 - t & 1 < t < 2 \\ 0 & t \ge 2 \end{cases}$$

VIBRATION MEASUREMENT AND TESTING

THERE WILL be applications where it is difficult to develop a differential equation model of the system from basic principles such as Newton's laws. It can even be difficult to develop an analytical model of the natural frequencies. In such cases we must resort to using measurements of the system response.

Vibration measurement and testing require a knowledge of the hardware available to produce vibration of the system under test and to measure the response. Knowledge of the algorithms and software available for processing the data is also necessary.

This chapter outlines the equipment requirements and the procedures that are available for these tasks.

LEARNING OBJECTIVES

After you have finished this chapter, you should be able to do the following:

■ Analyze the step and the frequency response to identify the form and the parameter values of a model.

■ Describe the commonly used equipment for stimulating a vibratory response and for collecting response data.

■ Perform a transform analysis of signal data to estimate natural frequencies.

■ Apply MATLAB and Simulink to the methods of the chapter.

9.1 SYSTEM IDENTIFICATION

In the previous chapters we have demonstrated how to apply the principles of dynamics to obtain a model. However, this is not always possible either because the system's dynamics are not understood well enough or because we may have a limited time in which to develop a model. In such circumstances engineers resort to testing the system. Note, however, that if the system is still under design, it cannot be tested, and a preliminary model must be developed to support the design process.

Assuming we can test the system, the basic information we often need to determine from a vibration test includes the following:

- The number of degrees of freedom of the system under test
- The natural or resonant frequencies of the system
- The mode shapes
- The damping associated with each mode

This information gives a general understanding of the system, but to design and analyze systems we need to be able to predict their response, and thus we may also need to have a specific model, including numerical values for the parameters in the model.

The process of obtaining a mathematical model from testing is sometimes called *system identification. Parameter identification* describes the situation where we already know the *form* of the model but need to obtain values for its parameters.

All real data will have some "scatter," and thus a perfect model fit will not be practical. However, to illustrate the methods clearly, in the following examples we use data that have very little scatter, and thus the derived models are unambiguous.

For the linear single-DOF model $m\ddot{x} + c\dot{x} + kx = f(t)$, whose frequency transfer function is

$$T(i\omega) = \frac{X(i\omega)}{F(i\omega)} = \frac{1}{k - m\omega^2 + c\omega i} \tag{9.1-1}$$

we need to know the values of m, c, and k. Throughout this text we have developed several methods for estimating parameter values. These include the following:

- Curve-fitting techniques using data from static tests to estimate the stiffness k and data from dynamic tests to estimate the damping c. See Chapter 1.
- Use of the logarithmic decrement to estimate c and k from measurements of oscillatory free response. See Chapter 3.
- Use of hysteresis data to estimate c. See Chapter 4.
- Use of measurements of the step response to estimate m, c, and k. See Example 9.1-1 in this section.
- Use of experimentally determined frequency response plots. See Examples 9.1-2 and 9.1-3 in this section.

EXAMPLE 9.1-1
Estimating Mass, Stiffness, and Damping from the Step Response

Figure 9.1-1 shows the response of a system to a step input of magnitude 6×10^3 N. The equation of motion is

$$m\ddot{x} + c\dot{x} + kx = f(t)$$

Estimate the values of m, c, and k.

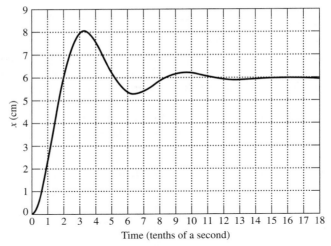

FIGURE 9.1-1 Plot of measured step response.

Solution

From the graph, we see that the steady-state response is $x_{ss} = 6$ cm. At steady state, $x_{ss} = f_{ss}/k$, and thus $k = 6 \times 10^3/6 \times 10^{-2} = 10^5$ N/m.

The peak value from the plot is $x = 8.1$ cm, so the maximum percent overshoot is $M_p = [(8.1 - 6)/6]100 = 35\%$. From Table 5.3-2, we may compute the damping ratio as follows:

$$A = \ln \frac{100}{35} = 1.0498 \qquad \zeta = \frac{A}{\sqrt{\pi^2 + A^2}} = 0.32$$

The peak occurs at $t_p = 0.32$ s. From Table 5.3-2,

$$t_p = 0.32 = \frac{\pi}{\omega_n \sqrt{1 - \zeta^2}} = \frac{3.316}{\omega_n}$$

Thus $\omega_n^2 = 107$ and

$$m = \frac{k}{\omega_n^2} = \frac{10^5}{107} = 930 \text{ kg}$$

From the expression for the damping ratio,

$$\zeta = \frac{c}{2\sqrt{mk}} = \frac{c}{2\sqrt{930(10^5)}} = 0.32$$

Thus $c = 6170$ N·s/m, and the model is

$$930\ddot{x} + 6170\dot{x} + 10^5 x = f(t) \qquad \blacksquare$$

Parameter Identification from Frequency Response

In cases where a transfer function or differential equation model is difficult to derive from general principles, or where the model's coefficient values are unknown, often an experimentally obtained frequency response plot can be used to determine the form of an appropriate model and the values of the model's coefficients.

Often a sinusoidal input is easier to apply to a system than a step input, because many devices, such as ac circuits and rotating machines, naturally produce a sinusoidal signal or motion. If a suitable apparatus can be devised to provide a sinusoidal input of adjustable frequency, then the system's output amplitude and phase shift relative to the input can be measured for various input frequencies. When these data are plotted on the logarithmic plot for a sufficient frequency range, the form of the model can often be determined. This procedure is easiest for systems with electrical inputs and outputs, because for these systems, variable-frequency oscillators and frequency response analyzers are commonly available. Some of these can automatically sweep through a range of frequencies and plot the decibel and phase angle data.

When using frequency response data for identification, it is important to understand how to reconstruct a transfer function form from the asymptotes and corner frequencies. The slopes of the asymptotes can be used to determine the orders of the numerator and the denominator. The corner frequencies can be used to estimate time constants and natural frequencies.

In theory we can determine the values of m, c, and k by identifying several features on the magnitude plot of $T(i\omega)$, given by Equation 9.1-1, as follows. Recall that the damping ratio is defined as $\zeta = c/2\sqrt{mk}$.

- The low-frequency asymptote of $|T(i\omega)|$ is $|T(0)| = 1/k$, from which we can obtain the value of k by measuring $|T(0)|$.

- The frequency corresponding to the peak value on the plot is

$$\omega_p = \sqrt{\frac{k}{m}}\sqrt{1 - 2\zeta^2} \qquad \text{if } 0 \le \zeta \le 0.707 \tag{9.1-2}$$

- The peak response is

$$X_p = \frac{F}{k}\frac{1}{2\zeta\sqrt{1 - \zeta^2}} \tag{9.1-3}$$

If ζ is small,

$$\frac{X_p}{F} \approx \frac{1}{2k\zeta} = \frac{1}{c}\sqrt{\frac{m}{k}} \tag{9.1-4}$$

So if we know the ratio X_p/F, we have a relation to use for c.

- At the natural frequency ($\omega = \omega_n$),

$$|T(i\omega_n)| = \frac{1}{c\omega_n} = \frac{1}{c}\sqrt{\frac{m}{k}} \tag{9.1-5}$$

So, if we know ω_n, this relation gives us a way of estimating c by measuring $|T(i\omega_n)|$.

- The bandwidth relation (Equation 4.2-18) may be manipulated to show that for ζ small,

$$\zeta \approx \frac{1}{2}\frac{\omega_2 - \omega_1}{\omega_p} \tag{9.1-6}$$

where ω_1 and ω_2 are the lower and upper bandwidth frequencies. Thus, by measuring the bandwidth and the peak frequency, we can estimate ζ.

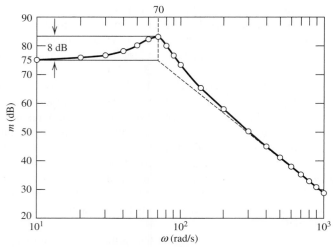

FIGURE 9.1-2 Plot of measured frequency response.

EXAMPLE 9.1-2	Measured response data are shown by the small circles in Figure 9.1-2. Determine the transfer function.
Identifying a System from Its Frequency Response	

Solution

After drawing the asymptotes shown by the dashed lines, we first note that the data have a low-frequency asymptote of zero slope and a high-frequency asymptote of slope -40 dB/decade. This suggests a second-order model without numerator dynamics, either of the overdamped form

$$T(s) = \frac{A}{(\tau_1 s + 1)(\tau_2 s + 1)}$$

or the underdamped form

$$T(s) = \frac{A}{s^2 + 2\zeta\omega_n s + \omega_n^2}$$

However, the peak in the data eliminates the overdamped form.

At low frequencies, $m \approx 20 \log A$. From the plot, at low frequency, $m = 75$ dB. Thus $75 = 20 \log A$, which gives

$$A = 10^{75/20} = 5623$$

The peak is estimated to be 83 dB. The peak when $A = 1$ is given by $m_p = -20 \log(2\zeta\sqrt{1 - \zeta^2})$. Thus, with $A = 5623$, the formula for the peak becomes

$$m_p = 20 \log 5623 - 20 \log(2\zeta\sqrt{1 - \zeta^2})$$

or

$$83 = 75 - 20 \log(2\zeta\sqrt{1 - \zeta^2})$$

Thus

$$\log(2\zeta\sqrt{1 - \zeta^2}) = \frac{75 - 83}{20} = -0.4$$

and

$$2\zeta\sqrt{1-\zeta^2} = 10^{-0.4}$$

Solve for ζ by squaring both sides:

$$4\zeta^2(1-\zeta^2) = 10^{-0.8}$$

$$4\zeta^4 - 4\zeta^2 + 10^{-0.8} = 0$$

This gives $\zeta^2 = 0.9587$ and 0.0413. The positive solutions are $\zeta = 0.98$ and 0.2. Because there is a resonance peak in the data, the first solution is not valid, and we obtain $\zeta = 0.2$.

Knowing ζ, we can now estimate ω_n from the peak frequency, which is estimated to be $\omega_p = 70$ rad/s. Thus, $\omega_p = \omega_n\sqrt{1-2\zeta^2}$, or

$$70 = \omega_n\sqrt{1 - 2(0.2)^2}$$

This gives $\omega_n = 73$ rad/s.

Thus the estimated model is

$$T(s) = \frac{5623}{s^2 + 29.2s + 5329}$$

∎

EXAMPLE 9.1-3
Identification of a Higher-Order System

Figure 9.1-3 shows the magnitude and phase plots determined from experiment. Estimate the DOF of the system, its resonant frequencies, and its modal damping ratios.

Solution

From the magnitude plot, we see that there are two peaks, one at 5 rad/s and the other at 50 rad/s. These are the resonant frequencies. The number of peaks does not necessarily equal the DOF. For example, a system with three DOF will be sixth order, and it could have two resonant peaks and either two real roots or a complex pair whose damping ratio is slightly less than 1. Here the phase angle is 180° at low frequencies and approaches −180° at high frequencies. This suggests that this system has a second-order numerator and a fourth-order denominator. Thus the system has two DOF.

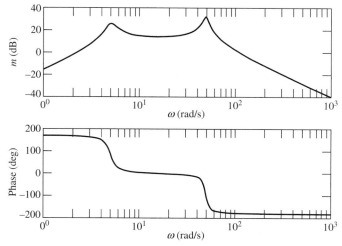

FIGURE 9.1-3 Magnitude and phase plots of measurements.

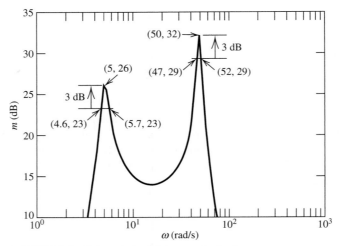

FIGURE 9.1-4 Enlarged view of the magnitude plot.

To estimate the modal damping ratios we use Equation 9.1-6. Figure 9.1-4 gives an enlarged view showing the two bandwidths of the system. For the peak at 5 rad/s we have

$$\zeta_1 = \frac{5.7 - 4.6}{2(5)} = 0.11$$

Similarly, for the second peak,

$$\zeta_2 = \frac{52 - 47}{2(50)} = 0.05 \qquad \blacksquare$$

The above methods, however, do not always produce useful results, and we need to be able to use more general methods. There are many reasons why this is so. For example, the model given by Equation 9.1-1 describes a linear, lumped-parameter, single-DOF system, but this model may not be sufficient to describe the real system under test for any of the following reasons:

- The system under test may not be accurately described by a single-DOF model. It may actually have more than one DOF.

- Real systems are not lumped-parameter systems; they are distributed-parameter systems because the mass, stiffness, and damping are not located at discrete points. For example, in Chapter 2 we developed a lumped-parameter, single-DOF model of a cantilever beam by using kinetic energy equivalence. This model predicts a single natural frequency and mode shape. As shown in Chapter 10, distributed-parameter systems have an infinite number of degrees of freedom and therefore an infinite number of natural frequencies and mode shapes.

- Real systems are nonlinear. Although many systems can be described well with linear models, this is not always the case.

In addition, it may be difficult to estimate m, c, and k with the above methods because of the following:

- In some systems the damping is so large that the frequency response plot has no peak and the free response dies out too quickly to obtain useful measurements or does not oscillate at all.

- It may be time consuming or difficult to apply a harmonic input at a large enough number of frequencies to obtain a good frequency response plot.

- Noise in the measurements may prevent us from obtaining a good frequency response plot.

In this chapter we present more general methods for obtaining the transfer function model of a system. The methods are based on random process theory, which we will develop. These methods are based on the fact that the transfer function is the ratio of the transform of the response to the transform of the forcing function.

Tests based on these methods can use a harmonic forcing function (often generated by a device called a *shaker*) to excite the system over a range of frequencies with a known input amplitude and phase. Another method uses an *impact hammer* to apply a sudden force to the structure, thus exciting what is equivalent to the free response. Both the response—which may be acceleration, velocity, or displacement—and the applied force are measured. The transforms of the response and the forcing function are calculated from these measurements, and the frequency transfer function is calculated from their ratio. Because most measurements contain noise, statistical methods are applied to the data.

In many applications we simply need to estimate the natural frequencies of the system, with no need for a more detailed model. In the remainder of this chapter we will develop methods for this purpose.

Vibration Exciters

The following components are required for a vibration test:

- An actuator called an *exciter* to create motion of or apply force to the object under test.

- *Sensors* to measure the required signals. Also called *transducers*, sensors convert the physical variable being measured (displacement, velocity, acceleration, or force) into another type of variable (usually an electrical signal).

- A data-acquisition system to acquire and process the signals from the sensors.

- An *analyzer* to analyze the processed signals and output the required information.

This section gives an overview of available means to produce motion of the test object or to apply a desired force to the object. The type of actuator used depends somewhat on what information is desired. For example, if we want to identify the transfer function of a system having a displacement input, such as a vehicle suspension, we would want to use an actuator that produces a base motion that is a prescribed function of time in order to reproduce the characteristics of a road surface. So the type of transfer function we are attempting to identify partly determines the type of actuator to be used. If we want to identify a transfer function $X(s)/F(s)$ that is a ratio of output displacement to input force, we would select an actuator that produces force and a sensor that measures displacement.

Impact Hammers

The most common exciters are the *impact hammer* and the *shaker*. Impact hammers are convenient and relatively inexpensive compared to other actuators. In addition, because they are not attached to the test object, they do not change its dynamics. The impact hammer (also called an *impulse hammer*) resembles an ordinary hammer, but it has a specially designed tip that contains a sensor for measuring the impact force. If the test object is struck crisply with the hammer, the applied force is a pulse that resembles an impulse. This is the advantage of using an impact hammer, because an impulse simplifies the analysis while exciting all of the object's natural frequencies.

By adding specially designed weights to the hammer, and by using tips with different hardness, we may adjust the applied force and the duration of the pulse. A harder tip generates a force having a higher frequency content (which enables you to analyze response at higher frequency), but it transfers less energy to the test object (which means the response may be too small to measure accurately).

Shakers

A shaker applies a time-varying force to the test object either by hydraulic, piezoelectric, or electromagnetic means. The electromagnetic type is based on the same principle as a speaker. A time-varying current is applied to an electromagnet, and the resulting magnetic force moves the shaker. Because a variety of time functions may be easily generated electronically under computer control, such a shaker is capable of generating a variety of time-varying forces, such as a sinusoidal force of constant frequency, a *swept sine* force whose frequency is gradually increased while keeping its amplitude constant, a randomly varying force, or a transient force that reproduces the effects of a shock input.

With some shakers the test object is mounted to the shaker, and thus the shaker may change the test object's dynamics by increasing its mass. Other shakers use a short, thin, but stiff rod called the stinger to transmit force to the object. The stinger helps to isolate the test object from the shaker mass and makes it easier to control the direction and point of application of the force. Small shakers are a little more expensive than impact hammers, while huge and expensive shakers exist for testing vehicles and even buildings.

9.2 TRANSFORM ANALYSIS OF SIGNALS

We have seen that a periodic signal may be expressed as a Fourier series of harmonic functions. The Fourier series coefficients, when plotted versus frequency, give a plot called the *spectrum* of the signal. The spectrum graphically displays the frequency content of the signal. If the signal is nonperiodic, however, the Fourier series cannot be used to represent the signal.

Nonperiodic Functions

Consider the series of rectangular pulses of unit height and width 2 shown in Figure 9.2-1. The period is $2p$ and the function is even, so no sine terms appear in its Fourier series. The series and its coefficients are

$$x(t) = \frac{a_0}{2} + a_1 \cos \frac{\pi t}{p} + a_2 \cos \frac{2\pi t}{p} + \cdots + a_n \cos \frac{n\pi t}{p} + \cdots$$

$$a_0 = \frac{2}{p}$$

$$a_n = \frac{2}{p} \frac{\sin(n\pi/p)}{n\pi/p} \quad n = 1, 2, \ldots$$

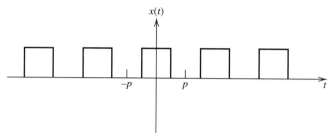

FIGURE 9.2-1 A series of rectangular pulses.

Consider what happens if the period increases. This is shown for four cases in Figure 9.2-2, in which $p = 2, 4, 8$, and ∞. The corresponding spectra, $|a_n|$ versus $n\pi/p$, are shown in Figure 9.2-3 for $p = 2, 4$, and 8. The difference between adjacent frequencies is

$$\Delta\omega = \frac{(n+1)\pi}{p} - \frac{n\pi}{p} = \frac{\pi}{p}$$

As we increase p, the frequencies get closer together and the height of the spectrum decreases. If we let $p \to \infty$, the series of pulses becomes one pulse, which is a nonperiodic function; there are an infinite number of frequencies, and the spectrum becomes a continuous function of frequency.

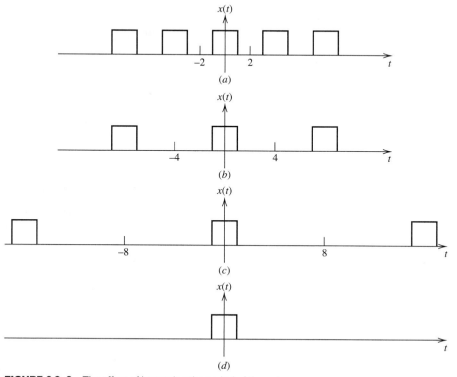

FIGURE 9.2-2 The effect of increasing the period of the pulse.

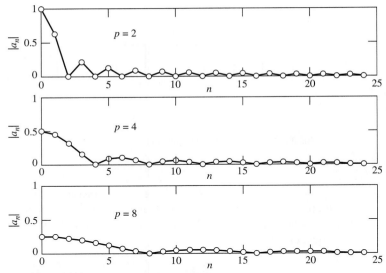

FIGURE 9.2-3   Spectra corresponding to Figure 9.2-2.

The Fourier Transform

An extension of the Fourier series to nonperiodic functions is the *Fourier transform pair*. We will not derive the following relations, but they can be shown to arise from the complex exponential form of the Fourier series as we let the period approach infinity. We express the time signal as

$$x(t) = \int_{-\infty}^{\infty} a(\omega) \cos \omega t \, d\omega + \int_{-\infty}^{\infty} b(\omega) \sin \omega t \, d\omega \tag{9.2-1}$$

where

$$a(\omega) = \frac{1}{2\pi} \int_{-\infty}^{\infty} x(t) \cos \omega t \, dt \tag{9.2-2}$$

$$b(\omega) = \frac{1}{2\pi} \int_{-\infty}^{\infty} x(t) \sin \omega t \, dt \tag{9.2-3}$$

The Fourier transform of $x(t)$ is

$$X(\omega) = a(\omega) - ib(\omega) \tag{9.2-4}$$

These equations are extensions of the Fourier series summation formulas. Because $x(t)$ is real, $a(\omega)$ and $b(\omega)$ will be real, but Equation 9.2-4 shows that $X(\omega)$ will be complex if $b(\omega) \neq 0$.

An equivalent form of the Fourier transform uses complex notation. It is given by

$$x(t) = \int_{-\infty}^{\infty} X(\omega) e^{i\omega t} \, d\omega \tag{9.2-5}$$

and

$$X(\omega) = \frac{1}{2\pi} \int_{-\infty}^{\infty} x(t) e^{-i\omega t} \, dt \tag{9.2-6}$$

The two forms are equivalent because $e^{-i\omega t} = \cos \omega t - i \sin \omega t$.

Signal Spectrum

The spectrum of a nonperiodic signal is the magnitude of its Fourier transform, that is, $|X(\omega)|$. The transform has a number of properties, including linearity and the *time-shifting* property. This property states that the transform of the time-shifted function $x(t + D)$ is $X(\omega)e^{i\omega D}$. An immediate implication of this property is that the spectrum of the shifted signal $x(t + D)$ is the same as the spectrum of $x(t)$. This is true because

$$\left|X(\omega)e^{i\omega D}\right| = |X(\omega)|\left|e^{i\omega D}\right| = |X(\omega)|$$

where $\left|e^{i\omega D}\right| = \sqrt{\cos^2 \omega D + \sin^2 \omega D} = 1$. The time shift D affects only the phase angle of the transform.

Note also that Equation 9.2-6 implies that the transform is symmetric about $\omega = 0$. That is,

$$X(-\omega) = X(\omega)$$

EXAMPLE 9.2-1
Fourier Transform of a Rectangular Pulse

Obtain the Fourier transform and the spectrum of the rectangular pulse function of height H and duration T, shown in Figure 9.2-4a. The pulse is defined by

$$x(t) = \begin{cases} 0 & t < 0 \\ H & 0 \le t \le T \\ 0 & t > T \end{cases}$$

Solution

For the complex form of the transform, Equation 9.2-6 gives

$$X(\omega) = \frac{1}{2\pi}\int_{-\infty}^{0} 0 e^{-i\omega t}\, dt + \frac{1}{2\pi}\int_{0}^{T} H e^{-i\omega t}\, dt + \frac{1}{2\pi}\int_{T}^{\infty} 0 e^{-i\omega t}\, dt$$

$$= -\frac{H}{2\pi i\omega}\left(e^{-i\omega T} - 1\right)$$

Applying the identity $e^{-i\omega T} = \cos \omega T - i \sin \omega T$, we obtain

$$X(\omega) = \frac{H}{2\pi\omega}\left[\sin \omega T + i\left(\cos \omega T - 1\right)\right]$$

The spectrum of the pulse is given by

$$|X(\omega)| = \frac{H}{2\pi\omega}\sqrt{\sin^2 \omega T + (\cos \omega T - 1)^2} = \frac{\sqrt{2}H}{2\pi\omega}\sqrt{1 - \cos \omega T} = \frac{H}{\pi\omega}\left|\sin \frac{\omega T}{2}\right|$$

where we have used the identities $\sin^2 \omega T + \cos^2 \omega T = 1$ and $1 - \cos \omega T = 2\sin^2(\omega T/2)$. The spectrum is shown in Figure 9.2-4b. From the shifting theorem, we see that this is also the spectrum of the shifted pulse shown in Figure 9.2-4c. ∎

Relationship of the Fourier and Laplace Transforms

The Fourier transform of $x(t)$ is denoted as $X(\omega) = \mathcal{F}[x(t)]$ and is related to the Laplace transform $X(s) = \mathcal{L}[x(t)]$ as follows, if $x(t) = 0$ for $t < 0$:

$$\mathcal{F}[x(t)] = \frac{1}{2\pi}\mathcal{L}[x(t)]\Big|_{s=i\omega} \tag{9.2-7}$$

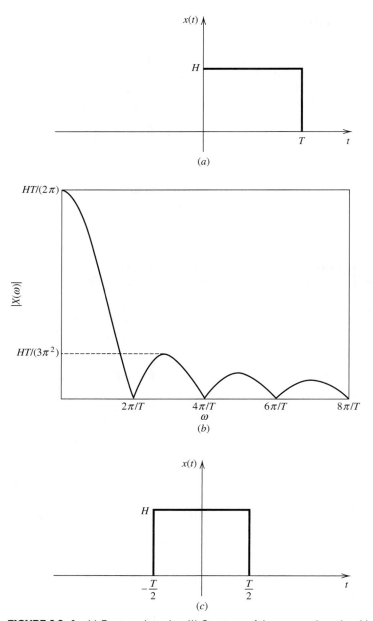

FIGURE 9.2-4 (*a*) Rectangular pulse. (*b*) Spectrum of the rectangular pulse. (*c*) Shifted pulse.

That is, we may obtain the Fourier transform by replacing s with $i\omega$ in the Laplace transform. Once you understand this, the following simpler notation should cause no confusion:

$$X(\omega) = \frac{1}{2\pi} X(s)\bigg|_{s=i\omega} \tag{9.2-8}$$

where $X(\omega)$ is the Fourier transform and $X(s)$ is the Laplace transform. In this notation the Fourier transform $X(\omega)$ does not contain i in the argument and is thereby distinguished from the Laplace transform with s replaced by $i\omega$, which is $X(i\omega)$. For example, the Laplace transform of the rectangular pulse shown in Figure 9.2-4a is

$$X(s) = \frac{H}{s}\left(1 - e^{-sT}\right)$$

Thus its Fourier transform is

$$X(\omega) = \frac{H}{2\pi i\omega}\left(1 - e^{-i\omega T}\right) = \frac{H}{2\pi\omega}\left[\sin \omega T + i(\cos \omega T - 1)\right]$$

which is the same result we obtained by integration in Example 9.2-1.

EXAMPLE 9.2-2 *Fourier Transform* *of an Exponential* *Signal*	Obtain the Fourier transform and the spectrum of the exponentially decaying signal defined by $$x(t) = \begin{cases} 0 & t < 0 \\ Be^{-at} & t \geq 0 \end{cases}$$
Solution	The Laplace transform of this function is $$X(s) = B\frac{1}{s + a}$$ Thus the Fourier transform is $$X(\omega) = \frac{B}{2\pi}\frac{1}{i\omega + a} = \frac{B}{2\pi}\frac{a - i\omega}{a^2 + \omega^2}$$ The spectrum is $$|X(\omega)| = \frac{B}{2\pi}\frac{1}{\sqrt{a^2 + \omega^2}}$$ Note that the time constant of the signal is $\tau = 1/a$, so the spectrum may be expressed as $$|X(\omega)| = \frac{B}{2\pi}\frac{\tau}{\sqrt{1 + \omega^2\tau^2}}$$ The spectrum is graphed in Figure 9.2-5 for $B = 1/2\pi$ and for several values of the parameter τ. The peak occurs at $\omega = 0$ and equals τ. A signal that disappears slowly (having a large time constant) has a spectrum with a higher and more narrow peak. The spectrum of a quickly decaying signal is flatter. ∎
EXAMPLE 9.2-3 *Fourier Transform* *of a Decaying* *Sinusoid*	Obtain the Fourier transform of the following signal: $$x(t) = \begin{cases} 0 & t < 0 \\ De^{-at}\sin bt & t \geq 0 \end{cases}$$
Solution	The Laplace transform of $x(t)$ is $$X(s) = D\frac{b}{(s + a)^2 + b^2}$$ So, replacing s with $i\omega$ gives the Fourier transform: $$X(\omega) = \frac{D}{2\pi}\frac{b}{(i\omega + a)^2 + b^2} = \frac{D}{2\pi}\frac{b}{a^2 + b^2 - \omega^2 + 2a\omega i}$$

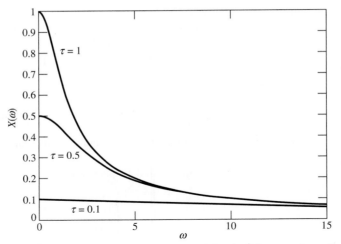

FIGURE 9.2-5 Spectrum of an exponential signal $e^{t/\tau}$ for two values of τ.

The magnitude gives the spectrum of the signal:

$$|X(\omega)| = \frac{D}{2\pi} \frac{b}{\sqrt{(a^2 + b^2 - \omega^2)^2 + 4a^2\omega^2}} \tag{1}$$

This is plotted in Figure 9.2-6, with $D = 1/2\pi$ for three cases:

1. $x(t) = e^{-t} \sin 50t$ $(a = 1, \quad b = 50)$
2. $x(t) = e^{-2t} \sin 50t$ $(a = 2, \quad b = 50)$
3. $x(t) = e^{-2t} \sin 20t$ $(a = 2, \quad b = 20)$

Note that each curve has a peak and that the height of the peak is identical for identical values of a. We can use calculus to find the height and location of the peak in terms of the parameters a and b. Solving $d|X(\omega)|/d\omega = 0$ for ω shows that the peak occurs at the frequency

$$\omega = \sqrt{b^2 - a^2} \tag{2}$$

and has a height of $0.5D/2\pi a$.   ■

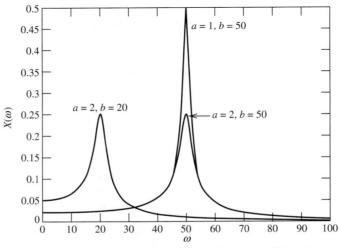

FIGURE 9.2-6 Spectrum of a decaying sinusoid $x(t) = e^{-at}\sin bt$ for two values of a and b.

Estimating Frequencies from the Spectrum

The function $x(t) = De^{-at} \sin bt$ has the form of the free response of a system with one DOF, for nonzero initial velocity and zero initial displacement. The system can be excited in this way by an impulse hammer. An impulsive force has the effect of creating the initial conditions $x(0) = 0$ and $m\dot{x}(0) = A$, where A is the strength of the impulse (the area under its force-versus-time curve). The impulse response may be expressed as

$$x(t) = De^{-t/\tau} \sin \omega_d t = (\dot{x}(0)/\omega_d)e^{-t/\tau} \sin \omega_d t$$

where τ is the time constant and ω_d is the damped natural frequency. A large time constant means that the oscillations take a while to decay and thus a large time constant is associated with a lightly damped system.

From Equation 1 of Example 9.2-3, the peak height of the signal spectrum is $0.5D/2\pi a$. Here the time constant is $1/a$, and thus we see that the height of the peak is $0.5D/2\pi a = 0.5D\tau/2\pi$ and so is proportional to the time constant. Thus a lightly damped system will have a large peak.

The location of the peak may be expressed as

$$\omega = \sqrt{b^2 - a^2} = \sqrt{\omega_d^2 - 1/\tau^2} = \omega_n\sqrt{1 - 2\zeta^2}$$

since $\tau = 1/\zeta\omega_n$. Thus, if the system is lightly damped (ζ is small), then the spectrum peak is located approximately at the undamped natural frequency ω_n.

This example shows that we can identify the natural frequency ω_n of a lightly damped signal from the location of the peak in the Fourier transform magnitude computed from the measured impulse response. Later in this chapter we will show how the transform can be computed in a data-acquisition system. Although the theory seems straightforward, in practice, identifying the time constant τ from the height of the peak is not usually done. Not only is the height of the peak difficult to determine due to noise in the measurements, but the height depends on D, which is a function of the strength of the impulsive input. If the strength is unknown, then we cannot use a measurement of the height of the peak to estimate a and the time constant τ.

The complete free response of a single-DOF underdamped system has the form

$$x(t) = D_1 e^{-t/\tau} \sin \omega_d t + D_2 e^{-t/\tau} \cos \omega_d t$$

where D_1 and D_2 depend on the specific values of the initial conditions. The Fourier transform of the cosine function may be derived in a manner similar to that used in Example 9.2-3. Its transform also has a peak at approximately $\omega = \omega_n$ if the system is slightly damped. Since $x(t)$ is a linear combination of the sine and cosine terms, the transform of $x(t)$ will also have a peak near $\omega = \omega_n$. So this method of identifying ω_n may be used even if the initial displacement is nonzero.

For a linear multi-DOF system with underdamped modes, each having the form $A_n e^{-t/\tau_n} \sin \omega_{dn} t + B_n e^{-t/\tau_n} \cos \omega_{dn} t$, the linearity property of the Fourier transform means that the spectrum will be the sum of the spectra for each mode. Thus the overall spectrum will display peaks nears the modal frequencies.

An advantage of using the Fourier transform method to identify frequencies is illustrated in Figure 9.2-7. The top graph shows the following decaying sinusoidal signal, whose frequency is 20 Hz:

$$x(t) = 2e^{-t} \sin 40\pi t$$

The middle graph shows the noise present in the measured signal. The bottom graph shows the measured signal, which is the sum of $x(t)$ and the noise. Observe that it would be

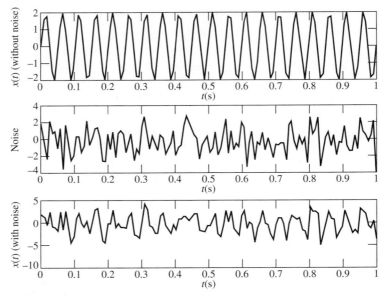

FIGURE 9.2-7 Sinusoidal signal and noise. (*a*) The pure signal. (*b*) Noise present in the signal. (*c*) The signal with additive noise present.

difficult to estimate parameters such as ω_n from these data. However, the spectrum, shown in Figure 9.2-8, clearly displays the peak at 20 Hz. In a data-acquisition application the transform is computed in software or by special-purpose computers called *spectrum analyzers*. Later in this chapter we will present a numerical method for doing this.

Power Spectrum

Some software packages and signal analyzers use vibration data to produce a plot called the *power spectrum* or a similar name, such as *power spectral density*. It is a plot, versus frequency, of a term that is proportional to the square of the spectrum amplitude $X(\omega)$.

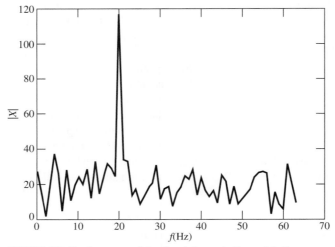

FIGURE 9.2-8 Spectrum of the signal shown in Figure 9.2-7*c*.

One such plot is a plot of $X^*(\omega)X(\omega)$ versus ω, where $X^*(\omega)$ is the complex conjugate of the Fourier transform $X(\omega)$. The term *power* derives from the fact that in some applications, the power transmitted is proportional to the square of the signal, such as with an electrical resistor in which the power dissipated is i^2R. Depending on the signal being processed, the power spectrum might or might not have units of power.

9.3 RANDOM PROCESSES

Until now we have treated only the response caused by inputs that are completely determined; for example, step functions, impulse functions, and harmonic functions. There is nothing uncertain about them. In many applications, however, the inputs are not well known and can be described only in terms of statistical measures such as their mean value. Examples include (1) wind gusts acting on buildings, bridges, aircraft, and antennas, (2) wave forces, and (3) ground motion during an earthquake. While these inputs are not truly random, it turns out that it is useful to treat them as if they were random.

A "random" input will generate a "random" response, and the output signals from any sensors measuring the input and the response will appear to be random. Random signals, like the example shown in the middle plot of Figure 9.2-7, have no apparent pattern and never repeat. Just because a signal does not repeat, however, does not mean it is random. For example, a signal behaving like the function $x(t) = t^2$ does not repeat but is not random. Another characteristic of a random signal is that it is impossible to predict what the signal will be in the future, even if we have the past values of the signal.

Some nonperiodic signals may appear to be random but are not. For example, a signal consisting of the sum of harmonic functions the ratio of whose frequencies is an irrational number will be nonperiodic, but it is not random because the function's value at any time t can be exactly predicted from the formula. An example is the function $x(t) = \sin 2t + \sin \sqrt{3}t$. Note also that the fact that a random signal does not repeat means that it cannot be expressed as a Fourier series.

Although we speak of "random" signals, it is a random *process* that produces the signal. Once measured and recorded, the signal is no longer random because it is now completely determined.

Expected Value

The *average* is also called the *mean* or *expected value*. For discrete values, such as those produced by digital data acquisition, the mean is defined as

$$E(x) = \mu = \frac{1}{n}\sum_{j=1}^{n} x_j \tag{9.3-1}$$

The mean is often represented by the symbol μ. For a continuous function $x(t)$, the mean is

$$E[x(T_d)] = \frac{1}{T_d}\int_0^{T_d} x(t)\,dt \tag{9.3-2}$$

where the time duration of the data sample is T_d.

Variance

Two signals may have the same mean value but one may fluctuate with greater amplitude about the mean. So we also need a measure of the range of fluctuation. Simply specifying the minimum and maximum values is insufficient because a single large fluctuation (above or below the mean) can be misleading. A measure that indicates the spread about the mean without regard for whether the fluctuations are above or below the mean is the *variance*, which is defined as the average value of the square of the difference between the signal and its mean. The variance is calculated for a discrete signal as follows:

$$\text{var}(x) = \sigma^2 = \frac{1}{n} \sum_{j=1}^{n} (x_j - \mu)^2 \tag{9.3-3}$$

where σ^2 is the variance and σ is called the *standard deviation*.

For a signal continuous in time, the variance is calculated from

$$\text{var}(x, T_d) = \frac{1}{T_d} \int_0^{T_d} [x(t) - \mu]^2 \, dt \tag{9.3-4}$$

The *mean-square* value of x is the expected value of x^2 and is denoted $E(x^2)$. The *root-mean-square (rms)* value of x is $\sqrt{E(x^2)}$. The relation between the variance, the mean square, and the mean is as follows:

$$\sigma^2 = E(x^2) - [E(x)]^2 \tag{9.3-5}$$

If the mean of x is zero, the standard deviation σ_x of x is calculated from

$$\sigma_x = \sqrt{E(x^2)} \tag{9.3-6}$$

and is thus the same as the rms value.

Stationary Processes

A special case of a random process is one that is *stationary*, which means that its statistical properties (such as its mean and variance) are constant. To simplify the mathematics, much of what follows depends on the truth of the stationarity assumption. This assumption is usually made in practice; however, real processes are not truly stationary.

In our applications we will apply statistical techniques to a single, random, *time* variable $x(t)$, instead of to an *ensemble*, $x_1(t)$, $x_2(t)$, $x_3(t)$, ..., which is a collection of possible outcomes. This means that we will use averaging processes over *time* instead of over an ensemble. Therefore, we will assume that the ensemble average equals the time average. This assumption is called the *ergodic hypothesis*, and a process that generates a time signal satisfying this hypothesis is said to be an *ergodic* process. The hypothesis implies that an ergodic process must be stationary. Therefore this assumption simplifies our calculations.

9.4 SPECTRAL ANALYSIS

We will want to examine an acquired signal to determine any periodic behavior because this will indicate resonant frequencies, for example. The amplitude spectrum based on the Fourier transform can often be used to do this, especially when the measured signal is not too noisy. If, however, we also want to develop a model of the system under test, rather than just determine its resonant frequencies, then we will need more than just the amplitude spectrum. The methods developed in this section can be used to compute the magnitude and phase angle plots of the system transfer function. The methods are also useful for situations in which signal noise is significant.

Note that a sine wave with a period P repeats itself every P seconds; that is, the function value at time t is identical to its value at time $t + P$. We say that the sine wave is perfectly *correlated* with itself with a period P. Thus, for a general signal, it will be useful to know if the signal contains some correlation. A perfectly random signal will not be correlated with itself for any positive time span τ because such correlation would indicate some periodicity.[1]

The Autocorrelation Function

The *autocorrelation function*, $R_{xx}(\tau)$, was developed as a measure of self-correlation; it is defined as

$$R_{xx}(\tau) = \lim_{T \to \infty} \frac{1}{T} \int_{-T/2}^{T/2} x(t)x(t + \tau)\, dt \tag{9.4-1}$$

In practice, when computing $R_{xx}(\tau)$ from data, we have a limited set of data to use and $x(t)$ will be zero for t greater than some value. So the time duration T in the formula cannot approach ∞ and the integral will have finite limits.

If $\tau = 0$ in Equation 9.4-1, we have

$$R_{xx}(0) = \lim_{T \to \infty} \frac{1}{T} \int_{-T/2}^{T/2} x^2(t)\, dt = E(x^2) \tag{9.4-2}$$

Thus the mean-square value of a signal is its autocorrelation when $\tau = 0$.

EXAMPLE 9.4-1 *Autocorrelation* *of the Sine* *Function*	Determine the autocorrelation of the sine function $x(t) = A \sin bt$.
Solution	The autocorrelation is found from $$R_{xx}(\tau) = \lim_{T \to \infty} \frac{1}{T} \int_{-T/2}^{T/2} (A \sin bt)A \sin[b(t + \tau)]\, dt$$ Using the identity $\sin[b(t + \tau)] = \sin bt \cos b\tau + \cos bt \sin b\tau$, we obtain $$R_{xx}(\tau) = \lim_{T \to \infty} \frac{A^2 \cos b\tau}{T} \int_{-T/2}^{T/2} (\sin bt)(\sin bt)\, dt$$ $$+ \lim_{T \to \infty} \frac{A^2 \sin b\tau}{T} \int_{-T/2}^{T/2} (\sin bt)(\cos bt)\, dt$$ Because $\sin bt$ is an odd function while $\cos bt$ is even, the limit of the second integral is zero, and we have $$R_{xx}(\tau) = \lim_{T \to \infty} \frac{A^2 \cos b\tau}{T} \int_{-T/2}^{T/2} \sin^2 bt\, dt = \frac{A^2 \cos b\tau}{2} \tag{1}$$

---

[1] We have used the symbol τ for the correlation parameter in order to be consistent with the standard usage in the signal-processing literature. Note that it is not the same as the time constant τ, although both have units of time.

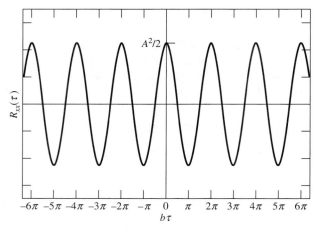

FIGURE 9.4-1 Autocorrelation $R_{xx}(\tau)$ of the sine function $A \sin bt$.

This function is plotted in Figure 9.4-1. It has a maximum when $\cos b\tau = 1$; that is, when $\tau = 2\pi n/b$, $n = 0, 1, 2, \ldots$. In addition, $R_{xx}(\tau) = 0$ when $\cos b\tau = 0$; that is, when $\tau = n\pi/2b$, $n = 1, 3, 5, \ldots$. The meaning of these results becomes clear if we think of τ as a time shift. A time shift of $2\pi/b$ converts a sine function into an identical sine function, while a time shift of $\pi/2b$ converts a sine function into a cosine function, which is independent of and thus uncorrelated with the sine function.

The mean-square value is obtained from $R_{xx}(0)$ and is equal to $A^2/2$. So the root-mean-square (rms) value of the function $A \sin bt$ is $A/\sqrt{2}$. ∎

Some of the following properties of the autocorrelation should be noted and kept in mind. If a process has a zero mean but is noisy, then $R_{xx}(\tau)$ will have a peak at $\tau = 0$ and will have small values for $\tau \neq 0$. If the process is periodic such that $x(t + \tau) = x(t)$, then $R_{xx}(\tau)$ will be a constant.

From the basic meaning of autocorrelation, we see that $R_{xx}(-\tau) = R_{xx}(\tau)$. In addition, it can be shown that (see [Newland, 1993]):

$$\mu^2 - \sigma^2 \leq R_{xx}(\tau) \leq \mu^2 + \sigma^2 \tag{9.4-3}$$

$$\lim_{\tau \to \infty} R_{xx}(\tau) = \mu^2 \tag{9.4-4}$$

The Cross-Correlation Function

We will also need to compute the correlation between two different signals, such as the measured input and the measured response signals, or response measurements at two different locations in a structure. For this application the *cross-correlation* function for two signals $x(t)$ and $y(t)$ is defined as follows:

$$R_{xy}(\tau) = \lim_{T \to \infty} \frac{1}{T} \int_{-T/2}^{T/2} x(t)y(t + \tau)\, dt \tag{9.4-5}$$

A significant application of the cross-correlation function is in sonar. Active sonar works by projecting a sound wave $x(t)$ and looking for its reflection $y(t)$ from an object. In this case the value of the parameter τ that maximizes $R_{xy}(\tau)$ can be related to the distance to the object, knowing the speed of sound in water.

The Spectral Density

We will see that the *spectral density*, $S_{xx}(\omega)$, is one of the most useful functions in vibration testing. It is defined as

$$S_{xx}(\omega) = \frac{1}{2\pi} \int_{-\infty}^{\infty} R_{xx}(\tau) e^{-i\omega\tau} \, d\tau \tag{9.4-6}$$

Although it is sometimes called the *power* spectral density, which suggests that $S_{xx}(\omega)$ is a measure of the power in a signal, this is true only for certain types of inputs and responses. Therefore a better name is simply *spectral density*.

Equation 9.4-6 has the form of the Fourier transform of the autocorrelation function. Therefore we can write $R_{xx}(\tau)$ as the inverse Fourier transform of $S_{xx}(\omega)$:

$$R_{xx}(\tau) = \int_{-\infty}^{\infty} S_{xx}(\omega) e^{i\omega\tau} \, d\omega \tag{9.4-7}$$

We can also write

$$S_{xy}(\omega) = \frac{1}{2\pi} \int_{-\infty}^{\infty} R_{xy}(\tau) e^{-i\omega\tau} \, d\tau \tag{9.4-8}$$

This is called the *cross-spectral density*.

From Equations 9.4-2 and 9.4-5, we obtain a very useful relation:

$$R_{xx}(0) = \int_{-\infty}^{\infty} S_{xx}(\omega) \, d\omega = E(x^2) \tag{9.4-9}$$

This says that the mean-square value can be computed from $R_{xx}(0)$.

EXAMPLE 9.4-2
Spectral Density of the Sine Function

Obtain the spectral density of the sine function $x(t) = A \sin bt$.

Solution

From Equation 9.4-4 and Equation 1 of Example 9.4-1,

$$S_{xx}(\omega) = \frac{1}{2\pi} \int_{-\infty}^{\infty} R_{xx}(\tau) e^{-i\omega\tau} \, d\tau = \frac{1}{2\pi} \int_{-\infty}^{\infty} \frac{A^2}{2} \cos(b\tau) \, e^{-i\omega\tau} \, d\tau$$

Care must be taken because of the infinite limits, so we express the integral as

$$S_{xx}(\omega) = \frac{A^2}{4\pi} \lim_{n \to \infty} \int_{-n\pi/b}^{n\pi/b} \cos(b\tau) \, e^{-i\omega\tau} \, d\tau$$

Evaluating the integral gives

$$S_{xx}(\omega) = \frac{A^2}{4\pi} \lim_{n \to \infty} \left[\frac{\sin n\pi(1 - \omega/b)}{b - \omega} + \frac{\sin n\pi(1 + \omega/b)}{b + \omega} \right]$$

We omit the details of carrying out the limit and simply give the result, which is

$$S_{xx}(\omega) = \frac{A^2}{4} [\delta(\omega + b) + \delta(\omega - b)] \tag{1}$$

where $\delta(x)$ is the *Dirac delta function*, which is defined such that

$$\int_{-\infty}^{\infty} \delta(x) \, dx = 1 \tag{2}$$

From Equation 2 we see that the area under the spectral density plot is $A^2/2$, which is the average "power" in the signal $A \sin bt$. Note that the rms value of $x(t)$ is $A/\sqrt{2}$, so the spectral density of the sine function is the square of its rms value. ∎

Loosely speaking, $\delta(0) = \infty$ and $\delta(x) = 0$ for $x \neq 0$. Applying this interpretation to Equation 1 of Example 9.4-2, we see that the spectral density plot of the sine function $A \sin bt$ consists of two "spikes," one at $\omega = -b$ and one at $\omega = b$. The plot is zero at all other frequencies. Because of the linearity property of the transform, we can conclude that a signal that contains two or more harmonics will have "spikes" at each harmonic frequency. Although a real, experimentally determined spectral density plot will not have infinite "spikes," the presence of high, narrow peaks in such a plot indicates that the signal contains distinct harmonic components at the frequencies corresponding to the peaks.

Although Equation 1 of Example 9.4-2 indicates that we could use the height of the peak to determine the amplitude contribution A of the harmonic, this is difficult to do in practice because of noise effects. For this reason, we will not concern ourselves with the numerical value of $S_{xx}(\omega)$ but only with the locations of its peaks. This will be of importance later when we discuss how to compute the spectral density in software. For example, we saw in Section 9.3 that the power spectrum is proportional to $X^*(\omega)X(\omega)$, where $X(\omega)$ is the Fourier transform of the signal and $X^*(\omega)$ is its complex conjugate. This transform can be computed with an algorithm called the *fast Fourier transform*. Thus we can use the fact that

$$S_{xx}(\omega) \propto X^*(\omega)X(\omega) \tag{9.4-10}$$

to obtain the *shape* but not the precise *amplitude* of the plot of the spectral density.

Response to Noise

The methods of this section are helpful for analyzing the effects of random inputs. Although we have not given a definition of *noise*, the term refers to a random and *unwanted* component in a measured signal. Although random, the noise produced by some sources may be characterized by a probability distribution, such as the Gaussian distribution or the Rayleigh distribution. Thus, although random, we may be able to describe such noisy signals by their mean and variance, for example.

The most common description of noise is the *white noise* model. The name derives from the fact that white light contains (theoretically, at least) a uniform distribution of light waves of different colors (that is, of different frequencies). Our formal definition of a white noise signal is that its autocorrelation function is an impulse function; that is,

$$R_{xx}(\tau) = W_0 \delta(\tau) \tag{9.4-11}$$

where W_0 is a constant. The signal is completely correlated for $\tau = 0$ and completely uncorrelated for any $\tau \neq 0$. Consequently, the spectral density of white noise is a constant, usually denoted S_0, and given by

$$S_{xx}(\omega) = \frac{W_0}{2\pi} = S_0 \tag{9.4-12}$$

Thus we may express Equation 9.4-11 as

$$R_{xx}(\tau) = 2\pi S_0 \delta(\tau) \tag{9.4-13}$$

Because of the constant-density function, the white noise model leads to simpler mathematics, and so it is often used as an approximate description of noisy real signals.

Note that, because of Equations 9.4-2 and 9.4-9, the mean-square value of white noise is infinite. This makes sense because a white noise signal can take on any number of values, including infinite values.

Figure 9.4-2 compares the autocorrelation function $R_{xx}(\tau)$ and the spectral density $S_{xx}(\omega)$ for a pure sinusoidal signal $x(t) = A \sin 2\pi t/T$ with those of an ideal white noise signal.

Band-Limited Noise

We note that a white noise signal cannot exist physically because it would contain infinite energy. We can see this by noting that the area under the spectral density curve is proportional to the energy content of the signal. But since the density is constant for $-\infty \le \omega \le \infty$, the area is infinite (see Figure 9.4-3a). So for real signals, the spectral density must decrease for higher frequencies.

In Figure 9.4-3b, the spectral density is approximately constant over some frequency range. Figure 9.4-3c shows an idealized case, which is easier to handle mathematically. The spectral densities in Figure 9.4-3 (b) and (c) are symmetric about $\omega = 0$ but are plotted for $\omega \ge 0$ only. They are examples of *band-limited noise*, which is so named because the density is nonzero only over a limited frequency band. Figure 9.4-4 illustrates the difference between an ideal white noise signal and a band-limited signal, which does not contain high frequencies.

The unrealistic behavior of the white noise model at high frequencies actually does not cause problems in practice because the magnitude of the frequency transfer function of mechanical systems is very small for frequencies above the highest natural frequency. The response to an ideal white noise input will therefore be very small at high frequencies and will thus be negligible. Thus we can use the white noise model as long as the spectral density of the real signal is approximately constant up to a frequency a little greater than the highest natural frequency of the system.

The autocorrelation of the band-limited noise illustrated in Figure 9.4-3c is

$$R_{xx}(\tau) = \int_{-\infty}^{\infty} S_{xx}(\omega)e^{i\omega\tau}\, d\omega = \frac{2S_0}{\tau}(\sin \omega_2\tau - \sin \omega_1\tau) \tag{9.4-14}$$

This function is plotted in Figure 9.4-3d. The correlation becomes very small for higher correlation times τ. Note that the correlation at $\tau = 0$ is now *finite*, whereas it is *infinite* for ideal white noise.

If the real signal may be modeled as consisting of a deterministic component plus a noise component, we may add the autocorrelations of each component to obtain the autocorrelation of the signal; similarly, we may add the spectral densities of the components to obtain the total density. So, for example, if we model the signal as consisting of a sine function $A \sin 2\pi t/T$ plus ideal white noise, then

$$R_{xx}(\tau) = 2\pi S_0\delta(\tau) + \frac{A^2}{2}\cos\frac{2\pi\tau}{T}$$

and

$$S_{xx}(\omega) = S_0 + \frac{A^2}{4}\delta\left(\omega + \frac{2\pi}{T}\right) + \frac{A^2}{4}\delta\left(\omega - \frac{2\pi}{T}\right)$$

The noise contributes a "spike" in the autocorrelation function, while the sine function contributes two spikes in the spectral density. The white noise raises the level of the spectral density by a constant amount S_0.

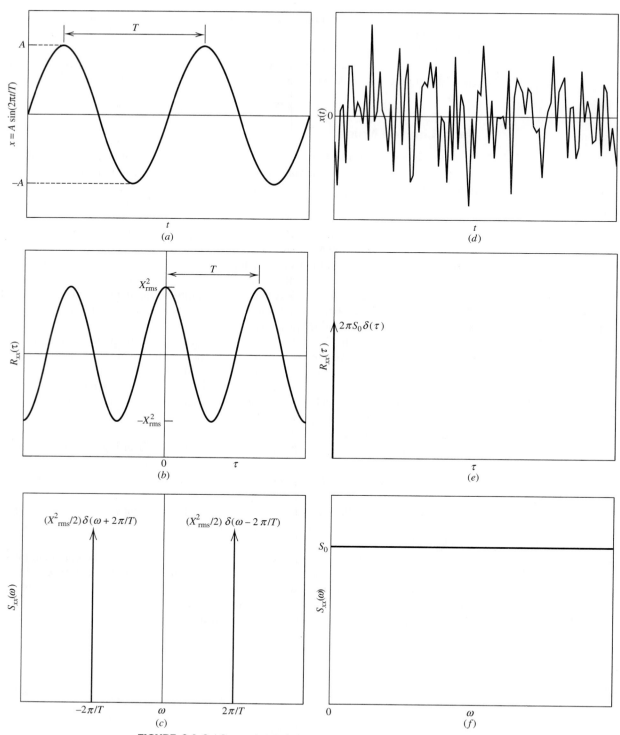

FIGURE 9.4-2 Comparison of autocorrelation $R_{xx}(\tau)$ and the spectral density $S_{xx}(\omega)$ of (a) a pure sinusoidal signal $A \sin \omega t$ with those of (d) ideal white noise.

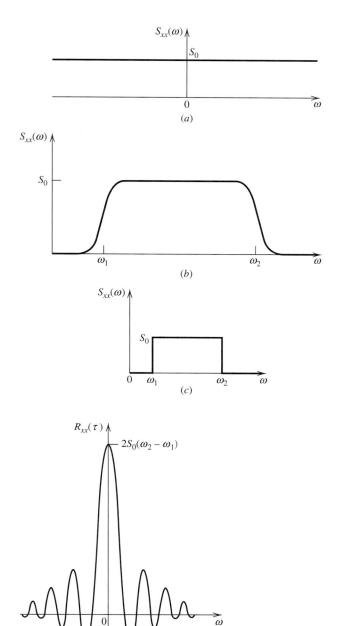

FIGURE 9.4-3 Characteristics of band-limited noise. (*a*) Spectral density of white noise. (*b*) Spectral density of band-limited noise. (*c*) Idealized band-limited noise. (*d*) Autocorrelation $R_{xx}(\tau)$ of idealized band-limited noise.

We may apply this additive property if more than one harmonic component is present in the signal. The autocorrelation will look more complicated because it contains the sum of two harmonic functions of different frequencies, but the density will have a pair of spikes for each harmonic. This is one reason the measured spectral density is

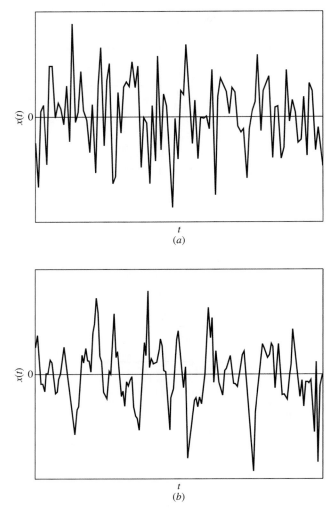

FIGURE 9.4-4   A white noise signal and a band-limited noisy signal.

useful; its peaks indicate the frequencies of the harmonic components. The presence of white noise merely shifts the plot upward.

Broad-Band and Narrow-Band Processes

Consider the two spectral density plots shown in Figure 9.4-5. The density in Figure 9.4-5a describes a *broad-band process*, which is one composed of components containing frequencies over a wide, or broad, frequency range. Figure 9.4-5b shows a density containing frequencies over a narrow band, where the frequencies ω_1 and ω_2 shown in Figure 9.4-3(b) are close to one another. The frequency ω_0 can be considered to be the average of the frequencies ω_1 and ω_2; that is, $\omega_0 = (\omega_1 + \omega_2)/2$. The corresponding process that generates such a density is called a *narrow-band process*. Figure 9.4-6 shows typical time histories for broad-band and narrow-band processes.

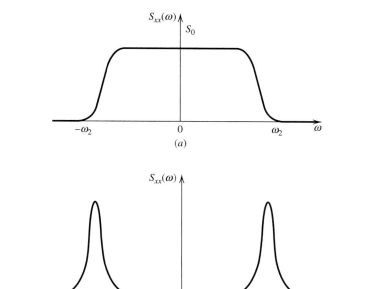

FIGURE 9.4-5 Broad-band versus narrow-band processes. (a) Spectral density of a broad-band process. (b) Spectral density of a narrow-band process.

For the density shown in Figure 9.4-5a,

$$E(x^2) = \int_{-\infty}^{\infty} S_{xx}(\omega)\, d\omega = 2S_0(\omega_2 - \omega_1) = 2S_0\, \omega_2 \qquad (9.4\text{-}15)$$

Since $\omega_1 = 0$. Thus we see that the mean-square value increases with the frequency range; for white noise, this range is infinite and so is the mean-square value.

Also for Figure 9.4-5a, with $\omega_1 = 0$ in Equation 9.4-14, we obtain

$$R_{xx}(\tau) = \frac{2S_0}{\tau} \sin \omega_2 \tau \qquad (9.4\text{-}16)$$

Identifying Transfer Functions from the Spectral Density

The spectral density can be used to determine the frequency response plot of a system under test. As we saw in Examples 9.1-2 and 9.1-3, this plot can be used to identify the number of modes, the modal damping ratios, and other information about the system's transfer function.

The following fundamental results are useful for this purpose. Their derivation will be omitted but can be found in references on signal processing. Suppose that the signal $x(t)$ is the measured output from a test and that $f(t)$ is the forcing function that generated the test results. Then

$$S_{xx}(\omega) = |T(i\omega)|^2 S_{ff}(\omega) \qquad (9.4\text{-}17)$$

$$S_{xf}(\omega) = |T(i\omega)| S_{ff}(\omega) \qquad (9.4\text{-}18)$$

$$S_{xx}(\omega) = |T(i\omega)| S_{xf}(\omega) \qquad (9.4\text{-}19)$$

where $T(i\omega)$ is the frequency transfer function of the system under test.

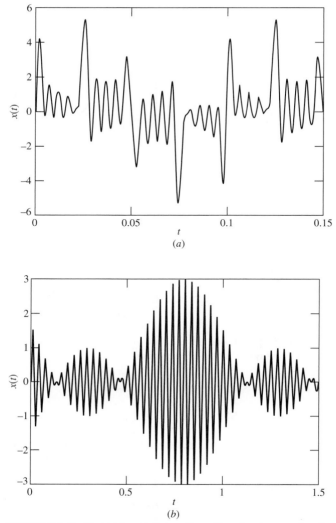

FIGURE 9.4-6 Typical time histories of broad-band and narrow-band processes. The frequency content of the broad-band signal is in the range $10 \leq f \leq 200$ Hz. The frequency content of the narrow-band signal is in the range $29 \leq f \leq 31$ Hz.

The meaning of Equation 9.4-17 is illustrated in Figure 9.4-7. The system shown in Figure 9.4-7a has a broad-band input whose spectral density is shown in Figure 9.4-7b. The system's frequency response plot is given in Figure 9.4-7c. The response, whose spectral density is plotted in Figure 9.4-7d, is a narrow-band process because of the filtering properties of the system (indicated by the fact that for a real system $|T(i\omega)|$ is small for high frequencies). Equation 9.4-17 implies that if we can compute $S_{xx}(\omega)$ and $S_{ff}(\omega)$, then we can compute

$$|T(i\omega)| = \sqrt{\frac{S_{xx}(\omega)}{S_{ff}(\omega)}} \qquad (9.4\text{-}20)$$

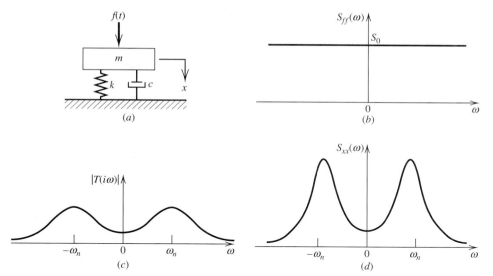

FIGURE 9.4-7 Illustration of how system dynamics can produce a narrow-band response from a broad-band input.

Equations 9.4-9 and 9.4-17 can be combined to give

$$E(x^2) = \int_{-\infty}^{\infty} |T(i\omega)|^2 S_{ff}(\omega)\, d\omega \tag{9.4-21}$$

This can be used to compute the mean-square response if we know the spectral density of the forcing function. For an ideal white noise with the constant spectral density S_0, the mean-square response reduces to

$$E(x^2) = S_0 \int_{-\infty}^{\infty} |T(i\omega)|^2\, d\omega \tag{9.4-22}$$

This integral has been tabulated for several common transfer functions in [James, 1947] and has been reproduced in more recent references, such as [Crandall, 1963] and [Newland, 1993].

$$\int_{-\infty}^{\infty} \left| \frac{B_0}{A_0 + i\omega A_1} \right|^2 d\omega = \pi \frac{B_0^2}{A_0 A_1} \tag{9.4-23}$$

$$\int_{-\infty}^{\infty} \left| \frac{B_0 + i\omega B_1}{A_0 - \omega^2 A_2 + i\omega A_1} \right|^2 d\omega = \pi \frac{(B_0^2/A_0)A_2 + B_1^2}{A_1 A_2} \tag{9.4-24}$$

$$\int_{-\infty}^{\infty} \left| \frac{B_0 - \omega^2 B_2 + i\omega B_1}{A_0 - \omega^2 A_2 + i\omega(A_1 - \omega^2 A_3)} \right|^2 d\omega$$
$$= \pi \frac{(B_0^2/A_0)A_2 A_3 + A_3(B_1^2 - 2B_0 B_2) + A_1 B_2^2}{A_1 A_2 A_3 - A_0 A_3^2} \tag{9.4-25}$$

$$\int_{-\infty}^{\infty} \left| \frac{B_0 - \omega^2 B_2 + i\omega(B_1 - \omega^2 B_3)}{A_0 - \omega^2 A_2 + \omega^4 A_4 + i\omega(A_1 - \omega^2 A_3)} \right|^2 d\omega$$
$$= \pi \frac{N}{A_1(A_2 A_3 - A_1 A_4) - A_0 A_3^2} \tag{9.4-26}$$

$$N = (B_0^2/A_0)(A_2A_3 - A_1A_4) + A_3(B_1^2 - 2B_0B_2)$$
$$+ A_1(B_2^2 - 2B_1B_3) + (B_3^2/A_4)(A_1A_2 - A_0A_3) \tag{9.4-27}$$

EXAMPLE 9.4-3
A Damped
System

The standard vibration model with damping is

$$m\ddot{x} + c\dot{x} + kx = f(t)$$

(**a**) Derive the expression for the spectral density $S_{xx}(\omega)$ of the response of the system in terms of the spectral density $S_{ff}(\omega)$ of the forcing function, assuming that the spectral density of the forcing function is the constant S_0.

(**b**) Derive the expression for the mean-square response.

Solution

(**a**) The transfer function is

$$T(s) = \frac{X(s)}{F(s)} = \frac{1}{ms^2 + cs + k}$$

Its frequency transfer function is

$$T(i\omega) = \frac{1}{k - m\omega^2 + c\omega i}$$

Thus

$$|T(i\omega)|^2 = \frac{1}{(k - m\omega^2)^2 + (c\omega)^2}$$

and

$$S_{xx}(\omega) = |T(i\omega)|^2 S_{ff}(\omega) = S_0 \frac{1}{(k - m\omega^2)^2 + (c\omega)^2}$$

Therefore the response, which is random, will have a spectral density that is a function of frequency, even though the spectral density of the input is constant.

(**b**) The mean-square response is given by

$$E(x^2) = S_0 \int_{-\infty}^{\infty} |T(i\omega)|^2 \, d\omega = S_0 \int_{-\infty}^{\infty} \left| \frac{1}{k - m\omega^2 + c\omega i} \right|^2 d\omega$$

This integral can be evaluated from Equation 9.4-24 with $B_0 = 1$, $B_1 = 0$, $A_0 = k$, $A_2 = m$, and $A_1 = c$.

$$E(x^2) = \frac{\pi S_0}{ck}$$

∎

EXAMPLE 9.4-4
Determining
Required Stiffness
for a Random
Input

The standard vibration model with damping and an applied force $f(t)$ is

$$m\ddot{x} + c\dot{x} + kx = f(t)$$

Suppose that the spectral density of the forcing function is the constant S_0. Determine the expression for the mean value of the stiffness k so that the standard deviation σ_x of the response does not exceed a specified limit D.

Solution

From the results of Example 9.4-3, we know that

$$E(x^2) = \frac{\pi S_0}{ck}$$

and thus, since the mean of x is zero, the standard deviation is

$$\sigma_x = \sqrt{E(x^2)} = \sqrt{\frac{\pi S_0}{ck}} \leq D$$

Therefore, k must be such that

$$k \geq \frac{\pi S_0}{cD}$$

∎

Identifying Transfer Functions with the Cross-Spectral Densities

The cross-spectral densities provide another way to obtain information about the transfer function. Because there will always be noise effects and instrument inaccuracies, the estimate of $|T(i\omega)|$ given by Equation 9.4-18 will usually differ from that given by Equation 9.4-19. Denote these estimates by $T_1(i\omega)$ and $T_2(i\omega)$. Then,

$$|T_1(i\omega)| = \frac{S_{xf}(\omega)}{S_{ff}(\omega)} \tag{9.4-28}$$

and

$$|T_2(i\omega)| = \frac{S_{xx}(\omega)}{S_{xf}(\omega)} \tag{9.4-29}$$

The *coherence function* γ^2 is defined as

$$\gamma^2 = \frac{|T_1(i\omega)|}{|T_2(i\omega)|} = \frac{S_{xf}(\omega)}{S_{ff}(\omega)} \frac{S_{xf}(\omega)}{S_{xx}(\omega)} \tag{9.4-30}$$

The coherence function is used as an estimate of the correctness of the estimates for $|T(i\omega)|$. If γ^2 is near 1, then the two estimates are close and thus more dependable than if γ^2 is near 0.

Estimating the Spectral Density

To use methods such as the one based on Equation 9.4-17, we must be able either to specify the input spectral density $S_{ff}(\omega)$ or to estimate it from test data. Figure 9.4-8a illustrates an analog system for doing this, assuming that the input $f(t)$ is a stationary random process. The input is processed by a narrow-band filter, whose transfer function $|T_f(i\omega)|$ is plotted in Figure 9.4-8b. The output of this filter, called $y(t)$, is then squared and time averaged using an averaging time T_{av} to produce $z(t)$, so that

$$z(t) = \frac{1}{T_{av}} \int_0^{T_{av}} y^2(t) \, dt$$

The averaging is a continual process, and T_{av} is finite, so z will be a function of time that fluctuates about a mean value.

The fluctuations will be small if T_{av} is large, and because $y(t)$ is stationary, $E(y^2) = E(z)$. Thus,

$$E(y^2) = \int_{-\infty}^{\infty} |T_f(i\omega)|^2 S_{xx}(\omega) \, d\omega$$

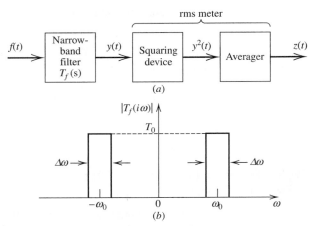

FIGURE 9.4-8   (a) Analog system for estimating the spectral density $S_{ff}(\omega)$ of the input $f(t)$. (b) Frequency response $|T_f(i\omega)|$ of the narrow-band filter.

Carrying out the integral for the transfer function shown in Figure 9.4-8b, for which $\Delta\omega << \omega_0$, we obtain

$$E(y^2) \approx 2T_0^2 \Delta\omega S_{ff}(\omega_0)$$

This gives

$$S_{ff}(\omega_0) \approx \frac{E(z)}{2T_0^2 \Delta\omega} \tag{9.4-31}$$

Therefore we can use the output of the averager to estimate the spectral density of the input $f(t)$.

We have illustrated this scheme with analog devices. Today, however, most spectral density estimation is done with digital filtering, often using, for example, the fast Fourier transform (FFT), which is discussed in Section 9.6. Theoretical development of digital filtering is beyond the scope of this text. An excellent reference for this development, which emphasizes mechanical vibration applications, is [Newland, 1993].

9.5   DATA ACQUISITION AND SIGNAL PROCESSING

A system for collecting and analyzing vibration data contains several hardware and software components. These are briefly discussed in this section. Detailed and up-to-date information on such components is best obtained from the vendors.

Sensors

A variety of sensors are available for vibration applications. Perhaps the most common are the *strain gauge* and the *accelerometer*. A strain gauge may be made of material, such as wire, whose electrical resistance changes when subjected to mechanical strain. Other gauges may be made of piezoelectric material that generates a charge when under strain. So that the gauge will experience the same strain as the test object, it must be bonded to the object, and this may be a disadvantage in some applications. The wire gauge is connected to a Wheatstone bridge circuit that measures the resistance change. Calibrating the gauge will give a relation between the resistance change and the strain. Because strain is a function of displacement, the gauge can be used as a displacement sensor. Force transducers, also called *load cells*, can be constructed with strain gauges. An advantage of strain gauges is that they are inexpensive, but they can produce a noisy signal and they have a limited *dynamic range*, which is the ratio of the minimum and maximum output levels over which the sensor can operate without significant error.

A model of an accelerometer was developed in Chapter 4, Section 2. It consists of a mass and stiffness element inside a case that is mounted to the test object. Often the stiffness element is piezoelectric. Accelerometers come in a variety of stiffnesses and masses. The accelerometer natural frequency must be greater than the highest frequency we want to measure, and its mass should be small compared to the mass of the test object.

Other types of sensors are available. One new type is a *laser vibrometer*, in which a laser beam is projected at the test object. The reflected beam is compared to the projected beam, and the result is processed to produce a measurement of the object's motion at the point where the beam strikes. Laser vibrometers do not affect the dynamics of the test object, so they are especially suitable for small objects if a suitable target point can be found. They are at present, however, rather expensive.

Analyzers

The signals from the sensors must be amplified and perhaps *conditioned* before being used either by an *analyzer* or by a computer. A wide variety of analyzers are commercially available. Their cost depends partly on the number of channels. The purpose of the analyzer is to help determine the natural frequencies and the transfer function between a given input-output pair of signals. This will also be the desired transfer function of the test object. For example, if a shaker provides a sinusoidal base displacement $y(t)$ whose frequency is varied, and if the sensor measures the displacement $x(t)$ of the test object, then the analyzer would display the plot of the frequency transfer function, in this case $X(i\omega)/Y(i\omega)$, and compute natural frequencies and modal damping ratios, as described in Example 9.1-3. Analyzers can also compute spectral densities, so they can handle response due to a random input.

A special type of analyzer that can also compute mode shapes is called a *modal analyzer*. It does this by using measured responses from several locations on the test object.

Sampling

Digital devices can handle mathematical relations and operations only when expressed as a finite set of numbers rather than as functions having an infinite number of possible values. Thus any continuous measurement signal must be converted into a set of pulses by *sampling*—the process by which a continuous-time variable is measured at distinct, separated instants of time. The sequence of measurements replaces the smooth curve of the measured variable versus time, and the infinite set of numbers represented by the smooth curve is replaced by a finite set of numbers. This is done because a digital device cannot store a continuous signal. Each pulse amplitude is then rounded off to one of a finite number of levels depending on the characteristics of the machine. This process is called *quantization*. Thus a digital device is one in which the signals are quantized in both time and amplitude. In an analog device, the signals are analog; that is, they are continuous in time and are not quantized in amplitude. The device that performs the sampling, quantization, and converting to binary form is an *analog-to-digital (A/D) converter*.

The number of binary digits carried by the machine is its *word length*, and this is obviously an important characteristic related to the device's resolution—the smallest change in the input signal that will produce a change in the output signal. If an A/D converter has a word length of 10 bits or more (a bit is one binary digit), an input signal can be resolved to one part in 2^{10}, or 1 in 1024. If the input signal has a range of 10 volts, the resolution is 10/1024, or approximately 0.01 volt. Thus the input must change by at least 0.01 volt in order to produce a change in the output.

Sampling extracts a discrete-time signal from a continuous-time signal. If the sampling frequency is not selected properly, the resulting sample sequence will not accurately represent the original signal. Fortunately, the proper frequency is readily determined in many cases by means of the *sampling theorem*, which we present shortly. *Uniform sampling* occurs when the sampling period T is constant.

Frequency Content of Signals

The proper value of the sampling period T depends on the nature of the signal being sampled. This is easily shown with the sinusoid of period P shown in Figure 9.5-1. If the sampling period T is slightly greater than the half-period $P/2$, it is possible to miss completely one lobe of the sinusoid. If $T < P/2$, each lobe will always be sampled at least once, and the oscillation will be detected. Stated in terms of frequencies, the sampling frequency $1/T$ must be at least twice the sinusoidal frequency $1/P$.

The issue is not so clear when the signal does not consist of sinusoids. The spectrum, which shows the frequency content of a signal, can be obtained from the Fourier series if the signal is periodic and from the Fourier transform if the signal is nonperiodic. For some physical signals, the energy content of a sinusoidal signal is proportional to the square of its amplitude. The energy contained in a general aperiodic signal $y(t)$ is found by integrating the square of the spectrum function $|Y(\omega)|$ over all frequencies and is given by

$$\text{Energy Content} = \int_0^\infty |Y(\omega)|^2 \, d\omega \tag{9.5-1}$$

For example, for the decaying exponential signal $e^{-t/\tau}$, $t \geq 0$, the frequency range $0 \leq \omega \leq \omega_u$ that contains 99% of the signal's energy is found as follows. We have the spectrum function:

$$|Y(\omega)| = \frac{\tau}{2\pi\sqrt{1 + \tau^2\omega^2}}$$

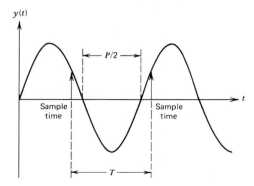

FIGURE 9.5-1. Sampling a sinusoid. The adjacent samples shown are too far apart to detect the oscillation in $y(t)$.

The upper frequency ω_u is found from

$$\int_0^{\omega_u} |Y(\omega)|^2 \, d\omega = 0.99 \int_0^{\infty} |Y(\omega)|^2 \, d\omega$$

With the given expression for $Y(\omega)$, this becomes

$$\int_0^{\omega_u} \frac{\tau^2}{1 + \tau^2 \omega^2} \, d\omega = 0.99 \int_0^{\infty} \frac{\tau^2}{1 + \tau^2 \omega^2} \, d\omega$$

$$\tan^{-1}(\tau\omega_u) - \tan^{-1}(0) = 0.99[\tan^{-1}(\infty) - \tan^{-1}(0)]$$

or

$$\omega_u = \frac{1}{\tau} \tan\left(\frac{0.99\pi}{2}\right) = \frac{63.657}{\tau}$$

Thus 99% of the energy lies in the frequency band $0 < \omega < 63.657/\tau$.

Aliasing and the Sampling Theorem

Now that we are able to determine the frequency content of a signal, we need to know what effects the higher frequencies have on the sampling process. Uniform sampling of the sinusoid

$$y(t) = \sin(\omega t + \phi)$$

produces the sequence

$$y(kT) = \sin(k\omega T + \phi)$$

and no loss of information is incurred in sampling if T is selected so that $0 \leq \omega \leq \pi/T$, as shown earlier. Uniform sampling cannot distinguish between two sinusoidal signals when their circular frequencies have a sum or difference equal to $2\pi n/T$, where n is any positive integer. This means that the only effective frequency range for uniform sampling is $0 \leq \omega \leq \pi/T$. That is, the signal being sampled must have no circular frequency content above π/T if uniform sampling is not to distort the signal's information. The frequency π/T is called the *Nyquist frequency* or the *folding frequency*, because all frequencies in the signal are folded into the interval $0 \leq \omega \leq \pi/T$ by uniform sampling. This phenomenon is called *aliasing*. It is seen in motion pictures of a rotating spoked wheel or aircraft propeller. As the object rotates faster, it appears to slow down and then stop or even rotate backward. The sampling process produced by the picture frames "aliases" the high rotation speed into the lower frequency interval defined by the Nyquist frequency.

The concept of aliasing leads directly to the Nyquist sampling theorem.

Sampling Theorem A continuous-time signal $y(t)$ can be reconstructed from its uniformly sampled values $y(kT)$ if the sampling period T satisfies

$$T \leq \frac{\pi}{\omega_u}$$

where ω_u is the highest frequency contained in the signal; that is, $|Y(\omega)| = 0$ for $\omega > \omega_u$.

With the principal exception of a pure sinusoid, most physical signals have no finite upper frequency ω_u. Their spectra $|Y(i\omega)|$ approach zero only as $\omega \to \infty$. In such cases, we estimate ω_u by finding the frequency range containing most of the signal's energy, as we did with the decaying exponential.

With a conservative engineering design philosophy, a safety factor between 2 and 10 is applied to determine the sampling rate. This factor is also necessary because we do not have in practice an infinite sequence of impulses as required by the sampling theorem. Because of aliasing, a low-pass filter, called a *guard filter*, is sometimes inserted before the sampler to eliminate frequencies above the Nyquist frequency π/T. By definition, these frequencies do not contribute significantly to the signal's energy. Thus they are equivalent to noise in the system and should be filtered out anyway.

For the decaying exponential, the upper frequency for the 99% energy criterion is $\omega_u = 63.657/\tau$. The sampling theorem requires that $T \le \pi/\omega_u = \pi\tau/63.657 = 0.049\tau$. Thus our sampling period should be less than 5% of the time constant τ.

9.6 MATLAB APPLICATIONS

Computer calculation of the Fourier transform to obtain the amplitude spectrum and the spectral density plots must use the discrete versions of the Fourier transform formulas. These are

$$x_k = \frac{1}{N}\sum_{k=1}^{N} X_k e^{i[2\pi(k-1)(n-1)/N]} \qquad 1 \le n \le N$$

$$X_k = \sum_{n=1}^{N} e^{-i2\pi(k-1)(n-1)/N} x_k \qquad 1 \le k \le N$$

The theory of the discrete Fourier transform (DFT) has been known for some time, but these formulas involve numerous calculations for typical data sets, and so practical application was not possible. In the early 1970s a fast algorithm, called the Cooley-Tukey algorithm, was developed that made practical applications feasible. The MATLAB implementation of the algorithm is the fft function, which stands for "fast Fourier transform" (FFT). Its basic syntax is fft(x), where x is a vector containing the data.

Most efficient operation of the algorithm is attained when the number of data points is a power of 2 (64, 128, 256, and 512 being commonly used values). If not, then MATLAB "pads" the data set with zeros to obtain the proper length. Proper use of the FFT requires an understanding of the Nyquist sampling theorem (see Section 9.5). Here we give some basic examples of its use.

Note that

$$\cos \omega t = \frac{e^{i\omega t} + e^{-i\omega t}}{2}$$

Thus the cosine actually consists of two complex functions, one having a frequency ω and the other having a frequency $-\omega$. Thus, since the Fourier transform is defined for complex functions over positive and negative frequencies, the spectrum of the cosine will have two "spikes," one at ω and the other at $-\omega$. In addition, the spectrum will have false spikes at frequency intervals equal to the period of the cosine. However, the function $e^{i\omega t}$ has only one frequency and thus only one "spike." This is illustrated in the following example.

EXAMPLE 9.6-1
Spectrum of the
Cosine Function

Obtain the spectrum plots of the functions $y_1(t) = e^{i\omega t}$ and $y_2(t) = \cos \omega t$, defined to be periodic over $-1 \le t \le 1$, for $\omega = 20\pi$, using 64 points.

Solution

The M-file is

```
n = 63;dt = 2/n;
t = [-1:dt:1];
f = 10;
omega = 2*pi*f;
y1 = exp(i*omega*t);
y2 = cos(omega*t);
Y1 = fft(y1);
Y2 = fft(y2);
subplot(2,1,1)
plot([0:n],abs(Y1)),ylabel('|Y|'),...
    title('Frequency content of y')
subplot(2,1,2)
plot([0:n],abs(Y2)),ylabel('|Y|'),xlabel('n')
```

The results are shown in Figure 9.6-1.

Effects of Noise

We can use the `randn` function to investigate the effects of noise in the signal. This function generates normally distributed random numbers of zero mean and unit variance. The syntax `randn(size(t))` generates an array of normally distributed random numbers the same size as the array `t`.

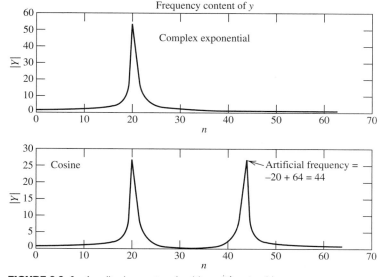

FIGURE 9.6-1   Amplitude spectra of $y_1(t) = e^{i\omega t}$ and $y_2(t) = \cos \omega t$ versus n.

EXAMPLE 9.6-2
Spectrum of the Sine Function with Noise

Obtain the amplitude spectrum and the spectral density of the function $y(t) = \sin 40\pi t$, using 128 points. The additive noise has an amplitude of 1.6.

Solution

Recall that

$$S_{yy}(\omega) \propto Y^*(\omega)Y(\omega)$$

Because of this, the vertical scale on the spectral density plot will be arbitrary. The following program generates the plots.

```
n = 127;dt = 1/n;
t = [0:dt:1];
noise = 1.6*randn(size(t));
y = 2*sin(40*pi*t) + noise;
Y = fft(y);
N = length(Y);
M = floor(N/2);
fHz = (1/max(t))*[0:M-1];
sd = conj(Y).*Y;
subplot(2,1,1)
plot(fHz,abs(Y(1:M))),ylabel('|Y|'),xlabel('f (Hz)')
subplot(2,1,2)
plot(fHz,sd(1:M)),ylabel('sd'),xlabel('f (Hz)')
```

The plots are shown in Figure 9.6-2. ■

EXAMPLE 9.6-3
Effects of Time Constant and Noise on the Spectrum

Obtain the amplitude spectrum of the function $y(t) = e^{-t/\tau} \sin 7\pi t$, for $\tau = 0.5$ and $\tau = 0.1$. The additive noise has an amplitude of 0.3.

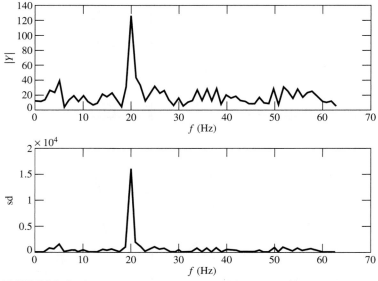

FIGURE 9.6-2 Amplitude spectrum and spectral density for $y(t) = \sin 40\pi t$ with additive noise present.

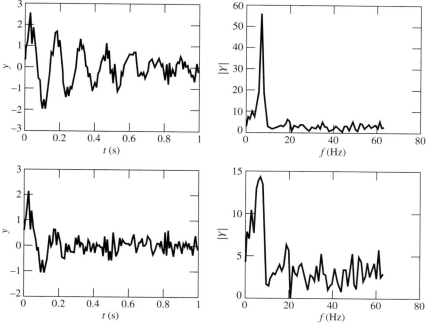

FIGURE 9.6-3 Effects of time constant and noise on the spectrum of $y(t) = e^{-t/\tau} \sin 7\pi t$, for $\tau = 0.5$ and $\tau = 0.1$.

Solution

The required program is a slight modification of that used in Example 9.6-2. The plots are shown in Figure 9.6-3. They show that the smaller time constant produces a less pronounced peak in the spectrum. The signal with the small time constant decays rapidly to a small level where the noise term predominates. This adversely affects the quality of the spectrum.

■

Duffing's Equation

In most applications the `fft` function is used with measured data. Here, in place of such data, we can solve an equation of motion numerically to generate some meaningful numerical values to use to demonstrate the use of the transform.

Duffing's equation describes the response of a mass subjected to a cosinusoidal forcing function and having a nonlinear spring whose spring force is described by $k_1 x + k_3 x^3$. The equation is

$$m\ddot{x} + c\dot{x} + k_1 x + k_3 x^3 = F \cos \omega_f t \qquad (9.6\text{-}1)$$

This is a nonlinear model because of the term x^3. As opposed to *linear* models, which oscillate only at the frequency of the forcing function, the response of this nonlinear model will contain more than one frequency. Note that it is difficult, if not impossible, to determine the frequencies by looking at a plot of the response versus time. The following example shows how to use the fast Fourier transform to obtain the spectrum plot of the response and to identify the response frequencies.

EXAMPLE 9.6-4
Spectrum of
Duffing's
Equation

Consider Duffing's equation, Equation 9.6-1, with the following parameter values, in SI units.

$$m = 1 \quad c = 0 \quad k_1 = 2 \quad k_3 = 0.1$$
$$F = 1 \quad \omega_f = 6$$

Obtain the spectrum plot of the response $x(t)$, and identify the dominant frequencies in the response.

Solution

The main program is as follows.

```
% Program Duffing Spectrum.m
% Plots the spectrum of the Duffing equation.
global m c k1 k3 omf F
m = 1;c = 0;k1 = 2;k3 = 0.1;F = 1;omf = 6;
[t, x] = ode45('duffing', [0 20], [0, 0]);
X = fft(x(:,1));
N = length(X);
M = floor(N/2);
f = (1/max(t))*[1:M];
subplot(2,1,1)
plot(t,x(:,1)),ylabel('x'),xlabel('t')
subplot(2,1,2)
plot(f,abs(X(1:M))),ylabel('|X|'),xlabel('f'),
[x,y]=ginput(2)
```

The program called by the `ode45` function is

```
function xdot = duffing(t,x)
global m c k1 k3 omf F
xdot = [x(2);(1/m)*(-k1*x(1)-k3*x(1)^3-c*x(2)+F*cos(omf*t))];
```

The plots are shown in Figure 9.6-4. The upper plot is the response of Duffing's equation. The lower plot is the spectrum of the response. The results from the `ginput` function

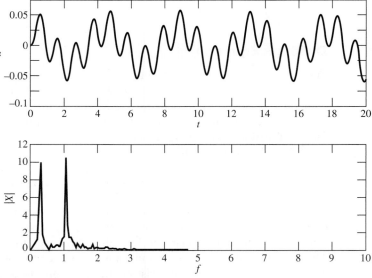

FIGURE 9.6-4   Time response of Duffing's equation and the amplitude spectrum.

show that the two peaks in the spectrum occur at $f = 0.239$ Hz and $f = 1.03$ Hz. Note that neither of these frequencies equals the forcing frequency $6/(2\pi) = 0.95$ Hz, although one is close. ∎

9.7 SIMULINK APPLICATIONS

Simulink provides the Power Spectral Density block to compute the spectral density. This block displays a plot of the data and plots of the spectral density amplitude and phase.

Figure 9.7-1 shows a Simulink model of Duffing's equation:

$$m\ddot{x} + c\dot{x} + k_1 x + k_3 x^3 = F \cos \omega_f t \tag{9.7-1}$$

In the Sine Wave block, set the Amplitude to F, the frequency to omf, and the phase to pi/2 to create a cosine input. In the Fcn block, type the expression u^3. In the Gain blocks, set the gains to 1/m, k1, and k3, respectively. Use the parameter values given in Example 9.6-1. Set the Stop time to 20, then run the model. The top two plots in the Power Spectral Density output window should look like the plots in Figure 9.6-4.

9.8 CHAPTER REVIEW

Sometimes it is difficult to develop a differential equation model of the system from basic principles such as Newton's laws. It can even be difficult to obtain an analytical model of the natural frequencies. In such cases we must resort to using measurements of the system response.

Vibration measurement and testing require a knowledge of the hardware available to produce vibration of the system under test and to measure the response. You must also be familiar with the algorithms and software available for processing the data.

This chapter described the equipment requirements and the procedures that are available for this task.

Now that you have finished this chapter, you should be able to do the following:

1. Analyze the step and the frequency response to identify the form and the parameter values of a model.

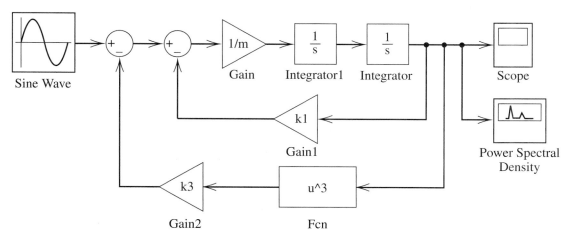

FIGURE 9.7-1   Simulink model of Duffing's equation, showing the Power Spectral Density block.

2. Describe the commonly used equipment for stimulating a vibratory response and for collecting response data.

3. Perform a transform analysis of signal data to estimate natural frequencies.

4. Apply MATLAB and Simulink to the methods of the chapter.

PROBLEMS

SECTION 9.1 SYSTEM IDENTIFICATION

9.1 Figure P9.1 shows the response of a system to a step input of magnitude 1000 N. The equation of motion is

$$m\ddot{x} + c\dot{x} + kx = f(t)$$

Estimate the values of m, c, and k.

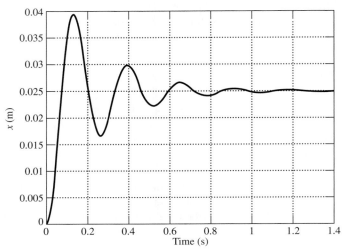

FIGURE P9.1

9.2 An input $15 \sin \omega t$ N was applied to a certain mechanical system for various values of the frequency ω, and the amplitude $|x|$ of the steady-state output was recorded. The data are shown in the following table. Determine the transfer function.

ω (rad/s)	$|x|$ (mm)
0.1	209
0.4	52
0.7	28
1	19
2	7
4	2
6	1

9.3 The following data were taken by driving a machine on its support with a rotating unbalance force at various frequencies. The machine's mass is 50 kg, but the stiffness and damping in the support are unknown. The frequency of the driving force is f Hz. The measured steady-state displacement of the machine is $|x|$ mm. Estimate the stiffness and damping in the support.

f (Hz)	|x| (mm)	f (Hz)	|x| (mm)
0.2	2	3.8	26
1	4	4	22
2	8	5	16
2.6	24	6	14
2.8	36	7	12
3	50	8	12
3.4	36	9	12
3.6	30	10	10

9.4 Figure P9.4 shows the magnitude and phase plots determined from experiment. Estimate the DOF of the system, its resonant frequencies, and its modal damping ratios.

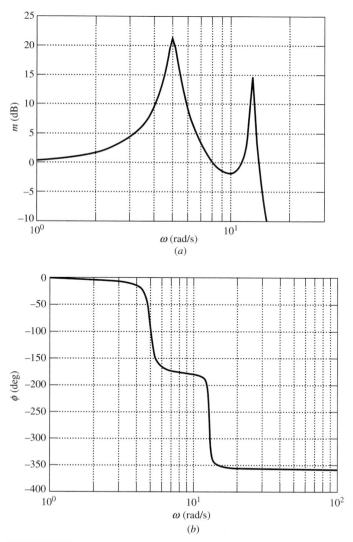

FIGURE P9.4

SECTION 9.2 TRANSFORM ANALYSIS OF SIGNALS

9.5 Use integration to obtain the Fourier transform and the spectrum of the triangular pulse of height H, width T, and centered at $t = T/2$. It is defined by

$$x(t) = \begin{cases} 0 & t < 0 \\ H\dfrac{2t}{T} & 0 \le t \le \dfrac{T}{2} \\ 2H\left(1 - \dfrac{t}{T}\right) & \dfrac{T}{2} \le t \le T \\ 0 & t > T \end{cases}$$

9.6 Use the results of Problem 9.5 and the shifting theorem to obtain the Fourier transform and the spectrum of the triangular pulse of height H, width T, and centered at $t = 0$.

9.7 Use the Laplace transform to obtain the Fourier transform and the spectrum of the triangular pulse of height H, width T, and centered at $t = T/2$.

9.8 (a) Obtain the Fourier transform of the following signal:

$$x(t) = \begin{cases} 0 & t < 0 \\ De^{-at}\cos bt & t \ge 0 \end{cases}$$

(b) Compare the results with those of Example 9.2-3.

SECTION 9.3 RANDOM PROCESSES

9.9 Determine the mean and variance of the function

$$x(t) = A + B\sin \omega t$$

9.10 Determine the mean and variance of the rectified sine wave

$$x(t) = \begin{cases} A\sin \omega t & 0 \le t \le \pi/\omega \\ -A\sin \omega t & \pi/\omega \le t \le 2\pi/\omega \end{cases}$$

SECTION 9.4 SPECTRAL ANALYSIS

9.11 Derive the autocorrelation function of $x(t) = A\cos bt$.

9.12 Derive the autocorrelation function of a rectangular pulse of height H and width T, centered at $t = 0$.

9.13 Derive the autocorrelation function of a periodic train of rectangular pulses of height H and width T, defined over the basic interval $0 \le t \le 2\pi$ as follows:

$$x(t) = \begin{cases} H & 0 < t < \pi \\ -H & \pi < t < 2\pi \end{cases}$$

9.14 A certain experimentally determined spectral density plot can be approximated by the function

$$S_{xx}(\omega) = 10\delta(\omega - 4) + 10\delta(\omega + 4)$$

Identify the corresponding time function, and compute the autocorrelation and the mean-square value.

9.15 A certain experimentally determined spectral density plot can be approximated by the function

$$S_{xx}(\omega) = 5\delta(\omega - 3) + 5\delta(\omega + 3) + 9\delta(\omega - 6) + 9\delta(\omega + 6)$$

Identify the corresponding time function and compute the autocorrelation and the mean-square value.

9.16 Derive the expression for the mean-square response of the system

$$100\ddot{x} + 4 \times 10^4 \dot{x} + 1.6 \times 10^5 x = f(t)$$

in terms of the constant spectral density $S_{ff}(\omega) = S_0$ of the forcing function.

9.17 The following is a model of a system having base motion. Suppose that the base motion $y(t)$ is random white noise with a constant spectral density S_0. Derive the expression for the mean-square response of the system

$$100\ddot{x} + 4 \times 10^4 \dot{x} + 1.6 \times 10^5 x = 4 \times 10^4 \dot{y} + 1.6 \times 10^5 y$$

9.18 Suppose an undamped oscillator is subjected to a forcing function having a constant spectral density S_0. The model is

$$m\ddot{x} + kx = f(t)$$

What happens if we try to obtain the expressions for the spectral density $S_{xx}(\omega)$ and the mean-square response? Explain the meaning of the results in physical terms.

9.19 Suppose an undamped oscillator is subjected to a forcing function consisting of band-limited white noise such that its spectral density is given by

$$S_{ff}(\omega) = \begin{cases} S_0 & |\omega| \leq \omega_n/2 \\ 0 & |\omega| > \omega_n/2 \end{cases}$$

where $\omega_n = \sqrt{k/m}$. The model is

$$m\ddot{x} + kx = f(t)$$

(a) Obtain the expression for the spectral density $S_{xx}(\omega)$ and compute the mean-square response.

(b) Discuss the results in light of the results of Problem 9.18.

9.20 The model of an undamped oscillator is

$$m\ddot{x} + kx = f(t)$$

Suppose it is subjected to a forcing function having an experimentally determined spectral density that can be approximated by the function

$$S_{ff}(\omega) = 10\delta(\omega - 4) + 10\delta(\omega + 4)$$

Obtain the expression for the spectral density $S_{xx}(\omega)$ and compute the mean-square response in two ways:
(a) from $S_{xx}(\omega)$ and
(b) by evaluating the time response.

SECTION 9.5 DATA ACQUISITION AND SIGNAL PROCESSING

9.21 Consider the triangular pulse defined by

$$x(t) = \begin{cases} 0 & t < 0 \\ \dfrac{2t}{T} H & 0 \leq t \leq \dfrac{T}{2} \\ 2H\left(1 - \dfrac{t}{T}\right) & \dfrac{T}{2} \leq t \leq T \\ 0 & t > T \end{cases}$$

Its Fourier transform and spectrum were obtained in Problem 9.5. Use these results to derive the expression for ω_u, the upper frequency for which 99% of the signal's energy lies in the range $0 \leq \omega \leq \omega_u$.

9.22 Determine the Fourier series for the function shown in Figure P9.22, and compute the frequency range that contains 99% of the signal's energy. Estimate the required sampling frequency.

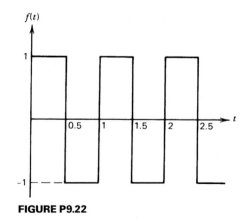

FIGURE P9.22

9.23 We suspect that the natural frequencies of a certain system with two degrees of freedom are approximately $\omega_1 = 10$ rad/s and $\omega_2 = 28$ rad/s. What is the minimum recommended sampling frequency to measure the motion?

9.24 We suspect that the time constant and the natural frequency of a certain system with one degree of freedom are approximately $\tau = 2$ s and $\omega = 6$ rad/s. What is the minimum recommended sampling frequency to measure the motion?

9.25 We suspect that the time constant and the natural frequency of a certain system with one degree of freedom are approximately $\tau = 2$ s and $\omega = 1/3$ rad/s. What is the minimum recommended sampling frequency to measure the motion?

SECTION 9.6 MATLAB APPLICATIONS

9.26 Modify the MATLAB program given in Example 9.6-1 to obtain the spectrum plots of the functions $y_1(t) = e^{i\omega t}$ and $y_2(t) = \sin \omega t$. Discuss the meaning of the "spikes" in the plots.

9.27 Obtain the amplitude spectrum and the spectral density of the function $y(t) = 10 \cos 30\pi t$ with additive noise having an amplitude of 15, using 128 points.

9.28 Obtain the amplitude spectrum of the function $y(t) = 10e^{-t/\tau} \sin 5\pi t$, for $\tau = 2$ and $\tau = 0.4$, with additive noise having an amplitude of 3.

9.29 Obtain the plot of the response $x(t)$ and the spectrum plot of Duffing's equation (Equation 9.6-1) using the parameter values given in Example 9.6-4, except that $c = 0.5$. Identify the dominant frequencies in the response, and explain any differences from the plot shown in Figure 9.6-4.

9.30 Obtain the plot of the response $x(t)$ and the spectrum plot of Duffing's equation (Equation 9.6-1), using the following parameter values.

$$m = 1 \quad c = 0 \quad k_1 = 20 \quad k_3 = 0.1$$

$$F = 1 \quad \omega_f = 10$$

Identify the dominant frequencies in the response

SECTION 9.7 SIMULINK APPLICATIONS

9.31 Create the Simulink model of Duffing's equation shown in Figure 9.7-1. Use the parameter values given in Example 9.6-4, except that $c = 0.06$. Identify the dominant frequencies in the response.

9.32 Create the Simulink model of Duffing's equation shown in Figure 9.7-1. Use the parameter values given in Problem 9.30. Identify the dominant frequencies in the response.

DISTRIBUTED SYSTEMS

CHAPTER OUTLINE

THIS CHAPTER treats the vibration of systems that cannot be described adequately with lumped-parameter models consisting of ordinary differential equations. If the model is linear, the number of its natural frequencies and the number of its modes equal its degrees of freedom. For example, a system with four degrees of freedom has four natural frequencies and four modes. In some systems the mass and elasticity cannot be considered to be concentrated at discrete points. These are called distributed systems. This distribution of the mass and elasticity throughout the system requires partial differential equations to describe the vibration. Such systems have an infinite number of natural frequencies and an infinite number of mode shapes.

We begin in Section 10.1 by considering how to model the simplest distributed system, a cable or string under tension, such as a guitar string. Torsional and longitudinal vibrations of rods are also described by such a model. The partial differential equation model is second order, and we introduce two methods for solving such an equation in Sections 10.2 and 10.3. We will see that to solve such models, we must specify not only the initial conditions, as we must do with ordinary differential equations, but also "boundary conditions" at the endpoints.

The second solution method is also useful for solving the fourth-order model that describes beam vibration. This model is derived in Section 10.4. Section 10.5 shows how to use MATLAB to solve the resulting transcendental equations for the natural frequencies.

LEARNING OBJECTIVES

After you have finished this chapter, you should be able to do the following:

- Derive equations of motion of common distributed systems.
- Solve the partial differential equation models.
- Interpret the solutions physically.
- Apply MATLAB to compute the natural frequencies.

10.1 INTRODUCTION

In our models considered thus far, we modeled the system as one or more lumped, or "discrete," masses, connected by discrete elastic and damping elements. Such models are called lumped-parameter models or discrete models. For example, we treated a cantilever beam as equivalent to a single mass and a single spring by equating its kinetic energy to that of a concentrated mass and by comparing its static deflection to that of a concentrated spring.

We can solve many practical problems with such models, but phenomena occur in real systems that are not predicted by or easily accounted for in lumped-parameter models. One example is wave motion in a cable. Another type of model, called a *distributed-parameter model*, treats the mass, elasticity, and damping as being distributed throughout the system. The mathematical form of such models is called a *partial differential equation* because it contains partial derivatives. Systems whose mass, elasticity, and damping are distributed parameters are called *distributed systems*. Examples of such systems are beams, shafts, and cables.

We now consider three typical examples of distributed systems, whose models have identical mathematical form. We will make three assumptions that are commonly used to reduce the complexity of a distributed-parameter model. We assume that the material is *homogeneous*, which means that its properties are uniform; for example, the mass density must be constant. The second assumption is that the material is *perfectly elastic*, which means that it obeys Hooke's law: stress is directly proportional to strain. The third assumption is that the material must be *isotropic*, which means that its properties are the same in every direction.

Transverse Vibration of a Cable

Consider the cable shown in in Figure 10.1-1. We assume that initially the cable is stretched along the x axis to a length L and that its ends are fixed at $x = 0$ and at $x = L$. We want to obtain an equation that can be solved for the deflection y at any point x, as a function of

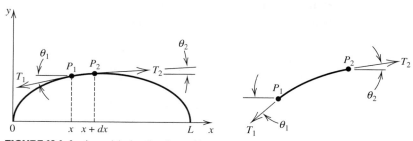

FIGURE 10.1-1 A model of a vibrating cable.

time t. We can see that y is a function of both x and t that describes the shape of the cable at any time, and thus we must use the *partial* derivatives of y with respect to x and t.

We assume the following:

1. The mass of the cable per unit length v is constant. (This is the "homogeneous" assumption.)

2. The cable is perfectly elastic, and it does not offer any resistance to bending.

3. The tension produced by stretching the cable is so large that the effect of gravity can be ignored.

4. The motions of the cable are only vertical, and they are small. This means that the slope is small.

Consider the infinitesimal cable element of length dx shown in Figure 10.1-1. The tension forces T_1 and T_2 at points P_1 and P_2 result from the cable being stretched between its two endpoints. We might or might not know the values of T_1 and T_2. The cable has a mass per unit length v, so the mass of an infinitesimal element of length dx is $v\,dx$. Because the cable does not move horizontally, the sum of forces in the x direction must be zero. This gives

$$T_2 \cos \theta_2 - T_1 \cos \theta_1 = 0$$

If we define $T = T_1 \cos \theta_1$, then we see that $T = T_2 \cos \theta_2$ also.

Summing forces in the y direction, we obtain

$$v\,dx\,\frac{\partial^2 y}{\partial t^2} = T_2 \sin \theta_2 - T_1 \sin \theta_1$$

where the acceleration $\partial^2 y/\partial t^2$ is taken to be evaluated at some point between x and $x + dx$. Dividing both sides by T gives

$$\frac{v}{T}\,dx\,\frac{\partial^2 y}{\partial t^2} = \frac{T_2 \sin \theta_2}{T} - \frac{T_1 \sin \theta_1}{T}$$

Since $T = T_1 \cos \theta_1 = T_2 \cos \theta_2$, this becomes

$$\frac{v}{T}\,dx\,\frac{\partial^2 y}{\partial t^2} = \frac{T_2 \sin \theta_2}{T_2 \cos \theta_2} - \frac{T_1 \sin \theta_1}{T_1 \cos \theta_1} = \tan \theta_2 - \tan \theta_1 \tag{10.1-1}$$

But $\tan \theta_1$ and $\tan \theta_2$ are the slopes of the cable at x and $x + dx$:

$$\tan \theta_1 = \left(\frac{\partial y}{\partial x}\right)\bigg|_x \qquad \tan \theta_2 = \left(\frac{\partial y}{\partial x}\right)\bigg|_{x+dx}$$

Substituting these expressions into Equation 10.1-1 and dividing by dx, we obtain

$$\frac{v}{T}\frac{\partial^2 y}{\partial t^2} = \frac{1}{dx}\left[\left(\frac{\partial y}{\partial x}\right)\bigg|_{x+dx} - \left(\frac{\partial y}{\partial x}\right)\bigg|_x\right]$$

The expression on the right-hand side becomes $\partial^2 y/\partial x^2$ if we let $dx \to 0$. Thus the model becomes

$$\frac{v}{T}\frac{\partial^2 y}{\partial t^2} = \frac{\partial^2 y}{\partial x^2} \tag{10.1-2}$$

Defining the constant c such that $c^2 = T/v$, the model becomes

$$\frac{\partial^2 y}{\partial t^2} = c^2 \frac{\partial^2 y}{\partial x^2} \tag{10.1-3}$$

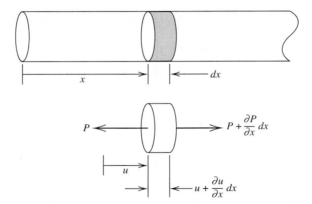

FIGURE 10.1-2   Longitudinal vibration of a rod.

This is the *one-dimensional wave equation*. We will see that c is the speed of a transverse wave traveling along the length of the cable.

Longitudinal Vibration of a Rod

Consider an elastic rod that has a uniform cross-sectional area A and an elastic modulus E. Figure 10.1-2 shows a cross-sectional element, which is a slice of infinitesimal thickness dx. The elastic forces on the left-hand and right-hand sides of the element are P and $P + (\partial P/\partial x)dx$. Under the action of these forces, the element has moved a distance u from its equilibrium position at x, as measured from some arbitrary point. The element also undergoes elastic strain, so that its width has increased by $(\partial u/\partial x)dx = \epsilon_x\, dx$, where ϵ_x is the strain in the x direction. Thus

$$\epsilon_x = \frac{\partial u}{\partial x}$$

Thus the force P is

$$P = EA\epsilon_x$$

Note that the displacement u is a function of both x and t.

The mass of the element is $\rho A\, dx$, where ρ is the mass density of the material (mass per unit volume). Summing forces on the element in the x direction, we obtain

$$\rho A\, dx\, \frac{\partial^2 u}{\partial t^2} = P + \frac{\partial P}{\partial x}\, dx - P = \frac{\partial P}{\partial x}\, dx$$

Since

$$\frac{\partial P}{\partial x} = \frac{\partial}{\partial x}\left(EA\frac{\partial u}{\partial x}\right)$$

we obtain

$$\rho A\, \frac{\partial^2 u}{\partial t^2} = \frac{\partial}{\partial x}\left(EA\frac{\partial u}{\partial x}\right)$$

Note that this model applies even if the area depends on x; that is, if $A = A(x)$. If A is constant and if we assume that the rod material is uniform, then A and E are constant, and the equation becomes

$$\frac{\partial^2 u}{\partial t^2} = \frac{E}{\rho}\frac{\partial^2 u}{\partial x^2} \tag{10.1-4}$$

This equation describes the propagation of a longitudinal stress wave. We will see that the propagation speed is $\sqrt{E/\rho}$.

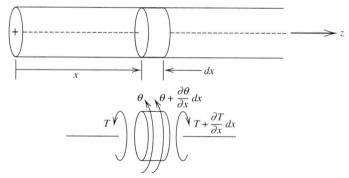

FIGURE 10.1-3 Torsional vibration of a rod.

Torsional Vibration of a Rod

We now consider the propagation of shear stress in a rod whose mass per unit volume is ρ. The cross-sectional area is A. The shear modulus of elasticity is G. Figure 10.1-3 shows a cross-sectional element that has been rotated through an angle θ from its equilibrium position and twisted by an amount $\gamma = R(\partial\theta/\partial x)$, where R is the radius of the rod.

Summing moments about the rod axis gives

$$\rho J \, dx \frac{\partial^2 \theta}{\partial t^2} = T + \frac{\partial T}{\partial x} \, dx - T = \frac{\partial T}{\partial x} \, dx$$

where J is the polar moment of inertia of the rod. The torque T is given by

$$T = GJ \frac{\gamma}{R} = GJ \frac{\partial\theta}{\partial x}$$

Substituting this for T, we obtain

$$\rho J \, dx \frac{\partial^2 \theta}{\partial t^2} = \frac{\partial T}{\partial x} \, dx = \frac{\partial}{\partial x} \left(GJ \frac{\partial\theta}{\partial x} \right)$$

Note that this model applies even if the cross-sectional area changes with x so that the polar moment depends on x; that is, if $J = J(x)$. If we assume that J and G are constant, the equation becomes

$$\frac{\partial^2 \theta}{\partial t^2} = \frac{G}{\rho} \frac{\partial^2 \theta}{\partial x^2} \tag{10.1-5}$$

This equation describes the propagation of a shear stress wave. We will see that the propagation speed is $\sqrt{G/\rho}$.

10.2 SOLUTION OF THE WAVE EQUATION

The three models developed in Section 10.1 all have the same mathematical form, the one-dimensional wave equation, which is

$$\frac{\partial^2 u}{\partial t^2} = c^2 \frac{\partial^2 u}{\partial x^2} \tag{10.2-1}$$

where $u = u(x, t)$.

d'Alembert's Solution

The solution of the wave equation is best visualized by considering the transverse vibration of a cable. With the cable under tension and fixed at both ends, if we "pluck" it by pulling it in the vertical direction, and then release it, two waves of identical shape move toward the opposite ends of the cable with a speed c, and each wave retains its shape as it moves. We will now prove this statement.

Suppose we set up a coordinate system that moves with the wave moving in the negative x direction, such that

$$v = x + ct \qquad (10.2\text{-}2)$$

Similarly for the wave moving to the right:

$$z = x - ct \qquad (10.2\text{-}3)$$

Then $u(x, t)$ becomes a function of v and z, and we can we can rewrite Equation 10.2-1 in terms of v and z as follows. Note that

$$\frac{\partial v}{\partial x} = \frac{\partial z}{\partial x} = 1$$

and

$$\frac{\partial u}{\partial x} = \frac{\partial u}{\partial v}\frac{\partial v}{\partial x} + \frac{\partial u}{\partial z}\frac{\partial z}{\partial x} = \frac{\partial u}{\partial v} + \frac{\partial u}{\partial z}$$

Assuming that all the partial derivatives are continuous, and applying the chain rule, we obtain

$$\frac{\partial^2 u}{\partial x^2} = \frac{\partial}{\partial x}\left(\frac{\partial u}{\partial v} + \frac{\partial u}{\partial z}\right)$$

$$= \frac{\partial}{\partial v}\left(\frac{\partial u}{\partial v} + \frac{\partial u}{\partial z}\right)\frac{\partial v}{\partial x} + \frac{\partial}{\partial z}\left(\frac{\partial u}{\partial v} + \frac{\partial u}{\partial z}\right)\frac{\partial z}{\partial x}$$

$$= \frac{\partial^2 u}{\partial v^2} + 2\frac{\partial^2 u}{\partial v\,\partial z} + \frac{\partial^2 u}{\partial z^2}$$

Applying the same procedure to $\partial^2 u / \partial t^2$, we obtain

$$\frac{\partial^2 u}{\partial t^2} = \frac{\partial^2 u}{\partial v^2} - 2\frac{\partial^2 u}{\partial v\,\partial z} + \frac{\partial^2 u}{\partial z^2}$$

Substitute these two expressions into Equation 10.2-1 to obtain

$$\frac{\partial^2 u}{\partial v\,\partial z} = 0 \qquad (10.2\text{-}4)$$

This equation can be solved directly with two integrations. Integrating with respect to z gives

$$\frac{\partial u}{\partial v} = h(v) \qquad (10.2\text{-}5)$$

where $h(v)$ is an arbitrary function of v. Integrating this equation with respect to v gives

$$u = \int h(v)\,dv + \psi(z) \qquad (10.2\text{-}6)$$

where $\psi(z)$ is an arbitrary function of z. Denote the integral, which is a function of v, by $\phi(v)$. Then

$$u(x, t) = \phi(v) + \psi(z) = \phi(x + ct) + \psi(x - ct) \qquad (10.2\text{-}7)$$

This procedure for solving the wave equation is due to d'Alembert, and the result is known as d'Alembert's solution.

Initial Conditions

Suppose that the initial shape of the cable (at $t = 0$) is given by

$$u(x, 0) = f(x) \qquad (10.2\text{-}8)$$

where $f(x)$ is a given function that describes the initial cable shape. Suppose also that we release the cable from rest so that the initial velocity is zero. Then

$$\left.\frac{\partial u}{\partial t}\right|_{t=0} = 0 \qquad (10.2\text{-}9)$$

Differentiate Equation 10.2-7 to obtain

$$\frac{\partial u}{\partial t} = \frac{\partial \phi}{\partial v}\frac{\partial v}{\partial t} + \frac{\partial \phi}{\partial z}\frac{\partial z}{\partial t} = c\phi'(x + ct) - c\psi'(x - ct)$$

where the prime $'$ denotes differentiation with respect to the arguments $x + ct$ and $x - ct$. Thus

$$u(x, 0) = \phi(x) + \psi(x) = f(x) \qquad (10.2\text{-}10)$$

and

$$\left.\frac{\partial u}{\partial t}\right|_{t=0} = c\phi'(x) - c\psi'(x) = 0 \qquad (10.2\text{-}11)$$

Divide the latter equation by c and integrate with respect to x to obtain

$$\phi(x) - \psi(x) = \phi(x_0) - \psi(x_0)$$

Adding this to Equation 10.2-10 gives

$$\phi(x) = \frac{1}{2}[f(x) + \phi(x_0) - \psi(x_0)]$$

Subtracting from Equation 10.2-10 gives

$$\psi(x) = \frac{1}{2}[f(x) - \phi(x_0) + \psi(x_0)]$$

If we replace x with $x + ct$ in $\phi(x)$ and x with $x - ct$ in $\psi(x)$, then from Equation 10.2-7 we see that

$$u(x, t) = \frac{1}{2}[f(x + ct) + f(x - ct)] \qquad (10.2\text{-}12)$$

Since $c > 0$, the function $f(x - ct)$ represents a wave that is traveling to the right with a speed c. Its shape does not change. Similarly, the function $f(x + ct)$ represents a wave that is traveling to the left with a speed c. The complete shape of the cable at any time is the *average* of these two functions.

If the initial velocity is not zero but some given function $g(x)$, we can use a similar procedure to show that

$$u(x,t) = \frac{1}{2}[f(x+ct) + f(x-ct)] + \frac{1}{2c}\int_{x-ct}^{x+ct} g(s)\,ds \qquad (10.2\text{-}13)$$

EXAMPLE 10.2-1
Wave Propagation
Speed

A string of length 4 m has a tension of 200 N. The string weighs 1.6 N. Compute the propagation speed of any transverse disturbances in the string.

Solution

The mass per unit length is

$$v = \frac{1.6/9.81}{4} = 0.0408\,\text{kg/m}$$

The propagation speed is

$$c = \sqrt{\frac{T}{v}} = \sqrt{\frac{200}{0.0408}} = 70\,\text{m/s}$$

■

EXAMPLE 10.2-2
Half-Sine Cable
Shape

Suppose that $L = 4$ and $c = 10$, and also assume that the cable is released from rest with the initial shape $f(x) = 4 \times 10^{-3} \sin \pi x/L$. Determine the cable shape $y(x,t)$.

Solution

The equation of motion is given by Equation 10.1-3 and its solution is given by Equation 10.2-12 with $u(x,t) = y(x,t)$. Thus

$$y(x,t) = \frac{4 \times 10^{-3}}{2}\left\{\sin\left[\frac{\pi}{4}(x+10t)\right] + \sin\left[\frac{\pi}{4}(x-10t)\right]\right\}$$

Using the identity for the sine of the sum of two angles and combining terms, we obtain

$$y(x,t) = 4 \times 10^{-3} \sin\left(\frac{\pi}{4}x\right)\cos\left(\frac{10\pi}{4}t\right)$$

$$= 4 \times 10^{-3} \sin\left(\frac{\pi}{4}x\right)\cos\left(\frac{5\pi}{2}t\right)$$

At any given time t, the cable has the same shape, $\sin(\pi x/4)$, but its amplitude varies with time as $4 \times 10^{-3} \cos(5\pi t/2)$. The frequency of vibration is thus $5\pi/2$ radians per unit time.

■

EXAMPLE 10.2-3
Full-Sine Cable
Shape

Suppose that $L = 4$ and $c = 10$ and that the cable is released from rest with the initial shape $f(x) = 4 \times 10^{-3} \sin 2\pi x/L$. Determine the cable shape $y(x,t)$.

Solution

The equation of motion is given by Equation 10.1-3 and its solution is given by Equation 10.2-12 with $u(x,t) = y(x,t)$. Thus

$$y(x,t) = \frac{4 \times 10^{-3}}{2}\left\{\sin\left[\frac{2\pi}{4}(x+10t)\right] + \sin\left[\frac{2\pi}{4}(x-10t)\right]\right\}$$

$$= \frac{4 \times 10^{-3}}{2}\left\{\sin\left[\frac{\pi}{2}(x+10t)\right] + \sin\left[\frac{\pi}{2}(x-10t)\right]\right\}$$

Using the identity for the sine of the sum of two angles and combining terms, we obtain

$$y(x, t) = 4 \times 10^{-3} \sin\left(\frac{\pi}{2}x\right) \cos(5\pi t)$$

At any given time t, the cable has the same shape, $\sin(\pi x/2)$, but its amplitude varies with time as $4 \times 10^{-3} \cos(5\pi t)$. The frequency of vibration is thus 5π radians per unit time. There is a node at $x = L/2 = 2$, so the midpoint of the cable never moves. ∎

10.3 SEPARATION OF VARIABLES

The d'Alembert solution of the wave equation, although elegant, does not provide a means for solving other distributed-parameter models. We now present a more general method of solution, known as *separation of variables*. We will use the wave equation as an example of this method. It is

$$\frac{\partial^2 u}{\partial t^2} = c^2 \frac{\partial^2 u}{\partial x^2} \tag{10.3-1}$$

Initial Conditions and Boundary Conditions

To solve this model, we must be given information about the initial values of u and $\partial u/\partial t$. These are the *initial conditions*. In addition, we must also be given information about the constraints, or lack thereof, at the endpoints, say where $x = 0$ and $x = L$. These are the *boundary conditions*. They can take on many forms, and this is what makes solving partial differential equations more complicated than solving ordinary differential equations. Considering the cable problem as an example, the two endpoints may be specified as fixed, or one end may be fixed and the other end free to move. We will see other examples of boundary conditions in later examples.

Obtaining Two Ordinary Differential Equations

We can always attempt a solution of a differential equation by trying a function form and substituting it into the equation. If it solves the equation, then we can evaluate the function for the given initial and boundary conditions. Suppose we try a solution of the form of the product of two functions, one a function only of x and one a function only of t.

$$u(x, t) = F(x)G(t) \tag{10.3-2}$$

From this we obtain

$$\frac{\partial^2 u}{\partial t^2} = F(x)\ddot{G}(t) \qquad \frac{\partial^2 u}{\partial x^2} = F''(x)G(t)$$

where the dot denotes d/dt and the prime denotes d/dx. Substituting these into Equation 10.3-1 gives

$$F(x)\ddot{G}(t) = c^2 F''(x)G(t)$$

Divide both sides by $c^2 F(x)G(t)$:

$$\frac{\ddot{G}(t)}{c^2 G(t)} = \frac{F''(x)}{F(x)}$$

Because the left-hand side is a function only of t and the right-hand side is a function only of x, the only way this equation can be true is if each side is equal to the same constant, which we will denote by α. Thus

$$\frac{\ddot{G}(t)}{c^2 G(t)} = \frac{F''(x)}{F(x)} = \alpha$$

and we have two *ordinary* differential equations:

$$\ddot{G} - c^2 \alpha G = 0 \tag{10.3-3}$$

$$F'' - \alpha F = 0 \tag{10.3-4}$$

Satisfying the Boundary Conditions

Let us take the following specific set of boundary conditions; namely, that the endpoints are fixed. Thus we take $u(x, t) = u(L, t) = 0$, where L is the cable length. In terms of our assumed solution (Equation 10.3-2), this implies that

$$u(0, t) = F(0)G(t) = 0 \qquad \text{for all } t \tag{10.3-5}$$

and

$$u(L, t) = F(L)G(t) = 0 \qquad \text{for all } t \tag{10.3-6}$$

One possible solution to these equations is $G(t) = 0$ for all t, but this implies that $u(x, t) = 0$ for all t, so we discard this trivial solution. So from Equations 10.3-5 and 10.3-6 we have

$$F(0) = 0 \qquad \text{and} \qquad F(L) = 0 \tag{10.3-7}$$

We do not as yet know the value of α, so now we must consider three cases:

1. $\alpha = 0$. In this case the solution of Equation 10.3-4 is $F(x) = ax + b$, where a and b are constants. From Equation 10.3-7, however, we see that $a = b = 0$, which gives a trivial solution, $F(x) = 0$, for all x and implies that $u(x, t) = 0$ for all x.
2. $\alpha > 0$. Let $r^2 = \alpha$. Then Equation 10.3-4 becomes $F'' - r^2 F = 0$, which has the solution

$$F(x) = Ae^{rx} + Be^{-rx}$$

Applying Equation 10.3-7, we find that $A = B = 0$, and thus again $F(x) = 0$ for all x.
3. $\alpha < 0$. Let $p^2 = -\alpha$. Then Equation 10.3-4 becomes $F'' + p^2 F = 0$, which has the solution

$$F(x) = A \cos px + B \sin px$$

Thus the solution to the wave equation has the form

$$u(x, t) = (A \cos px + B \sin px)G(t) \tag{10.3-8}$$

Applying the specific boundary condition given by Equation 10.3-7, we obtain $F(0) = A = 0$ and $F(L) = B \sin pL = 0$. If $B = 0$, the solution is again the trivial one, $F(x) = 0$. So the only possibility left to us is that $\sin pL = 0$.

Requiring that $\sin pL = 0$ implies that

$$p = \frac{n\pi}{L} \qquad n \text{ integer} \tag{10.3-9}$$

Thus there are an infinite number of solutions for $F(x)$, all of the form

$$F_n(x) = B_n \sin\left(\frac{n\pi}{L}x\right) \qquad n \text{ integer}$$

We will soon see that the constants B_n can be absorbed into another set of constants, so we can set $B_n = 1$ without loss of generality. Also, because $n = 0$ gives a trivial solution and because negative values of n give the same form of solution, we can ignore the nonpositive values and express the solution as

$$F_n(x) = \sin\left(\frac{n\pi}{L}x\right) \qquad n = 1, 2, 3, \cdots \tag{10.3-10}$$

The values $p = n\pi/L$ are called the *spatial frequencies* because they describe the wave shape of the cable. In this case, the wave shape is sinusoidal, $\sin n\pi x/L$, and the spatial wavelength is $2L/n$.

Not all applications will have the specific boundary conditions given by Equation 10.3-7, so the constant A will not always be zero in such problems. For example, the left-hand end of the cable could be free to move, so that $u(0, t) \neq 0$. So the general solution is given by Equation 10.3-8.

Computing the Temporal Frequencies with Both Ends Fixed

We now have values for α, which are $\alpha = -p^2 = -(n\pi/L)^2$. Thus Equation 10.3-3 becomes

$$\ddot{G} + c^2\left(\frac{n\pi}{L}\right)^2 G = 0$$

If we define

$$\lambda_n = \frac{cn\pi}{L} \tag{10.3-11}$$

then we can write

$$\ddot{G} + \lambda_n^2 G = 0 \tag{10.3-12}$$

The solution of this equation has the form

$$G_n(t) = C_n \cos \lambda_n t + D_n \sin \lambda_n t \qquad n = 1, 2, 3, \cdots$$

The λ_n give the frequencies of oscillation, which are sometimes called the *temporal frequencies* to distinguish them from the spatial frequencies. The period of oscillation is $2\pi/\lambda_n$.

The form of the solution to the wave equation (Equation 10.3-1) is thus

$$u_n(x, t) = G_n(t)F_n(x)$$
$$= (C_n \cos \lambda_n t + D_n \sin \lambda_n t)\left(A_n \cos \frac{n\pi}{L}x + B_n \sin \frac{n\pi}{L}x\right) \qquad n = 1, 2, 3, \cdots \tag{10.3.13}$$

The form of the solution satisfying the wave equation and the specific boundary conditions (Equations 10.3-5 and 10.3-6) is

$$u_n(x, t) = G_n(x)F_n(t)$$
$$= (C_n \cos \lambda_n t + D_n \sin \lambda_n t) \sin \frac{n\pi}{L}x \qquad n = 1, 2, 3, \cdots \tag{10.3-14}$$

Modes

The function $u_n(x,t) = G_n(x)F_n(t)$ is the nth mode of vibration. It consists of a *mode shape*, described by $G_n(x)$, and a time function $F_n(t)$ that describes how the mode shape oscillates in time. The function $G_n(x)$ is called the nth normal mode. The *fundamental mode* corresponds to $n = 1$, and λ_1 is the *fundamental frequency*. In musical terminology, the other modes are the *overtones*.

The *nodes* are the locations (the values of x) at which $G_n(x) = 0$. Thus $u_n(x,t) = 0$ for all times t at a node. Thus for the case where $G_n(x) = \sin(n\pi x/L)$, the first mode has two nodes, one at each endpoint (at $x = 0$ and $x = L$). The second mode has three nodes, one at each endpoint and one at $x = L/2$. The nth mode has $n+1$ nodes. Figure 10.3-1 shows the first four normal modes. Figure 10.3-2 shows the second mode at various values of t. It has the form of the sine wave $G_2(x) = \sin(2\pi x/L)$ at any instant t.

EXAMPLE 10.3-1 *Vibration of a* *Guitar String*	A certain guitar string has a mass of 0.25 g and a length of 38 cm. Compute the tension needed to obtain a temporal frequency of 196 Hz (which is a little below the note called "middle C").
Solution	We are give $m = 2.5 \times 10^{-4}$ kg and $L = 0.38$ m. The string actually has an infinite number of frequencies, λ_n, $n = 1, 2, 3, \cdots$. The implication here is that 196 Hz should be the "fundamental frequency," which is the lowest temporal frequency. Thus we require that $\lambda_1 = 2\pi(196)$ rad/s. From Equations 10.1-2 and 10.3-11, the tension is

$$T = vc^2 = v\left(\frac{\lambda_1 L}{\pi}\right)^2 = \frac{2.5 \times 10^{-4}}{0.38}\left(\frac{2\pi(196)(0.38)}{\pi}\right)^2 = 14.6 \text{ N} \qquad \blacksquare$$

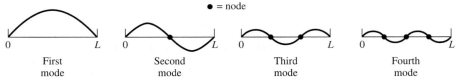

$\bullet$ = node

First mode   Second mode   Third mode   Fourth mode

FIGURE 10.3-1   First four modes of transverse vibration of a cable with both ends fixed.

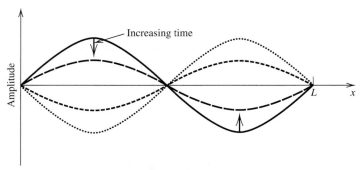

FIGURE 10.3-2   Time-dependent behavior of the second mode of transverse vibration of a cable with both ends fixed.

Satisfying the Initial Conditions

Each of the functions given by Equation 10.3-13 is a solution of Equation 10.3-1 but does not satisfy the initial condition:

$$u(x, 0) = f(x) \qquad (10.3\text{-}15)$$

where $f(x)$ is a given function that describes the initial cable shape. Suppose also that we release the cable from rest so that the initial velocity is zero. Then

$$\frac{\partial u}{\partial t}\bigg|_{t=0} = g(x) \qquad (10.3\text{-}16)$$

Because Equation 10.3-1 is linear and homogeneous, we can obtain a general solution by adding the functions given by Equation 10.3-13 as follows:

$$u(x, t) = \sum_{n=1}^{\infty} u_n(x, t) = \sum_{n=1}^{\infty} (C_n \cos \lambda_n t + D_n \sin \lambda_n t) \sin \frac{n\pi}{L} x \qquad (10.3\text{-}17)$$

Evaluating this at $t = 0$ gives

$$u(x, 0) = \sum_{n=1}^{\infty} C_n \sin \frac{n\pi}{L} x = f(x)$$

From this we see that the coefficients C_n are the coefficients of a Fourier sine series representing $f(x)$. These are given by

$$C_n = \frac{2}{L} \int_0^L f(x) \sin \frac{n\pi x}{L} \, dx \qquad n = 1, 2, 3, \cdots \qquad (10.3\text{-}18)$$

The initial condition on the velocity can be satisfied as follows. Differentiating Equation 10.3-17 with respect to t and using Equation 10.3-16, we obtain

$$\frac{\partial u}{\partial t}\bigg|_{t=0} = \sum_{n=1}^{\infty} D_n \lambda_n \sin \frac{n\pi x}{L} = g(x)$$

From this we see that the coefficients $D_n \lambda_n$ are the coefficients of a Fourier sine series representing $g(x)$. These are given by

$$D_n = \frac{2}{\lambda_n L} \int_0^L g(x) \sin \frac{n\pi x}{L} \, dx = \frac{2}{nc\pi} \int_0^L g(x) \sin \frac{n\pi x}{L} \, dx \qquad n = 1, 2, 3, \cdots \quad (10.3\text{-}19)$$

If the initial velocity is zero, then $D_n = 0$.

There are some mathematical details we have glossed over that may disqualify this method of obtaining a solution. These pertain to the convergence of the Fourier series, but for most practical problems, the nature of the functions $f(x)$ and $g(x)$ is such that there are no difficulties.

In summary, the solution of the wave equation (Equation 10.3-1) for the initial conditions (Equations 10.3-15 and 10.3-16) and the boundary conditions (Equations 10.3-5 and 10.3-6) is given by Equation 10.3-17, where C_n and D_n are computed from Equations 10.3-18 and 10.3-19.

EXAMPLE 10.3-2
Half-Sine Cable Shape

Obtain the solution to the wave equation when the cable is released from rest with the initial shape given by $f(x) = A \sin(\pi x / L)$.

Solution

Because the initial velocity is zero, $g(x) = 0$ and thus $D_n = 0$. For C_n, note that here $f(x)$ is already a Fourier sine series having only one term. Thus

$$C_1 = A \qquad \text{and} \qquad C_n = 0 \qquad \text{for } n > 1$$

Thus the solution is

$$u(x, t) = A \cos \lambda_1 t \sin \frac{n\pi x}{L} = A \cos \frac{c\pi t}{L} \sin \frac{n\pi x}{L}$$

The cable retains its original shape while oscillating at a radian frequency of $c\pi/L$. ∎

EXAMPLE 10.3-3
Triangular Cable Shape

Obtain the solution to the wave equation when the cable is released from rest with the initial shape given by

$$f(x) = \begin{cases} \dfrac{2x}{L} & 0 \le x \le \dfrac{L}{2} \\[2mm] \dfrac{2}{L}(L - x) & \dfrac{L}{2} < x \le L \end{cases}$$

Solution

Because the initial velocity is zero, $g(x) = 0$ and thus $D_n = 0$. For C_n,

$$C_n = \frac{2}{L} \int_0^{L/2} \frac{2x}{L} \sin \frac{n\pi x}{L} \, dx + \int_{L/2}^L \frac{2}{L}(L - x) \sin \frac{n\pi x}{L} \, dx$$

Using a table of integrals, we find that

$$C_n = \frac{8}{n^2 \pi^2} \sin \frac{n\pi}{2}$$

and thus

$$u(x, t) = \frac{8}{\pi^2} \left(\frac{1}{1^2} \sin \frac{\pi x}{L} \cos \frac{c\pi t}{L} - \frac{1}{3^2} \sin \frac{3\pi x}{L} \cos \frac{3c\pi t}{L} \right.$$
$$\left. + \frac{1}{5^2} \sin \frac{5\pi x}{L} \cos \frac{5c\pi t}{L} - \cdots \right) \tag{1}$$

Consider the case where $L = 2$ and $c = 4$. For this case the oscillation frequency is $c\pi/L = 2\pi$, and the period is 1. Figure 10.3-3 shows the cable shape for one-half a period, at the times $t = 0, 0.05, 0.1, \ldots 0.5$. This plot is based on the first seven nonzero terms in the series given by Equation 1. This corresponds to $n = 13$. Section 10.5 gives the MATLAB program used to generate this plot. ∎

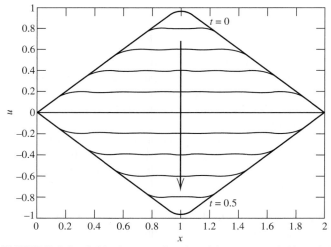

FIGURE 10.3-3   Cable shape as a function of time over one-half a period.

Rod with Concentrated Mass

Consider the rod with a concentrated mass m attached at its end (Figure 10.3-4a). Earlier we modeled the rod as a spring of stiffness $k = EA/L$, where A is the cross-sectional area of the rod and L is its length. Using kinetic energy equivalence, we then lumped one-third of the rod mass m_r with the concentrated mass to obtain the following lumped-parameter model:

$$\left(m + \frac{m_r}{3}\right)\ddot{x} + \frac{EA}{L}x = 0$$

The natural frequency is

$$\omega_n = \sqrt{\frac{EA}{L(m + m_r/3)}} \qquad (10.3\text{-}20)$$

Now we obtain the expressions for the natural frequencies of the system, treating the rod as a distributed-parameter system. The wave equation describes the longitudinal motion of the rod:

$$\frac{\partial^2 u}{\partial t^2} = c^2 \frac{\partial^2 u}{\partial x^2}$$

Figure 10.3-4b shows the axial forces acting on the rod. Assuming that the deflection u is measured from the equilibrium produced by the static load mg, then the weight mg will not appear in the equation of motion. The force f acting at the end of the rod is due to the rod deflection as follows (using Hooke's law):

$$f = EA\left(\frac{\partial u}{\partial x}\right)\bigg|_{x=L}$$

where $\partial u/\partial x$ is the unit strain.

Using separation of variables, $u(x,t) = F(x)G(t)$, we obtain $F'' + p^2 F = 0$, which has the solution

$$F(x) = A_1 \cos px + B_1 \sin px$$

The boundary conditions are as follows. The upper end of the rod is fixed, so $u(0,t) = 0$, which implies that $F(0) = 0$. Newton's law applied to mass m gives

$$m\left(\frac{\partial^2 u}{\partial t^2}\right)_{x=L} = -f = -EA\left(\frac{\partial u}{\partial x}\right)\bigg|_{x=L}$$

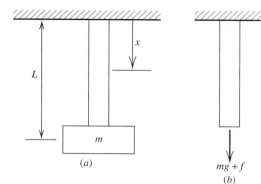

(a)

$mg + f$

(b)

FIGURE 10.3-4   A rod with a concentrated mass.

Thus

$$mF(L)\ddot{G} = -EAF'(L)G(t)$$

or

$$-\frac{EA}{m}\frac{F'(L)}{F(L)} = \frac{\ddot{G}}{G(t)} = -c^2p^2$$

Applying the condition $F(0) = 0$ to the solution form for $F(x)$ we obtain $A_1 = 0$ and thus $F(x) = B_1 \sin px$. The second boundary condition gives

$$-\frac{EAp \cos pL}{m \sin pL} = -c^2p^2$$

which can be expressed as

$$\tan pL = \frac{EA}{mpc^2} = \frac{\rho A}{mp} \tag{10.3-21}$$

since $c^2 = E/\rho$.

This equation defines the spatial frequencies p that identify the nodes. The temporal frequencies ω (the oscillation frequencies) are found from

$$\omega = pc = p\sqrt{\frac{E}{\rho}}$$

Letting $y = pL$, we can write

$$\tan y = \left(\frac{\rho AL}{m}\right)\frac{1}{y} = \frac{\beta}{y}$$

where $\beta = \rho AL/m$. This equation has an infinite number of roots and must be solved numerically. This requires that a numerical value for β be given. Many engineering calculators have a *solve* function that will compute the roots of such an equation. Computer software such as MATLAB can also be used, as discussed in Section 10.5.

However, if y is "small" enough, then $\tan y \approx y$ and we have $y \approx \beta/y$ or $y = \sqrt{\beta}$. This enables us to solve for the fundamental frequency as follows:

$$p \approx \sqrt{\frac{\rho A}{mL}}$$

Thus

$$\omega = \sqrt{\frac{E}{\rho}}p = \sqrt{\frac{E}{\rho}}\sqrt{\frac{\rho A}{mL}} = \sqrt{\frac{EA}{mL}}$$

This is the fundamental (lowest) oscillation frequency if y is small, which is true if the rod length L is small (and thus the rod mass is negligible relative to the concentrated mass). This expression is identical to Equation 10.3-20, obtained from the lumped-parameter analysis if the rod mass $m_r = 0$.

If the rod mass m_r equals the concentrated mass m, then $\beta = 1$ and the first three solutions are $y = 0.8603, 3.4256$, and 6.4373. These give the following spatial frequencies:

$$p = \frac{0.8603}{L}, \quad \frac{3.4256}{L}, \quad \frac{6.4373}{L}$$

If the rod mass is 1/10 of the concentrated mass, then $\beta = 0.1$ and the first three solutions are $p = 0.3111/L$, $p = 3.1731/L$, and $p = 6.2991/L$.

Torsional Vibration of a Rod and Disk

Consider the rod shown in Figure 10.3-5. The top end is fixed, and the bottom end of the rod is attached to a disk whose mass moment of inertia is I_d. The inertia of the rod about its long axis is $I_r = \rho L J_r$, where J_r is the polar moment of inertia of the rod.

The equation of motion is

$$\frac{\partial^2\theta}{\partial t^2} = c^2 \frac{\partial^2\theta}{\partial x^2}$$

where

$$c^2 = \frac{G}{\rho}$$

The general solution form is

$$\theta(t) = [A\ \sin(\omega cx) + B\ \cos(\omega cx)](C\ \sin \omega t + D\ \cos \omega t)$$

In this case the upper end of the rod is fixed. Thus the boundary condition is

$$\theta(0, t) = 0$$

which gives $B = 0$ and

$$\theta(t) = A\ \sin(\omega cx)(C\ \sin \omega t + D\ \cos \omega t)$$

This gives

$$\frac{\partial\theta}{\partial x} = A\omega c\ \cos(\omega cx)(C\ \sin \omega t + D\ \cos \omega t) \tag{10.3-22}$$

and

$$\frac{\partial^2\theta}{\partial t^2} = A\ \sin(\omega cx)\left(-C\omega^2\ \sin \omega t - D\omega^2\ \cos \omega t\right) \tag{10.3-23}$$

Summing moments at the lower end of the rod gives

$$I_d\left(\frac{\partial^2\theta}{\partial t^2}\right)_{x=L} = -GJ_r\left(\frac{\partial\theta}{\partial x}\right)_{x=L}$$

From Equations 10.3-22 and 10.3-23,

$$I_d\omega^2\ \sin\left(\omega L\sqrt{\frac{\rho}{G}}\right) = GJ_r\left(\omega\sqrt{\frac{\rho}{G}}\right)\cos\left(\omega L\sqrt{\frac{\rho}{G}}\right)$$

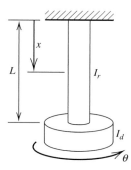
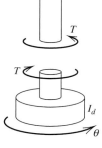

FIGURE 10.3-5 Torsional vibration of a rod and disk.

or

$$\tan \omega L \sqrt{\frac{\rho}{G}} = \frac{J_r}{\omega I_d} \sqrt{G\rho} = \frac{\rho L J_r}{\omega L I_d} \sqrt{\frac{G}{\rho}} = \frac{I_r}{\omega L I_d} \sqrt{\frac{G}{\rho}}$$

If we define

$$\beta = \omega L \sqrt{\frac{\rho}{G}} \tag{10.3-24}$$

then we obtain

$$\beta \tan \beta = \frac{I_r}{I_d} \tag{10.3-25}$$

For a cylinder of length L and radius R, the expression for its mass moment of inertia is

$$I = \frac{1}{2} \rho \pi L R^4$$

Thus if the rod and disk in Figure 10.3-5 are cylinders of length L_r and L_d and radius R_r and R_d, then Equation 10.3-25 becomes

$$\beta \tan \beta = \frac{I_r}{I_d} = \left(\frac{L_r}{L_d}\right) \left(\frac{R_r}{R_d}\right)^4 \tag{10.3-26}$$

EXAMPLE 10.3-4
Vibration of a
Router Bit

A representation of a router bit is given in Figure 10.3-6. Consider the upper end of the shank to be fixed to the motor shaft. The router speed is 25,000 rpm, and the bit has two cutters. The dimensions are

$$L_1 = 25 \text{ mm} \qquad L_2 = 6 \text{ mm}$$
$$R_1 = 3 \text{ mm} \qquad R_2 = 10 \text{ mm}$$

Use the following data (for steel):

$$G = 8 \times 10^{10} \ \frac{\text{N}}{\text{m}^2}$$

$$\rho = 7800 \ \frac{\text{kg}}{\text{m}^3}$$

Compute the natural frequencies and determine whether or not they are close to the forcing frequency.

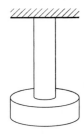
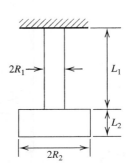

FIGURE 10.3-6   Representation of a router bit.

Solution From Equation 10.3-26,

$$\beta \tan \beta = \left(\frac{L_1}{L_2}\right)\left(\frac{R_1}{R_2}\right)^4 = 0.03375 \tag{1}$$

If β is small, then $\tan \beta \approx \beta$, so the previous equation gives

$$\beta^2 = 0.03375$$

or $\beta = 0.184$. Since $\tan 0.184 = 0.186$, our assumption that $\tan \beta \approx \beta$ is verified. The frequencies are given by Equation 10.3-24 to be

$$\omega = \frac{\beta}{L_1}\sqrt{\frac{G}{\rho}} = 1.28 \times 10^5 \beta$$

Thus the lowest frequency is

$$\omega = 1.28 \times 10^5 (0.184) = 2.36 \times 10^4 \text{ rad/s}$$

The remaining roots are found numerically to be

$$\beta = 3.1523, \quad 6.28855, \quad 59.6908,\ldots$$

and the corresponding frequencies are

$$\omega = 4.03 \times 10^5, \quad 8.813 \times 10^5, \quad 7.64 \times 10^6,\ldots \text{ rad/s}$$

The forcing frequency, which is twice the rotation speed because there are two cutters, is

$$\omega_f = 2\pi \frac{2(25,000)}{60} = 5236 \text{ rad/s}$$

Since this is not near any of the natural frequencies, we would not expect a resonant condition to exist. ∎

Table 10.3-1 summarizes the temporal frequencies and mode shapes of longitudinal and torsional vibrations for three common sets of boundary conditions.

10.4 TRANSVERSE VIBRATIONS OF BEAMS

Transverse beam vibrations, in which the beam vibrates in a direction perpendicular to its length, are also called *flexural* vibrations or *bending* vibrations. Figure 10.4-1 shows a section of a beam. The vibrations are assumed to be in the y direction, and the deflection is denoted by $u(x,t)$. The beam is assumed to be rectangular, of height h and width w. For now we allow the possibility that the beam cross-sectional area A changes along the beam length, so that $A = A(x)$. Let $I(x)$ be the area moment of inertia of the beam about the z axis. Let $f(x,t)$ be an externally applied load with dimensions of force per length.

We make the following assumptions (these assumptions constitute *Euler-Bernoulli beam theory*):

- The beam has uniform properties.
- The beam is slender (h/L and w/L are small).
- The beam material is linear, homogeneous, and isotropic; additionally, it obeys Hooke's law.
- There is no axial load.

TABLE 10.3-1   Longitudinal and Torsional Vibrations of a Uniform Rod of Length L

Longitudinal equation of motion:

$$\frac{\partial^2 u(x,t)}{\partial t^2} = c^2 \frac{\partial^2 u(x,t)}{\partial x^2} \qquad c^2 = \frac{E}{\rho}$$

Torsional equation of motion:

$$\frac{\partial^2 \theta(x,t)}{\partial t^2} = c^2 \frac{\partial^2 \theta(x,t)}{\partial x^2} \qquad c^2 = \frac{G}{\rho}$$

Boundary conditions	Temporal frequencies	Mode shape
Free-free	$\omega_n = \dfrac{n\pi c}{L}$ $n = 0,1,2,\cdots$	$\cos \dfrac{n\pi x}{L}$
Fixed-free	$\omega_n = \dfrac{(2n-1)\pi c}{2L}$ $n = 1,2,3,\cdots$	$\sin \dfrac{(2n-1)\pi x}{2L}$
Fixed-fixed	$\omega_n = \dfrac{n\pi c}{L}$ $n = 1,2,3,\cdots$	$\sin \dfrac{n\pi x}{L}$

- Plane sections remain plane during motion.
- The plane of motion is the same as the beam symmetry plane (so rotation and translation are independent).
- The mass moment of inertia about the z axis is negligible.
- Shear deformation is negligible.

Consider the infinitesimal beam element shown in Figure 10.4-1. Let V be the shear force and M the bending moment, which is described by

$$M(x,t) = EI(x)\frac{\partial^2 u}{\partial x^2} \tag{10.4-1}$$

The sides of the element remain straight and vertical during motion if the shear deformation is much smaller than $u(x,t)$. Thus Newton's law in the y direction gives

$$\rho A\, dx \frac{\partial^2 u}{\partial t^2} = \left(V + \frac{\partial V}{\partial x}\, dx\right) - V + f\, dx$$

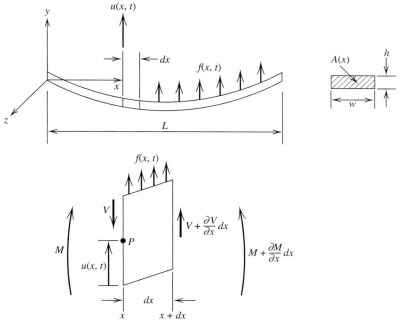

FIGURE 10.4-1 A model of transverse beam vibration.

This becomes

$$\rho A \, dx \frac{\partial^2 u}{\partial t^2} = \left(f + \frac{\partial V}{\partial x} \right) dx \qquad (10.4\text{-}2)$$

Summing moments about an axis passing through the point P and parallel to the z axis, we obtain

$$\left(M + \frac{\partial M}{\partial x} \, dx \right) - M + \left(V + \frac{\partial V}{\partial x} \, dx \right) dx + (f \, dx)\frac{dx}{2} = 0$$

This becomes

$$\left(V + \frac{\partial M}{\partial x} \right) dx + \left(\frac{f}{2} + \frac{\partial V}{\partial x} \right)(dx)^2 = 0$$

If $(dx)^2$ is small compared to dx, then the preceding equation gives

$$\left(V + \frac{\partial M}{\partial x} \right) dx = 0$$

or

$$V = -\frac{\partial M}{\partial x}$$

Substitute this into Equation 10.4-2 to obtain

$$\rho A \, dx \frac{\partial^2 u}{\partial t^2} = \left(f - \frac{\partial^2 M}{\partial x^2} \right) dx$$

From Equation 10.4-1,

$$\rho A \frac{\partial^2 u}{\partial t^2} = f - \frac{\partial^2 M}{\partial x^2} = f - \frac{\partial^2}{\partial x^2}\left[EI(x)\frac{\partial^2 u}{\partial x^2}\right]$$

or

$$\rho A(x)\frac{\partial^2 u}{\partial t^2} + \frac{\partial^2}{\partial x^2}\left[EI(x)\frac{\partial^2 u}{\partial x^2}\right] = f(x,t) \tag{10.4-3}$$

Note that this model allows both A and E to be functions of x.

If I is a constant, then

$$\rho A(x)\frac{\partial^2 u}{\partial t^2} + EI\frac{\partial^4 u}{\partial x^4} = f(x,t) \tag{10.4-4}$$

Solution of the Beam Equation

If A and I are constant and if the applied load $f(x,t)$ is zero, then the model becomes

$$\frac{\partial^2 u}{\partial t^2} + c^2\frac{\partial^4 u}{\partial x^4} = 0 \tag{10.4-5}$$

where

$$c^2 = \frac{EI}{\rho A} \tag{10.4-6}$$

To solve this model, we need two initial conditions: $u(x,0)$ and $(\partial u/\partial t)_{t=0}$. We also need four boundary conditions (because the model contains a fourth-order derivative with respect to x). These include information about the following at the ends of the beam:

- The deflection u
- The slope $\partial u/\partial x$
- The bending moment $EI(\partial^2 u/\partial x^2)$
- The shear force $\partial(EI\partial^2 u/\partial x^2)/\partial x$

These boundary conditions depend on the nature of the beam support at each end. The cases are summarized in Table 10.4-1.

We can solve Equation 10.4-5 by separation of variables. Let $u(x,t) = F(x)G(t)$ to obtain

$$F(x)\ddot{G}(t) + c^2 F''''(x)G(t) = 0$$

or

$$c^2\frac{F''''(x)}{F(x)} = -\frac{\ddot{G}(t)}{G(t)} = \omega^2$$

TABLE 10.4-1   Types of Beam Supports

End condition	Deflection	Slope	Bending moment	Shear force
Free	Unconstrained	Unconstrained	0	0
Clamped (fixed)	0	0	Unconstrained	Unconstrained
Simply supported (pinned)	0	Unconstrained	0	Unconstrained
Sliding	Unconstrained	0	Unconstrained	0

where ω is as yet unknown. This gives two equations. The first is

$$\ddot{G}(t) + \omega^2 G(t) = 0$$

which has the solution

$$G(t) = A \cos \omega t + B \sin \omega t$$

The second equation is

$$F''''(x) - \left(\frac{\omega}{c}\right)^2 F(x) = 0$$

Let

$$\beta^4 = \frac{\omega^2}{c^2} \tag{10.4-7}$$

Then the equation becomes

$$F''''(x) - \beta^4 F(x) = 0 \tag{10.4-8}$$

To solve Equation 10.4-8, we try the form

$$F(x) = Ce^{\lambda x}$$

Substituting this into Equation 10.4-8 gives

$$\lambda^4 Ce^{\lambda x} - \beta^4 Ce^{\lambda x} = 0$$

or $\lambda^4 = \beta^4$. This has four solutions: two real and two imaginary:

$$\lambda = \pm\beta, \qquad \pm i\beta$$

So the solution for $F(x)$ consists of the sum of four terms, each corresponding to a value for λ:

$$F(x) = C_1 e^{\beta x} + C_2 e^{-\beta x} + C_3 e^{i\beta x} + C_4 e^{-i\beta x}$$

or

$$F(x) = D_1 \cosh \beta x + D_2 \sinh \beta x + D_3 \cos \beta x + D_4 \sin \beta x \tag{10.4-9}$$

Since $\beta^4 = \omega^2/c^2$, the spatial frequencies are given by

$$\omega = c\beta^2 = \beta^2 \sqrt{\frac{EI}{\rho A}}$$

EXAMPLE 10.4-1
Natural Frequencies of a Cantilever Beam

Obtain the expression for the natural frequencies of the cantilever beam shown in Figure 10.4-2.

Solution

The natural frequencies are found from Equation 10.4-9:

$$F(x) = D_1 \cosh \beta x + D_2 \sinh \beta x + D_3 \cos \beta x + D_4 \sin \beta x$$

In this case the left-hand end of the beam is clamped and the right-hand end is free. Thus the boundary conditions are

$$u(0, t) = 0 \qquad \text{which gives } F(0) = 0$$

$$\left.\frac{\partial u}{\partial x}\right|_{x=0} = 0 \qquad \text{which gives } F'(0) = 0$$

$$\left.\frac{\partial^2 u}{\partial x^2}\right|_{x=L} = 0 \qquad \text{which gives } F''(L) = 0$$

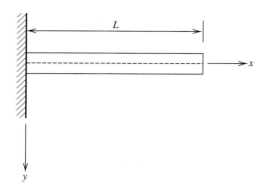

FIGURE 10.4-2 A cantilever beam.

and

$$\left.\frac{\partial^3 u}{\partial x^3}\right|_{x=L} = 0 \qquad \text{which gives } F'''(L) = 0$$

Application of the first two conditions gives

$$D_1 + D_3 = 0 \qquad \text{and} \qquad D_2 + D_4 = 0$$

and thus

$$F(x) = D_1(\cosh \beta x - \cos \beta x) + D_2(\sinh \beta x - \sin \beta x)$$

This gives

$$F'(x) = \beta D_1(\sinh \beta x + \sin \beta x) + \beta D_2(\cosh \beta x - \cos \beta x)$$
$$F''(x) = \beta^2 D_1(\cosh \beta x + \cos \beta x) + \beta^2 D_2(\sinh \beta x + \sin \beta x)$$
$$F'''(x) = \beta^3 D_1(\sinh \beta x - \sin \beta x) + \beta^3 D_2(\cosh \beta x + \cos \beta x)$$

Thus the third and fourth boundary conditions give

$$(\cosh \beta L + \cos \beta L)D_1 + (\sinh \beta L + \sin \beta L)D_2 = 0$$
$$(\sinh \beta L - \sin \beta L)D_1 + (\cosh \beta L + \cos \beta L)D_2 = 0$$

The determinant of these two equations is

$$\begin{vmatrix} (\cosh \beta L + \cos \beta L) & (\sinh \beta L + \sin \beta L) \\ (\sinh \beta L - \sin \beta L) & (\cosh \beta L + \cos \beta L) \end{vmatrix} = 0$$

Using the identity $\cosh^2 \beta L - \sinh^2 \beta L = 1$ and canceling terms, we obtain

$$\cosh \beta L \cos \beta L = -1$$

This equation has the following solutions for the five lowest spatial frequencies:

$$\beta = \frac{1.875}{L}, \frac{4.694}{L}, \frac{7.855}{L}, \frac{10.996}{L}, \frac{14.137}{L}$$

Thus the natural (temporal) frequencies are given by

$$\omega = c\beta^2 = \beta^2 \sqrt{\frac{EI}{\rho A}} = \frac{\gamma}{L^2} \sqrt{\frac{EI}{\rho A}}$$

where

$$\gamma = 3.516, \quad 22.034, \quad 61.701, \quad 120.91, \quad 199.85$$

Thus the fundamental frequency is

$$\omega_1 = \frac{3.516}{L^2} \sqrt{\frac{EI}{\rho A}}$$

which can be expressed as

$$\omega_1 = 3.516\sqrt{\frac{EI}{mL^3}} \tag{1}$$

since the beam mass is $m = \rho AL$.

In Chapter 1 we obtained the stiffness k from the static load-deflection curve of the beam:

$$k = \frac{3EI}{L^3}$$

The lumped-parameter analysis in earlier chapters took 23% of the beam mass m as an approximation for the system mass, and thus the natural frequency was estimated to be

$$\omega_n = \sqrt{\frac{k}{m}} = \sqrt{\frac{3EI}{0.23mL^3}} = 3.612\sqrt{\frac{EI}{mL^3}} \tag{2}$$

Comparing Equations 1 and 2, we see that the static deflection curve and the 23% approximation based on kinetic energy equivalence give a high but very accurate estimate of the fundamental frequency. The error is

$$100\frac{|3.516 - 3.612|}{3.516} = 2.7\% \qquad \blacksquare$$

Table 10.4-2 summarizes the temporal frequencies and mode shapes of transverse beam vibrations for five common sets of boundary conditions.

10.5 MATLAB APPLICATIONS

MATLAB can be of assistance with the material in this chapter in two ways: (1) for solving the transcendental equations to find the frequencies and (2) for plotting the solutions.

Solving Transcendental Equations

We can use the `fzero` function to find the zero of a function of a single variable, which we denote by x. One form of its syntax is

```
fzero('function',x0)
```

where `function` is a string containing the name of the function and `x0` is a numerical value supplied by the user. The value of `x0` is an initial guess for the location of the zero. The `fzero` function returns a value of x that is near `x0`. It identifies only points where the function crosses the x axis, not points where the function just touches the axis. For example, `fzero('cos',2)` returns the value $x = 1.5708$.

If the function is discontinuous, the `fzero` function may return values that are points of discontinuity. For example, typing x = `fzero('tan',1)` returns x = `1.5708`, a discontinuous point in tan x.

If `x0` is a *vector* of length two, the `fzero` algorithm assumes that `x0` is an interval where the function changes sign. An error occurs if this is not true. Calling `fzero` with such an interval guarantees that `fzero` will return a value near a point where the function changes sign.

TABLE 10.4-2 Transverse Vibrations of a Uniform Euler-Bernoulli Beam of Length *L*

Equation of motion:

$$\frac{\partial^2 u(x,t)}{\partial t^2} + c^2 \frac{\partial^4 u(x,t)}{\partial x^4} = 0 \qquad c^2 = \frac{EI}{\rho A} \qquad \omega_n = c\beta_n^2$$

Definitions:

$$P_n(x) = \cosh \beta_n x + \cos \beta_n x \qquad Q_n(x) = \cosh \beta_n x - \cos \beta_n x$$

$$R_n(x) = \sinh \beta_n x + \sin \beta_n x \qquad S_n(x) = \sinh \beta_n x - \sin \beta_n x$$

Boundary conditions	Lowest normalized frequencies $\beta_n L$ and frequency equation	Mode shape
Free-free	0 (rigid body mode) 4.73004 7.85320 10.99561 $\cos \beta_n L \cosh \beta_n L = 1$	$P_n(x) - \dfrac{Q_n(L)}{S_n(L)} R_n(x)$
Fixed-free	1.87510 4.69409 7.85476 10.99554 $\cos \beta_n L \cosh \beta_n L = -1$	$Q_n(x) - \dfrac{S_n(L)}{P_n(L)} S_n(x)$
Fixed-pinned	3.92660 7.06858 10.21018 13.35177 $\tan \beta_n L - \tanh \beta_n L = 0$	$Q_n(x) - \dfrac{Q_n(L)}{S_n(L)} S_n(x)$
Fixed-fixed	4.73004 7.85320 10.99561 14.13717 $\cos \beta_n L \cosh \beta_n L = 1$	$Q_n(x) - \dfrac{Q_n(L)}{S_n(L)} S_n(x)$
Pinned-pinned	$n\pi$ $\sin \beta_n L = 0$	$\sin \dfrac{n\pi x}{L}$

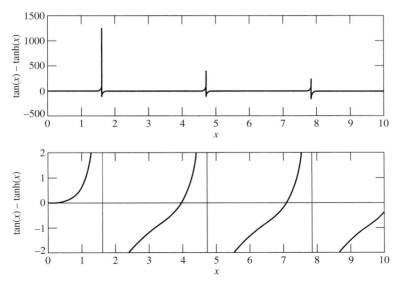

FIGURE 10.5-1 Plot of the function $y = \tan x - \tanh x$ for $0 \leq x \leq 10$.

To use the `fzero` function to find the zeros of more complicated functions, it is more convenient to define the function in an function M-file. For example, if $y = \tan x - \tanh x$, you create and save the following function file:

```
function y=fn1(x)
y=tan(x)-tanh(x);
```

Functions can have more than one zero, so it helps to plot the function first, and then use `fzero` to obtain an answer that is more accurate than the one read off the plot. The upper plot in Figure 10.5-1 shows the plot of this function for $0 \leq x \leq 10$. Note that there are discontinuities where $\tan x$ becomes infinite and changes sign. The lower plot has an expanded scale to display the discontinuities more clearly. These points are $x = \pi/2, 3\pi/2, 5\pi/2, \ldots$.

To find the zero near $x = 4$, you type `x = fzero('fn1',4)`. The answer to four decimal places is $x = 3.9266$.

To find the zero near $x = 7$, you type `x=fzero('f1',7)`. The answer is $x = 7.0686$.

However, if you type `x = fzero('f1',1)`, the answer returned is $x = 1.5708$, which is incorrect since this value corresponds to a point of discontinuity where $\tan x$ changes sign. The key point here is that the `fzero` function interpolates between computed points to estimate where the function crosses zero. When the $\tan x$ function changes sign at a discontinuity, the `fzero` function interprets this point as a zero of the function, which it is not. This is another reason to plot the function before using `fzero`.

In this way we have found the first two roots of the equation:

$$\tan x - \tanh x = 0$$

The plots shown in Figure 10.5-1 can be created with the following code.

```
x=[0:0.01:10];
subplot(2,1,1),plot(x,fn1(x)),xlabel('x')
subplot(2,1,2),plot(x,fn1(x),x,zeros(size(x))),
    xlabel('x'),...
axis([0 10 -2 2])
```

As another example, in Example 10.4-1 we found that the frequencies of a cantilever beam are the solutions of the equation

$$\cosh x \cos x + 1 = 0$$

The function file is

```
function y = fn2(x)
y = cosh(x)*cos(x) + 1;
```

For example, typing x = fzero('fn2',2) returns $x = 1.8751$.

Plotting Responses

Consider the cable response derived in Example 10.3-3 for an initial cable shape that is triangular. The equations are

$$C_n = \frac{8}{n^2\pi^2} \sin \frac{n\pi}{2}$$

$$u(x,t) = \frac{8}{\pi^2} \left(\frac{1}{1^2} \sin \frac{\pi x}{L} \cos \frac{c\pi t}{L} - \frac{1}{3^2} \sin \frac{3\pi x}{L} \cos \frac{3c\pi t}{L} + \frac{1}{5^2} \sin \frac{5\pi x}{L} \cos \frac{5c\pi t}{L} - \cdots \right)$$

For the case where $L = 2$ and $c = 4$, the oscillation period is 1. To plot this response for one-half period using a time increment of 0.05 and the first seven nonzero terms in the series ($n = 1, 3, 5, \ldots, 13$), the MATLAB code is

```
L = 2; c = 4;
for n = 1:2:13
  C(n) = (8/(n^2*pi^2))*sin(n*pi/2);
end
x = [0:0.01:L];
t = [0:0.05:0.5];
u = zeros(length(t),length(x));
k = 0;
for t = 0:0.05:0.5
  k = k+1;
  for n = 1:2:13
    u(k,:) = u(k,:) + C(n)*sin(n*pi*x/L).*cos(n*c*pi*t/L);
  end
end
plot(x,u(:,:)),xlabel('x'),ylabel('u')
```

The plot was given in Figure 10.3-1. The MATLAB Plot Editor was used to add the text and the arrow.

10.6 CHAPTER REVIEW

This chapter treated the vibration of systems that cannot be described adequately with lumped-parameter models. If a lumped-parameter model is linear, the number of its natural frequencies equals its degrees of freedom. Distributed systems, however, have an infinite number of natural frequencies and an infinite number of mode shapes.

We began by considering how to model the simplest distributed system, a cable or string under tension. Torsional and longitudinal vibrations of rods are also described by such a model. We introduced two methods for solving the equation, d'Alembert's method and the separation of variables method. The latter method can also be used to solve other partial differential equations, such as the beam equation, which is fourth order. The vibrational motion can be interpreted in terms of nodes and modes.

We then showed how to use MATLAB to solve the resulting transcendental equations for the natural frequencies.

Now that you have finished this chapter, you should be able to do the following:

1. Derive equations of motion of common distributed systems.

2. Apply the appropriate initial conditions and boundary conditions to solve the partial differential equation models.

3. Interpret the solutions physically in terms of nodes and mode shapes.

4. Apply MATLAB to compute the natural frequencies.

PROBLEMS

SECTION 10.1   INTRODUCTION

10.1 Derive the equation of motion for a string or cable vibrating in a viscous fluid. The cable length is L, and the damping constant c has units of (force/velocity)/length.

SECTION 10.2   SOLUTION OF THE WAVE EQUATION

10.2 A cable of length 5 m has a tension of 300 N. The cable weighs 3 N. Compute the propagation speed of any transverse disturbances in the cable.

10.3 Compute the velocity of transverse wave propagation in a cable of mass density 4 kg/m with a tension of 4500 N.

10.4 A specific cable, whose endpoints are fixed, has a length of 5 m and $c = \sqrt{T/v} = 15$. If the cable is released from rest with the initial shape $f(x) = 6 \times 10^{-3} \sin \pi x/L$, determine the cable shape $y(x, t)$.

10.5 A specific cable, whose endpoints are fixed, has a length of 5 m and $c = \sqrt{T/v} = 15$. If the cable is released from rest with the initial shape $f(x) = 6 \times 10^{-3} \sin 2\pi x/L$, determine the cable shape $y(x, t)$.

10.6 Compute the propagation speed of longitudinal stress waves in a steel bar for which $E = 2 \times 10^{11}$ N/m^2 and $\rho = 7800$ kg/m^3.

10.7 Compute the propagation speed of shear stress waves in a steel bar for which $G = 8 \times 10^{10}$ N/m^2 and $\rho = 7800$ kg/m^3.

SECTION 10.3   SEPARATION OF VARIABLES

10.8 Compute the wave propagation velocity and the fundamental frequency of a steel wire of diameter 1 mm and length 1 m. The tension in the wire is 200 N.

10.9 A certain guitar string has a mass of 0.3 g and a length of 45 cm. Compute the tension needed to obtain a temporal frequency of 196 Hz (which is a little below the note called "middle C").

10.10 Compute the three lowest temporal frequencies for transverse vibration of a steel cable of length 2 m, fixed at both ends. For steel, $E = 2 \times 10^{11}$ N/m^2 and $\rho = 7800$ kg/m^3.

10.11 Compute the three lowest temporal frequencies for torsional vibration of a solid steel shaft of length 2 m and diameter 30 mm, fixed at one end and free at the other. For steel, $G = 8 \times 10^{10}$ N/m^2 and $\rho = 7800$ kg/m^3.

10.12 Compute the three lowest temporal frequencies for torsional vibration of a solid steel shaft of length 2 m and diameter 30 mm, with both ends fixed. For steel, $G = 8 \times 10^{10}$ N/m^2 and $\rho = 7800$ kg/m^3.

10.13 Derive the temporal frequency equation for longitudinal vibration of the system shown in Figure P10.13.

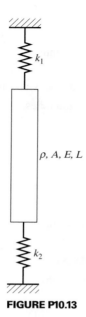

ρ, A, E, L

FIGURE P10.13

10.14 Derive the temporal frequency equation for longitudinal vibration of the system shown in Figure P10.14.

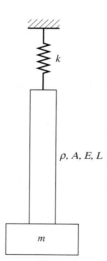

ρ, A, E, L

m

FIGURE P10.14

10.15 Derive the temporal frequency equation for longitudinal vibration of the system shown in Figure P10.15.

10.16 Derive the frequency equation for longitudinal vibration of the stepped bar shown in Figure P10.16. Both segments are made from identical material.

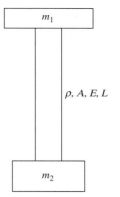

FIGURE P10.15

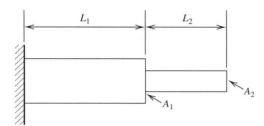

FIGURE P10.16

10.17 Derive the frequency equation for torsional vibration of the system shown in Figure P10.17.

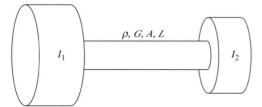

FIGURE P10.17

10.18 A motor of inertia $I = 150$ kg·m^2 is attached to one end of a solid steel shaft of length 2 m and diameter 30 mm. Taking the other end to be fixed, compute the three lowest temporal frequencies for torsional vibration of the system. For steel, $G = 8 \times 10^{10}$ N/m^2 and $\rho = 7800$ kg/m^3.

10.19 A solid steel shaft of length 2 m and diameter 30 mm has rotors attached to each end. The rotor inertias are 200 kg·m^2 and 100 kg·m^2. Compute the lowest temporal frequency for torsional vibration of the system. Compare this answer with the one obtained from a lumped-parameter model having two degrees of freedom that ignores the inertia of the shaft. For steel, $G = 8 \times 10^{10}$ N/m^2 and $\rho = 7800$ kg/m^3.

10.20 Derive the expressions for the frequency equation and mode shapes for the free-free case in Table 10.3-1.

10.21 Derive the expressions for the frequency equation and mode shapes for the fixed-free case in Table 10.3-1.

SECTION 10.4 TRANSVERSE VIBRATIONS OF BEAMS

10.22 Obtain the frequency equation for the transverse vibration of a simply supported beam of length L with a mass m attached at its center. Neglect the dimensions of the mass m.

10.23 **(a)** Derive the frequency equation for the system shown in Figure P10.23.

(b) Compute the three lowest temporal frequencies for the case where $L = 1$ m, $m = 100$ kg, $I = 10^{-5}$ m^4, $k = 9 \times 10^5$ N/m, and $E = 2 \times 10^{11}$ N/m^2.

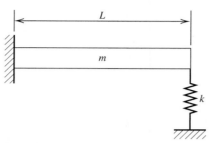

FIGURE P10.23

10.24 A certain steel rectangular beam is 50 mm by 150 mm, with a length of 1 m. For steel, $E = 2 \times 10^{11}$ N/m^2 and $\rho = 7800$ kg/m^3. Compare the three lowest temporal frequencies and their mode shapes for this beam supported four different ways:

(a) Free-free

(b) Fixed-free

(c) Fixed-fixed

(d) Pinned-pinned

10.25 Derive the expressions for the frequency equation and mode shapes for the free-free case in Table 10.4-2.

10.26 Derive the expressions for the frequency equation and mode shapes for the fixed-pinned case in Table 10.4-2.

10.27 Derive the expressions for the frequency equation and mode shapes for the fixed-fixed case in Table 10.4-2.

10.28 Derive the expressions for the frequency equation and mode shapes for the pinned-pinned case in Table 10.4-2.

SECTION 10.5   MATLAB APPLICATIONS

10.29 Use MATLAB to verify the three lowest nonzero temporal frequencies for the free-free case given in Table 10.4-2.

10.30 Use MATLAB to verify the three lowest temporal frequencies for the fixed-free case given in Table 10.4-2.

10.31 Use MATLAB to verify the three lowest temporal frequencies for the fixed-pinned case given in Table 10.4-2.

10.32 Use MATLAB to plot the first three mode shapes for the free-free case given in Table 10.4-2.

10.33 Use MATLAB to plot the first three mode shapes for the fixed-free case given in Table 10.4-2.

10.34 Use MATLAB to plot the first three mode shapes for the fixed-fixed case given in Table 10.4-2.

DYNAMIC FINITE ELEMENT ANALYSIS

UP TO now we have been limited to modeling a vibratory system either as a lumped-parameter system, in which the masses, stiffness, and damping are treated as being concentrated at discrete locations within the system, or as a distributed-parameter system, which accurately reflects the real situation. Distributed-parameter models, however, are difficult to derive and to solve for complex situations, and so we have available to us solutions for only relatively simple cases.

The partial differential equations that comprise a distributed-parameter model can be solved numerically using *finite-difference* methods, which we have not discussed. These methods approximate the partial differential equations by representing the partial derivatives as ratios of small differences in the variables, thus converting the partial differential equations into a large number of difference equations. Although this approach has had success in solving the equations of heat transfer and fluid mechanics, solving the equations for vibrating systems is not always easy or possible, even with a computer, because in some cases they do not possess stable solutions [Kreyszig, 1997].

In this chapter we introduce a method that provides a more accurate system description than that used to develop a lumped-parameter model but that for complex systems is easier to solve than partial differential equations and more accurate than their finite-difference approximations. This method is called the *finite element method*, sometimes abbreviated *FEM*, and sometimes called finite element *analysis* (FEA). It is particularly useful for irregular geometries, such as bars that have variable cross sections, and for systems such as trusses that are made up of several bars.

In the finite element method, the system is modeled as consisting of several pieces, and each piece is treated as a *continuous* element, which is called a *finite element*. Thus, in contrast to the finite-difference method, finite elements are not necessarily small.

The first step in analyzing the vibration of a structure using the finite element method is to divide the structure into a number of simple structural parts, such as bars, beams, or plates, whose equations of motion are easily derived. Once the equations have been obtained for each element, the equations are assembled and combined according to how the elements are connected in the system. These connection points are called *joints* or *nodes*.[1] The resulting collection of finite elements and nodes is called a *mesh*.

To solve the equation of motion for each finite element, we approximate the solution with low-order polynomials. These solutions are then assembled to obtain the mass and stiffness matrices for the entire structure as a whole. Once the equations have been assembled, the modal analysis techniques of Chapter 8 are used to compute the natural frequencies and mode shapes. The displacements obtained in the solution are the displacements of the nodes in the mesh.

The elements commonly used in FEM and treated here are the axial (bar) element, the transverse (beam) element, and the torsional element, which are one-dimensional. The plate and shell elements are treated in more specialized texts.

When developing a model of a structural element, we must use an approximation to the vibration mode shape. When only one finite element is used between structural joints or corners, we usually obtain an accurate result only for the lowest mode. To estimate the higher modes, we must use several elements between structural joints. As we will see, this results in a large number of equations that must be solved for the eigenvalues and eigenvectors of the system.

Thus, in most practical problems, a large number of elements is used, and the task of specifying the locations of the elements and of assembling the equations is very tedious if done by hand. Fortunately, the finite element method has now been widely implemented in many powerful commercially available computer programs. These programs provide a graphical interface for specifying the locations of the nodes; they assemble the equations and often provide graphical displays of the results.

The subject of finite elements is a deep one, and many schools devote an entire course to it. Here we provide a simple introduction that will familiarize you with the basic concepts and terminology. If you plan to continue your studies in vibrations, you undoubtedly will need to acquire a deeper understanding of the finite element method.

---

[1]Note that the term *node* in a finite element model does not have the same meaning as in modal analysis, where a node is a point of no motion.

LEARNING OBJECTIVES

After you have finished this chapter, you should be able to do the following:

- Write the equations for the bar, torsional, and beam elements.
- Assemble the equations, using the given boundary conditions, for a system consisting of multiple elements.
- Solve the assembled equations to determine the mode shapes and mode frequencies.

11.1 BAR ELEMENT

Recall that the force-deflection relation of a uniform bar or rod having an applied axial force f is

$$f = \frac{EA}{L} u = ku$$

where u is the resulting axial displacement. The relation is identical to that of a linear spring of stiffness k. For the case where forces are applied at both ends of the bar (Figure 11.1-1),

$$\begin{bmatrix} f_1 \\ f_2 \end{bmatrix} = \begin{bmatrix} k_{11} & k_{12} \\ k_{21} & k_{22} \end{bmatrix} \begin{bmatrix} u_1 \\ u_2 \end{bmatrix}$$

To determine the elements of the matrix, note that the first column contains the endpoint forces when $u_1 = 1$ and $u_2 = 0$. That is, $f_1 = ku_1$ and $f_2 = -ku_1$, so the first column is $[k, -k]^T$. Similarly, the second column contains the endpoint forces when $u_1 = 0$ and $u_2 = 1$. That is, $f_1 = -ku_2$ and $f_2 = ku_2$, so the second column is $[-k, k]^T$. Thus we have

$$\begin{bmatrix} f_1 \\ f_2 \end{bmatrix} = \begin{bmatrix} k & -k \\ -k & k \end{bmatrix} \begin{bmatrix} u_1 \\ u_2 \end{bmatrix} = \frac{EA}{L} \begin{bmatrix} 1 & -1 \\ -1 & 1 \end{bmatrix} \begin{bmatrix} u_1 \\ u_2 \end{bmatrix}$$

So the stiffness matrix $\mathbf{K}$ of the bar element is

$$\mathbf{K} = \frac{EA}{L} \begin{bmatrix} 1 & -1 \\ -1 & 1 \end{bmatrix} \tag{11.1-1}$$

The Mass Matrix

We now derive the mass matrix $\mathbf{M}$ for the bar element. Consider the bar shown in Figure 11.1-2a, which is a generalization of Figure 11.1-1 to show the variation of the displacement u as a function of time and location. Now consider the simpler situation

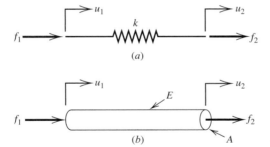

FIGURE 11.1-1 Force and axial displacement for a bar.

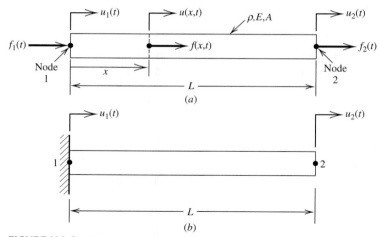

FIGURE 11.1-2 Variation of bar displacement u as a function of time and location.

shown in Figure 11.1-2b. From Chapter 10 we know that the *static* displacement of the bar is described by

$$EA\frac{d^2u(x)}{dx^2} = 0 \qquad 0 \le x \le L$$

This has the solution

$$u(x) = a + bx$$

where a and b are independent of x. Taking this *static* mode shape to be an approximation of the *dynamic* mode shape, we write

$$u(x,t) = a(t) + b(t)x \tag{11.1-2}$$

There are two nodes, whose displacements are $u_1(t)$ and $u_2(t)$. At $x = 0$, Equation 11.1-2 gives

$$u(0,t) = u_1(t)$$

$$u(L,t) = u_2(t)$$

where

$$a(t) = u_1(t)$$

$$b(t) = \frac{u_2(t) - u_1(t)}{L}$$

Thus

$$u(x,t) = \left(1 - \frac{x}{L}\right)u_1(t) + \frac{x}{L}u_2(t) = S_1(x)u_1(t) + S_2(x)u_2(t) \tag{11.1-3}$$

The functions S_1 and S_2 are called the *shape functions*. Equation 11.1-3 is an approximate solution of the equation of motion of the bar.

With the assumed mode shape specified by Equation 11.1-3, we can express the kinetic energy of the bar element as follows:

$$KE = \int_0^L \frac{1}{2}\rho A\left(\frac{\partial u}{\partial t}\right)^2 dx = \frac{1}{2}\rho A\int_0^L [S_1(x)\dot{u}_1(t) + S_2(x)\dot{u}_2(t)]^2 dx$$

or

$$KE = \frac{1}{2}\rho A \dot{u}_1^2(t) \int_0^L S_1^2(x)\,dx + \rho A \dot{u}_1(t)\dot{u}_2(t) \int_0^L S_1(x)S_2(x)\,dx$$
$$+ \frac{1}{2}\rho A \dot{u}_2^2(t) \int_0^L S_2^2(x)\,dx$$

Evaluation of the integrals gives

$$KE = \frac{1}{6}\rho A L \left[\dot{u}_1^2(t) + \dot{u}_1(t)\dot{u}_2(t) + \dot{u}_2^2(t) \right]$$

or

$$KE = \frac{1}{2}\rho A L \left[\frac{\dot{u}_1^2(t)}{3} + \frac{\dot{u}_1(t)\dot{u}_2(t)}{3} + \frac{\dot{u}_2^2(t)}{3} \right] = \frac{1}{2}\dot{\mathbf{u}}^T(t)\mathbf{M}\dot{\mathbf{u}}(t)$$

where

$$\dot{\mathbf{u}}(t) = \begin{bmatrix} \dot{u}_1(t) \\ \dot{u}_2(t) \end{bmatrix}$$

Thus the mass matrix of the bar element is

$$\mathbf{M} = \frac{\rho A L}{6} \begin{bmatrix} 2 & 1 \\ 1 & 2 \end{bmatrix} \tag{11.1-4}$$

Note that this result is dependent on the assumed mode shape specified by Equation 11.1-3.

Potential Energy Approach

We could have followed a similar approach to derive the stiffness matrix $\mathbf{K}$ from the potential energy expression for the bar element, as follows:

$$PE = \int_0^L \frac{1}{2} EA \left(\frac{\partial u}{\partial x} \right)^2 dx = \frac{1}{2}EA \int_0^L \left[\frac{dS_1(x)}{dx}u_1(t) + \frac{dS_2(x)}{dx}u_2(t) \right]^2 dx$$

or

$$PE = \frac{1}{2}\frac{EA}{L^2} \int_0^L [u_2(t) - u_1(t)]^2\,dx = \frac{1}{2}\frac{EA}{L} \left[u_1^2(t) - 2u_1(t)u_2(t) + u_2^2(t) \right]$$

or

$$PE = \frac{1}{2}\mathbf{u}^T(t)\mathbf{K}\mathbf{u}(t)$$

where $\mathbf{K}$ is given by Equation 11.1-1. The shape functions used in Equation 11.1-3 are equivalent to the linear force-deflection relation used to derive the stiffness matrix $\mathbf{K}$. So it is expected that the derivation of $\mathbf{K}$ based on potential energy should give the same result as Equation 11.1-3.

A Fixed-Free Bar

If we had focused on the specific situation shown in Figure 11.1-2b, in which the left-hand end of the bar is fixed, we would have observed that $u_1(t) = 0$ and that the kinetic and potential energy expressions would reduce to

$$KE = \frac{1}{2}\frac{\rho AL}{3}\dot{u}_2^2$$

$$PE = \frac{1}{2}\frac{EA}{L}u_2^2$$

From conservation of mechanical energy, $KE + PE =$ constant, so taking the time derivative of this expression yields

$$\frac{\rho AL}{3}\dot{u}_2\ddot{u}_2 + \frac{EA}{L}u_2\dot{u}_2 = 0$$

Canceling A and $\dot{u}_2$, we obtain

$$\frac{\rho L}{3}\ddot{u}_2 + \frac{E}{L}u_2 = 0 \tag{11.1-5}$$

This is the dynamic model of the bar element with no applied force and with the left-hand end fixed. The natural frequency is

$$\omega_n = \frac{1}{L}\sqrt{\frac{3E}{\rho}} = \frac{1.7321}{L}\sqrt{\frac{E}{\rho}} \tag{11.1-6}$$

Given the initial displacement $u_2(0)$ and initial velocity $\dot{u}_2(0)$ of the second node point, we can solve Equation 11.1-5 for $u_2(t)$. Then, using Equation 11.1-3 with $u_1(t) = 0$, we can obtain the expression for the axial displacement $u(x,t)$ of the bar element as a function of location x and time t:

$$u(x,t) = \frac{x}{L}u_2(t) \tag{11.1-7}$$

Comparison with Other Models

In Chapter 2 we derived the following expression for the natural frequency of a concentrated mass attached to a spring element, such as an elastic bar:

$$\omega_n = \sqrt{\frac{k}{m_c + m_s/3}}$$

where m_c is the concentrated mass at the end of the bar and m_s is the spring mass. Taking $m_c = 0$, using the fact that for the bar, $k = EA/L$, and noting that $m_s = \rho AL$, we obtain

$$\omega_n = \sqrt{\frac{EA}{L\rho AL/3}} = \frac{1}{L}\sqrt{\frac{3E}{\rho}}$$

which is identical to the frequency of the bar element given by Equation 11.1-6. This result should be expected because the derivation in Chapter 2 assumed the bar shape to be its static deflection shape, which is identical to the shape function used in this section.

We can also compare the result with that obtained from the solution of the partial differential equation (Equation 10.1-4) in Chapter 10, which is

$$\frac{\partial^2 u}{\partial t^2} = \frac{E}{\rho}\frac{\partial^2 u}{\partial x^2}$$

For the case where the left-hand end is fixed and the right-hand end is free, the method of separation of variables used in Chapter 10 results in the following expressions for the temporal frequencies and the mode shapes:

$$\omega_n = \frac{2n-1}{2L}\pi\sqrt{\frac{E}{\rho}} \qquad n = 1, 2, 3, \dots \tag{11.1-8}$$

$$F_n(x) = A_n \sin\frac{(2n-1)\pi x}{2L} \qquad n = 1, 2, 3, \dots \tag{11.1-9}$$

The first frequency and mode shape are

$$\omega_1 = \frac{\pi}{2L}\sqrt{\frac{E}{\rho}} = \frac{1.5708}{L}\sqrt{\frac{E}{\rho}}$$

$$F_1(x) = A_1 \sin\frac{\pi x}{2L}$$

Thus the finite element model predicts a frequency that is 10% higher than that predicted by the distributed-parameter model. The static deflection shape used by the finite element model is close to the first mode shape, as shown in Figure 11.1-3a.

The second frequency and mode shape are

$$\omega_2 = \frac{3\pi}{2L}\sqrt{\frac{E}{\rho}} = \frac{4.7124}{L}\sqrt{\frac{E}{\rho}}$$

$$F_2(x) = A_2 \sin\frac{3\pi x}{2L}$$

The finite element model predicts a frequency that is 63% lower than that predicted by the second mode of the distributed-parameter model. The second mode shape is quite different from the shape function used by the finite element model, as shown in

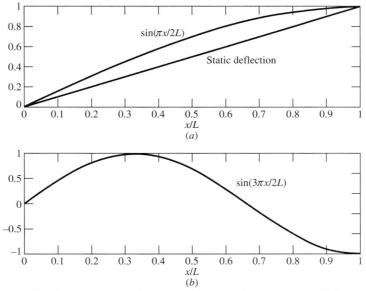

FIGURE 11.1-3   (a) First mode shape and the static deflection curve. (b) Second mode shape.

Figure 11.1-3*b*. Note that the second mode has a node (a point of no motion) at $x/L = 2/3$. The higher modes all have higher frequencies and more complex mode shapes.

So the manner in which the bar is excited (either by the initial conditions or by an applied force) will determine whether or not this finite element model will give reasonable results. If any mode higher than the first mode is excited, the model will not give accurate results.

Applying the Boundary Conditions

We now apply the boundary conditions to the equation of motion for a finite element model. The dynamic model of the bar element has the standard matrix form:

$$\mathbf{M\ddot{u}} + \mathbf{Ku} = \mathbf{0}$$

where we have assumed for now that the applied forces are zero. For the bar element, this equation becomes (after canceling A from all terms)

$$\frac{\rho L}{6} \begin{bmatrix} 2 & 1 \\ 1 & 2 \end{bmatrix} \begin{bmatrix} \ddot{u}_1 \\ \ddot{u}_2 \end{bmatrix} + \frac{E}{L} \begin{bmatrix} 1 & -1 \\ -1 & 1 \end{bmatrix} \begin{bmatrix} u_1 \\ u_2 \end{bmatrix} = \begin{bmatrix} 0 \\ 0 \end{bmatrix}$$

If the left-hand end of the element is fixed, $u_1(t) = \dot{u}_1(t) = 0$, so the first equation in the set is not applicable. This means we can strike out the first row and the first column of the $\mathbf{M}$ and $\mathbf{K}$ matrices and immediately obtain

$$\frac{\rho L}{3} \ddot{u}_2 + \frac{E}{L} u_2 = 0$$

which is identical to Equation 11.1-5 obtained directly from the energy expressions.

Striking out the row and column corresponding to the displacement of a fixed node is a procedure often used as a quick way to reduce the number of equations.

If the bar has an axial force $f(t)$ applied at its free end, its equation of motion is easily seen to be

$$\frac{\rho A L}{3} \ddot{u}_2 + \frac{EA}{L} u_2 = f(t) \tag{11.1-10}$$

11.2 MODELS WITH MULTIPLE BAR ELEMENTS

The next step in developing a finite element model is to decide how many elements to use. We have seen that a single-element model will give inaccurate results if higher modes are excited. To represent the higher modes, more elements must be used to model the entire bar. If more than one element is used, the equations for all the elements must be assembled into a model of the entire structure as a whole. The following example demonstrates how this is done.

EXAMPLE 11.2-1
A Two-Element Bar Model

Figure 11.2-1 shows a bar having a step change in cross section. It is fixed at the left-hand end and the bar material is uniform.

(a) Develop a finite element model of the longitudinal vibration of the bar.

(b) Solve the model for the frequencies and mode shapes of the node points in longitudinal vibration for the case where the cross section is constant (i.e., $A_1 = A_2 = A$ and $L_1 = L_2 = L/2$). Compare the results with those predicted by the distributed-parameter model.

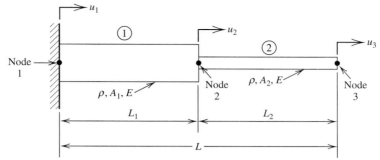

FIGURE 11.2-1   Bar having a step change in cross section.

Solution   (a) The step change in cross section suggests the two-element model shown in Figure 11.2-1. The mass and stiffness matrices for each element are the following:

$$\mathbf{M}_1 = \frac{m_1}{6}\begin{bmatrix} 2 & 1 \\ 1 & 2 \end{bmatrix} \qquad \mathbf{K}_1 = k_1\begin{bmatrix} 1 & -1 \\ -1 & 1 \end{bmatrix}$$

$$\mathbf{M}_2 = \frac{m_2}{6}\begin{bmatrix} 2 & 1 \\ 1 & 2 \end{bmatrix} \qquad \mathbf{K}_2 = k_2\begin{bmatrix} 1 & -1 \\ -1 & 1 \end{bmatrix}$$

where $m_i = \rho A_i L_i$ and $k_i = EA_i/L_i$, $i = 1, 2$.

The dynamic model of each element has the standard matrix form

$$\mathbf{M}_i\ddot{\mathbf{u}}_i + \mathbf{K}_i\mathbf{u}_i = \mathbf{0} \qquad i = 1, 2$$

For the first element, this gives

$$\frac{m_1}{6}\begin{bmatrix} 2 & 1 \\ 1 & 2 \end{bmatrix}\begin{bmatrix} \ddot{u}_1 \\ \ddot{u}_2 \end{bmatrix} + k_1\begin{bmatrix} 1 & -1 \\ -1 & 1 \end{bmatrix}\begin{bmatrix} u_1 \\ u_2 \end{bmatrix} = \begin{bmatrix} 0 \\ 0 \end{bmatrix}$$

Since the left-hand end is fixed, $u_1(t) = \dot{u}_1(t) = 0$, and so we strike out the first row and first column of the matrices. This leaves us with a single equation for the first element:

$$\frac{m_1}{3}\ddot{u}_2 + k_1 u_2 = 0 \tag{1}$$

For the second element,

$$\frac{m_2}{6}\begin{bmatrix} 2 & 1 \\ 1 & 2 \end{bmatrix}\begin{bmatrix} \ddot{u}_2 \\ \ddot{u}_3 \end{bmatrix} + k_2\begin{bmatrix} 1 & -1 \\ -1 & 1 \end{bmatrix}\begin{bmatrix} u_2 \\ u_3 \end{bmatrix} = \begin{bmatrix} 0 \\ 0 \end{bmatrix}$$

This gives the following equations:

$$\frac{m_2}{6}(2\ddot{u}_2 + \ddot{u}_3) + k_2(u_2 - u_3) = 0 \tag{2}$$

$$\frac{m_2}{6}(\ddot{u}_2 + 2\ddot{u}_3) + k_2(u_3 - u_2) = 0 \tag{3}$$

We now have three equations, but only two unknowns, u_2 and u_3. To solve the problem as an eigenvalue problem, we must reduce the set to just two equations. We can do this by adding Equations 1 and 2 to obtain

$$\left(\frac{m_1}{3} + \frac{m_2}{3}\right)\ddot{u}_2 + \frac{m_2}{6}\ddot{u}_3 + (k_1 + k_2)u_2 - k_2 u_3 = 0 \tag{4}$$

The model now consists of Equations 3 and 4, which can be put into matrix form as follows:

$$\frac{1}{6}\begin{bmatrix} 2(m_1+m_2) & m_2 \\ m_2 & 2m_2 \end{bmatrix}\begin{bmatrix} \ddot{u}_2 \\ \ddot{u}_3 \end{bmatrix} + \begin{bmatrix} (k_1+k_2) & -k_2 \\ -k_2 & k_2 \end{bmatrix}\begin{bmatrix} u_2 \\ u_3 \end{bmatrix} = \begin{bmatrix} 0 \\ 0 \end{bmatrix} \tag{5}$$

This is the finite element model of the bar, using two elements.

(b) With $A_1 = A_2 = A$, $L_1 = L_2 = L/2$, $m_1 = m_2 = m/2$, and $k_1 = k_2 = 2EA/L = k$, Equation 5 becomes

$$\frac{m}{12}\begin{bmatrix} 4 & 1 \\ 1 & 2 \end{bmatrix}\begin{bmatrix} \ddot{u}_2 \\ \ddot{u}_3 \end{bmatrix} + k\begin{bmatrix} 2 & -1 \\ -1 & 1 \end{bmatrix}\begin{bmatrix} u_2 \\ u_3 \end{bmatrix} = \begin{bmatrix} 0 \\ 0 \end{bmatrix} \tag{6}$$

Let $\lambda = m\omega^2/12k = \rho L^2\omega^2/24E$ to obtain

$$\begin{bmatrix} 2 & -1 \\ -1 & 1 \end{bmatrix}\begin{bmatrix} u_2 \\ u_3 \end{bmatrix} = \lambda\begin{bmatrix} 4 & 1 \\ 1 & 2 \end{bmatrix}\begin{bmatrix} u_2 \\ u_3 \end{bmatrix}$$

This is an eigenvalue problem whose characteristic equation is

$$7\lambda^2 - 10\lambda + 1 = 0$$

This has the roots $\lambda = 0.1082$ and $\lambda = 1.3204$. Thus the frequencies are

$$\omega = \sqrt{\frac{24EA\lambda}{\rho L^2}} = \frac{1.611}{L}\sqrt{\frac{E}{\rho}}, \qquad \frac{5.629}{L}\sqrt{\frac{E}{\rho}}$$

The partial differential equation (Equation 10.1-4) in Chapter 10 is

$$\frac{\partial^2 u}{\partial t^2} = \frac{E}{\rho}\frac{\partial^2 u}{\partial x^2}$$

For the case where the left-hand end is fixed and the right-hand end is free, the method of separation of variables used in Chapter 10 results in the following expressions for the temporal frequencies and the mode shapes:

$$\omega_n = \frac{2n-1}{2L}\pi\sqrt{\frac{E}{\rho}} \qquad n = 1, 2, 3, \ldots$$

$$F_n(x) = A_n \sin\frac{(2n-1)\pi x}{2L} \qquad n = 1, 2, 3, \ldots \tag{7}$$

The first frequency and mode shape are

$$\omega_1 = \frac{\pi}{2L}\sqrt{\frac{E}{\rho}} = \frac{1.5708}{L}\sqrt{\frac{E}{\rho}}$$

$$F_1(x) = A_1 \sin\frac{\pi x}{2L}$$

Thus the finite element model predicts a first mode frequency that is only 2.6% higher than that predicted by the distributed-parameter model. The second frequency

$$\omega_2 = \frac{3\pi}{2L}\sqrt{\frac{E}{\rho}} = \frac{4.7124}{L}\sqrt{\frac{E}{\rho}}$$

The finite element model predicts a frequency that is 19% higher than that predicted by the second mode of the distributed-parameter model.

Let us now compare the mode shapes predicted by the two models. For the finite element model the mode shape for the lowest frequency is $u_3/u_2 = 1.414$, and for the highest frequency, $u_3/u_2 = -1.414$.

From Equation (7), for the first mode,

$$\frac{F_1(L)}{F_1(L/2)} = \frac{\sin \pi/2}{\sin \pi/4} = 1.414$$

For the second mode,

$$\frac{F_2(L)}{F_2(L/2)} = \frac{\sin 3\pi/2}{\sin 3\pi/4} = -1.414$$

Thus both models predict the same mode shapes, for the first two modes. ∎

Extension to Multiple Elements

Consider the stepped bar shown in Figure 11.2-2. If the left-hand end is fixed and the right-hand end is free, application of the finite element procedure results in the following stiffness and mass matrices, where

$$k_i = \frac{E_i A_i}{L_i} \tag{11.2-1}$$

$$m_i = \rho A_i L_i \tag{11.2-2}$$

$$\mathbf{K} = \begin{bmatrix} k_1 + k_2 & -k_2 & 0 & 0 & 0 \\ -k_2 & k_2 + k_3 & -k_3 & 0 & 0 \\ 0 & -k_3 & k_3 + k_4 & -k_4 & 0 \\ 0 & 0 & -k_4 & k_4 + k_5 & -k_5 \\ 0 & 0 & 0 & -k_5 & k_5 \end{bmatrix} \tag{11.2-3}$$

$$\mathbf{M} = \frac{1}{6} \begin{bmatrix} 2m_1 + 2m_2 & m_2 & 0 & 0 & 0 \\ m_2 & 2m_2 + 2m_3 & m_3 & 0 & 0 \\ 0 & m_3 & 2m_3 + 2m_4 & m_4 & 0 \\ 0 & 0 & m_4 & 2m_4 + 2m_5 & m_5 \\ 0 & 0 & 0 & m_5 & 2m_5 \end{bmatrix} \tag{11.2-4}$$

The eigenvalue problem is

$$\mathbf{M}^{-1}\mathbf{K}\mathbf{u} = \omega^2 \mathbf{u}$$

Note the banded structure of these matrices, which is due to the fact that a given element is affected only by its adjacent elements. This structure, which consists of a main diagonal, one upper subdiagonal, and one lower subdiagonal, enables us to write a

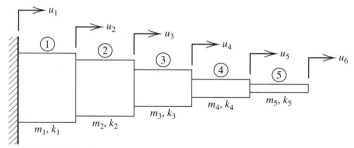

FIGURE 11.2-2 Stepped bar with five segments.

general-purpose program that can be applied to analyze a bar with an arbitrary number of elements.

11.3   CONSISTENT AND LUMPED-MASS MATRICES

The mass matrices we have used thus far were derived from the kinetic energy expression of the element, assuming that the velocity is the time derivative of the displacement function specified by the shape functions. The functions we have used actually specify the *static* deflection shape, and thus the kinetic energy expression derived from them is only an approximation.

The stiffness matrix derived from these shape functions is, however, accurate because stiffness is associated with static deflection. Mass matrices derived in this manner are said to be *consistent* mass matrices because they are consistent with the derivation of the stiffness matrix.

An easier way of calculating a mass matrix is to place an appropriate mass value at each node. For example, if the total mass is uniformly distributed throughout the system and if the kinematics are simple, then an appropriate mass value for a given node would be proportional to the dimensions of the element. For example, a bar element of length L, density ρ, and area A has a total mass of $\rho A L$. Placing one-half of the total mass at each node gives the following mass matrix:

$$\mathbf{M} = \frac{\rho A L}{2} \begin{bmatrix} 1 & 0 \\ 0 & 1 \end{bmatrix} \tag{11.3-1}$$

A mass matrix derived in this way is said to be a *lumped-mass* matrix.

An advantage of lumped-mass matrices derived in this manner is that they are diagonal and thus easily inverted to compute $\mathbf{M}^{-1}$. A disadvantage is that it is not always obvious how the mass should be apportioned to each node (the truss application treated in Section 11.4 is an example of this difficulty).

Another disadvantage of the lumped-mass approach occurs with beam elements (see Section 11.6), which have coordinates equal to the slope of the beam deflection curve. With no mass lumped for that coordinate, the mass matrix will have some zero elements along its diagonal and thus will not have an inverse. Such problems require special consideration.

EXAMPLE 11.3-1
Lumped versus Consistent Mass Matrices for a Fixed-Fixed Bar

Use a two-element bar model to obtain the frequencies of the fixed-fixed bar shown in Figure 11.3-1. Compare the results obtained from the consistent mass matrix, the lumped-mass matrix, and the distributed-parameter model

Solution

The stiffness matrix for a bar element is

$$\mathbf{K} = \frac{AE}{L} \begin{bmatrix} 1 & -1 \\ -1 & 1 \end{bmatrix}$$

The consistent mass matrix for a bar element is

$$\mathbf{M}_C = \frac{\rho A L}{6} \begin{bmatrix} 2 & 1 \\ 1 & 2 \end{bmatrix}$$

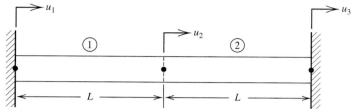

FIGURE 11.3-1 A fixed-fixed bar.

The lumped-mass matrix for a bar element is

$$\mathbf{M}_L = \frac{\rho A L}{2} \begin{bmatrix} 1 & 0 \\ 0 & 1 \end{bmatrix}$$

The assembled matrices are

$$\mathbf{K} = \frac{AE}{L} \begin{bmatrix} 1 & -1 & 0 \\ -1 & (1+1) & -1 \\ 0 & -1 & 1 \end{bmatrix} = \frac{AE}{L} \begin{bmatrix} 1 & -1 & 0 \\ -1 & 2 & -1 \\ 0 & -1 & 1 \end{bmatrix}$$

$$\mathbf{M}_C = \frac{\rho A L}{6} \begin{bmatrix} 2 & 1 & 0 \\ 1 & (2+2) & 1 \\ 0 & 1 & 2 \end{bmatrix} = \frac{\rho A L}{6} \begin{bmatrix} 2 & 1 & 0 \\ 1 & 4 & 1 \\ 0 & 1 & 2 \end{bmatrix}$$

$$\mathbf{M}_L = \frac{\rho A L}{2} \begin{bmatrix} 1 & 0 & 0 \\ 0 & (1+1) & 0 \\ 0 & 0 & 1 \end{bmatrix} = \frac{\rho A L}{2} \begin{bmatrix} 1 & 0 & 0 \\ 0 & 2 & 0 \\ 0 & 0 & 1 \end{bmatrix}$$

For the boundary conditions $u_1 = u_3 = 0$, we strike out the first and third rows and columns to obtain the following stiffness matrix:

$$K = \frac{2AE}{L}$$

Doing the same with the consistent mass matrix, we obtain

$$M_C = \frac{4\rho A L}{6}$$

Thus the equation of motion with the consistent matrix is

$$\frac{\rho L}{3} \ddot{u}_2 + \frac{E}{L} u_2 = 0$$

The frequency is

$$\omega = \frac{1}{L}\sqrt{\frac{3E}{\rho}} = \frac{1.73}{L}\sqrt{\frac{E}{\rho}}$$

Striking out the first and third rows and columns of the lumped-mass matrix, we obtain

$$M_L = \rho A L$$

Thus the equation of motion with the lumped-mass matrix is

$$\rho A L \ddot{u}_2 + \frac{2AE}{L} u_2 = 0$$

The frequency is

$$\omega = \frac{1}{L}\sqrt{\frac{2E}{\rho}} = \frac{1.41}{L}\sqrt{\frac{E}{\rho}}$$

For a uniform bar of length $2L$, the lowest frequency obtained from the distributed-parameter model is (see Table 10.3-1)

$$\omega_1 = \frac{1.57}{L}\sqrt{\frac{E}{\rho}}$$

The consistent mass matrix gives a frequency about 10% higher than that of the distributed-parameter model, while the lumped-mass matrix gives a value that is about 10% lower. ∎

The usefulness of the lumped-mass approximation, which leads to a diagonal mass matrix, is more evident in models having a large number of elements. In such models a diagonal mass matrix is inverted more easily without excessive numerical error. However, the process of obtaining a lumped-mass approximation introduces modeling error, and so the choice between the use of consistent versus lumped-mass matrices is usually not clear.

11.4 ANALYSIS OF TRUSSES

The purpose of the finite element method is to enable us to analyze structures that are difficult to analyze with distributed- or lumped-parameter models. A truss provides a good example of how this method can be applied. A truss is made up of bars or beams joined at their ends in a framework to form a rigid structure. The framework consists of a number of triangles. An example is shown in Figure 11.4-1. Cranes, bridges, and roof supports are common examples of trusses.

Practical trusses can have many members, so they are difficult to analyze by hand. Commercial finite element software is designed to assist the user in specifying the truss geometry. The software then assmbles the equations and performs the static analysis to determine the static stresses and the dynamic analysis to determine the vibrational characteristics of the structure.

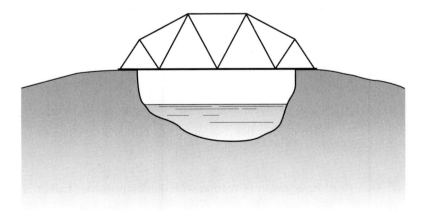

FIGURE 11.4-1 A bridge is an example of an application of the truss.

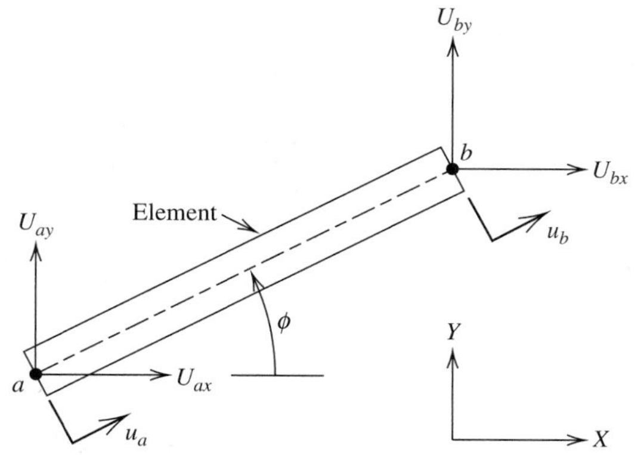

FIGURE 11.4-2 Coordinate systems used to represent a truss element.

Because the basic unit of a truss is a triangle, in this section we illustrate the finite element method for trusses by applying it to a simple triangular structure.

Coordinate Systems

Consider the element shown in Figure 11.4-2. The coordinates u_a and u_b are called *local* coordinates, and the coordinates U_{ax} and U_{bx} are called *global* coordinates. The two coordinate systems are related as follows:

$$u_a = U_{ax} \cos \phi + U_{ay} \sin \phi$$

$$u_b = U_{bx} \cos \phi + U_{by} \sin \phi$$

or

$$\begin{bmatrix} u_a \\ u_b \end{bmatrix} = \begin{bmatrix} \cos \phi & \sin \phi & 0 & \\ 0 & 0 & \cos \phi & \sin \phi \end{bmatrix} \begin{bmatrix} U_{ax} \\ U_{ay} \\ U_{bx} \\ U_{by} \end{bmatrix} = \mathbf{RU}$$

$$\mathbf{u} = \mathbf{RU}$$

where

$$\mathbf{u} = \begin{bmatrix} u_a \\ u_b \end{bmatrix} \tag{11.4-1}$$

$$\mathbf{R} = \begin{bmatrix} \cos \phi & \sin \phi & 0 & 0 \\ 0 & 0 & \cos \phi & \sin \phi \end{bmatrix} \tag{11.4-2}$$

and

$$\mathbf{U} = \begin{bmatrix} U_{ax} \\ U_{ay} \\ U_{bx} \\ U_{by} \end{bmatrix} \tag{11.4.3}$$

First consider element 1. See Figure 11.4-3. Note that $\phi = 2\pi - \psi$ and thus

$$\sin \phi = - \sin \psi \qquad \cos \phi = \cos \psi$$

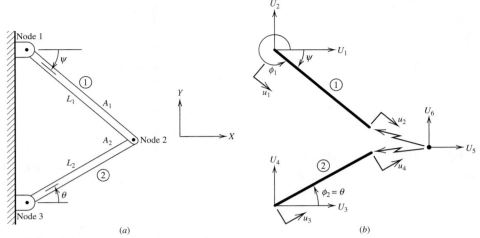

FIGURE 11.4-3 Analysis of a truss.

Thus

$$\mathbf{R}_1 = \begin{bmatrix} \cos\phi & \sin\phi & 0 & 0 \\ 0 & 0 & \cos\phi & \sin\phi \end{bmatrix} = \begin{bmatrix} \cos\psi & -\sin\psi & 0 & 0 \\ 0 & 0 & \cos\psi & -\sin\psi \end{bmatrix}$$

So, for element 1,

$$\mathbf{u}_1 = \mathbf{R}_1 \mathbf{U}_1$$

where

$$\mathbf{u}_1 = \begin{bmatrix} u_1 \\ u_2 \end{bmatrix} \qquad \mathbf{U}_1 = \begin{bmatrix} U_1 \\ U_2 \\ U_5 \\ U_6 \end{bmatrix}$$

Now consider element 2. See Figure 11.4-3. Note that for this element $\phi = \theta$. Thus

$$\mathbf{R}_2 = \begin{bmatrix} \cos\theta & \sin\theta & 0 & 0 \\ 0 & 0 & \cos\theta & \sin\theta \end{bmatrix}$$

and

$$\mathbf{u}_2 = \mathbf{R}_2 \mathbf{U}_2$$

where

$$\mathbf{u}_2 = \begin{bmatrix} u_3 \\ u_4 \end{bmatrix} \qquad \mathbf{U}_2 = \begin{bmatrix} U_3 \\ U_4 \\ U_5 \\ U_6 \end{bmatrix}$$

Mass and Stiffness Matrices

The kinetic energy of the system is the same regardless of what coordinate system is used to describe it. Thus, expressing the kinetic energy in both the **u** and the **U** coordinates, we have

$$KE = \frac{1}{2}\dot{\mathbf{u}}^T\mathbf{M}\dot{\mathbf{u}} = \frac{1}{2}\dot{\mathbf{U}}^T\mathbf{M}_G\dot{\mathbf{U}}$$

where $\mathbf{M}$ is the mass matrix of an element expressed in the $\mathbf{u}$ coordinate system and $\mathbf{M}_G$ is the mass matrix of the same element expressed in the global coordinate system $\mathbf{U}$. Since $\mathbf{u} = \mathbf{R}\mathbf{U}$, we have $\dot{\mathbf{u}} = \mathbf{R}\dot{\mathbf{U}}$, and thus $\dot{\mathbf{u}}^T = \dot{\mathbf{U}}^T\mathbf{R}^T$. Substituting this into the kinetic energy expression gives

$$\dot{\mathbf{u}}^T\mathbf{M}\dot{\mathbf{u}} = \dot{\mathbf{U}}^T\mathbf{R}^T\mathbf{M}\mathbf{R}\dot{\mathbf{U}} = \dot{\mathbf{U}}^T\mathbf{M}_G\dot{\mathbf{U}}$$

Thus the two mass matrices expressed in the different coordinate systems are related as follows:

$$\mathbf{M}_G = \mathbf{R}^T\mathbf{M}\mathbf{R} \tag{11.4-4}$$

We can use the same approach for the stiffness matrices. The potential energy of the system is the same regardless of what coordinate system is used to describe it. Thus, expressing the potential energy in both the $\mathbf{u}$ and the $\mathbf{U}$ coordinates, we have

$$PE = \frac{1}{2}\mathbf{u}^T\mathbf{K}\mathbf{u} = \frac{1}{2}\mathbf{U}^T\mathbf{K}_G\mathbf{U}$$

where $\mathbf{K}$ is the mass matrix of an element expressed in the $\mathbf{u}$ coordinate system and $\mathbf{K}_G$ is the mass matrix of the same element expressed in the global coordinate system $\mathbf{U}$. Thus the two stiffness matrices expressed in the different coordinate systems are related as follows:

$$\mathbf{K}_G = \mathbf{R}^T\mathbf{K}\mathbf{R} \tag{11.4-5}$$

The mass matrix for bar element i is

$$\mathbf{M} = \frac{\rho_i A_i L_i}{6}\begin{bmatrix} 2 & 1 \\ 1 & 2 \end{bmatrix} \tag{11.4-6}$$

and the stiffness matrix is

$$\mathbf{K} = \frac{E_i A_i}{L_i}\begin{bmatrix} 1 & -1 \\ -1 & 1 \end{bmatrix} \tag{11.4-7}$$

Thus, using Equations 11.4-2 and 11.4-6, we have

$$\mathbf{M}_G = \mathbf{R}^T\mathbf{M}\mathbf{R} = \begin{bmatrix} \cos\phi & 0 \\ \sin\phi & 0 \\ 0 & \cos\phi \\ 0 & \sin\phi \end{bmatrix} \frac{\rho_i A_i L_i}{6}\begin{bmatrix} 2 & 1 \\ 1 & 2 \end{bmatrix}\begin{bmatrix} \cos\phi & \sin\phi & 0 & 0 \\ 0 & 0 & \cos\phi & \sin\phi \end{bmatrix}$$

This evaluates to the following:

$$\mathbf{M}_G = \frac{\rho_i A_i L_i}{6}\begin{bmatrix} 2\cos^2\phi & 2\sin\phi\cos\phi & \vdots & \cos^2\phi & \sin\phi\cos\phi \\ 2\sin\phi\cos\phi & 2\sin^2\phi & \vdots & \sin\phi\cos\phi & \sin^2\phi \\ \cdots & \cdots & \cdots & \cdots & \cdots \\ \cos^2\phi & \sin\phi\cos\phi & \vdots & 2\cos^2\phi & 2\sin\phi\cos\phi \\ \sin\phi\cos\phi & \sin^2\phi & \vdots & 2\sin\phi\cos\phi & 2\sin^2\phi \end{bmatrix}$$

$$= \begin{bmatrix} \mathbf{M}_1 & \vdots & \mathbf{M}_2 \\ \cdots & \cdots & \cdots \\ \mathbf{M}_2 & \vdots & \mathbf{M}_1 \end{bmatrix} \tag{11.4-8}$$

where we have partitioned $\mathbf{M}_G$ into four (2×2) submatrices:

$$\mathbf{M}_1 = \frac{\rho_i A_i L_i}{6} \begin{bmatrix} 2\cos^2\phi & 2\sin\phi\cos\phi \\ 2\sin\phi\cos\phi & 2\sin^2\phi \end{bmatrix} \qquad (11.4\text{-}9)$$

$$\mathbf{M}_2 = \frac{\rho_i A_i L_i}{6} \begin{bmatrix} \cos^2\phi & \sin\phi\cos\phi \\ \sin\phi\cos\phi & \sin^2\phi \end{bmatrix} \qquad (11.4\text{-}10)$$

Using Equations 11.4-2 and 11.4-7, we have

$$\mathbf{K}_G = \mathbf{R}^T\mathbf{K}\mathbf{R} = \begin{bmatrix} \cos\phi & 0 \\ \sin\phi & 0 \\ 0 & \cos\phi \\ 0 & \sin\phi \end{bmatrix} \frac{E_i A_i}{L_i} \begin{bmatrix} 1 & -1 \\ -1 & 1 \end{bmatrix} \begin{bmatrix} \cos\phi & \sin\phi & 0 & 0 \\ 0 & 0 & \cos\phi & \sin\phi \end{bmatrix}$$

This evaluates to the following:

$$\mathbf{K}_G = \frac{E_i A_i}{L_i} \begin{bmatrix} \cos^2\phi & \sin\phi\cos\phi & \vdots & -\cos^2\phi & -\sin\phi\cos\phi \\ \sin\phi\cos\phi & \sin^2\phi & \vdots & -\sin\phi\cos\phi & -\sin^2\phi \\ \cdots & \cdots & \cdots & \cdots & \cdots \\ -\cos^2\phi & -\sin\phi\cos\phi & \vdots & \cos^2\phi & \sin\phi\cos\phi \\ -\sin\phi\cos\phi & -\sin^2\phi & \vdots & \sin\phi\cos\phi & \sin^2\phi \end{bmatrix}$$

$$= \begin{bmatrix} \mathbf{K}_1 & \vdots & \mathbf{K}_2 \\ \cdots & \cdots & \cdots \\ \mathbf{K}_2 & \vdots & \mathbf{K}_1 \end{bmatrix} \qquad (11.4\text{-}11)$$

where we have partitioned $\mathbf{K}_G$ into four (2×2) submatrices:

$$\mathbf{K}_1 = \frac{E_i A_i}{L_i} \begin{bmatrix} \cos^2\phi & \sin\phi\cos\phi \\ \sin\phi\cos\phi & \sin^2\phi \end{bmatrix} \qquad (11.4\text{-}12)$$

$$\mathbf{K}_2 = \frac{E_i A_i}{L_i} \begin{bmatrix} -\cos^2\phi & -\sin\phi\cos\phi \\ -\sin\phi\cos\phi & -\sin^2\phi \end{bmatrix} \qquad (11.4.13)$$

Assembling the Equations

The mass and stiffness matrices given by Equations 11.4-8 and 11.4-11 are for the global coordinate vector

$$\mathbf{U} = \begin{bmatrix} U_{ax} \\ U_{ay} \\ U_{bx} \\ U_{by} \end{bmatrix}$$

which contains only the four global coordinates attached to the specific bar element. The equations of motion for the entire truss must be obtained by expressing the mass and stiffness matrices in terms of the global coordinates of the entire truss. This coordinate vector is

$$\mathbf{U}_T = \begin{bmatrix} U_1 \\ U_2 \\ U_3 \\ U_4 \\ U_5 \\ U_6 \end{bmatrix} \tag{11.4-14}$$

Thus we must expand the stiffness matrices for each element as follows. For the first element,

$$\mathbf{K}_{G1} = \begin{bmatrix} \mathbf{K}_{11} & \mathbf{0}_2 & \mathbf{K}_{21} \\ \mathbf{0}_2 & \mathbf{0}_2 & \mathbf{0}_2 \\ \mathbf{K}_{21} & \mathbf{0}_2 & \mathbf{K}_{11} \end{bmatrix}$$

where $\mathbf{0}_2$ is a (2×2) matrix having all zero elements:

$$\mathbf{0}_2 = \begin{bmatrix} 0 & 0 \\ 0 & 0 \end{bmatrix}$$

Note that the location of the zero elements in $\mathbf{K}_{G1}$ were chosen in light of the fact that U_3 and U_4 are not associated with the first element. Note also that $\mathbf{K}_{11}$ is $\mathbf{K}_1$ evaluated for the first element and $\mathbf{K}_{21}$ is $\mathbf{K}_2$ evaluated for the first element.

For the second element,

$$\mathbf{K}_{G2} = \begin{bmatrix} \mathbf{0}_2 & \mathbf{0}_2 & \mathbf{0}_2 \\ \mathbf{0}_2 & \mathbf{K}_{12} & \mathbf{K}_{22} \\ \mathbf{0}_2 & \mathbf{K}_{22} & \mathbf{K}_{22} \end{bmatrix}$$

Note that the locations of the zero elements in $\mathbf{K}_{G2}$ were chosen in light of the fact that U_1 and U_2 are not associated with the second element. Note also that $\mathbf{K}_{12}$ is $\mathbf{K}_1$ evaluated for the second element and $\mathbf{K}_{22}$ is $\mathbf{K}_2$ evaluated for the second element.

The equation of motion for the entire truss in terms of the global coordinate vector $\mathbf{U}_T$ given by Equation 11.4-14 has the form

$$\mathbf{M}\ddot{\mathbf{U}}_T + \mathbf{K}\mathbf{U}_T = \mathbf{0} \tag{11.4-15}$$

where, by superimposing $\mathbf{M}_{G1}$ and $\mathbf{M}_{G2}$, we obtain

$$\begin{aligned} \mathbf{M} &= \mathbf{M}_{G1} + \mathbf{M}_{G2} \\ &= \begin{bmatrix} \mathbf{M}_{11} & \mathbf{0}_2 & \mathbf{M}_{21} \\ \mathbf{0}_2 & \mathbf{M}_{12} & \mathbf{M}_{22} \\ \mathbf{M}_{21} & \mathbf{M}_{22} & \mathbf{M}_{11} + \mathbf{M}_{12} \end{bmatrix} \end{aligned} \tag{11.4-16}$$

Superimposing $\mathbf{K}_{G1}$ and $\mathbf{K}_{G2}$, we obtain

$$\begin{aligned} \mathbf{K} &= \mathbf{K}_{G1} + \mathbf{K}_{G2} \\ &= \begin{bmatrix} \mathbf{K}_{11} & \mathbf{0}_2 & \mathbf{K}_{21} \\ \mathbf{0}_2 & \mathbf{K}_{12} & \mathbf{K}_{22} \\ \mathbf{K}_{21} & \mathbf{K}_{22} & \mathbf{K}_{21} + \mathbf{K}_{12} \end{bmatrix} \end{aligned} \tag{11.4-17}$$

Applying the Boundary Conditions

So far we have not imposed any boundary conditions on the truss. Suppose now that we are given that points 1 and 3 are pinned. This means that

$$U_1 = U_2 = U_3 = U_4 = 0$$

and thus the global displacement vector (Equation 11.4-14) for the truss reduces to $[U_5, U_6]^T$. This means that we may strike out the first four rows and the first four columns of $\mathbf{M}$ and $\mathbf{K}$ to obtain

$$\mathbf{M} = \mathbf{M}_{11} + \mathbf{M}_{12}$$

$$= \frac{\rho_1 A_1 L_1}{6} \begin{bmatrix} 2\cos^2\psi & -2\sin\psi\cos\psi \\ -2\sin\psi\cos\psi & 2\sin^2\psi \end{bmatrix}$$

$$+ \frac{\rho_2 A_2 L_2}{6} \begin{bmatrix} 2\cos^2\theta & 2\sin\theta\cos\theta \\ 2\sin\theta\cos\theta & 2\sin^2\theta \end{bmatrix} \tag{11.4-18}$$

$$\mathbf{K} = \mathbf{K}_{11} + \mathbf{K}_{12}$$

$$= \frac{E_1 A_1}{L_1} \begin{bmatrix} \cos^2\psi & -\sin\psi\cos\psi \\ -\sin\psi\cos\psi & \sin^2\psi \end{bmatrix}$$

$$+ \frac{E_2 A_2}{L_2} \begin{bmatrix} \cos^2\theta & \sin\theta\cos\theta \\ \sin\theta\cos\theta & \sin^2\theta \end{bmatrix} \tag{114-19}$$

The equation of motion of the truss becomes

$$\mathbf{M} \begin{bmatrix} \ddot{U}_5 \\ \ddot{U}_6 \end{bmatrix} + \mathbf{K} \begin{bmatrix} U_5 \\ U_6 \end{bmatrix} = \mathbf{0} \tag{11.4-20}$$

where $\mathbf{M}$ and $\mathbf{K}$ are given by Equations 11.4-18 and 11.4-19. These matrices can then be used to compute the natural frequencies and mode shapes. The eigenvalue problem is

$$\mathbf{M}^{-1}\mathbf{K} \begin{bmatrix} U_5 \\ U_6 \end{bmatrix} = \omega^2 \begin{bmatrix} U_6 \\ U_5 \end{bmatrix} \tag{11.4-21}$$

The displacements u_2 and u_4 can then be found by inverting the kinematic relations, noting that $u_1 = u_3 = 0$ because of the boundary conditions:

$$u_2 = U_5 \cos\psi - U_6 \sin\psi \tag{11.4-22}$$

$$u_4 = U_5 \cos\theta + U_6 \sin\theta \tag{11.4-23}$$

EXAMPLE 11.4-1
Vibration of a Nonsymmetric Truss

Consider the specific truss shown in Figure 11.4-4. The mass density and area are the same for both elements. Obtain the mode shapes and frequencies.

Solution

From the figure we see that $L_1 = 4L/5$, $L_2 = L$, and that

$$\sin\psi = 0 \qquad \cos\psi = 1$$

$$\sin\theta = \frac{3}{5} \qquad \cos\theta = \frac{4}{5}$$

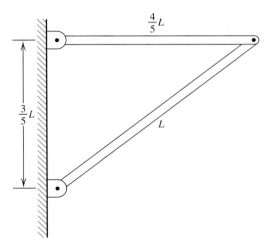

FIGURE 11.4-4   A nonsymmetric truss.

The mass matrix becomes

$$\mathbf{M} = \frac{\rho A L_1}{6}\begin{bmatrix} 2 & 0 \\ 0 & 0 \end{bmatrix} + \frac{\rho A L_2}{6}\begin{bmatrix} \dfrac{32}{25} & \dfrac{24}{25} \\[2mm] \dfrac{24}{25} & \dfrac{18}{25} \end{bmatrix}$$

$$= \frac{\rho A L}{25}\begin{bmatrix} 12 & 2 \\ 2 & 3 \end{bmatrix}$$

The stiffness matrix is

$$\mathbf{K} = \frac{EA}{L_1}\begin{bmatrix} 1 & 0 \\ 0 & 0 \end{bmatrix} + \frac{EA}{25 L_2}\begin{bmatrix} 16 & 12 \\ 12 & 9 \end{bmatrix}$$

$$= \frac{EA}{100L}\begin{bmatrix} 189 & 48 \\ 48 & 36 \end{bmatrix}$$

The modes and frequencies are found from

$$-\omega^2 \frac{\rho A L}{25}\begin{bmatrix} 12 & 2 \\ 2 & 3 \end{bmatrix}\begin{bmatrix} U_5 \\ U_6 \end{bmatrix} + \frac{EA}{100L}\begin{bmatrix} 189 & 48 \\ 48 & 36 \end{bmatrix}\begin{bmatrix} U_5 \\ U_6 \end{bmatrix} = \begin{bmatrix} 0 \\ 0 \end{bmatrix}$$

Let $\lambda = 4\rho L^2 \omega^2 / E$. Then these two equations become

$$\begin{bmatrix} (189 - 12\lambda) & (48 - 2\lambda) \\ (48 - 2\lambda) & (36 - 3\lambda) \end{bmatrix}\begin{bmatrix} U_5 \\ U_6 \end{bmatrix} = \begin{bmatrix} 0 \\ 0 \end{bmatrix}$$

The characeristic equation is $32\lambda^2 - 903\lambda + 4500 = 0$, which give $\lambda = 6.4642$ and $\lambda = 21.7546$. Thus the frequencies are

$$\omega = \frac{1.2712}{L}\sqrt{\frac{E}{\rho}}, \qquad \frac{2.3321}{L}\sqrt{\frac{E}{\rho}}$$

The mode shapes are

$$\frac{U_6}{U_5} = -3.1772, \qquad 16.045$$

∎

Special Cases

Note that if the truss elements are identical—that is, if each element has the same values for A, E, and L—then the $\mathbf{M}$ and $\mathbf{K}$ matrices will be diagonal if

$$\sin\psi\cos\psi = \sin\theta\cos\theta \tag{11.4-24}$$

One obvious and common case where this condition is satisfied is the case where $\psi = \theta$. Many trusses form an isosceles triangle (one whose side lengths are equal), for which $\psi = \theta = 30°$.

If $\mathbf{M}$ and $\mathbf{K}$ are diagonal, then Equation 11.4-20 has the form

$$m_5\ddot{U}_5 + k_5 U_5 = 0 \tag{11.4-25}$$

$$m_6\ddot{U}_6 + k_6 U_6 = 0 \tag{11.4-26}$$

where m_5, m_6, k_5, and k_6 are the diagonal elements of $\mathbf{M}$ and $\mathbf{K}$.

$$m_5 = \frac{\rho A L}{3}(\cos^2\psi + \cos^2\theta)$$

$$m_6 = \frac{\rho A L}{3}(\sin^2\psi + \sin^2\theta)$$

$$k_5 = \frac{EA}{L}(\cos^2\psi + \cos^2\theta)$$

$$k_6 = \frac{EA}{L}(\sin^2\psi + \sin^2\theta)$$

The frequencies are immediately found to be

$$\omega_1 = \sqrt{\frac{k_5}{m_5}}$$

$$\omega_2 = \sqrt{\frac{k_6}{m_6}}$$

If $\psi = \theta$, then

$$m_5 = \frac{2\rho A L}{3}\cos^2\theta$$

$$m_6 = \frac{2\rho A L}{3}\sin^2\theta$$

$$k_5 = \frac{2EA}{L}\cos^2\theta$$

$$k_6 = \frac{2EA}{L}\sin^2\theta$$

and

$$\ddot{U}_i + \frac{3E}{\rho L^2}U_i = 0 \qquad i = 5,\ 6 \tag{11.4-27}$$

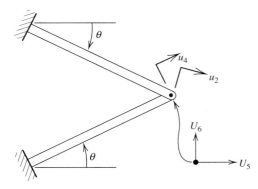

FIGURE 11.4-5   Two bars fixed at one end and pinned at the other end.

So the frequencies are identical and equal to

$$\omega = \frac{1}{L}\sqrt{\frac{3E}{\rho}}$$

and the two modes are uncoupled.

This perhaps unexpected result can be explained with Figure 11.4-5, which shows two bars, each having a mass m and stiffness k, fixed at one end and pinned at the other end. Suppose we assume that the vibration of the bars is in the longitudinal direction only (that is, the bars do not bend). If we apply the result of Chapter 2, which is based on the static deflection curve and says that the equivalent mass of each bar is $m/3$, we obtain the following equations of motion:

$$\frac{1}{3}m\ddot{u}_2 + ku_2 = 0 \qquad\qquad (11.4\text{-}28)$$

$$\frac{1}{3}m\ddot{u}_4 + ku_4 = 0 \qquad\qquad (11.4\text{-}29)$$

The natural frequency of each bar is

$$\omega = \sqrt{\frac{3k}{m}}$$

But $m = \rho AL$ and $k = EA/L$, so

$$\omega = \sqrt{\frac{3k}{m}} = \frac{1}{L}\sqrt{\frac{3E}{\rho}}$$

which is identical to the frequency predicted by the finite element model. The reason is that the finite element model is based on the same assumptions; namely, a bar model (longitudinal motion only) and the static deflection curve.

The equations of motion of the system in Figure 11.4-5 are identical to those of the truss equations (Equations 11.4-27) when expressed in the same coordinates. If we substitute the coordinate transformation

$$u_2 = U_5 \cos\theta - U_6 \sin\theta$$

$$u_4 = U_5 \cos\theta + U_6 \sin\theta$$

into Equations 11.4-28 and 11.4-29, we obtain, after some manipulation,

$$\frac{1}{3}m\ddot{U}_5 + kU_5 = 0$$

$$\frac{1}{3}m\ddot{U}_6 + kU_6 = 0$$

which are identical to the truss equations since $m = \rho AL$ and $k = EA/L$.

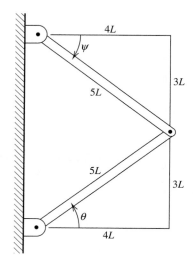

FIGURE 11.4-6 A symmetric truss.

EXAMPLE 11.4-2
Vibration of a Symmetric Truss

Consider the specific truss shown in Figure 11.4-6. The mass density, area, and length are the same for both elements. Obtain the mode shapes and frequencies.

Solution

From the figure we see that $L_1 = L_2 = 5L$ and that

$$\sin \psi = \sin \theta = \frac{3}{5} \qquad \cos \psi = \cos \theta = \frac{4}{5}$$

The mass matrix becomes

$$\mathbf{M} = \frac{5\rho AL}{6} \left\{ \begin{bmatrix} 2(4/5)^2 & -2(3/5)(4/5) \\ -2(3/5)(4/5) & 2(3/5)^2 \end{bmatrix} + \begin{bmatrix} 2(4/5)^2 & 2(3/5)(4/5) \\ 2(3/5)(4/5) & 2(3/5)^2 \end{bmatrix} \right\}$$

$$= \frac{2\rho AL}{15} \begin{bmatrix} 16 & 0 \\ 0 & 9 \end{bmatrix}$$

The stiffness matrix is

$$\mathbf{K} = \frac{EA}{5L} \left\{ \begin{bmatrix} (4/5)^2 & -2(3/5)(4/5) \\ -2(3/5)(4/5) & (3/5)^2 \end{bmatrix} + \begin{bmatrix} (4/5)^2 & 2(3/5)(4/5) \\ 2(3/5)(4/5) & (3/5)^2 \end{bmatrix} \right\}$$

$$= \frac{EA}{5L} \begin{bmatrix} 16 & 0 \\ 0 & 9 \end{bmatrix}$$

The modes and frequencies are found from

$$-\omega^2 \frac{2\rho AL}{15} \begin{bmatrix} 16 & 0 \\ 0 & 9 \end{bmatrix} \begin{bmatrix} U_5 \\ U_6 \end{bmatrix} + \frac{EA}{5L} \begin{bmatrix} 16 & 0 \\ 0 & 9 \end{bmatrix} \begin{bmatrix} U_5 \\ U_6 \end{bmatrix} = \begin{bmatrix} 0 \\ 0 \end{bmatrix}$$

Each of these two equations gives the same frequency

$$\omega = \frac{1}{L} \sqrt{\frac{3E}{2\rho}}$$

and the two modes are uncoupled. ∎

11.5 TORSIONAL ELEMENTS

In Section 11.1 we treated the longitudinal vibration of a bar. We now develop a finite element model of a rod in torsion. Recall that the force-deflection relation of a uniform rod having a net applied torque T is

$$T = \frac{GJ}{L}\theta = k\theta$$

where θ is the resulting twist, J is the polar moment of inertia, and k is the torsional stiffness, given by

$$k = \frac{GJ}{L} \tag{11.5-1}$$

For the case where torques are applied at both end of the bar (Figure 11.5-1),

$$\begin{bmatrix} T_1 \\ T_2 \end{bmatrix} = \begin{bmatrix} k_{11} & k_{12} \\ k_{21} & k_{22} \end{bmatrix} \begin{bmatrix} \theta_1 \\ \theta_2 \end{bmatrix}$$

To determine the elements of the matrix, note that the first column contains the endpoint torques when $\theta_1 = 1$ and $\theta_2 = 0$. That is, $T_1 = k\theta_1$ and $T_2 = -k\theta_1$, so the first column is $[k, -k]^T$. Similarly, the second column contains the endpoint torques when $\theta_1 = 0$ and $\theta_2 = 1$. That is, $T_1 = -k\theta_2$ and $T_2 = k\theta_2$, so the second column is $[-k, k]^T$. Thus we have

$$\begin{bmatrix} T_1 \\ T_2 \end{bmatrix} = \begin{bmatrix} k & -k \\ -k & k \end{bmatrix} \begin{bmatrix} \theta_1 \\ \theta_2 \end{bmatrix} = \frac{EA}{L} \begin{bmatrix} 1 & -1 \\ -1 & 1 \end{bmatrix} \begin{bmatrix} \theta_1 \\ \theta_2 \end{bmatrix}$$

So the torsional stiffness matrix $\mathbf{K}$ of the bar element is

$$\mathbf{K} = \frac{GJ}{L} \begin{bmatrix} 1 & -1 \\ -1 & 1 \end{bmatrix} \tag{11.5-2}$$

which is identical to the stiffness matrix for longitudinal vibration, except that GJ replaces EA.

The Mass Matrix

We now derive the mass matrix $\mathbf{M}$ for the torsional element. Actually, it is a matrix composed of inertias, and so should be called the inertia matrix and denoted by $\mathbf{I}$, but this symbol is universally reserved for the *identity* matrix.

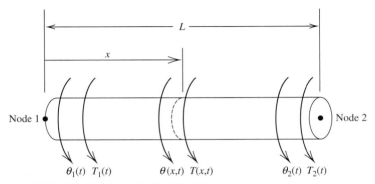

FIGURE 11.5-1 Torsional vibration element.

For the rod shown in Figure 11.5-1, from Chapter 10 we know that the *static* displacement of the rod is described by

$$GJ\frac{d^2\theta(x)}{dx^2} = 0 \qquad 0 \le x \le L$$

This has the solution

$$\theta(x) = a + bx$$

where a and b are independent of x. Taking this *static* mode shape to be an approximation of the *dynamic* mode shape, we write

$$\theta(x,t) = a(t) + b(t)x$$

There are two nodes, whose displacements are $\theta_1(t)$ and $\theta_2(t)$. At $x = 0$,

$$\theta(0,t) = \theta_1(t)$$

At $x = L$,

$$\theta(L,t) = \theta_2(t)$$

where

$$a(t) = \theta_1(t)$$

$$b(t) = \frac{\theta_2(t) - \theta_1(t)}{L}$$

Thus

$$\theta(x,t) = \left(1 - \frac{x}{L}\right)\theta_1(t) + \frac{x}{L}\theta_2(t) = S_1(x)\theta_1(t) + S_2(x)\theta_2(t) \tag{11.5-3}$$

The functions S_1 and S_2 are the shape functions for the torsional displacement of the bar. Equation 11.5-3 is an approximate solution of the equation of motion of the bar.

With the assumed mode shape specified by Equation 11.5-3, we can express the kinetic energy of the bar element as follows:

$$KE = \int_0^L \frac{1}{2}\rho J\left(\frac{\partial\theta}{\partial t}\right)^2 dx = \frac{1}{2}\rho A\int_0^L \left[S_1(x)\dot\theta_1(t) + S_2(x)\dot\theta_2(t)\right]^2 dx$$

Evaluation of this integral is identical to the process developed for the longitudinal vibration, and so we simply state the result:

$$KE = \frac{1}{2}\rho J\left[\frac{\dot\theta_1^2(t)}{3} + \frac{\dot\theta_1(t)\dot\theta_2(t)}{3} + \frac{\dot\theta_2^2(t)}{3}\right] = \frac{1}{2}\dot{\boldsymbol{\theta}}^T(t)\mathbf{M}\dot{\boldsymbol{\theta}}(t)$$

where

$$\dot{\boldsymbol{\theta}}(t) = \begin{bmatrix} \dot\theta_1(t) \\ \dot\theta_2(t) \end{bmatrix}$$

Thus the mass (or inertia) matrix of the bar element is

$$\mathbf{M} = \frac{\rho JL}{6}\begin{bmatrix} 2 & 1 \\ 1 & 2 \end{bmatrix} \tag{11.5.4}$$

which is identical to the mass matrix for longitudinal vibration, except that J replaces A. Note that this result is dependent on the assumed mode shape specified by Equation 11.5-3.

The equation of motion for the torsional element is

$$\mathbf{M}\ddot{\boldsymbol{\theta}} + \mathbf{K}\boldsymbol{\theta} = 0 \qquad (11.5\text{-}5)$$

where $\mathbf{M}$ and $\mathbf{K}$ are given by Equations 11.5-2 and 11.5-4. Thus the forms of the equation of motion and of the expressions for $\mathbf{M}$ and $\mathbf{K}$ are identical to those for the longitudinal bar vibration, and so we can apply the results of prior examples to the torsional vibration case if the boundary conditions have the same form.

EXAMPLE 11.5-1
*A Fixed-Free
Bar in Torsion*

Consider a model of a bar in torsion, for which the left-hand end of the bar is fixed and the right-hand end is free. Develop a model of the torsional vibration using one element.

Solution

Equation 11.5-4 becomes (after canceling J from all terms)

$$\frac{\rho L}{6}\begin{bmatrix} 2 & 1 \\ 1 & 2 \end{bmatrix}\begin{bmatrix} \ddot{\theta}_1 \\ \ddot{\theta}_2 \end{bmatrix} + \frac{G}{L}\begin{bmatrix} 1 & -1 \\ -1 & 1 \end{bmatrix}\begin{bmatrix} \theta_1 \\ \theta_2 \end{bmatrix} = \begin{bmatrix} 0 \\ 0 \end{bmatrix}$$

Since the left-hand end of the element is fixed, $\theta_1(t) = \dot{\theta}_1(t) = 0$; so the first equation in the set is not applicable. This means we can strike out the first row and the first column of the $\mathbf{M}$ and $\mathbf{K}$ matrices and immediately obtain

$$\frac{\rho L}{3}\ddot{\theta}_2 + \frac{G}{L}\theta_2 = 0 \qquad (1)$$

The natural frequency is

$$\omega_n = \frac{1}{L}\sqrt{\frac{3G}{\rho}} = \frac{1.7321}{L}\sqrt{\frac{G}{\rho}} \qquad (2)$$

Given the initial displacement $\theta_2(0)$ and initial velocity $\dot{\theta}_2(0)$ of the second node point, we can solve Equation 1 for $\theta_2(t)$. Then, using Equation 11.5-3 with $\theta_1(t) = 0$, we can obtain the expression for the displacement $\theta(x, t)$ of the torsional element as a function of location x and time t:

$$\theta(x, t) = \frac{x}{L}\theta_2(t)$$

In Chapter 2 we derived the following expression for the natural frequency of a concentrated inertia attached to the end of a rod:

$$\omega_n = \sqrt{\frac{k}{I_c + I_r/3}}$$

where I_c is the concentrated inertia at the end of the bar and I_r is the rod inertia. Taking $I_c = 0$, using the fact that for the rod, $k = GJ/L$, and noting that $I_r = \rho JL$, we obtain

$$\omega_n = \sqrt{\frac{GJ}{L\rho JL/3}} = \frac{1}{L}\sqrt{\frac{3G}{\rho}}$$

which is identical to the frequency of the finite element model of the rod. This result should be expected because the derivation in Chapter 2 assumed the rod's angular displacement to be its static deflection shape, which is identical to the shape function used in this section.

We can also compare the result with that obtained from the solution of the distributed-parameter model of the rod, which is

$$\frac{\partial^2 \theta}{\partial t^2} = \frac{G}{\rho}\frac{\partial^2 \theta}{\partial x^2}$$

For the case where the left-hand end is fixed and the right-hand end is free, the method of separation of variables used in Chapter 10 results in the following expressions for the temporal frequencies and the mode shapes:

$$\omega_n = \frac{2n-1}{2L}\pi\sqrt{\frac{G}{\rho}} \qquad n = 1, 2, 3, \ldots \tag{3}$$

$$F_n(x) = A_n \sin \frac{(2n-1)\pi x}{2L} \qquad n = 1, 2, 3, \ldots \tag{4}$$

The first frequency and mode shape are

$$\omega_1 = \frac{\pi}{2L}\sqrt{\frac{G}{\rho}} = \frac{1.5708}{L}\sqrt{\frac{G}{\rho}}$$

$$F_1(x) = A_1 \sin \frac{\pi x}{2L}$$

Thus the finite element model predicts a frequency that is 10% higher than that predicted by the distributed-parameter model. The static deflection shape used by the finite element model is close to the first mode shape.

The second frequency and mode shape are

$$\omega_2 = \frac{3\pi}{2L}\sqrt{\frac{G}{\rho}} = \frac{4.7124}{L}\sqrt{\frac{G}{\rho}}$$

$$F_2(x) = A_2 \sin \frac{3\pi x}{2L}$$

The finite element model predicts a frequency that is 63% lower than that predicted by the second mode of the distributed-parameter model. The second mode shape is quite different from the shape function used by the finite element model. The second mode has a node (a point of no motion) at $x/L = 2/3$. The higher modes all have higher frequencies and more complex mode shapes. ∎

EXAMPLE 11.5-2
Torsional Vibration of a Stepped Shaft

Figure 11.5-2 shows a shaft having a step change in cross-section area. It is fixed at the left-hand end, and the shaft material is uniform. Develop a finite element model of the torsional vibration of the shaft.

Solution

The step change in cross-section area suggests the two-element model shown in Figure 11.5-2. This model is identical in form to the two-element model of the longitudinal vibration of a stepped bar developed in Example 11.2-1, and the boundary conditions are

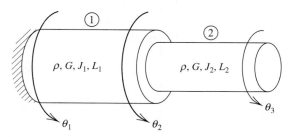

FIGURE 11.5-2 Shaft having a step change in cross-section area.

equivalent, with u_1, u_2, and u_3 replaced with θ_1, θ_2, and θ_3. Thus we may immediately use those results, with A replaced with J and E replaced with G. This gives

$$\frac{1}{6}\begin{bmatrix} 2(I_1 + I_2) & I_2 \\ I_2 & 2I_2 \end{bmatrix}\begin{bmatrix} \ddot{\theta}_2 \\ \ddot{\theta}_3 \end{bmatrix} + \begin{bmatrix} (k_1 + k_2) & -k_2 \\ -k_2 & k_2 \end{bmatrix}\begin{bmatrix} \theta_2 \\ \theta_3 \end{bmatrix} = \begin{bmatrix} 0 \\ 0 \end{bmatrix}$$

where $I_i = \rho_i J_i L_i$ and $k_i = G_i J_i / L_i$, $i = 1, 2$. This is the finite element model of the torsional vibration of the shaft, using two elements. ∎

11.6   BEAM ELEMENTS

The element equations for a beam element (called the Euler-Bernoulli beam element) are derived in a fashion similar to that used for the longitudinal vibration of a bar. The beam element is shown in Figure 11.6-1. The beam displacement normal to its length is denoted $v(x,t)$. The displacements v_1 and v_2 are the displacements of the endpoints. The variables θ_1 and θ_2 are the rotational displacements at the ends of the beam element. Thus we define the displacement vector of the beam element to be

$$\mathbf{v} = \begin{bmatrix} v_1 \\ \theta_1 \\ v_2 \\ \theta_2 \end{bmatrix} \tag{11.6-1}$$

The boundary conditions are

$$v(0, t) = v_1(t)$$

$$\frac{\partial v(0, t)}{\partial x} = \theta_1(t)$$

$$v(L, t) = v_2(t)$$

$$\frac{\partial v(L, t)}{\partial x} = \theta_2(t)$$

Here we assume that $v(x, t)$, the solution of the distributed-parameter model, is approximated by

$$v(x, t) = a(t) + b(t)x + c(t)x^2 + d(t)x^3$$

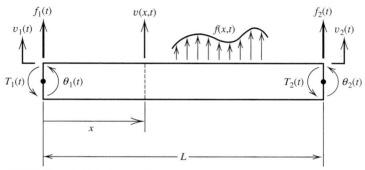

FIGURE 11.6-1   Euler-Bernoulli beam element.

Applying the boundary conditions, we find that

$$a(t) = v_1(t)$$

$$b(t) = \theta_1(t)$$

$$c(t) = \frac{1}{L^2}[-3v_1(t) - 2L\theta_1(t) + 3v_2(t) - L\theta_2(t)]$$

$$d(t) = \frac{1}{L^3}[2v_1(t) + L\theta_1(t) - 2v_2(t) + L\theta_2(t)]$$

Thus $v(x, t)$ has the form

$$v(x, t) = S_1(x)v_1(t) + S_2(x)\theta_2(t) + S_3(x)v_2(t) + S_4(x)\theta_2(t) \tag{11.6-2}$$

where the $S_i(x)$ terms are the following shape functions.

$$S_1(x) = 1 - 3\left(\frac{x}{L}\right)^2 + 2\left(\frac{x}{L}\right)^3 \tag{11.6-3}$$

$$S_2(x) = x - 2L\left(\frac{x}{L}\right)^2 + L\left(\frac{x}{L}\right)^3 \tag{11.6-4}$$

$$S_3(x) = 3\left(\frac{x}{L}\right)^2 - 2\left(\frac{x}{L}\right)^3 \tag{11.6-5}$$

$$S_4(x) = -L\left(\frac{x}{L}\right)^2 + L\left(\frac{x}{L}\right)^3 \tag{11.6-6}$$

Mass Matrix

With the assumed mode shape, we can express the kinetic energy of the beam element as follows:

$$KE = \int_0^L \frac{1}{2}\rho A\left(\frac{\partial v}{\partial t}\right)^2 dx$$

$$= \frac{1}{2}\rho A \int_0^L \left[S_1(x)\dot{v}_1(t) + S_2(x)\dot{\theta}_1(t) + S_3(x)\dot{v}_2(t) + S_4(x)\dot{\theta}_2(t)\right]^2 dx$$

Evaluation of the integrals gives

$$KE = \frac{\rho AL}{420}\dot{\mathbf{v}}^T \mathbf{M}\dot{\mathbf{v}}$$

where

$$\dot{\mathbf{v}} = \begin{bmatrix} \dot{v}_1 \\ \dot{\theta}_1 \\ \dot{v}_2 \\ \dot{\theta}_2 \end{bmatrix}$$

and the mass matrix of the beam element is

$$\mathbf{M} = \frac{\rho AL}{420}\begin{bmatrix} 156 & 22L & 54 & -13L \\ 22L & 4L^2 & 13L & -3L^2 \\ 54 & 13L & 156 & -22L \\ -13L & -3L^2 & -22L & 4L^2 \end{bmatrix} \tag{11.6-7}$$

Note that this result is dependent on the assumed mode shape specified by Equation 11.6-2.

Stiffness Matrix

We can derive the stiffness matrix **K** from the potential energy expression for the beam element, as follows:

$$PE = \frac{1}{2}EI \int_0^L \left(\frac{\partial^2 v}{\partial x^2}\right)^2 dx = \mathbf{v}^T \mathbf{K} \mathbf{v}$$

where I is the area moment of inertia of the section. Using Equation 11.6-2 and carrying out the integrals gives

$$\mathbf{K} = \frac{EI}{L^3}\begin{bmatrix} 12 & 6L & -12 & 6L \\ 6L & 4L^2 & -6L & 2L^2 \\ -12 & -6L & 12 & -6L \\ 6L & 2L^2 & -6L & 4L^2 \end{bmatrix} \tag{11.6-8}$$

The equation of motion of the beam element is

$$\mathbf{M}\ddot{\mathbf{v}} + \mathbf{K}\mathbf{v} = \mathbf{0} \tag{11.6-9}$$

For future reference, when we treat beams having multiple elements, we note that **M** and **K** have the following structure:

$$\mathbf{M} = \begin{bmatrix} \mathbf{M}_a & \vdots & \mathbf{M}_b \\ \cdots & \cdots & \cdots \\ \mathbf{M}_b^T & \vdots & \mathbf{M}_c \end{bmatrix} \tag{11.6-10}$$

where

$$\mathbf{M}_a = \frac{\rho AL}{420}\begin{bmatrix} 156 & 22L \\ 22L & 4L^2 \end{bmatrix} \tag{11.6-11}$$

$$\mathbf{M}_b = \frac{\rho AL}{420}\begin{bmatrix} 54 & -13L \\ 13L & -3L^2 \end{bmatrix} \tag{11.6-12}$$

$$\mathbf{M}_c = \frac{\rho AL}{420}\begin{bmatrix} 156 & -22L \\ -22L & 4L^2 \end{bmatrix} \tag{11.6-13}$$

$$\mathbf{K} = \begin{bmatrix} \mathbf{K}_a & \vdots & \mathbf{K}_b \\ \cdots & \cdots & \cdots \\ \mathbf{K}_b^T & \vdots & \mathbf{K}_c \end{bmatrix} \tag{11.6-14}$$

where

$$\mathbf{K}_a = \frac{EI}{L^3}\begin{bmatrix} 12 & 6L \\ 6L & 4L^2 \end{bmatrix} \tag{11.6-15}$$

$$\mathbf{K}_b = \frac{EI}{L^3}\begin{bmatrix} -12 & 6L \\ -6L & 2L^2 \end{bmatrix} \tag{11.6-16}$$

$$\mathbf{K}_c = \frac{EI}{420}\begin{bmatrix} 12 & -6L \\ -6L & 4L^2 \end{bmatrix} \tag{11.6-17}$$

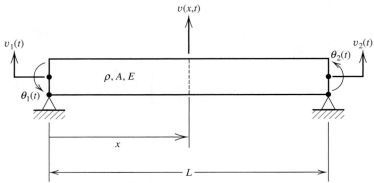

FIGURE 11.6-2 A pinned-pinned beam.

EXAMPLE 11.6-1
A Pinned-Pinned
Beam

Derive the expressions for the frequencies of the pinned-pinned beam shown in Figure 11.6-2. Use one element.

Solution

The displacement vector is

$$\mathbf{v} = \begin{bmatrix} v_1 \\ \theta_1 \\ v_2 \\ \theta_2 \end{bmatrix}$$

The boundary conditions imply that θ_1 and θ_2 are unrestrained but $v_1 = v_2 = 0$. Thus we strike out the first and third columns and rows of $\mathbf{M}$ and $\mathbf{K}$ given by Equations 11.6-10 and 11.6-14 to obtain

$$\mathbf{M} = \frac{\rho A L}{420} \begin{bmatrix} 4L^2 & -3L^2 \\ -3L^2 & 4L^2 \end{bmatrix}$$

$$\mathbf{K} = \frac{EI}{L^3} \begin{bmatrix} 4L^2 & 2L^2 \\ 2L^2 & 4L^2 \end{bmatrix}$$

The equation of motion is thus

$$\frac{\rho A L}{420} \begin{bmatrix} 4L^2 & -3L^2 \\ -3L^2 & 4L^2 \end{bmatrix} \begin{bmatrix} \ddot{\theta}_1 \\ \ddot{\theta}_2 \end{bmatrix} + \frac{EI}{L^3} \begin{bmatrix} 4L^2 & 2L^2 \\ 2L^2 & 4L^2 \end{bmatrix} \begin{bmatrix} \theta_1 \\ \theta_2 \end{bmatrix} = \begin{bmatrix} 0 \\ 0 \end{bmatrix}$$

or

$$\begin{bmatrix} 4 & -3 \\ -3 & 4 \end{bmatrix} \begin{bmatrix} \ddot{\theta}_1 \\ \ddot{\theta}_2 \end{bmatrix} + \frac{840EI}{\rho A L^4} \begin{bmatrix} 2 & 1 \\ 1 & 2 \end{bmatrix} \begin{bmatrix} \theta_1 \\ \theta_2 \end{bmatrix} = \begin{bmatrix} 0 \\ 0 \end{bmatrix}$$

Letting $\lambda = 840EI/\rho A L^4$ and $\theta_j = \Theta_j e^{i\omega t}$ for $j = 1, 2$, we obtain

$$(2\lambda - 4\omega^2)\Theta_1 + (3\omega^2 + \lambda)\Theta_2 = 0$$

$$(3\omega^2 + \lambda)\Theta_1 + (2\lambda - 4\omega^2)\Theta_2 = 0$$

The frequency equation is

$$7\omega^4 - 22\lambda\omega^2 + 3\lambda^2 = 0$$

which gives $\omega^2 = 0.142857\lambda$ and $\omega^2 = 3\lambda$, or

$$\omega_1 = \frac{10.9544}{L^2} \sqrt{\frac{EI}{\rho A}}$$

$$\omega_2 = \frac{50.9544}{L^2} \sqrt{\frac{EI}{\rho A}}$$

From the distributed-parameter model,

$$\omega_n = \left(\frac{n\pi}{L}\right)^2 \sqrt{\frac{EI}{\rho A}}$$

which gives

$$\omega_1 = \frac{9.8696}{L^2} \sqrt{\frac{EI}{\rho A}}$$

and

$$\omega_2 = \frac{39.4784}{L^2} \sqrt{\frac{EI}{\rho A}}$$

So the frequencies predicted by the finite element model are 11% and 27% higher than those predicted by the distributed-parameter model. ∎

EXAMPLE 11.6-2
Fixed-Free Beam with Two Elements

(a) Develop a two-element model of the fixed-free beam shown in Figure 11.6-3a.

(b) Discuss how the results can be extended to an arbitrary number of elements.

Solution

(a) For each element, where $i = 1, 2$,

$$\mathbf{M}_i = \frac{\rho_i A_i L_i}{420} \begin{bmatrix} 156 & 11L_i & \vdots & 54 & -13L_i/2 \\ 11L & L_i^2 & \vdots & 13L_i/2 & -3L_i^2/4 \\ \cdots & \cdots & \cdots & \cdots & \cdots \\ 54 & 13L_i/2 & \vdots & 156 & -11L_i \\ -13L_i/2 & -3L_i^2/4 & \vdots & -11L_i & L_i^2 \end{bmatrix}$$

$$\mathbf{K}_i = \frac{8E_i I_i}{L_i^3} \begin{bmatrix} 12 & 3L_i & \vdots & -12 & 3L_i \\ 3L & L_i^2 & \vdots & -3L_i & L_i^2/2 \\ \cdots & \cdots & \cdots & \cdots & \cdots \\ -12 & -3L_i & \vdots & 12 & -3L_i \\ 3L & L_i^2/2 & \vdots & -3L_i & L_i^2 \end{bmatrix}$$

The displacement vector for the first element is

$$\begin{bmatrix} v_1 \\ \theta_1 \\ v_2 \\ \theta_2 \end{bmatrix}$$

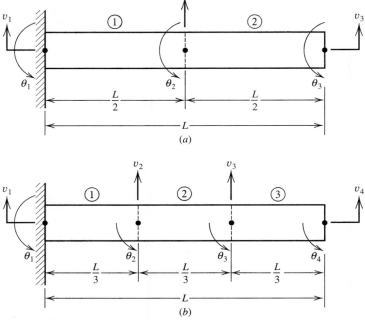

FIGURE 11.6-3 A fixed-free beam modeled with (a) two elements and (b) three elements.

Utilizing the structure of $\mathbf{M}$ and $\mathbf{K}$ shown in Equations 11.6-10 and 11.6-14, and applying the boundary conditions $v_1 = \theta_1 = 0$, we obtain

$$\mathbf{M}_{c1}\begin{bmatrix} \ddot{v}_2 \\ \ddot{\theta}_2 \end{bmatrix} + \mathbf{K}_{c1}\begin{bmatrix} v_2 \\ \theta_2 \end{bmatrix} = \begin{bmatrix} 0 \\ 0 \end{bmatrix}$$

where the submatrices $\mathbf{M}_{c1}$ and $\mathbf{K}_{c1}$ are evaluated from the expressions in Equations 11.6-13 and 11.6-17, with L replaced by $L/2$.

For element 2, the displacement vector is

$$\begin{bmatrix} v_2 \\ \theta_2 \\ v_3 \\ \theta_3 \end{bmatrix}$$

Adding the equations that share common elements gives

$$\begin{bmatrix} \mathbf{M}_{c1}+\mathbf{M}_{a2} & \vdots & \mathbf{M}_{b2} \\ \cdots & \cdots & \cdots \\ \mathbf{M}_{b2}^{T} & \vdots & \mathbf{M}_{c2} \end{bmatrix}\begin{bmatrix} \ddot{v}_2 \\ \ddot{\theta}_2 \\ \ddot{v}_3 \\ \ddot{\theta}_3 \end{bmatrix} + \begin{bmatrix} \mathbf{K}_{c1}+\mathbf{K}_{a2} & \vdots & \mathbf{K}_{b2} \\ \cdots & \cdots & \cdots \\ \mathbf{K}_{b2}^{T} & \vdots & \mathbf{K}_{c2} \end{bmatrix}\begin{bmatrix} v_2 \\ \theta_2 \\ v_3 \\ \theta_3 \end{bmatrix} = \begin{bmatrix} 0 \\ 0 \\ 0 \\ 0 \end{bmatrix}$$

where the submatrices are evaluated from the expressions in Equations 11.6-13 and 11.6-17, with L replaced by $L/2$.

The result for the mass matrix is

$$
\mathbf{M} = \frac{\rho AL}{840}
\begin{bmatrix}
156+156 & 11L-11L & \vdots & 54 & -13L/2 \\
11L-11L & L^2+L^2 & \vdots & 13L/2 & -3L^2/4 \\
\cdots & \cdots & \cdots & \cdots & \cdots \\
54 & 13L/2 & \vdots & 156 & -11L \\
-13L/2 & -3L^2/4 & \vdots & -11L & L^2
\end{bmatrix}
$$

$$
= \frac{\rho AL}{840}
\begin{bmatrix}
312 & 0 & \vdots & 54 & -13L/2 \\
0 & 2L^2 & \vdots & 13L/2 & -3L^2/4 \\
\cdots & \cdots & \cdots & \cdots & \cdots \\
54 & 13L/2 & \vdots & 156 & -11L \\
-13L/2 & -3L^2/4 & \vdots & -11L & L^2
\end{bmatrix}
$$

The stiffness matrix is

$$
\mathbf{K} = \frac{8EI}{L^3}
\begin{bmatrix}
12+12 & 3L-3L & \vdots & -12 & 3L \\
3L-3L & L^2+L^2 & \vdots & -13L & L^2/2 \\
\cdots & \cdots & \cdots & \cdots & \cdots \\
-12 & -3L & \vdots & 12 & -3L \\
3L & L^2/2 & \vdots & -3L & L^2
\end{bmatrix}
$$

$$
= \frac{8EI}{L^3}
\begin{bmatrix}
24 & 0 & \vdots & -12 & 3L \\
0 & 2L^2 & \vdots & -13L & L^2/2 \\
\cdots & \cdots & \cdots & \cdots & \cdots \\
-12 & -3L & \vdots & 12 & -3L \\
3L & L^2/2 & \vdots & -3L & L^2
\end{bmatrix}
$$

Because $\mathbf{M}$ and $\mathbf{K}$ are (4×4) matrices, there will be four frequencies and four mode shapes.

(b) With three elements, we evaluate the expressions for the submatrices in Equations 11.6-10 and 11.6-14, with L replaced by $L/3$. See Figure 11.6-3b. The general structure of the equations of motion is

$$
\mathbf{M}
\begin{bmatrix}
\ddot{v}_2 \\
\ddot{\theta}_2 \\
\ddot{v}_3 \\
\ddot{\theta}_3 \\
\ddot{v}_4 \\
\ddot{\theta}_4
\end{bmatrix}
+ \mathbf{K}
\begin{bmatrix}
v_2 \\
\theta_2 \\
v_3 \\
\theta_3 \\
v_4 \\
\theta_4
\end{bmatrix}
=
\begin{bmatrix}
0 \\
0 \\
0 \\
0
\end{bmatrix}
$$

where $\mathbf{O}_2$ is a (2×2) matrix of zeros and

$$\mathbf{M} = \begin{bmatrix} \mathbf{M}_{c1} + \mathbf{M}_{a2} & \vdots & \mathbf{M}_{b2} & \vdots & \mathbf{0}_2 \\ \cdots & \cdots & \cdots & \cdots & \cdots \\ \mathbf{M}_{b2}^T & \vdots & \mathbf{M}_{c2} + \mathbf{M}_{a3} & \vdots & \mathbf{M}_{b3} \\ \cdots & \cdots & \cdots & \cdots & \cdots \\ \mathbf{0}_2 & \vdots & \mathbf{M}_{b3}^T & \vdots & \mathbf{M}_{c3} \end{bmatrix}$$

$$\mathbf{K} = \begin{bmatrix} \mathbf{K}_{c1} + \mathbf{K}_{a2} & \vdots & \mathbf{K}_{b2} & \vdots & \mathbf{0}_2 \\ \cdots & \cdots & \cdots & \cdots & \cdots \\ \mathbf{K}_{b2}^T & \vdots & \mathbf{K}_{c2} + \mathbf{K}_{a3} & \vdots & \mathbf{K}_{b3} \\ \cdots & \cdots & \cdots & \cdots & \cdots \\ \mathbf{0}_2 & \vdots & \mathbf{K}_{b3}^T & \vdots & \mathbf{K}_{c3} \end{bmatrix}$$

It is straightforward to extend this structure to any number of elements.

The method is not limited to identical elements. If elements having dissimilar properties or lengths are chosen, then the expressions for the submatrices in Equations 11.6-10 and 11.6-14 are evaluated using each element's properties and lengths. ∎

Concentrated Beam Forces

When a beam is acted upon by an external concentrated force, such as that due to a point load or to an attached spring, we can simply include the force in the equations of motion provided that we choose a node to coincide with the location of the force and that the line of action of the force coincides with the axis of that node's rectilinear displacement. The effect of the spring force is to add a term to the stiffness matrix. The following example illustrates this procedure.

EXAMPLE 11.6-3
Concentrated Forces on a Fixed-Free Beam

Consider the two-element model of the fixed-free beam treated in Example 11.6-2. Now suppose that the beam is subjected to a point load P at its free end and a spring force located in the middle of the beam, as shown in Figure 11.6-4. Obtain the equations of motion.

Solution

We can use the mass and stiffness matrices obtained in Example 11.6-2. These are

$$\mathbf{M} = \frac{\rho A L}{840} \begin{bmatrix} 312 & 0 & \vdots & 54 & -13L/2 \\ 0 & 2L^2 & \vdots & 13L/2 & -3L^2/4 \\ \cdots & \cdots & \cdots & \cdots & \cdots \\ 54 & 13L/2 & \vdots & 156 & -11L \\ -13L/2 & -3L^2/4 & \vdots & -11L & L^2 \end{bmatrix} \quad (1)$$

$$\mathbf{K} = \frac{8EI}{L^3} \begin{bmatrix} 24 & 0 & \vdots & -12 & 3L \\ 0 & 2L^2 & \vdots & -13L & L^2/2 \\ \cdots & \cdots & \cdots & \cdots & \cdots \\ -12 & -3L & \vdots & 12 & -3L \\ 3L & L^2/2 & \vdots & -3L & L^2 \end{bmatrix}$$

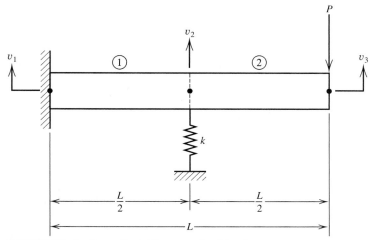

FIGURE 11.6-4   Concentrated force on a fixed-free beam.

The equations of motion have the form

$$\mathbf{M}\begin{bmatrix}\ddot{v}_2\\\ddot{\theta}_2\\\ddot{v}_3\\\ddot{\theta}_3\end{bmatrix}+\mathbf{K}\begin{bmatrix}v_2\\\theta_2\\v_3\\\theta_3\end{bmatrix}=\begin{bmatrix}f_2\\T_2\\f_3\\T_3\end{bmatrix}$$

There are no externally applied torques, so $T_2 = T_3 = 0$. The force f_3 at node 3 is due to the load P and acts in the opposite direction of v_3. Thus $f_3 = -P$. The spring force kv_2 acts at node 2. Thus, $f_2 = -kv_2$. Substituting these into the equations of motion and bringing the term kv_2 to the left-hand side, we see that the mass matrix remains unchanged but that the stiffness matrix becomes

$$\mathbf{K}=\frac{8EI}{L^3}\begin{bmatrix}\left(24+\dfrac{kL^3}{8EI}\right) & 0 & \vdots & -12 & 3L\\[2mm] 0 & 2L^2 & \vdots & -13L & L^2/2\\ \cdots & \cdots & \cdots & \cdots & \cdots\\ -12 & -3L & \vdots & 12 & -3L\\ 3L & L^2/2 & \vdots & -3L & L^2\end{bmatrix} \tag{2}$$

The only change in $\mathbf{K}$ is the addition of $kL^3/8EI$ to the element in the first row and first column. The equations of motion become

$$\mathbf{M}\begin{bmatrix}\ddot{v}_2\\\ddot{\theta}_2\\\ddot{v}_3\\\ddot{\theta}_3\end{bmatrix}+\mathbf{K}\begin{bmatrix}v_2\\\theta_2\\v_3\\\theta_3\end{bmatrix}=\begin{bmatrix}0\\0\\-P\\0\end{bmatrix}$$

where $\mathbf{M}$ is given by Equation 1 and $\mathbf{K}$ is given by Equation 2.   ∎

Distributed Beam Forces

When a distributed force acts on a beam, such as the force $f(x, t)$ as shown in Figure 11.6-1, the virtual work done by the force is

$$\delta W(t) = \int_0^L f(x, t)\,\delta v(x, t)\,dx \tag{11.6-18}$$

which can be expressed as

$$\delta W(t) = \delta \mathbf{v}^T(t)\mathbf{F}(t) \tag{11.6-19}$$

where we have defined the following vector of forces and torques (see Figure 11.6-1):

$$\mathbf{F} = \begin{bmatrix} F_1 \\ F_2 \\ F_3 \\ F_4 \end{bmatrix} = \begin{bmatrix} f_1 \\ T_1 \\ f_2 \\ T_2 \end{bmatrix} \tag{11.6-20}$$

Substituting $v(x, t)$ from Equation 11.6-2 into Equation 11.6-18, we obtain

$$\delta W(t) = \int_0^L f(x, t)S_1(x)\delta v_1(t)\, dx + \int_0^L f(x, t)S_2(x)\delta \theta_1(t)\, dx$$
$$+ \int_0^L f(x, t)S_3(x)\delta v_2(t)\, dx + \int_0^L f(x, t)S_4(x)\delta \theta_2(t)\, dx$$

Comparing this with Equation 11.6-19, we see that

$$F_i(t) = \int_0^L S_i(x)f(x, t)\, dx \qquad i = 1,\ 2,\ 3,\ 4 \tag{11.6-21}$$

where $F_i(t)$ is the force or torque at node i.
Specifically,

$$F_1(t) = f_1(t) = \int_0^L S_1(x)f(x, t)\, dx \tag{11.6-22}$$

$$F_2(t) = T_1(t) = \int_0^L S_2(x)f(x, t)\, dx \tag{11.6-23}$$

$$F_3(t) = f_2(t) = \int_0^L S_3(x)f(x, t)\, dx \tag{11.6-24}$$

$$F_4(t) = T_2(t) = \int_0^L S_4(x)f(x, t)\, dx \tag{11.6-25}$$

EXAMPLE 11.6-4
*Distributed Force
on a Fixed-Free
Beam*

Figure 11.6-5 shows a cantilever beam with a distributed load having a force density of w N/m. Use one element and obtain the equations of motion.

Solution

Applying the boundary conditions $v_1 = \theta_1 = 0$ gives the following matrices:

$$\mathbf{M} = \frac{\rho A L}{420} \begin{bmatrix} 156 & -22L \\ -22L & 4L^2 \end{bmatrix}$$

$$\mathbf{K} = \frac{EI}{L^3} \begin{bmatrix} 12 & -6L \\ -6L & 4L^2 \end{bmatrix}$$

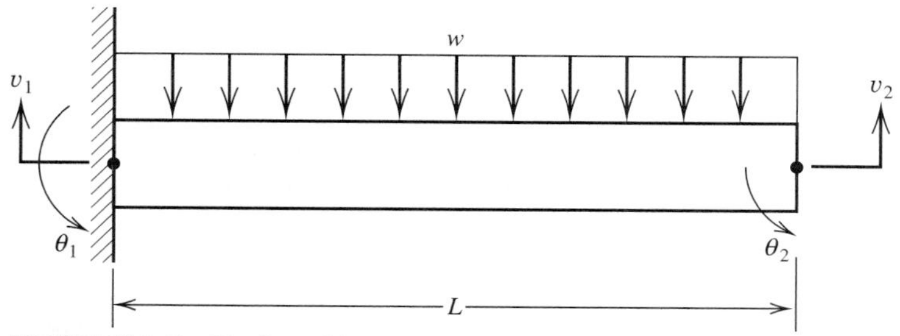

FIGURE 11.6-5 Distributed force on a fixed-free beam.

Here $f(x, t) = -w$, and the integrals in Equations 11.6-24 and 11.6-25 are evaluated as follows:

$$f_2 = -\int_0^L w \left[3\left(\frac{x}{L}\right)^2 - 2\left(\frac{x}{L}\right)^3 \right] dx = -\frac{wL}{2}$$

$$T_2 = -\int_0^L w \left[-L\left(\frac{x}{L}\right)^2 + L\left(\frac{x}{L}\right)^3 \right] dx = \frac{wL^2}{12}$$

Note that we need not evaluate the integrals for f_1 and T_1 since they do not appear in the equations of motion because of the particular boundary conditions.

The equations of motion are

$$\frac{\rho A L}{420} \begin{bmatrix} 156 & -22L \\ -22L & 4L^2 \end{bmatrix} \begin{bmatrix} \ddot{v}_2 \\ \ddot{\theta}_2 \end{bmatrix} + \frac{EI}{L^3} \begin{bmatrix} 12 & -6L \\ -6L & 4L^2 \end{bmatrix} \begin{bmatrix} v_2 \\ \theta_2 \end{bmatrix} = \begin{bmatrix} -\dfrac{wL}{2} \\ \dfrac{wL^2}{12} \end{bmatrix} \qquad \blacksquare$$

11.7 MATLAB APPLICATIONS

As shown in Chapter 8, MATLAB provides the `eig` function to support modal analysis. Here we illustrate some applications of this function to finite element analysis.

Consider the system shown in Figure 11.7-1 for the case where $m_1 = m_2 = m$, $m_3 = 2m$, and $k_1 = k_2 = k_3 = k$. The equations of motion are

$$m_1 \ddot{x}_1 = -k_1 x_1 - k_2(x_1 - x_2)$$

$$m_2 \ddot{x}_2 = k_2(x_1 - x_2) - k_3(x_2 - x_3)$$

$$m_3 \ddot{x}_3 = k_3(x_2 - x_3)$$

The mass and stiffness matrices are

$$\mathbf{M} = m \begin{bmatrix} 1 & 0 & 0 \\ 0 & 1 & 0 \\ 0 & 0 & 2 \end{bmatrix}$$

$$\mathbf{K} = k \begin{bmatrix} 2 & -1 & 0 \\ -1 & 2 & -1 \\ 0 & -1 & 1 \end{bmatrix}$$

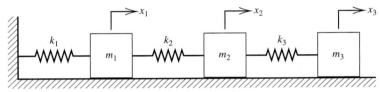

FIGURE 11.7-1 A system having three degrees of freedom.

Substituting $\mathbf{x}(t) = \mathbf{X}e^{i\omega t}$, canceling the $e^{i\omega t}$ terms, and rearranging gives

$$-m\omega^2 \begin{bmatrix} 1 & 0 & 0 \\ 0 & 1 & 0 \\ 0 & 0 & 2 \end{bmatrix} \mathbf{X} + k \begin{bmatrix} 2 & -1 & 0 \\ -1 & 2 & -1 \\ 0 & -1 & 1 \end{bmatrix} \mathbf{X} = \mathbf{0}$$

Frequently we can simplify such expressions by introducing the parameter $r = m\omega^2/k$. So dividing this equation by k and rearranging gives

$$\begin{bmatrix} 1 & 0 & 0 \\ 0 & 1 & 0 \\ 0 & 0 & 2 \end{bmatrix}^{-1} \begin{bmatrix} 2 & -1 & 0 \\ -1 & 2 & -1 \\ 0 & -1 & 1 \end{bmatrix} \mathbf{X} = r\mathbf{X}$$

for the eigenvalue problem.

The eigenvalues can be found with MATLAB by using the `eig` function. The session is

```
≫M = [1,0,0;0,1,0;0,0,2];
≫K = [2, - 1,0; - 1,2, - 1;0, - 1,1];
≫A = inv(M)*K;
≫[v,D] = eig(A)
```

The syntax [v, D] = eig(A) computes the arrays v and D. The eigenvectors of **A** are stored in the columns of v. The eigenvalues are the diagonal elements of D. In this example the results displayed on the screen after typing [v, D] = eig(A) are

v=

```
-0.6657  - 0.7557  0.3042
0.7328   - 0.5498  0.5699
-0.1409  0.3558    0.7633
```

D=

```
3.1007  0       0
0       1.2725  0
0       0       0.1267
```

The values of r are 3.1007, 1.2725, and 0.1267. Thus the frequencies are $\omega = \sqrt{rk/m}$, or

$$\omega = 1.7609\sqrt{\frac{k}{m}} \quad 1.1281\sqrt{\frac{k}{m}} \quad 0.3560\sqrt{\frac{k}{m}}$$

and the eigenvectors corresponding to the three modes are

$$\mathbf{X}_1 = \begin{bmatrix} -0.6657 \\ 0.7328 \\ -0.1409 \end{bmatrix} \quad \mathbf{X}_2 = \begin{bmatrix} -0.7557 \\ -0.5498 \\ 0.3558 \end{bmatrix} \quad \mathbf{X}_3 = \begin{bmatrix} 0.3042 \\ 0.5699 \\ 0.7633 \end{bmatrix}$$

Note that the scalar factor has been computed so that the length of each vector is 1. For example, $|\mathbf{X}_1| = \sqrt{(-0.6657)^2 + (0.7328)^2 + (-0.1409)^2} = 1$.

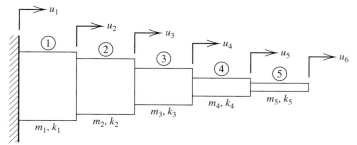

FIGURE 11.7-2 A fixed-free stepped bar.

Application to a Stepped Bar

Consider the stepped bar shown in Figure 11.7-2. If the left-hand end is fixed and the right-hand end is free, application of the finite element procedure results in the following stiffness and mass matrices, where

$$k_i = \frac{E_i A_i}{L_i} \qquad (11.7\text{-}1)$$

$$m_i = \rho A_i L_i \qquad (11.7\text{-}2)$$

$$\mathbf{K} = \begin{bmatrix} k_1 + k_2 & -k_2 & 0 & 0 & 0 \\ -k_2 & k_2 + k_3 & -k_3 & 0 & 0 \\ 0 & -k_3 & k_3 + k_4 & -k_4 & 0 \\ 0 & 0 & -k_4 & k_4 + k_5 & -k_5 \\ 0 & 0 & 0 & -k_5 & k_5 \end{bmatrix} \qquad (11.7\text{-}3)$$

$$\mathbf{M} = \frac{1}{6} \begin{bmatrix} 2m_1 + 2m_2 & m_2 & 0 & 0 & 0 \\ m_2 & 2m_2 + 2m_3 & m_3 & 0 & 0 \\ 0 & m_3 & 2m_3 + 2m_4 & m_4 & 0 \\ 0 & 0 & m_4 & 2m_4 + 2m_5 & m_5 \\ 0 & 0 & 0 & m_5 & 2m_5 \end{bmatrix} \qquad (11.7\text{-}4)$$

The eigenvalue problem is

$$\mathbf{M}^{-1}\mathbf{K}\mathbf{u} = \omega^2 \mathbf{u}$$

Note the banded structure of these matrices, which is due to the fact that a given element is affected only by its adjacent elements. This structure, which consists of a main diagonal, one upper subdiagonal, and one lower subdiagonal, enables us to write a general-purpose program that can be applied to analyze a bar with an arbitrary number of elements.

We can create a MATLAB program that takes advantage of this structure by using the MATLAB `diag` function as follows. If **S** is a $(1 \times n)$ vector, then `R1 = diag(S)` creates an $(n \times n)$ matrix **R1** whose main diagonal is **S**, with zeros in all other elements.

If **b** is a $[1 \times (n-1)]$ vector, then `R2 = diag(b,1)` creates an $(n \times n)$ matrix **R2** whose upper subdiagonal is **b**, with zeros in all other elements, and `R3 = diag(b,-1)` creates an $(n \times n)$ matrix **R3** whose lower subdiagonal is **b**, with zeros in all other elements. Adding the three matrices, as R = R1+R2+R3, produces a banded matrix **R** of the form shown in Equations 11.7-3 and 11.7-4. This is how we create the mass and stiffness matrices.

The following user-defined function creates the banded matrices:

```
function R = banded(S,d,q)
% Creates a banded matrix.
R1 = diag(S);
R2 = diag(d(2:q),1);
R3 = diag(d(2:q),-1);
R = R1 + R2 + R3;
```

The following program is for an aluminum bar having three sections, whose lengths are 0.5, 0.3, and 0.2 m, and whose radii are 0.1, 0.06, and 0.04 m. The program can easily be modified to handle other cases. Except for the first five lines, the program can be used without modification for any stepped bar of any material and dimensions, having any number of sections, as long as one end is fixed and the other is free.

```
% Define the parameter values.
E = 7e+10;density = 2700;
L = [0.5,0.3,0.2];
q = length(L);
A = pi*[0.1,0.06,0.04].^2;
rho = density*ones(size(L));
k = A*E./L;
m=rho.*A.*L;
%
% Create the stiffness matrix K.
V = zeros(1,q);
V(q) = k(q);
for n = 1:q-1
  for p = n:n+1
    V(n) = V(n) + k(p);
  end
end
K = banded(V,-k,q);
%
% Create the mass matrix M.
W = zeros(1,q);
W(1) = 2*m(1) + 2*m(2);
W(q) = 2*m(q);
for n = 2:q-1
  for p = n:n+1
    W(n) = W(n) + 2*m(p);
  end
end
M = banded(W,m,q);
M = M/6
% Compute the eigenvectors and eigenvalues.
B = inv(M)*K;
[U,D] = eig(B)
% Compute the frequencies.
omega = sqrt(diag(D))
```

The results for the mode shapes are

$$\mathbf{U} = \begin{bmatrix} -0.0574 & 0.3846 & -0.2962 \\ 0.3797 & 0.6149 & 0.4423 \\ -0.9233 & 0.6884 & 0.8465 \end{bmatrix}$$

and for the eigenvalues,

$$\mathbf{D} = 10^9 \begin{bmatrix} 3.4542 & 0 & 0 \\ 0 & 0.1435 & 0 \\ 0 & 0 & 0.7362 \end{bmatrix}$$

Thus the frequencies in rad/s are

$$\omega = 5.8772 \times 10^4$$

$$\omega = 1.1981 \times 10^4$$

$$\omega = 2.7132 \times 10^4$$

11.8 CHAPTER REVIEW

In this chapter we introduced the finite element method, which provides a more accurate system description than that used to develop a lumped-parameter model but that for complex systems is easier to solve than partial differential equations and more accurate than their finite-difference approximations. This method is particularly useful for irregular geometries, such as bars that have variable cross sections, and for systems such as trusses that are made up of several bars.

To use the finite element method we first divide the structure up into a number of simple structural parts, such as bars, beams, or plates, whose equations of motion are easily derived. After obtaining the equations for each element, we assemble the equations and combine them according to how the elements are connected in the system. The resulting collection of finite elements and nodes forms the mesh.

We then approximate the solution of the equation of motion for each finite element with low-order polynomials. These solutions are then assembled to obtain the mass and stiffness matrices for the entire structure as a whole, and the modal analysis techniques of Chapter 8 are used to compute the natural frequencies and mode shapes. The displacements obtained in the solution are the displacements of the nodes in the mesh.

The elements commonly used in FEM and treated here are the axial (bar) element, the transverse (beam) element, and the torsional element, which are one-dimensional. The plate and shell elements are treated in more specialized texts.

When we use several elements between structural joints to estimate the higher modes, we obtain a large number of equations to solve for the eigenvalues and eigenvectors of the system. When a large number of elements is used, the task of specifying the locations of the elements and of assembling the equations should be done with a computer program. Commercial programs provide a graphical interface for specifying the locations of the nodes and they often provide graphical displays of the results. To continue the study and applications of the finite element method, you will need to become familiar with such a program.

Now that you have finished this chapter, you should be able to do the following:

1. Write the equations for the bar, torsional, and beam elements.

2. Assemble the equations, using the given boundary conditions, for a system consisting of multiple elements.

3. Solve the assembled equations to determine the mode shapes and mode frequencies.

PROBLEMS

SECTION 11.1 BAR ELEMENT

11.1 Obtain the stiffness and the mass matrices for the longitudinal vibration of the tapered bar shown in Figure P11.1. The cross section is circular, and its diameter decreases linearly with x.

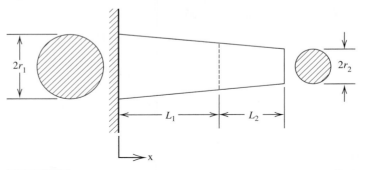

FIGURE P11.1

11.2 Obtain the stiffness and mass matrices for the longitudinal vibration of the tapered bar shown in Figure P11.2. The cross section is circular, and its area decreases with x as follows: $A(x) = A_0 e^{-x/L}$, where A_0 is the area at $x = 0$.

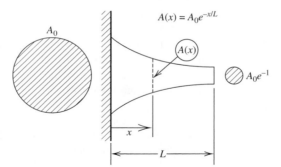

FIGURE P11.2

11.3 Obtain the frequencies and modes for the longitudinal vibration of a free-free bar having a single element. Compare the results with those obtained from the distributed-parameter model.

11.4 Figure P11.4 shows a spring connected to the free end of a fixed bar. Obtain the equation of longitudinal motion and frequencies using one element for the bar.

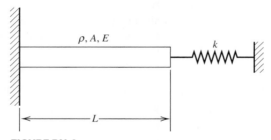

FIGURE P11.4

SECTION 11.2 MODELS WITH MULTIPLE BAR ELEMENTS

11.5 Obtain the stiffness and mass matrices for a two-element model of the longitudinal vibration of the fixed-fixed bar shown in Figure P11.5. Use elements of equal length.

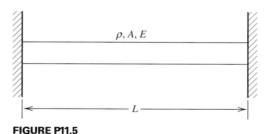

FIGURE P11.5

11.6 Obtain the stiffness and mass matrices, and the natural frequencies, for a two-element model of the longitudinal vibration of the fixed-free bar shown in Figure P11.6. Note that the elements have the same length, $L/2$.

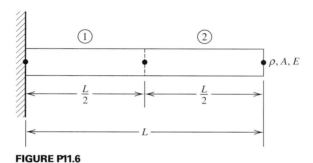

FIGURE P11.6

11.7 Obtain the stiffness and mass matrices, and the natural frequencies, for a two-element model of the longitudinal vibration of the fixed-free bar shown in Figure P11.7. Note that the elements have unequal lengths.

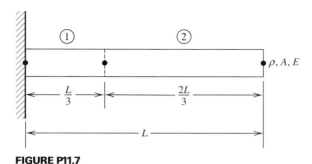

FIGURE P11.7

11.8 Compare the results of Problems 11.6 and 11.7, and discuss the effects of using unequal element lengths.

11.9 Obtain the stiffness and mass matrices, and the natural frequencies, for a three-element model of the longitudinal vibration of the fixed-free bar shown in Figure P11.9.

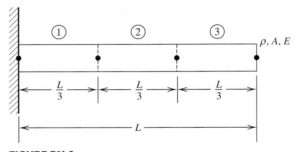

FIGURE P11.9

11.10 Obtain the stiffness and mass matrices for a three-element model of the longitudinal vibration of the fixed-fixed bar shown in Figure P11.5. Use elements of equal length.

11.11 Use a two-element model to obtain the mass and stiffness matrices for the longitudinal vibration of the tapered fixed-free bar shown in Figure P11.1. The cross section is circular and its radius varies linearly with x. (To solve this problem, you must use the appropriate shape functions and derive the expressions for the kinetic and potential energies, taking the variable cross section into account.)

11.12 Obtain the natural frequencies of the bar treated in Problem 11.11, for the case where $L = 0.5$ m, $r_1 = 5$ cm, $r_2 = 2.5$ cm, and the bar is made of steel.

11.13 Supposedly, the tradition of marching troops breaking step when crossing a bridge originated in England after a bridge collapsed due to the resonance caused by the marching cadence. Computing the resonant frequenies of a complex bridge structure as shown in Figure P11.13 is not easy, but we can start by determining the natural frequencies of its various elements. Suppose the vertical supports of a certain bridge are 10-m-long hollow steel cylinders whose inner and outer radii are 0.2 and 0.3 m, respectively. Use a three-element model of the longitudinal vibration to determine the natural frequencies.

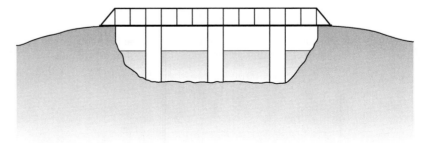

FIGURE P11.13

SECTION 11.3   CONSISTENT AND LUMPED-MASS MATRICES

11.14 Consider the two-element model of longitudinal vibration of the tapered fixed-free bar treated in Problem 11.11.

(a) Obtain the lumped-mass matrix by distributing the bar mass equally at each node.

(b) Obtain the lumped-mass matrix by distributing the bar mass at each node proportionally to the area.

11.15 Using the results of Problem 11.14 and the values from Problem 11.12, compute the natural frequencies for each lumped-mass matrix and compare with the results in Problem 11.12.

SECTION 11.4 ANALYSIS OF TRUSSES

11.16 Obtain the equations of motion in global coordinates and the expressions for the natural frequencies of the endpoint of the truss shown in Figure P11.16. Use one bar element for each truss member. Assume the truss members have the same density ρ, area A, and modulus E.

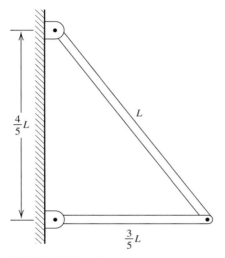

FIGURE P11.16

11.17 Obtain the equations of motion in global coordinates and the expressions for the natural frequencies of the endpoint of the truss shown in Figure P11.17. Use one bar element for each truss member. Assume the truss members have the same density ρ, area A, and modulus E.

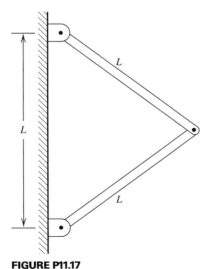

FIGURE P11.17

11.18 Obtain the equations of motion of the frame shown in Figure P11.18. Use one bar element for each frame member. Assume the members have the same density ρ, area A, and modulus E. The lower ends are fixed.

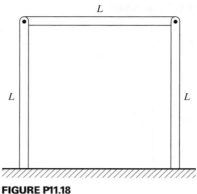

FIGURE P11.18

11.19 Obtain the equations of motion of the frame shown in Figure P11.19. Use one bar element for each frame member. Assume the members have the same density ρ, area A, and modulus E.

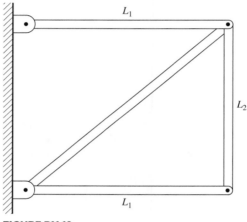

FIGURE P11.19

SECTION 11.5   TORSIONAL ELEMENTS

11.20 Consider Example 11.5-2. Obtain the expressions for the natural frequencies for the case where both elements are identical.

11.21 Consider Example 11.5-2. Obtain the expressions for the natural frequencies for the case where both elements are identical except that $J_2 = 2J_1$.

11.22 Consider a fixed-free bar in torsion, modeled with n elements that are not necessarily identical. Develop the general expression for the mass and stiffness matrices.

SECTION 11.6   BEAM ELEMENTS

11.23 Use two identical elements to obtain the expressions for the frequencies and transverse modes of the uniform fixed-fixed beam shown in Figure P11.5. The area moment of the beam cross section is I.

11.24 A steel beam 1 m long with rectangular cross section supports a 50-kg motor that rotates at 3450 rpm (Figure P11.24). The beam height is 50 mm and its width is 25 mm. Use a two-element model to compute the natural frequencies. If the motor has a rotating unbalance, will it cause resonance?

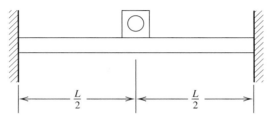

FIGURE P11.24

11.25 Use a one-element model to obtain the expressions for the frequencies of transverse vibration for the cantilever beam shown in Figure P11.25. The concentrated mass is fives times larger than the beam mass.

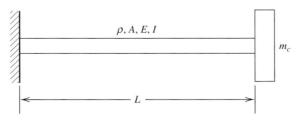

FIGURE P11.25

11.26 Use two elements to obtain the expressions for the frequencies of transverse vibration for the stepped fixed-fixed beam shown in Figure P11.26. The beam material is uniform.

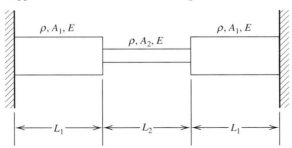

FIGURE P11.26

11.27 Use one element to obtain the expressions for the frequencies of transverse vibration for the fixed-pinned beam shown in Figure P11.27.

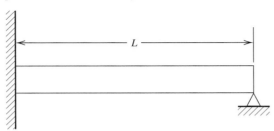

FIGURE P11.27

11.28 Derive the equations of motion for the multi-leaf spring shown in Figure P11.28. Treat the bolted connection as fixed and use symmetry. Each leaf has a width of $w = 4$ cm, has a thickness of $h = 0.6$ cm, and is made of steel. The dimensions are $L_1 = 20$ cm, $L_2 = 30$ cm, and $L_3 = 40$ cm.

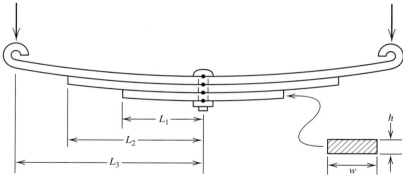

FIGURE P11.28

11.29 Use three identical elements to obtain the equations of transverse motion of the uniform fixed-fixed beam shown in Figure P11.29.

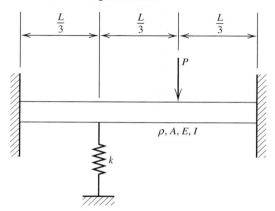

FIGURE P11.29

11.30 Use a one-element model to obtain the expressions for the frequencies of transverse vibration for the cantilever beam shown in Figure P11.30.

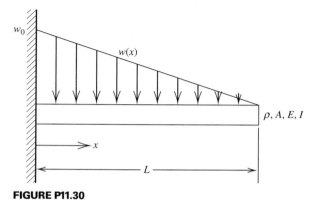

FIGURE P11.30

11.31 Use a two-element model to obtain the expressions for the frequencies of transverse vibration for the cantilever beam shown in Figure P11.31.

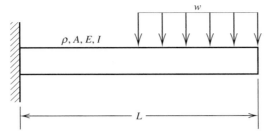

FIGURE P11.31

11.32 Figure P11.32 is a representation of a beam bending under the action of fluid pressure. An example is a car radio antenna subject to aerodynamic drag. Use a one-element model to obtain the expressions for the frequencies of the beam.

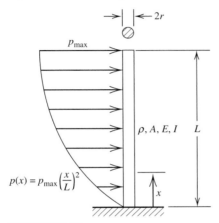

FIGURE P11.32

SECTION 11.7   MATLAB APPLICATIONS

11.33 Consider a steel bar having three sections, whose lengths are 0.5, 0.3, and 0.2 m and whose radii are 0.1, 0.06, and 0.04 m. Use MATLAB to compute the longitudinal mode shapes and frequencies. Compare the results with those found in Section 11.7 for the same bar but made of aluminum.

11.34 Use MATLAB to compute the longitudinal mode shapes and frequencies of the bar shown in Figure 11.7-2. The bar is made of steel. The section lengths are 0.7, 0.5, 0.3, 0.2, and 0.1 m, and the radii are 0.2, 0.1, 0.06, 0.04, and 0.02 m.

11.35 Use MATLAB to compute the torsional mode shapes and frequencies of the bar shown in Figure 11.7-2. The bar is made of steel. The section lengths are 0.7, 0.5, 0.3, 0.2, and 0.1 m, and the radii are 0.2, 0.1, 0.06, 0.04, and 0.02 m.

NUMERICAL SOLUTION METHODS

This appendix introduces the basics of solving ordinary differential equations with a numerical method. The algorithms presented here are simplified versions of the ones used by MATLAB and Simulink, and so an understanding of these methods will improve your understanding of these two software packages. The numerical solution algorithms used by the MATLAB `ode` solvers, covered in Section 3.8, are very complicated. Therefore, we will limit our presentation in this section to simple algorithms so that we may highlight the important issues to be considered when using numerical methods. Numerical methods require that the derivatives in the model be described by finite-difference expressions and that the resulting difference equations be solved in a step-by-step procedure. The issues related to these methods are the following:

- What finite-difference expressions provide the best approximations for derivatives?
- What are the effects of the step size used to obtain the approximations?
- What are the effects of round-off error when solving the finite-difference equations?

We will explore these issues and make you aware of some pitfalls that can be encountered. Such difficulties are more likely to occur when the solution is rapidly changing with time, and they can happen if the step size is not small compared to the smallest time constant of the system or the smallest oscillation period.

Test Cases

We now develop two test cases, whose solution can be found analytically, to use for checking the results of our numerical methods.

1. The following equation is used to illustrate the effect of step size relative to the system time constant, which is $\tau = 1/10$:

$$\frac{dy}{dt} + 10y = 0 \qquad y(0) = 2$$

   The solution is $y(t) = 2e^{-10t}$.

2. The following equation is used to illustrate the effect of step size relative to the solution's oscillation period:

$$\frac{dy}{dt} = \sin \omega t \qquad y(0) = y_0$$

   The solution is $y(t) = y_0 + (1 - \cos \omega t)/\omega$ and the period is $2\pi/\omega$.

The Euler Method

The essence of a numerical method is to convert the differential equation into a difference equation that can be programmed on a calculator or digital computer. Numerical algorithms differ partly as a result of the specific procedure used to obtain the difference

685

equations. In general, as the accuracy of the approximation is increased, so is the complexity of the programming involved. It is important to understand the concept of "step size" and its effects on solution accuracy. In order to provide a simple introduction to these issues, we begin with the simplest numerical method, the *Euler method*.

Consider the equation

$$\frac{dy}{dt} = f(t, y) \tag{B-1}$$

where $f(t, y)$ is a known function. From the definition of the derivative,

$$\frac{dy}{dt} = \lim_{\Delta t \to 0} \frac{y(t + \Delta t) - y(t)}{\Delta t}$$

If the time increment Δt is chosen small enough, the derivative can be replaced by the approximate expression

$$\frac{dy}{dt} \approx \frac{y(t + \Delta t) - y(t)}{\Delta t} \tag{B-2}$$

Assume that the right-hand side of Equation B-1 remains constant over the time interval $(t, t + \Delta t)$, and replace Equation B-1 by the following approximation:

$$\frac{y(t + \Delta t) - y(t)}{\Delta t} = f(t, y)$$

or

$$y(t + \Delta t) = y(t) + f(t, y)\Delta t \tag{B-3}$$

The smaller Δt is, the more accurate are our two assumptions leading to Equation B-3. This technique for replacing a differential equation with a difference equation is the *Euler method*. The increment Δt is called the *step size*.

A more concise representation is obtained by using the following notation:

$$y_k = y(t_k) \qquad y_{k+1} = y(t_{k+1}) = y(t_k + \Delta t)$$

where $t_{k+1} = t_k + \Delta t$. In this notation, the Euler algorithm (Equation B-3) is expressed as

$$y_{k+1} = y_k + f(t_k, y_k)\Delta t \tag{B-4}$$

EXAMPLE B-1
The Euler Method and an Exponential Solution

Use the Euler method to solve our first test case, which is

$$\frac{dy}{dt} + 10y = 0 \qquad y(0) = 2$$

which has the exact solution $y(t) = 2e^{-10t}$. Use a step size of $\Delta t = 0.02$, which is one-fifth of the time constant.

Solution

Here $f(t, y) = -10y$. Thus the Euler algorithm (Equation B-4) in this case becomes

$$y_{k+1} = y_k - (10y_k)\Delta t$$

or

$$y_{k+1} = y_k - (10y_k)0.02 = 0.8y_k$$

We show the computations for the first few steps, using four significant figures. The exact value, obtained from $y(t) = 2e^{-10t}$, and the percent error are shown in the following table.

Step	Numerical solution	Exact solution	Percent error
$k = 0$	$y_1 = 0.8y_0 = 1.6$	1.637	2.3%
$k = 1$	$y_2 = 0.8y_1 = 1.28$	1.341	4.5%
$k = 2$	$y_3 = 0.8y_2 = 1.024$	1.098	6.7%
$k = 3$	$y_4 = 0.8y_3 = 0.8192$	0.8987	8.8%
$k = 4$	$y_5 = 0.8y_4 = 0.6554$	0.7358	10.9%
$k = 5$	$y_6 = 0.8y_5 = 0.5243$	0.6024	13%

Notice how the percent error grows with each step. This is because the calculated result from the previous step is not exact. The numerical and exact solutions are shown in Figure B-1, where the numerical solution is shown by the small circles and the exact solution is shown by the solid curve.

Another observation here is that the step size should be much smaller than the time constant τ. A commonly used rule of thumb is that $\Delta t \leq \tau/20$. ∎

Round-Off Error   There is another reason the error increases with the number of steps. If we had retained six significant figures instead of four, we would have obtained $y_5 = 0.65536$ and $y_6 = 0.524288$. Even though computers can retain many more than four significant figures, they cannot represent numbers with infinite accuracy. Thus the calculated solution obtained by computer at each step is in effect rounded off to a finite number of significant figures. This rounded number is then used in the calculations for the next step, and so on, just as we rounded the value of y_5 to 0.6554 before using it to compute y_6. Therefore, the error in the numerical solution will increase with the number of steps required to obtain the solution.

Thus, because round-off error increases with each step, there is a trade-off between using a step size small enough to obtain an accurate solution yet not so small that many steps are required to obtain the solution.

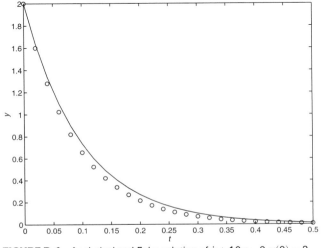

FIGURE B-1   Analytical and Euler solution of $\dot{y} + 10y = 0$, $y(0) = 2$.

Numerical methods have their greatest errors when trying to obtain solutions that are rapidly changing. The difficulties caused by an oscillating solution are illustrated in the following example.

EXAMPLE B-2
The Euler Method and an Oscillating Solution

Consider the following equation, which is our second test case:

$$\dot{y} = \sin t$$

for $y(0) = 0$ and $0 \leq t \leq 4\pi$. The exact solution is $y(t) = 1 - \cos t$, and its period is 2π. Solve this equation with Euler's method.

Solution

We choose a step size equal to one-thirteenth of the period, or $\Delta t = 2\pi/13$, so that we can compare the answer with that obtained by a method to be introduced later. The Euler algorithm (Equation B-4) becomes

$$y_{k+1} = y_k + (\sin t_k)\frac{2\pi}{13}$$

For successive values of $k = 0, 1, 2, \ldots$, we have $t_k = 0, 2\pi/13, 4\pi/13, \ldots$. Retaining four significant figures, we have

$$y_1 = y_0 + (\sin t_0)\frac{2\pi}{13} = 0 + (\sin 0)\frac{2\pi}{13} = 0$$

$$y_2 = y_1 + (\sin t_1)\frac{2\pi}{13} = 0 + \left(\sin\frac{2\pi}{13}\right)\frac{2\pi}{13} = 0.2246$$

$$y_3 = y_2 + (\sin t_2)\frac{2\pi}{13} = 0.2246 + \left(\sin\frac{4\pi}{13}\right)\frac{2\pi}{13} = 0.6224$$

and so on. The numerical and exact solutions are shown in Figure B-2, where the numerical solution is shown by the small circles and the exact solution is shown by the solid curve. There is noticeable error, especially near the peaks and valleys, where the solution is rapidly changing. ∎

The accuracy of the Euler method can be improved by using a smaller step size. However, very small step sizes require longer run times and can result in a large

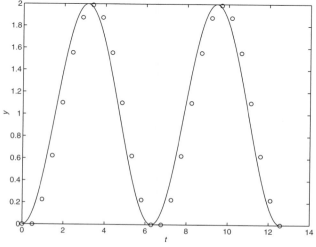

FIGURE B-2 Analytical and Euler solution of $\dot{y} = \sin t$, $y(0) = 0$.

accumulated error due to round-off effects. So we seek better algorithms to use for more challenging applications.

Predictor-Corrector Methods

We now consider *predictor-corrector* methods, which serve as the basis for many powerful algorithms. The Euler method can have a serious deficiency in problems where the variables are rapidly changing, because the method assumes the variables are constant over the time interval Δt. One way of improving the method is to use a better approximation to the right-hand side of the equation:

$$\frac{dy}{dt} = f(t, y) \tag{B-5}$$

The Euler approximation is

$$y(t_{k+1}) = y(t_k) + \Delta t f[t_k, y(t_k)] \tag{B-6}$$

Suppose instead we use the average of the right-hand side of Equation B-5 on the interval (t_k, t_{k+1}). This gives

$$y(t_{k+1}) = y(t_k) + \frac{\Delta t}{2}(f_k + f_{k+1}) \tag{B-7}$$

where

$$f_k = f[t_k, y(t_k)] \tag{B-8}$$

with a similar definition for f_{k+1}. Equation B-8 is equivalent to integrating Equation B-5 with the trapezoidal rule, whereas the Euler method is equivalent to integrating with the rectangular rule.

The difficulty with Equation B-8 is that f_{k+1} cannot be evaluated until $y(t_{k+1})$ is known, but this is precisely the quantity being sought. A way out of this difficulty is to use the Euler formula (Equation B-6) to obtain a preliminary estimate of $y(t_{k+1})$. This estimate is then used to compute f_{k+1} for the use in Equation B-8 to obtain the required value of $y(t_{k+1})$.

The notation can be changed to clarify the method. Let $h = \Delta t$ and $y_k = y(t_k)$, and let x_{k+1} be the estimate of $y(t_{k+1})$ obtained from the Euler formula (Equation B-6). Then, by omitting the t_k notation from the other equations, we obtain the following description of the predictor-corrector process:

$$\text{Euler predictor: } \quad x_{k+1} = y_k + h f(t_k, y_k) \tag{B-9}$$

$$\text{Trapezoidal corrector: } \quad y_{k+1} = y_k + \frac{h}{2}[f(t_k, y_k) + f(t_{k+1}, x_{k+1})] \tag{B-10}$$

This version of a predictor-corrector algorithm is sometimes called the *modified-Euler method*. However, note that any algorithm can be tried as a predictor or a corrector. Thus many other methods can also be classified as predictor-corrector.

EXAMPLE B-3
The Modified-Euler Method and an Exponential Solution

Use the modified-Euler method to solve our first test case:

$$\dot{y} = -10y \qquad y(0) = 2$$

for $0 \le t \le 0.5$. The exact solution is $y(t) = 2e^{-10t}$.

Solution

To illustrate the effect of the step size on the solution's accuracy, we use a step size $h = 0.02$, the same size used with the Euler method. The modified-Euler algorithm for this case has the following form:

$$x_{k+1} = y_k + h(-10y_k) = (1 - 10h)y_k = 0.8y_k$$

$$y_{k+1} = y_k + \frac{h}{2}(-10y_k - 10x_{k+1}) = 0.9y_k - 0.1x_{k+1}$$

The following table shows the numerical and exact solutions, rounded to four significant figures, and the percent error, for a few steps.

Step	Numerical solution	Exact solution	Percent error
$k = 0$	$x_1 = 0.8y_0 = 1.6$		
	$y_1 = 0.9y_0 - 0.1x_1 = 1.64$	1.637	0.2%
$k = 1$	$x_2 = 0.8y_1 = 1.312$		
	$y_2 = 0.9y_1 - 0.1x_2 = 1.3448$	1.341	0.3%
$k = 2$	$x_3 = 0.8y_2 = 1.07584$		
	$y_3 = 0.9y_2 - 0.1x_3 = 1.102736$	1.098	0.4%

There is much less error than with the Euler method using the same step size. Figure B-3 shows the results, with the numerical solution shown by the small circles and the exact solution by the solid line.   ∎

The modified-Euler method is a special case of the Runge-Kutta family of algorithms, to be discussed next. For purposes of comparison with the Runge-Kutta methods, we can express the modified-Euler method as follows:

$$y_{k+1} = y_k + \frac{1}{2}(g_1 + g_2) \tag{B-11}$$

$$g_1 = hf(t_k, y_k) \tag{B-12}$$

$$g_2 = hf(t_k + h, y_k + g_1) \tag{B-13}$$

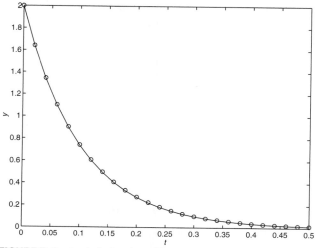

FIGURE B-3   Analytical and modified-Euler solution of $\dot{y} + 10y = 0$, $y(0) = 2$.

The Taylor series representation forms the basis of several methods for solving differential equations, including the Runge-Kutta methods. The Taylor series may be used to represent the solution $y(t + h)$ in terms of $y(t)$ and its derivatives, as follows.

$$y(t + h) = y(t) + h\,\dot{y}(t) + \frac{1}{2}h^2\ddot{y}(t) + \cdots \tag{B-14}$$

The number of terms kept in the series determines its accuracy. The required derivatives are calculated from the differential equation. If these derivatives can be found, Equation B-14 can be used to march forward in time. In practice, the high-order derivatives can be difficult to calculate, and the series (Equation B-14) is truncated at some term. The Runge-Kutta methods were developed because of the difficulty in computing the derivatives. These methods use several evaluations of the function $f(t, y)$ in a way that approximates the Taylor series. The number of terms in the series that are duplicated determines the order of the Runge-Kutta method. Thus, a fourth-order Runge-Kutta algorithm duplicates the Taylor series through the term involving h^4.

The second-order Runge-Kutta methods express y_{k+1} as

$$y_{k+1} = y_k + w_1 g_1 + w_2 g_2 \tag{B-15}$$

where w_1 and w_2 are constant weighting factors, and

$$g_1 = hf(t_k, y_k) \tag{B-16}$$

$$g_2 = hf(t_k + \alpha h, y_k + \beta h f_k) \tag{B-17}$$

The family of second-order Runge-Kutta algorithms is categorized by the parameters α, β, w_1, and w_2. To duplicate the Taylor series through the h^2 term, these coefficients must satisfy the following:

$$w_1 + w_2 = 1 \qquad w_1\alpha = \frac{1}{2} \qquad w_2\beta = \frac{1}{2} \tag{B-18}$$

Thus one of the parameters can be chosen independently.

The modified-Euler algorithm, Equations B-11 through B-13, is thus seen to be a second-order Runge-Kutta algorithm with $\alpha = \beta = 1$ and $w_1 = w_2 = 1/2$.

Fourth-Order Runge-Kutta Methods

The family of fourth-order Runge-Kutta algorithms expresses y_{k+1} as

$$y_{k+1} = y_k + w_1 g_1 + w_2 g_2 + w_3 g_3 + w_4 g_4 \tag{B-19}$$

$$g_1 = hf(t_k, y_k) \tag{B-20}$$
$$g_2 = hf(t_k + \alpha_1 h, y_k + \alpha_1 g_1)$$
$$g_3 = hf[t_k + \alpha_2 h, y_k + \beta_2 g_2 + (\alpha_2 - \beta_2)g_1]$$
$$g_4 = hf[t_k + \alpha_3 h, y_k + \beta_3 g_2 + \gamma_3 g_3 + (\alpha_3 - \beta_3 - \gamma_3)g_1]$$

Comparison with the Taylor series yields eight equations for the 10 parameters. Thus two parameters can be chosen in light of other considerations. A number of different choices have been used. For example, the *classical* Runge-Kutta method, which reduces to

Simpson's rule for integration if $f(t, y)$ is a function of only t, uses the following set of parameters:

$$w_1 = w_4 = 1/6 \qquad w_2 = w_3 = 1/3 \qquad \text{(B-21)}$$
$$\alpha_1 = \alpha_2 = 1/2 \qquad \beta_2 = 1/2$$
$$\gamma_3 = \alpha_3 = 1 \qquad \beta_3 = 0$$

The Runge-Kutta algorithms are very tedious to compute by hand, so we do not show the steps involved in the next example. The algorithms are easily programmed, however.

EXAMPLE B-4
Runge-Kutta Method for an Oscillating Solution

Illustrate how the fourth-order Runge-Kutta method works with an oscillating solution by using the method to solve our second test case:

$$\dot{y} = \sin t \qquad y(0) = 0$$

for $0 \le t \le 4\pi$. Use the parameter values given by Equation B-21.

Solution

To compare the results with those obtained with the Euler method, we will use the same step size $\Delta t = 2\pi/13$. The results are shown in Figure B-4, with the numerical solution shown by the small circles and the exact solution by the solid line. There is much less error than with the Euler method using the same step size.

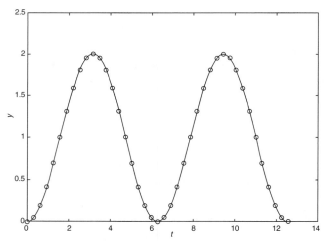

FIGURE B-4 Analytical and Runge-Kutta solution of $\dot{y} = \sin t$, $y(0) = 0$.

MECHANICAL PROPERTIES OF COMMON MATERIALS

Material	Mass density ρ (kg/m³)	Young's modulus of elasticity E (N/m²)	Shear modulus of elasticity G (N/m²)
Aluminum alloys*	2700	7.5×10^{10}	2.8×10^{10}
Brass*	8500	1×10^{11}	3.8×10^{10}
Bronze*	8500	1.1×10^{11}	4×10^{10}
Copper	8900	1.1×10^{11}	4.3×10^{10}
Rubber	960–1300	$7–40 \times 10^{5}$	$2–10 \times 10^{5}$
Steel*	7800	2×10^{11}	8×10^{10}
Titanium	4500	1.1×10^{11}	4.2×10^{10}

*These are alloys containing a primary metal plus various amounts of other substances. Thus their properties vary with the particular composition, but the values given here are representative.

REFERENCES

[**Bosch, 1986**] R. Bosch, *Automotive Handbook*, 2nd ed., Society of Automotive Engineers, Warrendale, PA, 1986.

[**Crandall, 1963**] S. Crandall and W. Mark, *Random Vibration in Mechanical Systems,* Academic Press, New York, 1963.

[**Den Hartog, 1985**] J. P. Den Hartog, *Mechanical Vibrations,* Dover Publications, New York, 1985 (slightly corrected republication of the 4th edition, originally published by McGraw-Hill, 1956).

[**Gillespie, 1992**] T. Gillespie, *Fundamentals of Vehicle Dynamics,* Society of Automotive Engineers, Warrendale, PA, 1992.

[**Greenberg, 1998**] M. D. Greenberg, *Advanced Engineering Mathematics,* 2nd ed., Prentice Hall, Upper Saddle River, NJ, 1998.

[**Harris, 2002**] C. M. Harris and A. G. Piersol, eds., *Shock and Vibration Handbook,* McGraw-Hill, New York, 2002.

[**James, 1947**] H. James, N. Nichols, and R. Phillips, *Theory of Servomechanisms,* McGraw-Hill, New York, 1947.

[**Kreyszig, 1997**] E. Kreyszig, *Advanced Engineering Mathematics,* 5th ed., John Wiley & Sons, New York, 1997.

[**Meriam, 2002**] J. L. Meriam and L. G. Kraige, *Engineering Mechanics, Volume 2, Dynamics,* John Wiley & Sons, New York, 2002.

[**Minorsky, 1962**] N. Minorsky, *Nonlinear Oscillations,* Van Nostrand, Princeton, NJ, 1962.

[**Newland, 1993**] D. Newland, *An Introduction to Random Vibration and Spectral Analysis,* Longman, New York, 1993.

[**Palm, 2000**] W. Palm III, *Modeling, Analysis, and Control of Dynamic Systems,* 2nd ed., John Wiley & Sons, New York, 2000.

[**Palm, 2005**] W. Palm III, *Introduction to MATLAB 7 for Engineers,* McGraw-Hill, New York, 2005.

[**Roark, 2001**] R. Roark, R. Budynas, and W. Young, *Roark's Formulas for Stess and Strain,* 7th ed., McGraw-Hill, New York, 2001.

INDEX